华章程序员书库

The Python 3
Standard Library by Example

Python 3标准库

[美] 道格·赫尔曼（Doug Hellmann）著

苏金国 李璜 等译

图书在版编目（CIP）数据

Python 3 标准库 /（美）道格·赫尔曼（Doug Hellmann）著；苏金国等译. —北京：机械工业出版社，2018.9（2024.7 重印）

（华章程序员书库）

书名原文：The Python 3 Standard Library by Example

ISBN 978-7-111-60895-0

I. P… II. ①道… ②苏… III. 软件工具–程序设计 IV. TP311.561

中国版本图书馆 CIP 数据核字（2018）第 211415 号

北京市版权局著作权合同登记　图字：01-2017-7499 号。

Authorized translation from the English language edition, entitled The Python 3 Standard Library by Example, ISBN: 9780134291055 by Doug Hellmann, published by Pearson Education, Inc., Copyright © 2017 Pearson Education, Inc.

All rights reserved. No part of this book may be reproduced or transmitted in any form or by any means, electronic or mechanical, including photocopying, recording or by any information storage retrieval system, without permission from Pearson Education, Inc.

Chinese simplified language edition published by China Machine Press, Copyright © 2018.

本书中文简体字版由 Pearson Education（培生教育出版集团）授权机械工业出版社在中国大陆地区（不包括香港、澳门特别行政区及台湾地区）独家出版发行。未经出版者书面许可，不得以任何方式抄袭、复制或节录本书中的任何部分。

本书封底贴有 Pearson Education（培生教育出版集团）激光防伪标签，无标签者不得销售。

Python 3 标准库

出版发行：机械工业出版社（北京市西城区百万庄大街 22 号　邮政编码：100037）	
责任编辑：王春华	责任校对：殷　虹
印　　刷：北京建宏印刷有限公司	版　　次：2024 年 7 月第 1 版第 11 次印刷
开　　本：186mm×240mm　1/16	印　　张：69.25
书　　号：ISBN 978-7-111-60895-0	定　　价：199.00 元

客服电话：(010) 88361066　68326294

版权所有·侵权必究
封底无防伪标均为盗版

译 者 序

自 1991 年首次发布以来，Python 的用户群体便开始不断增长，Python 也逐步成为开源开发中最受欢迎的编程语言之一。实际上，Python 语言的最大优势并不是语言本身的特性，而是它拥有脚本语言中最丰富的第三方模块，这是其他任何语言都无法比拟的巨大优势。很多人学习和使用 Python 正是因为 Python 拥有某个非常适合其工作领域的简单易用的模块。

标准库中包含数百个模块，为常见任务提供了丰富的工具，可以用来作为应用开发的起点。学习这些模块时，简短的例子要比详尽的手册文档更有帮助。这也正是写作本书的出发点。

作为 Python Software Foundation 的成员，作者道格·赫尔曼（Doug Hellmann）从 1.4 版本开始就一直在做 Python 编程工作，曾在大量 UNIX 和非 UNIX 平台上参与项目开发，涉及众多领域，积累了丰富的经验。他全面研究了标准库的模块，并在他的博客"Python Module of the Week"中利用实际例子介绍各个模块应该如何使用。为满足人们迫切的需求，他将这些博客文章进一步整理完善，并在 2011 年出版了针对 Python 2 的《Python 标准库》。

当前，Python 社区正在从 Python 2 向 Python 3 过渡。Python 2 与 Python 3 之间存在很多不兼容性，特别是很多曾经的标准库模块在 Python 3 中已经改名或者已经重新组织。鉴于此，时隔 7 年之后，终于有了这本《Python 3 标准库》，它主要强调 Python 3，同样沿袭《Python 标准库》的风格，通过轻松的方式，帮助你从具体的例子、具体的实践中了解技术细节，在知道"怎样做"的同时还能理解"为什么这样做"。所有例子都已经在 Python 3.5 上通过测试。

本书由苏金国、李璜主译，杨健康、乔会东、仝磊、王少轩、程芳、宋旭民、黄小钰等分别对全书各章进行了审阅，另外姚曜、程龙、吴忠望、张练达、陈峰、江健、姚勇、卢鋆、张莹参与了全书的修改整理工作，林琪、刘亮、刘跃邦、高强和王志淋统一全书术语，并完善了关键部分的翻译。全体人员共同完成了本书的翻译工作。由于译者水平有限，译文肯定有不当之处，敬请批评指正。

前　言

标准库会随每一版 Python 的发布而发布，其中包含数百个模块，为操作系统、解释器和互联网之间的交互提供了丰富的工具——所有这些模块都得到充分测试，可以用来作为应用开发的起点。本书会提供一些精选的例子，向你展示如何使用这些模块中最常用的一些特性，正是这些特性使 Python 有了"内含动力"（batteries included）的座右铭。这些例子均取自颇受关注的"Python Module of the Week（PyMOTW）"博客系列。

本书读者对象

本书的读者应该是中等程度的 Python 程序员，所以尽管书中对所有源代码都做了讨论，却也只有少数情况会逐行给出解释。每一节都强调了模块的特性，并通过源代码以及完全独立的示例程序的输出来具体说明。本书还尽可能简洁地介绍了各个特性，使读者能够把重点放在所展示的模块或函数上，而不会因支持代码而分心。

熟悉其他语言的有经验的程序员可以利用本书来了解 Python，但本书并不是关于 Python 语言的入门读物。研究这些例子时，如果之前有编写 Python 程序的经验，那么一定会很有帮助。

很多章节（比如介绍套接字网络编程或 hmac 加密的章节）还需要一些领域特定的知识。书中会提供解释这些例子所需的基本信息，不过由于标准库中模块涵盖的主题如此宽泛，所以不可能在一本书中全面地介绍每一个主题。在每个模块的讨论之后，还提供了一个推荐资源列表，大家可以进一步阅读这些资源，从中了解更多信息。推荐资源包括在线资源、RFC 标准文档以及相关图书。

Python 3 与 Python 2

Python 社区目前正在从 Python 2 向 Python 3 过渡。从主版本号可以看出，Python 2 和 Python 3 有很多不兼容之处，而且这种不兼容不只是存在于语言中。Python 3 中很多标准库模块都已经改名或者重新组织。

Python 开发社区认识到这种不兼容可能需要一个很长的过渡期，最终 Python 库和工具的生态系统会更新为使用 Python 3。尽管很多项目仍依赖于 Python 2，但 Python 2 目前只接受安全更新，并且还计划在 2020 年前完全废弃。所有新特性都只能在 Python 3 版本中使用。

编写可以同时用于这两个版本的程序可能很有难度，但并不是全无可能。这样做通常要求检查程序在哪个 Python 版本下运行，并且在导入时使用不同的模块名，或者在调用类或函数时使用不同的参数。在标准库之外，已经有大量工具可以简化这个过程。为了保证本书中的例子尽可能简洁，同时仍然只依赖于标准库，我们将主要强调 Python 3。所有例子已经在 Python 3.5 上通过测试（写作本书时 3.x 系列的当前版本），如果不加修改，可能无法用于 Python 2。要查看专门为使用 Python 2 而设计的例子，请参考本书的 Python 2 版本《Python 标准库》。

为了保证为每个例子提供清晰而简洁的描述，每一章不会过分强调 Python 2 和 Python 3 的差别。关于移植说明的附录会总结这两个版本之间一些最大的区别，这部分内容会合理组织，以便能有效地帮助从 Python 2 到 Python 3 的移植。

本书组织结构

http://docs.python.org 上提供了详尽的参考指南，而本书可以作为补充，提供功能完备的示例程序来展示这里介绍的特性。模块被分组为不同章节，以便轻松查找单个模块作为参考，并且可以按主题浏览进行更深层次的探讨。尽管可能不会一页页地从头到尾阅读本书，但如果你确实想要这么做，那么为了预备这种情况，本书也做了合理的组织，尽可能不要求你"提前参考"还没有介绍过的模块，不过要想完全避免这种情况是不可能的。

下载示例代码

原来的博客文章和示例代码可以在 https://pymotw.com/3/ 找到。本书勘误可以从作者的网站（https://doughellmann.com/blog/the-python-3-standard-library-by-example/）下载。

致谢

如果没有大家的贡献和支持，这本书绝无可能问世。

1997 年 Dick Wall 让我第一次接触到 Python，那时我们正在 ERDAS 一起合作开发 GIS 软件。记得在发现这样一个如此简便易用的新的工具语言时，我便立刻喜欢上了它，而且还对公司不让我们用它来完成"实际工作"颇有不满。在接下来的所有工作中我大量使用了 Python，而这正是因为 Dick，我要感谢从那以后软件开发给我带来的快乐时光。

Python 核心开发小组创建了一个由语言、工具和库共同构建的健壮的生态系统，这些库在日益普及，也在不断发现新的应用领域。如果没有他们付出的宝贵时间，没有他们提供的丰富资源，我们可能还得花时间一次又一次地从头开始。

本书中的材料最初是一系列博客帖子。如果没有博客读者们异常积极的响应，这些文章不会更新成使用 Python 3，这本新书也不可能出现。每个帖子都得到了 Python 社区成员的审阅和评论，有纠正，有建议，也有问题，这些评论促使我做出修改，这才有了你手上这本书。感谢大家日复一日地花时间来阅读我的博客，谢谢大家投入的时间和精力。

本书的技术审校人员——Diana Clarke、Ian Cordasco、Mark McClain、Paul McLanahan 和 Ryan Petrello——花了大量时间查找示例代码和相关解释中存在的问题。感谢他们的辛勤工作，最终的作品远比我靠一人之力得到的结果好得多。

Jim Baker 描述 readline 模块时提供了很有帮助的观点，特别是为那些 GNU 库很老或者默认未安装 GNU 库的平台提出了 gnureadline 包。

Patrick Kettner 帮助我收集了 Windows 上 platform 模块示例的输出。

还要特别感谢 Addison-Wesley 的编辑、制作人员和营销团队，感谢大家辛苦的工作和一贯的支持，帮助我明确这本书的目标，并取得成功。

最后，我要感谢我的妻子，Theresa Flynn。在完成这个新项目的过程中，很多个夜晚和周末我都无法相陪，她总能欣然接受。感谢她的建议、鼓励和支持。

目　　录

译者序
前言

第1章　文本 ··· 1

1.1　string：文本常量和模板 ············ 1
1.1.1　函数 ······································· 1
1.1.2　模板 ······································· 2
1.1.3　高级模板 ······························· 3
1.1.4　Formatter ······························· 5
1.1.5　常量 ······································· 5

1.2　textwrap：格式化文本段落 ········ 6
1.2.1　示例数据 ······························· 6
1.2.2　填充段落 ······························· 7
1.2.3　去除现有的缩进 ··················· 7
1.2.4　结合dedent和fill ················· 8
1.2.5　缩进块 ··································· 8
1.2.6　悬挂缩进 ····························· 10
1.2.7　截断长文本 ························· 10

1.3　re：正则表达式 ··························· 11
1.3.1　查找文本中的模式 ············· 11
1.3.2　编译表达式 ························· 12
1.3.3　多重匹配 ····························· 13
1.3.4　模式语法 ····························· 14
1.3.5　限制搜索 ····························· 22
1.3.6　用组解析匹配 ····················· 24
1.3.7　搜索选项 ····························· 29

1.3.8　前向或后向 ························· 35
1.3.9　自引用表达式 ····················· 38
1.3.10　用模式修改字符串 ··········· 42
1.3.11　利用模式拆分 ··················· 44

1.4　difflib：比较序列 ······················· 46
1.4.1　比较文本体 ························· 47
1.4.2　无用数据 ····························· 49
1.4.3　比较任意类型 ····················· 50

第2章　数据结构 ································· 52

2.1　enum：枚举类型 ·························· 53
2.1.1　创建枚举 ····························· 53
2.1.2　迭代 ····································· 53
2.1.3　比较Enum ··························· 54
2.1.4　唯一枚举值 ························· 55
2.1.5　通过编程创建枚举 ············· 56
2.1.6　非整数成员值 ····················· 58

2.2　collections：容器数据类型 ······ 60
2.2.1　ChainMap：搜索多个字典 ··· 60
2.2.2　Counter：统计可散列的
对象 ······································· 63
2.2.3　defaultdict：缺少的键返回
一个默认值 ··························· 66
2.2.4　deque：双端队列 ················ 67
2.2.5　namedtuple：带命名字段的
元组子类 ······························· 70

2.2.6 OrderedDict：记住向字典中
增加键的顺序 ················ 74
2.2.7 collections.abc：容器的
抽象基类 ···················· 76
2.3 数组：固定类型数据序列 ········ 78
2.3.1 初始化 ······················ 78
2.3.2 处理数组 ···················· 79
2.3.3 数组和文件 ·················· 79
2.3.4 候选字节顺序 ················ 80
2.4 heapq：堆排序算法 ············ 81
2.4.1 示例数据 ···················· 81
2.4.2 创建堆 ······················ 82
2.4.3 访问堆的内容 ················ 83
2.4.4 堆的数据极值 ················ 85
2.4.5 高效合并有序序列 ············ 85
2.5 bisect：维护有序列表 ········· 86
2.5.1 有序插入 ···················· 86
2.5.2 处理重复 ···················· 87
2.6 queue：线程安全的 FIFO 实现 ··· 88
2.6.1 基本 FIFO 队列 ·············· 88
2.6.2 LIFO 队列 ··················· 89
2.6.3 优先队列 ···················· 89
2.6.4 构建一个多线程播客客户
程序 ························· 90
2.7 struct：二进制数据结构 ······· 93
2.7.1 函数与 Struct 类 ············ 93
2.7.2 打包和解包 ·················· 93
2.7.3 字节序 ······················ 94
2.7.4 缓冲区 ······················ 95
2.8 weakref：对象的非永久引用 ···· 96
2.8.1 引用 ························ 96
2.8.2 引用回调 ···················· 97

2.8.3 最终化对象 ·················· 98
2.8.4 代理 ······················· 100
2.8.5 缓存对象 ··················· 101
2.9 copy：复制对象 ·············· 103
2.9.1 浅副本 ····················· 103
2.9.2 深副本 ····················· 104
2.9.3 定制复制行为 ··············· 105
2.9.4 深副本中的递归 ············· 106
2.10 pprint：美观打印数据结构 ···· 107
2.10.1 打印 ······················ 108
2.10.2 格式化 ···················· 108
2.10.3 任意类 ···················· 109
2.10.4 递归 ······················ 110
2.10.5 限制嵌套输出 ·············· 110
2.10.6 控制输出宽度 ·············· 111

第 3 章 算法 ························· 113
3.1 functools：管理函数的工具 ···· 113
3.1.1 修饰符 ····················· 113
3.1.2 比较 ······················· 119
3.1.3 缓存 ······················· 122
3.1.4 缩减数据集 ················· 125
3.1.5 泛型函数 ··················· 127
3.2 itertools：迭代器函数 ········ 129
3.2.1 合并和分解迭代器 ··········· 129
3.2.2 转换输入 ··················· 132
3.2.3 生成新值 ··················· 133
3.2.4 过滤 ······················· 135
3.2.5 数据分组 ··················· 138
3.2.6 合并输入 ··················· 139
3.3 operator：内置操作符的函数
接口 ························· 144

- 3.3.1 逻辑操作 ········· 144
- 3.3.2 比较操作符 ········· 145
- 3.3.3 算术操作符 ········· 145
- 3.3.4 序列操作符 ········· 146
- 3.3.5 原地操作符 ········· 148
- 3.3.6 属性和元素"获取方法"········· 148
- 3.3.7 结合操作符和定制类 ········· 150
- 3.4 contextlib：上下文管理器工具 ········· 151
  - 3.4.1 上下文管理器 API ········· 151
  - 3.4.2 上下文管理器作为函数修饰符 ········· 153
  - 3.4.3 从生成器到上下文管理器 ········· 154
  - 3.4.4 关闭打开的句柄 ········· 156
  - 3.4.5 忽略异常 ········· 157
  - 3.4.6 重定向输出流 ········· 158
  - 3.4.7 动态上下文管理器栈 ········· 159

第 4 章 日期和时间 ········· 166

- 4.1 time：时钟时间 ········· 166
  - 4.1.1 比较时钟 ········· 166
  - 4.1.2 墙上时钟时间 ········· 167
  - 4.1.3 单调时钟 ········· 168
  - 4.1.4 处理器时钟时间 ········· 169
  - 4.1.5 性能计数器 ········· 170
  - 4.1.6 时间组成 ········· 170
  - 4.1.7 处理时区 ········· 171
  - 4.1.8 解析和格式化时间 ········· 172
- 4.2 datetime：日期和时间值管理 ········· 174
  - 4.2.1 时间 ········· 174
  - 4.2.2 日期 ········· 175
  - 4.2.3 timedelta ········· 177
- 4.2.4 日期算术运算 ········· 178
- 4.2.5 比较值 ········· 179
- 4.2.6 结合日期和时间 ········· 179
- 4.2.7 格式化和解析 ········· 180
- 4.2.8 时区 ········· 182
- 4.3 calendar：处理日期 ········· 183
  - 4.3.1 格式化示例 ········· 183
  - 4.3.2 本地化环境 ········· 185
  - 4.3.3 计算日期 ········· 186

第 5 章 数学运算 ········· 188

- 5.1 decimal：定点数和浮点数的数学运算 ········· 188
  - 5.1.1 Decimal ········· 188
  - 5.1.2 格式化 ········· 189
  - 5.1.3 算术运算 ········· 190
  - 5.1.4 特殊值 ········· 191
  - 5.1.5 上下文 ········· 192
- 5.2 fractions：有理数 ········· 196
  - 5.2.1 创建 Fraction 实例 ········· 197
  - 5.2.2 算术运算 ········· 198
  - 5.2.3 近似值 ········· 199
- 5.3 random：伪随机数生成器 ········· 199
  - 5.3.1 生成随机数 ········· 200
  - 5.3.2 指定种子 ········· 200
  - 5.3.3 保存状态 ········· 201
  - 5.3.4 随机整数 ········· 202
  - 5.3.5 选择随机元素 ········· 203
  - 5.3.6 排列 ········· 203
  - 5.3.7 采样 ········· 205
  - 5.3.8 多个并发生成器 ········· 205
  - 5.3.9 SystemRandom ········· 206

5.3.10 非均匀分布 207
5.4 math：数学函数 208
　5.4.1 特殊常量 208
　5.4.2 测试异常值 208
　5.4.3 比较 210
　5.4.4 将浮点值转换为整数 212
　5.4.5 浮点值的其他表示 213
　5.4.6 正号和负号 214
　5.4.7 常用计算 215
　5.4.8 指数和对数 218
　5.4.9 角 222
　5.4.10 三角函数 224
　5.4.11 双曲函数 226
　5.4.12 特殊函数 227
5.5 statistics：统计计算 228
　5.5.1 平均值 228
　5.5.2 方差 230

第6章 文件系统 232

6.1 os.path：平台独立的文件名管理 233
　6.1.1 解析路径 233
　6.1.2 建立路径 236
　6.1.3 规范化路径 237
　6.1.4 文件时间 238
　6.1.5 测试文件 238
6.2 pathlib：文件系统路径作为对象 240
　6.2.1 路径表示 240
　6.2.2 建立路径 240
　6.2.3 解析路径 242
　6.2.4 创建具体路径 243
　6.2.5 目录内容 244
　6.2.6 读写文件 246
　6.2.7 管理目录和符号链接 246
　6.2.8 文件类型 247
　6.2.9 文件属性 248
　6.2.10 权限 250
　6.2.11 删除 250
6.3 glob：文件名模式匹配 252
　6.3.1 示例数据 252
　6.3.2 通配符 252
　6.3.3 单字符通配符 253
　6.3.4 字符区间 253
　6.3.5 转义元字符 254
6.4 fnmatch：UNIX 式 glob 模式匹配 254
　6.4.1 简单匹配 254
　6.4.2 过滤 255
　6.4.3 转换模式 256
6.5 linecache：高效读取文本文件 257
　6.5.1 测试数据 257
　6.5.2 读取特定行 257
　6.5.3 处理空行 258
　6.5.4 错误处理 258
　6.5.5 读取 Python 源文件 259
6.6 tempfile：临时文件系统对象 260
　6.6.1 临时文件 260
　6.6.2 命名文件 262
　6.6.3 假脱机文件 262
　6.6.4 临时目录 263
　6.6.5 预测名 264

6.6.6 临时文件位置 …… 264
6.7 shutil：高层文件操作 …… 265
　6.7.1 复制文件 …… 265
　6.7.2 复制文件元数据 …… 268
　6.7.3 处理目录树 …… 269
　6.7.4 查找文件 …… 271
　6.7.5 归档 …… 272
　6.7.6 文件系统空间 …… 275
6.8 filecmp：比较文件 …… 276
　6.8.1 示例数据 …… 276
　6.8.2 比较文件 …… 278
　6.8.3 比较目录 …… 279
　6.8.4 在程序中使用差异 …… 280
6.9 mmap：内存映射文件 …… 283
　6.9.1 读文件 …… 284
　6.9.2 写文件 …… 285
　6.9.3 正则表达式 …… 286
6.10 codecs：字符串编码和解码 …… 287
　6.10.1 Unicode 入门 …… 287
　6.10.2 处理文件 …… 289
　6.10.3 字节序 …… 291
　6.10.4 错误处理 …… 293
　6.10.5 编码转换 …… 295
　6.10.6 非 Unicode 编码 …… 296
　6.10.7 增量编码 …… 297
　6.10.8 Unicode 数据和网络通信 …… 299
　6.10.9 定义定制编码 …… 301
6.11 io：文本、十进制和原始流 I/O 工具 …… 307
　6.11.1 内存中的流 …… 307
　6.11.2 为文本数据包装字节流 …… 308

第 7 章　数据持久存储与交换 …… 310
7.1 pickle：对象串行化 …… 311
　7.1.1 编码和解码字符串中的数据 …… 311
　7.1.2 处理流 …… 312
　7.1.3 重构对象的问题 …… 313
　7.1.4 不可腌制的对象 …… 314
　7.1.5 循环引用 …… 316
7.2 shelve：对象的持久存储 …… 318
　7.2.1 创建一个新 shelf …… 318
　7.2.2 写回 …… 319
　7.2.3 特定 shelf 类型 …… 320
7.3 dbm：UNIX 键–值数据库 …… 320
　7.3.1 数据库类型 …… 321
　7.3.2 创建一个新数据库 …… 321
　7.3.3 打开一个现有数据库 …… 322
　7.3.4 错误情况 …… 322
7.4 sqlite3：嵌入式关系数据库 …… 323
　7.4.1 创建数据库 …… 323
　7.4.2 获取数据 …… 326
　7.4.3 查询元数据 …… 327
　7.4.4 行对象 …… 328
　7.4.5 在查询中使用变量 …… 329
　7.4.6 批量加载 …… 331
　7.4.7 定义新的列类型 …… 331
　7.4.8 确定列类型 …… 334
　7.4.9 事务 …… 336
　7.4.10 隔离级别 …… 338
　7.4.11 内存中的数据库 …… 341
　7.4.12 导出数据库内容 …… 341
　7.4.13 在 SQL 中使用 Python 函数 …… 342
　7.4.14 带正则表达式的查询 …… 344
　7.4.15 定制聚集 …… 345

7.4.16	线程和连接共享		346
7.4.17	限制对数据的访问		347

7.5 xml.etree.ElementTree：
XML 操纵 API ································ 349
 7.5.1 解析 XML 文档 ···················· 349
 7.5.2 遍历解析树 ·························· 350
 7.5.3 查找文档中的节点 ················ 351
 7.5.4 解析节点属性 ······················ 352
 7.5.5 解析时监视事件 ··················· 354
 7.5.6 创建一个定制树构造器 ·········· 356
 7.5.7 解析串 ································ 357
 7.5.8 用元素节点构造文档 ············· 359
 7.5.9 美观打印 XML ····················· 359
 7.5.10 设置元素属性 ···················· 360
 7.5.11 由节点列表构造树 ·············· 362
 7.5.12 将 XML 串行化至一个流 ····· 364

7.6 csv：逗号分隔值文件 ··················· 366
 7.6.1 读文件 ································ 366
 7.6.2 写文件 ································ 367
 7.6.3 方言 ··································· 368
 7.6.4 使用字段名 ·························· 373

第 8 章 数据压缩与归档 ················· 375

8.1 zlib：GNU zlib 压缩 ··················· 375
 8.1.1 处理内存中的数据 ················ 375
 8.1.2 增量压缩与解压缩 ················ 377
 8.1.3 混合内容流 ·························· 378
 8.1.4 校验和 ································ 378
 8.1.5 压缩网络数据 ······················ 379

8.2 gzip：读写 GNU zip 文件 ············ 382
 8.2.1 写压缩文件 ·························· 382
 8.2.2 读压缩数据 ·························· 384

 8.2.3 处理流 ································ 385

8.3 bz2：bzip2 压缩 ·························· 386
 8.3.1 内存中的一次性操作 ············· 386
 8.3.2 增量压缩和解压缩 ················ 388
 8.3.3 混合内容流 ·························· 388
 8.3.4 写压缩文件 ·························· 389
 8.3.5 读压缩文件 ·························· 390
 8.3.6 读写 Unicode 数据 ·············· 391
 8.3.7 压缩网络数据 ······················ 392

8.4 tarfile：tar 归档访问 ················ 395
 8.4.1 测试 tar 文件 ······················· 396
 8.4.2 从归档读取元数据 ················ 396
 8.4.3 从归档抽取文件 ··················· 397
 8.4.4 创建新归档 ·························· 399
 8.4.5 使用候选归档成员名 ············· 399
 8.4.6 从非文件源写数据 ················ 400
 8.4.7 追加到归档 ·························· 400
 8.4.8 处理压缩归档 ······················ 401

8.5 zipfile：ZIP 归档访问 ················ 402
 8.5.1 测试 ZIP 文件 ······················ 402
 8.5.2 从归档读取元数据 ················ 402
 8.5.3 从归档抽取归档文件 ············· 404
 8.5.4 创建新归档 ·························· 404
 8.5.5 使用候选归档成员名 ············· 406
 8.5.6 从非文件源写数据 ················ 406
 8.5.7 利用 ZipInfo 实例写数据 ····· 407
 8.5.8 追加到文件 ·························· 407
 8.5.9 Python ZIP 归档 ·················· 408
 8.5.10 限制 ·································· 410

第 9 章 加密 ···································· 411

9.1 hashlib：密码散列 ······················ 411

9.1.1	散列算法	411
9.1.2	示例数据	412
9.1.3	MD5 示例	412
9.1.4	SHA1 示例	412
9.1.5	按名创建散列	413
9.1.6	增量更新	413

9.2 hmac：密码消息签名与验证 414

9.2.1	消息签名	415
9.2.2	候选摘要类型	415
9.2.3	二进制摘要	416
9.2.4	消息签名的应用	416

第 10 章 使用进程、线程和协程提供并发性 420

10.1 subprocess：创建附加进程 420

10.1.1	运行外部命令	421
10.1.2	直接处理管道	425
10.1.3	连接管道段	427
10.1.4	与其他命令交互	428
10.1.5	进程间传递信号	430

10.2 signal：异步系统事件 434

10.2.1	接收信号	434
10.2.2	获取已注册的处理器	435
10.2.3	发送信号	436
10.2.4	闹铃	436
10.2.5	忽略信号	437
10.2.6	信号和线程	438

10.3 threading：进程中管理并发操作 440

10.3.1	Thread 对象	440
10.3.2	确定当前线程	441
10.3.3	守护与非守护线程	442
10.3.4	枚举所有线程	444
10.3.5	派生线程	445
10.3.6	定时器线程	447
10.3.7	线程间传送信号	447
10.3.8	控制资源访问	449
10.3.9	同步线程	453
10.3.10	限制资源的并发访问	456
10.3.11	线程特定的数据	457

10.4 multiprocessing：像线程一样管理进程 459

10.4.1	multiprocessing 基础	460
10.4.2	可导入的目标函数	461
10.4.3	确定当前进程	461
10.4.4	守护进程	462
10.4.5	等待进程	463
10.4.6	终止进程	465
10.4.7	进程退出状态	466
10.4.8	日志	467
10.4.9	派生进程	469
10.4.10	向进程传递消息	469
10.4.11	进程间信号传输	472
10.4.12	控制资源访问	473
10.4.13	同步操作	474
10.4.14	控制资源的并发访问	475
10.4.15	管理共享状态	476
10.4.16	共享命名空间	477
10.4.17	进程池	479
10.4.18	实现 MapReduce	480

10.5 asyncio：异步 I/O、事件循环和并发工具 484

10.5.1	异步并发概念	484
10.5.2	利用协程合作完成多任务	485

10.5.3 调度常规函数调用 …… 488
10.5.4 异步地生成结果 …… 490
10.5.5 并发地执行任务 …… 492
10.5.6 组合协程和控制结构 …… 495
10.5.7 同步原语 …… 499
10.5.8 提供协议类抽象的异步 I/O …… 505
10.5.9 使用协程和流的异步 I/O …… 510
10.5.10 使用 SSL …… 514
10.5.11 与域名服务交互 …… 516
10.5.12 使用子进程 …… 518
10.5.13 接收 UNIX 信号 …… 523
10.5.14 结合使用协程、线程与进程 …… 525
10.5.15 用 asyncio 调试 …… 527
10.6 concurrent.futures：管理并发任务池 …… 530
10.6.1 利用基本线程池使用 map() …… 531
10.6.2 调度单个任务 …… 532
10.6.3 按任意顺序等待任务 …… 532
10.6.4 Future 回调 …… 533
10.6.5 撤销任务 …… 534
10.6.6 任务中的异常 …… 535
10.6.7 上下文管理器 …… 536
10.6.8 进程池 …… 537

第 11 章 网络通信 …… 539

11.1 ipaddress：Internet 地址 …… 539
11.1.1 地址 …… 539
11.1.2 网络 …… 540
11.1.3 接口 …… 543
11.2 socket：网络通信 …… 544
11.2.1 寻址、协议簇和套接字类型 …… 544
11.2.2 TCP/IP 客户和服务器 …… 552
11.2.3 用户数据报客户和服务器 …… 558
11.2.4 UNIX 域套接字 …… 560
11.2.5 组播 …… 563
11.2.6 发送二进制数据 …… 566
11.2.7 非阻塞通信和超时 …… 568
11.3 selectors：I/O 多路复用抽象 …… 568
11.3.1 操作模型 …… 569
11.3.2 回送服务器 …… 569
11.3.3 回送客户 …… 570
11.3.4 服务器和客户 …… 571
11.4 select：高效等待 I/O …… 572
11.4.1 使用 select() …… 572
11.4.2 带超时的非阻塞 I/O …… 577
11.4.3 使用 poll() …… 579
11.4.4 平台特定的选项 …… 582
11.5 socketserver：创建网络服务器 …… 583
11.5.1 服务器类型 …… 583
11.5.2 服务器对象 …… 583
11.5.3 实现服务器 …… 584
11.5.4 请求处理器 …… 584
11.5.5 回送示例 …… 584
11.5.6 线程和进程 …… 588

第 12 章 互联网 …… 592

12.1 urllib.parse：分解 URL …… 592
12.1.1 解析 …… 593
12.1.2 反解析 …… 595

12.1.3 连接 …… 596	12.6.2 Morsel …… 622
12.1.4 解码查询参数 …… 597	12.6.3 编码的值 …… 624
12.2 urllib.request：网络资源访问 …… 599	12.6.4 接收和解析 Cookie 首部 …… 624
12.2.1 HTTP GET …… 599	12.6.5 候选输出格式 …… 625
12.2.2 编码参数 …… 600	12.7 webbrowser：显示 Web 页面 …… 626
12.2.3 HTTP POST …… 601	12.7.1 简单示例 …… 626
12.2.4 添加发出首部 …… 602	12.7.2 窗口与标签页 …… 626
12.2.5 从请求提交表单数据 …… 602	12.7.3 使用特定浏览器 …… 627
12.2.6 上传文件 …… 603	12.7.4 BROWSER 变量 …… 627
12.2.7 创建定制协议处理器 …… 606	12.7.5 命令行接口 …… 627
12.3 urllib.robotparser：Internet 蜘蛛访问控制 …… 608	12.8 uuid：全局唯一标识符 …… 628
12.3.1 robots.txt …… 608	12.8.1 UUID 1：IEEE 802 MAC 地址 …… 628
12.3.2 测试访问权限 …… 609	12.8.2 UUID 3 和 5：基于名字的值 …… 630
12.3.3 长寿命蜘蛛 …… 610	12.8.3 UUID 4：随机值 …… 631
12.4 base64：用 ASCII 编码二进制数据 …… 611	12.8.4 处理 UUID 对象 …… 631
12.4.1 Base64 编码 …… 611	12.9 json：JavaScript 对象记法 …… 632
12.4.2 Base64 解码 …… 612	12.9.1 编码和解码简单数据类型 …… 633
12.4.3 URL 安全的变种 …… 612	12.9.2 人类可读和紧凑输出 …… 633
12.4.4 其他编码 …… 613	12.9.3 编码字典 …… 635
12.5 http.server：实现 Web 服务器的基类 …… 615	12.9.4 处理定制类型 …… 636
12.5.1 HTTP GET …… 615	12.9.5 编码器和解码器类 …… 638
12.5.2 HTTP POST …… 616	12.9.6 处理流和文件 …… 640
12.5.3 线程和进程 …… 618	12.9.7 混合数据流 …… 641
12.5.4 处理错误 …… 619	12.9.8 命令行上处理 JSON …… 641
12.5.5 设置首部 …… 620	12.10 xmlrpc.client：XML-RPC 的客户库 …… 642
12.5.6 命令行用法 …… 621	12.10.1 连接服务器 …… 643
12.6 http.cookies：HTTP cookie …… 622	12.10.2 数据类型 …… 645
12.6.1 创建和设置 cookie …… 622	12.10.3 传递对象 …… 648
	12.10.4 二进制数据 …… 648

XVI

12.10.5 异常处理 ································ 650
12.10.6 将调用组合在一个消息中···· 650
12.11 xmlrpc.server：一个
       XML-RPC 服务器···················· 652
   12.11.1 一个简单的服务器 ············ 652
   12.11.2 候选 API 名 ···················· 653
   12.11.3 加点的 API 名 ················· 654
   12.11.4 任意 API 名 ···················· 655
   12.11.5 公布对象的方法 ··············· 656
   12.11.6 分派调用 ························ 657
   12.11.7 自省 API ························ 659

第 13 章 email ···································· 662
 13.1 smtplib：简单邮件传输协议
      客户···································· 662
   13.1.1 发送 email 消息················· 662
   13.1.2 认证和加密 ······················· 663
   13.1.3 验证 email 地址 ················· 666
 13.2 smtpd：示例邮件服务器··········· 667
   13.2.1 邮件服务器基类 ················· 667
   13.2.2 调试服务器 ······················· 669
   13.2.3 代理服务器 ······················· 670
 13.3 mailbox：管理 email 归档········ 670
   13.3.1 mbox ································· 671
   13.3.2 Maildir ····························· 673
   13.3.3 消息标志 ··························· 678
   13.3.4 其他格式 ··························· 680
 13.4 imaplib：IMAP4 客户库··········· 680
   13.4.1 变种 ································· 680
   13.4.2 连接服务器 ······················· 681
   13.4.3 示例配置 ··························· 682
   13.4.4 列出邮箱 ··························· 682

13.4.5 邮箱状态 ··························· 684
13.4.6 选择邮箱 ··························· 686
13.4.7 搜索消息 ··························· 686
13.4.8 搜索规则 ··························· 687
13.4.9 获取消息 ··························· 689
13.4.10 完整消息 ·························· 693
13.4.11 上传消息 ·························· 694
13.4.12 移动和复制消息················ 695
13.4.13 删除消息 ·························· 696

第 14 章 应用构建模块 ······················ 699
 14.1 argparse：命令行选项和参数
      解析···································· 700
   14.1.1 建立解析器 ······················· 700
   14.1.2 定义参数 ··························· 700
   14.1.3 解析命令行 ······················· 700
   14.1.4 简单示例 ··························· 701
   14.1.5 帮助输出 ··························· 707
   14.1.6 解析器组织 ······················· 711
   14.1.7 高级参数处理 ···················· 716
 14.2 getopt：命令行选项解析·········· 722
   14.2.1 函数参数 ··························· 722
   14.2.2 短格式选项 ······················· 723
   14.2.3 长格式选项 ······················· 723
   14.2.4 一个完整的例子·················· 723
   14.2.5 缩写长格式选项·················· 725
   14.2.6 GNU 式选项解析················ 725
   14.2.7 结束参数处理····················· 726
 14.3 readline：GNU readline 库····· 727
   14.3.1 配置 readline ····················· 727
   14.3.2 完成文本 ··························· 728
   14.3.3 访问完成缓冲区·················· 731

XVII

- 14.3.4 输入历史 ················ 733
- 14.3.5 hook ···················· 736
- 14.4 getpass：安全密码提示 ······ 737
  - 14.4.1 示例 ···················· 737
  - 14.4.2 无终端使用 getpass ····· 738
- 14.5 cmd：面向行的命令处理器 ···· 739
  - 14.5.1 处理命令 ················ 739
  - 14.5.2 命令参数 ················ 740
  - 14.5.3 现场帮助 ················ 741
  - 14.5.4 自动完成 ················ 742
  - 14.5.5 覆盖基类方法 ············ 744
  - 14.5.6 通过属性配置 Cmd ······· 745
  - 14.5.7 运行 shell 命令 ········· 746
  - 14.5.8 候选输入 ················ 747
  - 14.5.9 sys.argv 的命令 ········ 748
- 14.6 shlex：解析 shell 类语法 ···· 749
  - 14.6.1 解析加引号的字符串 ······ 749
  - 14.6.2 为 shell 建立安全的字符串 ···· 751
  - 14.6.3 嵌入注释 ················ 751
  - 14.6.4 将字符串分解为 token ···· 752
  - 14.6.5 包含其他 token 源 ······· 752
  - 14.6.6 控制解析器 ·············· 753
  - 14.6.7 错误处理 ················ 755
  - 14.6.8 POSIX 与非 POSIX 解析 ···· 756
- 14.7 configparser：处理配置文件 ···· 757
  - 14.7.1 配置文件格式 ············ 757
  - 14.7.2 读取配置文件 ············ 758
  - 14.7.3 访问配置设置 ············ 759
  - 14.7.4 修改设置 ················ 765
  - 14.7.5 保存配置文件 ············ 766
  - 14.7.6 选项搜索路径 ············ 767
- 14.7.7 用拼接合并值 ············ 768
- 14.8 logging：报告状态、错误和信息消息 ············ 772
  - 14.8.1 日志系统的组成 ·········· 773
  - 14.8.2 应用与库中的日志记录 ···· 773
  - 14.8.3 记入文件 ················ 773
  - 14.8.4 旋转日志文件 ············ 774
  - 14.8.5 详细级别 ················ 774
  - 14.8.6 命名日志记录器实例 ······ 776
  - 14.8.7 日志树 ··················· 776
  - 14.8.8 与 warnings 模块集成 ···· 777
- 14.9 fileinput：命令行过滤器框架 ···· 778
  - 14.9.1 将 m3u 文件转换为 RSS ···· 778
  - 14.9.2 进度元数据 ·············· 779
  - 14.9.3 原地过滤 ················ 781
- 14.10 atexit：程序关闭回调 ········ 782
  - 14.10.1 注册退出回调 ············ 782
  - 14.10.2 修饰符语法 ·············· 783
  - 14.10.3 撤销回调 ················ 784
  - 14.10.4 什么情况下不调用 atexit 函数 ···· 785
  - 14.10.5 处理异常 ················ 786
- 14.11 sched：定时事件调度器 ······ 787
  - 14.11.1 有延迟地运行事件 ········ 788
  - 14.11.2 重叠事件 ················ 788
  - 14.11.3 事件优先级 ·············· 789
  - 14.11.4 取消事件 ················ 790

第 15 章 国际化和本地化 ···· 791

- 15.1 gettext：消息编目 ············ 791
  - 15.1.1 转换工作流概述 ············ 791

15.1.2 由源代码创建消息编目 ……… 792
15.1.3 运行时查找消息编目 ……… 794
15.1.4 复数值 ……… 795
15.1.5 应用与模块本地化 ……… 797
15.1.6 切换转换 ……… 798
15.2 `locale`：文化本地化 API ……… 798
15.2.1 探查当前本地化环境 ……… 799
15.2.2 货币 ……… 803
15.2.3 格式化数字 ……… 804
15.2.4 解析数字 ……… 805
15.2.5 日期和时间 ……… 806

第 16 章 开发工具 ……… 807

16.1 `pydoc`：模块的联机帮助 ……… 808
16.1.1 纯文本帮助 ……… 808
16.1.2 HTML 帮助 ……… 809
16.1.3 交互式帮助 ……… 809
16.2 `doctest`：通过文档完成测试 ……… 810
16.2.1 起步 ……… 810
16.2.2 处理不可预测的输出 ……… 811
16.2.3 traceback ……… 814
16.2.4 避开空白符 ……… 815
16.2.5 测试位置 ……… 819
16.2.6 外部文档 ……… 822
16.2.7 运行测试 ……… 824
16.2.8 测试上下文 ……… 827
16.3 `unittest`：自动测试框架 ……… 829
16.3.1 基本测试结构 ……… 829
16.3.2 运行测试 ……… 829
16.3.3 测试结果 ……… 830
16.3.4 断言真值 ……… 831

16.3.5 测试相等性 ……… 832
16.3.6 几乎相等？ ……… 833
16.3.7 容器 ……… 833
16.3.8 测试异常 ……… 837
16.3.9 测试固件 ……… 838
16.3.10 用不同输入重复测试 ……… 840
16.3.11 跳过测试 ……… 842
16.3.12 忽略失败测试 ……… 842
16.4 `trace`：执行程序流 ……… 843
16.4.1 示例程序 ……… 843
16.4.2 跟踪执行 ……… 844
16.4.3 代码覆盖 ……… 845
16.4.4 调用关系 ……… 847
16.4.5 编程接口 ……… 848
16.4.6 保存结果数据 ……… 849
16.4.7 选项 ……… 850
16.5 `traceback`：异常和栈轨迹 ……… 850
16.5.1 支持函数 ……… 851
16.5.2 检查栈 ……… 851
16.5.3 traceback 异常 ……… 853
16.5.4 底层异常 API ……… 854
16.5.5 底层栈 API ……… 857
16.6 `cgitb`：详细的 traceback 报告 ……… 859
16.6.1 标准 traceback 转储 ……… 859
16.6.2 启用详细的 traceback ……… 860
16.6.3 traceback 中的局部变量 ……… 862
16.6.4 异常属性 ……… 864
16.6.5 HTML 输出 ……… 866
16.6.6 记录 traceback ……… 866
16.7 `pdb`：交互式调试工具 ……… 868
16.7.1 启动调试工具 ……… 869

- 16.7.2 控制调试工具 ⋯⋯⋯⋯⋯⋯ 871
- 16.7.3 断点 ⋯⋯⋯⋯⋯⋯⋯⋯⋯⋯ 881
- 16.7.4 改变执行流 ⋯⋯⋯⋯⋯⋯ 890
- 16.7.5 用别名定制调试工具 ⋯⋯⋯ 895
- 16.7.6 保存配置设置 ⋯⋯⋯⋯⋯ 897
- 16.8 profile 和 pstats：性能分析 ⋯⋯⋯⋯⋯⋯⋯⋯⋯⋯⋯ 898
  - 16.8.1 运行性能分析工具 ⋯⋯⋯⋯ 898
  - 16.8.2 在上下文中运行 ⋯⋯⋯⋯ 901
  - 16.8.3 pstats：保存和处理统计信息 ⋯⋯⋯⋯⋯⋯⋯⋯⋯⋯ 901
  - 16.8.4 限制报告内容 ⋯⋯⋯⋯⋯ 903
  - 16.8.5 调用者/被调用者图 ⋯⋯⋯ 903
- 16.9 timeit：测量小段 Python 代码执行的时间 ⋯⋯⋯⋯⋯⋯⋯ 905
  - 16.9.1 模块内容 ⋯⋯⋯⋯⋯⋯⋯ 905
  - 16.9.2 基本示例 ⋯⋯⋯⋯⋯⋯⋯ 905
  - 16.9.3 将值存储在字典中 ⋯⋯⋯ 906
  - 16.9.4 从命令行执行 ⋯⋯⋯⋯⋯ 908
- 16.10 tabnanny：缩进验证工具 ⋯⋯ 909
- 16.11 compileall：字节编译源文件 ⋯⋯⋯⋯⋯⋯⋯⋯⋯⋯⋯⋯ 910
  - 16.11.1 编译一个目录 ⋯⋯⋯⋯⋯ 910
  - 16.11.2 忽略文件 ⋯⋯⋯⋯⋯⋯ 911
  - 16.11.3 编译 sys.path ⋯⋯⋯⋯ 912
  - 16.11.4 编译单个文件 ⋯⋯⋯⋯ 912
  - 16.11.5 从命令行运行 ⋯⋯⋯⋯ 913
- 16.12 pyclbr：类浏览器 ⋯⋯⋯⋯⋯ 914
  - 16.12.1 扫描类 ⋯⋯⋯⋯⋯⋯⋯ 915
  - 16.12.2 扫描函数 ⋯⋯⋯⋯⋯⋯ 916
- 16.13 venv：创建虚拟环境 ⋯⋯⋯⋯ 917
  - 16.13.1 创建环境 ⋯⋯⋯⋯⋯⋯ 917
- 16.13.2 虚拟环境的内容 ⋯⋯⋯⋯ 917
- 16.13.3 使用虚拟环境 ⋯⋯⋯⋯ 918
- 16.14 ensurepip：安装 Python 包安装工具 ⋯⋯⋯⋯⋯⋯⋯⋯⋯ 920

第 17 章 运行时特性 ⋯⋯⋯⋯⋯⋯ 922

- 17.1 site：全站点配置 ⋯⋯⋯⋯⋯⋯ 922
  - 17.1.1 导入路径 ⋯⋯⋯⋯⋯⋯⋯ 922
  - 17.1.2 用户目录 ⋯⋯⋯⋯⋯⋯⋯ 923
  - 17.1.3 路径配置文件 ⋯⋯⋯⋯⋯ 924
  - 17.1.4 定制站点配置 ⋯⋯⋯⋯⋯ 926
  - 17.1.5 定制用户配置 ⋯⋯⋯⋯⋯ 927
  - 17.1.6 禁用 site 模块 ⋯⋯⋯⋯⋯ 929
- 17.2 sys：系统特定配置 ⋯⋯⋯⋯⋯ 929
  - 17.2.1 解释器设置 ⋯⋯⋯⋯⋯⋯ 929
  - 17.2.2 运行时环境 ⋯⋯⋯⋯⋯⋯ 935
  - 17.2.3 内存管理和限制 ⋯⋯⋯⋯ 937
  - 17.2.4 异常处理 ⋯⋯⋯⋯⋯⋯⋯ 942
  - 17.2.5 底层线程支持 ⋯⋯⋯⋯⋯ 944
  - 17.2.6 模块和导入 ⋯⋯⋯⋯⋯⋯ 947
  - 17.2.7 跟踪程序运行情况 ⋯⋯⋯ 963
- 17.3 os：可移植访问操作系统特定特性 ⋯⋯⋯⋯⋯⋯⋯⋯⋯⋯⋯ 968
  - 17.3.1 检查文件系统内容 ⋯⋯⋯ 968
  - 17.3.2 管理文件系统权限 ⋯⋯⋯ 971
  - 17.3.3 创建和删除目录 ⋯⋯⋯⋯ 973
  - 17.3.4 处理符号链接 ⋯⋯⋯⋯⋯ 973
  - 17.3.5 安全地替换现有文件 ⋯⋯ 974
  - 17.3.6 检测和改变进程所有者 ⋯ 975
  - 17.3.7 管理进程环境 ⋯⋯⋯⋯⋯ 976
  - 17.3.8 管理进程工作目录 ⋯⋯⋯ 977
  - 17.3.9 运行外部命令 ⋯⋯⋯⋯⋯ 977

17.3.10 用 `os.fork()` 创建
进程 ······ 979
17.3.11 等待子进程 ······ 980
17.3.12 Spawn 创建新进程 ······ 982
17.3.13 操作系统错误码 ······ 982
17.4 platform：系统版本信息 ······ 983
17.4.1 解释器 ······ 983
17.4.2 平台 ······ 984
17.4.3 操作系统和硬件信息 ······ 985
17.4.4 可执行程序体系结构 ······ 986
17.5 resource：系统资源管理 ······ 987
17.5.1 当前使用情况 ······ 987
17.5.2 资源限制 ······ 988
17.6 gc：垃圾回收器 ······ 990
17.6.1 跟踪引用 ······ 990
17.6.2 强制垃圾回收 ······ 992
17.6.3 查找无法回收的对象引用 ······ 993
17.6.4 回收阈值和代 ······ 995
17.6.5 调试 ······ 998
17.7 sysconfig：解释器编译时
配置 ······ 1002
17.7.1 配置变量 ······ 1002
17.7.2 安装路径 ······ 1004
17.7.3 Python 版本和平台 ······ 1007

第 18 章 语言工具 ······ 1009

18.1 warnings：非致命警告 ······ 1009
18.1.1 分类和过滤 ······ 1010
18.1.2 生成警告 ······ 1010
18.1.3 用模式过滤 ······ 1011
18.1.4 重复的警告 ······ 1013
18.1.5 候选消息传送函数 ······ 1013
18.1.6 格式化 ······ 1014
18.1.7 警告中的栈层次 ······ 1014
18.2 abc：抽象基类 ······ 1015
18.2.1 ABC 如何工作 ······ 1015
18.2.2 注册一个具体类 ······ 1016
18.2.3 通过派生实现 ······ 1017
18.2.4 辅助基类 ······ 1017
18.2.5 不完整的实现 ······ 1018
18.2.6 ABC 中的具体方法 ······ 1019
18.2.7 抽象属性 ······ 1020
18.2.8 抽象类和静态方法 ······ 1022
18.3 dis：Python 字节码反汇编
工具 ······ 1023
18.3.1 基本反汇编 ······ 1023
18.3.2 反汇编函数 ······ 1024
18.3.3 类 ······ 1025
18.3.4 源代码 ······ 1026
18.3.5 使用反汇编调试 ······ 1027
18.3.6 循环的性能分析 ······ 1028
18.3.7 编译器优化 ······ 1033
18.4 inspect：检查现场对象 ······ 1035
18.4.1 示例模块 ······ 1035
18.4.2 检查模块 ······ 1035
18.4.3 检查类 ······ 1036
18.4.4 检查实例 ······ 1038
18.4.5 文档串 ······ 1038
18.4.6 获取源代码 ······ 1039
18.4.7 方法和函数签名 ······ 1041
18.4.8 类层次体系 ······ 1043
18.4.9 方法解析顺序 ······ 1044
18.4.10 栈与帧 ······ 1045
18.4.11 命令行接口 ······ 1047

第 19 章　模块和包 ···········1048

19.1　importlib：Python 的导入机制 ···········1048
19.1.1　示例包 ···········1048
19.1.2　模块类型 ···········1049
19.1.3　导入模块 ···········1049
19.1.4　加载工具 ···········1051

19.2　pkgutil：包工具 ···········1052
19.2.1　包导入路径 ···········1052
19.2.2　包的开发版本 ···········1054
19.2.3　用 PKG 文件管理路径 ···········1055
19.2.4　嵌套包 ···········1056
19.2.5　包数据 ···········1058

19.3　zipimport：从 ZIP 归档加载 Python 代码 ···········1060
19.3.1　示例 ···········1060
19.3.2　查找模块 ···········1061
19.3.3　访问代码 ···········1061
19.3.4　源代码 ···········1062
19.3.5　包 ···········1063
19.3.6　数据 ···········1063

附录 A　移植说明 ···········1066

附录 B　标准库之外 ···········1081

第 1 章 文 本

对 Python 程序员来说，最显而易见的文本处理工具就是 `str` 类，不过除此以外，标准库还提供了大量其他的工具，可以帮助你轻松地完成高级文本处理。

程序可以使用 `string.Template` 作为一种简便方法来构建比 `str` 对象特性更丰富的字符串。与很多 Web 框架定义的模板或 Python Package Index 提供的扩展模块相比，尽管 `string.Template` 没有那么丰富的特性，但它确实能很好地支持用户可修改的模板，这些模板需要将动态值插入到静态文本中。

`textwrap` 模块包含格式化段落文本的工具，可以限制输出的宽度、增加缩进，以及插入换行符从而能一致地自动换行。

除了字符串对象支持的内置相等性和排序比较之外，标准库还包括两个与比较文本值有关的模块。`re` 提供了一个完整的正则表达式库，这个库使用 C 实现来保证速度。正则表达式非常适合在较大的数据集中查找子串，能够根据比固定字符串更复杂的模式来比较字符串，还可以完成一定程度的解析。

另一方面，`difflib` 则有所不同，它会根据增加、删除或修改的部分来计算不同文本序列之间的实际差别。`difflib` 中比较函数的输出可以用来为用户提供更详细的反馈，指出两个输入中出现变化的地方，文档随时间有哪些改变，等等。

1.1 string：文本常量和模板

`string` 模块在最早的 Python 版本中就已经有了。以前这个模块中提供的很多函数已经移植为 `str` 对象的方法，不过这个模块仍保留了很多有用的常量和类来处理 `str` 对象。这里将重点讨论这些常量和类。

1.1.1 函数

函数 `capwords()` 会把一个字符串中的所有单词首字母大写。

代码清单 1-1：`string_capwords.py`

```
import string

s = 'The quick brown fox jumped over the lazy dog.'
print(s)
print(string.capwords(s))
```

这个代码的结果等同于先调用 `split()`，把结果列表中的单词首字母大写，然后调用 `join()` 来合并结果。

```
$ python3 string_capwords.py

The quick brown fox jumped over the lazy dog.
The Quick Brown Fox Jumped Over The Lazy Dog.
```

1.1.2 模板

字符串模板是 **PEP 292**[①] 新增的部分，将作为内置拼接语法的替代做法。使用 `string.Template` 拼接时，要在名字前加前缀 `$` 来标识变量（例如，`$var`）。或者，如果有必要区分变量和周围的文本，可以用大括号包围变量（例如，`${var}`）。

下面这个例子对一个简单模板、使用 `%` 操作符的类似字符串拼接以及使用 `str.format()` 的新格式化字符串语法做了比较：

代码清单 1-2：**string_template.py**

```python
import string

values = {'var': 'foo'}

t = string.Template("""
Variable        : $var
Escape          : $$
Variable in text: ${var}iable
""")

print('TEMPLATE:', t.substitute(values))

s = """
Variable        : %(var)s
Escape          : %%
Variable in text: %(var)siable
"""

print('INTERPOLATION:', s % values)

s = """
Variable        : {var}
Escape          : {{}}
Variable in text: {var}iable
"""

print('FORMAT:', s.format(**values))
```

在前两种情况中，触发字符（`$` 或 `%`）要重复两次来进行转义。在格式化语法中，需要重复 `{` 和 `}` 来转义。

```
$ python3 string_template.py

TEMPLATE:
Variable        : foo
Escape          : $
Variable in text: fooiable
```

[①] www.python.org/dev/peps/pep-0292

```
INTERPOLATION:
Variable        : foo
Escape          : %
Variable in text: fooiable

FORMAT:
Variable        : foo
Escape          : {}
Variable in text: fooiable
```

模板与字符串拼接或格式化的一个关键区别是，它不考虑参数的类型。值会转换为字符串，再将字符串插入结果。这里没有提供格式化选项。例如，没有办法控制使用几位有效数字来表示一个浮点值。

不过，这也有一个好处，通过使用 safe_substitute() 方法，可以避免未能向模板提供所需的所有参数值时可能产生的异常。

代码清单 1-3：string_template_missing.py

```python
import string

values = {'var': 'foo'}

t = string.Template("$var is here but $missing is not provided")

try:
    print('substitute()     :', t.substitute(values))
except KeyError as err:
    print('ERROR:', str(err))

print('safe_substitute():', t.safe_substitute(values))
```

由于 values 字典中没有 missing 的值，所以 substitute() 会产生一个 KeyError。safe_substitute() 则不同，它不会抛出这个错误，而是会捕获这个错误并保留文本中的变量表达式。

```
$ python3 string_template_missing.py

ERROR: 'missing'
safe_substitute(): foo is here but $missing is not provided
```

1.1.3　高级模板

可以调整 string.Template 在模板体中查找变量名所使用的正则表达式模式，以改变它的默认语法。为此，一种简单的方法是修改 delimiter 和 idpattern 类属性。

代码清单 1-4：string_template_advanced.py

```python
import string

class MyTemplate(string.Template):
    delimiter = '%'
    idpattern = '[a-z]+_[a-z]+'


template_text = '''
```

```
    Delimiter : %%
    Replaced  : %with_underscore
    Ignored   : %notunderscored
'''

d = {
    'with_underscore': 'replaced',
    'notunderscored': 'not replaced',
}

t = MyTemplate(template_text)
print('Modified ID pattern:')
print(t.safe_substitute(d))
```

在这个例子中,替换规则已经改变,定界符是 % 而不是 $,而且变量名中间的某个位置必须包含一个下划线。模式 %notunderscored 不会被替换为任何字符串,因为它不包含下划线字符。

```
$ python3 string_template_advanced.py

Modified ID pattern:
Delimiter : %
Replaced  : replaced
Ignored   : %notunderscored
```

要完成更复杂的修改,可以覆盖 pattern 属性并定义一个全新的正则表达式。所提供的模式必须包含 4 个命名组,分别捕获转义定界符、命名变量、加括号的变量名和不合法的定界符模式。

代码清单 1-5：**string_template_defaultpattern.py**

```
import string

t = string.Template('$var')
print(t.pattern.pattern)
```

t.pattern 的值是一个已编译正则表达式,不过可以通过它的 pattern 属性得到原来的字符串。

```
\$(?:
  (?P<escaped>\$) |              # Two delimiters
  (?P<named>[_a-z][_a-z0-9]*)       |  # Identifier
  {(?P<braced>[_a-z][_a-z0-9]*)}    |  # Braced identifier
  (?P<invalid>)                     # Ill-formed delimiter exprs
)
```

下面这个例子定义了一个新模式以创建一个新的模板类型,这里使用 {{var}} 作为变量语法。

代码清单 1.6：**string_template_newsyntax.py**

```
import re
import string


class MyTemplate(string.Template):
```

```
        delimiter = '{{'
        pattern = r'''
        \{\{(?:
        (?P<escaped>\{\{)|
        (?P<named>[_a-z][_a-z0-9]*)\}\}|
        (?P<braced>[_a-z][_a-z0-9]*)\}\}|
        (?P<invalid>)
        )
        '''

t = MyTemplate('''
{{{{
{{var}}
''')

print('MATCHES:', t.pattern.findall(t.template))
print('SUBSTITUTED:', t.safe_substitute(var='replacement'))
```

必须分别提供 named 和 braced 模式，尽管它们实际上是一样的。运行这个示例程序会生成以下输出：

```
$ python3 string_template_newsyntax.py

MATCHES: [('{{', '', '', ''), ('', 'var', '', '')]
SUBSTITUTED:
{{
replacement
```

1.1.4 Formatter

Formatter 类实现了与 str 的 format() 方法同样的布局规范语言。它的特性包括类型强制转换、对齐、属性和域引用、命名和位置模板参数以及类型特定的格式化选项。大部分情况下 format() 方法都能更便利地访问这些特性，不过也可以利用 Formatter 构建子类，以备需要改动的情况。

1.1.5 常量

string 模块包括大量与 ASCII 和数值字符集相关的常量。

代码清单 1.7: string_constants.py

```
import inspect
import string


def is_str(value):
    return isinstance(value, str)


for name, value in inspect.getmembers(string, is_str):
    if name.startswith('_'):
        continue
    print('%s=%r\n' % (name, value))
```

这些常量在处理 ASCII 数据时很有用，不过由于采用某种 Unicode 的非 ASCII 文本越来越常见，这些常量的应用也变得很有限。

```
$ python3 string_constants.py

ascii_letters='abcdefghijklmnopqrstuvwxyzABCDEFGHIJKLMNOPQRSTUVW
XYZ'

ascii_lowercase='abcdefghijklmnopqrstuvwxyz'

ascii_uppercase='ABCDEFGHIJKLMNOPQRSTUVWXYZ'

digits='0123456789'

hexdigits='0123456789abcdefABCDEF'

octdigits='01234567'

printable='0123456789abcdefghijklmnopqrstuvwxyzABCDEFGHIJKLMNOPQ
RSTUVWXYZ!"#$%&\'()*+,-./:;<=>?@[\\]^_`{|}~ \t\n\r\x0b\x0c'

punctuation='!"#$%&\'()*+,-./:;<=>?@[\\]^_`{|}~'

whitespace=' \t\n\r\x0b\x0c'
```

提示：相关阅读材料
- `string`的标准库文档[一]。
- 字符串方法[二]：`str`对象的方法，取代`string`中已经废弃的函数。
- PEP 292[三]：更简单的字符串替换。
- 格式化字符串语法[四]：`Formatter`和`str.format()`中所用布局规范语言的形式化定义。

1.2 `textwrap`：格式化文本段落

需要美观打印（pretty-printing）的情况下，可以使用`textwrap`模块格式化要输出的文本。它提供了很多文本编辑器和字处理器中都有的段落自动换行或填充特性。

1.2.1 示例数据

本节中的例子会使用模块`textwrap_example.py`，其中包含一个字符串`sample_text`。

代码清单1.8：`textwrap_example.py`

```
sample_text = '''
    The textwrap module can be used to format text for output in
    situations where pretty-printing is desired.  It offers
    programmatic functionality similar to the paragraph wrapping
    or filling features found in many text editors.
    '''
```

[一] https://docs.python.org/3.5/library/string.html
[二] https://docs.python.org/3/library/stdtypes.html#string-methods
[三] www.python.org/dev/peps/pep-0292
[四] https://docs.python.org/3.5/library/string.html#format-string-syntax

1.2.2 填充段落

`fill()`函数取文本作为输入，生成格式化文本作为输出。

代码清单 1.9：**textwrap_fill.py**

```
import textwrap
from textwrap_example import sample_text

print(textwrap.fill(sample_text, width=50))
```

结果不是太让人满意。文本现在已经左对齐，不过只有第一行保留了缩进，后面各行前面的空格则嵌在段落中。

```
$ python3 textwrap_fill.py

    The textwrap module can be used to format
text for output in     situations where pretty-
printing is desired.    It offers    programmatic
functionality similar to the paragraph wrapping
or filling features found in many text editors.
```

1.2.3 去除现有的缩进

关于前面的例子，其输出中混合嵌入了制表符和额外的空格，所以格式不太美观。用`dedent()`可以去除示例文本中所有行前面的空白符，这会生成更好的结果，并且允许在Python代码中直接使用docstring或内嵌的多行字符串，同时去除代码本身的格式。示例字符串专门加入了一级缩进来展示这个特性。

代码清单 1-10：**textwrap_dedent.py**

```
import textwrap
from textwrap_example import sample_text

dedented_text = textwrap.dedent(sample_text)
print('Dedented:')
print(dedented_text)
```

现在结果好看多了。

```
$ python3 textwrap_dedent.py

Dedented:

The textwrap module can be used to format text for output in
situations where pretty-printing is desired.  It offers
programmatic functionality similar to the paragraph wrapping
or filling features found in many text editors.
```

由于"dedent"（去除缩进）与"indent"（缩进）正好相反，所以结果将得到一个文本块，其中每一行前面的空白符已经删除。如果一行比另一行缩进更多，有些空白符则不会被删除。

以下输入：

```
▫Line▫one.
▫▫▫Line▫two.
▫Line▫three.
```

将变成:

```
Line▫one.
▫▫Line▫two.
Line▫three.
```

1.2.4 结合 dedent 和 fill

接下来，可以把去除缩进的文本传入 `fill()`，并指定一些不同的 `width` 值。

代码清单 1-11：textwrap_fill_width.py

```python
import textwrap
from textwrap_example import sample_text

dedented_text = textwrap.dedent(sample_text).strip()
for width in [45, 60]:
    print('{} Columns:\n'.format(width))
    print(textwrap.fill(dedented_text, width=width))
    print()
```

这会生成指定宽度的输出。

```
$ python3 textwrap_fill_width.py

45 Columns:

The textwrap module can be used to format
text for output in situations where pretty-
printing is desired.  It offers programmatic
functionality similar to the paragraph
wrapping or filling features found in many
text editors.

60 Columns:

The textwrap module can be used to format text for output in
situations where pretty-printing is desired.  It offers
programmatic functionality similar to the paragraph wrapping
or filling features found in many text editors.
```

1.2.5 缩进块

可以使用 `indent()` 函数为一个字符串中的所有行增加一致的前缀文本。这个例子会格式化同样的示例文本，就好像它是回复邮件中所引用的原 email 的一部分，这里使用 > 作为每一行的前缀。

代码清单 1-12：textwrap_indent.py

```python
import textwrap
from textwrap_example import sample_text

dedented_text = textwrap.dedent(sample_text)
wrapped = textwrap.fill(dedented_text, width=50)
```

```
wrapped += '\n\nSecond paragraph after a blank line.'
final = textwrap.indent(wrapped, '> ')

print('Quoted block:\n')
print(final)
```

文本块按换行符分解,将为包含文本的各行增加前缀,然后再把这些行合并为一个新字符串并返回。

```
$ python3 textwrap_indent.py

Quoted block:
>  The textwrap module can be used to format text
> for output in situations where pretty-printing is
> desired.  It offers programmatic functionality
> similar to the paragraph wrapping or filling
> features found in many text editors.

> Second paragraph after a blank line.
```

为了控制哪些行接收新前缀,可以传入一个 callable 对象作为 `indent()` 的 `predicate` 参数。会依次为各行文本调用这个 callable,并为返回值为 true 的行添加前缀。

代码清单 1-13: **textwrap_indent_predicate.py**

```
import textwrap
from textwrap_example import sample_text


def should_indent(line):
    print('Indent {!r}?'.format(line))
    return len(line.strip()) % 2 == 0


dedented_text = textwrap.dedent(sample_text)
wrapped = textwrap.fill(dedented_text, width=50)
final = textwrap.indent(wrapped, 'EVEN ',
                        predicate=should_indent)

print('\nQuoted block:\n')
print(final)
```

这个例子会为包含偶数个字符的行添加前缀 EVEN。

```
$ python3 textwrap_indent_predicate.py

Indent ' The textwrap module can be used to format text\n'?
Indent 'for output in situations where pretty-printing is\n'?
Indent 'desired.  It offers programmatic functionality\n'?
Indent 'similar to the paragraph wrapping or filling\n'?
Indent 'features found in many text editors.'?

Quoted block:

EVEN  The textwrap module can be used to format text
for output in situations where pretty-printing is
EVEN desired.  It offers programmatic functionality
EVEN similar to the paragraph wrapping or filling
EVEN features found in many text editors.
```

1.2.6 悬挂缩进

不仅可以设置输出的宽度,还可以采用同样的方式单独控制首行的缩进,使首行的缩进不同于后续的各行。

代码清单 1-14:**textwrap_hanging_indent.py**

```
import textwrap
from textwrap_example import sample_text

dedented_text = textwrap.dedent(sample_text).strip()
print(textwrap.fill(dedented_text,
                    initial_indent='',
                    subsequent_indent=' ' * 4,
                    width=50,
                    ))
```

这就允许生成悬挂缩进,即首行缩进小于其他行的缩进。

```
$ python3 textwrap_hanging_indent.py

The textwrap module can be used to format text for
    output in situations where pretty-printing is
    desired.  It offers programmatic functionality
    similar to the paragraph wrapping or filling
    features found in many text editors.
```

缩进值也可以包含非空白字符。例如,悬挂缩进可以加前缀 * 来生成圆点项目符号。

1.2.7 截断长文本

可以使用 shorten() 截断文本来创建一个小结或预览。所有现有的空白符(如制表符、换行符以及多个空格组成的序列)都会被规范化为一个空格。然后文本将被截断为小于或等于指定的长度,而且会在单词边界之间截断,避免包含不完整的单词。

代码清单 1-15:**textwrap_shorten.py**

```
import textwrap
from textwrap_example import sample_text

dedented_text = textwrap.dedent(sample_text)
original = textwrap.fill(dedented_text, width=50)

print('Original:\n')
print(original)
shortened = textwrap.shorten(original, 100)
shortened_wrapped = textwrap.fill(shortened, width=50)

print('\nShortened:\n')
print(shortened_wrapped)
```

如果从原文本将非空白文本作为截断部分删除,那么它会被替换为一个占位值。可以为 shorten() 提供参数 placeholder 来替代默认值 [...]。

```
$ python3 textwrap_shorten.py

Original:

 The textwrap module can be used to format text
for output in situations where pretty-printing is
desired.  It offers programmatic functionality
similar to the paragraph wrapping or filling
features found in many text editors.

Shortened:

The textwrap module can be used to format text for
output in situations where pretty-printing [...]
```

提示：相关阅读材料
- `textwrap` 的标准库文档⊖。

1.3 re：正则表达式

正则表达式（regular expression）是用一种形式化语法描述的文本匹配模式。模式会被解释为一组指令，然后执行这些指令并提供一个字符串作为输入，将生成一个匹配子集或者生成原字符串的一个修改版本。"正则表达式"在讨论中经常简写为"regex"或"regexp"。表达式可以包含字面量文本匹配、重复、模式组合、分支和其他复杂的规则。与创建一个特定用途的词法分析器和解析器相比，利用正则表达式，很多解析问题可以更容易地得到解决。

正则表达式通常用于涉及大量文本处理的应用中。例如，在开发人员使用的文本编辑程序中，包括 vi、emacs 和其他现代 IDE，通常使用正则表达式作为搜索模式。正则表达式也是 UNIX 命令行工具的一个不可缺少的组成部分，如 sed、grep 和 awk。很多编程语言在语言语法中都包含了对正则表达式的支持（如 Perl、Ruby、Awk 和 Tcl）；另外一些语言（如 C、C++ 和 Python）则通过扩展库来支持正则表达式。

已经有很多开源的正则表达式实现，它们有一个共同的核心语法，不过提供了不同的扩展或修改来支持其高级特性。Python 的 `re` 模块中使用的语法以 Perl 使用的正则表达式语法为基础，另外还有一些 Python 特定的改进。

说明：尽管"正则表达式"的正式定义仅限于描述正则语言的表达式，不过 `re` 支持的一些扩展不只是能描述正则语言。这里使用的"正则表达式"有更一般的含义，是指可以用 Python 的 `re` 模块计算的所有表达式。

1.3.1 查找文本中的模式

`re` 最常见的用法是搜索文本中的模式。`search()` 函数取模式和要扫描的文本作为输入，找到这个模式时，就返回一个 `Match` 对象。如果没有找到模式，`search()` 将返回 `None`。每个 `Match` 对象包含有关匹配性质的信息，包括原输入字符串、所使用的正则表达式

⊖ https://docs.python.org/3.5/library/textwrap.html

以及模式在原字符串中出现的位置。

代码清单 1-16：`re_simple_match.py`

```python
import re

pattern = 'this'
text = 'Does this text match the pattern?'

match = re.search(pattern, text)

s = match.start()
e = match.end()

print('Found "{}"\nin "{}"\nfrom {} to {} ("{}")'.format(
    match.re.pattern, match.string, s, e, text[s:e]))
```

`start()` 和 `end()` 方法可以提供字符串中的相应索引，指示与模式匹配的文本在字符串中出现的位置。

```
$ python3 re_simple_match.py

Found "this"
in "Does this text match the pattern?"
from 5 to 9 ("this")
```

1.3.2 编译表达式

尽管 `re` 包括模块级函数，可以处理作为文本字符串的正则表达式，但是对于程序频繁使用的表达式而言，编译它们会更为高效。`compile()` 函数会把一个表达式字符串转换为一个 `RegexObject`。

代码清单 1-17：`re_simple_compiled.py`

```python
import re

# Precompile the patterns.
regexes = [
    re.compile(p)
    for p in ['this', 'that']
]
text = 'Does this text match the pattern?'

print('Text: {!r}\n'.format(text))

for regex in regexes:
    print('Seeking "{}" ->'.format(regex.pattern),
          end=' ')

    if regex.search(text):
        print('match!')
    else:
        print('no match')
```

模块级函数会维护一个包含已编译表达式的缓存，不过这个缓存的大小是有限的，另外直接使用已编译表达式可以避免与缓存查找相关的开销。使用已编译表达式的另一个好处是，通过在加载模块时预编译所有表达式，可以把编译工作转移到应用开始时，而不是当程序响应一个用户动作时才编译。

```
$ python3 re_simple_compiled.py

Text: 'Does this text match the pattern?'

Seeking "this" -> match!
Seeking "that" -> no match
```

1.3.3 多重匹配

到目前为止，示例模式都只是使用 `search()` 来查找字面量文本字符串的单个实例。`findall()` 函数会返回输入中与模式匹配而且不重叠的所有子串。

代码清单 1-18：**re_findall.py**

```python
import re

text = 'abbaaabbbbaaaaa'

pattern = 'ab'

for match in re.findall(pattern, text):
    print('Found {!r}'.format(match))
```

这个输入字符串中有 **ab** 的两个实例。

```
$ python3 re_findall.py

Found 'ab'
Found 'ab'
```

`finditer()` 返回一个迭代器，它会生成 Match 实例，而不是像 `findall()` 那样返回字符串。

代码清单 1.19：**re_finditer.py**

```python
import re

text = 'abbaaabbbbaaaaa'

pattern = 'ab'

for match in re.finditer(pattern, text):
    s = match.start()
    e = match.end()
    print('Found {!r} at {:d}:{:d}'.format(
        text[s:e], s, e))
```

这个例子同样会找到 **ab** 的两次出现，Match 实例显示了它们在原输入字符串中出现的位置。

```
$ python3 re_finditer.py

Found 'ab' at 0:2
Found 'ab' at 5:7
```

1.3.4 模式语法

除了简单的字面量文本字符串，正则表达式还支持更强大的模式。模式可以重复，可以锚定到输入中不同的逻辑位置，可以用紧凑的形式表述而不需要在模式中提供每一个字面量字符。可以结合字面量文本值和元字符来使用所有这些特性，元字符是 re 实现的正则表达式模式语法的一部分。

代码清单 1.20：re_test_patterns.py

```python
import re


def test_patterns(text, patterns):
    """Given source text and a list of patterns, look for
    matches for each pattern within the text and print
    them to stdout.
    """
    # Look for each pattern in the text and print the results.
    for pattern, desc in patterns:
        print("'{}' ({})\n".format(pattern, desc))
        print("  '{}'".format(text))
        for match in re.finditer(pattern, text):
            s = match.start()
            e = match.end()
            substr = text[s:e]
            n_backslashes = text[:s].count('\\')
            prefix = '.' * (s + n_backslashes)
            print("  {}'{}'".format(prefix, substr))
        print()
    return


if __name__ == '__main__':
    test_patterns('abbaaabbbbaaaaa',
                  [('ab', "'a' followed by 'b'"),
                   ])
```

下面的例子使用 test_patterns() 来研究模式的变化会如何改变与相同输入文本的匹配结果。输出显示了输入文本以及输入中与模式匹配的各个部分的子串区间。

```
$ python3 re_test_patterns.py

'ab' ('a' followed by 'b')

  'abbaaabbbbaaaaa'
  'ab'
  .....'ab'
```

1.3.4.1 重复

模式中有 5 种表示重复的方法。模式后面如果有元字符 *，则表示重复 0 次或多次（允许一个模式重复 0 次是指这个模式即使不出现也可以匹配）。如果把 * 替换为 +，那么模式必须至少出现 1 次才能匹配。使用 ? 表示模式出现 0 次或 1 次。如果要指定出现次数，需要在模式后面使用 {m}，这里 m 是模式应重复的次数。最后，如果要允许一个可变但有限的重复次数，那么可以使用 {m,n}，这里 m 是最小重复次数，n 是最大重复次数。如果省略 n（{m,}），则表示值必须至少出现 m 次，但没有最大限制。

代码清单 1-21：re_repetition.py

```
from re_test_patterns import test_patterns

test_patterns(
    'abbaabbba',
    [('ab*', 'a followed by zero or more b'),
     ('ab+', 'a followed by one or more b'),
     ('ab?', 'a followed by zero or one b'),
     ('ab{3}', 'a followed by three b'),
     ('ab{2,3}', 'a followed by two to three b')],
)
```

在这个例子中，ab∗ 和 ab? 的匹配要多于 ab+ 的匹配。

```
$ python3 re_repetition.py

'ab*' (a followed by zero or more b)

  'abbaabbba'
  'abb'
  ...'a'
  ....'abbb'
  ........'a'

'ab+' (a followed by one or more b)

  'abbaabbba'
  'abb'
  ....'abbb'

'ab?' (a followed by zero or one b)

  'abbaabbba'
  'ab'
  ...'a'
  ....'ab'
  ........'a'

'ab{3}' (a followed by three b)

  'abbaabbba'
  ....'abbb'

'ab{2,3}' (a followed by two to three b)

  'abbaabbba'
  'abb'
  ....'abbb'
```

处理重复指令时，re 在匹配模式时通常会尽可能多地消费输入。这种所谓的"贪心"行为可能会导致单个匹配减少，或者匹配结果可能包含比预想更多的输入文本。可以在重复指令后面加 ? 来关闭这种贪心行为。

代码清单 1-22：re_repetition_non_greedy.py

```
from re_test_patterns import test_patterns

test_patterns(
    'abbaabbba',
```

```
    [('ab*?', 'a followed by zero or more b'),
     ('ab+?', 'a followed by one or more b'),
     ('ab??', 'a followed by zero or one b'),
     ('ab{3}?', 'a followed by three b'),
     ('ab{2,3}?', 'a followed by two to three b')],
)
```

对于允许 b 出现 0 次的模式，如果消费输入时禁用贪心行为，那么这意味着匹配的子串不会包含任何 b 字符。

```
$ python3 re_repetition_non_greedy.py

'ab*?' (a followed by zero or more b)

  'abbaabbba'
  'a'
  ...'a'
  ....'a'
  ........'a'

'ab+?' (a followed by one or more b)

  'abbaabbba'
  'ab'
  ....'ab'

'ab??' (a followed by zero or one b)

  'abbaabbba'
  'a'
  ...'a'
  ....'a'
  ........'a'

'ab{3}?' (a followed by three b)

  'abbaabbba'
  ....'abbb'

'ab{2,3}?' (a followed by two to three b)

  'abbaabbba'
  'abb'
  ....'abb'
```

1.3.4.2 字符集

字符集（character set）是一组字符，包含可以与模式中当前位置匹配的所有字符。例如，[ab] 可以匹配 a 或 b。

代码清单 1-23：re_charset.py

```
from re_test_patterns import test_patterns

test_patterns(
    'abbaabbba',
    [('[ab]', 'either a or b'),
     ('a[ab]+', 'a followed by 1 or more a or b'),
     ('a[ab]+?', 'a followed by 1 or more a or b, not greedy')],
)
```

贪心形式的表达式（a[ab]+）会消费整个字符串，因为第一个字母是a，而且后续的各个字符要么是a要么是b。

```
$ python3 re_charset.py

'[ab]' (either a or b)

  'abbaabbba'
  'a'
  .'b'
  ..'b'
  ...'a'
  ....'a'
  .....'b'
  ......'b'
  .......'b'
  ........'a'

'a[ab]+' (a followed by 1 or more a or b)

  'abbaabbba'
  'abbaabbba'

'a[ab]+?' (a followed by 1 or more a or b, not greedy)

  'abbaabbba'
  'ab'
  ...'aa'
```

字符集还可以用来排除特定的字符。尖字符（^）意味着要查找不在这个尖字符后面的集合中的字符。

代码清单1-24：re_charset_exclude.py

```
from re_test_patterns import test_patterns

test_patterns(
    'This is some text -- with punctuation.',
    [('[^-. ]+', 'sequences without -, ., or space')],
)
```

这个模式会查找不包含字符 -、. 或空格的所有子串。

```
$ python3 re_charset_exclude.py

'[^-. ]+' (sequences without -, ., or space)

  'This is some text -- with punctuation.'
  'This'
  .....'is'
  ........'some'
  .............'text'
  ....................'with'
  .........................'punctuation'
```

随着字符集变得更大，键入每一个应当（或不应当）匹配的字符会变得很麻烦。可以使用一种更简洁的格式，利用字符区间（character range）来定义一个字符集，包含指定的起点和终点之间所有连续的字符。

代码清单 1-25：`re_charset_ranges.py`

```python
from re_test_patterns import test_patterns

test_patterns(
    'This is some text -- with punctuation.',
    [('[a-z]+', 'sequences of lowercase letters'),
     ('[A-Z]+', 'sequences of uppercase letters'),
     ('[a-zA-Z]+', 'sequences of lower- or uppercase letters'),
     ('[A-Z][a-z]+', 'one uppercase followed by lowercase')],
)
```

这里区间 a-z 包含小写 ASCII 字母，区间 A-Z 包含大写 ASCII 字母。还可以把这些区间结合到一个字符集中。

```
$ python3 re_charset_ranges.py

'[a-z]+' (sequences of lowercase letters)

  'This is some text -- with punctuation.'
  .'his'
  .....'is'
  ........'some'
  .............'text'
  ....................'with'
  ........................'punctuation'

'[A-Z]+' (sequences of uppercase letters)

  'This is some text -- with punctuation.'
  'T'

'[a-zA-Z]+' (sequences of lower- or uppercase letters)

  'This is some text -- with punctuation.'
  'This'
  .....'is'
  ........'some'
  .............'text'
  ....................'with'
  ........................'punctuation'

'[A-Z][a-z]+' (one uppercase followed by lowercase)

  'This is some text -- with punctuation.'
  'This'
```

作为字符集的一种特殊情况，元字符点号（.）指示模式应当匹配该位置的单个字符。

代码清单 1-26：`re_charset_dot.py`

```python
from re_test_patterns import test_patterns

test_patterns(
    'abbaabbba',
    [('a.', 'a followed by any one character'),
     ('b.', 'b followed by any one character'),
     ('a.*b', 'a followed by anything, ending in b'),
     ('a.*?b', 'a followed by anything, ending in b')],
)
```

结合点号和重复可以得到非常长的匹配,除非使用了非贪心形式。

```
$ python3 re_charset_dot.py

'a.' (a followed by any one character)

  'abbaabbba'
  'ab'
  ...'aa'

'b.' (b followed by any one character)

  'abbaabbba'
  .'bb'
  .....'bb'
  .......'ba'

'a.*b' (a followed by anything, ending in b)

  'abbaabbba'
  'abbaabbb'

'a.*?b' (a followed by anything, ending in b)

  'abbaabbba'
  'ab'
  ...'aab'
```

1.3.4.3 转义码

一种更简洁的表示是对一些预定义的字符集使用转义码。re 可识别的转义码如表 1-1 所示。

表 1-1 正则表达式转义码

转义码	含义
\d	数字
\D	非数字
\s	空白符(制表符、空格、换行等)
\S	非空白符
\w	字母数字
\W	非字母数字

说明:可以在字符前加一个反斜线(\)来指示转义。遗憾的是,反斜线本身在正常的 Python 字符串中也必须转义,这就会带来很难读的表达式。通过使用原始(raw)字符串可以消除这个问题,要保证可读性,可以在字面值前加 r 前缀来创建原始字符串。

代码清单 1-27:re_escape_codes.py

```
from re_test_patterns import test_patterns

test_patterns(
    'A prime #1 example!',
    [(r'\d+', 'sequence of digits'),
     (r'\D+', 'sequence of non-digits'),
```

```
    (r'\s+', 'sequence of whitespace'),
    (r'\S+', 'sequence of non-whitespace'),
    (r'\w+', 'alphanumeric characters'),
    (r'\W+', 'non-alphanumeric')],
)
```

这些示例表达式结合了转义码和重复来查找输入字符串中的类似字符序列。

```
$ python3 re_escape_codes.py

'\d+' (sequence of digits)

  'A prime #1 example!'
  .........'1'

'\D+' (sequence of non-digits)

  'A prime #1 example!'
  'A prime #'
  ..........' example!'

'\s+' (sequence of whitespace)

  'A prime #1 example!'
  .' '
  .......' '
  ..........' '

'\S+' (sequence of non-whitespace)

  'A prime #1 example!'
  'A'
  ..'prime'
  ........'#1'
  ...........'example!'

'\w+' (alphanumeric characters)

  'A prime #1 example!'
  'A'
  ..'prime'
  .........'1'
  ...........'example'

'\W+' (non-alphanumeric)

  'A prime #1 example!'
  .' '
  .......' #'
  ..........' '
  .................'!'
```

要匹配正则表达式语法中包含的字符，需要转义搜索模式中的字符。

代码清单 1-28：re_escape_escapes.py

```
from re_test_patterns import test_patterns

test_patterns(
    r'\d+ \D+ \s+',
    [(r'\\.\+', 'escape code')],
)
```

这个例子中的模式对反斜线和加号字符进行转义,因为这两个字符都是元字符,在正则表达式中有特殊的含义。

```
$ python3 re_escape_escapes.py

'\\.\+' (escape code)

  '\d+ \D+ \s+'
  '\d+'
  .....'\D+'
  ..........'\s+'
```

1.3.4.4 锚定

除了描述要匹配的模式的内容之外,还可以使用锚定指令指定模式在输入文本中的相对位置。表 1-2 列出了合法的锚定码。

代码清单 1-29: `re_anchoring.py`

```
from re_test_patterns import test_patterns
test_patterns(
    'This is some text -- with punctuation.',
    [(r'^\w+', 'word at start of string'),
     (r'\A\w+', 'word at start of string'),
     (r'\w+\S*$', 'word near end of string'),
     (r'\w+\S*\Z', 'word near end of string'),
     (r'\w*t\w*', 'word containing t'),
     (r'\bt\w+', 't at start of word'),
     (r'\w+t\b', 't at end of word'),
     (r'\Bt\B', 't, not start or end of word')],
)
```

这个例子中,匹配字符串开头和末尾单词的模式是不同的,因为字符串末尾的单词后面有结束句子的标点符号。模式 \w+$ 不能匹配,因为 . 不能被认为是一个字母数字字符。

表 1-2 正则表达式锚定码

锚定码	含 义
^	字符串或行的开头
$	字符串或行末尾
\A	字符串开头
\Z	字符串末尾
\b	单词开头或末尾的空串
\B	不在单词开头或末尾的空串

```
$ python3 re_anchoring.py

'^\w+' (word at start of string)

  'This is some text -- with punctuation.'
  'This'

'\A\w+' (word at start of string)
```

```
'This is some text -- with punctuation.'
'This'
```

`'\w+\S*$'` (word near end of string)

```
'This is some text -- with punctuation.'
.........................'punctuation.'
```

`'\w+\S*\Z'` (word near end of string)

```
'This is some text -- with punctuation.'
.........................'punctuation.'
```

`'\w*t\w*'` (word containing t)

```
'This is some text -- with punctuation.'
.............'text'
....................'with'
.............................'punctuation'
```

`'\bt\w+'` (t at start of word)

```
'This is some text -- with punctuation.'
.............'text'
```

`'\w+t\b'` (t at end of word)

```
'This is some text -- with punctuation.'
.............'text'
```

`'\Bt\B'` (t, not start or end of word)

```
'This is some text -- with punctuation.'
.......................'t'
............................'t'
...............................'t'
```

1.3.5 限制搜索

有些情况下，可以提前知道只需要搜索整个输入的一个子集，在这些情况下，可以告诉 re 限制搜索范围从而进一步约束正则表达式匹配。例如，如果模式必须出现在输入开头，那么使用 `match()` 而不是 `search()` 会锚定搜索，而不必显式地在搜索模式中包含一个锚。

代码清单 1-30：**re_match.py**

```
import re

text = 'This is some text -- with punctuation.'
pattern = 'is'

print('Text    :', text)
print('Pattern:', pattern)

m = re.match(pattern, text)
print('Match  :', m)
s = re.search(pattern, text)
print('Search :', s)
```

由于字面量文本 `is` 未出现在输入文本的开头,因此使用 `match()` 时找不到它。不过,这个序列在文本中另外还出现了两次,所以 `search()` 能找到。

```
$ python3 re_match.py

Text    : This is some text -- with punctuation.
Pattern : is
Match   : None
Search  : <_sre.SRE_Match object; span=(2, 4), match='is'>
```

`fullmatch()` 方法要求整个输入字符串与模式匹配。

代码清单 1-31:<code>re_fullmatch.py</code>

```python
import re

text = 'This is some text -- with punctuation.'
pattern = 'is'

print('Text      :', text)
print('Pattern   :', pattern)

m = re.search(pattern, text)
print('Search    :', m)
s = re.fullmatch(pattern, text)
print('Full match :', s)
```

这里 `search()` 显示模式确实出现在输入中,但是它没能消费所有输入,所以 `fullmatch()` 不会报告匹配。

```
$ python3 re_fullmatch.py

Text       : This is some text -- with punctuation.
Pattern    : is
Search     : <_sre.SRE_Match object; span=(2, 4), match='is'>
Full match : None
```

已编译正则表达式的 `search()` 方法还接受可选的 `start` 和 `end` 位置参数,将搜索范围限制到输入的一个子串中。

代码清单 1-32:<code>re_search_substring.py</code>

```python
import re

text = 'This is some text -- with punctuation.'
pattern = re.compile(r'\b\w*is\w*\b')

print('Text:', text)
print()

pos = 0
while True:
    match = pattern.search(text, pos)
    if not match:
        break
    s = match.start()
    e = match.end()
    print('  {:>2d} : {:>2d} = "{}"'.format(
        s, e - 1, text[s:e]))
    # Move forward in text for the next search.
    pos = e
```

这个例子实现了 `iterall()` 的一种不太高效的形式。每次找到一个匹配时，这个匹配的结束位置将用于下一个搜索。

```
$ python3 re_search_substring.py

Text: This is some text -- with punctuation.

0 :  3 = "This"
5 :  6 = "is"
```

1.3.6 用组解析匹配

搜索模式匹配是正则表达式强大能力的基础。为模式提供组可以隔离匹配文本的各个部分，以扩展这些功能来创建一个解析器。可以用小括号包围模式来定义组。

代码清单 1-33：**re_groups.py**

```
from re_test_patterns import test_patterns

test_patterns(
    'abbaaabbbbaaaaa',
    [('a(ab)', 'a followed by literal ab'),
     ('a(a*b*)', 'a followed by 0-n a and 0-n b'),
     ('a(ab)*', 'a followed by 0-n ab'),
     ('a(ab)+', 'a followed by 1-n ab')],
)
```

可以把完整的正则表达式转换为一个组，并嵌入到一个更大的表达式中。所有重复修饰符都可以应用到整个组，要求整个组模式重复。

```
$ python3 re_groups.py

'a(ab)' (a followed by literal ab)

  'abbaaabbbbaaaaa'
  ....'aab'

'a(a*b*)' (a followed by 0-n a and 0-n b)

  'abbaaabbbbaaaaa'
  'abb'
  ...'aaabbbb'
  ..........'aaaaa'

'a(ab)*' (a followed by 0-n ab)

  'abbaaabbbbaaaaa'
  'a'
  ...'a'
  ....'aab'
  ..........'a'
  ...........'a'
  ............'a'
  .............'a'
  ..............'a'

'a(ab)+' (a followed by 1-n ab)

  'abbaaabbbbaaaaa'
  ....'aab'
```

要访问与模式中各个组匹配的子串，可以使用 match 对象的 groups() 方法。

代码清单 1-34：re_groups_match.py

```
import re

text = 'This is some text -- with punctuation.'
print(text)
print()

patterns = [
    (r'^(\w+)', 'word at start of string'),
    (r'(\w+)\S*$', 'word at end, with optional punctuation'),
    (r'(\bt\w+)\W+(\w+)', 'word starting with t, another word'),
    (r'(\w+t)\b', 'word ending with t'),
]
for pattern, desc in patterns:
    regex = re.compile(pattern)
    match = regex.search(text)
    print("'{}' ({})\n".format(pattern, desc))
    print('  ', match.groups())
    print()
```

match.groups() 按匹配字符串的组在表达式中的顺序返回一个字符串序列。

```
$ python3 re_groups_match.py

This is some text -- with punctuation.

'^(\w+)' (word at start of string)

   ('This',)

'(\w+)\S*$' (word at end, with optional punctuation)

   ('punctuation',)

'(\bt\w+)\W+(\w+)' (word starting with t, another word)

   ('text', 'with')

'(\w+t)\b' (word ending with t)

   ('text',)
```

要访问单个组的匹配，可以使用 group() 方法。当使用组查找字符串的各个部分时，有些部分尽管与组匹配但在结果中并不需要，此时 group() 方法就很有用。

代码清单 1-35：re_groups_individual.py

```
import re

text = 'This is some text -- with punctuation.'

print('Input text            :', text)

# Word starting with 't' then another word
regex = re.compile(r'(\bt\w+)\W+(\w+)')
```

```
print('Pattern              :', regex.pattern)
match = regex.search(text)
print('Entire match         :', match.group(0))
print('Word starting with "t":', match.group(1))
print('Word after "t" word   :', match.group(2))
```

组 0 表示与整个表达式匹配的字符串,子组按其左括号在表达式中出现的顺序编号,从 1 开始。

```
$ python3 re_groups_individual.py

Input text              : This is some text -- with punctuation.
Pattern                 : (\bt\w+)\W+(\w+)
Entire match            : text -- with
Word starting with "t": text
Word after "t" word     : with
```

Python 扩展了基本组语法,还增加了命名组。通过使用名字来指示组可以更容易地修改模式,而不必同时修改使用了匹配结果的代码。要设置一个组的名字,可以使用语法 (?P<name>pattern)。

代码清单 1-36:**re_groups_named.py**

```
import re

text = 'This is some text -- with punctuation.'

print(text)
print()

patterns = [
    r'^(?P<first_word>\w+)',
    r'(?P<last_word>\w+)\S*$',
    r'(?P<t_word>\bt\w+)\W+(?P<other_word>\w+)',
    r'(?P<ends_with_t>\w+t)\b',
]

for pattern in patterns:
    regex = re.compile(pattern)
    match = regex.search(text)
    print("'{}'".format(pattern))
    print('  ', match.groups())
    print('  ', match.groupdict())
    print()
```

可以使用 `groupdict()` 获取一个字典,它将组名映射为匹配的子串。命名模式也包含在 `groups()` 返回的有序序列中。

```
$ python3 re_groups_named.py

This is some text -- with punctuation.

'^(?P<first_word>\w+)'
   ('This',)
   {'first_word': 'This'}

'(?P<last_word>\w+)\S*$'
```

```
    ('punctuation',)
    {'last_word': 'punctuation'}

'(?P<t_word>\bt\w+)\W+(?P<other_word>\w+)'
    ('text', 'with')
    {'t_word': 'text', 'other_word': 'with'}

'(?P<ends_with_t>\w+t)\b'
    ('text',)
    {'ends_with_t': 'text'}
```

更新的 `test_patterns()` 会显示一个模式匹配的编号组和命名组，使后面的例子更容易理解。

代码清单 1-37: **re_test_patterns_groups.py**

```python
import re


def test_patterns(text, patterns):
    """Given source text and a list of patterns, look for
    matches for each pattern within the text and print
    them to stdout.
    """
    # Look for each pattern in the text and print the results.
    for pattern, desc in patterns:
        print('{!r} ({})\n'.format(pattern, desc))
        print('  {!r}'.format(text))
        for match in re.finditer(pattern, text):
            s = match.start()
            e = match.end()
            prefix = ' ' * (s)
            print(
                '  {}{!r}{} '.format(prefix,
                                     text[s:e],
                                     ' ' * (len(text) - e)),
                end=' ',
            )
            print(match.groups())
            if match.groupdict():
                print('{}{}'.format(
                    ' ' * (len(text) - s),
                    match.groupdict()),
                )
        print()
    return
```

组本身是一个完整的正则表达式，可以嵌套在其他组中来创建更复杂的表达式。

代码清单 1-38: **re_groups_nested.py**

```python
from re_test_patterns_groups import test_patterns

test_patterns(
    'abbaabbba',
    [(r'a((a*)(b*))', 'a followed by 0-n a and 0-n b')],
)
```

在这里，组 `(a*)` 匹配一个空串，所以 `groups()` 的返回值包含空串作为匹配值。

```
$ python3 re_groups_nested.py

'a((a*)(b*))' (a followed by 0-n a and 0-n b)

  'abbaabbba'
  'abb'           ('bb', '', 'bb')
    'aabbb'       ('abbb', 'a', 'bbb')
        'a'       ('', '', '')
```

组还可以用于指定替代模式。可以使用管道符号（|）指示应当匹配某一个模式。不过，要仔细考虑管道符号的放置位置。下面这个例子中的第一个表达式匹配一个 a 序列，该序列后面跟着一个完全由某一个字母（a 或 b）组成的序列。第二个表达式匹配 a，其后面跟着一个可能包含 a 或 b 的序列。模式很相似，但是得到的匹配完全不同。

代码清单 1-39：re_groups_alternative.py

```
from re_test_patterns_groups import test_patterns

test_patterns(
    'abbaabbba',
    [(r'a((a+)|(b+))', 'a then seq. of a or seq. of b'),
     (r'a((a|b)+)', 'a then seq. of [ab]')],
)
```

如果一个替代组不匹配，但是整个模式确实匹配，那么 groups() 的返回值会在序列中本应出现替代组的位置包含一个 None 值。

```
$ python3 re_groups_alternative.py

'a((a+)|(b+))' (a then seq. of a or seq. of b)

  'abbaabbba'
  'abb'           ('bb', None, 'bb')
    'aa'          ('a', 'a', None)

'a((a|b)+)' (a then seq. of [ab])

  'abbaabbba'
  'abbaabbba'    ('bbaabbba', 'a')
```

如果匹配子模式的字符串不必从整个文本中抽取出来，那么在这种情况下，定义包含子模式的组也很有用。这些组被称为非捕获组（non-capturing）。非捕获组可以用来描述重复模式或替代，而不会隔离返回值中字符串的匹配部分。可以使用语法 (?:pattern) 创建一个非捕获组。

代码清单 1-40：re_groups_noncapturing.py

```
from re_test_patterns_groups import test_patterns

test_patterns(
    'abbaabbba',
    [(r'a((a+)|(b+))', 'capturing form'),
     (r'a((?:a+)|(?:b+))', 'noncapturing')],
)
```

尽管一个模式的捕获和非捕获形式会匹配得到相同的结果，但它们会返回不同的组，下面来做一个比较。

```
$ python3 re_groups_noncapturing.py

'a((a+)|(b+))' (capturing form)

  'abbaabbba'
  'abb'          ('bb', None, 'bb')
     'aa'        ('a', 'a', None)

'a((?:a+)|(?:b+))' (noncapturing)

  'abbaabbba'
  'abb'          ('bb',)
     'aa'        ('a',)
```

1.3.7 搜索选项

选项标志用来改变匹配引擎处理表达式的方式。可以使用位或（OR）操作结合这些标志，然后传递到 compile()、search()、match() 和其他接受搜索模式的函数。

1.3.7.1 大小写无关的匹配

IGNORECASE 会使模式中的字面量字符以及字符区间与大小写字符都匹配。

代码清单 1-41：**re_flags_ignorecase.py**

```
import re

text = 'This is some text -- with punctuation.'
pattern = r'\bT\w+'
with_case = re.compile(pattern)
without_case = re.compile(pattern, re.IGNORECASE)

print('Text:\n  {!r}'.format(text))
print('Pattern:\n  {}'.format(pattern))
print('Case-sensitive:')
for match in with_case.findall(text):
    print('  {!r}'.format(match))
print('Case-insensitive:')
for match in without_case.findall(text):
    print('  {!r}'.format(match))
```

由于这个模式包括字面量 T，因此若没有设置 IGNORECASE，则唯一的匹配是单词 This。忽略大小写时，text 也会匹配。

```
$ python3 re_flags_ignorecase.py

Text:
  'This is some text -- with punctuation.'
Pattern:
  \bT\w+
Case-sensitive:
  'This'
Case-insensitive:
  'This'
  'text'
```

1.3.7.2 多行输入

有两个标志会影响如何完成多行输入中的搜索：MULTILINE 和 DOTALL。MULTILINE

标志控制模式匹配代码如何对包含换行符的文本处理锚定指令。打开多行模式时，对 ^ 和 $ 的锚定规则除了应用于整个字符串之外，还会在各行的开头和末尾应用。

代码清单 1-42：`re_flags_multiline.py`

```python
import re

text = 'This is some text -- with punctuation.\nA second line.'
pattern = r'(^\w+)|(\w+\S*$)'
single_line = re.compile(pattern)
multiline = re.compile(pattern, re.MULTILINE)

print('Text:\n  {!r}'.format(text))
print('Pattern:\n  {}'.format(pattern))
print('Single Line :')
for match in single_line.findall(text):
    print('  {!r}'.format(match))
print('Multline    :')
for match in multiline.findall(text):
    print('  {!r}'.format(match))
```

这个例子中的模式可以匹配输入的第一个单词或最后一个单词。它会匹配字符串末尾的 line.，尽管这里没有换行符。

```
$ python3 re_flags_multiline.py

Text:
  'This is some text -- with punctuation.\nA second line.'
Pattern:
  (^\w+)|(\w+\S*$)
Single Line :
  ('This', '')
  ('', 'line.')
Multline    :
  ('This', '')
  ('', 'punctuation.')
  ('A', '')
  ('', 'line.')
```

DOTALL 是另一个与多行文本有关的标志。正常情况下，点字符（.）可以匹配输入文本中除换行符以外的所有字符。这个标志允许点字符还可以匹配换行符。

代码清单 1-43：`re_flags_dotall.py`

```python
import re

text = 'This is some text -- with punctuation.\nA second line.'
pattern = r'.+'
no_newlines = re.compile(pattern)
dotall = re.compile(pattern, re.DOTALL)

print('Text:\n  {!r}'.format(text))
print('Pattern:\n  {}'.format(pattern))
print('No newlines :')
for match in no_newlines.findall(text):
    print('  {!r}'.format(match))
print('Dotall      :')
for match in dotall.findall(text):
    print('  {!r}'.format(match))
```

如果没有这个标志，输入文本的各行会单独匹配模式。增加了这个标志则会消费整个字符串。

```
$ python3 re_flags_dotall.py

Text:
  'This is some text -- with punctuation.\nA second line.'
Pattern:
  .+
No newlines :
  'This is some text -- with punctuation.'
  'A second line.'
Dotall       :
  'This is some text -- with punctuation.\nA second line.'
```

1.3.7.3 Unicode

在 Python 3 中，str 对象使用完整的 Unicode 字符集，str 的正则表达式处理会假设模式和输入文本都是 Unicode。之前描述的转义码默认的也是按 Unicode 定义。这些假设意味着模式 \w+ 对单词"French"和"Français"都能匹配。要像 Python 2 中的默认假设那样将转义码限制到 ASCII 字符集，编译模式或者调用模块级函数 search() 和 match() 时要使用 ASCII 标志。

代码清单 1-44：**re_flags_ascii.py**

```
import re

text = u'Français łzoty Österreich'
pattern = r'\w+'
ascii_pattern = re.compile(pattern, re.ASCII)
unicode_pattern = re.compile(pattern)

print('Text    :', text)
print('Pattern :', pattern)
print('ASCII   :', list(ascii_pattern.findall(text)))
print('Unicode :', list(unicode_pattern.findall(text)))
```

其他转义序列（\W、\b、\B、\d、\D、\s 和 \S）对 ASCII 文本也会采用不同的处理方式。re 不再通过查询 Unicode 数据库来查找各个字符的属性，而是使用转义序列标识的字符集的 ASCII 定义。

```
$ python3 re_flags_ascii.py

Text    : Français łzoty Österreich
Pattern : \w+
ASCII   : ['Fran', 'ais', 'z', 'oty', 'sterreich']
Unicode : ['Français', 'łzoty', 'Österreich']
```

1.3.7.4 详细表达式语法

随着表达式变得越来越复杂，紧凑格式的正则表达式语法可能会变成障碍。随着表达式中组数的增加，需要做更多的工作来明确为什么需要各个元素，以及表达式的各部分究竟如何交互。使用命名组可以帮助缓解这些问题，不过更好的解决方案是使用详细模式（verbose mode）表达式，它允许在模式中嵌入注释和额外的空白符。

可以用一个验证 email 地址的模式来展示详细模式会让正则表达式的处理更加容易。第一个版本会识别以 3 个顶级域名之一结尾的地址：.com、.org 或 .edu。

代码清单 1-45：re_email_compact.py

```
import re

address = re.compile('[\w\d.+-]+@([\w\d.]+\.)+(com|org|edu)')
candidates = [
    u'first.last@example.com',
    u'first.last+category@gmail.com',
    u'valid-address@mail.example.com',
    u'not-valid@example.foo',
]

for candidate in candidates:
    match = address.search(candidate)
    print('{:<30}   {}'.format(
        candidate, 'Matches' if match else 'No match')
    )
```

这个表达式已经很复杂了。其中有多个字符类、组和重复表达式。

```
$ python3 re_email_compact.py

first.last@example.com           Matches
first.last+category@gmail.com    Matches
valid-address@mail.example.com   Matches
not-valid@example.foo            No match
```

将这个表达式转换为一种更详细的格式，使它更容易扩展。

代码清单 1-46：re_email_verbose.py

```
import re

address = re.compile(
    '''
    [\w\d.+-]+       # Username
    @
    ([\w\d.]+\.)+    # Domain name prefix
    (com|org|edu)    # TODO: support more top-level domains
    ''',
    re.VERBOSE)

candidates = [
    u'first.last@example.com',
    u'first.last+category@gmail.com',
    u'valid-address@mail.example.com',
    u'not-valid@example.foo',
]

for candidate in candidates:
    match = address.search(candidate)
    print('{:<30}   {}'.format(
        candidate, 'Matches' if match else 'No match'),
    )
```

这个表达式会匹配同样的输入，但是采用这种扩展格式更易读。注释也有助于识别模式的不同部分，从而能扩展以匹配更多输入。

```
$ python3 re_email_verbose.py

first.last@example.com              Matches
first.last+category@gmail.com       Matches
valid-address@mail.example.com      Matches
not-valid@example.foo               No match
```

这个扩展的版本会解析包含一个人名和 email 地址的输入（可能在 email 首部出现）。名字在前，后面是 email 地址，并用尖括号（< 和 >）包围。

代码清单 1-47：**re_email_with_name.py**

```
import re

address = re.compile(
    '''
    # A name is made up of letters, and may include "."
    # for title abbreviations and middle initials.
    ((?P<name>
       ([\w.,]+\s+)*[\w.,]+)
       \s*
       # Email addresses are wrapped in angle
       # brackets < >, but only if a name is
       # found, so keep the start bracket in this
       # group.
       <
    )? # The entire name is optional.

    # The address itself: username@domain.tld
    (?P<email>
      [\w\d.+-]+       # Username
      @
      ([\w\d.]+\.)+    # Domain name prefix
      (com|org|edu)    # Limit the allowed top-level domains.
    )

    >? # Optional closing angle bracket.
    ''',
    re.VERBOSE)

candidates = [
    u'first.last@example.com',
    u'first.last+category@gmail.com',
    u'valid-address@mail.example.com',
    u'not-valid@example.foo',
    u'First Last <first.last@example.com>',
    u'No Brackets first.last@example.com',
    u'First Last',
    u'First Middle Last <first.last@example.com>',
    u'First M. Last <first.last@example.com>',
    u'<first.last@example.com>',
]

for candidate in candidates:
    print('Candidate:', candidate)
    match = address.search(candidate)
    if match:
        print('  Name :', match.groupdict()['name'])
        print('  Email:', match.groupdict()['email'])
    else:
        print('  No match')
```

与其他编程语言一样，在详细正则表达式中插入注释有助于提高它的可维护性。最后这个版本包含了为将来的维护者提供的实现说明，另外还包括一些空白符以使各个组分开，并突出显示嵌套层次。

```
$ python3 re_email_with_name.py

Candidate: first.last@example.com
  Name : None
  Email: first.last@example.com
Candidate: first.last+category@gmail.com
  Name : None
  Email: first.last+category@gmail.com
Candidate: valid-address@mail.example.com
  Name : None
  Email: valid-address@mail.example.com
Candidate: not-valid@example.foo
  No match
Candidate: First Last <first.last@example.com>
  Name : First Last
  Email: first.last@example.com
Candidate: No Brackets first.last@example.com
  Name : None
  Email: first.last@example.com
Candidate: First Last
  No match
Candidate: First Middle Last <first.last@example.com>
  Name : First Middle Last
  Email: first.last@example.com
Candidate: First M. Last <first.last@example.com>
  Name : First M. Last
  Email: first.last@example.com
Candidate: <first.last@example.com>
  Name : None
  Email: first.last@example.com
```

1.3.7.5 在模式中嵌入标志

有些情况下，编译表达式时不能增加标志，如将一个模式作为参数传入一个库函数，这个库函数将在以后编译这个模式，在这种情况下，标志可以嵌入到表达式字符串本身。例如，要打开大小写无关匹配，可以在表达式开头增加 (?i)。

代码清单 1-48：`re_flags_embedded.py`

```
import re

text = 'This is some text -- with punctuation.'
pattern = r'(?i)\bT\w+'
regex = re.compile(pattern)

print('Text    :', text)
print('Pattern :', pattern)
print('Matches :', regex.findall(text))
```

由于这些选项会控制如何计算或解析整个表达式，所以它们总是要出现在表达式的最前面。

```
$ python3 re_flags_embedded.py

Text    : This is some text -- with punctuation.
Pattern : (?i)\bT\w+
Matches : ['This', 'text']
```

所有标志的缩写如表 1-3 所列。

嵌入标志可以放在同一个组中结合使用。例如，(?im) 会打开对多行字符串的大小写无关匹配。

表 1-3　正则表达式标志缩写

标志	缩写
ASCII	a
IGNORECASE	i
MULTILINE	m
DOTALL	s
VERBOSE	x

1.3.8　前向或后向

很多情况下，只有当模式中其他部分也匹配时才会匹配模式的某个部分。例如，在 email 解析表达式中，尖括号被标志为可选。在实际中，尖括号应当是成对的，所以只有当两个尖括号都出现或者都不出现时表达式才能匹配。这个修改版本的表达式使用一个肯定前向（positive look-ahead）断言来匹配尖括号对。前向断言语法为 (?=pattern)。

代码清单 1-49：**re_look_ahead.py**

```
import re

address = re.compile(
    '''
    # A name is made up of letters, and may include "."
    # for title abbreviations and middle initials.
    ((?P<name>
       ([\w.,]+\s+)*[\w.,]+
     )
     \s+
    ) # The name is no longer optional.

    # LOOKAHEAD
    # Email addresses are wrapped in angle brackets, but only
    # if both are present or neither is.
    (?= (<.*>$)       # Remainder wrapped in angle brackets
        |
        ([^<].*[^>]$) # Remainder *not* wrapped in angle brackets
      )

    <? # Optional opening angle bracket

    # The address itself: username@domain.tld
    (?P<email>
      [\w\d.+-]+       # Username
      @
      ([\w\d]+\.)+     # Domain name prefix
      (com|org|edu)    # Limit the allowed top-level domains.
    )

    >? # Optional closing angle bracket
    ''',
```

```
        re.VERBOSE)

candidates = [
    u'First Last <first.last@example.com>',
    u'No Brackets first.last@example.com',
    u'Open Bracket <first.last@example.com',
    u'Close Bracket first.last@example.com>',
]

for candidate in candidates:
    print('Candidate:', candidate)
    match = address.search(candidate)
    if match:
        print('  Name :', match.groupdict()['name'])
        print('  Email:', match.groupdict()['email'])
    else:
        print('  No match')
```

这个版本的表达式中有很多重要的变化。首先，name 部分不再是可选的。这说明，单独的地址将不能匹配，而且还会避免匹配那些格式不正确的"名 / 地址"组合。"name"组后面的肯定前向规则断言字符串的余下部分要么包围在一对尖括号中，要么不存在不匹配的尖括号，也就是尖括号要么都出现，要么都不出现。这个前向规则表述为一个组，不过前向组的匹配并不消费任何输入文本，所以这个模式的其余部分会从前向匹配之后的同一位置取字符。

```
$ python3 re_look_ahead.py

Candidate: First Last <first.last@example.com>
  Name : First Last
  Email: first.last@example.com
Candidate: No Brackets first.last@example.com
  Name : No Brackets
  Email: first.last@example.com
Candidate: Open Bracket <first.last@example.com
  No match
Candidate: Close Bracket first.last@example.com>
  No match
```

否定前向（negative look-ahead）断言（`(?!pattern)`）要求模式不匹配当前位置后面的文本。例如，email 识别模式可以修改为忽略自动系统常用的 `noreply` 邮件地址。

代码清单 1-50：re_negative_look_ahead.py

```
import re

address = re.compile(
    '''
    ^

    # An address: username@domain.tld

    # Ignore noreply addresses.
    (?!noreply@.*$)

    [\w\d.+-]+       # Username
    @
    ([\w\d.]+\.)+    # Domain name prefix
    (com|org|edu)    # Limit the allowed top-level domains.
```

```
        $
        ''',
        re.VERBOSE)

candidates = [
    u'first.last@example.com',
    u'noreply@example.com',
]

for candidate in candidates:
    print('Candidate:', candidate)
    match = address.search(candidate)
    if match:
        print('  Match:', candidate[match.start():match.end()])
    else:
        print('  No match')
```

以 noreply 开头的地址与这个模式不匹配，因为前向断言失败。

```
$ python3 re_negative_look_ahead.py

Candidate: first.last@example.com
  Match: first.last@example.com
Candidate: noreply@example.com
  No match
```

不用前向检查 email 地址 username 部分中的 noreply，可以借助语法 (?<!pattern) 在匹配 username 之后使用一个否定后向（negative look-behind）断言来改写这个模式。

代码清单 1-51：re_negative_look_behind.py

```
import re

address = re.compile(
    '''
    ^

    # An address: username@domain.tld

    [\w\d.+-]+       # Username

    # Ignore noreply addresses.
    (?<!noreply)

    @
    ([\w\d.]+\.)+    # Domain name prefix
    (com|org|edu)    # Limit the allowed top-level domains.

    $
    ''',
    re.VERBOSE)

candidates = [
    u'first.last@example.com',
    u'noreply@example.com',
]

for candidate in candidates:
    print('Candidate:', candidate)
    match = address.search(candidate)
    if match:
```

```
            print('  Match:', candidate[match.start():match.end()])
        else:
            print('  No match')
```

后向匹配与前向匹配的做法稍有不同，其表达式必须使用一个定长的模式。只要字符数固定（没有通配符或区间），后向匹配也允许重复。

```
$ python3 re_negative_look_behind.py

Candidate: first.last@example.com
  Match: first.last@example.com
Candidate: noreply@example.com
  No match
```

可以借助语法（**?<=pattern**）使用肯定后向（positive look-behind）断言查找符合某个模式的文本。例如，以下表达式可以查找推特用户名。

代码清单 1-52：re_look_behind.py

```
import re

twitter = re.compile(
    '''
    # A twitter handle: @username
    (?<=@)
    ([\w\d_]+)       # Username
    ''',
    re.VERBOSE)
text = '''This text includes two Twitter handles.
One for @ThePSF, and one for the author, @doughellmann.
'''

print(text)
for match in twitter.findall(text):
    print('Handle:', match)
```

这个模式会匹配能构成一个推特用户名的字符序列，只要字符序列前面有一个 @。

```
$ python3 re_look_behind.py

This text includes two Twitter handles.
One for @ThePSF, and one for the author, @doughellmann.

Handle: ThePSF
Handle: doughellmann
```

1.3.9 自引用表达式

还可以在表达式后面的部分中使用匹配的值。例如，前面的 email 例子可以更新为只匹配由人名和姓氏组成的地址，为此要包含这些组的反向引用。要达到这个目的，最容易的办法就是使用 \num 按 ID 编号引用先前匹配的组。

代码清单 1-53：re_refer_to_group.py

```
import re

address = re.compile(
    r'''
```

```
    # The regular name
    (\w+)                   # First name
    \s+
    (([\w.]+)\s+)?          # Optional middle name or initial
    (\w+)                   # Last name

    \s+

    <

    # The address: first_name.last_name@domain.tld
    (?P<email>
      \1                    # First name
      \.
      \4                    # Last name
      @
      ([\w\d.]+\.)+         # Domain name prefix
      (com|org|edu)         # Limit the allowed top-level domains.
    )

    >
    ''',
    re.VERBOSE | re.IGNORECASE)

candidates = [
    u'First Last <first.last@example.com>',
    u'Different Name <first.last@example.com>',
    u'First Middle Last <first.last@example.com>',
    u'First M. Last <first.last@example.com>',
]

for candidate in candidates:
    print('Candidate:', candidate)
    match = address.search(candidate)
    if match:
        print('  Match name :', match.group(1), match.group(4))
        print('  Match email:', match.group(5))
    else:
        print('  No match')
```

尽管这个语法很简单，按数字 ID 创建反向引用也依旧有几个缺点。从实用角度讲，表达式改变时，这些组就必须重新编号，每个引用可能都需要更新。另一个缺点是，采用标准反向引用语法 \n 只能创建 99 个引用，因为如果 ID 编号有 3 位，那么其便会被解释为一个 8 进制字符值而不是一个组引用。当然，如果一个表达式有超过 99 个组，那么问题就不仅仅是无法引用表达式中的所有组，这说明还存在一些更严重的维护问题。

```
$ python3 re_refer_to_group.py

Candidate: First Last <first.last@example.com>
  Match name : First Last
  Match email: first.last@example.com
Candidate: Different Name <first.last@example.com>
  No match
Candidate: First Middle Last <first.last@example.com>
  Match name : First Last
  Match email: first.last@example.com
Candidate: First M. Last <first.last@example.com>
  Match name : First Last
  Match email: first.last@example.com
```

Python 的表达式解析器包括一个扩展，可以使用 (?P=name) 来指示表达式中先前匹配的一个命名组的值。

代码清单 1-54：re_refer_to_named_group.py

```
import re

address = re.compile(
    '''

    # The regular name
    (?P<first_name>\w+)
    \s+
    (([\w.]+)\s+)?         # Optional middle name or initial
    (?P<last_name>\w+)

    \s+

    <

    # The address: first_name.last_name@domain.tld
    (?P<email>
      (?P=first_name)
      \.
      (?P=last_name)
      @
      ([\w\d.]+\.)+        # Domain name prefix
      (com|org|edu)        # Limit the allowed top-level domains.
    )

    >
    ''',
    re.VERBOSE | re.IGNORECASE)

candidates = [
    u'First Last <first.last@example.com>',
    u'Different Name <first.last@example.com>',
    u'First Middle Last <first.last@example.com>',
    u'First M. Last <first.last@example.com>',
]

for candidate in candidates:
    print('Candidate:', candidate)
    match = address.search(candidate)
    if match:
        print('  Match name :', match.groupdict()['first_name'],
              end=' ')
        print(match.groupdict()['last_name'])
        print('  Match email:', match.groupdict()['email'])
    else:
        print('  No match')
```

编译地址表达式时打开了 **IGNORECASE** 标志，因为正确的名字通常首字母会大写，而 email 地址往往不会大写首字母。

```
$ python3 re_refer_to_named_group.py

Candidate: First Last <first.last@example.com>
  Match name : First Last
  Match email: first.last@example.com
```

```
Candidate: Different Name <first.last@example.com>
  No match
Candidate: First Middle Last <first.last@example.com>
  Match name : First Last
  Match email: first.last@example.com
Candidate: First M. Last <first.last@example.com>
  Match name : First Last
  Match email: first.last@example.com
```

在表达式中使用反向引用还有一种机制，即根据前一个组是否匹配来选择不同的模式。可以修正这个 email 模式，使得如果出现名字就需要有尖括号，而如果只有 email 地址本身就不需要尖括号。查看一个组是否匹配的语法是 `(?(id)yes-expression|no-expression)`，这里 `id` 是组名或编号，`yes-expression` 是组有值时使用的模式，`no-expression` 则是组没有值时使用的模式。

代码清单 1-55：re_id.py

```
import re

address = re.compile(
    '''
    ^

    # A name is made up of letters, and may include "."
    # for title abbreviations and middle initials.
    (?P<name>
       ([\w.]+\s+)*[\w.]+
     )?
    \s*

    # Email addresses are wrapped in angle brackets, but
    # only if a name is found.
    (?(name)
      # Remainder wrapped in angle brackets because
      # there is a name
      (?P<brackets>(?=(<.*>$)))
      |
      # Remainder does not include angle brackets without name
      (?=([^<].*[^>]$))
     )

    # Look for a bracket only if the look-ahead assertion
    # found both of them.
    (?(brackets)<|\s*)

    # The address itself: username@domain.tld
    (?P<email>
      [\w\d.+-]+       # Username
      @
      ([\w\d.]+\.)+    # Domain name prefix
      (com|org|edu)    # Limit the allowed top-level domains.
     )

    # Look for a bracket only if the look-ahead assertion
    # found both of them.
    (?(brackets)>|\s*)

    $
    ''',
    re.VERBOSE)
```

```
candidates = [
    u'First Last <first.last@example.com>',
    u'No Brackets first.last@example.com',
    u'Open Bracket <first.last@example.com',
    u'Close Bracket first.last@example.com>',
    u'no.brackets@example.com',
]

for candidate in candidates:
    print('Candidate:', candidate)
    match = address.search(candidate)
    if match:
        print('  Match name :', match.groupdict()['name'])
        print('  Match email:', match.groupdict()['email'])
    else:
        print('  No match')
```

这个版本的 email 地址解析器使用了两个测试。如果 name 组匹配，则前向断言要求两个尖括号都出现，并建立 brackets 组。如果 name 不匹配，则这个断言要求余下文本不能用尖括号包围。接下来，如果设置了 brackets 组，那么具体的模式匹配代码会使用字面量模式消费输入中的尖括号；否则，它会消费所有空格。

```
$ python3 re_id.py

Candidate: First Last <first.last@example.com>
  Match name : First Last
  Match email: first.last@example.com
Candidate: No Brackets first.last@example.com
  No match
Candidate: Open Bracket <first.last@example.com
  No match
Candidate: Close Bracket first.last@example.com>
  No match
Candidate: no.brackets@example.com
  Match name : None
  Match email: no.brackets@example.com
```

1.3.10　用模式修改字符串

除了搜索文本之外，re 还支持使用正则表达式作为搜索机制来修改文本，而且替换（replacement）可以引用模式中的匹配组作为替代文本的一部分。使用 sub() 可以将一个模式的所有出现替换为另一个字符串。

代码清单 1-56：**re_sub.py**

```
import re

bold = re.compile(r'\*{2}(.*?)\*{2}')

text = 'Make this **bold**.  This **too**.'

print('Text:', text)
print('Bold:', bold.sub(r'<b>\1</b>', text))
```

可以使用向后引用的 \num 语法插入与模式匹配的文本的引用。

```
$ python3 re_sub.py

Text: Make this **bold**.  This **too**.
Bold: Make this <b>bold</b>.  This <b>too</b>.
```

要在替换中使用命名组,可以使用语法 \g<name>。

代码清单 1-57:re_sub_named_groups.py

```
import re

bold = re.compile(r'\*{2}(?P<bold_text>.*?)\*{2}')

text = 'Make this **bold**.  This **too**.'

print('Text:', text)
print('Bold:', bold.sub(r'<b>\g<bold_text></b>', text))
```

\g<name> 语法还适用于编号引用,使用这个语法可以消除组编号和外围字面量数字之间的模糊性。

```
$ python3 re_sub_named_groups.py

Text: Make this **bold**.  This **too**.
Bold: Make this <b>bold</b>.  This <b>too</b>.
```

向 count 传入一个值可以限制完成的替换数。

代码清单 1-58:re_sub_count.py

```
import re

bold = re.compile(r'\*{2}(.*?)\*{2}')

text = 'Make this **bold**.  This **too**.'

print('Text:', text)
print('Bold:', bold.sub(r'<b>\1</b>', text, count=1))
```

由于 count 为 1,所以只完成了第一个替换。

```
$ python3 re_sub_count.py

Text: Make this **bold**.  This **too**.
Bold: Make this <b>bold</b>.  This **too**.
```

subn() 的工作与 sub() 很相似,只是它会同时返回修改后的字符串和完成的替换数。

代码清单 1-59:re_subn.py

```
import re

bold = re.compile(r'\*{2}(?P<bold_text>.*?)\*{2}')

text = 'Make this **bold**.  This **too**.'

print('Text:', text)
print('Bold:', bold.sub(r'<b>\g<bold_text></b>', text))
```

这个例子中搜索模式有两次匹配。

```
$ python3 re_subn.py

Text: Make this **bold**.  This **too**.
Bold: ('Make this <b>bold</b>.  This <b>too</b>.', 2)
```

1.3.11 利用模式拆分

`str.split()`是分解字符串来完成解析的最常用的方法之一。不过，它只支持使用字面量值作为分隔符。有时，如果输入没有一致的格式，那么就需要有一个正则表达式。例如，很多纯文本标记语言都把段落分隔符定义为两个或多个换行符（\n）。在这种情况下，就不能使用`str.split()`，因为这个定义中提到了"或多个"。

通过`findall()`标识段落的一种策略是使用类似`(.+?)\n{2,}`的模式。

代码清单 1-60：`re_paragraphs_findall.py`

```
import re

text = '''Paragraph one
on two lines.

Paragraph two.


Paragraph three.'''

for num, para in enumerate(re.findall(r'(.+?)\n{2,}',
                                      text,
                                      flags=re.DOTALL)
                           ):
    print(num, repr(para))
    print()
```

对于输入文本末尾的段落，这个模式会失败，原因在于"Paragraph three."不是输出的一部分。

```
$ python3 re_paragraphs_findall.py

0 'Paragraph one\non two lines.'

1 'Paragraph two.'
```

可以扩展这个模式，指出段落以两个或多个换行符结束或者以输入末尾结束，这就能修正这个问题，但也会让模式变得更为复杂。可以转而使用`re.split()`而非`re.findall()`，这便能自动地处理边界条件，并保证模式更简单。

代码清单 1-61：`re_split.py`

```
import re

text = '''Paragraph one
on two lines.
Paragraph two.


Paragraph three.'''
```

```
print('With findall:')
for num, para in enumerate(re.findall(r'(.+?)(\n{2,}|$)',
                                      text,
                                      flags=re.DOTALL)):
    print(num, repr(para))
    print()

print()
print('With split:')
for num, para in enumerate(re.split(r'\n{2,}', text)):
    print(num, repr(para))
    print()
```

split()的模式参数更准确地表述了标记规范。由两个或多个换行符标记输入字符串中段落之间的分隔点。

```
$ python3 re_split.py

With findall:
0 ('Paragraph one\non two lines.', '\n\n')

1 ('Paragraph two.', '\n\n\n')

2 ('Paragraph three.', '')


With split:
0 'Paragraph one\non two lines.'

1 'Paragraph two.'

2 'Paragraph three.'
```

可以将表达式包围在括号里来定义一个组，这使得split()的工作更类似于str.partition()，因此它会返回分隔符值以及字符串的其他部分。

代码清单1-62：**re_split_groups.py**

```
import re

text = '''Paragraph one
on two lines.

Paragraph two.


Paragraph three.'''

print('With split:')
for num, para in enumerate(re.split(r'(\n{2,})', text)):
    print(num, repr(para))
    print()
```

现在输出包括各个段落，以及分隔这些段落的换行符序列。

```
$ python3 re_split_groups.py

With split:
0 'Paragraph one\non two lines.'

1 '\n\n'
```

```
2 'Paragraph two.'

3 '\n\n\n'

4 'Paragraph three.'
```

提示：相关阅读材料
- re 的标准库文档[一]。
- Regular Expression HOWTO[二]：Andrew Kuchling 为 Python 开发人员提供的正则表达式介绍。
- Kodos[三]：这是一个交互式的正则表达式测试工具，由 Phil Schwartz 创建。
- pythex[四]：这是一个用来测试正则表达式的基于 Web 的工具，由 Gabriel Rodríguez 创建，受到 Rubular 的启发。
- Wikipedia: Regular expression[五]：对正则表达式概念和技术的一般介绍。
- locale：处理 Unicode 文本时可以使用 locale 模块设置语言配置。
- unicodedata：通过程序访问 Unicode 字符属性数据库。

1.4 difflib：比较序列

difflib 模块包含一些计算和处理序列之间差异的工具。它对于比较文本尤其有用，其中的函数可以使用多种常用的差异格式生成报告。

本节中的例子都会使用以下公共测试数据（在 difflib_data.py 模块中）：

代码清单 1-63：difflib_data.py

```
text1 = """Lorem ipsum dolor sit amet, consectetuer adipiscing
elit. Integer eu lacus accumsan arcu fermentum euismod. Donec
pulvinar porttitor tellus. Aliquam venenatis. Donec facilisis
pharetra tortor.  In nec mauris eget magna consequat
convalis. Nam sed sem vitae odio pellentesque interdum. Sed
consequat viverra nisl. Suspendisse arcu metus, blandit quis,
rhoncus ac, pharetra eget, velit. Mauris urna. Morbi nonummy
molestie orci. Praesent nisi elit, fringilla ac, suscipit non,
tristique vel, mauris. Curabitur vel lorem id nisl porta
adipiscing. Suspendisse eu lectus. In nunc. Duis vulputate
tristique enim. Donec quis lectus a justo imperdiet tempus."""

text1_lines = text1.splitlines()

text2 = """Lorem ipsum dolor sit amet, consectetuer adipiscing
elit. Integer eu lacus accumsan arcu fermentum euismod. Donec
pulvinar, porttitor tellus. Aliquam venenatis. Donec facilisis
pharetra tortor. In nec mauris eget magna consequat
convalis. Nam cras vitae mi vitae odio pellentesque interdum. Sed
```

[一] https://docs.python.org/3.5/library/re.html
[二] https://docs.python.org/3.5/howto/regex.html
[三] http://kodos.sourceforge.net
[四] http://pythex.org
[五] https://en.wikipedia.org/wiki/Regular_expression

```
consequat viverra nisl. Suspendisse arcu metus, blandit quis,
rhoncus ac, pharetra eget, velit. Mauris urna. Morbi nonummy
molestie orci. Praesent nisi elit, fringilla ac, suscipit non,
tristique vel, mauris. Curabitur vel lorem id nisl porta
adipiscing. Duis vulputate tristique enim. Donec quis lectus a
justo imperdiet tempus.  Suspendisse eu lectus. In nunc."""

text2_lines = text2.splitlines()
```

1.4.1 比较文本体

`Differ`类用于处理文本行序列，并生成人类可读的差异（deltas）或更改指令，包括各行中的差异。`Differ`生成的默认输出与UNIX下的`diff`命令行工具类似，包括两个列表的原始输入值（包含共同的值），以及指示做了哪些更改的标记数据。

- 有 - 前缀的行在第一个序列中，而非第二个序列。
- 有 + 前缀的行在第二个序列中，而非第一个序列。
- 如果某一行在不同版本之间存在增量差异，那么会使用一个加?前缀的额外行来强调新版本中的变更。
- 如果一行未改变，则会打印输出，而且其左列有一个额外的空格，使它与其他可能有差异的输出对齐。

将文本传入`compare()`之前先将其分解为由单个文本行构成的序列，与传入大字符串相比，这样可以生成更可读的输出。

代码清单1-64：**difflib_differ.py**

```
import difflib
from difflib_data import *

d = difflib.Differ()
diff = d.compare(text1_lines, text2_lines)
print('\n'.join(diff))
```

示例数据中两个文本段的开始部分是一样的，所以第1行会直接打印而没有任何额外标注。

```
  Lorem ipsum dolor sit amet, consectetuer adipiscing
  elit. Integer eu lacus accumsan arcu fermentum euismod. Donec
```

数据的第3行有变化，修改后的文本中包含有一个逗号。这两个版本的数据行都会打印，而且第5行上的额外信息会显示文本中哪一列有修改，这里显示增加了,字符。

```
- pulvinar porttitor tellus. Aliquam venenatis. Donec facilisis
+ pulvinar, porttitor tellus. Aliquam venenatis. Donec facilisis
?         +
```

输出中接下来几行显示删除了一个多余的空格。

```
- pharetra tortor.  In nec mauris eget magna consequat
?                 -
+ pharetra tortor. In nec mauris eget magna consequat
```

接下来有一个更复杂的改变，其替换了一个短语中的多个单词。

```
- convalis. Nam sed sem vitae odio pellentesque interdum. Sed
?                       - --
+ convalis. Nam cras vitae mi vitae odio pellentesque interdum. Sed
?                 +++ +++++    +
```

段落中最后一句变化很大，所以表示差异时完全删除了老版本，而增加了新版本。

```
  consequat viverra nisl. Suspendisse arcu metus, blandit quis,
  rhoncus ac, pharetra eget, velit. Mauris urna. Morbi nonummy
  molestie orci. Praesent nisi elit, fringilla ac, suscipit non,
  tristique vel, mauris. Curabitur vel lorem id nisl porta
- adipiscing. Suspendisse eu lectus. In nunc. Duis vulputate
- tristique enim. Donec quis lectus a justo imperdiet tempus.
+ adipiscing. Duis vulputate tristique enim. Donec quis lectus a
+ justo imperdiet tempus.  Suspendisse eu lectus. In nunc.
```

`ndiff()` 函数生成的输出基本上相同，通过特别"加工"来处理文本数据，并删除输入中的"噪声"。

其他输出格式

`Differ` 类会显示所有输入行，统一差异格式（unified diff）则不同，它只包含有修改的文本行和一些上下文。`unified_diff()` 函数会生成这种输出。

代码清单 1-65：`difflib_unified.py`

```
import difflib
from difflib_data import *

diff = difflib.unified_diff(
    text1_lines,
    text2_lines,
    lineterm='',
)
print('\n'.join(list(diff)))
```

`lineterm` 参数用来告诉 `unified_diff()` 不必为它返回的控制行追加换行符，因为输入行不包括这些换行符。打印时所有行都会增加换行符。对于很多常用版本控制工具的用户来说，输出看上去应该很熟悉。

```
$ python3 difflib_unified.py

---
+++
@@ -1,11 +1,11 @@
 Lorem ipsum dolor sit amet, consectetuer adipiscing
 elit. Integer eu lacus accumsan arcu fermentum euismod. Donec
-pulvinar porttitor tellus. Aliquam venenatis. Donec facilisis
-pharetra tortor.  In nec mauris eget magna consequat
-convalis. Nam sed sem vitae odio pellentesque interdum. Sed
+pulvinar, porttitor tellus. Aliquam venenatis. Donec facilisis
+pharetra tortor. In nec mauris eget magna consequat
+convalis. Nam cras vitae mi vitae odio pellentesque interdum. S
ed
 consequat viverra nisl. Suspendisse arcu metus, blandit quis,
 rhoncus ac, pharetra eget, velit. Mauris urna. Morbi nonummy
 molestie orci. Praesent nisi elit, fringilla ac, suscipit non,
 tristique vel, mauris. Curabitur vel lorem id nisl porta
-adipiscing. Suspendisse eu lectus. In nunc. Duis vulputate
```

```
-tristique enim. Donec quis lectus a justo imperdiet tempus.
+adipiscing. Duis vulputate tristique enim. Donec quis lectus a
+justo imperdiet tempus.  Suspendisse eu lectus. In nunc.
```

使用 `context_diff()` 会产生类似的可续输出。

1.4.2 无用数据

所有生成差异序列的函数都会接受一些参数来指示应当忽略哪些行，以及要忽略一行中的哪些字符。例如，这些参数可用于跳过文件两个版本中的标记或空白符改变。

代码清单 1-66：`difflib_junk.py`

```python
# This example is adapted from the source for difflib.py.

from difflib import SequenceMatcher


def show_results(match):
    print('  a    = {}'.format(match.a))
    print('  b    = {}'.format(match.b))
    print('  size = {}'.format(match.size))
    i, j, k = match
    print('  A[a:a+size] = {!r}'.format(A[i:i + k]))
    print('  B[b:b+size] = {!r}'.format(B[j:j + k]))


A = " abcd"
B = "abcd abcd"

print('A = {!r}'.format(A))
print('B = {!r}'.format(B))

print('\nWithout junk detection:')
s1 = SequenceMatcher(None, A, B)
match1 = s1.find_longest_match(0, len(A), 0, len(B))
show_results(match1)

print('\nTreat spaces as junk:')
s2 = SequenceMatcher(lambda x: x == " ", A, B)
match2 = s2.find_longest_match(0, len(A), 0, len(B))
show_results(match2)
```

默认 `Differ` 不会显式地忽略任何行或字符，但会依赖 `SequenceMatcher` 的能力检测噪声。`ndiff()` 的默认行为是忽略空格和制表符。

```
$ python3 difflib_junk.py

A = ' abcd'
B = 'abcd abcd'

Without junk detection:
  a    = 0
  b    = 4
  size = 5
  A[a:a+size] = ' abcd'
  B[b:b+size] = ' abcd'

Treat spaces as junk:
  a    = 1
```

```
b    = 0
size = 4
A[a:a+size] = 'abcd'
B[b:b+size] = 'abcd'
```

1.4.3 比较任意类型

SequenceMatcher 类可以比较任意类型的两个序列，只要它们的值是可散列的。这个类使用一个算法来标识序列中最长的连续匹配块，并删除对实际数据没有贡献的无用值。

函数 get_opcodes() 返回一个指令列表来修改第一个序列，使它与第二个序列匹配。这些指令被编码为 5 元素元组，包括一个字符串指令（"操作码"）和序列的两对开始及结束索引（表示为 i1、i2、j1 和 j2），如表 1-4 所示。

表 1-4 difflib.get_opcodes() 指令

操作码	定义
'replace'	将 a[i1:i2] 替换为 b[j1:j2]
'delete'	完全删除 a[i1:i2]
'insert'	将 b[j1:j2] 插入 a[i1:i1]
'equal'	两个序列已经相等

代码清单 1-67：difflib_seq.py

```python
import difflib

s1 = [1, 2, 3, 5, 6, 4]
s2 = [2, 3, 5, 4, 6, 1]

print('Initial data:')
print('s1 =', s1)
print('s2 =', s2)
print('s1 == s2:', s1 == s2)
print()

matcher = difflib.SequenceMatcher(None, s1, s2)
for tag, i1, i2, j1, j2 in reversed(matcher.get_opcodes()):

    if tag == 'delete':
        print('Remove {} from positions [{}:{}]'.format(
            s1[i1:i2], i1, i2))
        print('  before =', s1)
        del s1[i1:i2]

    elif tag == 'equal':
        print('s1[{}:{}] and s2[{}:{}] are the same'.format(
            i1, i2, j1, j2))

    elif tag == 'insert':
        print('Insert {} from s2[{}:{}] into s1 at {}'.format(
            s2[j1:j2], j1, j2, i1))
        print('  before =', s1)
        s1[i1:i2] = s2[j1:j2]

    elif tag == 'replace':
        print(('Replace {} from s1[{}:{}] '
               'with {} from s2[{}:{}]').format(
```

```
            s1[i1:i2], i1, i2, s2[j1:j2], j1, j2))
    print('  before =', s1)
    s1[i1:i2] = s2[j1:j2]
    print('  after =', s1, '\n')
print('s1 == s2:', s1 == s2)
```

这个例子比较了两个整数列表,并使用 `get_opcodes()` 得出将原列表转换为新列表的指令。这里以逆序应用所做的修改,以便增加和删除元素之后列表索引仍是正确的。

```
$ python3 difflib_seq.py

Initial data:
s1 = [1, 2, 3, 5, 6, 4]
s2 = [2, 3, 5, 4, 6, 1]
s1 == s2: False

Replace [4] from s1[5:6] with [1] from s2[5:6]
  before = [1, 2, 3, 5, 6, 4]
  after = [1, 2, 3, 5, 6, 1]

s1[4:5] and s2[4:5] are the same
  after = [1, 2, 3, 5, 6, 1]

Insert [4] from s2[3:4] into s1 at 4
  before = [1, 2, 3, 5, 6, 1]
  after = [1, 2, 3, 5, 4, 6, 1]

s1[1:4] and s2[0:3] are the same
  after = [1, 2, 3, 5, 4, 6, 1]

Remove [1] from positions [0:1]
  before = [1, 2, 3, 5, 4, 6, 1]
  after = [2, 3, 5, 4, 6, 1]

s1 == s2: True
```

SequenceMatcher 用于处理定制类以及内置类型,前提是它们必须是可散列的。

提示:相关阅读材料

- `difflib` 的标准库文档[⊖]。
- "Pattern Matching: The Gestalt Approach"[⊖]:对 John W. Ratcliff 和 D. E. Metzener 提出的一种类似算法的讨论,发表于 1988 年 7 月的 "Dr. Dobb's Journal"。

⊖ https://docs.python.org/3.5/library/difflib.html

⊖ www.drdobbs.com/database/pattern-matching-the-gestalt-approach/184407970

第 ② 章
数据结构

　　Python 包含很多标准编程数据结构，如列表（list）、元组（tuple）、字典（dict）和集合（set），这些都属于其内置类型。对很多应用来说这些结构已经足够，不再需要其他结构了，不过，如果确实需要其他结构也大可放心，标准库提供有功能强大且经过充分测试的结构。

　　enum 模块提供了一个枚举类型的实现，其带有迭代和比较功能。可以用这个模块为值创建明确定义的符号，而不是使用字面量字符串或整数。

　　collections 模块包含多种数据结构的实现，这些结构扩展了其他模块中的相应结构。例如，Deque 是一个双端队列，允许从任意一端增加或删除元素。defaultdict 是一个字典，如果找不到某个键，则会返回一个默认值作为响应。OrderedDict 会记住按什么序列增加元素。namedtuple 扩展了一般的 tuple，除了为每个成员元素提供一个数值索引外，还会提供一个属性名。

　　对于很大量的数据，array 会比 list 更高效地利用内存。由于 array 仅限于一种数据类型，所以与通用的 list 相比，它可以采用一种更紧凑的内存表示。不仅如此，array 实例可以同样地使用 list 的很多方法来处理，所以完全可以把一个应用中的 list 替换为 array 而无须太多修改。

　　对序列中的元素排序是数据处理的一项基本内容。Python 的 list 包含一个 sort() 方法，不过有时维护一个有序列表会更高效，这样就无须在每次改变列表内容时都重新排序。heapq 中的函数可以修改列表的内容，同时还能以很低的开销维护列表原来的顺序。

　　构建有序列表或数组还有一种选择，即 bisect。它使用一种二分查找算法来查找新元素的插入点，如果要反复对一个频繁改变的列表排序，则可以将它作为一种候选方法。

　　尽管内置的 list 可以使用 insert() 和 pop() 方法模拟队列，但这不是线程安全的。要完成线程间的实序通信，可以使用 Queue 模块。multiprocessing 包含 Queue 的一个版本，它会处理进程间的通信，从而能更容易地转换多线程程序以使用进程而不是线程。

　　解码来自另一个应用的数据时（可能来自一个二进制文件或数据流），struct 会很有用，可以将这些数据解码为 Python 的原生类型以便于处理。

　　这一章会介绍两个与内存管理有关的模块。对于高度互连的数据结构，如图和树，可以使用 weakref 维护引用，而当不再需要某些对象时仍允许垃圾回收器进行清理。copy 中的函数用于复制数据结构及其内容，包括用 deepcopy() 完成递归复制。

　　调试数据结构可能很耗费时间，特别是查看大序列或字典的打印输出。可以使用

pprint 创建易读的表示，从而能打印到控制台或写至一个日志文件以便于调试。

最后一点，如果现有的类型不能满足需求，则可以派生某个原生类型进行定制，或者使用 collections 中定义的某个抽象基类作为起点来构建一个新的容器类型。

2.1 enum：枚举类型

enum 模块定义了一个提供迭代和比较功能的枚举类型。可以用这个模块为值创建明确定义的符号，而不是使用字面量整数或字符串。

2.1.1 创建枚举

可以使用 class 语法派生 Enum 并增加描述值的类属性来定义一个新枚举。

代码清单 2-1：**enum_create.py**

```
import enum


class BugStatus(enum.Enum):

    new = 7
    incomplete = 6
    invalid = 5
    wont_fix = 4
    in_progress = 3
    fix_committed = 2
    fix_released = 1


print('\nMember name: {}'.format(BugStatus.wont_fix.name))
print('Member value: {}'.format(BugStatus.wont_fix.value))
```

解析这个类时，Enum 的成员会被转换为实例。每个实例有一个对应成员名的 name 属性，另外有一个 value 属性，对应为类定义中的名所赋的值。

```
$ python3 enum_create.py

Member name: wont_fix
Member value: 4
```

2.1.2 迭代

迭代处理 enum 类会生成枚举的各个成员。

代码清单 2-2：**enum_iterate.py**

```
import enum


class BugStatus(enum.Enum):

    new = 7
    incomplete = 6
```

```
    invalid = 5
    wont_fix = 4
    in_progress = 3
    fix_committed = 2
    fix_released = 1


for status in BugStatus:
    print('{:15} = {}'.format(status.name, status.value))
```

这些成员按它们在类定义中声明的顺序生成。不会用名和值来对它们排序。

```
$ python3 enum_iterate.py

new             = 7
incomplete      = 6
invalid         = 5
wont_fix        = 4
in_progress     = 3
fix_committed   = 2
fix_released    = 1
```

2.1.3 比较 Enum

由于枚举成员是无序的，所以它们只支持按同一性和相等性进行比较。

代码清单 2-3：**enum_comparison.py**

```
import enum


class BugStatus(enum.Enum):

    new = 7
    incomplete = 6
    invalid = 5
    wont_fix = 4
    in_progress = 3
    fix_committed = 2
    fix_released = 1


actual_state = BugStatus.wont_fix
desired_state = BugStatus.fix_released

print('Equality:',
      actual_state == desired_state,
      actual_state == BugStatus.wont_fix)
print('Identity:',
      actual_state is desired_state,
      actual_state is BugStatus.wont_fix)
print('Ordered by value:')
try:
    print('\n'.join('   ' + s.name for s in sorted(BugStatus)))
except TypeError as err:
    print('  Cannot sort: {}'.format(err))
```

大于和小于比较符会产生 **TypeError** 异常。

```
$ python3 enum_comparison.py

Equality: False True
Identity: False True
Ordered by value:
  Cannot sort: unorderable types: BugStatus() < BugStatus()
```

有些枚举中的成员要表现得更像数字，例如，要支持比较，对于这些枚举要使用 Int-Enum 类。

代码清单 2-4：**enum_intenum.py**

```
import enum

class BugStatus(enum.IntEnum):

    new = 7
    incomplete = 6
    invalid = 5
    wont_fix = 4
    in_progress = 3
    fix_committed = 2
    fix_released = 1

print('Ordered by value:')
print('\n'.join('  ' + s.name for s in sorted(BugStatus)))
```

```
$ python3 enum_intenum.py

Ordered by value:
  fix_released
  fix_committed
  in_progress
  wont_fix
  invalid
  incomplete
  new
```

2.1.4 唯一枚举值

有相同值的 Enum 成员会被处理为同一个成员对象的别名引用。别名可以避免 Enum 的迭代器中出现重复的值。

代码清单 2-5：**enum_aliases.py**

```
import enum

class BugStatus(enum.Enum):

    new = 7
    incomplete = 6
    invalid = 5
    wont_fix = 4
    in_progress = 3
    fix_committed = 2
    fix_released = 1

    by_design = 4
    closed = 1
```

```
for status in BugStatus:
    print('{:15} = {}'.format(status.name, status.value))

print('\nSame: by_design is wont_fix: ',
      BugStatus.by_design is BugStatus.wont_fix)
print('Same: closed is fix_released: ',
      BugStatus.closed is BugStatus.fix_released)
```

由于 `by_design` 和 `closed` 是其他成员的别名，迭代处理 Enum 时它们不会单独出现在输出中。一个成员的规范名是与这个值关联的第一个名字。

```
$ python3 enum_aliases.py

new             = 7
incomplete      = 6
invalid         = 5
wont_fix        = 4
in_progress     = 3
fix_committed   = 2
fix_released    = 1

Same: by_design is wont_fix:   True
Same: closed is fix_released:  True
```

如果要求所有成员有唯一的值，则要为 Enum 增加 @unique 修饰符。

代码清单 2-6：enum_unique_enforce.py

```
import enum


@enum.unique
class BugStatus(enum.Enum):

    new = 7
    incomplete = 6
    invalid = 5
    wont_fix = 4
    in_progress = 3
    fix_committed = 2
    fix_released = 1

    # This will trigger an error with unique applied.
    by_design = 4
    closed = 1
```

解释 Enum 类时，有重复值的成员会触发一个 ValueError 异常。

```
$ python3 enum_unique_enforce.py

Traceback (most recent call last):
  File "enum_unique_enforce.py", line 11, in <module>
    class BugStatus(enum.Enum):
  File ".../lib/python3.5/enum.py", line 573, in unique
    (enumeration, alias_details))
ValueError: duplicate values found in <enum 'BugStatus'>:
by_design -> wont_fix, closed -> fix_released
```

2.1.5 通过编程创建枚举

有些情况下，通过编程创建枚举会更方便，而不是在类定义中硬编码定义枚举。在这些情况下，Enum 还支持向类构造函数传递成员名和值。

代码清单 2-7：**enum_programmatic_create.py**

```python
import enum

BugStatus = enum.Enum(
    value='BugStatus',
    names=('fix_released fix_committed in_progress '
           'wont_fix invalid incomplete new'),
)

print('Member: {}'.format(BugStatus.new))

print('\nAll members:')
for status in BugStatus:
    print('{:15} = {}'.format(status.name, status.value))
```

value 参数是枚举名，用于构建成员的表示。names 参数会列出枚举的成员。传递单个字符串时，会按空白符和逗号拆分，所得到的 token 会被用作成员名，这些成员还会自动赋值，从 1 开始。

```
$ python3 enum_programmatic_create.py

Member: BugStatus.new

All members:
fix_released    = 1
fix_committed   = 2
in_progress     = 3
wont_fix        = 4
invalid         = 5
incomplete      = 6
new             = 7
```

要想更多地控制与成员关联的值，可以把 names 字符串替换为一个由两部分元组构成的序列或者一个将名映射到值的字典。

代码清单 2-8：**enum_programmatic_mapping.py**

```python
import enum

BugStatus = enum.Enum(
    value='BugStatus',
    names=[
        ('new', 7),
        ('incomplete', 6),
        ('invalid', 5),
        ('wont_fix', 4),
        ('in_progress', 3),
        ('fix_committed', 2),
        ('fix_released', 1),
    ],
)

print('All members:')
for status in BugStatus:
    print('{:15} = {}'.format(status.name, status.value))
```

在这个例子中，指定了一个两部分元组的列表而不是一个只包含成员名的字符串。这样就可以按 enum_create.py 中定义的同样的顺序利用成员重新构造 BugStatus 枚举。

```
$ python3 enum_programmatic_mapping.py

All members:
new             = 7
incomplete      = 6
invalid         = 5
wont_fix        = 4
in_progress     = 3
fix_committed   = 2
fix_released    = 1
```

2.1.6 非整数成员值

Enum 成员值并不仅限于整数。任何类型的对象都可以与成员关联。如果值是一个元组，那么成员会作为单个参数被传递到 __init__()。

代码清单 2-9：**enum_tuple_values.py**

```
import enum


class BugStatus(enum.Enum):

    new = (7, ['incomplete',
               'invalid',
               'wont_fix',
               'in_progress'])
    incomplete = (6, ['new', 'wont_fix'])
    invalid = (5, ['new'])
    wont_fix = (4, ['new'])
    in_progress = (3, ['new', 'fix_committed'])
    fix_committed = (2, ['in_progress', 'fix_released'])
    fix_released = (1, ['new'])

    def __init__(self, num, transitions):
        self.num = num
        self.transitions = transitions

    def can_transition(self, new_state):
        return new_state.name in self.transitions


print('Name:', BugStatus.in_progress)
print('Value:', BugStatus.in_progress.value)
print('Custom attribute:', BugStatus.in_progress.transitions)
print('Using attribute:',
      BugStatus.in_progress.can_transition(BugStatus.new))
```

在这个例子中，每个成员值分别是一个元组，包括数值 ID（如可能存储在一个数据库中的 ID）和从当前状态迁移的一个合法变迁列表。

```
$ python3 enum_tuple_values.py

Name: BugStatus.in_progress
Value: (3, ['new', 'fix_committed'])
Custom attribute: ['new', 'fix_committed']
Using attribute: True
```

对于更复杂的情况，元组可能变得很难用。因为成员值可以是任意类型的对象，如果对每个 enum 值都需要跟踪大量单独的属性，那么这些情况下可以使用字典。复杂的值可以直接作为唯一参数传递到 `__init__()` 而不是 `self`。

代码清单 2-10：enum_complex_values.py

```python
import enum


class BugStatus(enum.Enum):

    new = {
        'num': 7,
        'transitions': [
            'incomplete',
            'invalid',
            'wont_fix',
            'in_progress',
        ],
    }
    incomplete = {
        'num': 6,
        'transitions': ['new', 'wont_fix'],
    }
    invalid = {
        'num': 5,
        'transitions': ['new'],
    }
    wont_fix = {
        'num': 4,
        'transitions': ['new'],
    }
    in_progress = {
        'num': 3,
        'transitions': ['new', 'fix_committed'],
    }
    fix_committed = {
        'num': 2,
        'transitions': ['in_progress', 'fix_released'],
    }
    fix_released = {
        'num': 1,
        'transitions': ['new'],
    }

    def __init__(self, vals):
        self.num = vals['num']
        self.transitions = vals['transitions']

    def can_transition(self, new_state):
        return new_state.name in self.transitions


print('Name:', BugStatus.in_progress)
print('Value:', BugStatus.in_progress.value)
print('Custom attribute:', BugStatus.in_progress.transitions)
print('Using attribute:',
      BugStatus.in_progress.can_transition(BugStatus.new))
```

这个例子描述了与前例相同的数据，不过使用的是字典而非元组。

```
$ python3 enum_complex_values.py

Name: BugStatus.in_progress
Value: {'transitions': ['new', 'fix_committed'], 'num': 3}
Custom attribute: ['new', 'fix_committed']
Using attribute: True
```

提示：相关阅读材料
- **enum** 的标准库文档[一]。
- **PEP 435**[二]：向 Python 标准库增加了一个 Enum 类型。
- **flufl.enum**[三]：enum 最早就受此启发，由 Barry Warsaw 创建。

2.2 collections：容器数据类型

collections 模块包含除内置类型 list、dict 和 tuple 以外的其他容器数据类型。

2.2.1 ChainMap：搜索多个字典

ChainMap 类管理一个字典序列，并按其出现的顺序搜索以查找与键关联的值。ChainMap 提供了一个很好的"上下文"容器，因为可以把它看作一个栈，栈增长时发生变更，栈收缩时这些变更被丢弃。

2.2.1.1 访问值

ChainMap 支持与常规字典相同的 API 来访问现有的值。

代码清单 2-11：collections_chainmap_read.py

```python
import collections

a = {'a': 'A', 'c': 'C'}
b = {'b': 'B', 'c': 'D'}

m = collections.ChainMap(a, b)

print('Individual Values')
print('a = {}'.format(m['a']))
print('b = {}'.format(m['b']))
print('c = {}'.format(m['c']))
print()

print('Keys = {}'.format(list(m.keys())))
print('Values = {}'.format(list(m.values())))
print()

print('Items:')
for k, v in m.items():
    print('{} = {}'.format(k, v))
```

[一] https://docs.python.org/3.5/library/enum.html
[二] www.python.org/dev/peps/pep-0435
[三] http://pythonhosted.org/flufl.enum/

```
print()
print('"d" in m: {}'.format(('d' in m)))
```

按子映射传递到构造函数的顺序来搜索这些子映射,所以对应键 'c' 报告的值来自 a 字典。

```
$ python3 collections_chainmap_read.py

Individual Values
a = A
b = B
c = C

Keys = ['c', 'b', 'a']
Values = ['C', 'B', 'A']

Items:
c = C
b = B
a = A

"d" in m: False
```

2.2.1.2 重排

ChainMap 会在它的 maps 属性中存储要搜索的映射列表。这个列表是可变的,所以可以直接增加新映射,或者改变元素的顺序以控制查找和更新行为。

代码清单 2-12:collections_chainmap_reorder.py

```
import collections

a = {'a': 'A', 'c': 'C'}
b = {'b': 'B', 'c': 'D'}

m = collections.ChainMap(a, b)

print(m.maps)
print('c = {}\n'.format(m['c']))

# Reverse the list.
m.maps = list(reversed(m.maps))

print(m.maps)
print('c = {}'.format(m['c']))
```

逆置映射列表时,与 'c' 关联的值会改变。

```
$ python3 collections_chainmap_reorder.py

[{'c': 'C', 'a': 'A'}, {'c': 'D', 'b': 'B'}]
c = C

[{'c': 'D', 'b': 'B'}, {'c': 'C', 'a': 'A'}]
c = D
```

2.2.1.3 更新值

ChainMap 不会缓存子映射中的值。因此,如果它们的内容有修改,则访问 ChainMap 时会反映到结果中。

代码清单 2-13: **collections_chainmap_update_behind.py**

```python
import collections

a = {'a': 'A', 'c': 'C'}
b = {'b': 'B', 'c': 'D'}

m = collections.ChainMap(a, b)
print('Before: {}'.format(m['c']))
a['c'] = 'E'
print('After : {}'.format(m['c']))
```

改变与现有键关联的值与增加新元素的做法一样。

```
$ python3 collections_chainmap_update_behind.py

Before: C
After : E
```

也可以直接通过 ChainMap 设置值，不过实际上只有链中的第一个映射会被修改。

代码清单 2-14: **collections_chainmap_update_directly.py**

```python
import collections

a = {'a': 'A', 'c': 'C'}
b = {'b': 'B', 'c': 'D'}

m = collections.ChainMap(a, b)
print('Before:', m)
m['c'] = 'E'
print('After :', m)
print('a:', a)
```

使用 m 存储新值时，a 映射会更新。

```
$ python3 collections_chainmap_update_directly.py

Before: ChainMap({'c': 'C', 'a': 'A'}, {'c': 'D', 'b': 'B'})
After : ChainMap({'c': 'E', 'a': 'A'}, {'c': 'D', 'b': 'B'})
a: {'c': 'E', 'a': 'A'}
```

ChainMap 提供了一种便利方法，可以用一个额外的映射在 maps 列表的最前面创建一个新实例，这样就能轻松地避免修改现有的底层数据结构。

代码清单 2-15: **collections_chainmap_new_child.py**

```python
import collections

a = {'a': 'A', 'c': 'C'}
b = {'b': 'B', 'c': 'D'}

m1 = collections.ChainMap(a, b)
m2 = m1.new_child()

print('m1 before:', m1)
print('m2 before:', m2)

m2['c'] = 'E'

print('m1 after:', m1)
print('m2 after:', m2)
```

正是基于这种堆栈行为，可以很方便地使用 `ChainMap` 实例作为模板或应用上下文。具体地，可以很容易地在一次迭代中增加或更新值，然后在下一次迭代中丢弃这些改变。

```
$ python3 collections_chainmap_new_child.py

m1 before: ChainMap({'c': 'C', 'a': 'A'}, {'c': 'D', 'b': 'B'})
m2 before: ChainMap({}, {'c': 'C', 'a': 'A'}, {'c': 'D', 'b':
'B'})
m1 after: ChainMap({'c': 'C', 'a': 'A'}, {'c': 'D', 'b': 'B'})
m2 after: ChainMap({'c': 'E'}, {'c': 'C', 'a': 'A'}, {'c': 'D',
'b': 'B'})
```

如果新上下文已知或提前构建，还可以向 `new_child()` 传递一个映射。

代码清单 2-16：`collections_chainmap_new_child_explicit.py`

```python
import collections

a = {'a': 'A', 'c': 'C'}
b = {'b': 'B', 'c': 'D'}
c = {'c': 'E'}

m1 = collections.ChainMap(a, b)
m2 = m1.new_child(c)

print('m1["c"] = {}'.format(m1['c']))
print('m2["c"] = {}'.format(m2['c']))
```

这相当于：

```
m2 = collections.ChainMap(c, *m1.maps)
```

并且还会产生：

```
$ python3 collections_chainmap_new_child_explicit.py

m1["c"] = C
m2["c"] = E
```

2.2.2 `Counter`：统计可散列的对象

`Counter` 是一个容器，可以跟踪等效值增加的次数。这个类可以用来实现其他语言中常用包（bag）或多集合（multiset）数据结构实现的算法。

2.2.2.1 初始化

`Counter` 支持 3 种形式的初始化。调用 `Counter` 的构造函数时可以提供一个元素序列或者一个包含键和计数的字典，还可以使用关键字参数将字符串名映射到计数。

代码清单 2-17：`collections_counter_init.py`

```python
import collections

print(collections.Counter(['a', 'b', 'c', 'a', 'b', 'b']))
print(collections.Counter({'a': 2, 'b': 3, 'c': 1}))
print(collections.Counter(a=2, b=3, c=1))
```

这 3 种形式的初始化结果都是一样的。

```
$ python3 collections_counter_init.py

Counter({'b': 3, 'a': 2, 'c': 1})
Counter({'b': 3, 'a': 2, 'c': 1})
Counter({'b': 3, 'a': 2, 'c': 1})
```

如果不提供任何参数，则可以构造一个空 Counter，然后通过 update() 方法填充。

代码清单 2-18：**collections_counter_update.py**

```
import collections

c = collections.Counter()
print('Initial :', c)

c.update('abcdaab')
print('Sequence:', c)

c.update({'a': 1, 'd': 5})
print('Dict    :', c)
```

计数值只会根据新数据增加，替换数据并不会改变计数。在下面的例子中，a 的计数会从 3 增加到 4。

```
$ python3 collections_counter_update.py

Initial : Counter()
Sequence: Counter({'a': 3, 'b': 2, 'c': 1, 'd': 1})
Dict    : Counter({'d': 6, 'a': 4, 'b': 2, 'c': 1})
```

2.2.2.2 访问计数

一旦填充了 Counter，便可以使用字典 API 获取它的值。

代码清单 2-19：**collections_counter_get_values.py**

```
import collections

c = collections.Counter('abcdaab')

for letter in 'abcde':
    print('{} : {}'.format(letter, c[letter]))
```

对于未知的元素，Counter 不会产生 KeyError。如果在输入中没有找到某个值（如此例中的 e），则其计数为 0。

```
$ python3 collections_counter_get_values.py

a : 3
b : 2
c : 1
d : 1
e : 0
```

elements() 方法返回一个迭代器，该迭代器将生成 Counter 知道的所有元素。

代码清单 2-20：**collections_counter_elements.py**

```
import collections
```

```
c = collections.Counter('extremely')
c['z'] = 0
print(c)
print(list(c.elements()))
```

不能保证元素的顺序不变，另外计数小于或等于 0 的元素不包含在内。

```
$ python3 collections_counter_elements.py

Counter({'e': 3, 'x': 1, 'm': 1, 't': 1, 'y': 1, 'l': 1, 'r': 1,
'z': 0})
['x', 'm', 't', 'e', 'e', 'e', 'y', 'l', 'r']
```

使用 most_common() 可以生成一个序列，其中包含 n 个最常遇到的输入值及相应计数。

代码清单 2-21：**collections_counter_most_common.py**

```
import collections

c = collections.Counter()
with open('/usr/share/dict/words', 'rt') as f:
    for line in f:
        c.update(line.rstrip().lower())

print('Most common:')
for letter, count in c.most_common(3):
    print('{}: {:>7}'.format(letter, count))
```

这个例子要统计系统字典内所有单词中出现的字母，以生成一个频度分布，然后打印 3 个最常见的字母。如果不向 most_common() 提供参数，则会生成由所有元素构成的一个列表，按频度排序。

```
$ python3 collections_counter_most_common.py

Most common:
e:  235331
i:  201032
a:  199554
```

2.2.2.3　算术操作

Counter 实例支持用算术和集合操作来完成结果的聚集。下面这个例子展示了创建新 Counter 实例的标准操作符，不过也支持 +=, -=, &= 和 |= 等原地执行的操作符。

代码清单 2-22：**collections_counter_arithmetic.py**

```
import collections

c1 = collections.Counter(['a', 'b', 'c', 'a', 'b', 'b'])
c2 = collections.Counter('alphabet')

print('C1:', c1)
print('C2:', c2)

print('\nCombined counts:')
print(c1 + c2)

print('\nSubtraction:')
print(c1 - c2)

print('\nIntersection (taking positive minimums):')
```

```
print(c1 & c2)

print('\nUnion (taking maximums):')
print(c1 | c2)
```

每次通过一个操作生成一个新的 Counter 时，计数为 0 或负数的元素都会被删除。在 c1 和 c2 中 a 的计数相同，所以减法操作后它的计数为 0。

```
$ python3 collections_counter_arithmetic.py

C1: Counter({'b': 3, 'a': 2, 'c': 1})
C2: Counter({'a': 2, 'b': 1, 'p': 1, 't': 1, 'l': 1, 'e': 1, 'h': 1})

Combined counts:
Counter({'b': 4, 'a': 4, 'p': 1, 't': 1, 'c': 1, 'e': 1, 'l': 1, 'h': 1})

Subtraction:
Counter({'b': 2, 'c': 1})

Intersection (taking positive minimums):
Counter({'a': 2, 'b': 1})

Union (taking maximums):
Counter({'b': 3, 'a': 2, 'p': 1, 't': 1, 'c': 1, 'e': 1, 'l': 1, 'h': 1})
```

2.2.3 defaultdict：缺少的键返回一个默认值

标准字典包括一个 setdefault() 方法，该方法被用来获取一个值，如果这个值不存在则建立一个默认值。与之相反，初始化容器时 defaultdict 会让调用者提前指定默认值。

代码清单 2-23：collections_defaultdict.py

```
import collections


def default_factory():
    return 'default value'


d = collections.defaultdict(default_factory, foo='bar')
print('d:', d)
print('foo =>', d['foo'])
print('bar =>', d['bar'])
```

只要所有键都有相同的默认值，那么这个方法就可以被很好地使用。如果默认值是一种用于聚集或累加值的类型，如 list、set 或者 int，那么这个方法尤其有用。标准库文档提供了很多以这种方式使用 defaultdict 的例子。

```
$ python3 collections_defaultdict.py

d: defaultdict(<function default_factory at 0x101921950>,
{'foo': 'bar'})
foo => bar
bar => default value
```

提示：相关阅读材料

- defaultdict examples[⊖]：标准库文档中使用 defaultdict 的例子。

⊖ https://docs.python.org/3.5/library/collections.html#defaultdict-examples

- Evolution of Default Dictionaries in Python[⊖]：James Tauber 对 `defaultdict` 与初始化字典的其他方式相关关系所做的讨论。

2.2.4 deque：双端队列

双端队列或 `deque` 支持从任意一端增加和删除元素。更为常用的两种结构（即栈和队列）就是双端队列的退化形式，它们的输入和输出被限制在某一端。

代码清单 2-24：**collections_deque.py**

```
import collections

d = collections.deque('abcdefg')
print('Deque:', d)
print('Length:', len(d))
print('Left end:', d[0])
print('Right end:', d[-1])

d.remove('c')
print('remove(c):', d)
```

由于 `deque` 是一种序列容器，因此同样支持 list 的一些操作，如用 `__getitem__()` 检查内容，确定长度，以及通过匹配标识从队列中间删除元素。

```
$ python3 collections_deque.py

Deque: deque(['a', 'b', 'c', 'd', 'e', 'f', 'g'])
Length: 7
Left end: a
Right end: g
remove(c): deque(['a', 'b', 'd', 'e', 'f', 'g'])
```

2.2.4.1 填充

可以从任意一端填充 `deque`，其在 Python 实现中被称为"左端"和"右端"。

代码清单 2-25：**collections_deque_populating.py**

```
import collections

# Add to the right.
d1 = collections.deque()
d1.extend('abcdefg')
print('extend    :', d1)
d1.append('h')
print('append    :', d1)

# Add to the left.
d2 = collections.deque()
d2.extendleft(range(6))
print('extendleft:', d2)
d2.appendleft(6)
print('appendleft:', d2)
```

`extendleft()` 函数迭代处理其输入，对各个元素完成与 `appendleft()` 同样的处

[⊖] http://jtauber.com/blog/2008/02/27/evolution_of_default_dictionaries_in_python/

理。最终结果是 deque 将包含逆序的输入序列。

```
$ python3 collections_deque_populating.py

extend     : deque(['a', 'b', 'c', 'd', 'e', 'f', 'g'])
append     : deque(['a', 'b', 'c', 'd', 'e', 'f', 'g', 'h'])
extendleft : deque([5, 4, 3, 2, 1, 0])
appendleft : deque([6, 5, 4, 3, 2, 1, 0])
```

2.2.4.2 消费

类似地，可以从两端或任意一端消费 deque 的元素，这取决于所应用的算法。

代码清单 2-26：**collections_deque_consuming.py**

```python
import collections

print('From the right:')
d = collections.deque('abcdefg')
while True:
    try:
        print(d.pop(), end='')
    except IndexError:
        break
print

print('\nFrom the left:')
d = collections.deque(range(6))
while True:
    try:
        print(d.popleft(), end='')
    except IndexError:
        break
print
```

使用 pop() 可以从 deque 的右端删除一个元素，使用 popleft() 可以从左端取一个元素。

```
$ python3 collections_deque_consuming.py

From the right:
gfedcba
From the left:
012345
```

由于双端队列是线程安全的，所以甚至可以在不同线程中同时从两端消费队列的内容。

代码清单 2-27：**collections_deque_both_ends.py**

```python
import collections
import threading
import time

candle = collections.deque(range(5))


def burn(direction, nextSource):
    while True:
        try:
            next = nextSource()
        except IndexError:
```

```
                break
        else:
            print('{:>8}: {}'.format(direction, next))
            time.sleep(0.1)
    print('{:>8} done'.format(direction))
    return

left = threading.Thread(target=burn,
                        args=('Left', candle.popleft))
right = threading.Thread(target=burn,
                         args=('Right', candle.pop))

left.start()
right.start()

left.join()
right.join()
```

这个例子中的线程交替处理两端，删除元素，直至这个 deque 为空。

```
$ python3 collections_deque_both_ends.py

    Left: 0
   Right: 4
   Right: 3
    Left: 1
   Right: 2
    Left done
   Right done
```

2.2.4.3 旋转

deque 的另一个很有用的方面是可以按任意一个方向旋转，从而跳过一些元素。

代码清单 2-28：**collections_deque_rotate.py**

```
import collections

d = collections.deque(range(10))
print('Normal        :', d)

d = collections.deque(range(10))
d.rotate(2)
print('Right rotation:', d)

d = collections.deque(range(10))
d.rotate(-2)
print('Left rotation :', d)
```

将 deque 向右旋转（使用一个正旋转值）会从右端取元素，并且把它们移到左端。向左旋转（使用一个负值）则从左端将元素移至右端。可以形象地把 deque 中的元素看作是刻在拨号盘上，这对于理解双端队列很有帮助。

```
$ python3 collections_deque_rotate.py

Normal        : deque([0, 1, 2, 3, 4, 5, 6, 7, 8, 9])
Right rotation: deque([8, 9, 0, 1, 2, 3, 4, 5, 6, 7])
Left rotation : deque([2, 3, 4, 5, 6, 7, 8, 9, 0, 1])
```

2.2.4.4 限制队列大小

配置 deque 实例时可以指定一个最大长度，使它不会超过这个大小。队列达到指定的长度时，随着新元素的增加会删除现有的元素。如果要查找一个长度不确定的流中的最后 n 个元素，那么这种行为会很有用。

代码清单 2-29：collections_deque_maxlen.py

```
import collections
import random

# Set the random seed so we see the same output each time
# the script is run.
random.seed(1)

d1 = collections.deque(maxlen=3)
d2 = collections.deque(maxlen=3)

for i in range(5):
    n = random.randint(0, 100)
    print('n =', n)
    d1.append(n)
    d2.appendleft(n)
    print('D1:', d1)
    print('D2:', d2)
```

不论元素增加到哪一端，队列长度都保持不变。

```
$ python3 collections_deque_maxlen.py

n = 17
D1: deque([17], maxlen=3)
D2: deque([17], maxlen=3)
n = 72
D1: deque([17, 72], maxlen=3)
D2: deque([72, 17], maxlen=3)
n = 97
D1: deque([17, 72, 97], maxlen=3)
D2: deque([97, 72, 17], maxlen=3)
n = 8
D1: deque([72, 97, 8], maxlen=3)
D2: deque([8, 97, 72], maxlen=3)
n = 32
D1: deque([97, 8, 32], maxlen=3)
D2: deque([32, 8, 97], maxlen=3)
```

提示：相关阅读材料

- Wikipedia: Deque[⊖]：对双端队列数据结构的讨论。
- deque Recipes[⊖]：标准库文档的算法中使用双端队列的例子。

2.2.5 namedtuple：带命名字段的元组子类

标准 tuple 使用数值索引来访问其成员。

[⊖] https://en.wikipedia.org/wiki/Deque

[⊖] https://docs.python.org/3.5/library/collections.html#deque-recipes

代码清单 2-30：collections_tuple.py

```
bob = ('Bob', 30, 'male')
print('Representation:', bob)

jane = ('Jane', 29, 'female')
print('\nField by index:', jane[0])

print('\nFields by index:')
for p in [bob, jane]:
    print('{} is a {} year old {}'.format(*p))
```

对于简单的用途，`tuple` 是很方便的容器。

```
$ python3 collections_tuple.py

Representation: ('Bob', 30, 'male')

Field by index: Jane

Fields by index:
Bob is a 30 year old male
Jane is a 29 year old female
```

另一方面，使用 `tuple` 时需要记住对应各个值要使用哪个索引，这可能会导致错误，特别是当 `tuple` 有大量字段，而且构造元组和使用元组的位置相距很远时。`namedtuple` 除了为各个成员指定数值索引外，还为其指定名字。

2.2.5.1 定义

与常规的元组一样，`namedtuple` 实例在内存使用方面同样很高效，因为它们没有每一个实例的字典。各种 `namedtuple` 都由自己的类表示，这个类使用 `namedtuple()` 工厂函数来创建。参数就是新类名和一个包含元素名的字符串。

代码清单 2-31：collections_namedtuple_person.py

```
import collections

Person = collections.namedtuple('Person', 'name age')

bob = Person(name='Bob', age=30)
print('\nRepresentation:', bob)

jane = Person(name='Jane', age=29)
print('\nField by name:', jane.name)

print('\nFields by index:')
for p in [bob, jane]:
    print('{} is {} years old'.format(*p))
```

如这个例子所示，除了使用标准元组的位置索引外，还可以使用点记法（`obj.attr`）按名字访问 `namedtuple` 的字段。

```
$ python3 collections_namedtuple_person.py

Representation: Person(name='Bob', age=30)
```

```
Field by name: Jane

Fields by index:
Bob is 30 years old
Jane is 29 years old
```

与常规 tuple 类似，namedtuple 也是不可修改的。这个限制允许 tuple 实例具有一致的散列值，这使得可以把它们用作字典中的键并包含在集合中。

代码清单 2-32：**collections_namedtuple_immutable.py**

```
import collections

Person = collections.namedtuple('Person', 'name age')

pat = Person(name='Pat', age=12)
print('\nRepresentation:', pat)

pat.age = 21
```

如果试图通过命名属性改变一个值，那么这会导致一个 AttributeError。

```
$ python3 collections_namedtuple_immutable.py

Representation: Person(name='Pat', age=12)
Traceback (most recent call last):
  File "collections_namedtuple_immutable.py", line 17, in <module>
    pat.age = 21
AttributeError: can't set attribute
```

2.2.5.2 非法字段名

如果字段名重复或者与 Python 关键字冲突，那么其就是非法字段名。

代码清单 2-33：**collections_namedtuple_bad_fields.py**

```
import collections

try:
    collections.namedtuple('Person', 'name class age')
except ValueError as err:
    print(err)

try:
    collections.namedtuple('Person', 'name age age')
except ValueError as err:
    print(err)
```

解析字段名时，非法值会导致 ValueError 异常。

```
$ python3 collections_namedtuple_bad_fields.py

Type names and field names cannot be a keyword: 'class'
Encountered duplicate field name: 'age'
```

如果要基于程序控制之外的值创建一个 namedtuple（如表示一个数据库查询返回的记录行，而事先并不知道数据库模式），那么这种情况下应把 rename 选项设置为 True，以对非法字段重命名。

代码清单2-34：**collections_namedtuple_rename.py**

```
import collections

with_class = collections.namedtuple(
    'Person', 'name class age',
    rename=True)
print(with_class._fields)

two_ages = collections.namedtuple(
    'Person', 'name age age',
    rename=True)
print(two_ages._fields)
```

重命名字段的新名字取决于它在元组中的索引，所以名为 `class` 的字段会变成 `_1`，重复的 `age` 字段则变成 `_2`。

```
$ python3 collections_namedtuple_rename.py

('name', '_1', 'age')
('name', 'age', '_2')
```

2.2.5.3 指定属性

`namedtuple` 提供了很多有用的属性和方法来处理子类和实例。所有这些内置属性名都有一个下划线（`_`）前缀，按惯例在大多数 Python 程序中，这都会指示一个私有属性。不过，对于 `namedtuple`，这个前缀是为了防止这个名字与用户提供的属性名冲突。

传入 `namedtuple` 来定义新类的字段名会保存在 `_fields` 属性中。

代码清单2-35：**collections_namedtuple_fields.py**

```
import collections

Person = collections.namedtuple('Person', 'name age')

bob = Person(name='Bob', age=30)
print('Representation:', bob)
print('Fields:', bob._fields)
```

尽管参数是一个用空格分隔的字符串，但存储的值却是由各个名字组成的一个序列。

```
$ python3 collections_namedtuple_fields.py

Representation: Person(name='Bob', age=30)
Fields: ('name', 'age')
```

可以使用 `_asdict()` 将 `namedtuple` 实例转换为 `OrderedDict` 实例。

代码清单2-36：**collections_namedtuple_asdict.py**

```
import collections

Person = collections.namedtuple('Person', 'name age')

bob = Person(name='Bob', age=30)
print('Representation:', bob)
print('As Dictionary:', bob._asdict())
```

OrderedDict 的键与相应 namedtuple 的字段顺序相同。

```
$ python3 collections_namedtuple_asdict.py

Representation: Person(name='Bob', age=30)
As Dictionary: OrderedDict([('name', 'Bob'), ('age', 30)])
```

_replace() 方法构建一个新实例，在这个过程中会替换一些字段的值。

代码清单 2-37：**collections_namedtuple_replace.py**

```
import collections

Person = collections.namedtuple('Person', 'name age')

bob = Person(name='Bob', age=30)
print('\nBefore:', bob)
bob2 = bob._replace(name='Robert')
print('After:', bob2)
print('Same?:', bob is bob2)
```

尽管从名字上看似乎会修改现有的对象，但由于 namedtuple 实例是不可变的，所以实际上这个方法会返回一个新对象。

```
$ python3 collections_namedtuple_replace.py

Before: Person(name='Bob', age=30)
After: Person(name='Robert', age=30)
Same?: False
```

2.2.6 OrderedDict：记住向字典中增加键的顺序

OrderedDict 是一个字典子类，可以记住其内容增加的顺序。

代码清单 2-38：**collections_ordereddict_iter.py**

```
import collections

print('Regular dictionary:')
d = {}
d['a'] = 'A'
d['b'] = 'B'
d['c'] = 'C'

for k, v in d.items():
    print(k, v)

print('\nOrderedDict:')
d = collections.OrderedDict()
d['a'] = 'A'
d['b'] = 'B'
d['c'] = 'C'

for k, v in d.items():
    print(k, v)
```

常规 dict 并不跟踪插入顺序，迭代处理时会根据散列表中如何存储键来按顺序生成值，而散列表中键的存储会受一个随机值的影响，以减少冲突。OrderedDict 中则相反，它会记住元素插入的顺序，并在创建迭代器时使用这个顺序。

```
$ python3 collections_ordereddict_iter.py

Regular dictionary:
c C
b B
a A

OrderedDict:
a A
b B
c C
```

2.2.6.1 相等性

常规的 dict 在检查相等性时会查看其内容。OrderedDict 还会考虑元素增加的顺序。

代码清单 2-39：**collections_ordereddict_equality.py**

```
import collections

print('dict       :', end=' ')
d1 = {}
d1['a'] = 'A'
d1['b'] = 'B'
d1['c'] = 'C'

d2 = {}
d2['c'] = 'C'
d2['b'] = 'B'
d2['a'] = 'A'

print(d1 == d2)

print('OrderedDict:', end=' ')

d1 = collections.OrderedDict()
d1['a'] = 'A'
d1['b'] = 'B'
d1['c'] = 'C'

d2 = collections.OrderedDict()
d2['c'] = 'C'
d2['b'] = 'B'
d2['a'] = 'A'

print(d1 == d2)
```

在这个例子中，由于两个有序字典由不同顺序的值创建，所以认为这两个有序字典是不同的。

```
$ python3 collections_ordereddict_equality.py

dict       : True
OrderedDict: False
```

2.2.6.2 重排

在 OrderedDict 中可以使用 move_to_end() 将键移至序列的起始或末尾位置来改变键的顺序。

代码清单 2-40：`collections_ordereddict_move_to_end.py`

```python
import collections

d = collections.OrderedDict(
    [('a', 'A'), ('b', 'B'), ('c', 'C')]
)

print('Before:')
for k, v in d.items():
    print(k, v)

d.move_to_end('b')

print('\nmove_to_end():')
for k, v in d.items():
    print(k, v)

d.move_to_end('b', last=False)

print('\nmove_to_end(last=False):')
for k, v in d.items():
    print(k, v)
```

`last` 参数会告诉 `move_to_end()` 要把元素移动为键序列的最后一个元素（参数值为 `True`）或者第一个元素（参数值为 `False`）。

```
$ python3 collections_ordereddict_move_to_end.py

Before:
a A
b B
c C

move_to_end():
a A
c C
b B

move_to_end(last=False):
b B
a A
c C
```

提示：相关阅读材料

- PYTHONHASHSEED[⊖]：这个环境变量可以控制散列算法（用来确定字典中键的位置）增加的随机种子值。

2.2.7 collections.abc：容器的抽象基类

`collections.abc` 模块包含一些抽象基类，其为 Python 内置容器数据结构以及 collections 模块定义的容器数据结构定义了 API。表 2-1 给出了这些类及其用途的一个列表。

⊖ https://docs.python.org/3.5/using/cmdline.html#envvar-PYTHONHASHSEED

表 2-1 抽象基类

类	基类	API 用途
Container		基本容器特性，如 in 操作符
Hashable		增加了散列支持，可以为容器实例提供散列值
Iterable		可以在容器内容上创建一个迭代器
Iterator	Iterable	这是容器内容上的一个迭代器
Generator	Iterator	为迭代器扩展了 PEP 342 的生成器协议
Sized		为知道自己大小的容器增加方法
Callable		可以作为函数来调用的容器
Sequence	Sized, Iterable, Container	支持获取单个元素以及迭代和改变元素顺序
MutableSequence	Sequence	支持创建一个实例之后增加和删除元素
ByteString	Sequence	合并 bytes 和 bytearray 的 API
Set	Sized, Iterable, Container	支持集合操作，如交集和并集
MutableSet	Set	增加了创建集合后管理集合内容的方法
Mapping	Sized, Iterable, Container	定义 dict 使用的只读 API
MutableMapping	Mapping	定义创建映射后管理映射内容的方法
MappingView	Sized	定义从迭代器访问映射的视图 API
ItemsView	MappingView, Set	视图 API 的一部分
KeysView	MappingView, Set	视图 API 的一部分
ValuesView	MappingView	视图 API 的一部分
Awaitable		await 表达式中可用的对象的 API，如协程
Coroutine	Awaitable	实现协程协议的类的 API
AsyncIterable		与 async for（PEP 492 中定义）兼容的 iterable 的 API
AsyncIterator	AsyncIterable	异步迭代器的 API

除了明确地定义不同容器的 API，这些抽象基类还可以在调用对象前用 isinstance() 测试一个对象是否支持一个 API。有些类还提供了方法实现，它们可以作为"混入类"（mix-in）构造定制容器类型，而不必从头实现每一个方法。

提示：相关阅读材料

- collections 的标准库文档[一]。
- 关于 collections 的 Python 2 到 Python 3 移植说明。
- **PEP 342**[二]：通过改进生成器实现的协程。
- **PEP 492**[三]：采用 async 和 await 语法的协程。

[一] https://docs.python.org/3.5/library/collections.html
[二] www.python.org/dev/peps/pep-0342
[三] www.python.org/dev/peps/pep-0492

2.3 数组：固定类型数据序列

array模块定义了一个序列数据结构，看起来与list很相似，只不过所有成员都必须是相同的基本类型。支持的类型包括所有数值类型或其他固定大小的基本类型（如字节）。

表2-2给出了支持的一些类型。array的标准库文档提供了所有类型代码的完整列表。

表 2-2 array成员的类型代码

代码	类型	最小大小（字节）
b	Int	1
B	Int	1
h	Signed short	2
H	Unsigned short	2
i	Signed int	2
I	Unsigned int	2
l	Signed long	4
L	Unsigned long	4
q	Signed long long	8
Q	Unsigned long long	8
f	Float	4
d	Double float	8

2.3.1 初始化

array被实例化时可以提供一个参数来描述允许哪种数据类型，还可以有一个存储在数组中的初始数据序列。

代码清单 2-41：array_string.py

```
import array
import binascii

s = b'This is the array.'
a = array.array('b', s)

print('As byte string:', s)
print('As array      :', a)
print('As hex        :', binascii.hexlify(a))
```

在这个例子中，数组被配置为包含一个字节序列，并用一个简单的字符串初始化。

```
$ python3 array_string.py

As byte string: b'This is the array.'
As array      : array('b', [84, 104, 105, 115, 32, 105, 115, 32,
 116, 104, 101, 32, 97, 114, 114, 97, 121, 46])
As hex        : b'546869732069732074686520617272617972e'
```

2.3.2 处理数组

与其他 Python 序列类似，可以采用同样的方式扩展和处理 array。

代码清单 2-42：`array_sequence.py`

```python
import array
import pprint

a = array.array('i', range(3))
print('Initial :', a)

a.extend(range(3))
print('Extended:', a)

print('Slice   :', a[2:5])

print('Iterator:')
print(list(enumerate(a)))
```

目前支持的操作包括分片、迭代以及在末尾增加元素。

```
$ python3 array_sequence.py

Initial : array('i', [0, 1, 2])
Extended: array('i', [0, 1, 2, 0, 1, 2])
Slice   : array('i', [2, 0, 1])
Iterator:
[(0, 0), (1, 1), (2, 2), (3, 0), (4, 1), (5, 2)]
```

2.3.3 数组和文件

可以使用专门的高效读写文件的内置方法将数组的内容写入文件或从文件读出数组。

代码清单 2-43：`array_file.py`

```python
import array
import binascii
import tempfile

a = array.array('i', range(5))
print('A1:', a)

# Write the array of numbers to a temporary file.
output = tempfile.NamedTemporaryFile()
a.tofile(output.file)  # Must pass an *actual* file
output.flush()

# Read the raw data.
with open(output.name, 'rb') as input:
    raw_data = input.read()
    print('Raw Contents:', binascii.hexlify(raw_data))

    # Read the data into an array.
    input.seek(0)
    a2 = array.array('i')
    a2.fromfile(input, len(a))
    print('A2:', a2)
```

这个例子展示了直接从二进制文件"原样"读取数据，还展示了将数据读入一个新数组，并把字节转换为适当的类型。

```
$ python3 array_file.py

A1: array('i', [0, 1, 2, 3, 4])
Raw Contents: b'0000000001000000020000000300000004000000'
A2: array('i', [0, 1, 2, 3, 4])
```

`tofile()`使用`tobytes()`格式化数据，`fromfile()`使用`frombytes()`再把它转换回一个数组实例。

代码清单2-44：array_tobytes.py

```
import array
import binascii

a = array.array('i', range(5))
print('A1:', a)

as_bytes = a.tobytes()
print('Bytes:', binascii.hexlify(as_bytes))

a2 = array.array('i')
a2.frombytes(as_bytes)
print('A2:', a2)
```

`tobytes()`和`frombytes()`都处理字节串，而不是Unicode字符串。

```
$ python3 array_tobytes.py

A1: array('i', [0, 1, 2, 3, 4])
Bytes: b'0000000001000000020000000300000004000000'
A2: array('i', [0, 1, 2, 3, 4])
```

2.3.4 候选字节顺序

如果数组中的数据没有采用原生的字节顺序，或者在发送到一个采用不同字节顺序的系统（或在网络上发送）之前数据需要交换顺序，那么可以由Python转换整个数组而不必迭代处理每一个元素。

代码清单2-45：array_byteswap.py

```
import array
import binascii


def to_hex(a):
    chars_per_item = a.itemsize * 2  # 2 hex digits
    hex_version = binascii.hexlify(a)
    num_chunks = len(hex_version) // chars_per_item
    for i in range(num_chunks):
        start = i * chars_per_item
        end = start + chars_per_item
        yield hex_version[start:end]


start = int('0x12345678', 16)
end = start + 5
a1 = array.array('i', range(start, end))
```

```
a2 = array.array('i', range(start, end))
a2.byteswap()

fmt = '{:>12} {:>12} {:>12} {:>12}'
print(fmt.format('A1 hex', 'A1', 'A2 hex', 'A2'))
print(fmt.format('-' * 12, '-' * 12, '-' * 12, '-' * 12))
fmt = '{!r:>12} {:12} {!r:>12} {:12}'
for values in zip(to_hex(a1), a1, to_hex(a2), a2):
    print(fmt.format(*values))
```

byteswap()方法会用C交换数组中元素的字节顺序，这比用Python循环处理数据高效得多。

```
$ python3 array_byteswap.py

      A1 hex           A1       A2 hex           A2
------------  -----------  ------------  -----------
b'78563412'     305419896   b'12345678'   2018915346
b'79563412'     305419897   b'12345679'   2035692562
b'7a563412'     305419898   b'1234567a'   2052469778
b'7b563412'     305419899   b'1234567b'   2069246994
b'7c563412'     305419900   b'1234567c'   2086024210
```

提示：相关阅读材料
- array 的标准库文档[1]。
- struct：struct 模块。
- Numerical Python[2]：NumPy 是一个高效处理大数据集的 Python 库。
- 关于 array 的 Python 2 到 Python 3 移植说明。

2.4 heapq：堆排序算法

堆（heap）是一个树形数据结构，其中子节点与父节点有一种有序关系。二叉堆（binary heap）可以使用一个有组织的列表或数组表示，其中元素 N 的子元素位于 $2*N+1$ 和 $2*N+2$（索引从 0 开始）。这种布局允许原地重新组织堆，从而不必在增加或删除元素时重新分配大量内存。

最大堆（max-heap）确保父节点大于或等于其两个子节点。最小堆（min-heap）要求父节点小于或等于其子节点。Python 的 heapq 模块实现了一个最小堆。

2.4.1 示例数据

这一节中的示例将使用 heapq_heapdata.py 中的数据。

代码清单 2-46：heapq_heapdata.py

```
# This data was generated with the random module.

data = [19, 9, 4, 10, 11]
```

---

[1] https://docs.python.org/3.5/library/array.html

[2] www.scipy.org

堆输出使用 heapq_showtree.py 打印。

代码清单 2-47：**heapq_showtree.py**

```python
import math
from io import StringIO


def show_tree(tree, total_width=36, fill=' '):
    """Pretty-print a tree."""
    output = StringIO()
    last_row = -1
    for i, n in enumerate(tree):
        if i:
            row = int(math.floor(math.log(i + 1, 2)))
        else:
            row = 0
        if row != last_row:
            output.write('\n')
        columns = 2 ** row
        col_width = int(math.floor(total_width / columns))
        output.write(str(n).center(col_width, fill))
        last_row = row
    print(output.getvalue())
    print('-' * total_width)
    print()
```

2.4.2 创建堆

创建堆有两种基本方式：heappush() 和 heapify()。

代码清单 2-48：**heapq_heappush.py**

```python
import heapq
from heapq_showtree import show_tree
from heapq_heapdata import data

heap = []
print('random :', data)
print()

for n in data:
    print('add {:>3}:'.format(n))
    heapq.heappush(heap, n)
    show_tree(heap)
```

使用 heappush()，从数据源增加新元素时会保持元素的堆排序顺序。

```
$ python3 heapq_heappush.py

random : [19, 9, 4, 10, 11]

add  19:

                 19
------------------------------------

add   9:

                 9
```

```
                    19
-----------------------------------
        add  4:
                         4
                   19          9
-----------------------------------
        add  10:
                         4
                   10          9
              19
-----------------------------------
        add  11:
                         4
                   10          9
              19   11
-----------------------------------
```

如果数据已经在内存中,那么使用 `heapify()` 原地重新组织列表中的元素会更高效。

代码清单 2-49:`heapq_heapify.py`

```python
import heapq
from heapq_showtree import show_tree
from heapq_heapdata import data

print('random    :', data)
heapq.heapify(data)
print('heapified :')
show_tree(data)
```

如果按堆顺序一次一个元素地构建列表,那么结果与构建一个无序列表再调用 `heapify()` 是一样的。

```
$ python3 heapq_heapify.py

random    : [19, 9, 4, 10, 11]
heapified :

                         4
              9               19
         10      11
-----------------------------------
```

2.4.3 访问堆的内容

一旦堆已经被正确组织,则可以使用 `heappop()` 删除有最小值的元素。

代码清单 2-50:`heapq_heappop.py`

```python
import heapq
from heapq_showtree import show_tree
from heapq_heapdata import data

print('random    :', data)
heapq.heapify(data)
```

```
print('heapified :')
show_tree(data)
print

for i in range(2):
    smallest = heapq.heappop(data)
    print('pop    {:>3}:'.format(smallest))
    show_tree(data)
```

这个例子是由标准库文档改写的,其中使用heapify()和heappop()对一个数字列表进行排序。

```
$ python3 heapq_heappop.py

random    : [19, 9, 4, 10, 11]
heapified :

                    4
         9                    19
   10         11
------------------------------------

pop     4:
                    9
         10                   19
    11
------------------------------------

pop     9:
                    10
         11                   19
------------------------------------
```

如果希望在一个操作中删除现有元素并替换为新值,则可以使用heapreplace()。

代码清单2-51:heapq_heapreplace.py

```
import heapq
from heapq_showtree import show_tree
from heapq_heapdata import data

heapq.heapify(data)
print('start:')
show_tree(data)

for n in [0, 13]:
    smallest = heapq.heapreplace(data, n)
    print('replace {:>2} with {:>2}:'.format(smallest, n))
    show_tree(data)
```

通过原地替换元素,这样可以维持一个固定大小的堆,如按优先级排序的作业队列。

```
$ python3 heapq_heapreplace.py

start:
                    4
         9                    19
```

```
         10      11
---------------------------------

replace  4 with  0:
                 0
         9               19
10      11
---------------------------------

replace  0 with 13:
                 9
         10              19
13      11
---------------------------------
```

2.4.4　堆的数据极值

heapq 还包括两个检查可迭代对象（iterable）的函数，可以查找其中包含的最大或最小值的范围。

代码清单 2-52　`heapq_extremes.py`

```
import heapq
from heapq_heapdata import data

print('all        :', data)
print('3 largest :', heapq.nlargest(3, data))
print('from sort :', list(reversed(sorted(data)[-3:])))
print('3 smallest:', heapq.nsmallest(3, data))
print('from sort :', sorted(data)[:3])
```

只有当 n 值（$n>1$）相对小时使用 nlargest() 和 nsmallest() 才算高效，不过有些情况下这两个函数会很方便。

```
$ python3 heapq_extremes.py

all        : [19, 9, 4, 10, 11]
3 largest : [19, 11, 10]
from sort : [19, 11, 10]
3 smallest: [4, 9, 10]
from sort : [4, 9, 10]
```

2.4.5　高效合并有序序列

对于小数据集，将多个有序序列合并到一个新序列很容易。

list(sorted(itertools.chain(*data)))

对于较大的数据集，这个技术可能会占用大量内存。merge() 不是对整个合并后的序列排序，而是使用一个堆一次一个元素地生成一个新序列，利用固定大小的内存确定下一个元素。

代码清单 2-53：`heapq_merge.py`

```
import heapq
import random
```

```
random.seed(2016)

data = []
for i in range(4):
    new_data = list(random.sample(range(1, 101), 5))
    new_data.sort()
    data.append(new_data)

for i, d in enumerate(data):
    print('{}: {}'.format(i, d))

print('\nMerged:')
for i in heapq.merge(*data):
    print(i, end=' ')
print()
```

由于 `merge()` 的实现使用了一个堆，所以它会根据所合并的序列个数消费内存，而不是根据这些序列中的元素个数。

```
$ python3 heapq_merge.py

0: [33, 58, 71, 88, 95]
1: [10, 11, 17, 38, 91]
2: [13, 18, 39, 61, 63]
3: [20, 27, 31, 42, 45]

Merged:
10 11 13 17 18 20 27 31 33 38 39 42 45 58 61 63 71 88 91 95
```

提示：相关阅读材料
- `heapq` 的标准库文档[⊖]。
- Wikipedia: Heap (data structure)[⊖]：堆数据结构的一般描述。
- 2.6.3 节"优先队列"：基于标准库中 `Queue` 的一个优先队列实现。

2.5 `bisect`：维护有序列表

`bisect` 模块实现了一个算法来向列表中插入元素，同时仍保持列表有序。

2.5.1 有序插入

下面给出一个简单的例子，这里使用 `insort()` 按有序顺序向一个列表中插入元素。

代码清单 2-54：**bisect_example.py**

```
import bisect

# A series of random numbers
values = [14, 85, 77, 26, 50, 45, 66, 79, 10, 3, 84, 77, 1]

print('New  Pos  Contents')
print('---  ---  --------')
```

⊖ https://docs.python.org/3.5/library/heapq.html

⊖ https://en.wikipedia.org/wiki/Heap_(data_structure)

```
l = []
for i in values:
    position = bisect.bisect(l, i)
    bisect.insort(l, i)
    print('{:3}  {:3}'.format(i, position), l)
```

输出的第一列显示了新随机数。第二列显示了这个数将插入到列表的哪个位置。每一行余下的部分则是当前的有序列表。

```
$ python3 bisect_example.py

New  Pos  Contents
---  ---  --------
 14    0  [14]
 85    1  [14, 85]
 77    1  [14, 77, 85]
 26    1  [14, 26, 77, 85]
 50    2  [14, 26, 50, 77, 85]
 45    2  [14, 26, 45, 50, 77, 85]
 66    4  [14, 26, 45, 50, 66, 77, 85]
 79    6  [14, 26, 45, 50, 66, 77, 79, 85]
 10    0  [10, 14, 26, 45, 50, 66, 77, 79, 85]
  3    0  [3, 10, 14, 26, 45, 50, 66, 77, 79, 85]
 84    9  [3, 10, 14, 26, 45, 50, 66, 77, 79, 84, 85]
 77    8  [3, 10, 14, 26, 45, 50, 66, 77, 77, 79, 84, 85]
  1    0  [1, 3, 10, 14, 26, 45, 50, 66, 77, 77, 79, 84, 85]
```

这是一个很简单的例子，实际上，对于此例处理的数据量来说，如果直接构建列表然后完成一次排序，可能速度更快。不过对于长列表而言，使用类似这样的一个插入排序算法可以大大节省时间和内存，尤其是比较两个列表成员的操作需要开销很大的计算时。

2.5.2 处理重复

之前显示的结果集包括一个重复的值 77。bisect 模块提供了两种方法来处理重复。新值可以插入到原值的左边或右边。insort() 函数实际上是 insort_right() 的别名，这个函数会在原值之后插入新值。相应的 insort_left() 函数则在原值之前插入新值。

代码清单 2-55 **bisect_example2.py**

```
import bisect

# A series of random numbers
values = [14, 85, 77, 26, 50, 45, 66, 79, 10, 3, 84, 77, 1]

print('New  Pos  Contents')
print('---  ---  --------')

# Use bisect_left and insort_left.
l = []
for i in values:
    position = bisect.bisect_left(l, i)
    bisect.insort_left(l, i)
    print('{:3}  {:3}'.format(i, position), l)
```

使用 bisect_left() 和 insort_left() 处理同样的数据时，结果是相同的有序列表，不过重复值插入的位置有所不同。

```
$ python3 bisect_example2.py

New  Pos  Contents
---  ---  --------
 14    0  [14]
 85    1  [14, 85]
 77    1  [14, 77, 85]
 26    1  [14, 26, 77, 85]
 50    2  [14, 26, 50, 77, 85]
 45    2  [14, 26, 45, 50, 77, 85]
 66    4  [14, 26, 45, 50, 66, 77, 85]
 79    6  [14, 26, 45, 50, 66, 77, 79, 85]
 10    0  [10, 14, 26, 45, 50, 66, 77, 79, 85]
  3    0  [3, 10, 14, 26, 45, 50, 66, 77, 79, 85]
 84    9  [3, 10, 14, 26, 45, 50, 66, 77, 79, 84, 85]
 77    7  [3, 10, 14, 26, 45, 50, 66, 77, 77, 79, 84, 85]
  1    0  [1, 3, 10, 14, 26, 45, 50, 66, 77, 77, 79, 84, 85]
```

提示：相关阅读材料
- `bisect` 的标准库文档⊖。
- Wikipedia: Insertion Sort⊖：插入排序算法的描述。

2.6 `queue`：线程安全的 FIFO 实现

`queue` 模块提供了一个适用于多线程编程的先进先出（FIFO，first-in, first-out）数据结构，可以用来在生产者和消费者线程之间安全地传递消息或其他数据。它会为调用者处理锁定，使多个线程可以安全而容易地处理同一个 `Queue` 实例。`Queue` 的大小（其中包含的元素个数）可能受限，以限制内存使用或处理。

说明：这里的讨论假设你已经了解队列的一般性质。如果你还不太清楚，那么在学习下面的内容之前可能需要先阅读一些有关的参考资料。

2.6.1 基本 FIFO 队列

`Queue` 类实现了一个基本的先进先出容器。使用 `put()` 将元素增加到这个序列的一端，使用 `get()` 从另一端删除。

代码清单 2-56：`queue_fifo.py`

```
import queue

q = queue.Queue()

for i in range(5):
    q.put(i)
```

⊖ https://docs.python.org/3.5/library/bisect.html
⊖ https://en.wikipedia.org/wiki/Insertion_sort

```
while not q.empty():
    print(q.get(), end=' ')
print()
```

这个例子使用了一个线程来展示按插入元素的相同顺序从队列删除元素。

```
$ python3 queue_fifo.py

0 1 2 3 4
```

2.6.2 LIFO 队列

与 `Queue` 的标准 FIFO 实现相反，`LifoQueue` 使用了（通常与栈数据结构关联的）后进先出（LIFO，last-in, first-out）顺序。

代码清单 2-57：**queue_lifo.py**

```
import queue

q = queue.LifoQueue()

for i in range(5):
    q.put(i)

while not q.empty():
    print(q.get(), end=' ')
print()
```

`get` 将删除最近使用 `put` 插入到队列的元素。

```
$ python3 queue_lifo.py

4 3 2 1 0
```

2.6.3 优先队列

有些情况下，需要根据队列中元素的特性来决定这些元素的处理顺序，而不是简单地采用在队列中创建或插入元素的顺序。例如，工资部门的打印作业可能就优先于某个开发人员想要打印的代码清单。`PriorityQueue` 使用队列内容的有序顺序来决定获取哪一个元素。

代码清单 2-58：**queue_priority.py**

```
import functools
import queue
import threading


@functools.total_ordering
class Job:

    def __init__(self, priority, description):
        self.priority = priority
        self.description = description
        print('New job:', description)
        return
```

```python
    def __eq__(self, other):
        try:
            return self.priority == other.priority
        except AttributeError:
            return NotImplemented

    def __lt__(self, other):
        try:
            return self.priority < other.priority
        except AttributeError:
            return NotImplemented

q = queue.PriorityQueue()

q.put(Job(3, 'Mid-level job'))
q.put(Job(10, 'Low-level job'))
q.put(Job(1, 'Important job'))

def process_job(q):
    while True:
        next_job = q.get()
        print('Processing job:', next_job.description)
        q.task_done()

workers = [
    threading.Thread(target=process_job, args=(q,)),
    threading.Thread(target=process_job, args=(q,)),
]
for w in workers:
    w.setDaemon(True)
    w.start()

q.join()
```

这个例子有多个线程在处理作业，要根据调用 get() 时队列中元素的优先级来处理。运行消费者线程时，增加到队列的元素的处理顺序取决于线程上下文切换。

```
$ python3 queue_priority.py

New job: Mid-level job
New job: Low-level job
New job: Important job
Processing job: Important job
Processing job: Mid-level job
Processing job: Low-level job
```

2.6.4 构建一个多线程播客客户程序

这一节将构建一个播客客户程序，程序的源代码展示了如何利用多个线程使用 Queue 类。这个程序要读入一个或多个 RSS 提要，对每个提要的专辑排队，显示最新的五集以供下载，并使用线程并行地处理多个下载。这里没有提供完备的错误处理，所以不能在实际的生产环境中使用，不过这个框架实现可以作为一个很好的例子来说明如何使用 queue 模块。

首先要建立一些操作参数。一般情况下，这些参数来自用户输入（例如，首选项、数据库等）。不过在这个例子中，线程数和要获取的 URL 列表都使用了硬编码值。

代码清单 2-59：`fetch_podcasts.py`

```python
from queue import Queue
import threading
import time
import urllib
from urllib.parse import urlparse

import feedparser

# Set up some global variables.
num_fetch_threads = 2
enclosure_queue = Queue()
# A real app wouldn't use hard-coded data.
feed_urls = [
    'http://talkpython.fm/episodes/rss',
]


def message(s):
    print('{}: {}'.format(threading.current_thread().name, s))
```

函数 `download_enclosures()` 在工作线程中运行，使用 `urllib` 处理下载。

```python
def download_enclosures(q):
    """This is the worker thread function.
    It processes items in the queue one after
    another.  These daemon threads go into an
    infinite loop, and exit only when
    the main thread ends.
    """
    while True:
        message('looking for the next enclosure')
        url = q.get()
        filename = url.rpartition('/')[-1]
        message('downloading {}'.format(filename))
        response = urllib.request.urlopen(url)
        data = response.read()
        # Save the downloaded file to the current directory.
        message('writing to {}'.format(filename))
        with open(filename, 'wb') as outfile:
            outfile.write(data)
        q.task_done()
```

一旦定义了线程的目标函数，接下来便可以启动工作线程。`download_enclosures()` 处理语句 `url = q.get()` 时，会阻塞并等待，直到队列返回某个结果。这说明，即使队列中没有任何内容也可以安全地启动线程。

```python
# Set up some threads to fetch the enclosures.
for i in range(num_fetch_threads):
    worker = threading.Thread(
        target=download_enclosures,
        args=(enclosure_queue,),
        name='worker-{}'.format(i),
    )
    worker.setDaemon(True)
    worker.start()
```

下一步使用 `feedparser` 模块获取提要内容，并将这些专辑的 URL 入队。一旦第一个 URL 增加到队列，就会有某个工作线程提取这个 URL，并且开始下载。这个循环会继续增加元素,直到这个提要已被完全消费,工作线程会依次将URL出队以完成下载。

```
# Download the feed(s) and put the enclosure URLs into
# the queue.
for url in feed_urls:
    response = feedparser.parse(url, agent='fetch_podcasts.py')
    for entry in response['entries'][:5]:
        for enclosure in entry.get('enclosures', []):
            parsed_url = urlparse(enclosure['url'])
            message('queuing {}'.format(
                parsed_url.path.rpartition('/')[-1]))
            enclosure_queue.put(enclosure['url'])
```

还有一件事要做,要使用join()再次等待队列清空。

```
# Now wait for the queue to be empty, indicating that we have
# processed all of the downloads.
message('*** main thread waiting')
enclosure_queue.join()
message('*** done')
```

运行这个示例脚本可以生成类似下面的输出。

```
$ python3 fetch_podcasts.py

worker-0: looking for the next enclosure
worker-1: looking for the next enclosure
MainThread: queuing turbogears-and-the-future-of-python-web-framework
s.mp3
MainThread: queuing continuum-scientific-python-and-the-business-of-o
pen-source.mp3
MainThread: queuing openstack-cloud-computing-built-on-python.mp3
MainThread: queuing pypy.js-pypy-python-in-your-browser.mp3
MainThread: queuing machine-learning-with-python-and-scikit-learn.mp3
MainThread: *** main thread waiting
worker-0: downloading turbogears-and-the-future-of-python-web-framewo
rks.mp3
worker-1: downloading continuum-scientific-python-and-the-business-of
-open-source.mp3
worker-0: looking for the next enclosure
worker-0: downloading openstack-cloud-computing-built-on-python.mp3
worker-1: looking for the next enclosure
worker-1: downloading pypy.js-pypy-python-in-your-browser.mp3
worker-0: looking for the next enclosure
worker-0: downloading machine-learning-with-python-and-scikit-learn.m
p3
worker-1: looking for the next enclosure
worker-0: looking for the next enclosure
MainThread: *** done
```

具体的输出取决于所使用的RSS提要的内容。

提示:相关阅读材料
- queue 的标准库文档[一]。
- deque:collections 包含的双端队列。
- Queue data structures[二]:解释队列的一篇维基百科文章。

[一] https://docs.python.org/3.5/library/queue.html
[二] https://en.wikipedia.org/wiki/Queue_(abstract_data_type)

- FIFO[①]：维基百科文章，解释了先进先出数据结构。
- feedparser 模块[②]：一个解析 RSS 和 Atom 提要的模块，由 Mark Pilgrim 创建并由 Kurt McKee 维护。

2.7 struct：二进制数据结构

struct 模块包括一些函数，这些函数可以完成字节串与原生 Python 数据类型（如数字和字符串）之间的转换。

2.7.1 函数与 Struct 类

struct 提供了一组处理结构值的模块级函数，另外还有一个 Struct 类。格式指示符将由字符串格式转换为一种编译表示，这与处理正则表达式的方式类似。这个转换会耗费一些资源，所以创建一个 Struct 实例并在这个实例上调用方法时（不是使用模块级函数）只完成一次转换，这会更高效。下面的例子使用了 Struct 类。

2.7.2 打包和解包

Struct 支持使用格式指示符将数据打包（packing）为字符串，另外支持从字符串解包（unpacking）数据，格式指示符由表示数据类型的字符和可选的数量及字节序（endianness）指示符构成。要全面了解目前支持的格式指示符，请参考标准库文档。

在下面的例子中，指示符要求有一个整型或长整型值、一个两字节字符串以及一个浮点数。格式指示符中包含的空格用来分隔类型指示符，并且在编译格式时会被忽略。

代码清单 2-60：struct_pack.py

```
import struct
import binascii

values = (1, 'ab'.encode('utf-8'), 2.7)
s = struct.Struct('I 2s f')
packed_data = s.pack(*values)

print('Original values:', values)
print('Format string  :', s.format)
print('Uses           :', s.size, 'bytes')
print('Packed Value   :', binascii.hexlify(packed_data))
```

这个例子将打包的值转换为一个十六进制字节序列，以便用 binascii.hexlify() 打印，因为有些字符是 null。

```
$ python3 struct_pack.py

Original values: (1, b'ab', 2.7)
Format string  : b'I 2s f'
Uses           : 12 bytes
Packed Value   : b'0100000061620000cdcc2c40'
```

[①] https://en.wikipedia.org/wiki/FIFO_(computing_and_electronics)
[②] https://pypi.python.org/pypi/feedparser

使用unpack()可以从打包的表示中抽取数据。

代码清单2-61：**struct_unpack.py**

```
import struct
import binascii

packed_data = binascii.unhexlify(b'0100000061620000cdcc2c40')

s = struct.Struct('I 2s f')
unpacked_data = s.unpack(packed_data)
print('Unpacked Values:', unpacked_data)
```

将打包值传入unpack()，基本上会得到相同的值（注意浮点值中的微小差别）。

```
$ python3 struct_unpack.py

Unpacked Values: (1, b'ab', 2.700000047683716)
```

2.7.3 字节序

默认地，值会使用原生C库的字节序（endianness）来编码。只需在格式串中提供一个显式的字节序指令，就可以很容易地覆盖这个默认选择。

代码清单2-62：**struct_endianness.py**

```
import struct
import binascii

values = (1, 'ab'.encode('utf-8'), 2.7)
print('Original values:', values)

endianness = [
    ('@', 'native, native'),
    ('=', 'native, standard'),
    ('<', 'little-endian'),
    ('>', 'big-endian'),
    ('!', 'network'),
]

for code, name in endianness:
    s = struct.Struct(code + ' I 2s f')
    packed_data = s.pack(*values)
    print()
    print('Format string  :', s.format, 'for', name)
    print('Uses           :', s.size, 'bytes')
    print('Packed Value   :', binascii.hexlify(packed_data))
    print('Unpacked Value :', s.unpack(packed_data))
```

表2-3列出了 Struct 使用的字节序指示符。

表2-3　**Struct** 的字节序指示符

代码	含义
@	原生顺序
=	原生标准
<	小端
>	大端
!	网络顺序

```
$ python3 struct_endianness.py

Original values: (1, b'ab', 2.7)

Format string  : b'@ I 2s f' for native, native
Uses           : 12 bytes
Packed Value   : b'0100000061620000cdcc2c40'
Unpacked Value : (1, b'ab', 2.700000047683716)

Format string  : b'= I 2s f' for native, standard
Uses           : 10 bytes
Packed Value   : b'010000006162cdcc2c40'
Unpacked Value : (1, b'ab', 2.700000047683716)

Format string  : b'< I 2s f' for little-endian
Uses           : 10 bytes
Packed Value   : b'010000006162cdcc2c40'
Unpacked Value : (1, b'ab', 2.700000047683716)

Format string  : b'> I 2s f' for big-endian
Uses           : 10 bytes
Packed Value   : b'000000016162402ccccd'
Unpacked Value : (1, b'ab', 2.700000047683716)

Format string  : b'! I 2s f' for network
Uses           : 10 bytes
Packed Value   : b'000000016162402ccccd'
Unpacked Value : (1, b'ab', 2.700000047683716)
```

2.7.4 缓冲区

通常在强调性能的情况下或者向扩展模块传入或传出数据时才会处理二进制打包数据。通过避免为每个打包结构分配一个新缓冲区所带来的开销，这些情况可以得到优化。`pack_into()` 和 `unpack_from()` 方法支持直接写入预分配的缓冲区。

代码清单 2-63 **struct_buffers.py**

```
import array
import binascii
import ctypes
import struct

s = struct.Struct('I 2s f')
values = (1, 'ab'.encode('utf-8'), 2.7)
print('Original:', values)

print()
print('ctypes string buffer')

b = ctypes.create_string_buffer(s.size)
print('Before  :', binascii.hexlify(b.raw))
s.pack_into(b, 0, *values)
print('After   :', binascii.hexlify(b.raw))
print('Unpacked:', s.unpack_from(b, 0))

print()
print('array')

a = array.array('b', b'\0' * s.size)
```

```
print('Before   :', binascii.hexlify(a))
s.pack_into(a, 0, *values)
print('After    :', binascii.hexlify(a))
print('Unpacked:', s.unpack_from(a, 0))
```

Struct 的 size 属性指出缓冲区需要有多大。

```
$ python3 struct_buffers.py

Original: (1, b'ab', 2.7)

ctypes string buffer
Before   : b'000000000000000000000000'
After    : b'0100000061620000cdcc2c40'
Unpacked: (1, b'ab', 2.700000047683716)

array
Before   : b'000000000000000000000000'
After    : b'0100000061620000cdcc2c40'
Unpacked: (1, b'ab', 2.700000047683716)
```

提示：相关阅读材料
- struct 的标准库文档[○]。
- struct 的 Python 2 到 Python 3 移植说明。
- array：array 模块，用于处理固定类型值序列。
- binascii：binascii 模块，用于生成二进制数据的 ASCII 表示。
- WikiPedia: Endianness[○]：维基百科文章，提供了字节顺序以及编码中字节序的解释。

2.8 weakref：对象的非永久引用

weakref 模块支持对象的弱引用。正常的引用会增加对象的引用数，并避免它被垃圾回收。但结果并不总是如期望中的那样，比如有时可能会出现一个循环引用，或者有时需要内存时可能要删除对象的缓存。弱引用（weak reference）是一个不能避免对象被自动清理的对象句柄。

2.8.1 引用

对象的弱引用要通过 ref 类来管理。要获取原对象，可以调用引用对象。

代码清单 2-64：weakref_ref.py

```
import weakref


class ExpensiveObject:

    def __del__(self):
        print('(Deleting {})'.format(self))
```

○ https://docs.python.org/3.5/library/struct.html
○ https://en.wikipedia.org/wiki/Endianness

```
obj = ExpensiveObject()
r = weakref.ref(obj)

print('obj:', obj)
print('ref:', r)
print('r():', r())

print('deleting obj')
del obj
print('r():', r())
```

在这里,由于 obj 在第二次调用引用之前已经被删除,所以 ref 返回 None。

```
$ python3 weakref_ref.py

obj: <__main__.ExpensiveObject object at 0x1007b1a58>
ref: <weakref at 0x1007a92c8; to 'ExpensiveObject' at 0x1007b1a58>
r(): <__main__.ExpensiveObject object at 0x1007b1a58>
deleting obj
(Deleting <__main__.ExpensiveObject object at 0x1007b1a58>)
r(): None
```

2.8.2 引用回调

ref 构造函数接受一个可选的回调函数,删除所引用的对象时会调用这个函数。

代码清单 2-65:weakref_ref_callback.py

```
import weakref


class ExpensiveObject:

    def __del__(self):
        print('(Deleting {})'.format(self))


def callback(reference):
    """Invoked when referenced object is deleted"""
    print('callback({!r})'.format(reference))


obj = ExpensiveObject()
r = weakref.ref(obj, callback)

print('obj:', obj)
print('ref:', r)
print('r():', r())

print('deleting obj')
del obj
print('r():', r())
```

当引用已经"死亡"而且不再引用原对象时,这个回调会接受这个引用对象作为参数。这个特性的一种用法就是从缓存中删除弱引用对象。

```
$ python3 weakref_ref_callback.py

obj: <__main__.ExpensiveObject object at 0x1010b1978>
```

```
ref: <weakref at 0x1010a92c8; to 'ExpensiveObject' at
0x1010b1978>
r(): <__main__.ExpensiveObject object at 0x1010b1978>
deleting obj
(Deleting <__main__.ExpensiveObject object at 0x1010b1978>)
callback(<weakref at 0x1010a92c8; dead>)
r(): None
```

2.8.3 最终化对象

清理弱引用时要对资源完成更健壮的管理，可以使用 finalize 将回调与对象关联。finalize 实例会一直保留（直到所关联的对象被删除），即使应用并没有保留最终化对象的引用。

代码清单 2-66　weakref_finalize.py

```
import weakref


class ExpensiveObject:

    def __del__(self):
        print('(Deleting {})'.format(self))


def on_finalize(*args):
    print('on_finalize({!r})'.format(args))


obj = ExpensiveObject()
weakref.finalize(obj, on_finalize, 'extra argument')

del obj
```

finalize 的参数包括要跟踪的对象，对象被垃圾回收时要调用的 callable，以及传入这个 callable 的所有位置或命名参数。

```
$ python3 weakref_finalize.py

(Deleting <__main__.ExpensiveObject object at 0x1019b10f0>)
on_finalize(('extra argument',))
```

这个 finalize 实例有一个可写属性 atexit，用来控制程序退出时是否调用这个回调（如果还未调用）。

代码清单 2-67：weakref_finalize_atexit.py

```
import sys
import weakref


class ExpensiveObject:

    def __del__(self):
        print('(Deleting {})'.format(self))


def on_finalize(*args):
    print('on_finalize({!r})'.format(args))
```

```
obj = ExpensiveObject()
f = weakref.finalize(obj, on_finalize, 'extra argument')
f.atexit = bool(int(sys.argv[1]))
```

默认设置是调用这个回调。将 `atexit` 设置为 false 会禁用这种行为。

```
$ python3 weakref_finalize_atexit.py 1

on_finalize(('extra argument',))
(Deleting <__main__.ExpensiveObject object at 0x1007b10f0>)

$ python3 weakref_finalize_atexit.py 0
```

如果向 `finalize` 实例提供所跟踪对象的一个引用，这便会导致一个引用被保留，所以这个对象永远不会被垃圾回收。

代码清单 2-68：**weakref_finalize_reference.py**

```
import gc
import weakref


class ExpensiveObject:

    def __del__(self):
        print('(Deleting {})'.format(self))


def on_finalize(*args):
    print('on_finalize({!r})'.format(args))


obj = ExpensiveObject()
obj_id = id(obj)

f = weakref.finalize(obj, on_finalize, obj)
f.atexit = False

del obj

for o in gc.get_objects():
    if id(o) == obj_id:
        print('found uncollected object in gc')
```

如上例所示，尽管 `obj` 的显式引用已经删除，但是这个对象仍保留，通过 f 对垃圾回收器可见。

```
$ python3 weakref_finalize_reference.py

found uncollected object in gc
```

使用所跟踪对象的一个绑定方法作为 callable 也可以适当地避免对象最终化。

代码清单 2-69：**weakref_finalize_reference_method.py**

```
import gc
import weakref
```

```
class ExpensiveObject:

    def __del__(self):
        print('(Deleting {})'.format(self))

    def do_finalize(self):
        print('do_finalize')


obj = ExpensiveObject()
obj_id = id(obj)

f = weakref.finalize(obj, obj.do_finalize)
f.atexit = False

del obj

for o in gc.get_objects():
    if id(o) == obj_id:
        print('found uncollected object in gc')
```

由于为 `finalize` 提供的 callable 是实例 `obj` 的一个绑定方法，所以最终化方法保留了 `obj` 的一个引用，它不能被删除和被垃圾回收。

```
$ python3 weakref_finalize_reference_method.py

found uncollected object in gc
```

2.8.4 代理

有时使用代理比使用弱引用更方便。使用代理可以像使用原对象一样，而且不要求在访问对象之前先调用代理。这说明，可以将代理传递到一个库，而这个库并不知道它接收的是一个引用而不是真正的对象。

代码清单 2-70：`weakref_proxy.py`

```
import weakref


class ExpensiveObject:

    def __init__(self, name):
        self.name = name

    def __del__(self):
        print('(Deleting {})'.format(self))


obj = ExpensiveObject('My Object')
r = weakref.ref(obj)
p = weakref.proxy(obj)

print('via obj:', obj.name)
print('via ref:', r().name)
print('via proxy:', p.name)
del obj
print('via proxy:', p.name)
```

如果引用对象被删除后再访问代理，会产生一个 ReferenceError 异常。

```
$ python3 weakref_proxy.py

via obj: My Object
via ref: My Object
via proxy: My Object
(Deleting <__main__.ExpensiveObject object at 0x1007aa7b8>)
Traceback (most recent call last):
  File "weakref_proxy.py", line 30, in <module>
    print('via proxy:', p.name)
ReferenceError: weakly-referenced object no longer exists
```

2.8.5 缓存对象

ref 和 proxy 类被认为是"底层"的。尽管它们对于维护单个对象的弱引用很有用，并且还支持对循环引用的垃圾回收，但 WeakKeyDictionary 和 WeakValueDictionary 类为创建多个对象的缓存提供了一个更适合的 API。

WeakValueDictionary 类使用它包含的值的弱引用，当其他代码不再真正使用这些值时，则允许垃圾回收。利用垃圾回收器的显式调用，下面展示了使用常规字典和 WeakValueDictionary 完成内存处理的区别：

代码清单 2-71：weakref_valuedict.py

```python
import gc
from pprint import pprint
import weakref

gc.set_debug(gc.DEBUG_UNCOLLECTABLE)


class ExpensiveObject:

    def __init__(self, name):
        self.name = name

    def __repr__(self):
        return 'ExpensiveObject({})'.format(self.name)

    def __del__(self):
        print('    (Deleting {})'.format(self))


def demo(cache_factory):
    # Hold objects so any weak references
    # are not removed immediately.
    all_refs = {}
    # Create the cache using the factory.
    print('CACHE TYPE:', cache_factory)
    cache = cache_factory()
    for name in ['one', 'two', 'three']:
        o = ExpensiveObject(name)
        cache[name] = o
        all_refs[name] = o
        del o  # decref

    print('  all_refs =', end=' ')
    pprint(all_refs)
    print('\n  Before, cache contains:', list(cache.keys()))
    for name, value in cache.items():
        print('    {} = {}'.format(name, value))
        del value  # decref
```

```
    # Remove all references to the objects except the cache.
    print('\n  Cleanup:')
    del all_refs
    gc.collect()

    print('\n  After, cache contains:', list(cache.keys()))
    for name, value in cache.items():
        print('    {} = {}'.format(name, value))
    print('  demo returning')
    return


demo(dict)
print()

demo(weakref.WeakValueDictionary)
```

如果循环变量指示所缓存的值，那么这些循环变量必须被显式清除，以使对象的引用数减少。否则，垃圾回收器不会删除这些对象，它们仍然会保留在缓存中。类似地，all_refs 变量用来保存引用，以防止它们被过早地垃圾回收。

```
$ python3 weakref_valuedict.py
CACHE TYPE: <class 'dict'>
  all_refs = {'one': ExpensiveObject(one),
 'three': ExpensiveObject(three),
 'two': ExpensiveObject(two)}

  Before, cache contains: ['one', 'three', 'two']
    one = ExpensiveObject(one)
    three = ExpensiveObject(three)
    two = ExpensiveObject(two)

  Cleanup:

  After, cache contains: ['one', 'three', 'two']
    one = ExpensiveObject(one)
    three = ExpensiveObject(three)
    two = ExpensiveObject(two)
  demo returning
    (Deleting ExpensiveObject(one))
    (Deleting ExpensiveObject(three))
    (Deleting ExpensiveObject(two))

CACHE TYPE: <class 'weakref.WeakValueDictionary'>
  all_refs = {'one': ExpensiveObject(one),
 'three': ExpensiveObject(three),
 'two': ExpensiveObject(two)}

  Before, cache contains: ['one', 'three', 'two']
    one = ExpensiveObject(one)
    three = ExpensiveObject(three)
    two = ExpensiveObject(two)

  Cleanup:
    (Deleting ExpensiveObject(one))
    (Deleting ExpensiveObject(three))
    (Deleting ExpensiveObject(two))

  After, cache contains: []
  demo returning
```

WeakKeyDictionary 的工作与之类似，不过使用了字典中键的弱引用而不是值的弱引用。

警告：weakref 的库文档有以下警告：
由于 WeakValueDictionary 建立在 Python 字典之上，迭代处理时不能改变大小。WeakValueDictionary 可能很难保证这一点，因为程序在迭代中完成的动作可能会导致字典中的项"魔法般地"消失（作为垃圾回收的副作用）。

提示：相关阅读材料
- **weakref** 的标准库文档。[一]
- **gc**：gc 模块是解释器垃圾回收器的接口。
- **PEP 205**[二]：弱引用增强提案。

2.9 copy：复制对象

copy 模块包括两个函数 copy() 和 deepcopy()，用于复制现有的对象。

2.9.1 浅副本

copy() 创建的浅副本（shallow copy）是一个新容器，其中填充了原对象内容的引用。建立 list 对象的一个浅副本时，会构造一个新的 list，并将原对象的元素追加到这个 list。

代码清单 2-72：copy_shallow.py

```python
import copy
import functools

@functools.total_ordering
class MyClass:

    def __init__(self, name):
        self.name = name

    def __eq__(self, other):
        return self.name == other.name

    def __gt__(self, other):
        return self.name > other.name


a = MyClass('a')
my_list = [a]
dup = copy.copy(my_list)

print('             my_list:', my_list)
```

[一] https://docs.python.org/3.5/library/weakref.html
[二] www.python.org/dev/peps/pep-0205

```
print('            dup:', dup)
print('    dup is my_list:', (dup is my_list))
print('    dup == my_list:', (dup == my_list))
print('dup[0] is my_list[0]:', (dup[0] is my_list[0]))
print('dup[0] == my_list[0]:', (dup[0] == my_list[0]))
```

作为一个浅副本,并不会复制 MyClass 实例,所以 dup 列表中的引用会指向 my_list 中相同的对象。

```
$ python3 copy_shallow.py

            my_list: [<__main__.MyClass object at 0x1007a87b8>]
                dup: [<__main__.MyClass object at 0x1007a87b8>]
     dup is my_list: False
     dup == my_list: True
dup[0] is my_list[0]: True
dup[0] == my_list[0]: True
```

2.9.2 深副本

deepcopy() 创建的深副本是一个新容器,其中填充了原对象内容的副本。要建立一个 list 的深副本,会构造一个新的 list,复制原列表的元素,然后将这些副本追加到新列表。

将前例中的 copy() 调用替换为 deepcopy(),可以清楚地看出输出的不同。

代码清单 2-73:**copy_deep.py**

```python
import copy
import functools


@functools.total_ordering
class MyClass:

    def __init__(self, name):
        self.name = name

    def __eq__(self, other):
        return self.name == other.name

    def __gt__(self, other):
        return self.name > other.name


a = MyClass('a')
my_list = [a]
dup = copy.deepcopy(my_list)

print('            my_list:', my_list)
print('                dup:', dup)
print('    dup is my_list:', (dup is my_list))
print('    dup == my_list:', (dup == my_list))
print('dup[0] is my_list[0]:', (dup[0] is my_list[0]))
print('dup[0] == my_list[0]:', (dup[0] == my_list[0]))
```

列表的第一个元素不再是相同的对象引用,不过比较这两个对象时,仍认为它们是相等的。

```
$ python3 copy_deep.py
            my_list: [<__main__.MyClass object at 0x1018a87b8>]
                dup: [<__main__.MyClass object at 0x1018b1b70>]
     dup is my_list: False
     dup == my_list: True
dup[0] is my_list[0]: False
dup[0] == my_list[0]: True
```

2.9.3 定制复制行为

可以使用特殊方法 `__copy__()` 和 `__deepcopy__()` 来控制如何建立副本。
- 调用 `__copy__()` 而不提供任何参数，这会返回对象的一个浅副本。
- 调用 `__deepcopy__()`，并提供一个备忘字典，这会返回对象的一个深副本。所有需要深复制的成员属性都要连同备忘字典传递到 `copy.deepcopy()` 以控制递归（备忘字典将在后面更详细地解释）。

下面这个例子展示了如何调用这些方法。

代码清单 2-74：copy_hooks.py

```python
import copy
import functools


@functools.total_ordering
class MyClass:

    def __init__(self, name):
        self.name = name

    def __eq__(self, other):
        return self.name == other.name

    def __gt__(self, other):
        return self.name > other.name

    def __copy__(self):
        print('__copy__()')
        return MyClass(self.name)

    def __deepcopy__(self, memo):
        print('__deepcopy__({})'.format(memo))
        return MyClass(copy.deepcopy(self.name, memo))


a = MyClass('a')
sc = copy.copy(a)
dc = copy.deepcopy(a)
```

备忘字典用于跟踪已复制的值，以避免无限递归。

```
$ python3 copy_hooks.py
__copy__()
__deepcopy__({})
```

2.9.4 深副本中的递归

为了避免复制递归数据结构可能带来的问题，deepcopy()使用了一个字典来跟踪已复制的对象。将这个字典传入__deepcopy__()方法，这样在该方法中也可以检查这个字典。

下面的例子显示了通过实现__deepcopy__()方法可以帮助一个互连的数据结构（如有向图）避免递归。

代码清单2-75：copy_recursion.py

```python
import copy


class Graph:

    def __init__(self, name, connections):
        self.name = name
        self.connections = connections

    def add_connection(self, other):
        self.connections.append(other)

    def __repr__(self):
        return 'Graph(name={}, id={})'.format(
            self.name, id(self))

    def __deepcopy__(self, memo):
        print('\nCalling __deepcopy__ for {!r}'.format(self))
        if self in memo:
            existing = memo.get(self)
            print('  Already copied to {!r}'.format(existing))
            return existing
        print('  Memo dictionary:')
        if memo:
            for k, v in memo.items():
                print('    {}: {}'.format(k, v))
        else:
            print('    (empty)')
        dup = Graph(copy.deepcopy(self.name, memo), [])
        print('  Copying to new object {}'.format(dup))
        memo[self] = dup
        for c in self.connections:
            dup.add_connection(copy.deepcopy(c, memo))
        return dup


root = Graph('root', [])
a = Graph('a', [root])
b = Graph('b', [a, root])
root.add_connection(a)
root.add_connection(b)

dup = copy.deepcopy(root)
```

Graph类包含一些基本的有向图方法。可以利用一个名和一个列表（包含已连接的现有节点）初始化一个Graph实例。add_connection()方法用于建立双向连接。深复制操作符也用到了这个方法。

__deepcopy__()方法将打印消息来显示这个方法是如何调用的，并根据需要管理备忘字典内容。它不是复制整个连接列表，而是创建一个新列表，再把各个连接的副本追加到这个列表。这样可以确保复制各个新节点时会更新备忘字典，而避免递归问题或多余的节点副本。与前面一样，完成时会返回复制的对象。

图 2-1 中的图有几个环，不过利用备忘字典处理递归就可以避免遍历导致栈溢出错误。复制根节点 root 时，会生成以下输出：

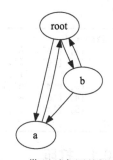

图 2-1　带环对象图的深副本

```
$ python3 copy_recursion.py

Calling __deepcopy__ for Graph(name=root, id=4314569528)
  Memo dictionary:
    (empty)
  Copying to new object Graph(name=root, id=4315093592)

Calling __deepcopy__ for Graph(name=a, id=4314569584)
  Memo dictionary:
    Graph(name=root, id=4314569528): Graph(name=root,
id=4315093592)
  Copying to new object Graph(name=a, id=4315094208)

Calling __deepcopy__ for Graph(name=root, id=4314569528)
  Already copied to Graph(name=root, id=4315093592)

Calling __deepcopy__ for Graph(name=b, id=4315092248)
  Memo dictionary:
    4314569528: Graph(name=root, id=4315093592)
    4315692808: [Graph(name=root, id=4314569528), Graph(name=a,
id=4314569584)]
    Graph(name=root, id=4314569528): Graph(name=root,
id=4315093592)
    4314569584: Graph(name=a, id=4315094208)
    Graph(name=a, id=4314569584): Graph(name=a, id=4315094208)
  Copying to new object Graph(name=b, id=4315177536)
```

第二次遇到 root 节点时，正在复制 a 节点，__deepcopy__()检测到递归，会重用备忘字典中现有的值，而不是创建一个新对象。

提示：相关阅读材料
- copy 的标准库文档⊖。

2.10　pprint：美观打印数据结构

pprint 模块包含一个"美观打印机"，用于生成数据结构的一个美观的视图。格式化工具会生成数据结构的一些表示，不仅能够由解释器正确地解析，还便于人阅读。输出会尽可能放在一行上，分解为多行时会缩进。

⊖ https://docs.python.org/3.5/library/copy.html

这一节中的例子都用到了 pprint_data.py，其中包含以下数据。

代码清单 2-76：**pprint_data.py**

```
data = [
    (1, {'a': 'A', 'b': 'B', 'c': 'C', 'd': 'D'}),
    (2, {'e': 'E', 'f': 'F', 'g': 'G', 'h': 'H',
         'i': 'I', 'j': 'J', 'k': 'K', 'l': 'L'}),
    (3, ['m', 'n']),
    (4, ['o', 'p', 'q']),
    (5, ['r', 's', 't''u', 'v', 'x', 'y', 'z']),
]
```

2.10.1 打印

要使用这个模块，最简单的方法就是利用 `pprint()` 函数。

代码清单 2-77：**pprint_pprint.py**

```
from pprint import pprint

from pprint_data import data

print('PRINT:')
print(data)
print()
print('PPRINT:')
pprint(data)
```

`pprint()` 格式化一个对象，并把它写至作为参数传入的一个数据流（或者是默认的 `sys.stdout`）。

```
$ python3 pprint_pprint.py

PRINT:
[(1, {'c': 'C', 'b': 'B', 'd': 'D', 'a': 'A'}), (2, {'k': 'K', 'i':
'I', 'g': 'G', 'f': 'F', 'e': 'E', 'h': 'H', 'l': 'L', 'j': 'J'}), (
3, ['m', 'n']), (4, ['o', 'p', 'q']), (5, ['r', 's', 'tu', 'v', 'x',
 'y', 'z'])]

PPRINT:
[(1, {'a': 'A', 'b': 'B', 'c': 'C', 'd': 'D'}),
 (2,
  {'e': 'E',
   'f': 'F',
   'g': 'G',
   'h': 'H',
   'i': 'I',
   'j': 'J',
   'k': 'K',
   'l': 'L'}),
 (3, ['m', 'n']),
 (4, ['o', 'p', 'q']),
 (5, ['r', 's', 'tu', 'v', 'x', 'y', 'z'])]
```

2.10.2 格式化

要格式化一个数据结构而不把它直接写至一个流（即用于日志），可以使用 `pformat()` 来构建一个字符串表示。

代码清单 2-78：`pprint_pformat.py`

```python
import logging
from pprint import pformat
from pprint_data import data

logging.basicConfig(
    level=logging.DEBUG,
    format='%(levelname)-8s %(message)s',
)

logging.debug('Logging pformatted data')
formatted = pformat(data)
for line in formatted.splitlines():
    logging.debug(line.rstrip())
```

然后可以单独地打印这个格式化的字符串或者记入日志。

```
$ python3 pprint_pformat.py

DEBUG    Logging pformatted data
DEBUG    [(1, {'a': 'A', 'b': 'B', 'c': 'C', 'd': 'D'}),
DEBUG     (2,
DEBUG      {'e': 'E',
DEBUG       'f': 'F',
DEBUG       'g': 'G',
DEBUG       'h': 'H',
DEBUG       'i': 'I',
DEBUG       'j': 'J',
DEBUG       'k': 'K',
DEBUG       'l': 'L'}),
DEBUG     (3, ['m', 'n']),
DEBUG     (4, ['o', 'p', 'q']),
DEBUG     (5, ['r', 's', 'tu', 'v', 'x', 'y', 'z'])]
```

2.10.3　任意类

如果一个定制类定义了一个 `__repr__()` 方法，那么 `pprint()` 使用的 PrettyPrinter 类还可以处理这样的定制类。

代码清单 2-79：`pprint_arbitrary_object.py`

```python
from pprint import pprint


class node:

    def __init__(self, name, contents=[]):
        self.name = name
        self.contents = contents[:]

    def __repr__(self):
        return (
            'node(' + repr(self.name) + ', ' +
            repr(self.contents) + ')'
        )


trees = [
    node('node-1'),
    node('node-2', [node('node-2-1')]),
```

```
      node('node-3', [node('node-3-1')]),
]
pprint(trees)
```

利用由 PrettyPrinter 组合的嵌套对象的表示来返回完整的字符串表示。

```
$ python3 pprint_arbitrary_object.py

[node('node-1', []),
 node('node-2', [node('node-2-1', [])]),
 node('node-3', [node('node-3-1', [])])]
```

2.10.4 递归

递归数据结构由指向原数据源的引用表示，形式为 <Recursion on typename with id=number>。

代码清单 2-80：pprint_recursion.py

```
from pprint import pprint

local_data = ['a', 'b', 1, 2]
local_data.append(local_data)

print('id(local_data) =>', id(local_data))
pprint(local_data)
```

在这个例子中，列表 local_data 增加到其自身，这会创建一个递归引用。

```
$ python3 pprint_recursion.py

id(local_data) => 4324368136
['a', 'b', 1, 2, <Recursion on list with id=4324368136>]
```

2.10.5 限制嵌套输出

对于非常深的数据结构，可能不要求输出中包含所有细节。数据有可能没有适当地格式化，也可能格式化文本过大而无法管理，或者有些数据可能是多余的。

代码清单 2-81：pprint_depth.py

```
from pprint import pprint

from pprint_data import data

pprint(data, depth=1)
pprint(data, depth=2)
```

使用 depth 参数可以控制美观打印机递归处理嵌套数据结构的深度。输出中未包含的层次用省略号表示。

```
$ python3 pprint_depth.py

[(...), (...), (...), (...), (...)]
[(1, {...}), (2, {...}), (3, [...]), (4, [...]), (5, [...])]
```

2.10.6 控制输出宽度

格式化文本的默认输出宽度为 80 列。要调整这个宽度，可以在 pprint() 中使用参数 width。

代码清单 2-82: **pprint_width.py**

```
from pprint import pprint

from pprint_data import data

for width in [80, 5]:
    print('WIDTH =', width)
    pprint(data, width=width)
    print()
```

当宽度太小而不能满足格式化数据结构时，倘若截断或转行会导致非法语法，那么便不会截断或转行。

```
$ python3 pprint_width.py

WIDTH = 80
[(1, {'a': 'A', 'b': 'B', 'c': 'C', 'd': 'D'}),
 (2,
  {'e': 'E',
   'f': 'F',
   'g': 'G',
   'h': 'H',
   'i': 'I',
   'j': 'J',
   'k': 'K',
   'l': 'L'}),
 (3, ['m', 'n']),
 (4, ['o', 'p', 'q']),
 (5, ['r', 's', 'tu', 'v', 'x', 'y', 'z'])]

WIDTH = 5
[(1,
  {'a': 'A',
   'b': 'B',
   'c': 'C',
   'd': 'D'}),
 (2,
  {'e': 'E',
   'f': 'F',
   'g': 'G',
   'h': 'H',
   'i': 'I',
   'j': 'J',
   'k': 'K',
   'l': 'L'}),
 (3,
  ['m',
   'n']),
 (4,
  ['o',
   'p',
   'q']),
 (5,
  ['r',
```

```
            's',
            'tu',
            'v',
            'x',
            'y',
            'z'])]
```

compact标志告诉pprint()尝试在每一行上放置更多数据，而不是把复杂数据结构分解为多行。

代码清单2-83：**pprint_compact.py**

```
from pprint import pprint

from pprint_data import data

print('DEFAULT:')
pprint(data, compact=False)
print('\nCOMPACT:')
pprint(data, compact=True)
```

这个例子展示了一个数据结构在一行上放不下时，它会分解（数据列表中的第二项也是如此）。如果多个元素可以放置在一行上（如第三个和第四个成员），那么便会把它们放在同一行上。

```
$ python3 pprint_compact.py

[(1, {'a': 'A', 'b': 'B', 'c': 'C', 'd': 'D'}),
 (2,
  {'e': 'E',
   'f': 'F',
   'g': 'G',
   'h': 'H',
   'i': 'I',
   'j': 'J',
   'k': 'K',
   'l': 'L'}),
 (3, ['m', 'n']),
 (4, ['o', 'p', 'q']),
 (5, ['r', 's', 'tu', 'v', 'x', 'y', 'z'])]
[(1, {'a': 'A', 'b': 'B', 'c': 'C', 'd': 'D'}),
 (2,
  {'e': 'E',
   'f': 'F',
   'g': 'G',
   'h': 'H',
   'i': 'I',
   'j': 'J',
   'k': 'K',
   'l': 'L'}),
 (3, ['m', 'n']), (4, ['o', 'p', 'q']),
 (5, ['r', 's', 'tu', 'v', 'x', 'y', 'z'])]
```

提示：相关阅读材料

- **pprint**的标准库文档[⊖]。

⊖ https://docs.python.org/3.5/library/pprint.html

第 3 章
算　法

Python 包含很多模块，可以采用最适合任务的方式来精巧而简洁地实现算法。它支持不同的编程方式，包括纯过程式、面向对象式和函数式。这 3 种方式经常在同一个程序的不同部分混合使用。

functools 包含的函数用于创建函数修饰符、启用面向方面（aspect-oriented）编程以及传统面向对象方法不能支持的代码重用。它还提供了一个类修饰符以使用一个快捷方式来实现所有富比较 API，另外提供了 partial 对象来创建函数（包含其参数）的引用。

itertools 模块包含的函数用于创建和处理函数式编程中使用的迭代器和生成器。通过提供基于函数的内置操作接口（如算术操作或元素查找），operator 模块在使用函数式编程时不再需要很多麻烦的 lambda 函数。

不论使用哪一种编程方式，contextlib 都会让资源管理更容易、更可靠且更简洁。结合上下文管理器和 with 语句，可以减少 try:finally 块的个数和所需的缩进层次，同时还能确保文件、套接字、数据库事务和其他资源在适当的时候关闭和释放。

3.1　functools：管理函数的工具

functools 模块提供了一些工具来调整或扩展函数和其他 callable 对象，从而不必完全重写。

3.1.1　修饰符

functools 模块提供的主要工具就是 partial 类，可以用来"包装"一个有默认参数的 callable 对象。得到的对象本身就是 callable，可以把它看作是原来的函数。它与原函数的参数完全相同，调用时还可以提供额外的位置或命名参数。可以使用 partial 而不是 lambda 为函数提供默认参数，有些参数可以不指定。

3.1.1.1　部分对象

第一个例子显示了函数 myfunc() 的两个简单 partial 对象。show_details() 的输出中包含这个部分对象（partial object）的 func、args 和 keywords 属性。

代码清单 3-1：functools_partial.py

```
import functools

def myfunc(a, b=2):
```

```
        "Docstring for myfunc()."
        print('  called myfunc with:', (a, b))


def show_details(name, f, is_partial=False):
    "Show details of a callable object."
    print('{}:'.format(name))
    print('  object:', f)
    if not is_partial:
        print('  __name__:', f.__name__)
    if is_partial:
        print('  func:', f.func)
        print('  args:', f.args)
        print('  keywords:', f.keywords)
    return


show_details('myfunc', myfunc)
myfunc('a', 3)
print()

# Set a different default value for 'b', but require
# the caller to provide 'a'.
p1 = functools.partial(myfunc, b=4)
show_details('partial with named default', p1, True)
p1('passing a')
p1('override b', b=5)
print()

# Set default values for both 'a' and 'b'.
p2 = functools.partial(myfunc, 'default a', b=99)
show_details('partial with defaults', p2, True)
p2()
p2(b='override b')
print()

print('Insufficient arguments:')
p1()
```

在这个例子的最后，调用了之前创建的第一个 **partial**，但没有为 a 传入一个值，这便会导致一个异常。

```
$ python3 functools_partial.py

myfunc:
  object: <function myfunc at 0x1007a6a60>
  __name__: myfunc
  called myfunc with: ('a', 3)

partial with named default:
  object: functools.partial(<function myfunc at 0x1007a6a60>,
b=4)
  func: <function myfunc at 0x1007a6a60>
  args: ()
  keywords: {'b': 4}
  called myfunc with: ('passing a', 4)
  called myfunc with: ('override b', 5)

partial with defaults:
  object: functools.partial(<function myfunc at 0x1007a6a60>,
'default a', b=99)
```

```
func: <function myfunc at 0x1007a6a60>
args: ('default a',)
keywords: {'b': 99}
called myfunc with: ('default a', 99)
called myfunc with: ('default a', 'override b')

Insufficient arguments:
Traceback (most recent call last):
  File "functools_partial.py", line 51, in <module>
    p1()
TypeError: myfunc() missing 1 required positional argument: 'a'
```

3.1.1.2 获取函数属性

默认地，partial 对象没有 __name__ 或 __doc__ 属性。如果没有这些属性，被修饰的函数将更难调试。使用 update_wrapper() 可以从原函数将属性复制或增加到 partial 对象。

代码清单 3-2：**functools_update_wrapper.py**

```
import functools


def myfunc(a, b=2):
    "Docstring for myfunc()."
    print('  called myfunc with:', (a, b))


def show_details(name, f):
    "Show details of a callable object."
    print('{}:'.format(name))
    print('  object:', f)
    print('  __name__:', end=' ')
    try:
        print(f.__name__)
    except AttributeError:
        print('(no __name__)')
    print('  __doc__', repr(f.__doc__))
    print()


show_details('myfunc', myfunc)

p1 = functools.partial(myfunc, b=4)
show_details('raw wrapper', p1)

print('Updating wrapper:')
print('  assign:', functools.WRAPPER_ASSIGNMENTS)
print('  update:', functools.WRAPPER_UPDATES)
print()

functools.update_wrapper(p1, myfunc)
show_details('updated wrapper', p1)
```

增加到包装器的属性在 WRAPPER_ASSIGNMENTS 中定义，另外 WRAPPER_UPDATES 列出了要修改的值。

```
$ python3 functools_update_wrapper.py

myfunc:
  object: <function myfunc at 0x1018a6a60>
```

```
    __name__: myfunc
    __doc__ 'Docstring for myfunc().'
raw wrapper:
  object: functools.partial(<function myfunc at 0x1018a6a60>, b=4)
    __name__: (no __name__)
    __doc__ 'partial(func, *args, **keywords) - new function with partial application\n   of the given arguments and keywords.\n'
Updating wrapper:
  assign: ('__module__', '__name__', '__qualname__', '__doc__', '__annotations__')
  update: ('__dict__',)
updated wrapper:
  object: functools.partial(<function myfunc at 0x1018a6a60>, b=4)
    __name__: myfunc
    __doc__ 'Docstring for myfunc().'
```

3.1.1.3 其他 callable

partial 适用于任何 callable 对象，而不只是独立的函数。

代码清单 3-3：**functools_callable.py**

```
import functools


class MyClass:
    "Demonstration class for functools"

    def __call__(self, e, f=6):
        "Docstring for MyClass.__call__"
        print('  called object with:', (self, e, f))


def show_details(name, f):
    "Show details of a callable object."
    print('{}:'.format(name))
    print('  object:', f)
    print('  __name__:', end=' ')
    try:
        print(f.__name__)
    except AttributeError:
        print('(no __name__)')
    print('  __doc__', repr(f.__doc__))
    return


o = MyClass()

show_details('instance', o)
o('e goes here')
print()

p = functools.partial(o, e='default for e', f=8)
functools.update_wrapper(p, o)
show_details('instance wrapper', p)
p()
```

这个例子从一个包含 __call__() 方法的类实例中创建部分对象。

```
$ python3 functools_callable.py
instance:
  object: <__main__.MyClass object at 0x1011b1cf8>
  __name__: (no __name__)
  __doc__ 'Demonstration class for functools'
  called object with: (<__main__.MyClass object at 0x1011b1cf8>,
'e goes here', 6)
instance wrapper:
  object: functools.partial(<__main__.MyClass object at
0x1011b1cf8>, f=8, e='default for e')
  __name__: (no __name__)
  __doc__ 'Demonstration class for functools'
  called object with: (<__main__.MyClass object at 0x1011b1cf8>,
'default for e', 8)
```

3.1.1.4 方法和函数

partial()返回一个可以直接使用的callable，partialmethod()返回的callable则可以用作对象的非绑定方法。在下面的例子中，这个独立函数两次被增加为MyClass的属性，一次使用partialmethod()作为method1()，另一次使用partial()作为method2()。

代码清单3-4：functools_partialmethod.py

```
import functools

def standalone(self, a=1, b=2):
    "Standalone function"
    print('  called standalone with:', (self, a, b))
    if self is not None:
        print('  self.attr =', self.attr)

class MyClass:
    "Demonstration class for functools"

    def __init__(self):
        self.attr = 'instance attribute'

    method1 = functools.partialmethod(standalone)
    method2 = functools.partial(standalone)

o = MyClass()
print('standalone')
standalone(None)
print()

print('method1 as partialmethod')
o.method1()
print()

print('method2 as partial')
try:
    o.method2()
except TypeError as err:
    print('ERROR: {}'.format(err))
```

method1()可以从MyClass的一个实例中调用，这个实例作为第一个参数传入，这与采用通常方式定义的方法是一样的。method2() 未被定义为绑定方法，所以必须显式传递 self 参数；否则，这个调用会导致 TypeError。

```
$ python3 functools_partialmethod.py

standalone
  called standalone with: (None, 1, 2)

method1 as partialmethod
  called standalone with: (<__main__.MyClass object at
0x1007b1d30>, 1, 2)
  self.attr = instance attribute

method2 as partial
ERROR: standalone() missing 1 required positional argument:
'self'
```

3.1.1.5 获取修饰符的函数属性

更新所包装 callable 的属性对修饰符尤其有用，因为转换后的函数最后会得到原 "裸" 函数的属性。

代码清单 3-5：**functools_wraps.py**

```python
import functools


def show_details(name, f):
    "Show details of a callable object."
    print('{}:'.format(name))
    print('  object:', f)
    print('  __name__:', end=' ')
    try:
        print(f.__name__)
    except AttributeError:
        print('(no __name__)')
    print('  __doc__', repr(f.__doc__))
    print()


def simple_decorator(f):
    @functools.wraps(f)
    def decorated(a='decorated defaults', b=1):
        print('  decorated:', (a, b))
        print('  ', end=' ')
        return f(a, b=b)
    return decorated


def myfunc(a, b=2):
    "myfunc() is not complicated"
    print('  myfunc:', (a, b))
    return


# The raw function
show_details('myfunc', myfunc)
myfunc('unwrapped, default b')
myfunc('unwrapped, passing b', 3)
print()

# Wrap explicitly.
```

```
wrapped_myfunc = simple_decorator(myfunc)
show_details('wrapped_myfunc', wrapped_myfunc)
wrapped_myfunc()
wrapped_myfunc('args to wrapped', 4)
print()

# Wrap with decorator syntax.
@simple_decorator
def decorated_myfunc(a, b):
    myfunc(a, b)
    return

show_details('decorated_myfunc', decorated_myfunc)
decorated_myfunc()
decorated_myfunc('args to decorated', 4)
```

functools 提供了一个修饰符 wraps()，它会对所修饰的函数应用 update_wrapper()。

```
$ python3 functools_wraps.py

myfunc:
  object: <function myfunc at 0x101241b70>
  __name__: myfunc
  __doc__ 'myfunc() is not complicated'

  myfunc: ('unwrapped, default b', 2)
  myfunc: ('unwrapped, passing b', 3)

wrapped_myfunc:
  object: <function myfunc at 0x1012e62f0>
  __name__: myfunc
  __doc__ 'myfunc() is not complicated'

  decorated: ('decorated defaults', 1)
     myfunc: ('decorated defaults', 1)
  decorated: ('args to wrapped', 4)
     myfunc: ('args to wrapped', 4)

decorated_myfunc:
  object: <function decorated_myfunc at 0x1012e6400>
  __name__: decorated_myfunc
  __doc__ None

  decorated: ('decorated defaults', 1)
     myfunc: ('decorated defaults', 1)
  decorated: ('args to decorated', 4)
     myfunc: ('args to decorated', 4)
```

3.1.2 比较

在 Python 2 中，类可以定义一个 __cmp__() 方法，它会根据这个对象小于、等于或者大于所比较的元素而分别返回 -1、0 或 1。Python 2.1 引入了富比较（rich comparison）方法 API（__lt__()、__le__()、__eq__()、__ne__()、__gt__() 和 __ge__()），可以完成一个比较操作并返回一个布尔值。Python 3 废弃了 __cmp__() 而代之以这些新方法，另外 functools 提供了一些工具，从而能更容易地编写符合新要求的类，即符合 Python 3 中新的比较需求。

3.1.2.1 富比较

设计富比较 API 是为了支持涉及复杂比较的类，以最高效的方式实现各个测试。不过，如果比较相对简单的类，就没有必要手动地创建各个富比较方法了。total_ordering() 类修饰符可以为一个提供了部分方法的类增加其余的方法。

代码清单 3-6：functools_total_ordering.py

```python
import functools
import inspect
from pprint import pprint


@functools.total_ordering
class MyObject:

    def __init__(self, val):
        self.val = val

    def __eq__(self, other):
        print('  testing __eq__({}, {})'.format(
            self.val, other.val))
        return self.val == other.val

    def __gt__(self, other):
        print('  testing __gt__({}, {})'.format(
            self.val, other.val))
        return self.val > other.val


print('Methods:\n')
pprint(inspect.getmembers(MyObject, inspect.isfunction))

a = MyObject(1)
b = MyObject(2)

print('\nComparisons:')
for expr in ['a < b', 'a <= b', 'a == b', 'a >= b', 'a > b']:
    print('\n{:<6}:'.format(expr))
    result = eval(expr)
    print('  result of {}: {}'.format(expr, result))
```

这个类必须提供 __eq__() 和另外一个富比较方法的实现。这个修饰符会增加其余方法的实现，它们会使用所提供的比较。如果无法完成一个比较，这个方法应当返回 NotImplemented，从而在另一个对象上使用逆比较操作符尝试比较，如果仍无法比较，便会完全失败。

```
$ python3 functools_total_ordering.py

Methods:

[('__eq__', <function MyObject.__eq__ at 0x10139a488>),
 ('__ge__', <function _ge_from_gt at 0x1012e2510>),
 ('__gt__', <function MyObject.__gt__ at 0x10139a510>),
 ('__init__', <function MyObject.__init__ at 0x10139a400>),
```

```
('__le__', <function _le_from_gt at 0x1012e2598>),
('__lt__', <function _lt_from_gt at 0x1012e2488>)]
Comparisons:

a < b :
  testing __gt__(1, 2)
  testing __eq__(1, 2)
  result of a < b: True

a <= b:
  testing __gt__(1, 2)
  result of a <= b: True

a == b :
  testing __eq__(1, 2)
  result of a == b: False

a >= b:
  testing __gt__(1, 2)
  testing __eq__(1, 2)
  result of a >= b: False

a > b :
  testing __gt__(1, 2)
  result of a > b: False
```

3.1.2.2　比对序

由于 Python 3 废弃了老式的比较函数，sort() 之类的函数中也不再支持 cmp 参数。对于使用了比较函数的较老的程序，可以用 cmp_to_key() 将比较函数转换为一个返回比对键 (collation key) 的函数，这个键用于确定元素在最终序列中的位置。

代码清单 3-7：functools_cmp_to_key.py

```
import functools


class MyObject:

    def __init__(self, val):
        self.val = val

    def __str__(self):
        return 'MyObject({})'.format(self.val)
def compare_obj(a, b):
    """Old-style comparison function.
    """
    print('comparing {} and {}'.format(a, b))
    if a.val < b.val:
        return -1
    elif a.val > b.val:
        return 1
    return 0

# Make a key function using cmp_to_key().
get_key = functools.cmp_to_key(compare_obj)


def get_key_wrapper(o):
    "Wrapper function for get_key to allow for print statements."
```

```
        new_key = get_key(o)
        print('key_wrapper({}) -> {!r}'.format(o, new_key))
        return new_key

objs = [MyObject(x) for x in range(5, 0, -1)]

for o in sorted(objs, key=get_key_wrapper):
    print(o)
```

正常情况下，可以直接使用 `cmp_to_key()`，不过这个例子中引入了一个额外的包装器函数，这样调用键函数时可以打印更多的信息。

如输出所示，`sorted()` 首先对序列中的每一个元素调用 `get_key_wrapper()` 以生成一个键。`cmp_to_key()` 返回的键是 `functools` 中定义的一个类的实例，这个类使用传入的老式比较函数实现富比较 API。所有键都创建之后，通过比较这些键来对序列排序。

```
$ python3 functools_cmp_to_key.py

key_wrapper(MyObject(5)) -> <functools.KeyWrapper object at
0x1011c5530>
key_wrapper(MyObject(4)) -> <functools.KeyWrapper object at
0x1011c5510>
key_wrapper(MyObject(3)) -> <functools.KeyWrapper object at
0x1011c54f0>
key_wrapper(MyObject(2)) -> <functools.KeyWrapper object at
0x1011c5390>
key_wrapper(MyObject(1)) -> <functools.KeyWrapper object at
0x1011c5710>
comparing MyObject(4) and MyObject(5)
comparing MyObject(3) and MyObject(4)
comparing MyObject(2) and MyObject(3)
comparing MyObject(1) and MyObject(2)
MyObject(1)
MyObject(2)
MyObject(3)
MyObject(4)
MyObject(5)
```

3.1.3 缓存

`lru_cache()` 修饰符将一个函数包装在一个"最近最少使用的"缓存中。函数的参数用来建立一个散列键，然后映射到结果。后续的调用如果有相同的参数，就会从这个缓存获取值而不会再次调用函数。这个修饰符还会为函数增加方法来检查缓存的状态（`cache_info()`）和清空缓存（`cache_clear()`）。

代码清单 3-8：`functools_lru_cache.py`

```
import functools


@functools.lru_cache()
def expensive(a, b):
    print('expensive({}, {})'.format(a, b))
    return a * b


MAX = 2
```

```
print('First set of calls:')
for i in range(MAX):
    for j in range(MAX):
        expensive(i, j)
print(expensive.cache_info())

print('\nSecond set of calls:')
for i in range(MAX + 1):
    for j in range(MAX + 1):
        expensive(i, j)
print(expensive.cache_info())

print('\nClearing cache:')
expensive.cache_clear()
print(expensive.cache_info())

print('\nThird set of calls:')
for i in range(MAX):
    for j in range(MAX):
        expensive(i, j)
print(expensive.cache_info())
```

这个例子在一组嵌套循环中执行了多个 expensive() 调用。第二次调用时有相同的参数值，结果在缓存中。清空缓存并再次运行循环时，这些值必须重新计算。

```
$ python3 functools_lru_cache.py

First set of calls:
expensive(0, 0)
expensive(0, 1)
expensive(1, 0)
expensive(1, 1)
CacheInfo(hits=0, misses=4, maxsize=128, currsize=4)

Second set of calls:
expensive(0, 2)
expensive(1, 2)
expensive(2, 0)
expensive(2, 1)
expensive(2, 2)
CacheInfo(hits=4, misses=9, maxsize=128, currsize=9)

Clearing cache:
CacheInfo(hits=0, misses=0, maxsize=128, currsize=0)

Third set of calls:
expensive(0, 0)
expensive(0, 1)
expensive(1, 0)
expensive(1, 1)
CacheInfo(hits=0, misses=4, maxsize=128, currsize=4)
```

为了避免一个长时间运行的进程导致缓存无限制地扩张，要指定一个最大大小。默认为 128 个元素，不过对于每个缓存可以用 maxsize 参数改变这个大小。

代码清单 3-9：functools_lru_cache_expire.py

```
import functools


@functools.lru_cache(maxsize=2)
```

```python
def expensive(a, b):
    print('called expensive({}, {})'.format(a, b))
    return a * b


def make_call(a, b):
    print('({}, {})'.format(a, b), end=' ')
    pre_hits = expensive.cache_info().hits
    expensive(a, b)
    post_hits = expensive.cache_info().hits
    if post_hits > pre_hits:
        print('cache hit')


print('Establish the cache')
make_call(1, 2)
make_call(2, 3)

print('\nUse cached items')
make_call(1, 2)
make_call(2, 3)

print('\nCompute a new value, triggering cache expiration')
make_call(3, 4)

print('\nCache still contains one old item')
make_call(2, 3)

print('\nOldest item needs to be recomputed')
make_call(1, 2)
```

在这个例子中，缓存大小设置为 2 个元素。使用第 3 组不同的参数 (3,4) 时，缓存中最老的元素会被清除，代之以这个新结果。

```
$ python3 functools_lru_cache_expire.py

Establish the cache
(1, 2) called expensive(1, 2)
(2, 3) called expensive(2, 3)

Use cached items
(1, 2) cache hit
(2, 3) cache hit

Compute a new value, triggering cache expiration
(3, 4) called expensive(3, 4)

Cache still contains one old item
(2, 3) cache hit

Oldest item needs to be recomputed
(1, 2) called expensive(1, 2)
```

`lru_cache()` 管理的缓存中键必须是可散列的，所以对于用缓存查找包装的函数，它的所有参数都必须是可散列的。

代码清单 3-10：**functools_lru_cache_arguments.py**

```python
import functools


@functools.lru_cache(maxsize=2)
```

```python
def expensive(a, b):
    print('called expensive({}, {})'.format(a, b))
    return a * b


def make_call(a, b):
    print('({}, {})'.format(a, b), end=' ')
    pre_hits = expensive.cache_info().hits
    expensive(a, b)
    post_hits = expensive.cache_info().hits
    if post_hits > pre_hits:
        print('cache hit')


make_call(1, 2)

try:
    make_call([1], 2)
except TypeError as err:
    print('ERROR: {}'.format(err))

try:
    make_call(1, {'2': 'two'})
except TypeError as err:
    print('ERROR: {}'.format(err))
```

如果将一个不能散列的对象传入这个函数,则会产生一个 TypeError。

```
$ python3 functools_lru_cache_arguments.py

(1, 2) called expensive(1, 2)
([1], 2) ERROR: unhashable type: 'list'
(1, {'2': 'two'}) ERROR: unhashable type: 'dict'
```

3.1.4 缩减数据集

reduce() 函数取一个 callable 和一个数据序列作为输入。它会用这个序列中的值调用这个 callable,并累加得到的输出来生成单个值作为输出。

代码清单 3-11:functools_reduce.py

```python
import functools


def do_reduce(a, b):
    print('do_reduce({}, {})'.format(a, b))
    return a + b


data = range(1, 5)
print(data)
result = functools.reduce(do_reduce, data)
print('result: {}'.format(result))
```

这个例子会累加输入序列中的数。

```
$ python3 functools_reduce.py

range(1, 5)
do_reduce(1, 2)
```

```
do_reduce(3, 3)
do_reduce(6, 4)
result: 10
```

可选的 initializer 参数放在序列最前面,像其他元素一样处理。可以利用这个参数以新输入更新前面计算的值。

代码清单 3-12:**functools_reduce_initializer.py**

```
import functools


def do_reduce(a, b):
    print('do_reduce({}, {})'.format(a, b))
    return a + b


data = range(1, 5)
print(data)
result = functools.reduce(do_reduce, data, 99)
print('result: {}'.format(result))
```

在这个例子中,使用前面的总和 99 来初始化 reduce() 计算的值。

```
$ python3 functools_reduce_initializer.py

range(1, 5)
do_reduce(99, 1)
do_reduce(100, 2)
do_reduce(102, 3)
do_reduce(105, 4)
result: 109
```

如果没有 initializer 参数,那么只有一个元素的序列会自动缩减为这个值。空列表会生成一个错误,除非提供了一个 initializer 参数。

代码清单 3-13:**functools_reduce_short_sequences.py**

```
import functools


def do_reduce(a, b):
    print('do_reduce({}, {})'.format(a, b))
    return a + b


print('Single item in sequence:',
      functools.reduce(do_reduce, [1]))

print('Single item in sequence with initializer:',
      functools.reduce(do_reduce, [1], 99))

print('Empty sequence with initializer:',
      functools.reduce(do_reduce, [], 99))

try:
    print('Empty sequence:', functools.reduce(do_reduce, []))
except TypeError as err:
    print('ERROR: {}'.format(err))
```

由于 `initializer` 参数相当于一个默认值，但也要与新值结合（如果输入序列不为空），所以必须仔细考虑这个参数的使用是否适当，这很重要。如果默认值与新值结合没有意义，那么最好是捕获 `TypeError` 而不是传入一个 `initializer` 参数。

```
$ python3 functools_reduce_short_sequences.py

Single item in sequence: 1
do_reduce(99, 1)
Single item in sequence with initializer: 100
Empty sequence with initializer: 99
ERROR: reduce() of empty sequence with no initial value
```

3.1.5 泛型函数

在类似 Python 的动态类型语言中，通常需要基于参数的类型完成稍有不同的操作，特别是在处理元素列表与单个元素的差别时。直接检查参数的类型固然很简单，但是有些情况下，行为差异可能被隔离到单个的函数中，对于这些情况，`functools` 提供了 `singledispatch()` 修饰符来注册一组泛型函数（generic function），可以根据函数第一个参数的类型自动切换。

代码清单 3-14：`functools_singledispatch.py`

```python
import functools


@functools.singledispatch
def myfunc(arg):
    print('default myfunc({!r})'.format(arg))


@myfunc.register(int)
def myfunc_int(arg):
    print('myfunc_int({})'.format(arg))


@myfunc.register(list)
def myfunc_list(arg):
    print('myfunc_list()')
    for item in arg:
        print('  {}'.format(item))


myfunc('string argument')
myfunc(1)
myfunc(2.3)
myfunc(['a', 'b', 'c'])
```

新函数的 `register()` 属性相当于另一个修饰符，用于注册替代实现。用 `singledispatch()` 包装的第一个函数是默认实现，在未指定其他类型特定函数时就使用这个默认实现，在这个例子中特定类型就是 `float`。

```
$ python3 functools_singledispatch.py

default myfunc('string argument')
myfunc_int(1)
default myfunc(2.3)
```

```
myfunc_list()
    a
    b
    c
```

没有找到这个类型的完全匹配时,会计算继承顺序,并使用最接近的匹配类型。

代码清单 3-15:**functools_singledispatch_mro.py**

```
import functools


class A:
    pass


class B(A):
    pass


class C(A):
    pass


class D(B):
    pass


class E(C, D):
    pass


@functools.singledispatch
def myfunc(arg):
    print('default myfunc({})'.format(arg.__class__.__name__))


@myfunc.register(A)
def myfunc_A(arg):
    print('myfunc_A({})'.format(arg.__class__.__name__))


@myfunc.register(B)
def myfunc_B(arg):
    print('myfunc_B({})'.format(arg.__class__.__name__))


@myfunc.register(C)
def myfunc_C(arg):
    print('myfunc_C({})'.format(arg.__class__.__name__))


myfunc(A())
myfunc(B())
myfunc(C())
myfunc(D())
myfunc(E())
```

在这个例子中,类 D 和 E 与已注册的任何泛型函数都不完全匹配,所选择的函数取决于类层次结构。

```
$ python3 functools_singledispatch_mro.py

myfunc_A(A)
myfunc_B(B)
myfunc_C(C)
myfunc_B(D)
myfunc_C(E)
```

提示：相关阅读材料
- `functools` 的标准库文档[一]。
- 富比较方法[二]：Python 参考指南中对富比较方法的描述。
- Isolated @memoize[三]：Ned Batchelder 的一篇文章，介绍如何创建适用于单元测试的 memoizing 修饰符。
- **PEP 443**[四]：single-dispatch 泛型函数。
- inspect：活动对象的自省 API。

3.2 itertools：迭代器函数

itertools 包括一组用于处理序列数据集的函数。这个模块提供的函数是受函数式编程语言（如 Clojure、Haskell、APL 和 SML）中类似特性的启发。其目的是要能快速处理，以及要高效地使用内存，而且可以联结在一起表述更复杂的基于迭代的算法。

与使用列表的代码相比，基于迭代器的代码可以提供更好的内存消费特性。在真正需要数据之前，并不从迭代器生成数据，由于这个原因，不需要把所有数据都同时存储在内存中。这种"懒"处理模式可以减少交换以及大数据集的其他副作用，从而改善性能。

除了 itertools 中定义的函数，这一节中的例子还会利用一些内置函数完成迭代。

3.2.1 合并和分解迭代器

chain() 函数取多个迭代器作为参数，最后返回一个迭代器，它会生成所有输入迭代器的内容，就好像这些内容来自一个迭代器一样。

代码清单 3-16：**itertools_chain.py**

```
from itertools import *

for i in chain([1, 2, 3], ['a', 'b', 'c']):
    print(i, end=' ')
print()
```

利用 chain()，可以轻松地处理多个序列而不必构造一个很大的列表。

```
$ python3 itertools_chain.py

1 2 3 a b c
```

[一] https://docs.python.org/3.5/library/functools.html
[二] https://docs.python.org/3/reference/datamodel.html#object.__lt__
[三] http://nedbatchelder.com/blog/201601/isolated_memoize.html
[四] www.python.org/dev/peps/pep-0443

如果不能提前知道所有要结合的迭代器（可迭代对象），或者如果需要采用懒方式计算，那么可以使用 `chain.from_iterable()` 来构造这个链。

代码清单 3-17：**itertools_chain_from_iterable.py**

```
from itertools import *

def make_iterables_to_chain():
    yield [1, 2, 3]
    yield ['a', 'b', 'c']

for i in chain.from_iterable(make_iterables_to_chain()):
    print(i, end=' ')
print()
```

```
$ python3 itertools_chain_from_iterable.py

1 2 3 a b c
```

内置函数 `zip()` 返回一个迭代器，它会把多个迭代器的元素结合到一个元组中。

代码清单 3-18：**itertools_zip.py**

```
for i in zip([1, 2, 3], ['a', 'b', 'c']):
    print(i)
```

与这个模块中的其他函数一样，返回值是一个可迭代对象，会一次生成一个值。

```
$ python3 itertools_zip.py

(1, 'a')
(2, 'b')
(3, 'c')
```

第一个输入迭代器处理完时 `zip()` 就会停止。要处理所有输入（即使迭代器生成的值个数不同），则要使用 `zip_longest()`。

代码清单 3-19：**itertools_zip_longest.py**

```
from itertools import *

r1 = range(3)
r2 = range(2)

print('zip stops early:')
print(list(zip(r1, r2)))

r1 = range(3)
r2 = range(2)

print('\nzip_longest processes all of the values:')
print(list(zip_longest(r1, r2)))
```

默认地，`zip_longest()` 会把所有缺少的值替换为 None。可以借助 fillvalue 参数来使用一个不同的替换值。

```
$ python3 itertools_zip_longest.py

zip stops early:
[(0, 0), (1, 1)]

zip_longest processes all of the values:
[(0, 0), (1, 1), (2, None)]
```

islice()函数返回一个迭代器,它按索引从输入迭代器返回所选择的元素。

代码清单 3-20: **itertools_islice.py**

```
from itertools import *

print('Stop at 5:')
for i in islice(range(100), 5):
    print(i, end=' ')
print('\n')

print('Start at 5, Stop at 10:')
for i in islice(range(100), 5, 10):
    print(i, end=' ')
print('\n')

print('By tens to 100:')
for i in islice(range(100), 0, 100, 10):
    print(i, end=' ')
print('\n')
```

islice()与列表的slice操作符参数相同,同样包括开始位置(start)、结束位置(stop)和步长(step)。start和step参数是可选的。

```
$ python3 itertools_islice.py

Stop at 5:
0 1 2 3 4

Start at 5, Stop at 10:
5 6 7 8 9

By tens to 100:
0 10 20 30 40 50 60 70 80 90
```

tee()函数根据一个原输入迭代器返回多个独立的迭代器(默认为2个)。

代码清单 3-21: **itertools_tee.py**

```
from itertools import *

r = islice(count(), 5)
i1, i2 = tee(r)

print('i1:', list(i1))
print('i2:', list(i2))
```

tee()的语义类似于UNIX tee工具,它会重复从输入读到的值,并把它们写至一个命名文件和标准输出。tee()返回的迭代器可以用来为并行处理的多个算法提供相同的数据集。

```
$ python3 itertools_tee.py

i1: [0, 1, 2, 3, 4]
i2: [0, 1, 2, 3, 4]
```

tee()创建的新迭代器会共享其输入迭代器,所以创建了新迭代器后,不应再使用原迭代器。

代码清单3-22:**itertools_tee_error.py**

```python
from itertools import *

r = islice(count(), 5)
i1, i2 = tee(r)

print('r:', end=' ')
for i in r:
    print(i, end=' ')
    if i > 1:
        break
print()

print('i1:', list(i1))
print('i2:', list(i2))
```

如果原输入迭代器的一些值已经消费,新迭代器不会再生成这些值。

```
$ python3 itertools_tee_error.py

r: 0 1 2
i1: [3, 4]
i2: [3, 4]
```

3.2.2 转换输入

内置的`map()`函数返回一个迭代器,它对输入迭代器中的值调用一个函数并返回结果。任何输入迭代器中的元素全部消费时,`map()`函数都会停止。

代码清单3-23:**itertools_map.py**

```python
def times_two(x):
    return 2 * x


def multiply(x, y):
    return (x, y, x * y)


print('Doubles:')
for i in map(times_two, range(5)):
    print(i)

print('\nMultiples:')
r1 = range(5)
r2 = range(5, 10)
for i in map(multiply, r1, r2):
    print('{:d} * {:d} = {:d}'.format(*i))

print('\nStopping:')
r1 = range(5)
r2 = range(2)
for i in map(multiply, r1, r2):
    print(i)
```

在第一个例子中，`lambda` 函数将输入值乘以 2。在第二个例子中，`lambda` 函数将两个参数相乘（这两个参数分别来自不同的迭代器），返回一个元组，其中包含原参数和计算得到的值。第三个例子会在生成两个元组后停止，因为第二个区间已经处理完。

```
$ python3 itertools_map.py

Doubles:
0
2
4
6
8

Multiples:
0 * 5 = 0
1 * 6 = 6
2 * 7 = 14
3 * 8 = 24
4 * 9 = 36

Stopping:
(0, 0, 0)
(1, 1, 1)
```

`starmap()` 函数类似于 `map()`，不过并不是由多个迭代器构造一个元组，它使用 `*` 语法分解一个迭代器中的元素作为映射函数的参数。

代码清单 3-24：**itertools_starmap.py**

```
from itertools import *

values = [(0, 5), (1, 6), (2, 7), (3, 8), (4, 9)]

for i in starmap(lambda x, y: (x, y, x * y), values):
    print('{} * {} = {}'.format(*i))
```

`map()` 的映射函数名为 `f(i1, i2)`，而传入 `starmap()` 的映射函数名为 `f(*i)`。

```
$ python3 itertools_starmap.py

0 * 5 = 0
1 * 6 = 6
2 * 7 = 14
3 * 8 = 24
4 * 9 = 36
```

3.2.3 生成新值

`count()` 函数返回一个迭代器，该迭代器能够无限地生成连续的整数。第一个数可以作为参数传入（默认为 0）。这里没有上界参数（参见内置的 `range()`，这个函数对结果集可以有更多控制）。

代码清单 3-25：**itertools_count.py**

```
from itertools import *

for i in zip(count(1), ['a', 'b', 'c']):
    print(i)
```

这个例子会停止，因为列表参数会被完全消费。

```
$ python3 itertools_count.py

(1, 'a')
(2, 'b')
(3, 'c')
```

count()的"开始位置"和"步长"参数可以是可相加的任意的数字值。

代码清单 3-26：itertools_count_step.py

```
import fractions
from itertools import *

start = fractions.Fraction(1, 3)
step = fractions.Fraction(1, 3)

for i in zip(count(start, step), ['a', 'b', 'c']):
    print('{}: {}'.format(*i))
```

在这个例子中，开始点和步长是来自fraction模块的Fraction对象。

```
$ python3 itertools_count_step.py

1/3: a
2/3: b
1: c
```

cycle()函数返回一个迭代器，它会无限地重复给定参数的内容。由于必须记住输入迭代器的全部内容，所以如果这个迭代器很长，则可能会耗费大量内存。

代码清单 3-27：itertools_cycle.py

```
from itertools import *

for i in zip(range(7), cycle(['a', 'b', 'c'])):
    print(i)
```

这个例子中使用了一个计数器变量，在数个周期后会中止循环。

```
$ python3 itertools_cycle.py

(0, 'a')
(1, 'b')
(2, 'c')
(3, 'a')
(4, 'b')
(5, 'c')
(6, 'a')
```

repeat()函数返回一个迭代器，每次访问时会生成相同的值。

代码清单 3-28：itertools_repeat.py

```
from itertools import *

for i in repeat('over-and-over', 5):
    print(i)
```

repeat() 返回的迭代器会一直返回数据，除非提供了可选的 times 参数来限制次数。

```
$ python3 itertools_repeat.py

over-and-over
over-and-over
over-and-over
over-and-over
over-and-over
```

如果既要包含来自其他迭代器的值，也要包含一些不变的值，那么可以结合使用 repeat() 以及 zip() 或 map()。

代码清单 3-29：**itertools_repeat_zip.py**

```
from itertools import *

for i, s in zip(count(), repeat('over-and-over', 5)):
    print(i, s)
```

这个例子中就结合了一个计数器值和 repeat() 返回的常量。

```
$ python3 itertools_repeat_zip.py

0 over-and-over
1 over-and-over
2 over-and-over
3 over-and-over
4 over-and-over
```

下面这个例子使用 map() 将 0 到 4 区间中的数乘以 2。

代码清单 3-30：**itertools_repeat_map.py**

```
from itertools import *

for i in map(lambda x, y: (x, y, x * y), repeat(2), range(5)):
    print('{:d} * {:d} = {:d}'.format(*i))
```

repeat() 迭代器不需要被显式限制，因为任何一个输入迭代器结束时 map() 就会停止处理，而且 range() 只返回 5 个元素。

```
$ python3 itertools_repeat_map.py

2 * 0 = 0
2 * 1 = 2
2 * 2 = 4
2 * 3 = 6
2 * 4 = 8
```

3.2.4 过滤

dropwhile() 函数返回一个迭代器，它会在条件第一次变为 false 之后生成输入迭代器的元素。

代码清单 3-31：**itertools_dropwhile.py**

```
from itertools import *
```

```
def should_drop(x):
    print('Testing:', x)
    return x < 1

for i in dropwhile(should_drop, [-1, 0, 1, 2, -2]):
    print('Yielding:', i)
```

dropwhile() 并不会过滤输入的每一个元素。第一次条件为 false 之后,输入迭代器的所有其余元素都会返回。

```
$ python3 itertools_dropwhile.py

Testing: -1
Testing: 0
Testing: 1
Yielding: 1
Yielding: 2
Yielding: -2
```

takewhile() 与 **dropwhile()** 正相反。它也返回一个迭代器,这个迭代器将返回输入迭代器中保证测试条件为 true 的元素。

代码清单 3-32: **itertools_takewhile.py**

```
from itertools import *


def should_take(x):
    print('Testing:', x)
    return x < 2

for i in takewhile(should_take, [-1, 0, 1, 2, -2]):
    print('Yielding:', i)
```

一旦 **should_take()** 返回 false, **takewhile()** 就停止处理输入。

```
$ python3 itertools_takewhile.py

Testing: -1
Yielding: -1
Testing: 0
Yielding: 0
Testing: 1
Yielding: 1
Testing: 2
```

内置函数 **filter()** 返回一个迭代器,它只包含测试条件返回 true 时所对应的元素。

代码清单 3-33: **itertools_filter.py**

```
from itertools import *


def check_item(x):
    print('Testing:', x)
    return x < 1

for i in filter(check_item, [-1, 0, 1, 2, -2]):
    print('Yielding:', i)
```

filter() 与 dropwhile() 和 takewhile() 不同,它在返回之前会测试每一个元素。

```
$ python3 itertools_filter.py

Testing: -1
Yielding: -1
Testing: 0
Yielding: 0
Testing: 1
Testing: 2
Testing: -2
Yielding: -2
```

filterfalse() 返回一个迭代器,其中只包含测试条件返回 false 时对应的元素。

代码清单 3-34:**itertools_filterfalse.py**

```
from itertools import *


def check_item(x):
    print('Testing:', x)
    return x < 1

for i in filterfalse(check_item, [-1, 0, 1, 2, -2]):
    print('Yielding:', i)
```

check_item() 中的测试表达式与前面相同,所以在这个使用 filterfalse() 的例子中,结果与上一个例子的结果正好相反。

```
$ python3 itertools_filterfalse.py

Testing: -1
Testing: 0
Testing: 1
Yielding: 1
Testing: 2
Yielding: 2
Testing: -2
```

compress() 提供了另一种过滤可迭代对象内容的方法。不是调用一个函数,而是使用另一个可迭代对象中的值指示什么时候接受一个值以及什么时候忽略一个值。

代码清单 3-35:**itertools_compress.py**

```
from itertools import *

every_third = cycle([False, False, True])
data = range(1, 10)

for i in compress(data, every_third):
    print(i, end=' ')
print()
```

第一个参数是要处理的数据迭代器。第二个参数是一个选择器迭代器,这个迭代器会生成布尔值指示从数据输入中取哪些元素(true 值说明将生成这个值;false 值表示这个值将被忽略)。

```
$ python3 itertools_compress.py

3 6 9
```

3.2.5 数据分组

`groupby()`函数返回一个迭代器,它会生成按一个公共键组织的值集。下面这个例子展示了如何根据一个属性对相关的值分组。

代码清单 3-36:**itertools_groupby_seq.py**

```python
import functools
from itertools import *
import operator
import pprint


@functools.total_ordering
class Point:

    def __init__(self, x, y):
        self.x = x
        self.y = y

    def __repr__(self):
        return '({}, {})'.format(self.x, self.y)

    def __eq__(self, other):
        return (self.x, self.y) == (other.x, other.y)

    def __gt__(self, other):
        return (self.x, self.y) > (other.x, other.y)


# Create a data set of Point instances.
data = list(map(Point,
                cycle(islice(count(), 3)),
                islice(count(), 7)))
print('Data:')
pprint.pprint(data, width=35)
print()

# Try to group the unsorted data based on X values.
print('Grouped, unsorted:')
for k, g in groupby(data, operator.attrgetter('x')):
    print(k, list(g))
print()

# Sort the data.
data.sort()
print('Sorted:')
pprint.pprint(data, width=35)
print()
# Group the sorted data based on X values.
print('Grouped, sorted:')
for k, g in groupby(data, operator.attrgetter('x')):
    print(k, list(g))
print()
```

输入序列要根据键值排序,以保证得到预期的分组。

```
$ python3 itertools_groupby_seq.py

Data:
[(0, 0),
 (1, 1),
 (2, 2),
 (0, 3),
 (1, 4),
 (2, 5),
 (0, 6)]

Grouped, unsorted:
0 [(0, 0)]
1 [(1, 1)]
2 [(2, 2)]
0 [(0, 3)]
1 [(1, 4)]
2 [(2, 5)]
0 [(0, 6)]

Sorted:
[(0, 0),
 (0, 3),
 (0, 6),
 (1, 1),
 (1, 4),
 (2, 2),
 (2, 5)]

Grouped, sorted:
0 [(0, 0), (0, 3), (0, 6)]
1 [(1, 1), (1, 4)]
2 [(2, 2), (2, 5)]
```

3.2.6 合并输入

accumulate() 函数处理输入迭代器，向一个函数传递第 n 和 $n+1$ 个元素，并且生成返回值而不是某个输入。合并两个值的默认函数会将两个值相加，所以 accumulate() 可以用来生成一个数值输入序列的累加和。

代码清单 3-37: **itertools_accumulate.py**

```python
from itertools import *

print(list(accumulate(range(5))))
print(list(accumulate('abcde')))
```

用于非整数值序列时，结果取决于将两个元素"相加"是什么含义。这个脚本中的第二个例子显示了当 accumulate() 接收到一个字符串输入时，每个响应都将是该字符串的一个前缀，而且长度不断增加。

```
$ python3 itertools_accumulate.py

[0, 1, 3, 6, 10]
['a', 'ab', 'abc', 'abcd', 'abcde']
```

accumulate() 可以与任何取两个输入值的函数结合来得到不同的结果。

代码清单 3-38: **itertools_accumulate_custom.py**

```
from itertools import *


def f(a, b):
    print(a, b)
    return b + a + b

print(list(accumulate('abcde', f)))
```

这个例子以一种特殊的方式合并字符串值，会生成一系列（无意义的）回文。每一步调用 `f()` 时，它都会打印 `accumulate()` 传入的输入值。

```
$ python3 itertools_accumulate_custom.py

a b
bab c
cbabc d
dcbabcd e
['a', 'bab', 'cbabc', 'dcbabcd', 'edcbabcde']
```

迭代处理多个序列的嵌套 `for` 循环通常可以被替换为 `product()`，它会生成一个迭代器，值为输入值集合的笛卡儿积。

代码清单 3-39: **itertools_product.py**

```
from itertools import *
import pprint

FACE_CARDS = ('J', 'Q', 'K', 'A')
SUITS = ('H', 'D', 'C', 'S')

DECK = list(
    product(
        chain(range(2, 11), FACE_CARDS),
        SUITS,
    )
)

for card in DECK:
    print('{:>2}{}'.format(*card), end=' ')
    if card[1] == SUITS[-1]:
        print()
```

`product()` 生成的值是元组，成员取自作为参数传入的各个迭代器（按其传入的顺序）。返回的第一个元组包含各个迭代器的第一个值。传入 `product()` 的最后一个迭代器最先处理，接下来处理倒数第二个迭代器，依此类推。结果是按第一个迭代器、下一个迭代器等的顺序得到的返回值。

在这个例子中，扑克牌首先按牌面大小排序，然后按花色排序。

```
$ python3 itertools_product.py

 2H  2D  2C  2S
 3H  3D  3C  3S
 4H  4D  4C  4S
 5H  5D  5C  5S
```

```
6H   6D   6C   6S
7H   7D   7C   7S
8H   8D   8C   8S
9H   9D   9C   9S
10H  10D  10C  10S
JH   JD   JC   JS
QH   QD   QC   QS
KH   KD   KC   KS
AH   AD   AC   AS
```

要改变这些扑克牌的顺序,需要改变传入 `product()` 的参数的顺序。

代码清单 3-40:**itertools_product_ordering.py**

```python
from itertools import *
import pprint
FACE_CARDS = ('J', 'Q', 'K', 'A')
SUITS = ('H', 'D', 'C', 'S')

DECK = list(
    product(
        SUITS,
        chain(range(2, 11), FACE_CARDS),
    )
)

for card in DECK:
    print('{:>2}{}'.format(card[1], card[0]), end=' ')
    if card[1] == FACE_CARDS[-1]:
        print()
```

这个例子中的打印循环会查找一个 A 而不是黑桃,然后增加一个换行使输出分行显示。

```
$ python3 itertools_product_ordering.py

2H  3H  4H  5H  6H  7H  8H  9H  10H  JH  QH  KH  AH
2D  3D  4D  5D  6D  7D  8D  9D  10D  JD  QD  KD  AD
2C  3C  4C  5C  6C  7C  8C  9C  10C  JC  QC  KC  AC
2S  3S  4S  5S  6S  7S  8S  9S  10S  JS  QS  KS  AS
```

要计算一个序列与自身的积,可以指定输入重复多少次。

代码清单 3-41:**itertools_product_repeat.py**

```python
from itertools import *


def show(iterable):
    for i, item in enumerate(iterable, 1):
        print(item, end=' ')
        if (i % 3) == 0:
            print()
    print()


print('Repeat 2:\n')
show(list(product(range(3), repeat=2)))

print('Repeat 3:\n')
show(list(product(range(3), repeat=3)))
```

由于重复一个迭代器就像把同一个迭代器传入多次,`product()` 生成的每个元组所包含的元素个数就等于重复计数器。

```
$ python3 itertools_product_repeat.py

Repeat 2:

(0, 0) (0, 1) (0, 2)
(1, 0) (1, 1) (1, 2)
(2, 0) (2, 1) (2, 2)

Repeat 3:

(0, 0, 0) (0, 0, 1) (0, 0, 2)
(0, 1, 0) (0, 1, 1) (0, 1, 2)
(0, 2, 0) (0, 2, 1) (0, 2, 2)
(1, 0, 0) (1, 0, 1) (1, 0, 2)
(1, 1, 0) (1, 1, 1) (1, 1, 2)
(1, 2, 0) (1, 2, 1) (1, 2, 2)
(2, 0, 0) (2, 0, 1) (2, 0, 2)
(2, 1, 0) (2, 1, 1) (2, 1, 2)
(2, 2, 0) (2, 2, 1) (2, 2, 2)
```

permutations() 函数从输入迭代器生成元素,这些元素以给定长度的排列形式组合。默认地它会生成所有排列的全集。

代码清单 3-42:**itertools_permutations.py**

```python
from itertools import *


def show(iterable):
    first = None
    for i, item in enumerate(iterable, 1):
        if first != item[0]:
            if first is not None:
                print()
            first = item[0]
        print(''.join(item), end=' ')
    print()

print('All permutations:\n')
show(permutations('abcd'))

print('\nPairs:\n')
show(permutations('abcd', r=2))
```

可以使用 r 参数限制返回的各个排列的长度和个数。

```
$ python3 itertools_permutations.py

All permutations:

abcd abdc acbd acdb adbc adcb
bacd badc bcad bcda bdac bdca
cabd cadb cbad cbda cdab cdba
dabc dacb dbac dbca dcab dcba

Pairs:

ab ac ad
ba bc bd
ca cb cd
da db dc
```

为了将值限制为唯一的组合而不是排列，可以使用 `combinations()`。只要输入的成员是唯一的，输出就不会包含任何重复的值。

代码清单 3-43：`itertools_combinations.py`

```python
from itertools import *

def show(iterable):
    first = None
    for i, item in enumerate(iterable, 1):
        if first != item[0]:
            if first is not None:
                print()
            first = item[0]
        print(''.join(item), end=' ')
    print()

print('Unique pairs:\n')
show(combinations('abcd', r=2))
```

与排列不同，`combinations()` 的 r 参数是必要参数。

```
$ python3 itertools_combinations.py

Unique pairs:

ab ac ad
bc bd
cd
```

尽管 `combinations()` 不会重复单个的输入元素，但有时可能也需要考虑包含重复元素的组合。对于这种情况，可以使用 `combinations_with_replacement()`。

代码清单 3-44：`itertools_combinations_with_replacement.py`

```python
from itertools import *

def show(iterable):
    first = None
    for i, item in enumerate(iterable, 1):
        if first != item[0]:
            if first is not None:
                print()
            first = item[0]
        print(''.join(item), end=' ')
    print()

print('Unique pairs:\n')
show(combinations_with_replacement('abcd', r=2))
```

在这个输出中，每个输入元素会与自身以及输入序列的所有其他成员配对。

```
$ python3 itertools_combinations_with_replacement.py

Unique pairs:

aa ab ac ad
bb bc bd
cc cd
dd
```

提示：相关阅读材料
- `itertools` 的标准库文档[一]。
- `itertools` 的 Python 2 到 Python 3 移植说明。
- The Standard ML Basis Library[二]：SML 库。
- Definition of Haskell and the Standard Libraries[三]：函数语言 Haskell 的标准库规范。
- Clojure[四]：Clojure 是一种在 Java 虚拟机上运行的动态函数语言。
- tee[五]：这是一个 UNIX 命令行工具，用于将一个输入分解为多个相同的输出流。
- Wikipedia: Cartesian product[六]：两个序列笛卡儿积的数学定义。

3.3 operator：内置操作符的函数接口

使用迭代器编程时，有时需要为简单的表达式创建小函数。有些情况下，尽管这确实可以被实现为 `lambda` 函数，但某些操作根本不需要新函数。`operator` 模块定义了一些函数，可以对应标准对象 API 中内置的算术、比较和其他操作。

3.3.1 逻辑操作

有些函数可以用来确定一个值的相应布尔值，将其取反以创建相反的布尔值，以及比较对象以查看它们是否相等。

代码清单 3-45：`operator_boolean.py`

```
from operator import *

a = -1
b = 5

print('a =', a)
print('b =', b)
print()

print('not_(a)      :', not_(a))
print('truth(a)     :', truth(a))
print('is_(a, b)    :', is_(a, b))
print('is_not(a, b):', is_not(a, b))
```

`not_()` 后面有下划线，因为 `not` 是一个 Python 关键字。在 `if` 语句中测试一个表达式或将一个表达式转换为一个 `bool` 时会使用某种逻辑，`truth()` 会应用与之相同的逻辑。`is_()` 实现了 `is` 关键字使用的检查，`is_not()` 完成同样的测试，不过返回相反的答案。

[一] https://docs.python.org/3.5/library/itertools.html
[二] www.standardml.org/Basis/
[三] www.haskell.org/definition/
[四] http://clojure.org
[五] http://man7.org/linux/man-pages/man1/tee.1.html
[六] https://en.wikipedia.org/wiki/Cartesian_product

```
$ python3 operator_boolean.py

a = -1
b = 5
not_(a)      : False
truth(a)     : True
is_(a, b)    : False
is_not(a, b): True
```

3.3.2 比较操作符

支持所有富比较操作符。

代码清单 3-46：**operator_comparisons.py**

```
from operator import *

a = 1
b = 5.0

print('a =', a)
print('b =', b)
for func in (lt, le, eq, ne, ge, gt):
    print('{}(a, b): {}'.format(func.__name__, func(a, b)))
```

这些函数等价于使用 <、<=、==、>= 和 > 的表达式语法。

```
$ python3 operator_comparisons.py

a = 1
b = 5.0
lt(a, b): True
le(a, b): True
eq(a, b): False
ne(a, b): True
ge(a, b): False
gt(a, b): False
```

3.3.3 算术操作符

也支持处理数字值的算术操作符。

代码清单 3-47：**operator_math.py**

```
from operator import *

a = -1
b = 5.0
c = 2
d = 6

print('a =', a)
print('b =', b)
print('c =', c)
print('d =', d)

print('\nPositive/Negative:')
print('abs(a):', abs(a))
print('neg(a):', neg(a))
print('neg(b):', neg(b))
```

```
print('pos(a):', pos(a))
print('pos(b):', pos(b))

print('\nArithmetic:')
print('add(a, b)      :', add(a, b))
print('floordiv(a, b):', floordiv(a, b))
print('floordiv(d, c):', floordiv(d, c))
print('mod(a, b)      :', mod(a, b))
print('mul(a, b)      :', mul(a, b))
print('pow(c, d)      :', pow(c, d))
print('sub(b, a)      :', sub(b, a))
print('truediv(a, b) :', truediv(a, b))
print('truediv(d, c) :', truediv(d, c))

print('\nBitwise:')
print('and_(c, d)  :', and_(c, d))
print('invert(c)   :', invert(c))
print('lshift(c, d):', lshift(c, d))
print('or_(c, d)   :', or_(c, d))
print('rshift(d, c):', rshift(d, c))
print('xor(c, d)   :', xor(c, d))
```

提供了两个不同的除法操作符：floordiv()（Python 3.0 版本之前实现的整数除法）和 truediv()（浮点数除法）。

```
$ python3 operator_math.py

a = -1
b = 5.0
c = 2
d = 6

Positive/Negative:
abs(a): 1
neg(a): 1
neg(b): -5.0
pos(a): -1
pos(b): 5.0

Arithmetic:
add(a, b)      : 4.0
floordiv(a, b): -1.0
floordiv(d, c): 3
mod(a, b)      : 4.0
mul(a, b)      : -5.0
pow(c, d)      : 64
sub(b, a)      : 6.0
truediv(a, b) : -0.2
truediv(d, c) : 3.0

Bitwise:
and_(c, d)  : 2
invert(c)   : -3
lshift(c, d): 128
or_(c, d)   : 6
rshift(d, c): 1
xor(c, d)   : 4
```

3.3.4 序列操作符

处理序列的操作符可以分为 4 组：建立序列、搜索元素、访问内容以及从序列删除元素。

代码清单 3-48: operator_sequences.py

```
from operator import *

a = [1, 2, 3]
b = ['a', 'b', 'c']

print('a =', a)
print('b =', b)

print('\nConstructive:')
print('  concat(a, b):', concat(a, b))

print('\nSearching:')
print('  contains(a, 1)   :', contains(a, 1))
print('  contains(b, "d"):', contains(b, "d"))
print('  countOf(a, 1)    :', countOf(a, 1))
print('  countOf(b, "d") :', countOf(b, "d"))
print('  indexOf(a, 5)    :', indexOf(a, 1))

print('\nAccess Items:')
print('  getitem(b, 1)                :',
      getitem(b, 1))
print('  getitem(b, slice(1, 3))      :',
      getitem(b, slice(1, 3)))
print('  setitem(b, 1, "d")           :', end=' ')
setitem(b, 1, "d")
print(b)
print('  setitem(a, slice(1, 3), [4, 5]):', end=' ')
setitem(a, slice(1, 3), [4, 5])
print(a)

print('\nDestructive:')
print('  delitem(b, 1)                :', end=' ')
delitem(b, 1)
print(b)
print('  delitem(a, slice(1, 3)):', end=' ')
delitem(a, slice(1, 3))
print(a)
```

其中一些操作（如 setitem() 和 delitem()）会原地修改序列, 而且不返回任何值。

```
$ python3 operator_sequences.py

a = [1, 2, 3]
b = ['a', 'b', 'c']
Constructive:
  concat(a, b): [1, 2, 3, 'a', 'b', 'c']

Searching:
  contains(a, 1)   : True
  contains(b, "d"): False
  countOf(a, 1)    : 1
  countOf(b, "d") : 0
  indexOf(a, 5)    : 0

Access Items:
  getitem(b, 1)                  : b
  getitem(b, slice(1, 3))        : ['b', 'c']
  setitem(b, 1, "d")             : ['a', 'd', 'c']
  setitem(a, slice(1, 3), [4, 5]): [1, 4, 5]
```

```
Destructive:
  delitem(b, 1)         : ['a', 'c']
  delitem(a, slice(1, 3)): [1]
```

3.3.5 原地操作符

除了标准操作符，很多对象类型还通过一些特殊操作符（如 +=）支持"原地"修改。这些原地修改也有相应的等价函数。

代码清单 3-49：**operator_inplace.py**

```
from operator import *

a = -1
b = 5.0
c = [1, 2, 3]
d = ['a', 'b', 'c']
print('a =', a)
print('b =', b)
print('c =', c)
print('d =', d)
print()

a = iadd(a, b)
print('a = iadd(a, b) =>', a)
print()

c = iconcat(c, d)
print('c = iconcat(c, d) =>', c)
```

这些例子只展示了部分函数。要全面了解相关详细内容，请参考标准库文档。

```
$ python3 operator_inplace.py

a = -1
b = 5.0
c = [1, 2, 3]
d = ['a', 'b', 'c']

a = iadd(a, b) => 4.0

c = iconcat(c, d) => [1, 2, 3, 'a', 'b', 'c']
```

3.3.6 属性和元素"获取方法"

operator 模块最特别的特性之一是获取方法（getter）的概念。获取方法是运行时构造的一些 callable 对象，用来获取对象的属性或序列的内容。获取方法在处理迭代器或生成器序列时特别有用，因为获取方法引入的开销会大大低于 lambda 或 Python 函数的开销。

代码清单 3-50：**operator_attrgetter.py**

```
from operator import *


class MyObj:
    """example class for attrgetter"""
    def __init__(self, arg):
        super().__init__()
        self.arg = arg
```

```python
    def __repr__(self):
        return 'MyObj({})'.format(self.arg)

l = [MyObj(i) for i in range(5)]
print('objects   :', l)

# Extract the 'arg' value from each object.
g = attrgetter('arg')
vals = [g(i) for i in l]
print('arg values:', vals)

# Sort using arg.
l.reverse()
print('reversed  :', l)
print('sorted    :', sorted(l, key=g))
```

属性获取方法类似于 `lambda x,n='attrname': getattr(x,n)`:

```
$ python3 operator_attrgetter.py

objects    : [MyObj(0), MyObj(1), MyObj(2), MyObj(3), MyObj(4)]
arg values : [0, 1, 2, 3, 4]
reversed   : [MyObj(4), MyObj(3), MyObj(2), MyObj(1), MyObj(0)]
sorted     : [MyObj(0), MyObj(1), MyObj(2), MyObj(3), MyObj(4)]
```

元素获取方法类似于 `lambda x,y=5: x[y]`:

代码清单 3-51: **operator_itemgetter.py**

```python
from operator import *

l = [dict(val=-1 * i) for i in range(4)]
print('Dictionaries:')
print(' original:', l)
g = itemgetter('val')
vals = [g(i) for i in l]
print('   values:', vals)
print('   sorted:', sorted(l, key=g))

print
l = [(i, i * -2) for i in range(4)]
print('\nTuples:')
print(' original:', l)
g = itemgetter(1)
vals = [g(i) for i in l]
print('   values:', vals)
print('   sorted:', sorted(l, key=g))
```

除了序列，元素获取方法还适用于映射。

```
$ python3 operator_itemgetter.py

Dictionaries:
 original: [{'val': 0}, {'val': -1}, {'val': -2}, {'val': -3}]
   values: [0, -1, -2, -3]
   sorted: [{'val': -3}, {'val': -2}, {'val': -1}, {'val': 0}]

Tuples:
 original: [(0, 0), (1, -2), (2, -4), (3, -6)]
   values: [0, -2, -4, -6]
   sorted: [(3, -6), (2, -4), (1, -2), (0, 0)]
```

3.3.7 结合操作符和定制类

operator 模块中的函数完成操作时会使用标准 Python 接口，所以它们不仅适用于内置类型，也适用于用户定义的类。

代码清单 3-52：`operator_classes.py`

```
from operator import *

class MyObj:
    """Example for operator overloading"""
    def __init__(self, val):
        super(MyObj, self).__init__()
        self.val = val

    def __str__(self):
        return 'MyObj({})'.format(self.val)

    def __lt__(self, other):
        """compare for less-than"""
        print('Testing {} < {}'.format(self, other))
        return self.val < other.val

    def __add__(self, other):
        """add values"""
        print('Adding {} + {}'.format(self, other))
        return MyObj(self.val + other.val)

a = MyObj(1)
b = MyObj(2)

print('Comparison:')
print(lt(a, b))

print('\nArithmetic:')
print(add(a, b))
```

要全面了解每个操作符使用的所有特殊方法，请参见 Python 参考指南。

```
$ python3 operator_classes.py

Comparison:
Testing MyObj(1) < MyObj(2)
True

Arithmetic:
Adding MyObj(1) + MyObj(2)
MyObj(3)
```

提示：相关阅读材料

- **operator** 的标准库文档[⊖]。
- **functools**：函数编程工具，包括为类增加富比较方法的 `total_ordering()` 修饰符。
- **itertools**：迭代器操作。

⊖ https://docs.python.org/3.5/library/operator.html

- collections：集合的抽象类型。
- numbers：数值的抽象类型。

3.4 contextlib：上下文管理器工具

contextlib 模块包含的工具用于处理上下文管理器和 with 语句。

3.4.1 上下文管理器 API

上下文管理器（context manager）负责管理一个代码块中的资源，会在进入代码块时创建资源，然后在退出代码块后清理这个资源。例如，文件就支持上下文管理器 API，可以确保完成文件读写后关闭文件。

代码清单 3-53：**contextlib_file.py**

```
with open('/tmp/pymotw.txt', 'wt') as f:
    f.write('contents go here')
# File is automatically closed
```

上下文管理器由 with 语句启用，这个 API 包括两个方法。执行流进入 with 中的代码块时会运行 __enter__() 方法。它会返回在这个上下文中使用的一个对象。执行流离开 with 块时，则调用这个上下文管理器的 __exit__() 方法来清理所使用的资源。

代码清单 3-54：**contextlib_api.py**

```
class Context:

    def __init__(self):
        print('__init__()')

    def __enter__(self):
        print('__enter__()')
        return self

    def __exit__(self, exc_type, exc_val, exc_tb):
        print('__exit__()')

with Context():
    print('Doing work in the context')
```

相对于 try:finally 块，结合上下文管理器和 with 语句是一种更紧凑的写法，因为总会调用上下文管理器的 __exit__() 方法，即使产生异常的情况下也会调用这个方法。

```
$ python3 contextlib_api.py

__init__()
__enter__()
Doing work in the context
__exit__()
```

如果 with 语句的 as 子句中指定了名，那么 __enter__() 方法可以返回与这个名关联的任何对象。在这个例子中，Context 会返回一个使用打开的上下文的对象。

代码清单 3-55：**contextlib_api_other_object.py**

```python
class WithinContext:

    def __init__(self, context):
        print('WithinContext.__init__({})'.format(context))

    def do_something(self):
        print('WithinContext.do_something()')

    def __del__(self):
        print('WithinContext.__del__')


class Context:

    def __init__(self):
        print('Context.__init__()')

    def __enter__(self):
        print('Context.__enter__()')
        return WithinContext(self)

    def __exit__(self, exc_type, exc_val, exc_tb):
        print('Context.__exit__()')


with Context() as c:
    c.do_something()
```

与变量 c 关联的值是 __enter__() 返回的对象，这不一定是 with 语句中创建的 Context 实例。

```
$ python3 contextlib_api_other_object.py

Context.__init__()
Context.__enter__()
WithinContext.__init__(<__main__.Context object at 0x1007b1c50>)
WithinContext.do_something()
Context.__exit__()
WithinContext.__del__
```

__exit__() 方法接收一些参数，其中包含 with 块中产生的所有异常的详细信息。

代码清单 3-56：**contextlib_api_error.py**

```python
class Context:

    def __init__(self, handle_error):
        print('__init__({})'.format(handle_error))
        self.handle_error = handle_error

    def __enter__(self):
        print('__enter__()')
        return self

    def __exit__(self, exc_type, exc_val, exc_tb):
        print('__exit__()')
        print('  exc_type =', exc_type)
        print('  exc_val  =', exc_val)
        print('  exc_tb   =', exc_tb)
        return self.handle_error
```

```
with Context(True):
    raise RuntimeError('error message handled')

print()

with Context(False):
    raise RuntimeError('error message propagated')
```

如果上下文管理器可以处理这个异常,那么 __exit__() 应当返回一个 true 值来指示这个异常不需要传播。如果返回 false,则会在 __exit__() 返回后再次抛出这个异常。

```
$ python3 contextlib_api_error.py

__init__(True)
__enter__()
__exit__()
  exc_type = <class 'RuntimeError'>
  exc_val  = error message handled
  exc_tb   = <traceback object at 0x10115cc88>

__init__(False)
__enter__()
__exit__()
  exc_type = <class 'RuntimeError'>
  exc_val  = error message propagated
  exc_tb   = <traceback object at 0x10115cc88>
Traceback (most recent call last):
  File "contextlib_api_error.py", line 33, in <module>
    raise RuntimeError('error message propagated')
RuntimeError: error message propagated
```

3.4.2 上下文管理器作为函数修饰符

类 ContextDecorator 增加了对常规上下文管理器类的支持,因此其不仅可以作为上下文管理器,也可以作为函数修饰符。

代码清单 3-57:**contextlib_decorator.py**

```
import contextlib


class Context(contextlib.ContextDecorator):

    def __init__(self, how_used):
        self.how_used = how_used
        print('__init__({})'.format(how_used))

    def __enter__(self):
        print('__enter__({})'.format(self.how_used))
        return self

    def __exit__(self, exc_type, exc_val, exc_tb):
        print('__exit__({})'.format(self.how_used))


@Context('as decorator')
def func(message):
    print(message)


print()
```

```
with Context('as context manager'):
    print('Doing work in the context')

print()
func('Doing work in the wrapped function')
```

使用上下文管理器作为修饰符时有一点不同：`__enter__()` 返回的值在被修饰的函数中不可用，这与使用 `with` 和 `as` 时不一样。传入被修饰函数的参数可以正常使用。

```
$ python3 contextlib_decorator.py

__init__(as decorator)

__init__(as context manager)
__enter__(as context manager)
Doing work in the context
__exit__(as context manager)

__enter__(as decorator)
Doing work in the wrapped function
__exit__(as decorator)
```

3.4.3 从生成器到上下文管理器

采用传统方式创建上下文管理器并不难，即编写一个包含 `__enter__()` 和 `__exit__()` 方法的类。不过有些时候，如果只有很少的上下文要管理，那么完整地写出所有代码便会成为额外的负担。在这些情况下，可以使用 `contextmanager()` 修饰符将一个生成器函数转换为上下文管理器。

代码清单 3-58：contextlib_contextmanager.py

```
import contextlib


@contextlib.contextmanager
def make_context():
    print('  entering')
    try:
        yield {}
    except RuntimeError as err:
        print('  ERROR:', err)
    finally:
        print('  exiting')


print('Normal:')
with make_context() as value:
    print('  inside with statement:', value)

print('\nHandled error:')
with make_context() as value:
    raise RuntimeError('showing example of handling an error')

print('\nUnhandled error:')
with make_context() as value:
    raise ValueError('this exception is not handled')
```

生成器要初始化上下文，调用一次 `yield`，然后清理上下文。所生成的值（如果有）会绑定到 `with` 语句 `as` 子句中的变量。`with` 块中抛出的异常会在生成器中再次抛出，从而可以在生成器中得到处理。

```
$ python3 contextlib_contextmanager.py

Normal:
  entering
  inside with statement: {}
  exiting

Handled error:
  entering
  ERROR: showing example of handling an error
  exiting

Unhandled error:
  entering
  exiting
Traceback (most recent call last):
  File "contextlib_contextmanager.py", line 32, in <module>
    raise ValueError('this exception is not handled')
ValueError: this exception is not handled
```

contextmanager()返回的上下文管理器派生自ContextDecorator，所以也可以被用作函数修饰符。

代码清单 3-59：**contextlib_contextmanager_decorator.py**

```python
import contextlib


@contextlib.contextmanager
def make_context():
    print('  entering')
    try:
        # Yield control, but not a value, because any value
        # yielded is not available when the context manager
        # is used as a decorator.
        yield
    except RuntimeError as err:
        print('  ERROR:', err)
    finally:
        print('  exiting')


@make_context()
def normal():
    print('  inside with statement')


@make_context()
def throw_error(err):
    raise err


print('Normal:')
normal()

print('\nHandled error:')
throw_error(RuntimeError('showing example of handling an error'))

print('\nUnhandled error:')
throw_error(ValueError('this exception is not handled'))
```

与前面的 `ContextDecorator` 例子一样，上下文管理器被用作修饰符时，生成器生成的值在被修饰的函数中不可用。传入被修饰函数的参数仍然可用，如这个例子中的 `throw_error()` 所示。

```
$ python3 contextlib_contextmanager_decorator.py

Normal:
  entering
  inside with statement
  exiting

Handled error:
  entering
  ERROR: showing example of handling an error
  exiting

Unhandled error:
  entering
  exiting
Traceback (most recent call last):
  File "contextlib_contextmanager_decorator.py", line 43, in <module>
    throw_error(ValueError('this exception is not handled'))
  File ".../lib/python3.5/contextlib.py", line 30, in inner
    return func(*args, **kwds)
  File "contextlib_contextmanager_decorator.py", line 33, in throw_error
    raise err
ValueError: this exception is not handled
```

3.4.4 关闭打开的句柄

`file` 类直接支持上下文管理器 API，但另外一些表示打开句柄的对象却并不支持。`contextlib` 的标准库文档中给出的示例是从 `urllib.urlopen()` 返回的对象。另外一些遗留的类会使用一个 `close()` 方法但不支持上下文管理器 API。为了确保关闭句柄，要使用 `closing()` 为它创建一个上下文管理器。

代码清单 3-60：`contextlib_closing.py`

```python
import contextlib


class Door:

    def __init__(self):
        print('  __init__()')
        self.status = 'open'

    def close(self):
        print('  close()')
        self.status = 'closed'


print('Normal Example:')
with contextlib.closing(Door()) as door:
    print('  inside with statement: {}'.format(door.status))
print('  outside with statement: {}'.format(door.status))

print('\nError handling example:')
try:
```

```
    with contextlib.closing(Door()) as door:
        print('  raising from inside with statement')
        raise RuntimeError('error message')
except Exception as err:
    print('  Had an error:', err)
```

不论 with 块中是否有错误,都会关闭这个句柄。

```
$ python3 contextlib_closing.py

Normal Example:
  __init__()
  inside with statement: open
  close()
  outside with statement: closed

Error handling example:
  __init__()
  raising from inside with statement
  close()
  Had an error: error message
```

3.4.5 忽略异常

很多情况下,忽略库产生的异常通常很有用,因为这个错误可能会显示期望的状态已经被实现,否则该错误可以被忽略。要忽略异常,最常用的方法是利用一个 try:except 语句,其在 except 块中只包含一个 pass 语句。

代码清单 3-61:contextlib_ignore_error.py

```
import contextlib


class NonFatalError(Exception):
    pass


def non_idempotent_operation():
    raise NonFatalError(
        'The operation failed because of existing state'
    )


try:
    print('trying non-idempotent operation')
    non_idempotent_operation()
    print('succeeded!')
except NonFatalError:
    pass

print('done')
```

在这种情况下,这个操作会失败,而错误将被忽略。

```
$ python3 contextlib_ignore_error.py

trying non-idempotent operation
done
```

try:except 也可以被替换为 contextlib.suppress(),以更显式地抑制 with 块中产生某一类异常。

代码清单3-62：`contextlib_suppress.py`

```
import contextlib


class NonFatalError(Exception):
    pass


def non_idempotent_operation():
    raise NonFatalError(
        'The operation failed because of existing state'
    )


with contextlib.suppress(NonFatalError):
    print('trying non-idempotent operation')
    non_idempotent_operation()
    print('succeeded!')

print('done')
```

在这个更新后的版本中，异常会被完全丢弃。

```
$ python3 contextlib_suppress.py

trying non-idempotent operation
done
```

3.4.6 重定向输出流

设计不当的库代码可能会直接写 `sys.stdout` 或 `sys.stderr`，而没有提供参数来配置不同的输出目标。可以用 `redirect_stdout()` 和 `redirect_stderr()` 上下文管理器从这些函数捕获输出，因为无法修改这些函数的源代码来接受新的输出参数。

代码清单3-63：`contextlib_redirect.py`

```
from contextlib import redirect_stdout, redirect_stderr
import io
import sys


def misbehaving_function(a):
    sys.stdout.write('(stdout) A: {!r}\n'.format(a))
    sys.stderr.write('(stderr) A: {!r}\n'.format(a))


capture = io.StringIO()
with redirect_stdout(capture), redirect_stderr(capture):
    misbehaving_function(5)

print(capture.getvalue())
```

在这个例子中，`misbehaving_function()` 同时写至 `stdout` 和 `stderr`，不过两个上下文管理器将这个输出发送到同一个 `io.StringIO` 实例，会在这里保存以备以后使用。

```
$ python3 contextlib_redirect.py

(stdout) A: 5
(stderr) A: 5
```

说明：redirect_stdout() 和 redirect_stderr() 会修改全局状态，替换 sys 模块中的对象；出于这个原因，要小心使用这两个函数。这些函数并不保证真正的线程安全，所以在多线程应用中调用这些函数可能会有不确定的结果。如果有其他操作希望标准输出流关联到终端设备，那么 redirect_stdout() 和 redirect_stderr() 将会干扰和影响那些操作。

3.4.7 动态上下文管理器栈

大多数上下文管理器都一次处理一个对象，如单个文件或数据库句柄。在这些情况下，对象是提前已知的，并且使用上下文管理器的代码可以建立这一个对象上。另外一些情况下，程序可能需要在一个上下文中创建未知数目的对象，控制流退出这个上下文时所有这些对象都要清理。ExitStack 就是用来处理这些更动态的情况。

ExitStack 实例会维护清理回调的一个栈数据结构。这些回调显式地填充在上下文中，在控制流退出上下文时会以逆序调用所有注册的回调。结果类似于有多个嵌套的 with 语句，只不过它们是动态建立的。

3.4.7.1 上下文管理器入栈

可以使用多种方法填充 ExitStack。下面这个例子使用 enter_context() 来为栈增加一个新的上下文管理器。

代码清单 3-64：contextlib_exitstack_enter_context.py

```
import contextlib


@contextlib.contextmanager
def make_context(i):
    print('{} entering'.format(i))
    yield {}
    print('{} exiting'.format(i))


def variable_stack(n, msg):
    with contextlib.ExitStack() as stack:
        for i in range(n):
            stack.enter_context(make_context(i))
        print(msg)


variable_stack(2, 'inside context')
```

enter_context() 首先在上下文管理器上调用 __enter__()。然后把它的 __exit__() 方法注册为一个回调，撤销栈时将调用这个回调。

```
$ python3 contextlib_exitstack_enter_context.py

0 entering
1 entering
inside context
1 exiting
0 exiting
```

提供给 ExitStack 的上下文管理器被当作出现在一系列嵌套的 with 语句中。这个上下文发生的错误会通过上下文管理器正常的错误处理来传播。下面的上下文管理器类展示了错误如何传播。

代码清单3-65：**contextlib_context_managers.py**

```
import contextlib


class Tracker:
    "Base class for noisy context managers."

    def __init__(self, i):
        self.i = i

    def msg(self, s):
        print('  {}({}): {}'.format(
            self.__class__.__name__, self.i, s))

    def __enter__(self):
        self.msg('entering')


class HandleError(Tracker):
    "If an exception is received, treat it as handled."

    def __exit__(self, *exc_details):
        received_exc = exc_details[1] is not None
        if received_exc:
            self.msg('handling exception {!r}'.format(
                exc_details[1]))
        self.msg('exiting {}'.format(received_exc))
        # Return a boolean value indicating whether the exception
        # was handled.
        return received_exc


class PassError(Tracker):
    "If an exception is received, propagate it."

    def __exit__(self, *exc_details):
        received_exc = exc_details[1] is not None
        if received_exc:
            self.msg('passing exception {!r}'.format(
                exc_details[1]))
        self.msg('exiting')
        # Return False, indicating any exception was not handled.
        return False


class ErrorOnExit(Tracker):
    "Cause an exception."

    def __exit__(self, *exc_details):
        self.msg('throwing error')
        raise RuntimeError('from {}'.format(self.i))


class ErrorOnEnter(Tracker):
    "Cause an exception."

    def __enter__(self):
        self.msg('throwing error on enter')
        raise RuntimeError('from {}'.format(self.i))

    def __exit__(self, *exc_info):
        self.msg('exiting')
```

下面的例子使用了这些类，它基于 variable_stack()，这个函数使用传入的上下文管理器来构造一个 ExitStack，继而逐步地构建整个上下文。这些例子传入了不同的上下文管理器来研究错误处理行为。第一个例子展示了没有异常的正常情况。

```
print('No errors:')
variable_stack([
    HandleError(1),
    PassError(2),
])
```

下一个例子展示了栈末尾的上下文管理器中的错误处理，撤销栈时，所有打开的上下文都会关闭。

```
print('\nError at the end of the context stack:')
variable_stack([
    HandleError(1),
    HandleError(2),
    ErrorOnExit(3),
])
```

下一个例子中，异常会在栈中间的上下文管理器内被处理。直到一些上下文已经关闭时这些错误才出现，所以那些上下文不会看到这个错误。

```
print('\nError in the middle of the context stack:')
variable_stack([
    HandleError(1),
    PassError(2),
    ErrorOnExit(3),
    HandleError(4),
])
```

最后一个例子展示了异常未处理并且继续向上传播到其调用代码的情况。

```
try:
    print('\nError ignored:')
    variable_stack([
        PassError(1),
        ErrorOnExit(2),
    ])
except RuntimeError:
    print('error handled outside of context')
```

如果栈中的任何上下文管理器接收到一个异常并返回一个 True 值，那么这会阻止该异常继续向上传播到任何其他上下文管理器。

```
$ python3 contextlib_exitstack_enter_context_errors.py

No errors:
  HandleError(1): entering
  PassError(2): entering
  PassError(2): exiting
  HandleError(1): exiting False
  outside of stack, any errors were handled

Error at the end of the context stack:
  HandleError(1): entering
  HandleError(2): entering
  ErrorOnExit(3): entering
  ErrorOnExit(3): throwing error
  HandleError(2): handling exception RuntimeError('from 3',)
```

```
    HandleError(2): exiting True
    HandleError(1): exiting False
    outside of stack, any errors were handled
Error in the middle of the context stack:
    HandleError(1): entering
    PassError(2): entering
    ErrorOnExit(3): entering
    HandleError(4): entering
    HandleError(4): exiting False
    ErrorOnExit(3): throwing error
    PassError(2): passing exception RuntimeError('from 3',)
    PassError(2): exiting
    HandleError(1): handling exception RuntimeError('from 3',)
    HandleError(1): exiting True
    outside of stack, any errors were handled

Error ignored:
    PassError(1): entering
    ErrorOnExit(2): entering
    ErrorOnExit(2): throwing error
    PassError(1): passing exception RuntimeError('from 2',)
    PassError(1): exiting
error handled outside of context
```

3.4.7.2 任意上下文回调

ExitStack 还支持关闭上下文的任意回调，从而可以很容易地清理不通过上下文管理器控制的资源。

代码清单 3-66：**contextlib_exitstack_callbacks.py**

```
import contextlib


def callback(*args, **kwds):
    print('closing callback({}, {})'.format(args, kwds))


with contextlib.ExitStack() as stack:
    stack.callback(callback, 'arg1', 'arg2')
    stack.callback(callback, arg3='val3')
```

与完整上下文管理器的 __exit__() 方法一样，这些回调会按其注册的逆序调用。

```
$ python3 contextlib_exitstack_callbacks.py

closing callback((), {'arg3': 'val3'})
closing callback(('arg1', 'arg2'), {})
```

不论是否出现错误，都会调用这些回调，而且不会为它们提供是否出现错误的任何信息。其返回值将被忽略。

代码清单 3-67：**contextlib_exitstack_callbacks_error.py**

```
import contextlib


def callback(*args, **kwds):
    print('closing callback({}, {})'.format(args, kwds))
```

```
try:
    with contextlib.ExitStack() as stack:
        stack.callback(callback, 'arg1', 'arg2')
        stack.callback(callback, arg3='val3')
        raise RuntimeError('thrown error')
except RuntimeError as err:
    print('ERROR: {}'.format(err))
```

由于回调不能访问错误,所以它们无法防止异常在上下文管理器栈中继续传播。

```
$ python3 contextlib_exitstack_callbacks_error.py

closing callback((), {'arg3': 'val3'})
closing callback(('arg1', 'arg2'), {})
ERROR: thrown error
```

回调提供了一种便捷的方式来清晰地定义清理逻辑,而没有创建新上下文管理器类的开销。为了提高代码的可读性,这个逻辑可以被封装到一个内联函数中,callback()可以作为一个修饰符。

代码清单 3-68:`contextlib_exitstack_callbacks_decorator.py`

```
import contextlib


with contextlib.ExitStack() as stack:

    @stack.callback
    def inline_cleanup():
        print('inline_cleanup()')
        print('local_resource = {!r}'.format(local_resource))

    local_resource = 'resource created in context'
    print('within the context')
```

没有办法为使用 `callback()` 修饰符注册的函数指定参数。不过,如果清理回调采用内联方式定义,那么根据作用域规则,它就可以访问调用代码中定义的变量。

```
$ python3 contextlib_exitstack_callbacks_decorator.py

within the context
inline_cleanup()
local_resource = 'resource created in context'
```

3.4.7.3 部分栈

在构建复杂的上下文时,如果上下文无法完整构建,那么能够中止操作会很有用,另外如果所有资源都能正确建立,那么可以把所有资源的清理延迟到之后某个时间。例如,如果一个操作需要多个持续时间很长的网络连接,那么倘若一个连接失败,最好就不要启动这个操作。不过,如果所有连接都可以打开,那么这些连接需要长时间保持打开,并且要比上下文管理器的持续时间还要长。对于这种情况可以使用 `ExitStack` 的 `pop_all()` 方法。

`pop_all()` 会从调用它的栈中清除所有上下文管理器和回调,并返回一个新的栈,其中预填充了同样的上下文管理器和回调。之后(原栈撤销之后)可以调用这个新栈的 `close()` 方法来清理这些资源。

代码清单 3-69: **contextlib_exitstack_pop_all.py**

```
import contextlib

from contextlib_context_managers import *


def variable_stack(contexts):
    with contextlib.ExitStack() as stack:
        for c in contexts:
            stack.enter_context(c)
        # Return the close() method of a new stack as a clean-up
        # function.
        return stack.pop_all().close
    # Explicitly return None, indicating that the ExitStack could
    # not be initialized cleanly but that cleanup has already
    # occurred.
    return None


print('No errors:')
cleaner = variable_stack([
    HandleError(1),
    HandleError(2),
])
cleaner()

print('\nHandled error building context manager stack:')
try:
    cleaner = variable_stack([
        HandleError(1),
        ErrorOnEnter(2),
    ])
except RuntimeError as err:
    print('caught error {}'.format(err))
else:
    if cleaner is not None:
        cleaner()
    else:
        print('no cleaner returned')

print('\nUnhandled error building context manager stack:')
try:
    cleaner = variable_stack([
        PassError(1),
        ErrorOnEnter(2),
    ])
except RuntimeError as err:
    print('caught error {}'.format(err))
else:
    if cleaner is not None:
        cleaner()
    else:
        print('no cleaner returned')
```

这个例子使用了之前定义的相同的上下文管理器类，不过 **ErrorOnEnter** 会在 **__enter__()** 而不是 **__exit__()** 中生成一个错误。在 **variable_stack()** 中，如果所有上下文都成功进入而没有任何错误，则会返回新 **ExitStack** 的 **close()** 方法。如果出现了一个已处理的错误，则 **variable_stack()** 返回 **None** 指示清理工作已经完成。如果出现一个未处理的错误,则清理这个部分栈,并继续传播这个错误。

```
$ python3 contextlib_exitstack_pop_all.py

No errors:
  HandleError(1): entering
  HandleError(2): entering
  HandleError(2): exiting False
  HandleError(1): exiting False

Handled error building context manager stack:
  HandleError(1): entering
  ErrorOnEnter(2): throwing error on enter
  HandleError(1): handling exception RuntimeError('from 2',)
  HandleError(1): exiting True
no cleaner returned

Unhandled error building context manager stack:
  PassError(1): entering
  ErrorOnEnter(2): throwing error on enter
  PassError(1): passing exception RuntimeError('from 2',)
  PassError(1): exiting
caught error from 2
```

提示:相关阅读材料

- **contextlib** 的标准库文档[一]。
- **PEP 343**[二]:with 语句。
- Context Manager Types[三]:标准库文档中关于上下文管理器 API 的描述。
- **with** Statement Context Managers[四]:Python 参考指南中关于上下文管理器 API 的描述。
- Resource management in Python 3.3 或者 **contextlib.ExitStack** FTW![五]: Barry Warsaw 提供的使用 **ExitStack** 部署安全代码的描述。

[一] https://docs.python.org/3.5/library/contextlib.html
[二] www.python.org/dev/peps/pep-0343
[三] https://docs.python.org/library/stdtypes.html#typecontextmanager
[四] https://docs.python.org/reference/datamodel.html#context-managers
[五] www.wefearchange.org/2013/05/resource-management-in-python-33-or.html

第 4 章
日期和时间

不同于 `int`、`float` 和 `str`，Python 没有包含对应日期和时间的原生类型，不过提供了 3 个相应的模块，可以采用多种表示来管理日期和时间值。

`time` 模块由底层 C 库提供与时间相关的函数。它包含一些函数，可以用于获取时钟时间和处理器运行时间，还提供了基本的解析和字符串格式化工具。

`datetime` 模块为日期、时间以及日期时间值提供了一个更高层接口。`datetime` 中的类支持算术、比较和时区配置。

`calendar` 模块可以创建周、月和年的格式化表示。它还可以用来计算重复事件，给定日期是星期几，以及其他基于日历的值。

4.1 time：时钟时间

`time` 模块允许访问多种类型的时钟，分别用于不同的用途。标准系统调用（如 `time()`）会报告系统"墙上时钟"时间。`monotonic()` 时钟可以用于测量一个长时间运行的进程的耗用时间（elapsed time），因为即使系统时间有改变，也能保证这个时钟不会逆转。对于性能测试，`perf_counter()` 允许访问有最高可用分辨率的时钟，这使得短时间测量更为准确。CPU 时间可以通过 `clock()` 得到，`process_time()` 会返回处理器时间和系统时间的组合结果。

说明：这些实现提供了一些用于管理日期和时间的 C 库函数。由于它们绑定到底层 C 实现，一些细节（如纪元开始时间和支持的最大日期值）会特定于具体的平台。要全面了解有关的详细信息，请参考库文档。

4.1.1 比较时钟

时钟的实现细节因平台而异。可以使用 `get_clock_info()` 获得当前实现的基本信息，包括时钟的分辨率。

代码清单 4-1：time_get_clock_info.py

```
import textwrap
import time

available_clocks = [
    ('clock', time.clock),
    ('monotonic', time.monotonic),
    ('perf_counter', time.perf_counter),
```

```
        ('process_time', time.process_time),
        ('time', time.time),
    ]

    for clock_name, func in available_clocks:
        print(textwrap.dedent('''\
        {name}:
            adjustable    : {info.adjustable}
            implementation: {info.implementation}
            monotonic     : {info.monotonic}
            resolution    : {info.resolution}
            current       : {current}
        ''').format(
            name=clock_name,
            info=time.get_clock_info(clock_name),
            current=func())
        )
```

下面在 Mac OS X 上的输出显示，monotonic 和 perf_counter 时钟是通过相同的底层系统调用来实现的。

```
$ python3 time_get_clock_info.py

clock:
    adjustable    : False
    implementation: clock()
    monotonic     : True
    resolution    : 1e-06
    current       : 0.028399

monotonic:
    adjustable    : False
    implementation: mach_absolute_time()
    monotonic     : True
    resolution    : 1e-09
    current       : 172336.002232467

perf_counter:
    adjustable    : False
    implementation: mach_absolute_time()
    monotonic     : True
    resolution    : 1e-09
    current       : 172336.002280763

process_time:
    adjustable    : False
    implementation: getrusage(RUSAGE_SELF)
    monotonic     : True
    resolution    : 1e-06
    current       : 0.028593

time:
    adjustable    : True
    implementation: gettimeofday()
    monotonic     : False
    resolution    : 1e-06
    current       : 1471198232.045526
```

4.1.2 墙上时钟时间

time 模块的核心函数之一是 time()，它会把从"纪元"开始以来的秒数作为一个浮点值返回。

代码清单 4-2：**time_time.py**

```
import time
print('The time is:', time.time())
```

纪元是时间测量的起始点，对于 UNIX 系统这个起始时间就是 1970 年 1 月 1 日 0:00。尽管这个值总是一个浮点数，但具体的精度依赖于具体的平台。

```
$ python3 time_time.py
The time is: 1471198232.091589
```

浮点数表示对于存储或比较日期很有用，但是对于生成人类可读的表示就有些差强人意了。要记录或打印时间，`ctime()` 可能是更好的选择。

代码清单 4-3：**time_ctime.py**

```
import time
print('The time is      :', time.ctime())
later = time.time() + 15
print('15 secs from now :', time.ctime(later))
```

这个例子中的第二个 `print()` 调用显示了如何使用 `ctime()` 格式化非当前时间的另一个时间值。

```
$ python3 time_ctime.py

The time is      : Sun Aug 14 14:10:32 2016
15 secs from now : Sun Aug 14 14:10:47 2016
```

4.1.3 单调时钟

由于 `time()` 查看系统时钟，并且用户或系统服务可能改变系统时钟来同步多个计算机上的时钟，所以反复调用 `time()` 所产生的值可能向前和向后。试图测量持续时间或者使用这些时间来完成计算时，这可能会导致意想不到的行为。为了避免这些情况，可以使用 `monotonic()`，它总是返回向前的值。

代码清单 4-4：**time_monotonic.py**

```
import time
start = time.monotonic()
time.sleep(0.1)
end = time.monotonic()
print('start : {:>9.2f}'.format(start))
print('end   : {:>9.2f}'.format(end))
print('span  : {:>9.2f}'.format(end - start))
```

单调时钟的起始点没有被定义，所以返回值只是在与其他时钟值完成计算时有用。在这个例子中，使用 `monotonic()` 来测量睡眠持续时间。

```
$ python3 time_monotonic.py

start : 172336.14
end   : 172336.24
span  :      0.10
```

4.1.4 处理器时钟时间

time()返回的是一个墙上时钟时间,而clock()返回处理器时钟时间。clock()返回的值反映了程序运行时使用的实际时间。

代码清单4-5:**time_clock.py**

```
import hashlib
import time
# Data to use to calculate md5 checksums
data = open(__file__, 'rb').read()

for i in range(5):
    h = hashlib.sha1()
    print(time.ctime(), ': {:0.3f} {:0.3f}'.format(
        time.time(), time.clock()))
    for i in range(300000):
        h.update(data)
    cksum = h.digest()
```

在这个例子中,每次循环迭代时,会打印格式化的ctime()时间,以及time()和clock()返回的浮点值。

说明: 如果想要在你的系统上运行这个例子,可能必须在内循环中增加更多周期,或者处理更大量的数据,这样才能真正看到时间的差异。

```
$ python3 time_clock.py

Sun Aug 14 14:10:32 2016 : 1471198232.327 0.033
Sun Aug 14 14:10:32 2016 : 1471198232.705 0.409
Sun Aug 14 14:10:33 2016 : 1471198233.086 0.787
Sun Aug 14 14:10:33 2016 : 1471198233.466 1.166
Sun Aug 14 14:10:33 2016 : 1471198233.842 1.540
```

一般情况下,如果程序什么也没有做,则处理器时钟不会"滴答"(tick)。

代码清单4-6:**time_clock_sleep.py**

```
import time

template = '{} - {:0.2f} - {:0.2f}'

print(template.format(
    time.ctime(), time.time(), time.clock())
)

for i in range(3, 0, -1):
    print('Sleeping', i)
    time.sleep(i)
    print(template.format(
        time.ctime(), time.time(), time.clock())
    )
```

在这个例子中,循环几乎不做什么工作,每次迭代后都会睡眠。应用睡眠时,time()值会增加,而clock()值不会增加。

```
$ python3 -u time_clock_sleep.py

Sun Aug 14 14:10:34 2016 - 1471198234.28 - 0.03
Sleeping 3
Sun Aug 14 14:10:37 2016 - 1471198237.28 - 0.03
Sleeping 2
Sun Aug 14 14:10:39 2016 - 1471198239.29 - 0.03
Sleeping 1
Sun Aug 14 14:10:40 2016 - 1471198240.29 - 0.03
```

调用`sleep()`会从当前线程交出控制，并要求这个线程等待系统再次将其唤醒。如果程序只有一个线程，则这个函数实际上会阻塞应用，使它不做任何工作。

4.1.5 性能计数器

在测量性能时，高分辨率时钟是必不可少的。要确定最好的时钟数据源，需要有平台特定的知识，Python通过`perf_counter()`来提供所需的这些知识。

代码清单 4-7：**time_perf_counter.py**

```python
import hashlib
import time

# Data to use to calculate md5 checksums
data = open(__file__, 'rb').read()

loop_start = time.perf_counter()

for i in range(5):
    iter_start = time.perf_counter()
    h = hashlib.sha1()
    for i in range(300000):
        h.update(data)
    cksum = h.digest()
    now = time.perf_counter()
    loop_elapsed = now - loop_start
    iter_elapsed = now - iter_start
    print(time.ctime(), ': {:0.3f} {:0.3f}'.format(
        iter_elapsed, loop_elapsed))
```

类似于`monotonic()`，`perf_counter()`的纪元未定义，所以返回值只用于比较和计算值，而不作为绝对时间。

```
$ python3 time_perf_counter.py

Sun Aug 14 14:10:40 2016 : 0.487 0.487
Sun Aug 14 14:10:41 2016 : 0.485 0.973
Sun Aug 14 14:10:41 2016 : 0.494 1.466
Sun Aug 14 14:10:42 2016 : 0.487 1.953
Sun Aug 14 14:10:42 2016 : 0.480 2.434
```

4.1.6 时间组成

有些情况下需要把时间存储为过去了多少秒（秒数），但是另外一些情况下，程序需要访问一个日期的各个字段（例如，年和月）。`time`模块定义了`struct_time`来保存日期和时间值，其中分解了各个组成部分以便于访问。很多函数都要处理`struct_time`值而不是浮点值。

代码清单 4-8：**time_struct.py**

```python
import time

def show_struct(s):
    print('  tm_year :', s.tm_year)
    print('  tm_mon  :', s.tm_mon)
    print('  tm_mday :', s.tm_mday)
    print('  tm_hour :', s.tm_hour)
    print('  tm_min  :', s.tm_min)
    print('  tm_sec  :', s.tm_sec)
    print('  tm_wday :', s.tm_wday)
    print('  tm_yday :', s.tm_yday)
    print('  tm_isdst:', s.tm_isdst)

print('gmtime:')
show_struct(time.gmtime())
print('\nlocaltime:')
show_struct(time.localtime())
print('\nmktime:', time.mktime(time.localtime()))
```

`gmtime()` 函数以 UTC 格式返回当前时间。`localtime()` 会返回应用了当前时区的当前时间。`mktime()` 取一个 `struct_time` 实例，将它转换为浮点数表示。

```
$ python3 time_struct.py

gmtime:
  tm_year : 2016
  tm_mon  : 8
  tm_mday : 14
  tm_hour : 18
  tm_min  : 10
  tm_sec  : 42
  tm_wday : 6
  tm_yday : 227
  tm_isdst: 0

localtime:
  tm_year : 2016
  tm_mon  : 8
  tm_mday : 14
  tm_hour : 14
  tm_min  : 10
  tm_sec  : 42
  tm_wday : 6
  tm_yday : 227
  tm_isdst: 1

mktime: 1471198242.0
```

4.1.7　处理时区

用于确定当前时间的函数有一个前提，即已经设置了时区，其可以由程序设置，也可以使用系统的默认时区。修改时区不会改变具体的时间，只会改变表示时间的方式。

要改变时区，需要设置环境变量 `TZ`，然后调用 `tzset()`。设置时区时可以指定很多细节，甚至细致到夏令时的开始和结束时间。不过，通常更容易的做法是使用时区名，由底层库推导出其他信息。

下面这个例子会将时区修改为一些不同的值，并展示这些改变对 time 模块中的其他设置有什么影响。

代码清单 4-9：time_timezone.py

```python
import time
import os


def show_zone_info():
    print('  TZ    :', os.environ.get('TZ', '(not set)'))
    print('  tzname:', time.tzname)
    print('  Zone  : {} ({})'.format(
        time.timezone, (time.timezone / 3600)))
    print('  DST   :', time.daylight)
    print('  Time  :', time.ctime())
    print()


print('Default :')
show_zone_info()

ZONES = [
    'GMT',
    'Europe/Amsterdam',
]

for zone in ZONES:
    os.environ['TZ'] = zone
    time.tzset()
    print(zone, ':')
    show_zone_info()
```

这些例子使用的系统默认时区是 U.S./Eastern（美国东部时间）。例子中的其他时区会改变 **tzname**、**daylight** 标志和 **timezone** 偏移值。

```
$ python3 time_timezone.py
Default :
  TZ    : (not set)
  tzname: ('EST', 'EDT')
  Zone  : 18000 (5.0)
  DST   : 1
  Time  : Sun Aug 14 14:10:42 2016

GMT :
  TZ    : GMT
  tzname: ('GMT', 'GMT')
  Zone  : 0 (0.0)
  DST   : 0
  Time  : Sun Aug 14 18:10:42 2016

Europe/Amsterdam :
  TZ    : Europe/Amsterdam
  tzname: ('CET', 'CEST')
  Zone  : -3600 (-1.0)
  DST   : 1
  Time  : Sun Aug 14 20:10:42 2016
```

4.1.8 解析和格式化时间

函数 strptime() 和 strftime() 可以在时间值的 struct_time 表示和字符串表

示之间转换。这两个函数支持大量格式化指令，允许不同方式的输入和输出。所有这些格式化指令的完整列表参见 time 模块的库文档。

下面的这个例子将当前时间从字符串转换为 struct_time 实例，然后再转换回字符串。

代码清单 4-10：time_strptime.py

```
import time

def show_struct(s):
    print('  tm_year :', s.tm_year)
    print('  tm_mon  :', s.tm_mon)
    print('  tm_mday :', s.tm_mday)
    print('  tm_hour :', s.tm_hour)
    print('  tm_min  :', s.tm_min)
    print('  tm_sec  :', s.tm_sec)
    print('  tm_wday :', s.tm_wday)
    print('  tm_yday :', s.tm_yday)
    print('  tm_isdst:', s.tm_isdst)

now = time.ctime(1483391847.433716)
print('Now:', now)

parsed = time.strptime(now)
print('\nParsed:')
show_struct(parsed)

print('\nFormatted:',
      time.strftime("%a %b %d %H:%M:%S %Y", parsed))
```

输出字符串与输入字符串并不完全相同，因为日期前面加了一个前缀 0（由"2"变为"02"）。

```
$ python3 time_strptime.py

Now: Mon Jan  2 16:17:27 2017

Parsed:
  tm_year : 2017
  tm_mon  : 1
  tm_mday : 2
  tm_hour : 16
  tm_min  : 17
  tm_sec  : 27
  tm_wday : 0
  tm_yday : 2
  tm_isdst: -1

Formatted: Mon Jan 02 16:17:27 2017
```

提示：相关阅读材料

- time 的标准库文档[○]。
- time 的 Python 2 到 Python 3 移植说明。
- datetime：datetime 模块包含一些类，用于完成关于日期和时间的计算。
- calendar：处理更高级的日期函数，以生成日历或计算重复事件。

○ https://docs.python.org/3.5/library/time.html

4.2 datetime：日期和时间值管理

datetime 包含一些函数和类，用于完成日期和时间的解析、格式化和算术运算。

4.2.1 时间

时间值用 time 类表示。time 实例包含 hour、minute、second 和 microsecond 属性，还可以包含时区信息。

代码清单 4-11：**datetime_time.py**

```
import datetime

t = datetime.time(1, 2, 3)
print(t)
print('hour         :', t.hour)
print('minute       :', t.minute)
print('second       :', t.second)
print('microsecond:', t.microsecond)
print('tzinfo       :', t.tzinfo)
```

初始化 time 实例的参数是可选的，不过默认值 0 通常都不会是正确的设置。

```
$ python3 datetime_time.py

01:02:03
hour        : 1
minute      : 2
second      : 3
microsecond: 0
tzinfo      : None
```

time 实例只包含时间值，而不包含与时间关联的日期值。

代码清单 4-12：**datetime_time_minmax.py**

```
import datetime

print('Earliest  :', datetime.time.min)
print('Latest    :', datetime.time.max)
print('Resolution:', datetime.time.resolution)
```

min 和 max 类属性可以反映一天中的合法时间范围。

```
$ python3 datetime_time_minmax.py

Earliest   : 00:00:00
Latest     : 23:59:59.999999
Resolution: 0:00:00.000001
```

time 的分辨率被限制为整微秒值。

代码清单 4-13：**datetime_time_resolution.py**

```
import datetime

for m in [1, 0, 0.1, 0.6]:
```

```
        try:
            print('{:02.1f} :'.format(m),
                  datetime.time(0, 0, 0, microsecond=m))
        except TypeError as err:
            print('ERROR:', err)
```

如果微秒为浮点值，则其会产生一个 TypeError。

```
$ python3 datetime_time_resolution.py

1.0 : 00:00:00.000001
0.0 : 00:00:00
ERROR: integer argument expected, got float
ERROR: integer argument expected, got float
```

4.2.2 日期

日历日期值用 date 类表示。date 实例包含 year、month 和 day 属性。使用 today() 类方法很容易创建一个表示当前日期的日期实例。

代码清单 4-14：**datetime_date.py**

```
import datetime

today = datetime.date.today()
print(today)
print('ctime  :', today.ctime())
tt = today.timetuple()
print('tuple  : tm_year =', tt.tm_year)
print('         tm_mon  =', tt.tm_mon)
print('         tm_mday =', tt.tm_mday)
print('         tm_hour =', tt.tm_hour)
print('         tm_min  =', tt.tm_min)
print('         tm_sec  =', tt.tm_sec)
print('         tm_wday =', tt.tm_wday)
print('         tm_yday =', tt.tm_yday)
print('         tm_isdst=', tt.tm_isdst)
print('ordinal:', today.toordinal())
print('Year   :', today.year)
print('Mon    :', today.month)
print('Day    :', today.day)
```

下面这个例子采用多种不同格式来打印当前日期。

```
$ python3 datetime_date.py

2016-07-10
ctime  : Sun Jul 10 00:00:00 2016
tuple  : tm_year  = 2016
         tm_mon   = 7
         tm_mday  = 10
         tm_hour  = 0
         tm_min   = 0
         tm_sec   = 0
         tm_wday  = 6
         tm_yday  = 192
         tm_isdst = -1
ordinal: 736155
Year   : 2016
Mon    : 7
Day    : 10
```

还有一些类方法可以由 POSIX 时间戳或 Gregorian 日历中表示日期值的整数（第 1 年的 1 月 1 日对应的值为 1，以后每天对应的值逐个加 1）来创建 date 实例。

代码清单 4-15：datetime_date_fromordinal.py

```python
import datetime
import time

o = 733114
print('o                :', o)
print('fromordinal(o)   :', datetime.date.fromordinal(o))

t = time.time()
print('t                :', t)
print('fromtimestamp(t):', datetime.date.fromtimestamp(t))
```

这个例子表明 fromordinal() 和 fromtimestamp() 使用了不同的值类型。

```
$ python3 datetime_date_fromordinal.py

o                : 733114
fromordinal(o)   : 2008-03-13
t                : 1468161894.788508
fromtimestamp(t): 2016-07-10
```

与 time 类类似，可以使用 min 和 max 属性确定所支持的日期值范围。

代码清单 4-16：datetime_date_minmax.py

```python
import datetime

print('Earliest  :', datetime.date.min)
print('Latest    :', datetime.date.max)
print('Resolution:', datetime.date.resolution)
```

日期的分辨率为整天。

```
$ python3 datetime_date_minmax.py

Earliest  : 0001-01-01
Latest    : 9999-12-31
Resolution: 1 day, 0:00:00
```

创建新的 date 实例还有一种方法，可以使用现有日期的 replace() 方法来创建。

代码清单 4-17：datetime_date_replace.py

```python
import datetime

d1 = datetime.date(2008, 3, 29)
print('d1:', d1.ctime())

d2 = d1.replace(year=2009)
print('d2:', d2.ctime())
```

下面这个例子会改变年，但日和月保持不变。

```
$ python3 datetime_date_replace.py

d1: Sat Mar 29 00:00:00 2008
d2: Sun Mar 29 00:00:00 2009
```

4.2.3 `timedelta`

通过对两个 `datetime` 对象完成算术运算，或者结合使用 `datetime` 和 `timedelta`，可以计算出将来和过去的日期。将两个日期相减可以生成一个 `timedelta`，还可以对某个日期增加或减去一个 `timedelta` 来生成另一个日期。`timedelta` 的内部值按日、秒和微秒存储。

代码清单 4-18：`datetime_timedelta.py`

```
import datetime

print('microseconds:', datetime.timedelta(microseconds=1))
print('milliseconds:', datetime.timedelta(milliseconds=1))
print('seconds      :', datetime.timedelta(seconds=1))
print('minutes      :', datetime.timedelta(minutes=1))
print('hours        :', datetime.timedelta(hours=1))
print('days         :', datetime.timedelta(days=1))
print('weeks        :', datetime.timedelta(weeks=1))
```

传入构造函数的中间值会被转换为日、秒和微秒。

```
$ python3 datetime_timedelta.py

microseconds: 0:00:00.000001
milliseconds: 0:00:00.001000
seconds      : 0:00:01
minutes      : 0:01:00
hours        : 1:00:00
days         : 1 day, 0:00:00
weeks        : 7 days, 0:00:00
```

一个 `timedelta` 的完整时间段可以使用 `total_seconds()` 得到，并作为一个秒数返回。

代码清单 4-19：`datetime_timedelta_total_seconds.py`

```
import datetime

for delta in [datetime.timedelta(microseconds=1),
              datetime.timedelta(milliseconds=1),
              datetime.timedelta(seconds=1),
              datetime.timedelta(minutes=1),
              datetime.timedelta(hours=1),
              datetime.timedelta(days=1),
              datetime.timedelta(weeks=1),
              ]:
    print('{:15} = {:8} seconds'.format(
        str(delta), delta.total_seconds())
    )
```

返回值是一个浮点数，因为有些时间段不到 1 秒。

```
$ python3 datetime_timedelta_total_seconds.py

0:00:00.000001  =     1e-06 seconds
0:00:00.001000  =     0.001 seconds
0:00:01         =       1.0 seconds
0:01:00         =      60.0 seconds
1:00:00         =    3600.0 seconds
1 day, 0:00:00  =   86400.0 seconds
7 days, 0:00:00 =  604800.0 seconds
```

4.2.4 日期算术运算

日期算术运算使用标准算术操作符来完成。

代码清单 4-20：**datetime_date_math.py**

```
import datetime

today = datetime.date.today()
print('Today    :', today)

one_day = datetime.timedelta(days=1)
print('One day  :', one_day)

yesterday = today - one_day
print('Yesterday:', yesterday)

tomorrow = today + one_day
print('Tomorrow :', tomorrow)

print()
print('tomorrow - yesterday:', tomorrow - yesterday)
print('yesterday - tomorrow:', yesterday - tomorrow)
```

这个处理日期对象的例子展示了如何使用 `timedelta` 对象计算新日期，另外，这里将日期实例相减来生成 `timedelta`（包括一个负差异值）。

```
$ python3 datetime_date_math.py

Today    : 2016-07-10
One day  : 1 day, 0:00:00
Yesterday: 2016-07-09
Tomorrow : 2016-07-11

tomorrow - yesterday: 2 days, 0:00:00
yesterday - tomorrow: -2 days, 0:00:00
```

`timedelta` 对象还支持与整数、浮点数和其他 `timedelta` 实例的算术运算。

代码清单 4-21：**datetime_timedelta_math.py**

```
import datetime

one_day = datetime.timedelta(days=1)
print('1 day    :', one_day)
print('5 days   :', one_day * 5)
print('1.5 days :', one_day * 1.5)
print('1/4 day  :', one_day / 4)

# Assume an hour for lunch.
work_day = datetime.timedelta(hours=7)
meeting_length = datetime.timedelta(hours=1)
print('meetings per day :', work_day / meeting_length)
```

在这个例子中，计算了一天的多个倍数，得到的 `timedelta` 包含相应的天数或小时数。

最后这个例子展示了如何结合两个 `timedelta` 对象来计算值。在这里，结果是一个浮点数。

```
$ python3 datetime_timedelta_math.py

1 day          : 1 day, 0:00:00
5 days         : 5 days, 0:00:00
1.5 days       : 1 day, 12:00:00
1/4 day        : 6:00:00
meetings per day : 7.0
```

4.2.5 比较值

日期和时间值都可以使用标准比较操作符来比较,从而确定哪一个在前,哪一个在后。

代码清单4-22: `datetime_comparing.py`

```python
import datetime
import time

print('Times:')
t1 = datetime.time(12, 55, 0)
print('  t1:', t1)
t2 = datetime.time(13, 5, 0)
print('  t2:', t2)
print('  t1 < t2:', t1 < t2)

print
print('Dates:')
d1 = datetime.date.today()
print('  d1:', d1)
d2 = datetime.date.today() + datetime.timedelta(days=1)
print('  d2:', d2)
print('  d1 > d2:', d1 > d2)
```

支持所有比较操作符。

```
$ python3 datetime_comparing.py

Times:
  t1: 12:55:00
  t2: 13:05:00
  t1 < t2: True
Dates:
  d1: 2016-07-10
  d2: 2016-07-11
  d1 > d2: False
```

4.2.6 结合日期和时间

使用`datetime`类可以存储由日期和时间分量构成的值。类似于`date`,可以使用很多便利的类方法来从其他常用值创建`datetime`实例。

代码清单4-23: `datetime_datetime.py`

```python
import datetime

print('Now     :', datetime.datetime.now())
print('Today   :', datetime.datetime.today())
print('UTC Now:', datetime.datetime.utcnow())
print

FIELDS = [
```

```
            'year', 'month', 'day',
            'hour', 'minute', 'second',
            'microsecond',
]

d = datetime.datetime.now()
for attr in FIELDS:
    print('{:15}: {}'.format(attr, getattr(d, attr)))
```

可以想见，`datetime` 实例包含 `date` 和 `time` 对象的所有属性。

```
$ python3 datetime_datetime.py

Now      : 2016-07-10 10:44:55.215677
Today    : 2016-07-10 10:44:55.215719
UTC Now  : 2016-07-10 14:44:55.215732
year           : 2016
month          : 7
day            : 10
hour           : 10
minute         : 44
second         : 55
microsecond    : 216198
```

与 `date` 类似，`datetime` 提供了便利的类方法来创建新实例。它还包括 `fromordinal()` 和 `fromtimestamp()`。

代码清单 4-24：**datetime_datetime_combine.py**

```
import datetime

t = datetime.time(1, 2, 3)
print('t :', t)

d = datetime.date.today()
print('d :', d)

dt = datetime.datetime.combine(d, t)
print('dt:', dt)
```

利用 `combine()`，可以由一个 `date` 实例和一个 `time` 实例创建一个 `datetime` 实例。

```
$ python3 datetime_datetime_combine.py

t : 01:02:03
d : 2016-07-10
dt: 2016-07-10 01:02:03
```

4.2.7 格式化和解析

`datetime` 对象的默认字符串表示使用 ISO-8601 格式（YYYY-MM-DDTHH:MM:SS.mmmmmm）。可以使用 `strftime()` 生成其他格式。

代码清单 4-25：**datetime_datetime_strptime.py**

```
import datetime

format = "%a %b %d %H:%M:%S %Y"
```

```
today = datetime.datetime.today()
print('ISO     :', today)

s = today.strftime(format)
print('strftime:', s)

d = datetime.datetime.strptime(s, format)
print('strptime:', d.strftime(format))
```

使用 `datetime.strptime()` 可以将格式化的字符串转换为 `datetime` 实例。

```
$ python3 datetime_datetime_strptime.py

ISO     : 2016-07-10 10:44:55.325247
strftime: Sun Jul 10 10:44:55 2016
strptime: Sun Jul 10 10:44:55 2016
```

还可以对 Python 的字符串格式化微语言[⊖]使用同样的格式化代码，只需要把这些格式化代码放在格式化字符串字段规范的 : 后面。

代码清单 4-26：`datetime_format.py`

```
import datetime

today = datetime.datetime.today()
print('ISO     :', today)
print('format(): {:%a %b %d %H:%M:%S %Y}'.format(today))
```

每个 `datetime` 格式化代码都必须有 `%` 前缀，后面的冒号会作为字面量字符包含在输出中。

```
$ python3 datetime_format.py

ISO     : 2016-07-10 10:44:55.389239
format(): Sun Jul 10 10:44:55 2016
```

表 4-1 给出了 U.S./Eastern 时区 2016 年 1 月 13 日 5:00PM 的所有格式化代码。

表 4-1 `strptime/strftime` 格式化代码

符号	含义	示例
%a	缩写的星期几	'Wed'
%A	完整的星期几	'Wednesday'
%w	星期几的编号：0（星期天）到 6（星期六）	'3'
%d	当月哪一天（补 0）	'13'
%b	缩写的月份名	'Jan'
%B	完整的月份名	'January'
%m	当年哪个月	'01'
%y	不加世纪编号的年份	'16'
%Y	加世纪编号的年份	'2016'
%H	24 小时制的小时数	'17'
%I	12 小时制的小时数	'05'
%p	AM/PM	'PM'

⊖ https://docs.python.org/3.5/library/string.html#formatspec

（续）

符号	含义	示例
%M	分钟	'00'
%S	秒	'00'
%f	微秒	'000000'
%z	区分时区的日期时间对象的 UTC 偏移	'-0500'
%Z	时区名	'EST'
%j	当年哪一天	'013'
%W	当年哪一周	'02'
%c	当前本地化环境的日期和时间表示	'Wed Jan 13 17:00:00 2016'
%x	当前本地化环境的日期表示	'01/13/16'
%X	当前本地化环境的时间表示	'17:00:00'
%%	字面量 % 字符	'%'

4.2.8 时区

在 datetime 中，时区由 tzinfo 的子类表示。由于 tzinfo 是一个抽象基类，实际使用时，应用需要定义它的一个子类，并为一些方法提供适当的实现。

datetime 在类 timezone 中确实包含一个原生实现，该类使用了一个固定的 UTC 偏移。这个实现不支持一年中不同日期有不同的偏移值，如有些地方采用夏令时，或者有些地方 UTC 偏移会随时间改变。

代码清单 4-27：**datetime_timezone.py**

```
import datetime

min6 = datetime.timezone(datetime.timedelta(hours=-6))
plus6 = datetime.timezone(datetime.timedelta(hours=6))
d = datetime.datetime.now(min6)

print(min6, ':', d)
print(datetime.timezone.utc, ':',
      d.astimezone(datetime.timezone.utc))
print(plus6, ':', d.astimezone(plus6))

# Convert to the current system timezone.
d_system = d.astimezone()
print(d_system.tzinfo, '        :', d_system)
```

要把一个 datetime 值从一个时区转换为另一个时区，可以使用 astimezone()。在前面的例子中，显示了 UTC 两侧正负 6 个小时的两个不同时区，另外还使用了由 datetime.timezone 得到的 utc 实例来作为参考。最后的输出行显示了系统时区的值，这是不提供任何参数调用 astimezone() 得到的。

```
$ python3 datetime_timezone.py

UTC-06:00 : 2016-07-10 08:44:55.495995-06:00
UTC+00:00 : 2016-07-10 14:44:55.495995+00:00
UTC+06:00 : 2016-07-10 20:44:55.495995+06:00
EDT       : 2016-07-10 10:44:55.495995-04:00
```

说明：第三方模块 pytz[①] 是一个更好的时区实现。它支持命名时区，另外当世界各地的政府机构做出改变时，它会及时更新时区偏移数据库。

提示：相关阅读材料
- datetime 的标准库文档[②]。
- datetime 的 Python 2 到 Python 3 移植说明。
- calendar：calendar 模块。
- time：time 模块。
- dateutil[③]：Labix 的 dateutil 为 datetime 模块扩展了一些额外特性。
- pytz：提供了一个世界时区数据库和一些类，使 datetime 对象能够区分时区。
- Wikipedia: Proleptic Gregorian calendar[④]：Gregorian 日历系统的一个描述。
- Wikipedia: ISO 8601[⑤]：日期和时间数值表示的标准。

4.3　calendar：处理日期

calendar 模块定义了 Calendar 类，其中封装了一些值的计算，如给定的一个月或一年中的周日期。另外，TextCalendar 和 HTMLCalendar 类可以生成经过预格式化的输出。

4.3.1　格式化示例

prmonth() 方法是一个简单的函数，可以生成月的格式化文本输出。

代码清单 4-28：**calendar_textcalendar.py**

```
import calendar

c = calendar.TextCalendar(calendar.SUNDAY)
c.prmonth(2017, 7)
```

这个例子按照美国的惯例，将 TextCalendar 配置为一周从星期日开始。而默认则会使用欧洲惯例，即一周从星期一开始。这个例子会生成以下输出。

```
$ python3 calendar_textcalendar.py

     July 2017
Su Mo Tu We Th Fr Sa
                   1
 2  3  4  5  6  7  8
```

---

① http://pytz.sourceforge.net/
② https://docs.python.org/3.5/library/datetime.html
③ http://labix.org/python-dateutil
④ https://en.wikipedia.org/wiki/Proleptic_Gregorian_calendar
⑤ https://en.wikipedia.org/wiki/ISO_8601

```
 9 10 11 12 13 14 15
16 17 18 19 20 21 22
23 24 25 26 27 28 29
30 31
```

利用 HTMLCalendar 和 formatmonth() 可以生成一个类似的 HTML 表格。显示的输出看起来与纯文本的版本大致是一样的，不过会用 HTML 标记包围。各个表单元格有一个类属性对应星期几，从而可以通过 CSS 指定 HTML 的样式。

除了可用的默认格式外，要想以其他的某种格式生成输出，可以使用 calendar 计算日期，并把这些值组织为周和月区间，然后迭代处理结果。对于这个任务，Calendar 的 weekheader()、monthcalendar() 和 yeardays2calendar() 方法尤其有用。

调用 yeardays2calendar() 会生成一个由"月行"列表构成的序列。每个月列表包含一些月，每个月是一个周列表。周是元组列表，元组则由日编号（1～31）和星期几编号（0～6）构成。当月以外的日编号为 0。

代码清单 4-29：calendar_yeardays2calendar.py

```python
import calendar
import pprint

cal = calendar.Calendar(calendar.SUNDAY)

cal_data = cal.yeardays2calendar(2017, 3)
print('len(cal_data)      :', len(cal_data))

top_months = cal_data[0]
print('len(top_months)    :', len(top_months))

first_month = top_months[0]
print('len(first_month)   :', len(first_month))

print('first_month:')
pprint.pprint(first_month, width=65)
```

调用 yeardays2calendar(2017,3) 会返回 2017 年的数据，按每行 3 个月组织。

```
$ python3 calendar_yeardays2calendar.py

len(cal_data)      : 4
len(top_months)    : 3
len(first_month)   : 5
first_month:
[[(1, 6), (2, 0), (3, 1), (4, 2), (5, 3), (6, 4), (7, 5)],
 [(8, 6), (9, 0), (10, 1), (11, 2), (12, 3), (13, 4), (14, 5)],
 [(15, 6), (16, 0), (17, 1), (18, 2), (19, 3), (20, 4), (21,
5)],
 [(22, 6), (23, 0), (24, 1), (25, 2), (26, 3), (27, 4), (28,
5)],
 [(29, 6), (30, 0), (31, 1), (0, 2), (0, 3), (0, 4), (0, 5)]]
```

这等价于 formatyear() 使用的数据。

代码清单 4-30：calendar_formatyear.py

```python
import calendar

cal = calendar.TextCalendar(calendar.SUNDAY)
print(cal.formatyear(2017, 2, 1, 1, 3))
```

如果给定相同的参数，则 formatyear() 会生成以下输出。

```
$ python3 calendar_formatyear.py

                              2017

      January               February                March
Su Mo Tu We Th Fr Sa  Su Mo Tu We Th Fr Sa  Su Mo Tu We Th Fr Sa
 1  2  3  4  5  6  7            1  2  3  4            1  2  3  4
 8  9 10 11 12 13 14   5  6  7  8  9 10 11   5  6  7  8  9 10 11
15 16 17 18 19 20 21  12 13 14 15 16 17 18  12 13 14 15 16 17 18
22 23 24 25 26 27 28  19 20 21 22 23 24 25  19 20 21 22 23 24 25
29 30 31              26 27 28              26 27 28 29 30 31

       April                   May                   June
Su Mo Tu We Th Fr Sa  Su Mo Tu We Th Fr Sa  Su Mo Tu We Th Fr Sa
                   1      1  2  3  4  5  6               1  2  3
 2  3  4  5  6  7  8   7  8  9 10 11 12 13   4  5  6  7  8  9 10
 9 10 11 12 13 14 15  14 15 16 17 18 19 20  11 12 13 14 15 16 17
16 17 18 19 20 21 22  21 22 23 24 25 26 27  18 19 20 21 22 23 24
23 24 25 26 27 28 29  28 29 30 31           25 26 27 28 29 30
30

        July                  August               September
Su Mo Tu We Th Fr Sa  Su Mo Tu We Th Fr Sa  Su Mo Tu We Th Fr Sa
                   1         1  2  3  4  5                  1  2
 2  3  4  5  6  7  8   6  7  8  9 10 11 12   3  4  5  6  7  8  9
 9 10 11 12 13 14 15  13 14 15 16 17 18 19  10 11 12 13 14 15 16
16 17 18 19 20 21 22  20 21 22 23 24 25 26  17 18 19 20 21 22 23
23 24 25 26 27 28 29  27 28 29 30 31        24 25 26 27 28 29 30
30 31

      October               November               December
Su Mo Tu We Th Fr Sa  Su Mo Tu We Th Fr Sa  Su Mo Tu We Th Fr Sa
 1  2  3  4  5  6  7            1  2  3  4                  1  2
 8  9 10 11 12 13 14   5  6  7  8  9 10 11   3  4  5  6  7  8  9
15 16 17 18 19 20 21  12 13 14 15 16 17 18  10 11 12 13 14 15 16
22 23 24 25 26 27 28  19 20 21 22 23 24 25  17 18 19 20 21 22 23
29 30 31              26 27 28 29 30        24 25 26 27 28 29 30
                                            31
```

day_name、day_abbr、month_name 和 month_abbr 模块属性对于生成定制格式的输出很有用（例如，在 HTML 输出中包含链接）。这些属性会针对当前本地化环境正确地自动配置。

4.3.2 本地化环境

如果不是为了当前本地化环境，而是要为另外一个本地化环境生成一个格式化的日历，那么可以使用 LocaleTextCalendar 或 LocaleHTMLCalendar。

代码清单 4-31：calendar_locale.py

```
import calendar

c = calendar.LocaleTextCalendar(locale='en_US')
c.prmonth(2017, 7)

print()

c = calendar.LocaleTextCalendar(locale='fr_FR')
c.prmonth(2017, 7)
```

一周的第一天不属于本地化环境的设置。这个值取自 `calendar` 类的参数，常规的 `TextCalendar` 类也是如此。

```
$ python3 calendar_locale.py

      July 2017
Mo Tu We Th Fr Sa Su
                1  2
 3  4  5  6  7  8  9
10 11 12 13 14 15 16
17 18 19 20 21 22 23
24 25 26 27 28 29 30
31

     juillet 2017
Lu Ma Me Je Ve Sa Di
                1  2
 3  4  5  6  7  8  9
10 11 12 13 14 15 16
17 18 19 20 21 22 23
24 25 26 27 28 29 30
31
```

4.3.3 计算日期

尽管 `calendar` 模块主要强调采用不同格式打印完整的日历，但它还是提供了另外一些函数，对采用其他方式处理日期很有用，如为一个重复事件计算日期。例如，Python Atlanta 用户组每月的第二个星期四会召开一次会议。要计算一年中的会议日期，可以使用 `monthcalendar()` 的返回值。

代码清单 4-32：`calendar_monthcalendar.py`

```python
import calendar
import pprint

pprint.pprint(calendar.monthcalendar(2017, 7))
```

有些日期的值为 0。这说明这几天对应的星期几尽管在给定的当前月份里，但它们实际上属于另一个月。

```
$ python3 calendar_monthcalendar.py

[[0, 0, 0, 0, 0, 1, 2],
 [3, 4, 5, 6, 7, 8, 9],
 [10, 11, 12, 13, 14, 15, 16],
 [17, 18, 19, 20, 21, 22, 23],
 [24, 25, 26, 27, 28, 29, 30],
 [31, 0, 0, 0, 0, 0, 0]]
```

一周中的第一天默认为星期一。可以通过调用 `setfirstweekday()` 来改变这个设置，不过由于 `calendar` 模块包含了一些常量来索引 `monthcalendar()` 返回的日期区间，所以在这种情况下更方便的做法是跳过这一步。

要计算一年的会议日期，假设是每个月的第二个星期四，那么可以查看 `monthcalendar()` 的输出，找到星期四对应的日期。一个月的第一周和最后一周都要填充 0 值作为占位符，分别表示相应日期实际上在前一个月或下一个月。例如，如果一个月从星期五开始，那么第一周星期四位置上的值就是 0。

代码清单 4-33：**calendar_secondthursday.py**

```
import calendar
import sys

year = int(sys.argv[1])

# Show every month.
for month in range(1, 13):

    # Compute the dates for each week that overlaps the month.
    c = calendar.monthcalendar(year, month)
    first_week = c[0]
    second_week = c[1]
    third_week = c[2]

    # If there is a Thursday in the first week,
    # the second Thursday is in the second week.
    # Otherwise, the second Thursday must be in
    # the third week.
    if first_week[calendar.THURSDAY]:
        meeting_date = second_week[calendar.THURSDAY]
    else:
        meeting_date = third_week[calendar.THURSDAY]

    print('{:>3}: {:>2}'.format(calendar.month_abbr[month],
                                meeting_date))
```

所以，这一年的会议日程为：

```
$ python3 calendar_secondthursday.py 2017

Jan: 12
Feb:  9
Mar:  9
Apr: 13
May: 11
Jun:  8
Jul: 13
Aug: 10
Sep: 14
Oct: 12
Nov:  9
Dec: 14
```

提示：相关阅读材料

- **calendar** 的标准库文档⊖。
- **time**：底层时间函数。
- **datetime**：管理日期值，包括时间戳和时区。
- **locale**：本地化环境设置。

⊖ https://docs.python.org/3.5/library/calendar.html

第 5 章
数学运算

作为一种通用的编程语言，Python 经常用来解决数学问题。它包含一些用于管理整数和浮点数的内置类型，这很适合完成一般应用中可能出现的基本的数学运算。标准库包含一些用于满足更高级需求的模块。

Python 的内置浮点数使用底层 double 表示。对于大多数有数学运算需求的程序来说，这已经足够精确，但是如果需要非整数值的更为精确的表示，那么 decimal 和 fractions 模块会很有用。小数和分数值的算术运算可以保证精度，但是不如原生 float 的运算速度快。

random 模块包含一个均匀分布伪随机数生成器，还提供了一些函数用于模拟很多常见的非均匀分布。

math 模块包含一些高级数学函数的快速实现，如对数和三角函数。这个模块对原生平台 C 库中常见的 IEEE 函数提供了全面的补充。

5.1 decimal：定点数和浮点数的数学运算

decimal 模块实现了定点和浮点算术运算，使用的是大多数人所熟悉的模型，而不是程序员熟悉的模式（即大多数计算机硬件实现的 IEEE 浮点数运算）。Decimal 实例可以准确地表示任何数，对其上或其下取整，还可以限制有效数字个数。

5.1.1 Decimal

小数值被表示为 Decimal 类的实例。构造函数取一个整数或字符串作为参数。在使用浮点数创建 Decimal 之前，可以先将浮点数转换为一个字符串，以使调用者能够显式地处理值的位数，因为如果使用硬件浮点数表示则可能无法准确地表述。或者，类方法 from_float() 可以把一个浮点数转换为精确的小数表示。

代码清单 5-1：decimal_create.py

```
import decimal

fmt = '{0:<25} {1:<25}'
print(fmt.format('Input', 'Output'))
print(fmt.format('-' * 25, '-' * 25))

# Integer
print(fmt.format(5, decimal.Decimal(5)))
```

```
# String
print(fmt.format('3.14', decimal.Decimal('3.14')))

# Float
f = 0.1
print(fmt.format(repr(f), decimal.Decimal(str(f))))
print('{:<0.23g} {:<25}'.format(
    f,
    str(decimal.Decimal.from_float(f))[:25])
)
```

浮点值 **0.1** 并没有被表示为一个精确的二进制值，所以 **float** 的表示与 **Decimal** 值不同。在这个输出的最后一行，完整的字符串表示被截断为 25 个字符。

```
$ python3 decimal_create.py

Input                          Output
------------------------       ------------------------
5                              5
3.14                           3.14
0.1                            0.1
0.10000000000000000555112      0.1000000000000000055511
```

Decimal 还可以由元组创建，其中包含一个符号标志（**0** 表示正，**1** 表示负）、由数位组成的一个 **tuple** 以及一个整数指数。

代码清单 5-2：decimal_tuple.py

```
import decimal

# Tuple
t = (1, (1, 1), -2)
print('Input   :', t)
print('Decimal:', decimal.Decimal(t))
```

基于元组的表示在创建时不太方便，不过它提供了一种可移植的方式，这样可以导出小数值而不损失精度。元组形式可以通过网络传输，或者在不支持精确小数值的数据库中存储，以后再转换回 **Decimal** 实例。

```
$ python3 decimal_tuple.py

Input   : (1, (1, 1), -2)
Decimal: -0.11
```

5.1.2 格式化

Decimal 对应 Python 的字符串格式化协议[⊖]，使用与其他数值类型一样的语法和选项。

代码清单 5-3：decimal_format.py

```
import decimal

d = decimal.Decimal(1.1)
print('Precision:')
print('{:.1}'.format(d))
```

⊖ https://docs.python.org/3.5/library/string.html#formatspec

```
print('{:.2}'.format(d))
print('{:.3}'.format(d))
print('{:.18}'.format(d))

print('\nWidth and precision combined:')
print('{:5.1f} {:5.1g}'.format(d, d))
print('{:5.2f} {:5.2g}'.format(d, d))
print('{:5.2f} {:5.2g}'.format(d, d))

print('\nZero padding:')
print('{:05.1}'.format(d))
print('{:05.2}'.format(d))
print('{:05.3}'.format(d))
```

格式字符串可以控制输出的宽度，精度（即有效数字个数），以及其填充值以占满宽度的方式。

```
$ python3 decimal_format.py

Precision:
1
1.1
1.10
1.10000000000000009

Width and precision combined:
  1.1     1
 1.10   1.1
 1.10   1.1
Zero padding:
00001
001.1
01.10
```

5.1.3 算术运算

Decimal 重载了简单的算术操作符，所以可以采用与内置数值类型相同的方式来处理 Decimal 实例。

代码清单 5-4：decimal_operators.py

```
import decimal

a = decimal.Decimal('5.1')
b = decimal.Decimal('3.14')
c = 4
d = 3.14

print('a     =', repr(a))
print('b     =', repr(b))
print('c     =', repr(c))
print('d     =', repr(d))
print()

print('a + b =', a + b)
print('a - b =', a - b)
print('a * b =', a * b)
print('a / b =', a / b)
print()
```

```
print('a + c =', a + c)
print('a - c =', a - c)
print('a * c =', a * c)
print('a / c =', a / c)
print()

print('a + d =', end=' ')
try:
    print(a + d)
except TypeError as e:
    print(e)
```

`Decimal` 操作符还接受整数参数，不过，在这些操作符使用浮点值之前必须把浮点值转换为 `Decimal` 实例。

```
$ python3 decimal_operators.py

a     = Decimal('5.1')
b     = Decimal('3.14')
c     = 4
d     = 3.14

a + b = 8.24
a - b = 1.96
a * b = 16.014
a / b = 1.624203821656050955414012739

a + c = 9.1
a - c = 1.1
a * c = 20.4
a / c = 1.275

a + d = unsupported operand type(s) for +: 'decimal.Decimal' and
 'float'
```

除了基本算术运算，`Decimal` 还包括一些方法来查找以 10 为底的对数和自然对数。`log10()` 和 `ln()` 返回的值都是 `Decimal` 实例，所以可以与其他值一样在公式中直接使用。

5.1.4 特殊值

除了期望的数字值，`Decimal` 还可以表示很多特殊值，包括正负无穷大值、"不是一个数"（NaN）和 0。

代码清单 5-5：decimal_special.py

```
import decimal

for value in ['Infinity', 'NaN', '0']:
    print(decimal.Decimal(value), decimal.Decimal('-' + value))
print()

# Math with infinity
print('Infinity + 1:', (decimal.Decimal('Infinity') + 1))
print('-Infinity + 1:', (decimal.Decimal('-Infinity') + 1))

# Print comparing NaN
print(decimal.Decimal('NaN') == decimal.Decimal('Infinity'))
print(decimal.Decimal('NaN') != decimal.Decimal(1))
```

与无穷大值相加会返回另一个无穷大值。与 NaN 比较相等性总会返回 false,而比较不等性总会返回 true。与 NaN 比较大小来确定排序顺序是未定义的,这会导致一个错误。

```
$ python3 decimal_special.py

Infinity -Infinity
NaN -NaN
0 -0
Infinity + 1: Infinity
-Infinity + 1: -Infinity
False
True
```

5.1.5 上下文

到目前为止,前面的所有例子使用的都是 decimal 模块的默认行为。还可以使用一个上下文(context)来覆盖某些设置,如保持的精度、如何完成取整、错误处理等。上下文可以应用于一个线程中的所有 Decimal 实例,或者在一个小代码区中本地应用。

5.1.5.1 当前上下文

要获取当前全局上下文,可以使用 getcontext()。

代码清单 5-6:decimal_getcontext.py

```
import decimal

context = decimal.getcontext()

print('Emax     =', context.Emax)
print('Emin     =', context.Emin)
print('capitals =', context.capitals)
print('prec     =', context.prec)
print('rounding =', context.rounding)
print('flags    =')
for f, v in context.flags.items():
    print('  {}: {}'.format(f, v))
print('traps    =')
for t, v in context.traps.items():
    print('  {}: {}'.format(t, v))
```

这个示例脚本显示了 Context 的公共属性。

```
$ python3 decimal_getcontext.py

Emax     = 999999
Emin     = -999999
capitals = 1
prec     = 28
rounding = ROUND_HALF_EVEN
flags    =
  <class 'decimal.InvalidOperation'>: False
  <class 'decimal.FloatOperation'>: False
  <class 'decimal.DivisionByZero'>: False
  <class 'decimal.Overflow'>: False
  <class 'decimal.Underflow'>: False
  <class 'decimal.Subnormal'>: False
  <class 'decimal.Inexact'>: False
  <class 'decimal.Rounded'>: False
  <class 'decimal.Clamped'>: False
```

```
traps        =
 <class 'decimal.InvalidOperation'>: True
 <class 'decimal.FloatOperation'>: False
 <class 'decimal.DivisionByZero'>: True
 <class 'decimal.Overflow'>: True
 <class 'decimal.Underflow'>: False
 <class 'decimal.Subnormal'>: False
 <class 'decimal.Inexact'>: False
 <class 'decimal.Rounded'>: False
 <class 'decimal.Clamped'>: False
```

5.1.5.2 精度

上下文的 `prec` 属性控制了作为算术运算结果创建的新值所要保持的精度。字面量值会按这个属性保持精度。

代码清单 5-7：`decimal_precision.py`

```python
import decimal

d = decimal.Decimal('0.123456')

for i in range(1, 5):
    decimal.getcontext().prec = i
    print(i, ':', d, d * 1)
```

要改变精度，可以直接为这个属性赋一个 1 到 `decimal.MAX_PREC` 之间的新值。

```
$ python3 decimal_precision.py

1 : 0.123456 0.1
2 : 0.123456 0.12
3 : 0.123456 0.123
4 : 0.123456 0.1235
```

5.1.5.3 取整

取整有多种选择，以保证值在所需的精度范围内。

`ROUND_CEILING`：总是趋向无穷大向上取整。

`ROUND_DOWN`：总是趋向 0 取整。

`ROUND_FLOOR`：总是趋向负无穷大向下取整。

`ROUND_HALF_DOWN`：如果最后一个有效数字大于或等于 5 则朝 0 反方向取整；否则，趋向 0 取整。

`ROUND_HALF_EVEN`：类似于 `ROUND_HALF_DOWN`，不过，如果最后一个有效数字为 5，则会检查前一位。偶数值会导致结果向下取整，奇数值导致结果向上取整。

`ROUND_HALF_UP`：类似于 `ROUND_HALF_DOWN`，不过如果最后一位有效数字为 5，则值会朝 0 的反方向取整。

`ROUND_UP`：朝 0 的反方向取整。

`ROUND_05UP`：如果最后一位是 0 或 5，则朝 0 的反方向取整；否则向 0 取整。

代码清单 5-8：`decimal_rounding.py`

```python
import decimal

context = decimal.getcontext()
```

```
ROUNDING_MODES = [
    'ROUND_CEILING',
    'ROUND_DOWN',
    'ROUND_FLOOR',
    'ROUND_HALF_DOWN',
    'ROUND_HALF_EVEN',
    'ROUND_HALF_UP',
    'ROUND_UP',
    'ROUND_05UP',
]

header_fmt = '{:10} ' + ' '.join(['{:^8}'] * 6)

print(header_fmt.format(
    ' ',
    '1/8 (1)', '-1/8 (1)',
    '1/8 (2)', '-1/8 (2)',
    '1/8 (3)', '-1/8 (3)',
))
for rounding_mode in ROUNDING_MODES:
    print('{0:10}'.format(rounding_mode.partition('_')[-1]),
          end=' ')
    for precision in [1, 2, 3]:
        context.prec = precision
        context.rounding = getattr(decimal, rounding_mode)
        value = decimal.Decimal(1) / decimal.Decimal(8)
        print('{0:^8}'.format(value), end=' ')
        value = decimal.Decimal(-1) / decimal.Decimal(8)
        print('{0:^8}'.format(value), end=' ')
    print()
```

这个程序显示了使用不同算法将同一个值取整为不同精度的效果。

```
$ python3 decimal_rounding.py
```

	1/8 (1)	-1/8 (1)	1/8 (2)	-1/8 (2)	1/8 (3)	-1/8 (3)
CEILING	0.2	-0.1	0.13	-0.12	0.125	-0.125
DOWN	0.1	-0.1	0.12	-0.12	0.125	-0.125
FLOOR	0.1	-0.2	0.12	-0.13	0.125	-0.125
HALF_DOWN	0.1	-0.1	0.12	-0.12	0.125	-0.125
HALF_EVEN	0.1	-0.1	0.12	-0.12	0.125	-0.125
HALF_UP	0.1	-0.1	0.13	-0.13	0.125	-0.125
UP	0.2	-0.2	0.13	-0.13	0.125	-0.125
05UP	0.1	-0.1	0.12	-0.12	0.125	-0.125

5.1.5.4 本地上下文

可以使用 with 语句对一个代码块应用上下文。

代码清单 5-9：decimal_context_manager.py

```
import decimal

with decimal.localcontext() as c:
```

```python
    c.prec = 2
    print('Local precision:', c.prec)
    print('3.14 / 3 =', (decimal.Decimal('3.14') / 3))

print()
print('Default precision:', decimal.getcontext().prec)
print('3.14 / 3 =', (decimal.Decimal('3.14') / 3))
```

Context 支持 with 使用的上下文管理器 API，所以这个设置只在块内应用。

```
$ python3 decimal_context_manager.py

Local precision: 2

3.14 / 3 = 1.0

Default precision: 28
3.14 / 3 = 1.0466666666666666666666666667
```

5.1.5.5 各实例的上下文

还可以用上下文构造 Decimal 实例，然后从这个上下文继承精度以及转换的取整参数。

代码清单 5-10：**decimal_instance_context.py**

```python
import decimal

# Set up a context with limited precision.
c = decimal.getcontext().copy()
c.prec = 3

# Create our constant.
pi = c.create_decimal('3.1415')

# The constant value is rounded off.
print('PI      :', pi)

# The result of using the constant uses the global context.
print('RESULT:', decimal.Decimal('2.01') * pi)
```

例如，这样一来，应用就可以选择与用户数据精度不同的常量值精度。

```
$ python3 decimal_instance_context.py

PI     : 3.14
RESULT: 6.3114
```

5.1.5.6 线程

"全局"上下文实际上是线程本地上下文，所以完全可以使用不同的值分别配置各个线程。

代码清单 5-11：**decimal_thread_context.py**

```python
import decimal
import threading
from queue import PriorityQueue


class Multiplier(threading.Thread):
    def __init__(self, a, b, prec, q):
```

```python
        self.a = a
        self.b = b
        self.prec = prec
        self.q = q
        threading.Thread.__init__(self)

    def run(self):
        c = decimal.getcontext().copy()
        c.prec = self.prec
        decimal.setcontext(c)
        self.q.put((self.prec, a * b))


a = decimal.Decimal('3.14')
b = decimal.Decimal('1.234')
# A PriorityQueue will return values sorted by precision,
# no matter in which order the threads finish.
q = PriorityQueue()
threads = [Multiplier(a, b, i, q) for i in range(1, 6)]
for t in threads:
    t.start()

for t in threads:
    t.join()

for i in range(5):
    prec, value = q.get()
    print('{}  {}'.format(prec, value))
```

这个例子使用指定的值来创建一个新的上下文,然后安装到各个线程中。

```
$ python3 decimal_thread_context.py

1   4
2   3.9
3   3.87
4   3.875
5   3.8748
```

提示:相关阅读材料

- decimal 的标准库文档[⊖]。
- decimal 的 Python 2 到 Python 3 移植说明。
- Wikipedia: Floating Point[⊜]:有关浮点数表示和算术运算的维基百科文章。
- Floating Point Arithmetic: Issues and Limitations[⊛]:Python 教程中的一篇文章,介绍了浮点数数学表示问题。

5.2 fractions:有理数

Fraction 类基于 numbers 模块中 Rational 定义的 API 来实现有理数的数值运算。

[⊖] https://docs.python.org/3.5/library/decimal.html
[⊜] https://en.wikipedia.org/wiki/Floating_point
[⊛] https://docs.python.org/tutorial/floatingpoint.html

5.2.1 创建 Fraction 实例

与 decimal 模块类似,可以采用多种方式创建新值。一种简便的方式是由单独的分子和分母值来创建。

代码清单 5-12:**fractions_create_integers.py**

```python
import fractions

for n, d in [(1, 2), (2, 4), (3, 6)]:
    f = fractions.Fraction(n, d)
    print('{}/{} = {}'.format(n, d, f))
```

计算新值时要保持最小公分母。

```
$ python3 fractions_create_integers.py

1/2 = 1/2
2/4 = 1/2
3/6 = 1/2
```

创建 Fraction 的另一种方法是使用 <numerator> / <denominator> 字符串表示:

代码清单 5-13:**fractions_create_strings.py**

```python
import fractions

for s in ['1/2', '2/4', '3/6']:
    f = fractions.Fraction(s)
    print('{} = {}'.format(s, f))
```

解析这个字符串,以找出分子和分母值。

```
$ python3 fractions_create_strings.py

1/2 = 1/2
2/4 = 1/2
3/6 = 1/2
```

字符串还可以使用更常用的小数或浮点数记法,即用一个小数点分隔的一系列数字。能够由 float() 解析而且不表示 NaN 或无穷大值的所有字符串都被支持。

代码清单 5-14:**fractions_create_strings_floats.py**

```python
import fractions

for s in ['0.5', '1.5', '2.0', '5e-1']:
    f = fractions.Fraction(s)
    print('{0:>4} = {1}'.format(s, f))
```

浮点值表示的分子和分母值会自动计算。

```
$ python3 fractions_create_strings_floats.py

 0.5 = 1/2
 1.5 = 3/2
 2.0 = 2
5e-1 = 1/2
```

还可以从有理数值的其他表示（如 float 或 Decimal）直接创建 Fraction 实例。

代码清单 5-15：fractions_from_float.py

```
import fractions

for v in [0.1, 0.5, 1.5, 2.0]:
    print('{} = {}'.format(v, fractions.Fraction(v)))
```

不能精确表示的浮点值可能会得到意料外的结果。

```
$ python3 fractions_from_float.py

0.1 = 3602879701896397/36028797018963968
0.5 = 1/2
1.5 = 3/2
2.0 = 2
```

使用值的 Decimal 表示则会给出期望的结果。

代码清单 5-16：fractions_from_decimal.py

```
import decimal
import fractions

values = [
    decimal.Decimal('0.1'),
    decimal.Decimal('0.5'),
    decimal.Decimal('1.5'),
    decimal.Decimal('2.0'),
]

for v in values:
    print('{} = {}'.format(v, fractions.Fraction(v)))
```

Decimal 的内部实现不存在标准浮点数表示的精度误差。

```
$ python3 fractions_from_decimal.py

0.1 = 1/10
0.5 = 1/2
1.5 = 3/2
2.0 = 2
```

5.2.2 算术运算

一旦分数被实例化，就可以在数学表达式中使用了。

代码清单 5-17：fractions_arithmetic.py

```
import fractions

f1 = fractions.Fraction(1, 2)
f2 = fractions.Fraction(3, 4)

print('{} + {} = {}'.format(f1, f2, f1 + f2))
print('{} - {} = {}'.format(f1, f2, f1 - f2))
print('{} * {} = {}'.format(f1, f2, f1 * f2))
print('{} / {} = {}'.format(f1, f2, f1 / f2))
```

分数运算支持所有标准操作符。

```
$ python3 fractions_arithmetic.py

1/2 + 3/4 = 5/4
1/2 - 3/4 = -1/4
1/2 * 3/4 = 3/8
1/2 / 3/4 = 2/3
```

5.2.3 近似值

`Fraction`有一个有用的特性，即能够将一个浮点数转换为一个近似的有理数值。

代码清单5-18：**fractions_limit_denominator.py**

```python
import fractions
import math

print('PI        =', math.pi)

f_pi = fractions.Fraction(str(math.pi))
print('No limit =', f_pi)

for i in [1, 6, 11, 60, 70, 90, 100]:
    limited = f_pi.limit_denominator(i)
    print('{0:8} = {1}'.format(i, limited))
```

可以通过限制分母大小来控制这个分数的值。

```
$ python3 fractions_limit_denominator.py

PI       = 3.141592653589793
No limit = 3141592653589793/1000000000000000
       1 = 3
       6 = 19/6
      11 = 22/7
      60 = 179/57
      70 = 201/64
      90 = 267/85
     100 = 311/99
```

提示：相关阅读材料

- `fractions`的标准库文档⊖。
- `decimal`：`decimal`模块提供了一个API来完成定点数和浮点数的数学运算。
- `numbers`：数值抽象基类。
- `fractions`的Python 2到Python 3移植说明。

5.3 `random`：伪随机数生成器

`random`模块基于Mersenne Twister算法提供了一个快速伪随机数生成器。原先开发这个生成器是为了向蒙特卡洛模拟生成输入，Mersenne Twister算法会生成大周期近均匀分布的数，因此适用于大量不同类型的应用。

⊖ https://docs.python.org/3.5/library/fractions.html

5.3.1 生成随机数

random()函数从所生成的序列返回下一个随机的浮点值。返回的所有值都落在 0 <= n < 1.0区间内。

代码清单5-19：**random_random.py**

```python
import random

for i in range(5):
    print('%04.3f' % random.random(), end=' ')
print()
```

重复运行这个程序会生成不同的数字序列。

```
$ python3 random_random.py

0.859 0.297 0.554 0.985 0.452

$ python3 random_random.py

0.797 0.658 0.170 0.297 0.593
```

要生成一个指定数值区间内的数，则要使用uniform()。

代码清单5-20：**random_uniform.py**

```python
import random

for i in range(5):
    print('{:04.3f}'.format(random.uniform(1, 100)), end=' ')
print()
```

传入最小值和最大值，uniform()会使用公式 min + (max - min) * random()来调整random()的返回值。

```
$ python3 random_uniform.py

12.428 93.766 95.359 39.649 88.983
```

5.3.2 指定种子

每次调用random()都会生成不同的值，并且在一个非常大的周期之后数字才会重复。这对于生成唯一值或变化的值很有用，不过有些情况下可能需要提供相同的数据集，从而以不同的方式处理。对此，一种技术是使用一个程序生成随机值，并保存这些随机值，以便在另一个步骤中再做处理。不过，这对于量很大的数据来说可能并不实用，所以random包含了一个seed()函数，可以用来初始化伪随机数生成器，使它能生成一个期望的值集。

代码清单5-21：**random_seed.py**

```python
import random

random.seed(1)

for i in range(5):
    print('{:04.3f}'.format(random.random()), end=' ')
print()
```

种子（seed）值会控制由公式生成的第一个值，该公式可用来生成伪随机数。由于公式是确定的，所以改变种子后便设置了将生成的整个序列。seed()的参数可以是任意的可散列对象。默认为使用一个平台特定的随机源（如果有的话）。但如果没有这样一个随机源，则使用当前时间。

```
$ python3 random_seed.py

0.134 0.847 0.764 0.255 0.495

$ python3 random_seed.py

0.134 0.847 0.764 0.255 0.495
```

5.3.3 保存状态

random()使用的伪随机算法的内部状态可以保存，并用于控制后续生成的随机数。如果在继续生成随机数之前恢复前一个状态，则会减少出现重复的可能性，即避免出现之前输入中重复的值或值序列。getstate()函数会返回一些数据，以后可以借助setstate()利用这些数据重新初始化伪随机数生成器。

代码清单 5-22：**random_state.py**

```python
import random
import os
import pickle
if os.path.exists('state.dat'):
    # Restore the previously saved state.
    print('Found state.dat, initializing random module')
    with open('state.dat', 'rb') as f:
        state = pickle.load(f)
    random.setstate(state)
else:
    # Use a well-known start state.
    print('No state.dat, seeding')
    random.seed(1)

# Produce random values.
for i in range(3):
    print('{:04.3f}'.format(random.random()), end=' ')
print()

# Save state for next time.
with open('state.dat', 'wb') as f:
    pickle.dump(random.getstate(), f)

# Produce more random values.
print('\nAfter saving state:')
for i in range(3):
    print('{:04.3f}'.format(random.random()), end=' ')
print()
```

getstate()返回的数据是一个实现细节，所以这个例子用pickle将数据保存到一个文件；否则，它会把伪随机数生成器当作一个黑盒。如果程序开始时这个文件存在，则加载原来的状态并继续。每次运行时都会在保存状态之前和之后生成一些数，以展示恢复状态会使生成器再次生成同样的值。

```
$ python3 random_state.py

No state.dat, seeding
0.134 0.847 0.764

After saving state:
0.255 0.495 0.449

$ python3 random_state.py

Found state.dat, initializing random module
0.255 0.495 0.449

After saving state:
0.652 0.789 0.094
```

5.3.4 随机整数

`random()`将生成浮点数。可以把结果转换为整数,不过直接使用`randint()`生成整数会更方便。

代码清单 5-23:`random_randint.py`

```python
import random

print('[1, 100]:', end=' ')

for i in range(3):
    print(random.randint(1, 100), end=' ')

print('\n[-5, 5]:', end=' ')
for i in range(3):
    print(random.randint(-5, 5), end=' ')
print()
```

`randint()`的参数是值的闭区间的两端。这些数可以是正数或负数,不过第一个值要小于第二个值。

```
$ python3 random_randint.py

[1, 100]: 98 75 34
[-5, 5]: 4 0 5
```

`randrange()`是从区间选择值的一种更一般的形式。

代码清单 5-24:`random_randrange.py`

```python
import random

for i in range(3):
    print(random.randrange(0, 101, 5), end=' ')
print()
```

除了开始值(start)和结束值(stop),`randrange()`还支持一个步长(step)参数,所以它完全等价于从`range(start, stop, step)`选择一个随机值。不过randrange更高效,因为它并没有真正构造区间。

```
$ python3 random_randrange.py

15 20 85
```

5.3.5 选择随机元素

随机数生成器有一种常见用法，即从一个枚举值序列中选择元素，即使这些值并不是数字。random 包括一个 choice() 函数，可以从一个序列中随机选择。下面这个例子模拟抛硬币 10 000 次来统计多少次面朝上，多少次面朝下。

代码清单 5-25：**random_choice.py**

```
import random
import itertools

outcomes = {
    'heads': 0,
    'tails': 0,
}
sides = list(outcomes.keys())

for i in range(10000):
    outcomes[random.choice(sides)] += 1

print('Heads:', outcomes['heads'])
print('Tails:', outcomes['tails'])
```

由于只允许两个结果，所以不必使用数字然后再进行转换，这里对 choice() 使用了单词"heads"（表示面朝上）和"tails"（表示面朝下）。结果以表格形式存储在一个字典中，使用结果名作为键。

```
$ python3 random_choice.py

Heads: 5091
Tails: 4909
```

5.3.6 排列

要模拟一个扑克牌游戏，需要把一副牌混起来，然后向玩家发牌，同一张牌不能多次使用。使用 choice() 可能导致同一张牌被发出两次，所以，可以用 shuffle() 来洗牌，然后在发各张牌时删除所发的牌。

代码清单 5-26：**random_shuffle.py**

```
import random
import itertools

FACE_CARDS = ('J', 'Q', 'K', 'A')
SUITS = ('H', 'D', 'C', 'S')
def new_deck():
    return [
        # Always use 2 places for the value, so the strings
        # are a consistent width.
        '{:>2}{}'.format(*c)
        for c in itertools.product(
            itertools.chain(range(2, 11), FACE_CARDS),
            SUITS,
        )
    ]
```

```python
def show_deck(deck):
    p_deck = deck[:]
    while p_deck:
        row = p_deck[:13]
        p_deck = p_deck[13:]
        for j in row:
            print(j, end=' ')
        print()


# Make a new deck, with the cards in order.
deck = new_deck()
print('Initial deck:')
show_deck(deck)

# Shuffle the deck to randomize the order.
random.shuffle(deck)
print('\nShuffled deck:')
show_deck(deck)

# Deal 4 hands of 5 cards each.
hands = [[], [], [], []]

for i in range(5):
    for h in hands:
        h.append(deck.pop())

# Show the hands.
print('\nHands:')
for n, h in enumerate(hands):
    print('{}:'.format(n + 1), end=' ')
    for c in h:
        print(c, end=' ')
    print()

# Show the remaining deck.
print('\nRemaining deck:')
show_deck(deck)
```

这些扑克牌被表示为字符串，包括面值和一个表示花色的字母。要创建发出的"一手牌"，可以一次向 4 个列表分别增加一张牌，然后从这副牌中将发出的牌删除，使这些牌不会再次发出。

```
$ python3 random_shuffle.py

Initial deck:
 2H  2D  2C  2S  3H  3D  3C  3S  4H  4D  4C  4S  5H
 5D  5C  5S  6H  6D  6C  6S  7H  7D  7C  7S  8H  8D
 8C  8S  9H  9D  9C  9S 10H 10D 10C 10S  JH  JD  JC
 JS  QH  QD  QC  QS  KH  KD  KC  KS  AH  AD  AC  AS

Shuffled deck:
 QD  8C  JD  2S  AC  2C  6S  6D  6C  7H  JC  QS  QC
 KS  4D 10C  KH  5S  9C 10S  5C  7C  AS  6H  3C  9H
 4S  7S 10H  2D  8S  AH  9S  8H  QH  5D  5H  KD  8D
10D  4C  3S  3H  7D  AD  4H  9D  3D  2H  KC  JH  JS

Hands:
1:  JS  3D  7D 10D  5D
2:  JH  9D  3H  8D  QH
3:  KC  4H  3S  KD  8H
4:  2H  AD  4C  5H  9S
```

```
Remaining deck:
QD  8C  JD  2S  AC  2C  6S  6D  6C  7H  JC  QS  QC
KS  4D  10C KH  5S  9C  10S 5C  7C  AS  6H  3C  9H
4S  7S  10H 2D  8S  AH
```

5.3.7 采样

很多模拟需要从大量输入值中得到随机样本。`sample()`函数可以生成无重复值的样本，并且不会修改输入序列。下面的例子会打印系统字典中单词的一个随机样本。

代码清单 5-27：`random_sample.py`

```python
import random

with open('/usr/share/dict/words', 'rt') as f:
    words = f.readlines()
words = [w.rstrip() for w in words]

for w in random.sample(words, 5):
    print(w)
```

生成结果集的算法会考虑输入的规模和所请求的样本，从而尽可能高效地生成结果。

```
$ python3 random_sample.py

streamlet
impestation
violaquercitrin
mycetoid
plethoretical

$ python3 random_sample.py

nonseditious
empyemic
ultrasonic
Kyurinish
amphide
```

5.3.8 多个并发生成器

除了模块级函数，random 还包括一个 Random 类以管理多个随机数生成器的内部状态。之前介绍的所有函数都可以作为 Random 实例的方法得到，并且每个实例都可以被单独初始化和使用，而不会干扰其他实例返回的值。

代码清单 5-28：`random_random_class.py`

```python
import random
import time

print('Default initializiation:\n')

r1 = random.Random()
r2 = random.Random()

for i in range(3):
    print('{:04.3f}  {:04.3f}'.format(r1.random(), r2.random()))
```

```python
print('\nSame seed:\n')

seed = time.time()
r1 = random.Random(seed)
r2 = random.Random(seed)

for i in range(3):
    print('{:04.3f}  {:04.3f}'.format(r1.random(), r2.random()))
```

如果系统上设置了很好的原生随机值种子,那么实例会有独特的初始状态。不过,如果没有一个好的平台随机值生成器,那么不同实例往往会以当前时间作为种子,并因此生成相同的值。

```
$ python3 random_random_class.py

Default initializiation:

0.862  0.390
0.833  0.624
0.252  0.080

Same seed:

0.466  0.466
0.682  0.682
0.407  0.407
```

5.3.9 SystemRandom

有些操作系统提供了一个随机数生成器,可以访问更多能引入生成器的信息源。random通过SystemRandom类提供了这个特性,该类与Random的API相同,不过使用os.urandom()生成值,该值会构成所有其他算法的基础。

代码清单5-29: random_system_random.py

```python
import random
import time

print('Default initializiation:\n')

r1 = random.SystemRandom()
r2 = random.SystemRandom()

for i in range(3):
    print('{:04.3f}  {:04.3f}'.format(r1.random(), r2.random()))

print('\nSame seed:\n')

seed = time.time()
r1 = random.SystemRandom(seed)
r2 = random.SystemRandom(seed)

for i in range(3):
    print('{:04.3f}  {:04.3f}'.format(r1.random(), r2.random()))
```

SystemRandom产生的序列是不可再生的,因为其随机性来自系统,而不是来自软件状态(实际上,seed()和setstate()根本不起作用)。

```
$ python3 random_system_random.py

Default initializiation:

0.110  0.481
0.624  0.350
0.378  0.056

Same seed:

0.634  0.731
0.893  0.843
0.065  0.177
```

5.3.10 非均匀分布

`random()`生成的值为均匀分布，这对于很多目的来说非常有用，不过，另外一些分布可以更准确地对特定情况建模。`random`模块还包含一些函数以生成满足这些分布的值。这里将列出这些分布，但是并不打算详细介绍，因为它们往往只在特定条件下使用，而且需要更复杂的例子来说明。

5.3.10.1 正态分布

正态分布（normal distribution）常用于非均匀的连续值，如梯度、高度、重量等。正态分布产生的曲线有一个独特形状，所以被昵称为"钟形曲线"。`random`包含两个函数，可以生成正态分布的值，分别是`normalvariate()`和稍快一些的`gauss()`（正态分布也被称为高斯分布）。

还有一个相关的函数`lognormvariate()`，它可以生成对数呈正态分布的伪随机值。对数正态分布适用于多个不交互随机变量的积。

5.3.10.2 近似分布

三角分布（triangular distribution）被用作小样本的近似分布。三角分布的"曲线"中，低点位于已知的最小和最大值，在模式值处有一个高点，这要根据"最接近"的结果（由`triangular()`的模式参数反映）来估计。

5.3.10.3 指数分布

`expovariate()`可以生成一个指数分布，这对于模拟齐次泊松过程的到达或间隔时间值很有用，如放射衰变速度或到达 Web 服务器的请求。

很多可观察的现象都适用于帕累托分布（或幂律分布），这个分布因 Chris Anderson 的"长尾效应"而得到普及。`paretovariate()`函数对于模拟资源分配很有用（如人的财产、对音乐家的需求以及对博客的关注等）。

5.3.10.4 角分布

冯·米塞斯分布（或循环正态分布，由`vonmisesvariate()`生成）用于计算周期值的概率，如角度、日历日期和时间。

5.3.10.5 大小分布

`betavariate()`生成 Beta 分布的值，常用于贝叶斯统计和应用，如任务持续时间建模。`gammavariate()`生成的伽马分布用于对事物的规模建模，如等待时间、雨量和计算误差。

  `weibullvariate()` 计算的韦伯分布用于故障分析、工业工程和天气预报。它描述了粒子或其他离散对象的大小分布。

提示：相关阅读材料
- `random` 的标准库文档[⊖]。
- Mersenne Twister: A 623-dimensionally equidistributed uniform pseudorandom number generator：M. Matsumoto 和 T. Nishimura 撰写的一篇文章，发表在 ACM Transactions on Modeling and Computer Simulation Vol. 8, No. 1, January 1998, pp. 3–30.
- Wikipedia: Mersenne Twister[⊖]：维基百科文章，关于 Python 使用的伪随机数生成器算法。
- Wikipedia: Uniform distribution[⊖]：维基百科文章，关于统计中的连续均匀分布。

5.4 `math`：数学函数

  `math` 模块实现了正常情况下原生平台 C 库中才有的很多专用 IEEE 函数，可以使用浮点值完成复杂的数学运算，包括对数和三角函数运算。

5.4.1 特殊常量

  很多数学运算依赖于一些特殊的常量。`math` 包含有 π (pi)、e、nan（不是一个数）和 infinity（无穷大）的值。

代码清单 5-30：<code>math_constants.py</code>

```
import math

print(' π : {:.30f}'.format(math.pi))
print(' e : {:.30f}'.format(math.e))
print('nan: {:.30f}'.format(math.nan))
print('inf: {:.30f}'.format(math.inf))
```

  π 和 e 的精度仅受平台的浮点数 C 库限制。

```
$ python3 math_constants.py
π : 3.141592653589793115997963468544
e : 2.718281828459045090795598298428
nan: nan
inf: inf
```

5.4.2 测试异常值

  浮点数计算可能导致两种类型的异常值。第一种是 `inf`（无穷大），当用 `double` 存储一个浮点值，而该值会从一个具有很大绝对值的值上溢出时，就会出现这个异常值。

[⊖] https://docs.python.org/3.5/library/random.html
[⊖] https://en.wikipedia.org/wiki/Mersenne_twister
[⊖] https://en.wikipedia.org/wiki/Uniform_distribution_(continuous)

代码清单 5-31：**math_isinf.py**

```
import math

print('{:^3} {:6} {:6} {:6}'.format(
    'e', 'x', 'x**2', 'isinf'))
print('{:-^3} {:-^6} {:-^6} {:-^6}'.format(
    '', '', '', ''))

for e in range(0, 201, 20):
    x = 10.0 ** e
    y = x * x
    print('{:3d} {:<6g} {:<6g} {!s:6}'.format(
        e, x, y, math.isinf(y),
    ))
```

当这个例子中的指数变得足够大时，x 的平方无法再存放于一个 double 中，这个值就会被记录为无穷大。

```
$ python3 math_isinf.py

 e  x      x**2   isinf
--- ------ ------ ------
  0 1      1      False
 20 1e+20  1e+40  False
 40 1e+40  1e+80  False
 60 1e+60  1e+120 False
 80 1e+80  1e+160 False
100 1e+100 1e+200 False
120 1e+120 1e+240 False
140 1e+140 1e+280 False
160 1e+160 inf    True
180 1e+180 inf    True
200 1e+200 inf    True
```

不过，并不是所有浮点数溢出都会导致 inf 值。具体地，用浮点值计算一个指数时，会生成 OverflowError 而不是保留 inf 结果。

代码清单 5-32：**math_overflow.py**

```
x = 10.0 ** 200

print('x    =', x)
print('x*x  =', x * x)
print('x**2 =', end=' ')
try:
    print(x ** 2)
except OverflowError as err:
    print(err)
```

这种差异是由 C 和 Python 所用库中的实现差别造成的。

```
$ python3 math_overflow.py

x    = 1e+200
x*x  = inf
x**2 = (34, 'Result too large')
```

使用无穷大值的除法运算未定义。将一个数除以无穷大值的结果是 nan（不是一个数）。

代码清单 5-33：`math_isnan.py`

```
import math

x = (10.0 ** 200) * (10.0 ** 200)
y = x / x
print('x =', x)
print('isnan(x) =', math.isnan(x))
print('y = x / x =', x / x)
print('y == nan =', y == float('nan'))
print('isnan(y) =', math.isnan(y))
```

nan 不等于任何值，甚至不等于其自身，所以要想检查 nan，需要使用 `isnan()`。

```
$ python3 math_isnan.py

x = inf
isnan(x) = False
y = x / x = nan
y == nan = False
isnan(y) = True
```

可以使用 `isfinite()` 检查其是普通的数还是特殊值 inf 或 nan。

代码清单 5-34：`math_isfinite.py`

```
import math

for f in [0.0, 1.0, math.pi, math.e, math.inf, math.nan]:
    print('{:5.2f} {!s}'.format(f, math.isfinite(f)))
```

如果是特殊值 inf 或 nan，则 `isfinite()` 返回 false，否则返回 true。

```
$ python3 math_isfinite.py

 0.00 True
 1.00 True
 3.14 True
 2.72 True
  inf False
  nan False
```

5.4.3 比较

涉及浮点值的比较很容易出错，每一步计算都可能由于数值表示而引入误差。`isclose()` 函数使用一种稳定的算法来尽可能减少这些误差，同时完成相对和绝对比较。所用的公式等价于

```
abs(a-b) <= max(rel_tol * max(abs(a), abs(b)), abs_tol)
```

默认地，`isclose()` 会完成相对比较，容差被设置为 `1e-09`，这表示两个值之差必须小于或等于 1e-09 乘以 a 和 b 中较大的绝对值。向 `isclose()` 传入关键字参数 `rel_tol` 可以改变这个容差。在这个例子中，值之间的差距必须在 10% 以内。

代码清单 5-35：`math_isclose.py`

```
import math
```

```
INPUTS = [
    (1000, 900, 0.1),
    (100, 90, 0.1),
    (10, 9, 0.1),
    (1, 0.9, 0.1),
    (0.1, 0.09, 0.1),
]

print('{:^8} {:^8} {:^8} {:^8} {:^8} {:^8}'.format(
    'a', 'b', 'rel_tol', 'abs(a-b)', 'tolerance', 'close')
)
print('{:-^8} {:-^8} {:-^8} {:-^8} {:-^8} {:-^8}'.format(
    '-', '-', '-', '-', '-', '-'),
)

fmt = '{:8.2f} {:8.2f} {:8.2f} {:8.2f} {:8.2f} {!s:>8}'

for a, b, rel_tol in INPUTS:
    close = math.isclose(a, b, rel_tol=rel_tol)
    tolerance = rel_tol * max(abs(a), abs(b))
    abs_diff = abs(a - b)
    print(fmt.format(a, b, rel_tol, abs_diff, tolerance, close))
```

0.1 和 0.09 之间的比较失败, 因为误差表示 0.1。

```
$ python3 math_isclose.py

   a        b      rel_tol  abs(a-b) tolerance  close
-------- -------- -------- -------- -------- --------
 1000.00   900.00     0.10   100.00   100.00     True
  100.00    90.00     0.10    10.00    10.00     True
   10.00     9.00     0.10     1.00     1.00     True
    1.00     0.90     0.10     0.10     0.10     True
    0.10     0.09     0.10     0.01     0.01    False
```

要使用一个固定或 "绝对" 容差, 可以传入 abs_tol 而不是 rel_tol。

代码清单 5-36: math_isclose_abs_tol.py

```
import math

INPUTS = [
    (1.0, 1.0 + 1e-07, 1e-08),
    (1.0, 1.0 + 1e-08, 1e-08),
    (1.0, 1.0 + 1e-09, 1e-08),
]

print('{:^8} {:^11} {:^8} {:^10} {:^8}'.format(
    'a', 'b', 'abs_tol', 'abs(a-b)', 'close')
)
print('{:-^8} {:-^11} {:-^8} {:-^10} {:-^8}'.format(
    '-', '-', '-', '-', '-'),
)

for a, b, abs_tol in INPUTS:
    close = math.isclose(a, b, abs_tol=abs_tol)
    abs_diff = abs(a - b)
    print('{:8.2f} {:11} {:8} {:0.9f} {!s:>8}'.format(
        a, b, abs_tol, abs_diff, close))
```

对于绝对容差, 输入值之差必须小于给定的容差。

```
$ python3 math_isclose_abs_tol.py

    a           b         abs_tol  abs(a-b)    close
--------  -----------    -------  ----------  -------
    1.00  1.0000001      1e-08    0.000000100  False
    1.00  1.00000001     1e-08    0.000000010  True
    1.00  1.000000001    1e-08    0.000000001  True
```

nan 和 inf 是特殊情况。

代码清单 5-37: **math_isclose_inf.py**

```
import math

print('nan, nan:', math.isclose(math.nan, math.nan))
print('nan, 1.0:', math.isclose(math.nan, 1.0))
print('inf, inf:', math.isclose(math.inf, math.inf))
print('inf, 1.0:', math.isclose(math.inf, 1.0))
```

nan 不接近任何值，包括它自身。inf 只接近它自身。

```
$ python3 math_isclose_inf.py

nan, nan: False
nan, 1.0: False
inf, inf: True
inf, 1.0: False
```

5.4.4 将浮点值转换为整数

math 模块中有 3 个函数用于将浮点值转换为整数。这 3 个函数分别采用不同的方法，并适用于不同的场合。

最简单的是 trunc()，其会截断小数点后的数字，只留下构成这个值整数部分的有效数字。floor() 将其输入转换为不大于它的最大整数，ceil()（上限）会生成按顺序排在这个输入值之后的最小整数。

代码清单 5-38: **math_integers.py**

```
import math

HEADINGS = ('i', 'int', 'trunk', 'floor', 'ceil')
print('{:^5} {:^5} {:^5} {:^5} {:^5}'.format(*HEADINGS))
print('{:-^5} {:-^5} {:-^5} {:-^5} {:-^5}'.format(
    '', '', '', '', '',
))

fmt = '{:5.1f} {:5.1f} {:5.1f} {:5.1f} {:5.1f}'

TEST_VALUES = [
    -1.5,
    -0.8,
    -0.5,
    -0.2,
    0,
    0.2,
    0.5,
    0.8,
    1,
```

```
]
for i in TEST_VALUES:
    print(fmt.format(
        i,
        int(i),
        math.trunc(i),
        math.floor(i),
        math.ceil(i),
    ))
```

trunc() 等价于直接转换为 int。

```
$ python3 math_integers.py

   i    int   trunk floor ceil
 ----- ----- ----- ----- -----
 -1.5  -1.0  -1.0  -2.0  -1.0
 -0.8   0.0   0.0  -1.0   0.0
 -0.5   0.0   0.0  -1.0   0.0
 -0.2   0.0   0.0  -1.0   0.0
  0.0   0.0   0.0   0.0   0.0
  0.2   0.0   0.0   0.0   1.0
  0.5   0.0   0.0   0.0   1.0
  0.8   0.0   0.0   0.0   1.0
  1.0   1.0   1.0   1.0   1.0
```

5.4.5 浮点值的其他表示

modf() 取一个浮点数,并返回一个元组,其中包含这个输入值的小数和整数部分。

代码清单 5-39:math_modf.py

```
import math

for i in range(6):
    print('{}/2 = {}'.format(i, math.modf(i / 2.0)))
```

返回值中的两个数都是浮点数。

```
$ python3 math_modf.py

0/2 = (0.0, 0.0)
1/2 = (0.5, 0.0)
2/2 = (0.0, 1.0)
3/2 = (0.5, 1.0)
4/2 = (0.0, 2.0)
5/2 = (0.5, 2.0)
```

frexp() 返回一个浮点数的尾数和指数,可以用这个函数创建值的一种更可移植的表示。

代码清单 5-40:math_frexp.py

```
import math

print('{:^7} {:^7} {:^7}'.format('x', 'm', 'e'))
print('{:-^7} {:-^7} {:-^7}'.format('', '', ''))

for x in [0.1, 0.5, 4.0]:
    m, e = math.frexp(x)
    print('{:7.2f} {:7.2f} {:7d}'.format(x, m, e))
```

frexp() 使用公式 x = m * 2**e，并返回值 m 和 e。

```
$ python3 math_frexp.py

   x        m        e
-------  -------  -------
  0.10     0.80      -3
  0.50     0.50       0
  4.00     0.50       3
```

ldexp() 与 frexp() 正好相反。

代码清单 5-41：math_ldexp.py

```
import math

print('{:^7} {:^7} {:^7}'.format('m', 'e', 'x'))
print('{:-^7} {:-^7} {:-^7}'.format('', '', ''))

INPUTS = [
    (0.8, -3),
    (0.5, 0),
    (0.5, 3),
]

for m, e in INPUTS:
    x = math.ldexp(m, e)
    print('{:7.2f} {:7d} {:7.2f}'.format(m, e, x))
```

ldexp() 使用与 frexp() 相同的公式，取尾数和指数值作为参数，并返回一个浮点数。

```
$ python3 math_ldexp.py

   m        e        x
-------  -------  -------
  0.80      -3      0.10
  0.50       0      0.50
  0.50       3      4.00
```

5.4.6 正号和负号

一个数的绝对值就是不带正负号的本值。使用 fabs() 可以计算一个浮点数的绝对值。

代码清单 5-42：math_fabs.py

```
import math

print(math.fabs(-1.1))
print(math.fabs(-0.0))
print(math.fabs(0.0))
print(math.fabs(1.1))
```

在实际中，float 的绝对值表示为一个正值。

```
$ python3 math_fabs.py

1.1
0.0
0.0
1.1
```

要确定一个值的符号，以便为一组值指定相同的符号或者比较两个值，可以使用 copysign() 来设置正确值的符号。

代码清单 5-43：math_copysign.py

```
import math

HEADINGS = ('f', 's', '< 0', '> 0', '= 0')
print('{:^5} {:^5} {:^5} {:^5} {:^5}'.format(*HEADINGS))
print('{:-^5} {:-^5} {:-^5} {:-^5} {:-^5}'.format(
    '', '', '', '', '',
))

VALUES = [
    -1.0,
    0.0,
    1.0,
    float('-inf'),
    float('inf'),
    float('-nan'),
    float('nan'),
]

for f in VALUES:
    s = int(math.copysign(1, f))
    print('{:5.1f} {:5d} {!s:5} {!s:5} {!s:5}'.format(
        f, s, f < 0, f > 0, f == 0,
    ))
```

还需要另一个类似 copysign() 的函数，因为不能将 nan 和 –nan 与其他值直接比较。

```
$ python3 math_copysign.py

  f      s     < 0   > 0   = 0
----- ----- ----- ----- -----
 -1.0    -1 True  False False
  0.0     1 False False True
  1.0     1 False True  False
 -inf    -1 True  False False
  inf     1 False True  False
  nan    -1 False False False
  nan     1 False False False
```

5.4.7 常用计算

在二进制浮点数内存中表示精确值很有难度。有些值无法准确地表示，而且如果通过反复计算来处理一个值，那么计算越频繁就越容易引入表示误差。math 包含一个函数来计算一系列浮点数的和，它使用一种高效的算法来尽量减少这种误差。

代码清单 5-44：math_fsum.py

```
import math

values = [0.1] * 10

print('Input values:', values)

print('sum()          : {:.20f}'.format(sum(values)))
```

```
س = 0.0
for i in values:
    s += i
print('for-loop   : {:.20f}'.format(s))

print('math.fsum() : {:.20f}'.format(math.fsum(values)))
```

给定一个包含10个值的序列，每个值都等于 `0.1`，这个序列总和的期望值为 `1.0`。不过，由于 `0.1` 不能精确地表示为一个浮点值，所以会在总和中引入误差，除非用 `fsum()` 来计算。

```
$ python3 math_fsum.py

Input values: [0.1, 0.1, 0.1, 0.1, 0.1, 0.1, 0.1, 0.1, 0.1, 0.1]
sum()       : 0.99999999999999988898
for-loop    : 0.99999999999999988898
math.fsum() : 1.00000000000000000000
```

`factorial()` 常用于计算一系列对象的排列和组合数。一个正整数 n 的阶乘（表示为 n!）被递归地定义为 `(n-1)!*n`，并在 `0!==1` 停止递归。

代码清单 5-45：math_factorial.py

```
import math

for i in [0, 1.0, 2.0, 3.0, 4.0, 5.0, 6.1]:
    try:
        print('{:2.0f} {:6.0f}'.format(i, math.factorial(i)))
    except ValueError as err:
        print('Error computing factorial({}): {}'.format(i, err))
```

`factorial()` 只能处理整数，不过它确实也接受 `float` 参数，只要这个参数可以转换为一个整数而不丢值。

```
$ python3 math_factorial.py

 0      1
 1      1
 2      2
 3      6
 4     24
 5    120
Error computing factorial(6.1): factorial() only accepts integral
 values
```

`gamma()` 类似于 `factorial()`，不过它可以处理实数，而且值会下移一个数（gamma 等于 `(n - 1)!`）。

代码清单 5-46：math_gamma.py

```
import math

for i in [0, 1.1, 2.2, 3.3, 4.4, 5.5, 6.6]:
    try:
        print('{:2.1f} {:6.2f}'.format(i, math.gamma(i)))
    except ValueError as err:
        print('Error computing gamma({}): {}'.format(i, err))
```

由于 0 会导致开始值为负，所以这是不允许的。

```
$ python3 math_gamma.py

Error computing gamma(0): math domain error
1.1    0.95
2.2    1.10
3.3    2.68
4.4   10.14
5.5   52.34
6.6  344.70
```

lgamma() 会返回对输入值求 gamma 所得结果的绝对值的自然对数。

代码清单 5-47：math_lgamma.py

```python
import math

for i in [0, 1.1, 2.2, 3.3, 4.4, 5.5, 6.6]:
    try:
        print('{:2.1f} {:.20f} {:.20f}'.format(
            i,
            math.lgamma(i),
            math.log(math.gamma(i)),
        ))
    except ValueError as err:
        print('Error computing lgamma({}): {}'.format(i, err))
```

使用 lgamma() 会比使用 gamma() 的结果单独计算对数更精确。

```
$ python3 math_lgamma.py

Error computing lgamma(0): math domain error
1.1 -0.04987244125984036103 -0.04987244125983997245
2.2 0.09694746679063825923 0.09694746679063866168
3.3 0.98709857789473387513 0.98709857789473409717
4.4 2.31610349142485727469 2.31610349142485727469
5.5 3.95781396761871651080 3.95781396761871606671
6.6 5.84268005527463252236 5.84268005527463252236
```

求模操作符（%）会计算一个除法表达式的余数（例如，5 % 2 = 1）。Python 语言内置的这个操作符可以很好地处理整数，但是与很多其他浮点数运算类似，中间计算可能带来表示问题，从而进一步造成数据丢失。fmod() 可以为浮点值提供一个更精确的实现。

代码清单 5-48：math_fmod.py

```python
import math

print('{:^4} {:^4} {:^5} {:^5}'.format(
    'x', 'y', '%', 'fmod'))
print('{:-^4} {:-^4} {:-^5} {:-^5}'.format(
    '-', '-', '-', '-'))

INPUTS = [
    (5, 2),
    (5, -2),
    (-5, 2),
]

for x, y in INPUTS:
```

```
    print('{:4.1f} {:4.1f} {:5.2f} {:5.2f}'.format(
        x,
        y,
        x % y,
        math.fmod(x, y),
    ))
```

还有一点可能经常产生混淆，即 `fmod()` 计算模所使用的算法与 `%` 使用的算法也有所不同，所以结果的符号不同。

```
$ python3 math_fmod.py

  x    y      %    fmod
---- ---- ----- -----
 5.0  2.0  1.00  1.00
 5.0 -2.0 -1.00  1.00
-5.0  2.0  1.00 -1.00
```

可以使用 `gcd()` 找出两个整数公约数中最大的整数——也就是最大公约数。

代码清单 5-49：**math_gcd.py**

```
import math

print(math.gcd(10, 8))
print(math.gcd(10, 0))
print(math.gcd(50, 225))
print(math.gcd(11, 9))
print(math.gcd(0, 0))
```

如果两个值都为 `0`，则结果为 `0`。

```
$ python3 math_gcd.py

2
10
25
1
0
```

5.4.8 指数和对数

指数生长曲线在经济学、物理学和其他科学中经常出现。Python 有一个内置的幂运算符（"**"），不过，如果需要将一个可调用函数作为另一个函数的参数，那么可能需要用到 `pow()`。

代码清单 5-50：**math_pow.py**

```
import math

INPUTS = [
    # Typical uses
    (2, 3),
    (2.1, 3.2),

    # Always 1
    (1.0, 5),
    (2.0, 0),
```

```
    # Not a number
    (2, float('nan')),

    # Roots
    (9.0, 0.5),
    (27.0, 1.0 / 3),
]

for x, y in INPUTS:
    print('{:5.1f} ** {:5.3f} = {:6.3f}'.format(
        x, y, math.pow(x, y)))
```

1 的任何次幂总返回 1.0，同样，任何值的指数为 0.0 时也总是返回 1.0。对于 nan 值（不是一个数），大多数运算都返回 nan。如果指数小于 1，pow() 会计算一个根。

```
$ python3 math_pow.py

  2.0 **  3.000 =   8.000
  2.1 **  3.200 =  10.742
  1.0 **  5.000 =   1.000
  2.0 **  0.000 =   1.000
  2.0 **    nan =     nan
  9.0 **  0.500 =   3.000
 27.0 **  0.333 =   3.000
```

由于平方根（指数为 1/2）被使用得非常频繁，所以有一个单独的函数来计算平方根。

代码清单 5-51：math_sqrt.py

```
import math

print(math.sqrt(9.0))
print(math.sqrt(3))
try:
    print(math.sqrt(-1))
except ValueError as err:
    print('Cannot compute sqrt(-1):', err)
```

计算负数的平方根需要用到复数，这不在 math 的处理范围内。试图计算一个负值的平方根时，会导致一个 ValueError。

```
$ python3 math_sqrt.py

3.0
1.7320508075688772
Cannot compute sqrt(-1): math domain error
```

对数函数查找满足条件 x = b ** y 的 y。默认 log() 计算自然对数（底数为 e）。如果提供了第二个参数，则使用这个参数值作为底数。

代码清单 5-52：math_log.py

```
import math

print(math.log(8))
print(math.log(8, 2))
print(math.log(0.5, 2))
```

x 小于 1 时，求对数会生成负数结果。

```
$ python3 math_log.py

2.0794415416798357
3.0
-1.0
```

log() 有三个变形。在给定浮点数表示和取整误差的情况下，由 log(x, b) 生成的计算值只有有限的精度（特别是对于某些底数）。log10() 完成 log(x, 10) 计算，但是会使用一种比 log() 更精确的算法。

代码清单 5-53：**math_log10.py**

```python
import math

print('{:2} {:^12} {:^10} {:^20} {:8}'.format(
    'i', 'x', 'accurate', 'inaccurate', 'mismatch',
))
print('{:-^2} {:-^12} {:-^10} {:-^20} {:-^8}'.format(
    '', '', '', '', '',
))

for i in range(0, 10):
    x = math.pow(10, i)
    accurate = math.log10(x)
    inaccurate = math.log(x, 10)
    match = '' if int(inaccurate) == i else '*'
    print('{:2d} {:12.1f} {:10.8f} {:20.18f} {:^5}'.format(
        i, x, accurate, inaccurate, match,
    ))
```

输出中末尾有 * 的行突出强调了不精确的值。

```
$ python3 math_log10.py

 i       x        accurate        inaccurate       mismatch
-- ------------ ---------- -------------------- --------
 0          1.0 0.00000000 0.000000000000000000
 1         10.0 1.00000000 1.000000000000000000
 2        100.0 2.00000000 2.000000000000000000
 3       1000.0 3.00000000 2.999999999999999556    *
 4      10000.0 4.00000000 4.000000000000000000
 5     100000.0 5.00000000 5.000000000000000000
 6    1000000.0 6.00000000 5.999999999999999112    *
 7   10000000.0 7.00000000 7.000000000000000000
 8  100000000.0 8.00000000 8.000000000000000000
 9 1000000000.0 9.00000000 8.999999999999998224    *
```

类似于 log10()，log2() 会完成等价于 math.log(x,2) 的计算。

代码清单 5-54：**math_log2.py**

```python
import math

print('{:>2} {:^5} {:^5}'.format(
    'i', 'x', 'log2',
))
print('{:-^2} {:-^5} {:-^5}'.format(
    '', '', '',
))
```

```
for i in range(0, 10):
    x = math.pow(2, i)
    result = math.log2(x)
    print('{:2d} {:5.1f} {:5.1f}'.format(
        i, x, result,
    ))
```

取决于底层平台，这个内置的特殊用途函数能提供更好的性能和精度，因为它利用了针对底数 2 的特殊用途算法，而在更一般用途的函数中没有使用这些算法。

```
$ python3 math_log2.py

 i    x    log2
--  -----  -----
 0    1.0    0.0
 1    2.0    1.0
 2    4.0    2.0
 3    8.0    3.0
 4   16.0    4.0
 5   32.0    5.0
 6   64.0    6.0
 7  128.0    7.0
 8  256.0    8.0
 9  512.0    9.0
```

log1p() 会计算 Newton-Mercator 序列（1+x 的自然对数）。

代码清单 5-55：**math_log1p.py**

```
import math

x = 0.0000000000000000000000001
print('x        :', x)
print('1 + x    :', 1 + x)
print('log(1+x):', math.log(1 + x))
print('log1p(x):', math.log1p(x))
```

对于非常接近于 0 的 x，log1p() 会更为精确，因为它使用的算法可以补偿由初始加法带来的取整误差。

```
$ python3 math_log1p.py

x        : 1e-25
1 + x    : 1.0
log(1+x): 0.0
log1p(x): 1e-25
```

exp() 会计算指数函数（e**x）。

代码清单 5-56：**math_exp.py**

```
import math

x = 2

fmt = '{:.20f}'
print(fmt.format(math.e ** 2))
print(fmt.format(math.pow(math.e, 2)))
print(fmt.format(math.exp(2)))
```

类似于其他特殊函数,与等价的通用函数 `math.pow(math.e, x)` 相比,`exp()` 使用的算法可以生成更精确的结果。

```
$ python3 math_exp.py

7.38905609893064951876
7.38905609893064951876
7.38905609893065040694
```

`expm1()` 是 `log1p()` 的逆运算,会计算 e**x − 1。

代码清单 5-57:**math_expm1.py**

```
import math

x = 0.0000000000000000000000001

print(x)
print(math.exp(x) - 1)
print(math.expm1(x))
```

类似于 `log1p()`,x 值很小时,如果单独完成减法,则可能会损失精度。

```
$ python3 math_expm1.py

1e-25
0.0
1e-25
```

5.4.9 角

尽管我们每天讨论角时更常用的是度,但弧度才是科学和数学领域中度量角度的标准单位。弧度是在圆心相交的两条线所构成的角,其终点落在圆的圆周上,终点之间相距一个弧度。

圆周长计算为 $2\pi r$,所以弧度与 π(这是三角函数计算中经常出现的一个值)之间存在一个关系。这个关系使得三角学和微积分中都使用了弧度,因为利用弧度可以得到更紧凑的公式。

要把度转换为弧度,可以使用 `radians()`。

代码清单 5-58:**math_radians.py**

```
import math

print('{:^7} {:^7} {:^7}'.format(
    'Degrees', 'Radians', 'Expected'))
print('{:-^7} {:-^7} {:-^7}'.format(
    '', '', ''))

INPUTS = [
    (0, 0),
    (30, math.pi / 6),
    (45, math.pi / 4),
    (60, math.pi / 3),
    (90, math.pi / 2),
    (180, math.pi),
    (270, 3 / 2.0 * math.pi),
    (360, 2 * math.pi),
]

for deg, expected in INPUTS:
```

```
    print('{:7d} {:7.2f} {:7.2f}'.format(
        deg,
        math.radians(deg),
        expected,
    ))
```

转换公式为 rad = deg * π / 180。

```
$ python3 math_radians.py

Degrees Radians Expected
------- ------- -------
      0    0.00    0.00
     30    0.52    0.52
     45    0.79    0.79
     60    1.05    1.05
     90    1.57    1.57
    180    3.14    3.14
    270    4.71    4.71
    360    6.28    6.28
```

要从弧度转换为度，可以使用 degrees()。

代码清单 5-59: math_degrees.py

```
import math

INPUTS = [
    (0, 0),
    (math.pi / 6, 30),
    (math.pi / 4, 45),
    (math.pi / 3, 60),
    (math.pi / 2, 90),
    (math.pi, 180),
    (3 * math.pi / 2, 270),
    (2 * math.pi, 360),
]

print('{:^8} {:^8} {:^8}'.format(
    'Radians', 'Degrees', 'Expected'))
print('{:-^8} {:-^8} {:-^8}'.format('', '', ''))
for rad, expected in INPUTS:
    print('{:8.2f} {:8.2f} {:8.2f}'.format(
        rad,
        math.degrees(rad),
        expected,
    ))
```

具体转换公式为 deg = rad * 180 / π。

```
$ python3 math_degrees.py

Radians  Degrees  Expected
-------- -------- --------
    0.00     0.00     0.00
    0.52    30.00    30.00
    0.79    45.00    45.00
    1.05    60.00    60.00
    1.57    90.00    90.00
    3.14   180.00   180.00
    4.71   270.00   270.00
    6.28   360.00   360.00
```

5.4.10 三角函数

三角函数将三角形中的角与其边长相关联。在有周期性质的公式中经常出现三角函数，如谐波或圆周运动；在处理角时也会经常用到三角函数。标准库中所有三角函数的角参数都被表示为弧度。

给定一个直角三角形中的角，其正弦是对边长度与斜边长度之比（sin A = 对边 / 斜边）。余弦是邻边长度与斜边长度之比（cos A = 邻边 / 斜边）。正切是对边与邻边之比（tan A = 对边 / 邻边）。

代码清单 5-60：**math_trig.py**

```
import math

print('{:^7} {:^7} {:^7} {:^7} {:^7}'.format(
    'Degrees', 'Radians', 'Sine', 'Cosine', 'Tangent'))
print('{:-^7} {:-^7} {:-^7} {:-^7} {:-^7}'.format(
    '-', '-', '-', '-', '-'))

fmt = '{:7.2f} {:7.2f} {:7.2f} {:7.2f} {:7.2f}'

for deg in range(0, 361, 30):
    rad = math.radians(deg)
    if deg in (90, 270):
        t = float('inf')
    else:
        t = math.tan(rad)
    print(fmt.format(deg, rad, math.sin(rad), math.cos(rad), t))
```

正切也可以被定义为角的正弦值与其余弦值之比，因为弧度 π/2 和 3π/2 的余弦是 0，所以相应的正切值为无穷大。

```
$ python3 math_trig.py

Degrees Radians  Sine   Cosine Tangent
------- ------- ------- ------- -------
   0.00    0.00    0.00    1.00    0.00
  30.00    0.52    0.50    0.87    0.58
  60.00    1.05    0.87    0.50    1.73
  90.00    1.57    1.00    0.00     inf
 120.00    2.09    0.87   -0.50   -1.73
 150.00    2.62    0.50   -0.87   -0.58
 180.00    3.14    0.00   -1.00   -0.00
 210.00    3.67   -0.50   -0.87    0.58
 240.00    4.19   -0.87   -0.50    1.73
 270.00    4.71   -1.00   -0.00     inf
 300.00    5.24   -0.87    0.50   -1.73
 330.00    5.76   -0.50    0.87   -0.58
 360.00    6.28   -0.00    1.00   -0.00
```

给定一个点 (x, y)，点 [(0, 0), (x, 0), (x, y)] 构成的三角形中斜边长度为 (x**2 + y**2) ** 1/2，可以用 hypot() 来计算。

代码清单 5-61：**math_hypot.py**

```
import math

print('{:^7} {:^7} {:^10}'.format('X', 'Y', 'Hypotenuse'))
```

```
print('{:-^7} {:-^7} {:-^10}'.format('', '', ''))

POINTS = [
    # Simple points
    (1, 1),
    (-1, -1),
    (math.sqrt(2), math.sqrt(2)),
    (3, 4),  # 3-4-5 triangle
    # On the circle
    (math.sqrt(2) / 2, math.sqrt(2) / 2),  # pi/4 rads
    (0.5, math.sqrt(3) / 2),  # pi/3 rads
]

for x, y in POINTS:
    h = math.hypot(x, y)
    print('{:7.2f} {:7.2f} {:7.2f}'.format(x, y, h))
```

对于圆上的点，其斜边总是等于 1。

```
$ python3 math_hypot.py

   X       Y    Hypotenuse
------- ------- ----------
  1.00    1.00    1.41
 -1.00   -1.00    1.41
  1.41    1.41    2.00
  3.00    4.00    5.00
  0.71    0.71    1.00
  0.50    0.87    1.00
```

还可以用这个函数查看两个点之间的距离。

代码清单 5-62：math_distance_2_points.py

```
import math

print('{:^8} {:^8} {:^8} {:^8} {:^8}'.format(
    'X1', 'Y1', 'X2', 'Y2', 'Distance',
))
print('{:-^8} {:-^8} {:-^8} {:-^8} {:-^8}'.format(
    '', '', '', '', '',
))

POINTS = [
    ((5, 5), (6, 6)),
    ((-6, -6), (-5, -5)),
    ((0, 0), (3, 4)),  # 3-4-5 triangle
    ((-1, -1), (2, 3)),  # 3-4-5 triangle
]

for (x1, y1), (x2, y2) in POINTS:
    x = x1 - x2
    y = y1 - y2
    h = math.hypot(x, y)
    print('{:8.2f} {:8.2f} {:8.2f} {:8.2f} {:8.2f}'.format(
        x1, y1, x2, y2, h,
    ))
```

使用 x 值之差和 y 值之差将一个端点移至原点，然后将结果传入 hypot()。

```
$ python3 math_distance_2_points.py
```

```
    X1        Y1        X2        Y2    Distance
--------  --------  --------  --------  --------
    5.00      5.00      6.00      6.00      1.41
   -6.00     -6.00     -5.00     -5.00      1.41
    0.00      0.00      3.00      4.00      5.00
   -1.00     -1.00      2.00      3.00      5.00
```

math 还定义了反三角函数。

代码清单 5-63：**math_inverse_trig.py**

```
import math

for r in [0, 0.5, 1]:
    print('arcsine({:.1f})    = {:5.2f}'.format(r, math.asin(r)))
    print('arccosine({:.1f})  = {:5.2f}'.format(r, math.acos(r)))
    print('arctangent({:.1f}) = {:5.2f}'.format(r, math.atan(r)))
    print()
```

1.57 大约等于 π/2，或 90 度，这个角的正弦为 1，余弦为 0。

```
$ python3 math_inverse_trig.py

arcsine(0.0)    =  0.00
arccosine(0.0)  =  1.57
arctangent(0.0) =  0.00

arcsine(0.5)    =  0.52
arccosine(0.5)  =  1.05
arctangent(0.5) =  0.46

arcsine(1.0)    =  1.57
arccosine(1.0)  =  0.00
arctangent(1.0) =  0.79
```

5.4.11 双曲函数

双曲函数经常出现在线性微分方程中，处理电磁场、流体力学、狭义相对论和其他高级物理和数学问题时常会用到。

代码清单 5-64：**math_hyperbolic.py**

```
import math

print('{:^6} {:^6} {:^6} {:^6}'.format(
    'X', 'sinh', 'cosh', 'tanh',
))
print('{:-^6} {:-^6} {:-^6} {:-^6}'.format('', '', '', ''))

fmt = '{:6.4f} {:6.4f} {:6.4f} {:6.4f}'

for i in range(0, 11, 2):
    x = i / 10.0
    print(fmt.format(
        x,
        math.sinh(x),
        math.cosh(x),
        math.tanh(x),
    ))
```

余弦函数和正弦函数构成一个圆,而双曲余弦函数和双曲正弦函数构成半个双曲线。

```
$ python3 math_hyperbolic.py

   X    sinh   cosh   tanh
------ ------ ------ ------
0.0000 0.0000 1.0000 0.0000
0.2000 0.2013 1.0201 0.1974
0.4000 0.4108 1.0811 0.3799
0.6000 0.6367 1.1855 0.5370
0.8000 0.8881 1.3374 0.6640
1.0000 1.1752 1.5431 0.7616
```

另外还提供了反双曲函数 acosh()、asinh() 和 atanh()。

5.4.12 特殊函数

统计学中经常用到高斯误差函数(Gauss error function)。

代码清单 5-65:math_erf.py

```python
import math

print('{:^5} {:7}'.format('x', 'erf(x)'))
print('{:-^5} {:-^7}'.format('', ''))

for x in [-3, -2, -1, -0.5, -0.25, 0, 0.25, 0.5, 1, 2, 3]:
    print('{:5.2f} {:7.4f}'.format(x, math.erf(x)))
```

对于误差函数,erf(-x) == -erf(x)。

```
$ python3 math_erf.py

  x    erf(x)
----- -------
-3.00 -1.0000
-2.00 -0.9953
-1.00 -0.8427
-0.50 -0.5205
-0.25 -0.2763
 0.00  0.0000
 0.25  0.2763
 0.50  0.5205
 1.00  0.8427
 2.00  0.9953
 3.00  1.0000
```

补余误差函数 erfc() 生成等价于 1 - erf(x) 的值。

代码清单 5-66:math_erfc.py

```python
import math

print('{:^5} {:7}'.format('x', 'erfc(x)'))
print('{:-^5} {:-^7}'.format('', ''))

for x in [-3, -2, -1, -0.5, -0.25, 0, 0.25, 0.5, 1, 2, 3]:
    print('{:5.2f} {:7.4f}'.format(x, math.erfc(x)))
```

如果 x 值很小,那么在从 1 做减法时 erfc() 实现便可以避免可能的精度误差。

```
$ python3 math_erfc.py

    x    erfc(x)
 -----   -------
 -3.00   2.0000
 -2.00   1.9953
 -1.00   1.8427
 -0.50   1.5205
 -0.25   1.2763
  0.00   1.0000
  0.25   0.7237
  0.50   0.4795
  1.00   0.1573
  2.00   0.0047
  3.00   0.0000
```

提示：相关阅读材料

- **math** 的标准库文档[一]。
- **IEEE floating-point arithmetic in Python**[二]：John Cook 写的博客文章，介绍特殊值如何产生，以及在 Python 中完成数学运算时如何处理特殊值。
- **SciPy**[三]：Python 中实现科学和数学计算的开源库。
- **PEP 485**[四]：测试近似相等性的函数。

5.5 **statistics**：统计计算

statistics 模块实现了很多常用的统计公式，允许使用 Python 的各种数值类型（**int**、**float**、**Decimal** 和 **Fraction**）来完成高效计算。

5.5.1 平均值

共支持 3 种形式的平均值：均值（mean），中值或中位数（median），以及众数（mode）。可以用 **mean()** 计算算术平均值。

代码清单 5-67：**statistics_mean.py**

```
from statistics import *
data = [1, 2, 2, 5, 10, 12]
print('{:0.2f}'.format(mean(data)))
```

对于整数和浮点数，这个函数的返回值总是 **float**。对于 **Decimal** 和 **Fraction** 输入数据，结果与输入的类型相同。

```
$ python3 statistics_mean.py

5.33
```

[一] https://docs.python.org/3.5/library/math.html
[二] www.johndcook.com/blog/2009/07/21/ieee-arithmetic-python/
[三] http://scipy.org
[四] www.python.org/dev/peps/pep-0485

可以使用 `mode()` 计算一个数据集中最常见的数据点。

代码清单 5-68：`statistics_mode.py`

```
from statistics import *
data = [1, 2, 2, 5, 10, 12]
print(mode(data))
```

其返回值总是输入数据集的一个成员。由于 `mode()` 把输入处理为一个离散值集合，并且统计出现次数，所以实际上输入不需要是数值。

```
$ python3 statistics_mode.py

2
```

计算中值（或中位数）有 4 种变形。前三种是一般算法的简单版本，只是在处理元素个数为偶数的数据集时采用了不同方法。

代码清单 5-69：`statistics_median.py`

```
from statistics import *
data = [1, 2, 2, 5, 10, 12]

print('median : {:0.2f}'.format(median(data)))
print('low    : {:0.2f}'.format(median_low(data)))
print('high   : {:0.2f}'.format(median_high(data)))
```

`median()` 会查找中间的值。如果数据集包含偶数个值，则取两个中间元素的平均值。`median_low()` 总是返回输入数据集中的一个值，对于有偶数个元素的数据集，会返回两个中间元素中较小的一个。`median_high()` 与之类似，不过会返回两个中间元素中较大的一个。

```
$ python3 statistics_median.py

median : 3.50
low    : 2.00
high   : 5.00
```

中值计算的第 4 个版本是 `median_grouped()`，它会把输入看作连续数据。这个函数计算 50% 百分位数（即中值）的做法是首先使用所提供的间隔宽度找出中值区间，然后使用落入该区间的数据集中的具体值位置在该区间中插值。

代码清单 5-70：`statistics_median_grouped.py`

```
from statistics import *
data = [10, 20, 30, 40]

print('1: {:0.2f}'.format(median_grouped(data, interval=1)))
print('2: {:0.2f}'.format(median_grouped(data, interval=2)))
print('3: {:0.2f}'.format(median_grouped(data, interval=3)))
```

随着间隔宽度的增加，为相同数据集计算的中值会改变。

```
$ python3 statistics_median_grouped.py

1: 29.50
2: 29.00
3: 28.50
```

5.5.2 方差

统计使用两个值描述一个值集相对于均值的分散度。方差（variance）是各个值与均值之差平方的平均，标准偏差或标准差（standard deviation）是方差的平方根（这很有用，因为取平方根可以使标准差与输入数据有相同的单位）。如果方差或标准差的值很大，这说明一个数据集是分散的，而如果这个值很小，则说明数据在靠近均值聚集。

代码清单 5-71: **statistics_variance.py**

```python
from statistics import *
import subprocess
def get_line_lengths():
    cmd = 'wc -l ../[a-z]*/*.py'
    out = subprocess.check_output(
        cmd, shell=True).decode('utf-8')
    for line in out.splitlines():
        parts = line.split()
        if parts[1].strip().lower() == 'total':
            break
        nlines = int(parts[0].strip())
        if not nlines:
            continue  # Skip empty files.
        yield (nlines, parts[1].strip())


data = list(get_line_lengths())

lengths = [d[0] for d in data]
sample = lengths[::2]

print('Basic statistics:')
print('  count     : {:3d}'.format(len(lengths)))
print('  min       : {:6.2f}'.format(min(lengths)))
print('  max       : {:6.2f}'.format(max(lengths)))
print('  mean      : {:6.2f}'.format(mean(lengths)))

print('\nPopulation variance:')
print('  pstdev    : {:6.2f}'.format(pstdev(lengths)))
print('  pvariance : {:6.2f}'.format(pvariance(lengths)))

print('\nEstimated variance for sample:')
print('  count     : {:3d}'.format(len(sample)))
print('  stdev     : {:6.2f}'.format(stdev(sample)))
print('  variance  : {:6.2f}'.format(variance(sample)))
```

Python 包括两组函数来计算方差和标准差，具体取决于数据集是表示总体还是表示总体中的一个样本。这个例子首先使用 wc 统计所有示例程序输入文件中的行数。然后使用 `pvariance()` 和 `pstdev()` 计算总体的方差和标准差。最后，它使用 `variance()` 和 `stddev()` 计算一个子集的样本方差和标准差，这个子集是由每隔一个文件的长度创建的。

```
$ python3 statistics_variance.py
Basic statistics:
    count    : 959
    min      :   4.00
    max      : 228.00
    mean     :  28.62
Population variance:
    pstdev   :  18.52
    pvariance: 342.95

Estimated variance for sample:
    count    : 480
    stdev    :  21.09
    variance : 444.61
```

提示：相关阅读材料

- statistics 的标准库文档[一]。
- Median for Discrete and Continuous Frequency Type Data (grouped data)[二]：对连续数据中值的讨论。
- **PEP 450**[三]：为标准库增加了一个统计模块。

[一] https://docs.python.org/3.5/library/statistics.html
[二] www.mathstips.com/statistics/median-for-discrete-and-continuous-frequency-type.html
[三] www.python.org/dev/peps/pep-0450

第 ⑥ 章
文件系统

　　Python 的标准库包括大量工具，可以处理文件系统中的文件，构造和解析文件名，还可以检查文件内容。

　　处理文件的第一步是确定要处理的文件的名字。Python 将文件名表示为简单的字符串，另外还提供了一些工具，可以由 `os.path` 中平台独立的标准组成部分构造文件名。

　　`pathlib` 模块提供了一个面向对象 API 来处理文件系统路径。使用这个模块而不是 `os.path` 可以提供更大的便利，因为它会在更高抽象层完成处理。

　　用 `os` 中的 `listdir()` 可以列出一个目录中的内容，或者使用 `glob` 由一个模式建立文件名列表。

　　`glob` 使用的文件名模式匹配还可以通过 `fnmatch` 直接提供，从而可以在其他上下文中使用。

　　明确文件名之后，可以用 `os.stat()` 和 `stat` 中的常量来检查其他特性，如权限或文件大小。

　　应用需要随机访问文件时，利用 `linecache` 可以很容易地按行号读取行。文件的内容在缓存中维护，所以要当心内存消耗。

　　有些情况下需要创建草稿文件来临时保存数据，或者将数据移动到一个永久位置之前需要用临时文件存储，此时 `tempfile` 就很有用。它提供了一些类，可以安全而稳妥地创建临时文件和目录。可以保证文件名和目录名是唯一的，其中包含随机的组成部分，因此不容易猜出。

　　程序经常需要把文件作为一个整体来处理，而不考虑其内容。`shutil` 模块包含了一些高级文件操作，如复制文件和目录，以及创建或解压缩文件归档。

　　`filecmp` 模块通过查看文件和目录包含的字节来完成文件和目录比较，不过不需要有关其格式的任何特殊知识。

　　内置的 `file` 类可以用于读写本地文件系统上可见的文件。不过，通过 `read()` 和 `write()` 接口访问大文件时，程序的性能可能会受影响，因为文件从磁盘移动到应用可见的内存时这两个操作都会涉及多次数据复制。使用 `mmap` 可以告诉操作系统使用其虚拟内存子系统，将文件的内容直接映射到程序可以访问的内存，从而避免操作系统与 `file` 对象内部缓冲区之间的复制步骤。

　　如果文本数据中使用了非 ASCII 字符，那么通常会采用一种 Unicode 数据格式保存。由于标准 `file` 句柄假设文本文件的各个字节分别表示一个字符，所以读取使用多字节编码的 Unicode 文本需要额外的处理。`codecs` 模块会自动处理编码和解码，所以在很多情况下，完全可以使用一个非 ASCII 文件而无须对程序做任何其他修改。

io 模块提供的类用来实现 Python 的基于文件的输入和输出。如果测试代码依赖于从文件读写数据，那么对于这些测试代码，io 提供了一个内存中流对象，它就像一个文件，不过不驻留在磁盘上。

6.1 `os.path`：平台独立的文件名管理

利用 `os.path` 模块中包含的函数，很容易编写代码来处理多个平台上的文件。即使程序不打算在平台之间移植，也应当使用 `os.path` 来完成可靠的文件名解析。

6.1.1 解析路径

`os.path` 中的第一组函数可以用来将表示文件名的字符串解析为文件名的各个组成部分。这些函数并不要求路径真正存在：它们只是处理字符串。

路径解析依赖于 `os` 中定义的一些变量：
- `os.sep`：路径各部分之间的分隔符（例如，"/"或"\"）。
- `os.extsep`：文件名与文件"扩展名"之间的分隔符（例如"."）。
- `os.pardir`：路径中表示目录树上一级的部分（例如".."）。
- `os.curdir`：路径中指示当前目录的部分（例如"."）。

`split()` 函数将路径分解为两个单独的部分，并返回包含这些结果的一个 `tuple`。这个 `tuple` 的第二个元素是路径的最后一部分，第一个元素则是此前的所有内容。

代码清单 6-1：`ospath_split.py`

```
import os.path

PATHS = [
    '/one/two/three',
    '/one/two/three/',
    '/',
    '.',
    '',
]

for path in PATHS:
    print('{!r:>17} : {}'.format(path, os.path.split(path)))
```

输入参数以 `os.sep` 结尾时，路径的最后一个元素是一个空串。

```
$ python3 ospath_split.py

  '/one/two/three' : ('/one/two', 'three')
 '/one/two/three/' : ('/one/two/three', '')
               '/' : ('/', '')
               '.' : ('', '.')
                '' : ('', '')
```

`basename()` 函数返回的值等价于 `split()` 值的第二部分。

代码清单 6-2: **ospath_basename.py**

```python
import os.path

PATHS = [
    '/one/two/three',
    '/one/two/three/',
    '/',
    '.',
    '',
]

for path in PATHS:
    print('{!r:>17} : {!r}'.format(path, os.path.basename(path)))
```

整个路径会剥除到只剩下最后一个元素，不论这指示的是一个文件还是一个目录。如果路径以目录分隔符结尾（**os.sep**），则认为基本部分为空。

```
$ python3 ospath_basename.py

 '/one/two/three' : 'three'
'/one/two/three/' : ''
              '/' : ''
              '.' : '.'
               '' : ''
```

dirname() 函数返回分解路径得到的第一部分。

代码清单 6-3: **ospath_dirname.py**

```python
import os.path

PATHS = [
    '/one/two/three',
    '/one/two/three/',
    '/',
    '.',
    '',
]

for path in PATHS:
    print('{!r:>17} : {!r}'.format(path, os.path.dirname(path)))
```

将 **basename()** 的结果与 **dirname()** 结合可以得到原来的路径。

```
$ python3 ospath_dirname.py

 '/one/two/three' : '/one/two'
'/one/two/three/' : '/one/two/three'
              '/' : '/'
              '.' : ''
               '' : ''
```

splitext() 的工作类似于 **split()**，不过它会根据扩展名分隔符而不是目录分隔符来分解路径。

代码清单 6-4: **ospath_splitext.py**

```python
import os.path
```

```
PATHS = [
    'filename.txt',
    'filename',
    '/path/to/filename.txt',
    '/',
    '',
    'my-archive.tar.gz',
    'no-extension.',
]

for path in PATHS:
    print('{!r:>21} : {!r}'.format(path, os.path.splitext(path)))
```

查找扩展名时，只使用 `os.extsep` 的最后一次出现，所以如果一个文件名有多个扩展名，那么分解这个文件名时，部分扩展名会留在前缀上。

```
$ python3 ospath_splitext.py

         'filename.txt' : ('filename', '.txt')
             'filename' : ('filename', '')
'/path/to/filename.txt' : ('/path/to/filename', '.txt')
                  '/' : ('/', '')
                   '' : ('', '')
    'my-archive.tar.gz' : ('my-archive.tar', '.gz')
        'no-extension.' : ('no-extension', '.')
```

`commonprefix()` 取一个路径列表作为参数，并且返回一个字符串，表示所有路径中都出现的公共前缀。这个值可能表示一个根本不存在的路径，而且并不考虑路径分隔符，所以这个前缀可能并不落在一个分隔符边界上。

代码清单 6-5：`ospath_commonprefix.py`

```
import os.path

paths = ['/one/two/three/four',
         '/one/two/threefold',
         '/one/two/three/',
         ]
for path in paths:
    print('PATH:', path)

print()
print('PREFIX:', os.path.commonprefix(paths))
```

在这个例子中，公共前缀字符串是 `/one/two/three`，尽管其中一个路径并不包括一个名为 `three` 的目录。

```
$ python3 ospath_commonprefix.py

PATH: /one/two/three/four
PATH: /one/two/threefold
PATH: /one/two/three/

PREFIX: /one/two/three
```

`commonpath()` 则要考虑路径分隔符。它返回的前缀不包括部分路径值。

代码清单 6-6：ospath_commonpath.py

```
import os.path

paths = ['/one/two/three/four',
         '/one/two/threefold',
         '/one/two/three/',
         ]
for path in paths:
    print('PATH:', path)

print()
print('PREFIX:', os.path.commonpath(paths))
```

由于 "threefold" 在 "three" 后面没有一个路径分隔符，所以公共前缀为 /one/two。

```
$ python3 ospath_commonpath.py

PATH: /one/two/three/four
PATH: /one/two/threefold
PATH: /one/two/three/

PREFIX: /one/two
```

6.1.2 建立路径

除了分解现有的路径，还经常需要从其他字符串建立路径。要将多个路径组成部分结合为一个值，可以使用 join()。

代码清单 6-7：ospath_join.py

```
import os.path

PATHS = [
    ('one', 'two', 'three'),
    ('/', 'one', 'two', 'three'),
    ('/one', '/two', '/three'),
]

for parts in PATHS:
    print('{} : {!r}'.format(parts, os.path.join(*parts)))
```

如果要连接的某个参数以 os.sep 开头，那么前面的所有参数都会被丢弃，并且这个新参数会成为返回值的开始部分。

```
$ python3 ospath_join.py

('one', 'two', 'three') : 'one/two/three'
('/', 'one', 'two', 'three') : '/one/two/three'
('/one', '/two', '/three') : '/three'
```

还可以处理包含"可变"部分的路径，这些"可变"部分可以自动扩展。例如，expanduser() 可以将波浪线（~）字符转换为用户主目录名。

代码清单 6-8：ospath_expanduser.py

```
import os.path
```

```
for user in ['', 'dhellmann', 'nosuchuser']:
    lookup = '~' + user
    print('{!r:>15} : {!r}'.format(
        lookup, os.path.expanduser(lookup)))
```

如果用户的主目录无法找到，那么字符串将不做任何改动并直接返回，如下面这个例子中的 ~nosuchuser。

```
$ python3 ospath_expanduser.py

            '~' : '/Users/dhellmann'
   '~dhellmann' : '/Users/dhellmann'
  '~nosuchuser' : '~nosuchuser'
```

expandvars() 更为通用，它会扩展路径中出现的所有 shell 环境变量。

代码清单 6-9：**ospath_expandvars.py**

```
import os.path
import os

os.environ['MYVAR'] = 'VALUE'

print(os.path.expandvars('/path/to/$MYVAR'))
```

这里不会完成任何验证来确保变量值能够得到真正存在的文件名。

```
$ python3 ospath_expandvars.py

/path/to/VALUE
```

6.1.3 规范化路径

使用 join() 或利用嵌入变量由单独的字符串组合路径时，得到的路径最后可能会有多余的分隔符或相对路径部分。使用 normpath() 可以清除这些内容。

代码清单 6-10：**ospath_normpath.py**

```
import os.path

PATHS = [
    'one//two//three',
    'one/./two/./three',
    'one/../alt/two/three',
]

for path in PATHS:
    print('{!r:>22} : {!r}'.format(path, os.path.normpath(path)))
```

这里会估算并折叠 os.curdir 和 os.pardir 构成的路径段。

```
$ python3 ospath_normpath.py

    'one//two//three' : 'one/two/three'
   'one/./two/./three' : 'one/two/three'
'one/../alt/two/three' : 'alt/two/three'
```

要把一个相对路径转换为一个绝对文件名，可以使用 abspath()。

代码清单 6-11: **ospath_abspath.py**

```
import os
import os.path

os.chdir('/usr')

PATHS = [
    '.',
    '..',
    './one/two/three',
    '../one/two/three',
]

for path in PATHS:
    print('{!r:>21} : {!r}'.format(path, os.path.abspath(path)))
```

结果是一个从文件系统树最顶层开始的完整的路径。

```
$ python3 ospath_abspath.py

                  '.' : '/usr'
                 '..' : '/'
    './one/two/three' : '/usr/one/two/three'
   '../one/two/three' : '/one/two/three'
```

6.1.4 文件时间

除了处理路径, `os.path` 还包括一些用于获取文件属性的函数,类似于 `os.stat()` 返回的结果。

代码清单 6-12: **ospath_properties.py**

```
import os.path
import time

print('File         :', __file__)
print('Access time  :', time.ctime(os.path.getatime(__file__)))
print('Modified time:', time.ctime(os.path.getmtime(__file__)))
print('Change time  :', time.ctime(os.path.getctime(__file__)))
print('Size         :', os.path.getsize(__file__))
```

`os.path.getatime()` 返回访问时间, `os.path.getmtime()` 返回修改时间, `os.path.getctime()` 返回创建时间。`os.path.getsize()` 返回文件中的数据量,以字节为单位表示。

```
$ python3 ospath_properties.py

File          : ospath_properties.py
Access time   : Fri Aug 26 16:38:05 2016
Modified time : Fri Aug 26 15:50:48 2016
Change time   : Fri Aug 26 15:50:49 2016
Size          : 481
```

6.1.5 测试文件

程序在遇到一个路径名时,通常需要知道这个路径指示的是一个文件、目录还是一个符号连接(symlink),另外还要知道它是否确实存在。`os.path` 包含了一些用于测试所有这些条件的函数。

代码清单 6-13：ospath_tests.py

```python
import os.path

FILENAMES = [
    __file__,
    os.path.dirname(__file__),
    '/',
    './broken_link',
]

for file in FILENAMES:
    print('File         : {!r}'.format(file))
    print('Absolute     :', os.path.isabs(file))
    print('Is File?     :', os.path.isfile(file))
    print('Is Dir?      :', os.path.isdir(file))
    print('Is Link?     :', os.path.islink(file))
    print('Mountpoint? :', os.path.ismount(file))
    print('Exists?      :', os.path.exists(file))
    print('Link Exists?:', os.path.lexists(file))
    print()
```

所有这些测试函数都返回布尔值。

```
$ ln -s /does/not/exist broken_link
$ python3 ospath_tests.py
File         : 'ospath_tests.py'
Absolute     : False
Is File?     : True
Is Dir?      : False
Is Link?     : False
Mountpoint? : False
Exists?      : True
Link Exists?: True

File         : ''
Absolute     : False
Is File?     : False
Is Dir?      : False
Is Link?     : False
Mountpoint? : False
Exists?      : False
Link Exists?: False

File         : '/'
Absolute     : True
Is File?     : False
Is Dir?      : True
Is Link?     : False
Mountpoint? : True
Exists?      : True
Link Exists?: True

File         : './broken_link'
Absolute     : False
Is File?     : False
Is Dir?      : False
Is Link?     : True
Mountpoint? : False
Exists?      : False
Link Exists?: True
```

提示：相关阅读材料
- `os.path` 的标准库文档[⊖]。
- `os.path` 的 Python 2 到 Python 3 移植说明。
- `pathlib`：路径作为对象。
- `os`：`os` 模块是 `os.path` 的父模块。
- `time`：`time` 模块包含有一些函数，可以在 `os.path` 中时间属性函数所用的表示与易读的字符串表示之间完成转换。

6.2 `pathlib`：文件系统路径作为对象

`pathlib` 模块提供了一个面向对象 API 来解析、建立、测试和处理文件名和路径，而不是使用底层字符串操作。

6.2.1 路径表示

`pathlib` 包含一些类来管理使用 POSIX 标准或 Microsoft Windows 语法格式化的文件系统路径。这个模块包含一些"纯"类，会处理字符串但不与实际的文件系统交互，另外还包含一些"具体"类，它们扩展了 API，以包含可以反映或修改本地文件系统上数据的操作。

纯类 `PurePosixPath` 和 `PureWindowsPath` 可以在任意操作系统上实例化和使用，因为它们只处理文件名和目录名。要实例化一个具体类来处理真正的文件系统，需要使用 `Path` 得到一个 `PosixPath` 或 `WindowsPath`，这取决于具体的平台。

6.2.2 建立路径

要实例化一个新路径，可以提供一个字符串作为第一个参数。路径对象的字符串表示就是这个名值。要创建一个新路径来指示相对于已有路径的一个值，可以使用 `/` 操作符扩展这个路径。这个操作符的参数可以是一个字符串，也可以是另一个路径对象。

代码清单 6-14：`pathlib_operator.py`

```
import pathlib

usr = pathlib.PurePosixPath('/usr')
print(usr)

usr_local = usr / 'local'
print(usr_local)

usr_share = usr / pathlib.PurePosixPath('share')
print(usr_share)

root = usr / '..'
print(root)

etc = root / '/etc/'
print(etc)
```

⊖ https://docs.python.org/3.5/library/os.path.html

如示例输出中 root 的值所示，这个操作符会组合所提供的路径值，但是在包含父目录引用 ".." 时没有对结果进行规范化。不过，如果一个路径段以路径分隔符开头，那么与 os.path.join() 中一样，会把它解释为一个新的"根"引用。会从路径值中间删除额外的路径分隔符，如这里的 etc 示例所示。

```
$ python3 pathlib_operator.py

/usr
/usr/local
/usr/share
/usr/..
/etc
```

具体路径类包括一个用于规范化路径 resolve() 方法，它会在文件系统中查找目录和符号链接，并生成一个名指示的绝对路径。

代码清单 6-15：**pathlib_resolve.py**

```python
import pathlib

usr_local = pathlib.Path('/usr/local')
share = usr_local / '..' / 'share'
print(share.resolve())
```

这里相对路径被转换为绝对路径 /usr/share。如果输入路径包含符号链接，那么它们也会展开，以使规范化的路径直接指示目标。

```
$ python3 pathlib_resolve.py

/usr/share
```

要想在路径段提前不可知时建立路径，可以使用 joinpath()，并传入各个路径段作为一个单独的参数。

代码清单 6-16：**pathlib_joinpath.py**

```python
import pathlib

root = pathlib.PurePosixPath('/')
subdirs = ['usr', 'local']
usr_local = root.joinpath(*subdirs)
print(usr_local)
```

与 / 操作符一样，调用 joinpath() 会创建一个新实例。

```
$ python3 pathlib_joinpath.py

/usr/local
```

给定一个现有的路径对象，可以很容易地建立一个与它稍有差别的新对象，如指示同一个目录中的另一个文件。可以使用 with_name() 创建一个新路径，将一个路径中的名部分替换为另一个不同的文件名。使用 with_suffix() 也可以创建一个新路径，将文件名的扩展名替换为一个不同的值。

代码清单 6-17：**pathlib_from_existing.py**

```
import pathlib

ind = pathlib.PurePosixPath('source/pathlib/index.rst')
print(ind)

py = ind.with_name('pathlib_from_existing.py')
print(py)

pyc = py.with_suffix('.pyc')
print(pyc)
```

这两个方法都返回新对象，原来的对象仍保持不变。

```
$ python3 pathlib_from_existing.py

source/pathlib/index.rst
source/pathlib/pathlib_from_existing.py
source/pathlib/pathlib_from_existing.pyc
```

6.2.3 解析路径

路径对象提供了一些方法和属性可以从路径名中抽取出部分值。例如，**parts** 属性可以生成根据路径分隔符值解析得到的一个路径段序列。

代码清单 6-18：**pathlib_parts.py**

```
import pathlib

p = pathlib.PurePosixPath('/usr/local')
print(p.parts)
```

这个序列是一个元组，反映了路径实例的不可变性。

```
$ python3 pathlib_parts.py

('/', 'usr', 'local')
```

有两种办法可以从一个给定的路径对象在文件系统层次结构中"向上"导航。**parent** 属性指示一个新的路径实例，对应包含给定路径（**os.path.dirname()** 返回的值）的目录。**parents** 属性是一个迭代器，会生成一系列父目录引用，在路径层次结构中不断"向上"，直到到达文件系统的根目录。

代码清单 6-19：**pathlib_parents.py**

```
import pathlib

p = pathlib.PurePosixPath('/usr/local/lib')

print('parent: {}'.format(p.parent))

print('\nhierarchy:')
for up in p.parents:
    print(up)
```

这个例子迭代处理 **parents** 属性并打印成员值。

```
$ python3 pathlib_parents.py

parent: /usr/local

hierarchy:
/usr/local
/usr
/
```

可以通过路径对象的属性来访问路径的其他部分。Name 属性包含路径的最后一部分，即最后一个路径分隔符后面的部分（与 os.path.basename() 生成的值相同）。suffix 属性包含扩展名分隔符后面的值，stem 属性包含名字中后缀之前的部分。

代码清单 6-20：**pathlib_name.py**

```python
import pathlib

p = pathlib.PurePosixPath('./source/pathlib/pathlib_name.py')
print('path  : {}'.format(p))
print('name  : {}'.format(p.name))
print('suffix: {}'.format(p.suffix))
print('stem  : {}'.format(p.stem))
```

尽管 suffix 和 stem 的值与 os.path.splitext() 生成的值类似，但这些值只是基于 name 的值，而不是完整路径。

```
$ python3 pathlib_name.py

path   : source/pathlib/pathlib_name.py
name   : pathlib_name.py
suffix : .py
stem   : pathlib_name
```

6.2.4 创建具体路径

可以由字符串参数创建具体 Path 类的实例，字符串参数可能指示文件系统中一个文件、目录或符号链接的名字（或可能的名字）。这个类还提供了很多便利方法，可以使用常用位置（如当前工作目录和用户的主目录）建立路径实例，这些常用位置可能会改变。

代码清单 6-21：**pathlib_convenience.py**

```python
import pathlib

home = pathlib.Path.home()
print('home: ', home)

cwd = pathlib.Path.cwd()
print('cwd : ', cwd)
```

这两个方法都会创建预填充一个绝对文件系统引用的 Path 实例。

```
$ python3 pathlib_convenience.py

home:  /Users/dhellmann
cwd :  /Users/dhellmann/PyMOTW
```

6.2.5 目录内容

可以使用 3 个方法来访问目录列表以及发现文件系统中的文件名。`iterdir()` 是一个生成器，会为包含目录中的每个元素生成一个新的 `Path` 实例。

代码清单 6-22：**pathlib_iterdir.py**

```
import pathlib

p = pathlib.Path('.')

for f in p.iterdir():
    print(f)
```

如果 `Path` 不指示一个目录，则 `iterdir()` 会产生 `NotADirectoryError`。

```
$ python3 pathlib_iterdir.py

example_link
index.rst
pathlib_chmod.py
pathlib_convenience.py
pathlib_from_existing.py
pathlib_glob.py
pathlib_iterdir.py
pathlib_joinpath.py
pathlib_mkdir.py
pathlib_name.py
pathlib_operator.py
pathlib_ownership.py
pathlib_parents.py
pathlib_parts.py
pathlib_read_write.py
pathlib_resolve.py
pathlib_rglob.py
pathlib_rmdir.py
pathlib_stat.py
pathlib_symlink_to.py
pathlib_touch.py
pathlib_types.py
pathlib_unlink.py
```

可以使用 `glob()` 找出与一个模式匹配的文件。

代码清单 6-23：**pathlib_glob.py**

```
import pathlib

p = pathlib.Path('..')

for f in p.glob('*.rst'):
    print(f)
```

这个例子展示了脚本父目录中的所有 reStructuredText[①]输入文件。

```
$ python3 pathlib_glob.py

../about.rst
```

---

① http://docutils.sourceforge.net/

```
../algorithm_tools.rst
../book.rst
../compression.rst
../concurrency.rst
../cryptographic.rst
../data_structures.rst
../dates.rst
../dev_tools.rst
../email.rst
../file_access.rst
../frameworks.rst
../i18n.rst
../importing.rst
../index.rst
../internet_protocols.rst
../language.rst
../networking.rst
../numeric.rst
../persistence.rst
../porting_notes.rst
../runtime_services.rst
../text.rst
../third_party.rst
../unix.rst
```

glob 处理器支持使用模式前缀 ** 或者通过调用 rglob() 而不是 glob() 来完成递归扫描。

代码清单 6-24：pathlib_rglob.py

```python
import pathlib

p = pathlib.Path('..')

for f in p.rglob('pathlib_*.py'):
    print(f)
```

由于这个例子从父目录开始，所以必须通过一个递归搜索来查找与 pathlib_*.py 匹配的示例文件。

```
$ python3 pathlib_rglob.py

../pathlib/pathlib_chmod.py
../pathlib/pathlib_convenience.py
../pathlib/pathlib_from_existing.py
../pathlib/pathlib_glob.py
../pathlib/pathlib_iterdir.py
../pathlib/pathlib_joinpath.py
../pathlib/pathlib_mkdir.py
../pathlib/pathlib_name.py
../pathlib/pathlib_operator.py
../pathlib/pathlib_ownership.py
../pathlib/pathlib_parents.py
../pathlib/pathlib_parts.py
../pathlib/pathlib_read_write.py
../pathlib/pathlib_resolve.py
../pathlib/pathlib_rglob.py
../pathlib/pathlib_rmdir.py
../pathlib/pathlib_stat.py
../pathlib/pathlib_symlink_to.py
../pathlib/pathlib_touch.py
../pathlib/pathlib_types.py
../pathlib/pathlib_unlink.py
```

6.2.6 读写文件

每个 Path 实例都包含一些方法来处理所指示文件的内容。要直接获取内容，可以使用 read_bytes() 或 read_text()。要写入文件，可以使用 write_bytes() 或 write_text()。可以使用 open() 方法打开文件并保留文件句柄，而不是向内置的 open() 函数传入文件名。

代码清单 6-25：**pathlib_read_write.py**

```python
import pathlib

f = pathlib.Path('example.txt')

f.write_bytes('This is the content'.encode('utf-8'))

with f.open('r', encoding='utf-8') as handle:
    print('read from open(): {!r}'.format(handle.read()))

print('read_text(): {!r}'.format(f.read_text('utf-8')))
```

这些便利方法会在打开文件和写入文件之前完成一些类型检查，除此之外，它们与直接操作是等价的。

```
$ python3 pathlib_read_write.py

read from open(): 'This is the content'
read_text(): 'This is the content'
```

6.2.7 管理目录和符号链接

可以用表示不存在的目录或符号链接的路径来创建关联的文件系统项。

代码清单 6-26：**pathlib_mkdir.py**

```python
import pathlib

p = pathlib.Path('example_dir')

print('Creating {}'.format(p))
p.mkdir()
```

如果这个路径已经存在，则 mkdir() 会产生一个 FileExistsError。

```
$ python3 pathlib_mkdir.py

Creating example_dir

$ python3 pathlib_mkdir.py

Creating example_dir
Traceback (most recent call last):
  File "pathlib_mkdir.py", line 16, in <module>
    p.mkdir()
  File ".../lib/python3.5/pathlib.py", line 1214, in mkdir
    self._accessor.mkdir(self, mode)
  File ".../lib/python3.5/pathlib.py", line 371, in wrapped
    return strfunc(str(pathobj), *args)
FileExistsError: [Errno 17] File exists: 'example_dir'
```

可以使用symlink_to()创建一个符号链接。这个链接根据路径的值命名，将指示symlink_to()参数给定的名字。

代码清单6-27：**pathlib_symlink_to.py**

```python
import pathlib

p = pathlib.Path('example_link')

p.symlink_to('index.rst')

print(p)
print(p.resolve().name)
```

这个例子首先创建一个符号链接，然后使用resolve()读取这个链接来找出它指示的名字，并打印这个名字。

```
$ python3 pathlib_symlink_to.py

example_link
index.rst
```

6.2.8 文件类型

Path实例包含一些方法来检查路径指示的文件的类型。下面这个例子会创建多个不同类型的文件并测试它们，还会测试本地操作系统上的其他设备特定的文件。

代码清单6-28：**pathlib_types.py**

```python
import itertools
import os
import pathlib
root = pathlib.Path('test_files')

# Clean up from previous runs.
if root.exists():
    for f in root.iterdir():
        f.unlink()
else:
    root.mkdir()

# Create test files.
(root / 'file').write_text(
    'This is a regular file', encoding='utf-8')
(root / 'symlink').symlink_to('file')
os.mkfifo(str(root / 'fifo'))

# Check the file types.
to_scan = itertools.chain(
    root.iterdir(),
    [pathlib.Path('/dev/disk0'),
     pathlib.Path('/dev/console')],
)
hfmt = '{:18s}' + ('  {:>5}' * 6)
print(hfmt.format('Name', 'File', 'Dir', 'Link', 'FIFO', 'Block',
                  'Character'))
print()

fmt = '{:20s}  ' + ('{!r:>5}  ' * 6)
```

```
    for f in to_scan:
        print(fmt.format(
            str(f),
            f.is_file(),
            f.is_dir(),
            f.is_symlink(),
            f.is_fifo(),
            f.is_block_device(),
            f.is_char_device(),
        ))
```

所有这些方法（is_dir()、is_file()、is_symlink()、is_socket()、is_fifo()、is_block_device() 和 is_char_device()）都不带参数。

```
$ python3 pathlib_types.py

Name                File    Dir     Link    FIFO    Block   Character
test_files/fifo     False   False   False   True    False   False
test_files/file     True    False   False   False   False   False
test_files/symlink  True    False   True    False   False   False
/dev/disk0          False   False   False   False   True    False
/dev/console        False   False   False   False   False   True
```

6.2.9 文件属性

可以使用方法 stat() 和 lstat() 来访问文件的有关详细信息（lstat() 用于检查一个可能是符号链接的目标的状态）。这些方法生成的结果分别与 os.stat() 和 os.lstat() 相同。

代码清单 6-29：**pathlib_stat.py**

```python
import pathlib
import sys
import time

if len(sys.argv) == 1:
    filename = __file__
else:
    filename = sys.argv[1]

p = pathlib.Path(filename)
stat_info = p.stat()

print('{}:'.format(filename))
print('  Size:', stat_info.st_size)
print('  Permissions:', oct(stat_info.st_mode))
print('  Owner:', stat_info.st_uid)
print('  Device:', stat_info.st_dev)
print('  Created      :', time.ctime(stat_info.st_ctime))
print('  Last modified:', time.ctime(stat_info.st_mtime))
print('  Last accessed:', time.ctime(stat_info.st_atime))
```

取决于在哪里安装这个示例代码，输出可能有变化。试着在命令行上向 pathlib_stat.py 传入不同的文件名。

```
$ python3 pathlib_stat.py

pathlib_stat.py:
  Size: 607
  Permissions: 0o100644
  Owner: 527
```

```
  Device: 16777218
  Created      : Thu Dec 29 12:25:25 2016
  Last modified: Thu Dec 29 12:25:25 2016
  Last accessed: Thu Dec 29 12:25:34 2016

$ python3 pathlib_stat.py index.rst
index.rst:
  Size: 19363
  Permissions: 0o100644
  Owner: 527
  Device: 16777218
  Created      : Thu Dec 29 11:27:58 2016
  Last modified: Thu Dec 29 11:27:58 2016
  Last accessed: Thu Dec 29 12:25:33 2016
```

如果想更简单地访问文件所有者的信息，则可以使用 `owner()` 和 `group()`。

代码清单 6-30：pathlib_ownership.py

```
import pathlib

p = pathlib.Path(__file__)

print('{} is owned by {}/{}'.format(p, p.owner(), p.group()))
```

`stat()` 返回系统 ID 值（数值），而这些方法则会查找与 ID 关联的名。

```
$ python3 pathlib_ownership.py

pathlib_ownership.py is owned by dhellmann/dhellmann
```

`touch()` 方法的做法与 UNIX 命令 touch 类似，可以创建一个文件，或者更新一个现有文件的修改时间和权限。

代码清单 6-31：pathlib_touch.py

```
import pathlib
import time

p = pathlib.Path('touched')
if p.exists():
    print('already exists')
else:
    print('creating new')

p.touch()
start = p.stat()

time.sleep(1)

p.touch()
end = p.stat()

print('Start:', time.ctime(start.st_mtime))
print('End  :', time.ctime(end.st_mtime))
```

多次运行这个例子时，后续的运行会更新现有文件。

```
$ python3 pathlib_touch.py

creating new
```

```
Start: Thu Dec 29 12:25:34 2016
End  : Thu Dec 29 12:25:35 2016

$ python3 pathlib_touch.py

already exists
Start: Thu Dec 29 12:25:35 2016
End  : Thu Dec 29 12:25:36 2016
```

6.2.10 权限

在类 UNIX 系统上，可以用 chmod() 更改文件权限，模式作为整数传入。模式值可以使用 stat 模块中定义的常量来构造。下面这个例子会反转用户的执行权限位。

代码清单 6-32：**pathlib_chmod.py**

```python
import os
import pathlib
import stat

# Create a fresh test file.
f = pathlib.Path('pathlib_chmod_example.txt')
if f.exists():
    f.unlink()
f.write_text('contents')

# Determine which permissions are already set using stat.
existing_permissions = stat.S_IMODE(f.stat().st_mode)
print('Before: {:o}'.format(existing_permissions))

# Decide which way to toggle them.
if not (existing_permissions & os.X_OK):
    print('Adding execute permission')
    new_permissions = existing_permissions | stat.S_IXUSR
else:
    print('Removing execute permission')
    # Use xor to remove the user execute permission.
    new_permissions = existing_permissions ^ stat.S_IXUSR

# Make the change and show the new value.
f.chmod(new_permissions)
after_permissions = stat.S_IMODE(f.stat().st_mode)
print('After: {:o}'.format(after_permissions))
```

这个脚本假设运行时其有修改文件模式所需的权限。

```
$ python3 pathlib_chmod.py

Before: 644
Adding execute permission
After: 744
```

6.2.11 删除

提供了两个方法来删除文件系统中的对象，使用哪一个方法取决于具体的类型。要删除一个空目录，可以使用 rmdir()。

代码清单 6-33：**pathlib_rmdir.py**

```python
import pathlib
```

```
p = pathlib.Path('example_dir')

print('Removing {}'.format(p))
p.rmdir()
```

如果后置条件已经满足而目录不存在,则会产生一个 FileNotFoundError 异常。如果试图删除一个不为空的目录,则也会出现错误。

```
$ python3 pathlib_rmdir.py

Removing example_dir

$ python3 pathlib_rmdir.py

Removing example_dir
Traceback (most recent call last):
  File "pathlib_rmdir.py", line 16, in <module>
    p.rmdir()
  File ".../lib/python3.5/pathlib.py", line 1262, in rmdir
    self._accessor.rmdir(self)
  File ".../lib/python3.5/pathlib.py", line 371, in wrapped
    return strfunc(str(pathobj), *args)
FileNotFoundError: [Errno 2] No such file or directory:
'example_dir'
```

对于文件、符号链接和大多数其他路径类型,可以使用 unlink()。

代码清单 6-34:**pathlib_unlink.py**

```
import pathlib

p = pathlib.Path('touched')

p.touch()

print('exists before removing:', p.exists())

p.unlink()

print('exists after removing:', p.exists())
```

用户必须有删除文件、符号链接、套接字或其他文件系统对象的权限。

```
$ python3 pathlib_unlink.py

exists before removing: True
exists after removing: False
```

提示:相关阅读材料

- pathlib 的标准库文档[○]。
- os.path:独立于平台的文件名管理。
- Managing File System Permissions:对 os.stat() 和 os.lstat() 的讨论。
- glob:UNIX shell 对文件名的模式匹配。
- PEP 428[○]:pathlib 模块。

[○] https://docs.python.org/3.5/library/pathlib.html
[○] www.python.org/dev/peps/pep-0428

6.3 glob：文件名模式匹配

尽管 glob API 很小，但这个模块的功能却很强大。只要程序需要查找文件系统中名字与某个模式匹配的一组文件，就可以使用这个模块。要创建一个文件名列表，要求其中各个文件名都有某个特定的扩展名、前缀或者中间都有某个共同的字符串，就可以使用 glob 而不用编写定制代码来扫描目录内容。

glob 的模式规则与 re 模块使用的正则表达式并不相同。实际上，glob 的模式遵循标准 UNIX 路径扩展规则。只使用几个特殊字符来实现两个不同的通配符和字符区间。模式规则应用于文件名中的段（在路径分隔符 / 处截止）。模式中的路径可以是相对路径或绝对路径。shell 变量名和波浪线（~）都不会被扩展。

6.3.1 示例数据

这一节中的例子假设当前工作目录中有以下测试文件。

```
$ python3 glob_maketestdata.py

dir
dir/file.txt
dir/file1.txt
dir/file2.txt
dir/filea.txt
dir/fileb.txt
dir/file?.txt
dir/file*.txt
dir/file[.txt
dir/subdir
dir/subdir/subfile.txt
```

如果这些文件不存在，那么在运行以下例子之前，请使用示例代码中的 glob_maketestdata.py 来创建这些文件。

6.3.2 通配符

星号（*）匹配一个文件名段中的 0 个或多个字符。例如，dir/*。

代码清单 6-35：glob_asterisk.py

```
import glob
for name in sorted(glob.glob('dir/*')):
    print(name)
```

这个模式会匹配目录 dir 中的所有路径名（文件或目录），但不会进一步递归搜索到子目录。glob() 返回的数据不会排序，所以这里的示例会进行排序以便研究结果。

```
$ python3 glob_asterisk.py

dir/file*.txt
dir/file.txt
dir/file1.txt
dir/file2.txt
dir/file?.txt
dir/file[.txt
dir/filea.txt
dir/fileb.txt
dir/subdir
```

要列出子目录中的文件，必须把子目录包含在模式中。

代码清单 6-36：`glob_subdir.py`

```python
import glob

print('Named explicitly:')
for name in sorted(glob.glob('dir/subdir/*')):
    print('  {}'.format(name))

print('Named with wildcard:')
for name in sorted(glob.glob('dir/*/*')):
    print('  {}'.format(name))
```

前面显示的第一种情况显式列出了子目录名，第二种情况则依赖一个通配符来查找目录。

```
$ python3 glob_subdir.py

Named explicitly:
  dir/subdir/subfile.txt
Named with wildcard:
  dir/subdir/subfile.txt
```

在这里，两种做法的结果是一样的。如果还有另一个子目录，则通配符会匹配这两个子目录，并且两个子目录中的文件名都会出现在结果中。

6.3.3 单字符通配符

问号（?）也是一个通配符。它会匹配文件名中该位置的单个字符。

代码清单 6-37：`glob_question.py`

```python
import glob

for name in sorted(glob.glob('dir/file?.txt')):
    print(name)
```

前面的例子会匹配以 `file` 开头，然后是另外一个任意字符，最后以 `.txt` 结尾的所有文件名。

```
$ python3 glob_question.py

dir/file*.txt
dir/file1.txt
dir/file2.txt
dir/file?.txt
dir/file[.txt
dir/filea.txt
dir/fileb.txt
```

6.3.4 字符区间

如果使用字符区间（[a-z]）而不是问号，则可以匹配多个字符中的一个字符。下面这个例子会查找名字中扩展名前有一个数字的所有文件。

代码清单 6-38：`glob_charrange.py`

```python
import glob
for name in sorted(glob.glob('dir/*[0-9].*')):
    print(name)
```

字符区间 [0-9] 会匹配所有单个数字。区间根据各字母 / 数字的字符码排序，短横线指示连续字符组成的一个不间断区间。这个区间值也可以写为 [0123456789]。

```
$ python3 glob_charrange.py

dir/file1.txt
dir/file2.txt
```

6.3.5 转义元字符

有时有必要搜索名字中包含一些特殊元字符的文件，glob 使用这些特殊元字符表示模式。escape() 函数会建立一个合适的模式，其中的特殊字符会被"转义"，使它们不会被 glob 扩展或解释为特殊字符。

代码清单 6-39：**glob_escape.py**

```python
import glob

specials = '?*['

for char in specials:
    pattern = 'dir/*' + glob.escape(char) + '.txt'
    print('Searching for: {!r}'.format(pattern))
    for name in sorted(glob.glob(pattern)):
        print(name)
    print()
```

可以通过构建一个包含单个元素的字符区间来转义各个特殊字符。

```
$ python3 glob_escape.py

Searching for: 'dir/*[?].txt'
dir/file?.txt

Searching for: 'dir/*[*].txt'
dir/file*.txt

Searching for: 'dir/*[[].txt'
dir/file[.txt
```

提示：相关阅读材料

- glob 的标准库文档[⊖]。
- Pattern Matching Notation[⊖]：Open Group 的 shell 命令语言规范中对文件名模式匹配的解释。
- fnmatch：文件名匹配实现。
- glob 的 Python 2 到 Python 3 移植说明。

6.4 fnmatch：UNIX 式 glob 模式匹配

fnmatch 模块用于根据 glob 模式（如 UNIX shell 所用的模式）比较文件名。

6.4.1 简单匹配

fnmatch() 将一个文件名与一个模式进行比较，并返回一个布尔值，指示二者是否匹

⊖ https://docs.python.org/3.5/library/glob.html
⊖ www.opengroup.org/onlinepubs/000095399/utilities/xcu_chap02.html#tag_02_13

配。如果操作系统使用一个区分大小写的文件系统，则这个比较就是区分大小写的。

代码清单6-40：**fnmatch_fnmatch.py**

```
import fnmatch
import os

pattern = 'fnmatch_*.py'
print('Pattern :', pattern)
print()

files = os.listdir('.')
for name in files:
    print('Filename: {:<25} {}'.format(
        name, fnmatch.fnmatch(name, pattern)))
```

在这个例子中，这个模式会匹配所有以 `'fnmatch_'` 开头并以 `'.py'` 结尾的文件。

```
$ python3 fnmatch_fnmatch.py

Pattern : fnmatch_*.py

Filename: fnmatch_filter.py         True
Filename: fnmatch_fnmatch.py        True
Filename: fnmatch_fnmatchcase.py    True
Filename: fnmatch_translate.py      True
Filename: index.rst                 False
```

要强制完成一个区分大小写的比较，而不论文件系统和操作系统如何设置，可以使用 `fnmatchcase()`。

代码清单6-41：**fnmatch_fnmatchcase.py**

```
import fnmatch
import os

pattern = 'FNMATCH_*.PY'
print('Pattern :', pattern)
print()

files = os.listdir('.')

for name in files:
    print('Filename: {:<25} {}'.format(
        name, fnmatch.fnmatchcase(name, pattern)))
```

由于测试这个程序所用的 OS X 系统使用的是区分大小写的文件系统，所以模式修改后不会匹配任何文件。

```
$ python3 fnmatch_fnmatchcase.py

Pattern : FNMATCH_*.PY

Filename: fnmatch_filter.py         False
Filename: fnmatch_fnmatch.py        False
Filename: fnmatch_fnmatchcase.py    False
Filename: fnmatch_translate.py      False
Filename: index.rst                 False
```

6.4.2 过滤

要测试一个文件名序列，可以使用 `filter()`，它会返回与模式参数匹配的文件名列表。

代码清单 6-42：**fnmatch_filter.py**

```
import fnmatch
import os
import pprint

pattern = 'fnmatch_*.py'
print('Pattern :', pattern)

files = os.listdir('.')

print('\nFiles   :')
pprint.pprint(files)

print('\nMatches :')
pprint.pprint(fnmatch.filter(files, pattern))
```

在这个例子中，filter() 返回了与这一节关联的示例源文件的文件名列表。

```
$ python3 fnmatch_filter.py

Pattern : fnmatch_*.py

Files   :
['fnmatch_filter.py',
 'fnmatch_fnmatch.py',
 'fnmatch_fnmatchcase.py',
 'fnmatch_translate.py',
 'index.rst']

Matches :
['fnmatch_filter.py',
 'fnmatch_fnmatch.py',
 'fnmatch_fnmatchcase.py',
 'fnmatch_translate.py']
```

6.4.3 转换模式

在内部，fnmatch 将 glob 模式转换为一个正则表达式，并使用 re 模块比较文件名和模式。translate() 函数是将 glob 模式转换为正则表达式的公共 API。

代码清单 6-43：**fnmatch_translate.py**

```
import fnmatch

pattern = 'fnmatch_*.py'
print('Pattern :', pattern)
print('Regex   :', fnmatch.translate(pattern))
```

要建立一个合法的表达式，需要对一些字符进行转义。

```
$ python3 fnmatch_translate.py

Pattern : fnmatch_*.py
Regex   : fnmatch_.*\.py\Z(?ms)
```

提示：相关阅读材料

- fnmatch 的标准库文档[⊖]。

⊖ https://docs.python.org/3.5/library/fnmatch.html

- glob：glob 模块结合使用 fnmatch 匹配和 os.listdir() 来生成与模式匹配的文件和目录列表。
- re：正则表达式模式匹配。

6.5 linecache：高效读取文本文件

处理 Python 源文件时，在 Python 标准库的其他部分中用到了 linecache 模块。缓存实现将在内存中保存文件的内容（解析为单独的行）。这个 API 通过索引一个 list 来返回所请求的行，与反复地读取文件并解析文本来查找所需文本行相比，这样可以节省时间。这个模块在查找同一个文件中的多行时尤其有用，比如为一个错误报告生成一个跟踪记录 (traceback)。

6.5.1 测试数据

以下文本由一个 Lorem Ipsum 生成器生成，该生成器将作为后面的示例输入。

代码清单 6-44：linecache_data.py

```
import os
import tempfile

lorem = '''Lorem ipsum dolor sit amet, consectetuer
adipiscing elit.  Vivamus eget elit. In posuere mi non
risus. Mauris id quam posuere lectus sollicitudin
varius. Praesent at mi. Nunc eu velit. Sed augue massa,
fermentum id, nonummy a, nonummy sit amet, ligula. Curabitur
eros pede, egestas at, ultricies ac, apellentesque eu,
tellus.

Sed sed odio sed mi luctus mollis. Integer et nulla ac augue
convallis accumsan. Ut felis. Donec lectus sapien, elementum
nec, condimentum ac, interdum non, tellus. Aenean viverra,
mauris vehicula semper porttitor, ipsum odio consectetuer
lorem, ac imperdiet eros odio a sapien. Nulla mauris tellus,
aliquam non, egestas a, nonummy et, erat. Vivamus sagittis
porttitor eros.'''

def make_tempfile():
    fd, temp_file_name = tempfile.mkstemp()
    os.close(fd)
    with open(temp_file_name, 'wt') as f:
        f.write(lorem)
    return temp_file_name


def cleanup(filename):
    os.unlink(filename)
```

6.5.2 读取特定行

linecache 模块读取的文件行号从 1 开始，不过通常列表的数组索引会从 0 开始。

代码清单 6-45：**linecache_getline.py**

```
import linecache
from linecache_data import *

filename = make_tempfile()

# Pick out the same line from source and cache.
# (Notice that linecache counts from 1.)
print('SOURCE:')
print('{!r}'.format(lorem.split('\n')[4]))
print()
print('CACHE:')
print('{!r}'.format(linecache.getline(filename, 5)))

cleanup(filename)
```

返回的各行包括末尾的一个换行符。

```
$ python3 linecache_getline.py

SOURCE:
'fermentum id, nonummy a, nonummy sit amet, ligula. Curabitur'

CACHE:
'fermentum id, nonummy a, nonummy sit amet, ligula. Curabitur\n'
```

6.5.3　处理空行

返回值总是在行末尾包含一个换行符，所以如果文本行为空，则返回值就是一个换行符。

代码清单 6-46：**linecache_empty_line.py**

```
import linecache
from linecache_data import *

filename = make_tempfile()

# Blank lines include the newline.
print('BLANK : {!r}'.format(linecache.getline(filename, 8)))

cleanup(filename)
```

输入文件的第 8 行不包含任何文本。

```
$ python3 linecache_empty_line.py

BLANK : '\n'
```

6.5.4　错误处理

如果所请求的行号超出了文件中合法行号的范围，则 `getline()` 会返回一个空串。

代码清单 6-47：**linecache_out_of_range.py**

```
import linecache
from linecache_data import *

filename = make_tempfile()
```

```
# The cache always returns a string, and uses
# an empty string to indicate a line that does
# not exist.
not_there = linecache.getline(filename, 500)
print('NOT THERE: {!r} includes {} characters'.format(
    not_there, len(not_there)))

cleanup(filename)
```

输入文件只有 15 行，所以请求第 500 行就类似于试图越过文件末尾继续读文件。

```
$ python3 linecache_out_of_range.py

NOT THERE: '' includes 0 characters
```

读取一个不存在的文件时，也采用同样的方式处理。

代码清单 6-48：`linecache_missing_file.py`

```
import linecache

# Errors are even hidden if linecache cannot find the file.
no_such_file = linecache.getline(
    'this_file_does_not_exist.txt', 1,
)
print('NO FILE: {!r}'.format(no_such_file))
```

调用者试图读取数据时，这个模块不会产生异常。

```
$ python3 linecache_missing_file.py

NO FILE: ''
```

6.5.5 读取 Python 源文件

由于生成 traceback 跟踪记录时 `linecache` 使用得非常频繁，其关键特性之一是能够指定模块的基名在导入路径中查找 Python 源模块。

代码清单 6-49：`linecache_path_search.py`

```
import linecache
import os

# Look for the linecache module using
# the built-in sys.path search.
module_line = linecache.getline('linecache.py', 3)
print('MODULE:')
print(repr(module_line))
# Look at the linecache module source directly.
file_src = linecache.__file__
if file_src.endswith('.pyc'):
    file_src = file_src[:-1]
print('\nFILE:')
with open(file_src, 'r') as f:
    file_line = f.readlines()[2]
print(repr(file_line))
```

如果 `linecache` 中的缓存填充代码在当前目录中无法找到指定名的文件，那么它会

在 `sys.path` 中搜索指定名的模块。这个例子要查找 `linecache.py`。由于当前目录中没有这个文件副本，所以会找到标准库中的相应文件。

```
$ python3 linecache_path_search.py

MODULE:
'This is intended to read lines from modules imported -- hence
if a filename\n'
FILE:
'This is intended to read lines from modules imported -- hence
if a filename\n'
```

提示：相关阅读材料

- `linecache` 的标准库文档⊖。

6.6　`tempfile`：临时文件系统对象

要想安全地创建名字唯一的临时文件，以防止被试图破坏应用或窃取数据的人猜出，这很有难度。`tempfile` 模块提供了多个函数来安全地创建临时文件系统资源。`Temporary-File()` 打开并返回一个未命名的文件，`NamedTemporaryFile()` 打开并返回一个命名文件，`SpooledTemporaryFile` 在将内容写入磁盘之前先将其保存在内存中，`Temporary-Directory` 是一个上下文管理器，上下文关闭时会删除这个目录。

6.6.1　临时文件

如果应用需要临时文件来存储数据，而不需要与其他程序共享这些文件，则应当使用 `TemporaryFile()` 函数创建文件。这个函数会创建一个文件，而且如果平台支持，它会立即断开这个新文件的链接。这样一来，其他程序就不可能找到或打开这个文件，因为文件系统表中根本没有这个文件的引用。对于 `TemporaryFile()` 创建的文件，不论通过调用 `close()` 还是结合使用上下文管理器 API 和 `with` 语句，关闭文件时都会自动删除这个文件。

代码清单 6-50：`tempfile_TemporaryFile.py`

```python
import os
import tempfile

print('Building a filename with PID:')
filename = '/tmp/guess_my_name.{}.txt'.format(os.getpid())
with open(filename, 'w+b') as temp:
    print('temp:')
    print('  {!r}'.format(temp))
    print('temp.name:')
    print('  {!r}'.format(temp.name))

# Clean up the temporary file yourself.
os.remove(filename)
```

⊖ https://docs.python.org/3.5/library/linecache.html

```
print()
print('TemporaryFile:')
with tempfile.TemporaryFile() as temp:
    print('temp:')
    print('   {!r}'.format(temp))
    print('temp.name:')
    print('   {!r}'.format(temp.name))

# Automatically cleans up the file
```

这个例子展示了采用不同方法创建临时文件的差别，一种做法是使用一个通用模式来构造临时文件的文件名，另一种做法是使用 TemporaryFile() 函数。TemporaryFile() 返回的文件没有文件名。

```
$ python3 tempfile_TemporaryFile.py

Building a filename with PID:
temp:
   <_io.BufferedRandom name='/tmp/guess_my_name.12151.txt'>
temp.name:
   '/tmp/guess_my_name.12151.txt'
TemporaryFile:
temp:
   <_io.BufferedRandom name=4>
temp.name:
   4
```

默认地，文件句柄是使用模式 'w+b' 创建的，以便它在所有平台上都表现一致，并允许调用者读写这个文件。

代码清单 6-51：tempfile_TemporaryFile_binary.py

```
import os
import tempfile

with tempfile.TemporaryFile() as temp:
    temp.write(b'Some data')

    temp.seek(0)
    print(temp.read())
```

写文件之后，必须使用 seek() "回转"文件句柄以便从文件读回数据。

```
$ python3 tempfile_TemporaryFile_binary.py

b'Some data'
```

要以文本模式打开文件，创建文件时要设置 mode 为 'w+t'。

代码清单 6-52：tempfile_TemporaryFile_text.py

```
import tempfile

with tempfile.TemporaryFile(mode='w+t') as f:
    f.writelines(['first\n', 'second\n'])

    f.seek(0)
    for line in f:
        print(line.rstrip())
```

这个文件句柄将把数据处理为文本。

```
$ python3 tempfile_TemporaryFile_text.py

first
second
```

6.6.2 命名文件

有些情况下,可能非常需要一个命名的临时文件。对于跨多个进程甚至主机的应用来说,为文件命名是在应用不同部分之间传递文件的最简单的方法。NamedTemporaryFile() 函数会创建一个文件,但不会断开它的链接,所以会保留它的文件名(用 name 属性访问)。

代码清单 6-53:**tempfile_NamedTemporaryFile.py**

```python
import os
import pathlib
import tempfile

with tempfile.NamedTemporaryFile() as temp:
    print('temp:')
    print('  {!r}'.format(temp))
    print('temp.name:')
    print('  {!r}'.format(temp.name))

    f = pathlib.Path(temp.name)

print('Exists after close:', f.exists())
```

句柄关闭后文件将被删除。

```
$ python3 tempfile_NamedTemporaryFile.py

temp:
  <tempfile._TemporaryFileWrapper object at 0x1011b2d30>
temp.name:
  '/var/folders/5q/8gk0wq888xlggz008k8dr7180000hg/T/tmps4qh5zde'
Exists after close: False
```

6.6.3 假脱机文件

如果临时文件中包含的数据相对较少,则使用 SpooledTemporaryFile 可能更高效,因为它使用一个 io.BytesIO 或 io.StringIO 缓冲区在内存中保存内容,直到数据达到一个阈值大小。当数据量超过这个阈值时,数据将"滚动"并写入磁盘,然后用常规的 TemporaryFile() 替换这个缓冲区。

代码清单 6-54:**tempfile_SpooledTemporaryFile.py**

```python
import tempfile

with tempfile.SpooledTemporaryFile(max_size=100,
                                   mode='w+t',
                                   encoding='utf-8') as temp:
    print('temp: {!r}'.format(temp))

    for i in range(3):
        temp.write('This line is repeated over and over.\n')
        print(temp._rolled, temp._file)
```

这个例子使用 `SpooledTemporaryFile` 的私有属性来确定何时滚动到磁盘。除非要调整缓冲区大小，否则很少需要检查这个状态。

```
$ python3 tempfile_SpooledTemporaryFile.py

temp: <tempfile.SpooledTemporaryFile object at 0x1007b2c88>
False <_io.StringIO object at 0x1007a3d38>
False <_io.StringIO object at 0x1007a3d38>
True <_io.TextIOWrapper name=4 mode='w+t' encoding='utf-8'>
```

要显式地将缓冲区写至磁盘，可以调用 `rollover()` 或 `fileno()` 方法。

代码清单 6-55：`tempfile_SpooledTemporaryFile_explicit.py`

```python
import tempfile

with tempfile.SpooledTemporaryFile(max_size=1000,
                                   mode='w+t',
                                   encoding='utf-8') as temp:
    print('temp: {!r}'.format(temp))

    for i in range(3):
        temp.write('This line is repeated over and over.\n')
        print(temp._rolled, temp._file)
    print('rolling over')
    temp.rollover()
    print(temp._rolled, temp._file)
```

在这个例子中，由于缓冲区非常大，远远大于实际的数据量，所以除非调用 `rollover()`，否则不会在磁盘上创建任何文件。

```
$ python3 tempfile_SpooledTemporaryFile_explicit.py

temp: <tempfile.SpooledTemporaryFile object at 0x1007b2c88>
False <_io.StringIO object at 0x1007a3d38>
False <_io.StringIO object at 0x1007a3d38>
False <_io.StringIO object at 0x1007a3d38>
rolling over
True <_io.TextIOWrapper name=4 mode='w+t' encoding='utf-8'>
```

6.6.4 临时目录

需要多个临时文件时，可能更方便的做法是用 `TemporaryDirectory` 创建一个临时目录，并打开该目录中的所有文件。

代码清单 6-56：`tempfile_TemporaryDirectory.py`

```python
import pathlib
import tempfile

with tempfile.TemporaryDirectory() as directory_name:
    the_dir = pathlib.Path(directory_name)
    print(the_dir)
    a_file = the_dir / 'a_file.txt'
    a_file.write_text('This file is deleted.')

print('Directory exists after?', the_dir.exists())
print('Contents after:', list(the_dir.glob('*')))
```

上下文管理器会生成目录名，可以在上下文块中用来建立其他文件名。

```
$ python3 tempfile_TemporaryDirectory.py

/var/folders/5q/8gk0wq888xlggz008k8dr7180000hg/T/tmp_urhiioj
Directory exists after? False
Contents after: []
```

6.6.5 预测名

虽然没有严格匿名的临时文件那么安全，但有时还需要在名字中包含一个可预测的部分，以便查找和检查文件来进行调试。目前为止介绍的所有函数都取 3 个参数，可以在某种程度上控制文件。文件名使用以下公式生成。

```
dir + prefix + random + suffix
```

除了 `random` 外，所有其他值都可以作为参数传递到这些函数以创建临时文件或目录。

代码清单 6-57：**tempfile_NamedTemporaryFile_args.py**

```python
import tempfile

with tempfile.NamedTemporaryFile(suffix='_suffix',
                                 prefix='prefix_',
                                 dir='/tmp') as temp:
    print('temp:')
    print('  ', temp)
    print('temp.name:')
    print('  ', temp.name)
```

前缀（`prefix`）和后缀（`suffix`）参数与一个随机的字符串结合来建立文件名，`dir` 参数保持不变，作为新文件的位置。

```
$ python3 tempfile_NamedTemporaryFile_args.py

temp:
   <tempfile._TemporaryFileWrapper object at 0x1018b2d68>
temp.name:
   /tmp/prefix_q6wd5czl_suffix
```

6.6.6 临时文件位置

如果没有使用 `dir` 参数指定明确的目标位置，则临时文件使用的路径会根据当前平台和设置而变化。`tempfile` 模块包括两个函数来查询运行时使用的设置。

代码清单 6-58：**tempfile_settings.py**

```python
import tempfile

print('gettempdir():', tempfile.gettempdir())
print('gettempprefix():', tempfile.gettempprefix())
```

`gettempdir()` 返回包含所有临时文件的默认目录，`gettempprefix()` 返回新文件和目录名的字符串前缀。

```
$ python3 tempfile_settings.py

gettempdir(): /var/folders/5q/8gk0wq888xlggz008k8dr7180000hg/T
gettempprefix(): tmp
```

gettempdir() 返回的值根据一个简单算法来设置，它会查找一个位置列表，寻找第一个允许当前进程创建文件的位置。搜索列表的顺序如下：
1. 环境变量 TMPDIR。
2. 环境变量 TEMP。
3. 环境变量 TMP。
4. 作为"后路"的位置，这取决于具体平台（Windows 使用 C:\temp、C:\tmp、\temp 或 \tmp 中第一个可用的位置。其他平台使用 /tmp、/var/tmp 或 /usr/tmp）。
5. 如果找不到其他目录，则使用当前工作目录。

代码清单 6-59：**tempfile_tempdir.py**

```
import tempfile

tempfile.tempdir = '/I/changed/this/path'
print('gettempdir():', tempfile.gettempdir())
```

如果程序需要对所有临时文件使用一个全局位置，但不使用以上任何环境变量，则应当直接设置 tempfile.tempdir，为该变量赋一个值。

```
$ python3 tempfile_tempdir.py

gettempdir(): /I/changed/this/path
```

提示：相关阅读材料
- tempfile 的标准库文档[⊖]。
- random：伪随机数生成器，用来在临时文件名中引入随机值。

6.7 shutil：高层文件操作

shutil 模块包括一些高层文件操作，如复制和归档。

6.7.1 复制文件

copyfile() 将源文件的内容复制到目标文件，如果没有权限写目标文件，则会产生 IOError。

代码清单 6-60：**shutil_copyfile.py**

```
import glob
import shutil

print('BEFORE:', glob.glob('shutil_copyfile.*'))

shutil.copyfile('shutil_copyfile.py', 'shutil_copyfile.py.copy')

print('AFTER:', glob.glob('shutil_copyfile.*'))
```

⊖ https://docs.python.org/3.5/library/tempfile.html

由于这个函数会打开输入文件进行读取，而不论其类型，所以某些特殊文件（如 UNIX 设备节点）不能用 `copyfile()` 复制为新的特殊文件。

```
$ python3 shutil_copyfile.py

BEFORE: ['shutil_copyfile.py']
AFTER: ['shutil_copyfile.py', 'shutil_copyfile.py.copy']
```

`copyfile()` 的实现使用了底层函数 `copyfileobj()`。`copyfile()` 的参数是文件名，但 `copyfileobj()` 的参数是打开的文件句柄。还可以有第三个参数（可选）：读入块使用的一个缓冲区长度。

代码清单 6-61：**shutil_copyfileobj.py**

```python
import io
import os
import shutil
import sys


class VerboseStringIO(io.StringIO):

    def read(self, n=-1):
        next = io.StringIO.read(self, n)
        print('read({}) got {} bytes'.format(n, len(next)))
        return next


lorem_ipsum = '''Lorem ipsum dolor sit amet, consectetuer
adipiscing elit.  Vestibulum aliquam mollis dolor. Donec
vulputate nunc ut diam. Ut rutrum mi vel sem. Vestibulum
ante ipsum.'''

print('Default:')
input = VerboseStringIO(lorem_ipsum)
output = io.StringIO()
shutil.copyfileobj(input, output)

print()

print('All at once:')
input = VerboseStringIO(lorem_ipsum)
output = io.StringIO()
shutil.copyfileobj(input, output, -1)

print()

print('Blocks of 256:')
input = VerboseStringIO(lorem_ipsum)
output = io.StringIO()
shutil.copyfileobj(input, output, 256)
```

默认行为是使用大数据块读取。使用 `-1` 会一次读入所有输入，或者也可以使用其他正数，这会设置特定的块大小。下面这个例子将使用多个不同的块大小来展示效果。

```
$ python3 shutil_copyfileobj.py

Default:
read(16384) got 166 bytes
```

```
read(16384) got 0 bytes

All at once:
read(-1) got 166 bytes
read(-1) got 0 bytes

Blocks of 256:
read(256) got 166 bytes
read(256) got 0 bytes
```

类似于 UNIX 命令行工具 cp，copy() 函数会用同样的方式解释输出名。如果指定的目标指示一个目录而不是一个文件，则会使用源文件的基名在该目录中创建一个新文件。

代码清单 6-62：shutil_copy.py

```python
import glob
import os
import shutil

os.mkdir('example')
print('BEFORE:', glob.glob('example/*'))

shutil.copy('shutil_copy.py', 'example')

print('AFTER :', glob.glob('example/*'))
```

一同复制文件的权限与内容。

```
$ python3 shutil_copy.py

BEFORE: []
AFTER : ['example/shutil_copy.py']
```

copy2() 的工作类似于 copy()，不过会在复制到新文件的元数据中包含访问和修改时间。

代码清单 6-63：shutil_copy2.py

```python
import os
import shutil
import time

def show_file_info(filename):
    stat_info = os.stat(filename)
    print('  Mode    :', oct(stat_info.st_mode))
    print('  Created :', time.ctime(stat_info.st_ctime))
    print('  Accessed:', time.ctime(stat_info.st_atime))
    print('  Modified:', time.ctime(stat_info.st_mtime))

os.mkdir('example')
print('SOURCE:')
show_file_info('shutil_copy2.py')

shutil.copy2('shutil_copy2.py', 'example')

print('DEST:')
show_file_info('example/shutil_copy2.py')
```

这个新文件的所有特性都与原文件完全相同。

```
$ python3 shutil_copy2.py

SOURCE:
  Mode    : 0o100644
  Created : Wed Dec 28 19:03:12 2016
```

```
    Accessed: Wed Dec 28 19:03:49 2016
    Modified: Wed Dec 28 19:03:12 2016
DEST:
    Mode    : 0o100644
    Created : Wed Dec 28 19:03:49 2016
    Accessed: Wed Dec 28 19:03:49 2016
    Modified: Wed Dec 28 19:03:12 2016
```

6.7.2 复制文件元数据

默认地，在 UNIX 下创建一个新文件时，它会根据当前用户的 umask 接受权限。要把权限从一个文件复制到另一个文件，可以使用 copymode()。

代码清单 6-64：**shutil_copymode.py**

```
import os
import shutil
import subprocess
with open('file_to_change.txt', 'wt') as f:
    f.write('content')
os.chmod('file_to_change.txt', 0o444)

print('BEFORE:', oct(os.stat('file_to_change.txt').st_mode))

shutil.copymode('shutil_copymode.py', 'file_to_change.txt')

print('AFTER :', oct(os.stat('file_to_change.txt').st_mode))
```

这个示例脚本创建了一个要修改的文件，然后使用 copymode() 将脚本的权限复制到示例文件。

```
$ python3 shutil_copymode.py

BEFORE: 0o100444
AFTER : 0o100644
```

要为文件复制其他元数据，可以使用 copystat()。

代码清单 6-65：**shutil_copystat.py**

```
import os
import shutil
import time

def show_file_info(filename):
    stat_info = os.stat(filename)
    print('  Mode    :', oct(stat_info.st_mode))
    print('  Created :', time.ctime(stat_info.st_ctime))
    print('  Accessed:', time.ctime(stat_info.st_atime))
    print('  Modified:', time.ctime(stat_info.st_mtime))

with open('file_to_change.txt', 'wt') as f:
    f.write('content')
os.chmod('file_to_change.txt', 0o444)

print('BEFORE:')
show_file_info('file_to_change.txt')

shutil.copystat('shutil_copystat.py', 'file_to_change.txt')

print('AFTER:')
show_file_info('file_to_change.txt')
```

使用 copystat() 只会复制与文件关联的权限和日期。

```
$ python3 shutil_copystat.py

BEFORE:
  Mode    : 0o100444
  Created : Wed Dec 28 19:03:49 2016
  Accessed: Wed Dec 28 19:03:49 2016
  Modified: Wed Dec 28 19:03:49 2016
AFTER:
  Mode    : 0o100644
  Created : Wed Dec 28 19:03:49 2016
  Accessed: Wed Dec 28 19:03:49 2016
  Modified: Wed Dec 28 19:03:46 2016
```

6.7.3 处理目录树

shutil 包含 3 个函数来处理目录树。要把一个目录从一个位置复制到另一个位置，可以使用 copytree()。这个函数会递归遍历源目录树，将文件复制到目标位置。目标目录必须不存在。

代码清单 6-66：**shutil_copytree.py**

```python
import glob
import pprint
import shutil

print('BEFORE:')
pprint.pprint(glob.glob('/tmp/example/*'))

shutil.copytree('../shutil', '/tmp/example')

print('\nAFTER:')
pprint.pprint(glob.glob('/tmp/example/*'))
```

symlinks 参数控制符号链接作为链接复制或文件复制。默认地会将内容复制到新文件。如果这个选项为 true，则会在目标树中创建新的符号链接。

```
$ python3 shutil_copytree.py

BEFORE:
[]

AFTER:
['/tmp/example/example',
 '/tmp/example/example.out',
 '/tmp/example/file_to_change.txt',
 '/tmp/example/index.rst',
 '/tmp/example/shutil_copy.py',
 '/tmp/example/shutil_copy2.py',
 '/tmp/example/shutil_copyfile.py',
 '/tmp/example/shutil_copyfile.py.copy',
 '/tmp/example/shutil_copyfileobj.py',
 '/tmp/example/shutil_copymode.py',
 '/tmp/example/shutil_copystat.py',
 '/tmp/example/shutil_copytree.py',
 '/tmp/example/shutil_copytree_verbose.py',
 '/tmp/example/shutil_disk_usage.py',
 '/tmp/example/shutil_get_archive_formats.py',
 '/tmp/example/shutil_get_unpack_formats.py',
```

```
    '/tmp/example/shutil_make_archive.py',
    '/tmp/example/shutil_move.py',
    '/tmp/example/shutil_rmtree.py',
    '/tmp/example/shutil_unpack_archive.py',
    '/tmp/example/shutil_which.py',
    '/tmp/example/shutil_which_regular_file.py']
```

`copytree()` 接受两个 callable 参数来控制它的行为。调用 `ignore` 参数时要提供复制的各个目录或子目录的名，以及一个目录内容列表。这个函数应当返回待复制元素的一个列表。另外调用 `copy_function` 参数来具体复制文件。

代码清单 6-67：**shutil_copytree_verbose.py**

```python
import glob
import pprint
import shutil


def verbose_copy(src, dst):
    print('copying\n {!r}\n to {!r}'.format(src, dst))
    return shutil.copy2(src, dst)


print('BEFORE:')
pprint.pprint(glob.glob('/tmp/example/*'))
print()

shutil.copytree(
    '../shutil', '/tmp/example',
    copy_function=verbose_copy,
    ignore=shutil.ignore_patterns('*.py'),
)

print('\nAFTER:')
pprint.pprint(glob.glob('/tmp/example/*'))
```

在这个例子中，使用了 `ignore_patterns()` 来创建一个 `ignore` 函数，要求不复制 Python 源文件。`verbose_copy()` 首先打印复制文件时的文件名；然后使用 `copy2()`（即默认的复制函数）来建立副本。

```
$ python3 shutil_copytree_verbose.py

BEFORE:
[]

copying
 '../shutil/example.out'
 to '/tmp/example/example.out'
copying
 '../shutil/file_to_change.txt'
 to '/tmp/example/file_to_change.txt'
copying
 '../shutil/index.rst'
 to '/tmp/example/index.rst'

AFTER:
['/tmp/example/example',
 '/tmp/example/example.out',
 '/tmp/example/file_to_change.txt',
 '/tmp/example/index.rst']
```

要删除一个目录以及其中的内容，可以使用 `rmtree()`。

代码清单 6-68：**shutil_rmtree.py**

```
import glob
import pprint
import shutil

print('BEFORE:')
pprint.pprint(glob.glob('/tmp/example/*'))

shutil.rmtree('/tmp/example')

print('\nAFTER:')
pprint.pprint(glob.glob('/tmp/example/*'))
```

默认地，如果出现错误，则会作为异常抛出，不过如果第二个参数为 true，则可以忽略这些错误。可以在第三个参数中提供一个特殊的错误处理函数。

```
$ python3 shutil_rmtree.py

BEFORE:
['/tmp/example/example',
 '/tmp/example/example.out',
 '/tmp/example/file_to_change.txt',
 '/tmp/example/index.rst']

AFTER:
[]
```

要把一个文件或目录从一个位置移动到另一个位置，可以使用 `move()`。

代码清单 6-69：**shutil_move.py**

```
import glob
import shutil

with open('example.txt', 'wt') as f:
    f.write('contents')

print('BEFORE: ', glob.glob('example*'))

shutil.move('example.txt', 'example.out')

print('AFTER : ', glob.glob('example*'))
```

其语义与 UNIX 命令 mv 类似。如果源和目标都在同一个文件系统中，则会重命名源文件。否则，源文件会被复制到目标文件，然后被删除。

```
$ python3 shutil_move.py

BEFORE: ['example.txt']
AFTER : ['example.out']
```

6.7.4 查找文件

`which()` 函数会扫描一个搜索路径以查找一个命名文件。典型的用法是在环境变量 PATH 定义的 shell 搜索路径中查找一个可执行程序。

代码清单 6-70：**shutil_which.py**

```
import shutil

print(shutil.which('virtualenv'))
print(shutil.which('tox'))
print(shutil.which('no-such-program'))
```

如果无法找到与搜索参数匹配的文件，则 `which()` 返回 `None`。

```
$ python3 shutil_which.py

/Users/dhellmann/Library/Python/3.5/bin/virtualenv
/Users/dhellmann/Library/Python/3.5/bin/tox
None
```

`which()` 接收参数，可以根据文件的权限和要检查的搜索路径来完成过滤。`path` 参数默认为 `os.environ('PATH')`，不过也可以是包含目录名并用 `os.pathsep` 分隔的任意字符串。`mode` 参数应当是与文件权限对应的一个位掩码。默认地，这个掩码会查找可执行文件，不过下面的例子使用了可读位掩码和另外一个搜索路径来查找一个配置文件。

代码清单 6-71：**shutil_which_regular_file.py**

```
import os
import shutil

path = os.pathsep.join([
    '.',
    os.path.expanduser('~/pymotw'),
])

mode = os.F_OK | os.R_OK

filename = shutil.which(
    'config.ini',
    mode=mode,
    path=path,
)

print(filename)
```

以这种方式搜索可读文件时，可能还会出现竞态条件，因为在找到文件和真正使用这个文件的间隙，这个文件可能被删除或者它的权限可能更改。

```
$ touch config.ini
$ python3 shutil_which_regular_file.py

./config.ini
```

6.7.5 归档

Python 的标准库包含很多模块来管理归档文件，如 `tarfile` 和 `zipfile`。另外，`shutil` 中还提供了很多更高层函数来创建和解压归档文件。`get_archive_formats()` 返回当前系统上支持的所有格式的名字和描述。

代码清单 6-72：**shutil_get_archive_formats.py**

```
import shutil

for format, description in shutil.get_archive_formats():
    print('{:<5}: {}'.format(format, description))
```

支持的格式取决于有哪些模块和底层库。因此，根据这个例子在哪里运行，它的输出可能会有变化。

```
$ python3 shutil_get_archive_formats.py

bztar: bzip2'ed tar-file
gztar: gzip'ed tar-file
tar  : uncompressed tar file
xztar: xz'ed tar-file
zip  : ZIP file
```

可以使用 make_archive() 创建一个新的归档文件。它的输入被设计为能最好地支持将整个目录及其所有内容递归地归档。默认地会使用当前工作目录，所以所有文件和子目录出现在归档文件的顶层。要改变这种行为，可以使用 root_dir 参数移至文件系统中的一个新的相对位置，另外可以用 base_dir 参数指定一个目录增加到归档。

代码清单 6-73：**shutil_make_archive.py**

```
import logging
import shutil
import sys
import tarfile

logging.basicConfig(
    format='%(message)s',
    stream=sys.stdout,
    level=logging.DEBUG,
)
logger = logging.getLogger('pymotw')

print('Creating archive:')
shutil.make_archive(
    'example', 'gztar',
    root_dir='..',
    base_dir='shutil',
    logger=logger,
)
print('\nArchive contents:')
with tarfile.open('example.tar.gz', 'r') as t:
    for n in t.getnames():
        print(n)
```

这个例子从 shutil 示例源文件所在的目录开始，在文件系统中上移一层；然后将这个 shutil 目录增加到一个 tar 归档，并用 gzip 压缩。logging 模块被配置为显示 make_archive() 中操作的有关消息。

```
$ python3 shutil_make_archive.py

Creating archive:
changing into '..'
```

```
Creating tar archive
changing back to '...'

Archive contents:
shutil
shutil/config.ini
shutil/example.out
shutil/file_to_change.txt
shutil/index.rst
shutil/shutil_copy.py
shutil/shutil_copy2.py
shutil/shutil_copyfile.py
shutil/shutil_copyfileobj.py
shutil/shutil_copymode.py
shutil/shutil_copystat.py
shutil/shutil_copytree.py
shutil/shutil_copytree_verbose.py
shutil/shutil_disk_usage.py
shutil/shutil_get_archive_formats.py
shutil/shutil_get_unpack_formats.py
shutil/shutil_make_archive.py
shutil/shutil_move.py
shutil/shutil_rmtree.py
shutil/shutil_unpack_archive.py
shutil/shutil_which.py
shutil/shutil_which_regular_file.py
```

shutil 维护了一个格式注册表，可以在当前系统上解包；可以通过 get_unpack_formats() 访问这个注册表。

代码清单 6-74：**shutil_get_unpack_formats.py**

```
import shutil

for format, exts, description in shutil.get_unpack_formats():
    print('{:<5}: {}, names ending in {}'.format(
        format, description, exts))
```

shutil 管理的注册表不同于创建归档所用的注册表，因为它还包括对应各种格式的常见文件扩展名。解压缩归档文件的函数使用这个注册表来根据文件扩展名猜测应当使用哪种格式。

```
$ python3 shutil_get_unpack_formats.py

bztar: bzip2'ed tar-file, names ending in ['.tar.bz2', '.tbz2']
gztar: gzip'ed tar-file, names ending in ['.tar.gz', '.tgz']
tar  : uncompressed tar file, names ending in ['.tar']
xztar: xz'ed tar-file, names ending in ['.tar.xz', '.txz']
zip  : ZIP file, names ending in ['.zip']
```

可以传入归档文件名用 unpack_archive() 解压缩归档文件，还可以传入要把它解压缩到哪个目录（可选）。如果没有指定目录，就使用当前目录。

代码清单 6-75：**shutil_unpack_archive.py**

```
import pathlib
import shutil
import sys
import tempfile
```

```
with tempfile.TemporaryDirectory() as d:
    print('Unpacking archive:')
    shutil.unpack_archive(
        'example.tar.gz',
        extract_dir=d,
    )

    print('\nCreated:')
    prefix_len = len(d) + 1
    for extracted in pathlib.Path(d).rglob('*'):
        print(str(extracted)[prefix_len:])
```

在这个例子中，`unpack_archive()` 能确定归档的格式，因为文件名以 `tar.gz` 结尾，并且这个值与 unpack 格式注册表中的 `gztar` 格式相关联。

```
$ python3 shutil_unpack_archive.py

Unpacking archive:

Created:
shutil
shutil/config.ini
shutil/example.out
shutil/file_to_change.txt
shutil/index.rst
shutil/shutil_copy.py
shutil/shutil_copy2.py
shutil/shutil_copyfile.py
shutil/shutil_copyfileobj.py
shutil/shutil_copymode.py
shutil/shutil_copystat.py
shutil/shutil_copytree.py
shutil/shutil_copytree_verbose.py
shutil/shutil_disk_usage.py
shutil/shutil_get_archive_formats.py
shutil/shutil_get_unpack_formats.py
shutil/shutil_make_archive.py
shutil/shutil_move.py
shutil/shutil_rmtree.py
shutil/shutil_unpack_archive.py
shutil/shutil_which.py
shutil/shutil_which_regular_file.py
```

6.7.6 文件系统空间

完成一个长时间运行的可能耗尽可用空间的操作之前，最好先检查本地文件系统，来看看有多少可用的空间，这会很有用。`disk_usage()` 会返回一个元组，包括总空间、当前正在使用的空间以及未用的空间（自由空间）。

代码清单 6-76：**shutil_disk_usage.py**

```
import shutil

total_b, used_b, free_b = shutil.disk_usage('.')

gib = 2 ** 30    # GiB == gibibyte
gb = 10 ** 9     # GB == gigabyte

print('Total: {:6.2f} GB   {:6.2f} GiB'.format(
```

```
    total_b / gb, total_b / gib))
print('Used  : {:6.2f} GB  {:6.2f} GiB'.format(
    used_b / gb, used_b / gib))
print('Free  : {:6.2f} GB  {:6.2f} GiB'.format(
    free_b / gb, free_b / gib))
```

`disk_usage()` 返回的值以字节为单位，所以这个示例程序在打印之前先把它转换为更可读的一些单位。

```
$ python3 shutil_disk_usage.py

Total: 499.42 GB  465.12 GiB
Used : 246.68 GB  229.73 GiB
Free : 252.48 GB  235.14 GiB
```

提示：相关阅读材料
- `shutil` 的标准库文档[⊖]。
- 第 8 章：处理归档和压缩格式的模块。

6.8 `filecmp`：比较文件

`filecmp` 模块提供了一些函数和一个类来比较文件系统上的文件和目录。

6.8.1 示例数据

以下讨论中的例子使用了 `filecmp_mkexamples.py` 创建的一组测试文件。

代码清单 6-77：`filecmp_mkexamples.py`

```python
import os


def mkfile(filename, body=None):
    with open(filename, 'w') as f:
        f.write(body or filename)
    return


def make_example_dir(top):
    if not os.path.exists(top):
        os.mkdir(top)
    curdir = os.getcwd()
    os.chdir(top)

    os.mkdir('dir1')
    os.mkdir('dir2')
    mkfile('dir1/file_only_in_dir1')
    mkfile('dir2/file_only_in_dir2')

    os.mkdir('dir1/dir_only_in_dir1')
    os.mkdir('dir2/dir_only_in_dir2')

    os.mkdir('dir1/common_dir')
```

⊖ https://docs.python.org/3.5/library/shutil.html

```
        os.mkdir('dir2/common_dir')

        mkfile('dir1/common_file', 'this file is the same')
        mkfile('dir2/common_file', 'this file is the same')

        mkfile('dir1/not_the_same')
        mkfile('dir2/not_the_same')

        mkfile('dir1/file_in_dir1', 'This is a file in dir1')
        os.mkdir('dir2/file_in_dir1')

        os.chdir(curdir)
        return

if __name__ == '__main__':
    os.chdir(os.path.dirname(__file__) or os.getcwd())
    make_example_dir('example')
    make_example_dir('example/dir1/common_dir')
    make_example_dir('example/dir2/common_dir')
```

运行这个脚本会在 example 目录下生成一个文件树。

```
$ find example

example
example/dir1
example/dir1/common_dir
example/dir1/common_dir/dir1
example/dir1/common_dir/dir1/common_dir
example/dir1/common_dir/dir1/common_file
example/dir1/common_dir/dir1/dir_only_in_dir1
example/dir1/common_dir/dir1/file_in_dir1
example/dir1/common_dir/dir1/file_only_in_dir1
example/dir1/common_dir/dir1/not_the_same
example/dir1/common_dir/dir2
example/dir1/common_dir/dir2/common_dir
example/dir1/common_dir/dir2/common_file
example/dir1/common_dir/dir2/dir_only_in_dir2
example/dir1/common_dir/dir2/file_in_dir1
example/dir1/common_dir/dir2/file_only_in_dir2
example/dir1/common_dir/dir2/not_the_same
example/dir1/common_file
example/dir1/dir_only_in_dir1
example/dir1/file_in_dir1
example/dir1/file_only_in_dir1
example/dir1/not_the_same
example/dir2
example/dir2/common_dir
example/dir2/common_dir/dir1
example/dir2/common_dir/dir1/common_dir
example/dir2/common_dir/dir1/common_file
example/dir2/common_dir/dir1/dir_only_in_dir1
example/dir2/common_dir/dir1/file_in_dir1
example/dir2/common_dir/dir1/file_only_in_dir1
example/dir2/common_dir/dir1/not_the_same
example/dir2/common_dir/dir2
example/dir2/common_dir/dir2/common_dir
example/dir2/common_dir/dir2/common_file
example/dir2/common_dir/dir2/dir_only_in_dir2
example/dir2/common_dir/dir2/file_in_dir1
example/dir2/common_dir/dir2/file_only_in_dir2
example/dir2/common_dir/dir2/not_the_same
```

```
example/dir2/common_file
example/dir2/dir_only_in_dir2
example/dir2/file_in_dir1
example/dir2/file_only_in_dir2
example/dir2/not_the_same
```

common_dir 目录下也有同样的目录结构，以提供有意思的递归比较选择。

6.8.2 比较文件

cmp() 用于比较文件系统上的两个文件。

代码清单6-78：**filecmp_cmp.py**

```python
import filecmp

print('common_file :', end=' ')
print(filecmp.cmp('example/dir1/common_file',
                  'example/dir2/common_file'),
      end=' ')
print(filecmp.cmp('example/dir1/common_file',
                  'example/dir2/common_file',
                  shallow=False))

print('not_the_same:', end=' ')
print(filecmp.cmp('example/dir1/not_the_same',
                  'example/dir2/not_the_same'),
      end=' ')
print(filecmp.cmp('example/dir1/not_the_same',
                  'example/dir2/not_the_same',
                  shallow=False))

print('identical   :', end=' ')
print(filecmp.cmp('example/dir1/file_only_in_dir1',
                  'example/dir1/file_only_in_dir1'),
      end=' ')
print(filecmp.cmp('example/dir1/file_only_in_dir1',
                  'example/dir1/file_only_in_dir1',
                  shallow=False))
```

shallow 参数告诉 cmp() 除了文件的元数据外，是否还要查看文件的内容。默认情况下，会使用由 os.stat() 得到的信息来完成一个浅比较。如果结果是一样的，则认为文件相同。因此，对于同时创建的相同大小的文件，即使它们的内容不同，也会报告为是相同的文件。当 shallow 为 False 时，则要比较文件的内容。

```
$ python3 filecmp_cmp.py

common_file : True True
not_the_same: True False
identical   : True True
```

如果非递归地比较两个目录中的一组文件，则可以使用 cmpfiles()。参数是目录名和两个位置上要检查的文件列表。传入的公共文件列表应当只包含文件名（目录会导致匹配不成功），而且这些文件在两个位置上都应当出现。下一个例子显示了构造公共列表的一种简单方法。与 cmp() 一样，这个比较也有一个 shallow 标志。

代码清单 6-79：`filecmp_cmpfiles.py`

```python
import filecmp
import os

# Determine the items that exist in both directories.
d1_contents = set(os.listdir('example/dir1'))
d2_contents = set(os.listdir('example/dir2'))
common = list(d1_contents & d2_contents)
common_files = [
    f
    for f in common
    if os.path.isfile(os.path.join('example/dir1', f))
]
print('Common files:', common_files)
# Compare the directories.
match, mismatch, errors = filecmp.cmpfiles(
    'example/dir1',
    'example/dir2',
    common_files,
)
print('Match       :', match)
print('Mismatch    :', mismatch)
print('Errors      :', errors)
```

`cmpfiles()` 返回 3 个文件名列表，分别包含匹配的文件、不匹配的文件和不能比较的文件（由于权限问题或出于其他原因）。

```
$ python3 filecmp_cmpfiles.py

Common files: ['not_the_same', 'file_in_dir1', 'common_file']
Match        : ['not_the_same', 'common_file']
Mismatch     : ['file_in_dir1']
Errors       : []
```

6.8.3 比较目录

前面介绍的函数适合完成相对简单的比较。对于大目录树的递归比较或者更完整的分析，`dircmp` 类会更有用。在最简单的用例中，`report()` 会打印比较两个目录的报告。

代码清单 6-80：`filecmp_dircmp_report.py`

```python
import filecmp

dc = filecmp.dircmp('example/dir1', 'example/dir2')
dc.report()
```

输出是一个纯文本报告，显示的结果只包括给定目录的内容，而不会递归比较其子目录。在这里，认为文件 `not_the_same` 是相同的，因为这里没有比较内容。无法让 `dircmp` 像 `cmp()` 那样比较文件的内容。

```
$ python3 filecmp_dircmp_report.py

diff example/dir1 example/dir2
Only in example/dir1 : ['dir_only_in_dir1', 'file_only_in_dir1']
Only in example/dir2 : ['dir_only_in_dir2', 'file_only_in_dir2']
Identical files : ['common_file', 'not_the_same']
Common subdirectories : ['common_dir']
Common funny cases : ['file_in_dir1']
```

为了更多的细节，也为了完成一个递归比较，可以使用report_full_closure()：

代码清单6-81：`filecmp_dircmp_report_full_closure.py`

```
import filecmp

dc = filecmp.dircmp('example/dir1', 'example/dir2')
dc.report_full_closure()
```

输出将包括所有同级子目录的比较。

```
$ python3 filecmp_dircmp_report_full_closure.py

diff example/dir1 example/dir2
Only in example/dir1 : ['dir_only_in_dir1', 'file_only_in_dir1']
Only in example/dir2 : ['dir_only_in_dir2', 'file_only_in_dir2']
Identical files : ['common_file', 'not_the_same']
Common subdirectories : ['common_dir']
Common funny cases : ['file_in_dir1']

diff example/dir1/common_dir example/dir2/common_dir
Common subdirectories : ['dir1', 'dir2']

diff example/dir1/common_dir/dir1 example/dir2/common_dir/dir1
Identical files : ['common_file', 'file_in_dir1',
 'file_only_in_dir1', 'not_the_same']
Common subdirectories : ['common_dir', 'dir_only_in_dir1']

diff example/dir1/common_dir/dir1/dir_only_in_dir1
example/dir2/common_dir/dir1/dir_only_in_dir1

diff example/dir1/common_dir/dir1/common_dir
example/dir2/common_dir/dir1/common_dir

diff example/dir1/common_dir/dir2 example/dir2/common_dir/dir2
Identical files : ['common_file', 'file_only_in_dir2',
 'not_the_same']
Common subdirectories : ['common_dir', 'dir_only_in_dir2',
 'file_in_dir1']

diff example/dir1/common_dir/dir2/common_dir
example/dir2/common_dir/dir2/common_dir

diff example/dir1/common_dir/dir2/file_in_dir1
example/dir2/common_dir/dir2/file_in_dir1

diff example/dir1/common_dir/dir2/dir_only_in_dir2
example/dir2/common_dir/dir2/dir_only_in_dir2
```

6.8.4 在程序中使用差异

除了生成打印报告，dircmp还能计算文件列表，可以在程序中直接使用。以下各个属性只在请求时才计算，所以对于未用的数据，创建dircmp实例不会带来开销。

代码清单6-82：`filecmp_dircmp_list.py`

```
import filecmp
import pprint

dc = filecmp.dircmp('example/dir1', 'example/dir2')
```

```
print('Left:')
pprint.pprint(dc.left_list)

print('\nRight:')
pprint.pprint(dc.right_list)
```

所比较目录中包含的文件和子目录分别列在 left_list 和 right_list 中。

```
$ python3 filecmp_dircmp_list.py

Left:
['common_dir',
 'common_file',
 'dir_only_in_dir1',
 'file_in_dir1',
 'file_only_in_dir1',
 'not_the_same']

Right:
['common_dir',
 'common_file',
 'dir_only_in_dir2',
 'file_in_dir1',
 'file_only_in_dir2',
 'not_the_same']
```

可以向构造函数传入一个要忽略的名字列表（该列表中指定的名字将被忽略）来对输入进行过滤。默认地，RCS、CVS 和 tags 等名字会被忽略。

代码清单 6-83：filecmp_dircmp_list_filter.py

```
import filecmp
import pprint

dc = filecmp.dircmp('example/dir1', 'example/dir2',
                    ignore=['common_file'])
print('Left:')
pprint.pprint(dc.left_list)

print('\nRight:')
pprint.pprint(dc.right_list)
```

在这里，将 common_file 从要比较的文件列表中去除。

```
$ python3 filecmp_dircmp_list_filter.py

Left:
['common_dir',
 'dir_only_in_dir1',
 'file_in_dir1',
 'file_only_in_dir1',
 'not_the_same']

Right:
['common_dir',
 'dir_only_in_dir2',
 'file_in_dir1',
 'file_only_in_dir2',
 'not_the_same']
```

两个输入目录中共有的文件名会保存在 common 内，各目录独有的文件会列在 left_

only 和 right_only 中。

代码清单 6-84：`filecmp_dircmp_membership.py`

```
import filecmp
import pprint

dc = filecmp.dircmp('example/dir1', 'example/dir2')
print('Common:')
pprint.pprint(dc.common)

print('\nLeft:')
pprint.pprint(dc.left_only)

print('\nRight:')
pprint.pprint(dc.right_only)
```

"左"目录是 `dircmp()` 的第一个参数，"右"目录是第二个参数。

```
$ python3 filecmp_dircmp_membership.py

Common:
['file_in_dir1', 'common_file', 'common_dir', 'not_the_same']

Left:
['dir_only_in_dir1', 'file_only_in_dir1']

Right:
['file_only_in_dir2', 'dir_only_in_dir2']
```

公共成员可以被进一步分解为文件、目录和"有趣"元素（两个目录中类型不同的内容，或者 `os.stat()` 指出的有错误的地方）。

代码清单 6-85：`filecmp_dircmp_common.py`

```
import filecmp
import pprint

dc = filecmp.dircmp('example/dir1', 'example/dir2')
print('Common:')
pprint.pprint(dc.common)

print('\nDirectories:')
pprint.pprint(dc.common_dirs)

print('\nFiles:')
pprint.pprint(dc.common_files)

print('\nFunny:')
pprint.pprint(dc.common_funny)
```

在示例数据中，`file_in_dir1` 元素在一个目录中是一个文件，而在另一个目录中是一个子目录，所以它会出现在"有趣"列表中。

```
$ python3 filecmp_dircmp_common.py

Common:
['file_in_dir1', 'common_file', 'common_dir', 'not_the_same']

Directories:
```

```
['common_dir']
Files:
['common_file', 'not_the_same']
Funny:
['file_in_dir1']
```

文件之间的差别也可以做类似的划分。

代码清单 6-86：filecmp_dircmp_diff.py

```
import filecmp

dc = filecmp.dircmp('example/dir1', 'example/dir2')
print('Same      :', dc.same_files)
print('Different :', dc.diff_files)
print('Funny     :', dc.funny_files)
```

文件 `not_the_same` 通过 `os.stat()` 比较，并且不检查内容，所以它包含在 `same_files` 列表中。

```
$ python3 filecmp_dircmp_diff.py

Same      : ['common_file', 'not_the_same']
Different : []
Funny     : []
```

最后一点，子目录也会被保存，以便容易地完成递归比较。

代码清单 6-87：filecmp_dircmp_subdirs.py

```
import filecmp

dc = filecmp.dircmp('example/dir1', 'example/dir2')
print('Subdirectories:')
print(dc.subdirs)
```

属性 `subdirs` 是一个字典，它将目录名映射到新的 `dircmp` 对象。

```
$ python3 filecmp_dircmp_subdirs.py

Subdirectories:
{'common_dir': <filecmp.dircmp object at 0x1019b2be0>}
```

提示：相关阅读材料
- `filecmp` 的标准库文档[⊖]。
- `difflib`：计算两个序列之间的差异。

6.9 `mmap`：内存映射文件

建立一个文件的内存映射将使用操作系统虚拟内存来直接访问文件系统上的数据，而不是使用常规的 I/O 函数访问数据。内存映射通常可以提高 I/O 性能，因为使用内存映射

⊖ https://docs.python.org/3.5/library/filecmp.html

时，不需要对每个访问都建立一个单独的系统调用，也不需要在缓冲区之间复制数据；实际上，内核和用户应用都能直接访问内存。

内存映射文件可以看作是可修改的字符串或类似文件的对象，这取决于具体的需要。映射文件支持一般的文件 API 方法，如 `close()`、`flush()`、`read()`、`readline()`、`seek()`、`tell()` 和 `write()`。它还支持字符串 API，提供分片等特性以及类似 `find()` 的方法。

下面的所有示例都会使用文本文件 `lorem.txt`，其中包含一些 Lorem Ipsum。为便于参考，下面的代码清单给出这个文件的文本。

代码清单 6-88：`lorem.txt`

```
Lorem ipsum dolor sit amet, consectetuer adipiscing elit.
Donec egestas, enim et consectetuer ullamcorper, lectus ligula
rutrum leo, a elementum elit tortor eu quam. Duis tincidunt nisi ut
ante. Nulla facilisi. Sed tristique eros eu libero. Pellentesque vel
arcu. Vivamus purus orci, iaculis ac, suscipit sit amet, pulvinar eu,
lacus. Praesent placerat tortor sed nisl. Nunc blandit diam egestas
dui. Pellentesque habitant morbi tristique senectus et netus et
malesuada fames ac turpis egestas. Aliquam viverra fringilla
leo. Nulla feugiat augue eleifend nulla. Vivamus mauris. Vivamus sed
mauris in nibh placerat egestas. Suspendisse potenti. Mauris
massa. Ut eget velit auctor tortor blandit sollicitudin. Suspendisse
imperdiet justo.
```

说明：UNIX 和 Windows 上 `mmap()` 的参数和行为有所不同，不过这里不会全面讨论这些差别。有关的更多细节，可以参考标准库文档。

6.9.1 读文件

使用 `mmap()` 函数可以创建一个内存映射文件。第一个参数是文件描述符，可能来自 `file` 对象的 `fileno()` 方法，也可能来自 `os.open()`。调用者在调用 `mmap()` 之前负责打开文件，不再需要文件时要负责将其关闭。

`mmap()` 的第二个参数是要映射的文件部分的大小（以字节为单位）。如果这个值为 `0`，则映射整个文件。如果这个大小大于文件的当前大小，则会扩展该文件。

说明：Windows 不支持创建长度为 `0` 的映射。

这两个平台都支持一个可选的关键字参数 `access`。使用 `ACCESS_READ` 表示只读访问；`ACCESS_WRITE` 表示"写通过"（write-through），即对内存的赋值直接写入文件；`ACCESS_COPY` 表示"写时复制"（copy-on-write），对内存的赋值不会写至文件。

代码清单 6-89：`mmap_read.py`

```
import mmap

with open('lorem.txt', 'r') as f:
    with mmap.mmap(f.fileno(), 0,
                   access=mmap.ACCESS_READ) as m:
        print('First 10 bytes via read :', m.read(10))
        print('First 10 bytes via slice:', m[:10])
        print('2nd   10 bytes via read :', m.read(10))
```

文件指针会跟踪通过一个分片操作访问的最后一个字节。在这个例子中，第一次读之后，指针向前移动 10 个字节。然后由分片操作将指针重置回文件的起始位置，并由分片使指针再次向前移动 10 个字节。分片操作之后，再调用 read() 会给出文件的 11～20 字节。

```
$ python3 mmap_read.py

First 10 bytes via read  : b'Lorem ipsu'
First 10 bytes via slice: b'Lorem ipsu'
2nd   10 bytes via read  : b'm dolor si'
```

6.9.2 写文件

要建立内存映射文件来接收更新，映射之前首先要使用模式 'r+'（而不是 'w'）打开文件以便完成追加。然后可以使用任何改变数据的 API 方法（例如 write() 或赋值到一个分片等）。

下面的例子使用了默认访问模式 ACCESS_WRITE，并赋值到一个分片，以原地修改某一行的一部分。

代码清单 6-90：mmap_write_slice.py

```
import mmap
import shutil

# Copy the example file.
shutil.copyfile('lorem.txt', 'lorem_copy.txt')

word = b'consectetuer'
reversed = word[::-1]
print('Looking for    :', word)
print('Replacing with :', reversed)

with open('lorem_copy.txt', 'r+') as f:
    with mmap.mmap(f.fileno(), 0) as m:
        print('Before:\n{}'.format(m.readline().rstrip()))
        m.seek(0)  # Rewind

        loc = m.find(word)
        m[loc:loc + len(word)] = reversed
        m.flush()

        m.seek(0)  # Rewind
        print('After :\n{}'.format(m.readline().rstrip()))

        f.seek(0)  # Rewind
        print('File  :\n{}'.format(f.readline().rstrip()))
```

内存和文件中第一行中间的单词"consectetuer"将被替换。

```
$ python3 mmap_write_slice.py

Looking for    : b'consectetuer'
Replacing with : b'reutetcesnoc'
Before:
b'Lorem ipsum dolor sit amet, consectetuer adipiscing elit.'
After :
b'Lorem ipsum dolor sit amet, reutetcesnoc adipiscing elit.'
File  :
Lorem ipsum dolor sit amet, reutetcesnoc adipiscing elit.
```

复制模式

使用访问设置 ACCESS_COPY 时不会把修改写入磁盘上的文件。

代码清单 6-91：**mmap_write_copy.py**

```python
import mmap
import shutil

# Copy the example file.
shutil.copyfile('lorem.txt', 'lorem_copy.txt')

word = b'consectetuer'
reversed = word[::-1]
with open('lorem_copy.txt', 'r+') as f:
    with mmap.mmap(f.fileno(), 0,
                   access=mmap.ACCESS_COPY) as m:
        print('Memory Before:\n{}'.format(
            m.readline().rstrip()))
        print('File Before  :\n{}\n'.format(
            f.readline().rstrip()))

        m.seek(0)  # Rewind
        loc = m.find(word)
        m[loc:loc + len(word)] = reversed

        m.seek(0)  # Rewind
        print('Memory After :\n{}'.format(
            m.readline().rstrip()))

        f.seek(0)
        print('File After   :\n{}'.format(
            f.readline().rstrip()))
```

在这个例子中，必须单独地回转文件句柄和 mmap 句柄，因为这两个对象的内部状态会单独维护。

```
$ python3 mmap_write_copy.py

Memory Before:
b'Lorem ipsum dolor sit amet, consectetuer adipiscing elit.'
File Before  :
Lorem ipsum dolor sit amet, consectetuer adipiscing elit.

Memory After :
b'Lorem ipsum dolor sit amet, reutetcesnoc adipiscing elit.'
File After   :
Lorem ipsum dolor sit amet, consectetuer adipiscing elit.
```

6.9.3 正则表达式

由于内存映射文件就类似于一个字符串，因此也常与其他处理字符串的模块一起使用，如正则表达式。下面的例子会找出所有包含"nulla"的句子。

代码清单 6-92：**mmap_regex.py**

```python
import mmap
import re

pattern = re.compile(rb'(\.\W+)?([^.]?nulla[^.]*?\.)',
                     re.DOTALL | re.IGNORECASE | re.MULTILINE)
```

```
with open('lorem.txt', 'r') as f:
    with mmap.mmap(f.fileno(), 0,
                   access=mmap.ACCESS_READ) as m:
        for match in pattern.findall(m):
            print(match[1].replace(b'\n', b' '))
```

由于这个模式包含两个组，所以 `findall()` 的返回值是一个元组序列。`print` 语句会找出匹配的句子，并用空格代替换行符，使各个结果都打印在同一行上。

```
$ python3 mmap_regex.py

b'Nulla facilisi.'
b'Nulla feugiat augue eleifend nulla.'
```

提示：相关阅读材料
- `mmap` 的标准库文档[⊖]。
- `mmap` 的 Python 2 到 Python 3 移植说明。
- `os`：os 模块。
- `re`：正则表达式。

6.10 codecs：字符串编码和解码

codecs 模块提供了流接口和文件接口来完成文本数据不同表示之间的转换。通常用于处理 Unicode 文本，不过也提供了其他编码来满足其他用途。

6.10.1 Unicode 入门

CPython 3.x 区分了文本（text）和字节（byte）串。`bytes` 实例使用一个 8 位字节值序列。与之不同，`str` 串在内部作为一个 Unicode 码点（code point）序列来管理。码点值使用 2 字节或 4 字节表示，这取决于编译 Python 时指定的选项。

输出 str 值时，会使用某种标准机制编码，以后可以将这个字节序列重构为同样的文本串。编码值的字节不一定与码点值完全相同，编码只是定义了两个值集之间转换的一种方式。读取 Unicode 数据时还需要知道编码，这样才能把接收到的字节转换为 `unicode` 类使用的内部表示。

西方语言最常用的编码是 UTF-8 和 UTF-16，这两种编码分别使用单字节和两字节值序列表示各个码点。对于其他语言，由于大多数字符都由超过两字节的码点表示，所以使用其他编码来存储可能更为高效。

提示：相关阅读材料
关于 Unicode 的更多介绍信息，请参考本节最后所列的参考资料。《The Python Unicode HOWTO》尤其有帮助。

⊖ https://docs.python.org/3.5/library/mmap.html

编码

要了解编码,最好的方法就是采用不同方式对相同的串进行编码,并查看所生成的不同的字节序列。下面的例子使用以下函数格式化字节串,使之更易读。

代码清单 6-93:`codecs_to_hex.py`

```python
import binascii


def to_hex(t, nbytes):
    """Format text t as a sequence of nbyte long values
    separated by spaces.
    """
    chars_per_item = nbytes * 2
    hex_version = binascii.hexlify(t)
    return b' '.join(
        hex_version[start:start + chars_per_item]
        for start in range(0, len(hex_version), chars_per_item)
    )


if __name__ == '__main__':
    print(to_hex(b'abcdef', 1))
    print(to_hex(b'abcdef', 2))
```

这个函数使用 `binascii` 得到输入字节串的十六进制表示,在返回这个值之前每隔 `nbytes` 字节就插入一个空格。

```
$ python3 codecs_to_hex.py

b'61 62 63 64 65 66'
b'6162 6364 6566'
```

第一个编码示例首先使用 `unicode` 类的原始表示来打印文本 `'français'`,后面是 Unicode 数据库中各个字符的名。接下来两行将这个字符串分别编码为 UTF-8 和 UTF-16,并显示编码得到的十六进制值。

代码清单 6-94:`codecs_encodings.py`

```python
import unicodedata
from codecs_to_hex import to_hex

text = 'français'

print('Raw    : {!r}'.format(text))
for c in text:
    print('  {!r}: {}'.format(c, unicodedata.name(c, c)))
print('UTF-8 : {!r}'.format(to_hex(text.encode('utf-8'), 1)))
print('UTF-16: {!r}'.format(to_hex(text.encode('utf-16'), 2)))
```

对一个 `str` 编码的结果是一个 `bytes` 对象。

```
$ python3 codecs_encodings.py

Raw    : 'français'
  'f': LATIN SMALL LETTER F
  'r': LATIN SMALL LETTER R
```

```
   'a':  LATIN SMALL LETTER A
   'n':  LATIN SMALL LETTER N
   'ç':  LATIN SMALL LETTER C WITH CEDILLA
   'a':  LATIN SMALL LETTER A
   'i':  LATIN SMALL LETTER I
   's':  LATIN SMALL LETTER S
UTF-8 : b'66 72 61 6e c3 a7 61 69 73'
UTF-16: b'fffe 6600 7200 6100 6e00 e700 6100 6900 7300'
```

给定一个编码字节序列（作为一个 bytes 实例），decode() 方法将其转换为码点，并作为一个 str 实例返回这个序列。

代码清单 6-95：codecs_decode.py

```
from codecs_to_hex import to_hex

text = 'français'
encoded = text.encode('utf-8')
decoded = encoded.decode('utf-8')

print('Original :', repr(text))
print('Encoded  :', to_hex(encoded, 1), type(encoded))
print('Decoded  :', repr(decoded), type(decoded))
```

选择使用哪一种编码不会改变输出类型。

```
$ python3 codecs_decode.py

Original : 'français'
Encoded  : b'66 72 61 6e c3 a7 61 69 73' <class 'bytes'>
Decoded  : 'français' <class 'str'>
```

说明：解释器启动过程中（即加载 site 时）会设置默认编码。关于默认编码设置的描述，可以参考 sys 相关讨论中的 17.2.1.4 节。

6.10.2 处理文件

处理 I/O 操作时，编码和解码字符串尤其重要。不论是写至一个文件、套接字还是其他流，数据都必须使用适当的编码。一般来讲，所有文本数据在读取时都需要由其字节表示解码，写数据时则需要从内部值编码为一种特定的表示。程序可以显式地编码和解码数据，不过取决于所用的编码，要想确定是否已经读取足够的字节来充分解码数据，这可能并不容易。codecs 提供了一些类来管理数据编码和解码，所以应用不再需要做这个工作。

codecs 提供的最简单的接口可以替代内置 open() 函数。这个新版本的函数与内置函数的做法很相似，不过增加了两个参数来指定编码和所需的错误处理技术。

代码清单 6-96：codecs_open_write.py

```
from codecs_to_hex import to_hex

import codecs
import sys

encoding = sys.argv[1]
filename = encoding + '.txt'
```

```
print('Writing to', filename)
with codecs.open(filename, mode='w', encoding=encoding) as f:
    f.write('français')

# Determine the byte grouping to use for to_hex().
nbytes = {
    'utf-8': 1,
    'utf-16': 2,
    'utf-32': 4,
}.get(encoding, 1)

# Show the raw bytes in the file.
print('File contents:')
with open(filename, mode='rb') as f:
    print(to_hex(f.read(), nbytes))
```

这个例子首先处理一个包含 ç 的 unicode 串，使用命令行上指定的编码将这个文本保存到一个文件。

```
$ python3 codecs_open_write.py utf-8

Writing to utf-8.txt
File contents:
b'66 72 61 6e c3 a7 61 69 73'

$ python3 codecs_open_write.py utf-16

Writing to utf-16.txt
File contents:
b'fffe 6600 7200 6100 6e00 e700 6100 6900 7300'

$ python3 codecs_open_write.py utf-32

Writing to utf-32.txt
File contents:
b'fffe0000 66000000 72000000 61000000 6e000000 e7000000 61000000
69000000 73000000'
```

用 open() 读数据很简单，但有一点要注意：必须提前知道编码才能正确地建立解码器。尽管有些数据格式（如 XML）会在文件中指定编码，但是通常都由应用来管理。codecs 只是取一个编码参数，并假设这个编码是正确的。

代码清单 6-97：codecs_open_read.py

```
import codecs
import sys

encoding = sys.argv[1]
filename = encoding + '.txt'

print('Reading from', filename)
with codecs.open(filename, mode='r', encoding=encoding) as f:
    print(repr(f.read()))
```

这个例子读取上一个程序创建的文件，并把得到的 unicode 对象的表示打印到控制台。

```
$ python3 codecs_open_read.py utf-8

Reading from utf-8.txt
'français'
```

```
$ python3 codecs_open_read.py utf-16

Reading from utf-16.txt
'français'

$ python3 codecs_open_read.py utf-32

Reading from utf-32.txt
'français'
```

6.10.3 字节序

在不同计算机系统之间传输数据时（可能直接复制一个文件，或者使用网络通信来完成传输），多字节编码（如 UTF-16 和 UTF-32）会带来一个问题。不同系统中使用的高字节和低字节的顺序不同。数据的这个特性被称为字节序（endianness），这取决于硬件体系结构等因素，还取决于操作系统和应用开发人员做出的选择。通常没有办法提前知道给定的一组数据要使用哪一个字节序，所以多字节编码还包含一个字节序标志（Byte-Order Marker, BOM），这个标志出现在编码输出的前几个字节。例如，UTF-16 定义 0xFFFE 和 0xFEFF 不是合法字符，可以用于指示字节序。codecs 定义了 UTF-16 和 UTF-32 所用字节序标志的相应常量。

代码清单 6-98：**codecs_bom.py**

```python
import codecs
from codecs_to_hex import to_hex

BOM_TYPES = [
    'BOM', 'BOM_BE', 'BOM_LE',
    'BOM_UTF8',
    'BOM_UTF16', 'BOM_UTF16_BE', 'BOM_UTF16_LE',
    'BOM_UTF32', 'BOM_UTF32_BE', 'BOM_UTF32_LE',
]

for name in BOM_TYPES:
    print('{:12} : {}'.format(
        name, to_hex(getattr(codecs, name), 2)))
```

取决于当前系统的原生字节序，BOM、BOM_UTF16 和 BOM_UTF32 会自动设置为适当的大端（big-endian）或小端（little-endian）值。

```
$ python3 codecs_bom.py

BOM          : b'fffe'
BOM_BE       : b'feff'
BOM_LE       : b'fffe'
BOM_UTF8     : b'efbb bf'
BOM_UTF16    : b'fffe'
BOM_UTF16_BE : b'feff'
BOM_UTF16_LE : b'fffe'
BOM_UTF32    : b'fffe 0000'
BOM_UTF32_BE : b'0000 feff'
BOM_UTF32_LE : b'fffe 0000'
```

可以由 codecs 中的解码器自动检测和处理字节序，也可以在编码时显式地指定字节序。

代码清单 6-99：`codecs_bom_create_file.py`

```python
import codecs
from codecs_to_hex import to_hex

# Pick the non-native version of UTF-16 encoding.
if codecs.BOM_UTF16 == codecs.BOM_UTF16_BE:
    bom = codecs.BOM_UTF16_LE
    encoding = 'utf_16_le'
else:
    bom = codecs.BOM_UTF16_BE
    encoding = 'utf_16_be'

print('Native order  :', to_hex(codecs.BOM_UTF16, 2))
print('Selected order:', to_hex(bom, 2))

# Encode the text.
encoded_text = 'français'.encode(encoding)
print('{:14}: {}'.format(encoding, to_hex(encoded_text, 2)))

with open('nonnative-encoded.txt', mode='wb') as f:
    # Write the selected byte-order marker.  It is not included
    # in the encoded text because the byte order was given
    # explicitly when selecting the encoding.
    f.write(bom)
    # Write the byte string for the encoded text.
    f.write(encoded_text)
```

`codecs_bom_create_file.py` 首先得出原生字节序，然后显式地使用替代形式，以便下一个例子可以在展示读取时自动检测字节序。

```
$ python3 codecs_bom_create_file.py

Native order   : b'fffe'
Selected order: b'feff'
utf_16_be     : b'0066 0072 0061 006e 00e7 0061 0069 0073'
```

`codecs_bom_detection.py` 打开文件时没有指定字节序，所以解码器会使用文件前两个字节中的 BOM 值来确定字节序。

代码清单 6-100：`codecs_bom_detection.py`

```python
import codecs
from codecs_to_hex import to_hex

# Look at the raw data.
with open('nonnative-encoded.txt', mode='rb') as f:
    raw_bytes = f.read()

print('Raw     :', to_hex(raw_bytes, 2))

# Reopen the file and let codecs detect the BOM.
with codecs.open('nonnative-encoded.txt',
                 mode='r',
                 encoding='utf-16',
                 ) as f:
    decoded_text = f.read()

print('Decoded:', repr(decoded_text))
```

由于文件的前两个字节用于字节序检测，所以它们并不包含在 `read()` 返回的数据中。

```
$ python3 codecs_bom_detection.py

Raw     : b'feff 0066 0072 0061 006e 00e7 0061 0069 0073'
Decoded: 'français'
```

6.10.4 错误处理

前面几节指出,读写 Unicode 文件时需要知道所使用的编码。正确地设置编码很重要,这有两个原因:首先,如果读文件时未能正确地配置编码,就无法正确地解释数据,数据有可能被破坏或者无法解码;其次,并不是所有 Unicode 字符都可以用所有编码表示,所以如果写文件时使用了错误的编码,就会产生一个错误,可能丢失数据。

类似于 `str` 的 `encode()` 方法和 `bytes` 的 `decode()` 方法,`codecs` 也使用了同样的 5 个错误处理选项,如表 6-1 所列。

表 6-1 Codec 错误处理模式

错误模式	描述
strict	如果数据无法转换,则产生一个异常
replace	将无法编码的数据替换为一个特殊的标志字符
ignore	跳过数据
xmlcharrefreplace	XML 字符(仅适用于编码)
backslashreplace	转义序列(仅适用于编码)

6.10.4.1 编码错误

最常见的错误是在向一个 ASCII 输出流(如一个常规文件或 `sys.stdout`)写 Unicode 数据时接收到一个 `UnicodeEncodeError`。代码清单 6-101 中的示例程序可以用来试验不同的错误处理模式。

代码清单 6-101:**codecs_encode_error.py**

```
import codecs
import sys

error_handling = sys.argv[1]

text = 'français'

try:
    # Save the data, encoded as ASCII, using the error
    # handling mode specified on the command line.
    with codecs.open('encode_error.txt', 'w',
                     encoding='ascii',
                     errors=error_handling) as f:
        f.write(text)

except UnicodeEncodeError as err:
    print('ERROR:', err)

else:
    # If there was no error writing to the file,
    # show the file's contents.
    with open('encode_error.txt', 'rb') as f:
        print('File contents: {!r}'.format(f.read()))
```

要确保一个应用显式地为所有 I/O 操作设置正确的编码,`strict` 模式是最安全的选择,但是产生一个异常时,这种模式可能导致程序崩溃。

```
$ python3 codecs_encode_error.py strict

ERROR: 'ascii' codec can't encode character '\xe7' in position
4: ordinal not in range(128)
```

另外一些错误模式更为灵活。例如,`replace` 确保不会产生任何错误,其代价是一些无法转换为所需编码的数据可能会丢失。pi(π) 的 Unicode 字符仍然无法用 ASCII 编码,但是采用这种错误处理模式时,并不是产生一个异常,而是会在输出中将这个字符替换为 ?。

```
$ python3 codecs_encode_error.py replace

File contents: b'fran?ais'
```

要完全跳过有问题的数据,可以使用 `ignore`。无法编码的数据都会被丢弃。

```
$ python3 codecs_encode_error.py ignore

File contents: b'franais'
```

还有两种无损的错误处理选项,这两种模式会把字符替换为标准中定义的一个与该编码不同的候选表示。`xmlcharrefreplace` 使用一个 XML 字符引用作为替代(W3C 文档"字符的 XML 实体定义"中指定了字符引用列表)。

```
$ python3 codecs_encode_error.py xmlcharrefreplace

File contents: b'fran&#231;ais'
```

另一种无损的错误处理机制是 `backslashreplace`,它生成的输出格式类似于打印 `unicode` 对象的 `repr()` 时返回的值。Unicode 字符会被替换为 \u 以及码点的十六进制值。

```
$ python3 codecs_encode_error.py backslashreplace

File contents: b'fran\\xe7ais'
```

6.10.4.2 解码错误

数据解码时也有可能遇到错误,特别是如果使用了错误的编码。

代码清单 6-102:`codecs_decode_error.py`

```python
import codecs
import sys

from codecs_to_hex import to_hex

error_handling = sys.argv[1]

text = 'français'
print('Original    :', repr(text))

# Save the data with one encoding.
with codecs.open('decode_error.txt', 'w',
                 encoding='utf-16') as f:
    f.write(text)

# Dump the bytes from the file.
with open('decode_error.txt', 'rb') as f:
    print('File contents:', to_hex(f.read(), 1))
```

```
    # Try to read the data with the wrong encoding.
    with codecs.open('decode_error.txt', 'r',
                     encoding='utf-8',
                     errors=error_handling) as f:
        try:
            data = f.read()
        except UnicodeDecodeError as err:
            print('ERROR:', err)
        else:
            print('Read            :', repr(data))
```

与编码一样，如果不能正确地解码字节流，则 strict 错误处理模式会产生一个异常。在这里，产生 UnicodeDecodeError 的原因是尝试使用 UTF-8 解码器将 UTF-16 BOM 部分转换为一个字符。

```
$ python3 codecs_decode_error.py strict

Original       : 'français'
File contents: b'ff fe 66 00 72 00 61 00 6e 00 e7 00 61 00 69 00
73 00'
ERROR: 'utf-8' codec can't decode byte 0xff in position 0:
invalid start byte
```

切换到 ignore 会让解码器跳过不合法的字节。不过，结果仍然不是原来期望的结果，因为其中包括嵌入的 null 字节。

```
$ python3 codecs_decode_error.py ignore

Original       : 'français'
File contents: b'ff fe 66 00 72 00 61 00 6e 00 e7 00 61 00 69 00
73 00'
Read           : 'f\x00r\x00a\x00n\x00\x00a\x00i\x00s\x00'
```

采用 replace 模式时，非法的字节会被替换为 \uFFFD，这是官方的 Unicode 替换字符，看起来像是一个有黑色背景的菱形，其中包含一个白色的问号。

```
$ python3 codecs_decode_error.py replace

Original       : 'français'
File contents: b'ff fe 66 00 72 00 61 00 6e 00 e7 00 61 00 69 00
73 00'
Read           : 'f\x00r\x00a\x00n\x00\x00a\x00i\x00s\x00'
```

6.10.5 编码转换

尽管大多数应用都在内部处理 str 数据，将数据解码或编码作为 I/O 操作的一部分，但有些情况下，可能需要改变文件的编码而不继续坚持这种中间数据格式，这可能很有用。EncodedFile() 取一个使用某种编码打开的文件句柄，用一个类包装这个文件句柄，有 I/O 操作时它会把数据转换为另一种编码。

代码清单 6-103：codecs_encodedfile.py

```
from codecs_to_hex import to_hex

import codecs
import io

# Raw version of the original data
```

```
data = 'français'

# Manually encode it as UTF-8.
utf8 = data.encode('utf-8')
print('Start as UTF-8   :', to_hex(utf8, 1))

# Set up an output buffer, then wrap it as an EncodedFile.
output = io.BytesIO()
encoded_file = codecs.EncodedFile(output, data_encoding='utf-8',
                                  file_encoding='utf-16')
encoded_file.write(utf8)

# Fetch the buffer contents as a UTF-16 encoded byte string.
utf16 = output.getvalue()
print('Encoded to UTF-16:', to_hex(utf16, 2))

# Set up another buffer with the UTF-16 data for reading,
# and wrap it with another EncodedFile.
buffer = io.BytesIO(utf16)
encoded_file = codecs.EncodedFile(buffer, data_encoding='utf-8',
                                  file_encoding='utf-16')

# Read the UTF-8 encoded version of the data.
recoded = encoded_file.read()
print('Back to UTF-8    :', to_hex(recoded, 1))
```

这个例子显示了如何读写 `EncodedFile()` 返回的不同句柄。不论这个句柄用于读还是写，`file_encoding` 总是指示打开文件句柄所用的编码（作为第一个参数传入），`data_encoding` 值则指示通过 `read()` 和 `write()` 调用传递数据时所用的编码。

```
$ python3 codecs_encodedfile.py

Start as UTF-8   : b'66 72 61 6e c3 a7 61 69 73'
Encoded to UTF-16: b'fffe 6600 7200 6100 6e00 e700 6100 6900
7300'
Back to UTF-8    : b'66 72 61 6e c3 a7 61 69 73'
```

6.10.6 非 Unicode 编码

尽管之前大多数例子都使用 Unicode 编码，但实际上 `codecs` 还可以用于很多其他数据转换。例如，Python 包含了处理 base-64、bzip2、ROT-13、ZIP 和其他数据格式的 `codecs`。

代码清单 6-104：<code>codecs_rot13.py</code>

```
import codecs
import io

buffer = io.StringIO()
stream = codecs.getwriter('rot_13')(buffer)

text = 'abcdefghijklmnopqrstuvwxyz'

stream.write(text)
stream.flush()

print('Original:', text)
print('ROT-13  :', buffer.getvalue())
```

如果转换可以被表述为有单个输入参数的函数，并且返回一个字节或 Unicode 串，那么这样的转换都可以注册为一个 codec。对于 `'rot_13'` codec，输入应当是一个 Unicode 串；输出也是一个 Unicode 串。

```
$ python3 codecs_rot13.py

Original: abcdefghijklmnopqrstuvwxyz
ROT-13  : nopqrstuvwxyzabcdefghijklm
```

使用 `codecs` 包装一个数据流，可以提供比直接使用 `zlib` 更简单的接口。

代码清单6-105：`codecs_zlib.py`

```
import codecs
import io

from codecs_to_hex import to_hex

buffer = io.BytesIO()
stream = codecs.getwriter('zlib')(buffer)

text = b'abcdefghijklmnopqrstuvwxyz\n' * 50

stream.write(text)
stream.flush()

print('Original length :', len(text))
compressed_data = buffer.getvalue()
print('ZIP compressed  :', len(compressed_data))

buffer = io.BytesIO(compressed_data)
stream = codecs.getreader('zlib')(buffer)

first_line = stream.readline()
print('Read first line :', repr(first_line))

uncompressed_data = first_line + stream.read()
print('Uncompressed    :', len(uncompressed_data))
print('Same            :', text == uncompressed_data)
```

并不是所有压缩或编码系统都支持使用 `readline()` 或 `read()` 通过流接口读取数据的一部分，因为这需要找到压缩段的末尾来完成解压缩。如果一个程序无法在内存中保存整个解压缩的数据集，那么可以使用压缩库的增量访问特性，而不是 `codecs`。

```
$ python3 codecs_zlib.py

Original length : 1350
ZIP compressed  : 48
Read first line : b'abcdefghijklmnopqrstuvwxyz\n'
Uncompressed    : 1350
Same            : True
```

6.10.7 增量编码

目前提供的一些编码（特别是 `bz2` 和 `zlib`）在处理数据流时可能会显著改变数据流的长度。对于大的数据集，这些编码采用增量方式可以更好地处理，即一次只处理一个小数据块。`IncrementalEncoder`/`IncrementalDecoder` API 就是为此而设计。

代码清单 6-106: **codecs_incremental_bz2.py**

```python
import codecs
import sys

from codecs_to_hex import to_hex

text = b'abcdefghijklmnopqrstuvwxyz\n'
repetitions = 50

print('Text length :', len(text))
print('Repetitions :', repetitions)
print('Expected len:', len(text) * repetitions)

# Encode the text several times to build up a
# large amount of data.
encoder = codecs.getincrementalencoder('bz2')()
encoded = []

print()
print('Encoding:', end=' ')
last = repetitions - 1
for i in range(repetitions):
    en_c = encoder.encode(text, final=(i == last))
    if en_c:
        print('\nEncoded : {} bytes'.format(len(en_c)))
        encoded.append(en_c)
    else:
        sys.stdout.write('.')

all_encoded = b''.join(encoded)
print()
print('Total encoded length:', len(all_encoded))
print()

# Decode the byte string one byte at a time.
decoder = codecs.getincrementaldecoder('bz2')()
decoded = []

print('Decoding:', end=' ')
for i, b in enumerate(all_encoded):
    final = (i + 1) == len(text)
    c = decoder.decode(bytes([b]), final)
    if c:
        print('\nDecoded : {} characters'.format(len(c)))
        print('Decoding:', end=' ')
        decoded.append(c)
    else:
        sys.stdout.write('.')
print()
restored = b''.join(decoded)

print()
print('Total uncompressed length:', len(restored))
```

每次将数据传递到编码器或解码器时,其内部状态都会更新。状态一致时(按照 codec 的定义),会返回数据并重置状态。在此之前,**encode()** 或 **decode()** 调用并不返回任何数据。传入最后一位数据时,参数 **final** 应当被设置为 **True**,这样 codec 就能知道需要刷新输出所有余下的缓冲数据。

```
$ python3 codecs_incremental_bz2.py

Text length : 27
Repetitions : 50
Expected len: 1350

Encoding: .........................................................
Encoded : 99 bytes

Total encoded length: 99

Decoding: .........................................................
................................
Decoded : 1350 characters
Decoding: .........

Total uncompressed length: 1350
```

6.10.8　Unicode 数据和网络通信

不同于标准输入和输出流，网络套接字是字节流，默认情况下不支持编码。因此，如果程序希望通过网络发送或接收 Unicode 数据，那么在将数据写至一个套接字之前必须先将数据编码为字节。下例中的服务器会把接收到的数据回送给发送者。

代码清单 6-107：**codecs_socket_fail.py**

```python
import sys
import socketserver


class Echo(socketserver.BaseRequestHandler):

    def handle(self):
        # Get some bytes and echo them back to the client.
        data = self.request.recv(1024)
        self.request.send(data)
        return


if __name__ == '__main__':
    import codecs
    import socket
    import threading

    address = ('localhost', 0)  # Let the kernel assign a port.
    server = socketserver.TCPServer(address, Echo)
    ip, port = server.server_address   # Which port was assigned?

    t = threading.Thread(target=server.serve_forever)
    t.setDaemon(True)   # Don't hang on exit.
    t.start()

    # Connect to the server.
    s = socket.socket(socket.AF_INET, socket.SOCK_STREAM)
    s.connect((ip, port))

    # Send the data.
    # WRONG: Not encoded first!
    text = 'français'
    len_sent = s.send(text)
```

```
    # Receive a response.
    response = s.recv(len_sent)
    print(repr(response))

    # Clean up.
    s.close()
    server.socket.close()
```

可以在每个 send() 调用之前对数据显式编码,不过如果少一个 send() 调用,则会导致编码错误。

```
$ python3 codecs_socket_fail.py

Traceback (most recent call last):
  File "codecs_socket_fail.py", line 43, in <module>
    len_sent = s.send(text)
TypeError: a bytes-like object is required, not 'str'
```

可以使用 makefile() 得到套接字的一个类似文件的句柄,然后用一个基于流的阅读器或书写器来包装这个句柄,确保 Unicode 串传入和传出套接字时会完成编码。

代码清单 6-108:codecs_socket.py

```python
import sys
import socketserver


class Echo(socketserver.BaseRequestHandler):

    def handle(self):
        """Get some bytes and echo them back to the client.

        There is no need to decode them, since they are not used.

        """
        data = self.request.recv(1024)
        self.request.send(data)


class PassThrough:

    def __init__(self, other):
        self.other = other

    def write(self, data):
        print('Writing :', repr(data))
        return self.other.write(data)

    def read(self, size=-1):
        print('Reading :', end=' ')
        data = self.other.read(size)
        print(repr(data))
        return data

    def flush(self):
        return self.other.flush()

    def close(self):
        return self.other.close()
```

```python
if __name__ == '__main__':
    import codecs
    import socket
    import threading

    address = ('localhost', 0)  # Let the kernel assign a port.
    server = socketserver.TCPServer(address, Echo)
    ip, port = server.server_address  # Which port was assigned?

    t = threading.Thread(target=server.serve_forever)
    t.setDaemon(True)  # Don't hang on exit.
    t.start()

    # Connect to the server.
    s = socket.socket(socket.AF_INET, socket.SOCK_STREAM)
    s.connect((ip, port))

    # Wrap the socket with a reader and a writer.
    read_file = s.makefile('rb')
    incoming = codecs.getreader('utf-8')(PassThrough(read_file))
    write_file = s.makefile('wb')
    outgoing = codecs.getwriter('utf-8')(PassThrough(write_file))

    # Send the data.
    text = 'français'
    print('Sending :', repr(text))
    outgoing.write(text)
    outgoing.flush()

    # Receive a response.
    response = incoming.read()
    print('Received:', repr(response))

    # Clean up.
    s.close()
    server.socket.close()
```

这个例子使用 `PassThrough` 来展示数据在发送之前会进行编码，另外客户端接收响应之后会进行解码。

```
$ python3 codecs_socket.py

Sending : 'français'
Writing : b'fran\xc3\xa7ais'
Reading : b'fran\xc3\xa7ais'
Reading : b''
Received: 'français'
```

6.10.9 定义定制编码

由于 Python 已经提供了大量标准 `codecs`，所以应用一般不太可能需要定义定制的编码器或解码器。不过，如果确实有必要，`codecs` 中的很多基类可以帮助你更容易地定义定制编码。

第一步是了解编码描述的转换性质。这一节中的例子将使用一个"invertcaps"编码，它把大写字母转换为小写，把小写字母转换为大写。下面是一个编码函数的简单定义，它会对输入字符串完成这个转换。

代码清单6-109: **codecs_invertcaps.py**

```python
import string


def invertcaps(text):
    """Return new string with the case of all letters switched.
    """
    return ''.join(
        c.upper() if c in string.ascii_lowercase
        else c.lower() if c in string.ascii_uppercase
        else c
        for c in text
    )


if __name__ == '__main__':
    print(invertcaps('ABCdef'))
    print(invertcaps('abcDEF'))
```

在这里，编码器和解码器都是同一个函数（与ROT-13类似）。

```
$ python3 codecs_invertcaps.py

abcDEF
ABCdef
```

尽管很容易理解，但这个实现效率不高，特别是对于非常大的文本串。幸运的是，**codecs**包含一些辅助函数，可以创建基于字符映射（character map）的**codecs**，如invertcaps。字符映射编码由两个字典构成。编码映射（encoding map）将输入串的字符值转换为输出中的字节值，解码映射（decoding map）则相反。首先创建解码映射，然后使用**make_encoding_map()**把它转换为一个编码映射。C函数**charmap_encode()**和**charmap_decode()**可以使用这些映射高效地转换输入数据。

代码清单6-110: **codecs_invertcaps_charmap.py**

```python
import codecs
import string

# Map every character to itself.
decoding_map = codecs.make_identity_dict(range(256))

# Make a list of pairs of ordinal values for the
# lowercase and uppercase letters.
pairs = list(zip(
    [ord(c) for c in string.ascii_lowercase],
    [ord(c) for c in string.ascii_uppercase],
))
# Modify the mapping to convert upper to lower and
# lower to upper.
decoding_map.update({
    upper: lower
    for (lower, upper)
    in pairs
})
decoding_map.update({
    lower: upper
    for (lower, upper)
    in pairs
```

```
})
# Create a separate encoding map.
encoding_map = codecs.make_encoding_map(decoding_map)

if __name__ == '__main__':
    print(codecs.charmap_encode('abcDEF', 'strict',
                                encoding_map))
    print(codecs.charmap_decode(b'abcDEF', 'strict',
                                decoding_map))
    print(encoding_map == decoding_map)
```

尽管 invertcaps 的编码和解码映射是一样的，但并不总是如此。有时会把多个输入字符编码为相同的输出字节，`make_encoding_map()` 会检测这些情况，并把编码值替换为 `None`，以标志编码为未定义。

```
$ python3 codecs_invertcaps_charmap.py

(b'ABCdef', 6)
('ABCdef', 6)
True
```

字符映射编码器和解码器支持前面介绍的所有标准错误处理方法，所以不需要做任何额外的工作来支持这部分 API。

代码清单 6-111: `codecs_invertcaps_error.py`

```
import codecs
from codecs_invertcaps_charmap import encoding_map

text = 'pi: \u03c0'

for error in ['ignore', 'replace', 'strict']:
    try:
        encoded = codecs.charmap_encode(
            text, error, encoding_map)
    except UnicodeEncodeError as err:
        encoded = str(err)
    print('{:7}: {}'.format(error, encoded))
```

由于 π 的 Unicode 码点不在编码映射中，所以采用 strict 错误处理模式时会产生一个异常。

```
$ python3 codecs_invertcaps_error.py

ignore : (b'PI: ', 5)
replace: (b'PI: ?', 5)
strict : 'charmap' codec can't encode character '\u03c0' in
position 4: character maps to <undefined>
```

定义了编码和解码映射之后，还需要建立一些额外的类，另外要注册编码。`register()` 向注册表增加一个搜索函数，使得当用户希望使用这种编码时，`codecs` 能够找到它。这个搜索函数必须有一个字符串参数，其中包含编码名，如果它知道这个编码则返回一个 `CodecInfo` 对象，否则返回 `None`。

代码清单 6-112: `codecs_register.py`

```
import codecs
import encodings
```

```
def search1(encoding):
    print('search1: Searching for:', encoding)
    return None


def search2(encoding):
    print('search2: Searching for:', encoding)
    return None

codecs.register(search1)
codecs.register(search2)

utf8 = codecs.lookup('utf-8')
print('UTF-8:', utf8)

try:
    unknown = codecs.lookup('no-such-encoding')
except LookupError as err:
    print('ERROR:', err)
```

可以注册多个搜索函数,每个搜索函数将依次调用,直到一个搜索函数返回一个 CodecInfo,或者所有搜索函数都已经调用。codecs 注册的内部搜索函数知道如何加载标准 codecs,如 encodings 的 UTF-8,所以这些编码名不会传递到定制搜索函数。

```
$ python3 codecs_register.py

UTF-8: <codecs.CodecInfo object for encoding utf-8 at
0x1007773a8>
search1: Searching for: no-such-encoding
search2: Searching for: no-such-encoding
ERROR: unknown encoding: no-such-encoding
```

搜索函数返回的 CodecInfo 实例告诉 codecs 如何使用所支持的各种不同机制来完成编码和解码,包括:无状态编码、增量式编码和流编码。codecs 包括一些基类来帮助建立字符映射编码。下面这个例子集成了所有内容,它会注册一个搜索函数,并返回为 invertcaps codec 配置的一个 CodecInfo 实例。

代码清单 6-113:codecs_invertcaps_register.py

```
import codecs

from codecs_invertcaps_charmap import encoding_map, decoding_map


class InvertCapsCodec(codecs.Codec):
    "Stateless encoder/decoder"

    def encode(self, input, errors='strict'):
        return codecs.charmap_encode(input, errors, encoding_map)

    def decode(self, input, errors='strict'):
        return codecs.charmap_decode(input, errors, decoding_map)


class InvertCapsIncrementalEncoder(codecs.IncrementalEncoder):
    def encode(self, input, final=False):
        data, nbytes = codecs.charmap_encode(input,
                                              self.errors,
```

```python
                                        encoding_map)
        return data

class InvertCapsIncrementalDecoder(codecs.IncrementalDecoder):
    def decode(self, input, final=False):
        data, nbytes = codecs.charmap_decode(input,
                                             self.errors,
                                             decoding_map)
        return data

class InvertCapsStreamReader(InvertCapsCodec,
                             codecs.StreamReader):
    pass


class InvertCapsStreamWriter(InvertCapsCodec,
                             codecs.StreamWriter):
    pass


def find_invertcaps(encoding):
    """Return the codec for 'invertcaps'.
    """
    if encoding == 'invertcaps':
        return codecs.CodecInfo(
            name='invertcaps',
            encode=InvertCapsCodec().encode,
            decode=InvertCapsCodec().decode,
            incrementalencoder=InvertCapsIncrementalEncoder,
            incrementaldecoder=InvertCapsIncrementalDecoder,
            streamreader=InvertCapsStreamReader,
            streamwriter=InvertCapsStreamWriter,
        )
    return None

codecs.register(find_invertcaps)

if __name__ == '__main__':

    # Stateless encoder/decoder
    encoder = codecs.getencoder('invertcaps')
    text = 'abcDEF'
    encoded_text, consumed = encoder(text)
    print('Encoded "{}" to "{}", consuming {} characters'.format(
        text, encoded_text, consumed))

    # Stream writer
    import io
    buffer = io.BytesIO()
    writer = codecs.getwriter('invertcaps')(buffer)
    print('StreamWriter for io buffer: ')
    print('  writing "abcDEF"')
    writer.write('abcDEF')
    print('  buffer contents: ', buffer.getvalue())

    # Incremental decoder
    decoder_factory = codecs.getincrementaldecoder('invertcaps')
    decoder = decoder_factory()
    decoded_text_parts = []
    for c in encoded_text:
        decoded_text_parts.append(
            decoder.decode(bytes([c]), final=False)
        )
```

```
decoded_text_parts.append(decoder.decode(b'', final=True))
decoded_text = ''.join(decoded_text_parts)
print('IncrementalDecoder converted {!r} to {!r}'.format(
    encoded_text, decoded_text))
```

无状态编码器/解码器的基类是 Codec，要用新实现来覆盖 encode() 和 decode() （在这里分别调用了 charmap_encode() 和 charmap_decode()）。这些方法必须分别返回一个元组，其中包含转换的数据和已消费的输入字节或字符数。charmap_encode() 和 charmap_decode() 已经返回了这个信息，所以很方便。

IncrementalEncoder 和 incrementalDecoder 可以作为增量式编码接口的基类。增量类的 encode() 和 decode() 方法被定义为只返回真正的转换数据。缓冲的有关信息都作为内部状态来维护。invertcaps 编码不需要缓冲数据（它使用一种一对一映射）。如果编码根据所处理的数据会生成不同数量的输出，如压缩算法，那么对于这些编码，BufferedIncrementalEncoder 和 BufferedIncrementalDecoder 将是更合适的基类，因为它们可以管理输入中未处理的部分。

StreamReader 和 StreamWriter 也需要 encode() 和 decode() 方法，而且因为它们往往返回与 Codec 中相应方法同样的值，所以实现时可以使用多重继承。

```
$ python3 codecs_invertcaps_register.py

Encoded "abcDEF" to "b'ABCdef'", consuming 6 characters
StreamWriter for io buffer:
  writing "abcDEF"
  buffer contents:  b'ABCdef'
IncrementalDecoder converted b'ABCdef' to 'abcDEF'
```

提示：相关阅读材料

- codecs 的标准库文档[一]。
- locale：访问和管理基于本地化的配置设置和行为。
- io：io 模块还包含处理编码和解码的文件和流包装器。
- socketserver：要了解一个更详细的回送服务器的例子，请参见 socketserver 模块。
- encodings：标准库中的一个包，其中包含 Python 提供的编码器/解码器实现。
- **PEP 100**[二]：Python Unicode 集成。
- Unicode HOWTO[三]：Python 中使用 Unicode 的官方指南。
- Text vs. Data Instead of Unicode vs. 8-bit[四]：Python 3.0 "What's New" 文章中的一节，介绍文本处理有哪些改变。
- Python Unicode Objects[五]：Fredrik Lundh 的一篇文章，介绍在 Python 2.0 中使用非 ASCII 字符。

[一] https://docs.python.org/3.5/library/codecs.html
[二] www.python.org/dev/peps/pep-0100
[三] https://docs.python.org/3/howto/unicode.html
[四] https://docs.python.org/3.0/whatsnew/3.0.html#text-vs-data-instead-of-unicode-vs-8-bit
[五] http://effbot.org/zone/unicode-objects.html

- How to Use UTF-8 with Python[一]：Evan Jones 编写的一个关于处理 Unicode 的快速指南，包括 XML 数据和字节序标志。
- On the Goodness of Unicode[二]：Tim Bray 对国际化和 Unicode 的介绍。
- On Character Strings[三]：Tim Bray 的一篇文章，介绍编程语言中字符串处理的历史。
- Characters vs. Bytes[四]：这是 Tim Bray 的论文"modern character string processing for computer programmers"的第一部分。这一部分介绍了除 ASCII 字节以外其余格式文本的内存表示。
- Wikipedia: Endianness[五]：维基百科中对字节序的解释。
- W3C XML Entity Definitions for Characters[六]：如果字符引用无法采用一种编码表示，那么这里给出了这些字符引用 XML 表示的规范。

6.11 io：文本、十进制和原始流 I/O 工具

io 模块在解释器的内置 open() 之上实现了一些类来完成基于文件的输入和输出操作。这些类得到了适当的分解，从而可以针对不同的用途重新组合——例如，支持向一个网络套接字写 Unicode 数据。

6.11.1 内存中的流

StringIO 提供了一种很便利的方式，可以使用文件 API（如 read()、write() 等）处理内存中的文本。有些情况下，与其他一些字符串连接技术相比，使用 StringIO 构造大字符串可以提供更好的性能。内存中的流缓冲区对测试也很有用，写入磁盘上真正的文件并不会减慢测试套件的速度。

下面是使用 StringIO 缓冲区的一些标准例子：

代码清单 6-114：io_stringio.py

```
import io

# Write to a buffer.
output = io.StringIO()
output.write('This goes into the buffer. ')
print('And so does this.', file=output)

# Retrieve the value written.
print(output.getvalue())

output.close()   # Discard buffer memory.
```

[一] http://evanjones.ca/python-utf8.html
[二] www.tbray.org/ongoing/When/200x/2003/04/06/Unicode
[三] www.tbray.org/ongoing/When/200x/2003/04/13/Strings
[四] www.tbray.org/ongoing/When/200x/2003/04/26/UTF
[五] https://en.wikipedia.org/wiki/Endianness
[六] www.w3.org/TR/xml-entity-names/

```
# Initialize a read buffer.
input = io.StringIO('Inital value for read buffer')

# Read from the buffer.
print(input.read())
```

这个例子使用了 read(),不过也可以用 readline() 和 readlines() 方法。StringIO 类还提供了一个 seek() 方法,读取文本时可以在缓冲区中跳转,如果使用一种前向解析算法,则这个方法对于回转很有用。

```
$ python3 io_stringio.py

This goes into the buffer. And so does this.

Inital value for read buffer
```

要处理原始字节而不是 Unicode 文本,可以使用 BytesIO。

代码清单 6-115:**io_bytesio.py**

```
import io

# Write to a buffer.
output = io.BytesIO()
output.write('This goes into the buffer. '.encode('utf-8'))
output.write('ÁÇÊ'.encode('utf-8'))

# Retrieve the value written.
print(output.getvalue())

output.close()  # Discard buffer memory.
# Initialize a read buffer.
input = io.BytesIO(b'Inital value for read buffer')

# Read from the buffer.
print(input.read())
```

写入 BytesIO 实例的值一定是 bytes 而不是 str。

```
$ python3 io_bytesio.py

b'This goes into the buffer. \xc3\x81\xc3\x87\xc3\x8a'
b'Inital value for read buffer'
```

6.11.2 为文本数据包装字节流

原始字节流(如套接字)可以被包装为一个层来处理串编码和解码,从而可以更容易地用于处理文本数据。TextIOWrapper 类支持读写。write_through 参数会禁用缓冲,并且立即将写至包装器的所有数据刷新输出到底层缓冲区。

代码清单 6-116:**io_textiowrapper.py**

```
import io

# Write to a buffer.
output = io.BytesIO()
wrapper = io.TextIOWrapper(
```

```
        output,
        encoding='utf-8',
        write_through=True,
    )
    wrapper.write('This goes into the buffer. ')
    wrapper.write('ÁÇÊ')

    # Retrieve the value written.
    print(output.getvalue())

    output.close()  # Discard buffer memory.

    # Initialize a read buffer.
    input = io.BytesIO(
        b'Inital value for read buffer with unicode characters ' +
        'ÁÇÊ'.encode('utf-8')
    )
    wrapper = io.TextIOWrapper(input, encoding='utf-8')

    # Read from the buffer.
    print(wrapper.read())
```

这个例子使用了一个 `BytesIO` 实例作为流。对应 `bz2`、`http.server` 和 `subprocess` 的例子展示了如何对其他类型的类似文件的对象使用 `TextIOWrapper`。

```
$ python3 io_textiowrapper.py

b'This goes into the buffer. \xc3\x81\xc3\x87\xc3\x8a'
Inital value for read buffer with unicode characters ÁÇÊ
```

提示：相关阅读材料

- `io` 的标准库文档[⊖]。
- 12.2.3 节 "HTTP POST"：使用 `TextIOWrapper` 的 `detach()` 独立于所包装的套接字单独管理包装器。
- Efficient String Concatenation in Python[⊖]：分析合并字符串的各种方法及相对优点。

⊖ https://docs.python.org/3.5/library/io.html
⊖ www.skymind.com/%7Eocrow/python_string/

第 7 章

数据持久存储与交换

持久存储数据以便长期使用包括两个方面：在对象的内存中表示和存储格式之间来回转换数据，以及处理转换后数据的存储区。标准库包含很多模块可以处理不同情况下的这两个方面。

有两个模块可以将对象转换为一种可传输或存储的格式（这个过程被称为串行化(serializing)）。最常用的是使用 pickle 持久存储，因为它可以与其他一些具体存储串行化数据的标准库模块集成，如 shelve。而对基于 Web 的应用，json 更为常用，因为它能更好地与现有 Web 服务存储工具集成。

一旦将内存中对象转换为一种可保存的格式，那么下一步就是确定如何存储这个数据。如果数据不需要以某种方式索引，则依序先后写入串行化对象的简单平面文件就很适用。Python 包括一组模块可以在一个简单的数据库中存储键 - 值对，需要索引查找时会使用某种 DBM 变形格式。

要利用 DBM 格式，最直接的方式就是使用 shelve。可以打开 shelve 文件，通过一个类字典的 API 来访问。保存到数据库的对象会自动"腌制"并保存，而无须调用者做任何额外的工作。

不过 shelve 有一个缺点，使用默认接口时，没有办法预测将使用哪一个 DBM 格式：因为 shelve 会根据创建数据库的系统上有哪些可用的库来选择一个格式。如果应用不需要在配置有不同的库的主机之间共享数据库文件，那么选择哪个格式并不重要；不过，如果必须保证可移植性，则可以使用这个模块中的某个类来确保选择一个特定的格式。

对于 Web 应用，由于这些应用已经在处理 JSON 格式的数据，因此可以使用 json 和 dbm 提供另一种持久存储机制。直接使用 dbm 会比使用 shelve 稍微多做一些工作，因为 DBM 数据库键和值必须是字符串，而且在数据库中访问值时不会自动地重新创建对象。

大多数 Python 发布版本都提供了 sqlite3 进程中关系数据库，可以采用比键 - 值对更复杂的组织来存储数据。它将数据库存储在内存中或者存储在一个本地文件中，所有访问都在同一个进程中，所以不存在网络通信延迟。sqlite3 的紧凑性使它尤其适合嵌入到桌面应用或 Web 应用的开发版本中。

还有一些模块可以用来解析这些定义得更正式的格式，这对于在 Python 程序和用其他语言编写的应用之间交换数据非常有用。xml.etree.ElementTree 可以解析 XML 文档，为不同应用提供多种操作模式。除了解析工具，ElementTree 还包括一个接口可以由内存中的对象创建良构的 XML 文档。csv 模块可以读写表格数据（采用由电子表格或数据库应用生成的格式），这对于批量加载数据或者将数据从一种格式转换为另一种格式非常有用。

7.1 pickle：对象串行化

pickle 模块实现了一个算法可以将一个任意的 Python 对象转换为一系列字节。这个过程也被称为串行化对象。可以传输或存储表示对象的字节流，然后再重新构造来创建有相同性质的新对象。

警告：pickle 的文档明确指出它不提供任何安全保证。实际上，对数据解除腌制可以执行任意的代码。使用 pickle 完成进程间通信或数据存储时要当心，另外不要相信未经过安全验证的数据。参见 hmac 模块，其中有一个例子展示了验证腌制数据源来源的一种安全方法。

7.1.1 编码和解码字符串中的数据

第一个例子使用 dumps() 将一个数据结构编码为一个字符串，然后把这个字符串打印到控制台。它使用了一个完全由内置类型构成的数据结构。任何类的实例都可以腌制，如后面的例子所示。

代码清单 7-1：**pickle_string.py**

```
import pickle
import pprint

data = [{'a': 'A', 'b': 2, 'c': 3.0}]
print('DATA:', end=' ')
pprint.pprint(data)

data_string = pickle.dumps(data)
print('PICKLE: {!r}'.format(data_string))
```

默认地，pickle 将以一种二进制格式写入，在 Python 3 程序之间共享时这种格式兼容性最好。

```
$ python3 pickle_string.py

DATA: [{'a': 'A', 'b': 2, 'c': 3.0}]
PICKLE: b'\x80\x03]q\x00}q\x01(X\x01\x00\x00\x00cq\x02G@\x08\x00
\x00\x00\x00\x00\x00X\x01\x00\x00\x00bq\x03K\x02X\x01\x00\x00\x0
0aq\x04X\x01\x00\x00\x00Aq\x05ua.'
```

数据串行化后，可以写到一个文件、套接字、管道或者其他位置。之后可以读取这个文件，将数据解除腌制，以便用同样的值构造一个新对象。

代码清单 7-2：**pickle_unpickle.py**

```
import pickle
import pprint

data1 = [{'a': 'A', 'b': 2, 'c': 3.0}]
print('BEFORE: ', end=' ')
pprint.pprint(data1)

data1_string = pickle.dumps(data1)

data2 = pickle.loads(data1_string)
```

```
            print('AFTER : ', end=' ')
            pprint.pprint(data2)

            print('SAME? :', (data1 is data2))
            print('EQUAL?:', (data1 == data2))
```

新构造的对象等于原来的对象,但并不是同一个对象。

```
$ python3 pickle_unpickle.py

BEFORE: [{'a': 'A', 'b': 2, 'c': 3.0}]
AFTER : [{'a': 'A', 'b': 2, 'c': 3.0}]
SAME? : False
EQUAL?: True
```

7.1.2 处理流

除了 `dumps()` 和 `loads()`,pickle 还提供了一些便利函数来处理类似文件的流。可以向一个流写多个对象,然后从流读取这些对象,而无须事先知道要写多少个对象或者这些对象有多大。

代码清单 7-3:`pickle_stream.py`

```python
import io
import pickle
import pprint


class SimpleObject:

    def __init__(self, name):
        self.name = name
        self.name_backwards = name[::-1]
        return


data = []
data.append(SimpleObject('pickle'))
data.append(SimpleObject('preserve'))
data.append(SimpleObject('last'))

# Simulate a file.
out_s = io.BytesIO()

# Write to the stream.
for o in data:
    print('WRITING : {} ({})'.format(o.name, o.name_backwards))
    pickle.dump(o, out_s)
    out_s.flush()

# Set up a readable stream.
in_s = io.BytesIO(out_s.getvalue())

# Read the data.
while True:
    try:
        o = pickle.load(in_s)
    except EOFError:
        break
    else:
        print('READ    : {} ({})'.format(
            o.name, o.name_backwards))
```

这个例子使用两个 BytesIO 缓冲区来模拟流。第一个缓冲区接收腌制的对象，它的值被填入第二个缓冲区，load() 将读取这个缓冲区。简单的数据库格式也可以使用 pickle 来存储对象。shelve 模块就是这样一个实现。

```
$ python3 pickle_stream.py

WRITING : pickle (elkcip)
WRITING : preserve (evreserp)
WRITING : last (tsal)
READ    : pickle (elkcip)
READ    : preserve (evreserp)
READ    : last (tsal)
```

除了存储数据，pickle 对于进程间通信也很方便。例如，os.fork() 和 os.pipe() 可以用来建立工作进程，从一个管道读取作业指令，并把结果写至另一个管道。管理工作线程池以及发送作业和接收响应的核心代码可以重用，因为作业和响应对象不必基于一个特定的类。使用管道或套接字时，在转储各个对象之后不要忘记刷新输出，以便将数据通过连接推送到另一端。参见 multiprocessing 模块来了解一个可重用的工作线程池管理器。

7.1.3 重构对象的问题

处理定制类时，腌制的类必须出现在读取 pickle 的进程所在的命名空间里。只会腌制这个实例的数据，而不是类定义。类名用于查找构造函数，以便在解除腌制时创建新对象。下面这个例子将一个类的实例写至一个文件。

代码清单 7-4：`pickle_dump_to_file_1.py`

```python
import pickle
import sys


class SimpleObject:

    def __init__(self, name):
        self.name = name
        l = list(name)
        l.reverse()
        self.name_backwards = ''.join(l)


if __name__ == '__main__':
    data = []
    data.append(SimpleObject('pickle'))
    data.append(SimpleObject('preserve'))
    data.append(SimpleObject('last'))

    filename = sys.argv[1]

    with open(filename, 'wb') as out_s:
        for o in data:
            print('WRITING: {} ({})'.format(
                o.name, o.name_backwards))
            pickle.dump(o, out_s)
```

运行这个脚本时，会根据作为命令行参数给定的名字来创建一个文件。

```
$ python3 pickle_dump_to_file_1.py test.dat

WRITING: pickle (elkcip)
WRITING: preserve (evreserp)
WRITING: last (tsal)
```

通过简单地尝试加载而得到的腌制对象将会失败。

代码清单 7-5：`pickle_load_from_file_1.py`

```python
import pickle
import pprint
import sys

filename = sys.argv[1]

with open(filename, 'rb') as in_s:
    while True:
        try:
            o = pickle.load(in_s)
        except EOFError:
            break
        else:
            print('READ: {} ({})'.format(
                o.name, o.name_backwards))
```

这个版本失败的原因在于并没有 `SimpleObject` 类。

```
$ python3 pickle_load_from_file_1.py test.dat

Traceback (most recent call last):
  File "pickle_load_from_file_1.py", line 15, in <module>
    o = pickle.load(in_s)
AttributeError: Can't get attribute 'SimpleObject' on <module '_
_main__' from 'pickle_load_from_file_1.py'>
```

修正后的版本从原脚本导入了 `SimpleObject`，这一次运行会成功。在导入列表的最后增加这个 `import` 语句后，现在脚本就能找到这个类并构造对象了。

```python
from pickle_dump_to_file_1 import SimpleObject
```

现在运行修改后的脚本会生成期望的结果。

```
$ python3 pickle_load_from_file_2.py test.dat

READ: pickle (elkcip)
READ: preserve (evreserp)
READ: last (tsal)
```

7.1.4 不可腌制的对象

并不是所有对象都是可腌制的。套接字、文件句柄、数据库连接以及其他运行时状态依赖于操作系统或其他进程的对象，其可能无法用一种有意义的方式保存。如果对象包含不可腌制的属性，则可以定义 `__getstate__()` 和 `__setstate__()` 来返回所腌制实例的状态的一个子集。

`__getstate__()` 方法必须返回一个对象，其中包含所腌制对象的内部状态。表示状态的一种便利方式是使用字典，不过值可以是任意的可腌制对象。保存状态，然后在从 `pickle` 加载对象时将所保存的状态传入 `__setstate__()`。

代码清单 7-6：`pickle_state.py`

```python
import pickle


class State:

    def __init__(self, name):
        self.name = name

    def __repr__(self):
        return 'State({!r})'.format(self.__dict__)


class MyClass:

    def __init__(self, name):
        print('MyClass.__init__({})'.format(name))
        self._set_name(name)

    def _set_name(self, name):
        self.name = name
        self.computed = name[::-1]

    def __repr__(self):
        return 'MyClass({!r}) (computed={!r})'.format(
            self.name, self.computed)

    def __getstate__(self):
        state = State(self.name)
        print('__getstate__ -> {!r}'.format(state))
        return state

    def __setstate__(self, state):
        print('__setstate__({!r})'.format(state))
        self._set_name(state.name)


inst = MyClass('name here')
print('Before:', inst)
dumped = pickle.dumps(inst)

reloaded = pickle.loads(dumped)
print('After:', reloaded)
```

这个例子使用了一个单独的 `State` 对象来保存 `MyClass` 的内部状态。从 `pickle` 加载 `MyClass` 的一个实例时，会向 `__setstate__()` 传入一个 `State` 实例，用来初始化这个对象。

```
$ python3 pickle_state.py

MyClass.__init__(name here)
Before: MyClass('name here') (computed='ereh eman')
__getstate__ -> State({'name': 'name here'})
__setstate__(State({'name': 'name here'}))
After: MyClass('name here') (computed='ereh eman')
```

警告：如果返回值为 false，则对象解除腌制时不会调用 `__setstate__()`。

7.1.5 循环引用

pickle协议会自动处理对象之间的循环引用，所以复杂数据结构不需要任何特殊的处理。考虑图7-1中的有向图。这个图中包含几个环，不过仍然可以腌制所得的正确结构然后重新加载。

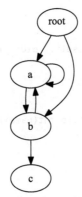

图 7-1 腌制带环的数据结构

代码清单 7-7：**pickle_cycle.py**

```
import pickle


class Node:
    """A simple digraph
    """
    def __init__(self, name):
        self.name = name
        self.connections = []

    def add_edge(self, node):
        "Create an edge between this node and the other."
        self.connections.append(node)

    def __iter__(self):
        return iter(self.connections)


def preorder_traversal(root, seen=None, parent=None):
    """Generator function to yield the edges in a graph.
    """
    if seen is None:
        seen = set()
    yield (parent, root)
    if root in seen:
        return
    seen.add(root)
    for node in root:
        recurse = preorder_traversal(node, seen, root)
        for parent, subnode in recurse:
            yield (parent, subnode)


def show_edges(root):
```

```
        "Print all the edges in the graph."
        for parent, child in preorder_traversal(root):
            if not parent:
                continue
            print('{:>5} -> {:>2} ({})'.format(
                parent.name, child.name, id(child)))

# Set up the nodes.
root = Node('root')
a = Node('a')
b = Node('b')
c = Node('c')

# Add edges between them.
root.add_edge(a)
root.add_edge(b)
a.add_edge(b)
b.add_edge(a)
b.add_edge(c)
a.add_edge(a)

print('ORIGINAL GRAPH:')
show_edges(root)

# Pickle and unpickle the graph to create
# a new set of nodes.
dumped = pickle.dumps(root)
reloaded = pickle.loads(dumped)

print('\nRELOADED GRAPH:')
show_edges(reloaded)
```

重新加载的节点并不是同一个对象，但保持了节点之间的关系，而且如果对象有多个引用，那么只会重新加载这个对象的一个副本。要验证这两点，可以在通过 `pickle` 传递节点之前和之后检查节点的 `id()` 值。

```
$ python3 pickle_cycle.py

ORIGINAL GRAPH:
 root ->  a (4315798272)
    a ->  b (4315798384)
    b ->  a (4315798272)
    b ->  c (4315799112)
    a ->  a (4315798272)
 root ->  b (4315798384)

RELOADED GRAPH:
 root ->  a (4315904096)
    a ->  b (4315904152)
    b ->  a (4315904096)
    b ->  c (4315904208)
    a ->  a (4315904096)
 root ->  b (4315904152)
```

提示：相关阅读材料
- `pickle` 的标准库文档[⊖]。

[⊖] https://docs.python.org/3.5/library/pickle.html

- **PEP 3154**[⊖]: pickle 协议版本 4。
- **shelve**:shelve 模块使用 pickle 在 DBM 数据库中存储数据。
- Pickle: An interesting stack language[⊖]:Alexandre Vassalotti 提供的一个教程。

7.2 shelve:对象的持久存储

不需要关系数据库时,可以用 shelve 模块作为持久存储 Python 对象的一个简单的选择。类似于字典,shelf 按键访问。值将被腌制并写至由 dbm 创建和管理的数据库。

7.2.1 创建一个新 shelf

使用 shelve 最简单的方法就是利用 DbfilenameShelf 类。它使用 dbm 存储数据。这个类可以直接使用,也可以通过调用 shelve.open() 来使用。

代码清单 7-8:**shelve_create.py**

```
import shelve

with shelve.open('test_shelf.db') as s:
    s['key1'] = {
        'int': 10,
        'float': 9.5,
        'string': 'Sample data',
    }
```

要再次访问这个数据,可以打开 shelf,并像字典一样使用它。

代码清单 7-9:**shelve_existing.py**

```
import shelve

with shelve.open('test_shelf.db') as s:
    existing = s['key1']

print(existing)
```

运行这两个示例脚本会生成以下输出。

```
$ python3 shelve_create.py
$ python3 shelve_existing.py

{'string': 'Sample data', 'int': 10, 'float': 9.5}
```

dbm 模块不支持多个应用同时写同一个数据库,不过它支持并发的只读客户。如果一个客户没有修改 shelf,则可以通过传入 flag='r' 来告诉 shelve 以只读方式打开数据库。

代码清单 7-10:**shelve_readonly.py**

```
import dbm
import shelve
```

[⊖] www.python.org/dev/peps/pep-3154

[⊖] http://peadrop.com/blog/2007/06/18/pickle-an-interesting-stack-language/

```
with shelve.open('test_shelf.db', flag='r') as s:
    print('Existing:', s['key1'])
    try:
        s['key1'] = 'new value'
    except dbm.error as err:
        print('ERROR: {}'.format(err))
```

如果数据库作为只读数据源打开，并且程序试图修改数据库，那么便会生成一个访问错误异常。具体的异常类型取决于创建数据库时 **dbm** 选择的数据库模块。

```
$ python3 shelve_readonly.py

Existing: {'string': 'Sample data', 'int': 10, 'float': 9.5}
ERROR: cannot add item to database
```

7.2.2 写回

默认地，shelf 不会跟踪对可变对象的修改。这说明，如果存储在 shelf 中的一个元素的内容有变化，那么 shelf 必须再次存储整个元素来显式地更新。

代码清单 7-11: shelve_withoutwriteback.py

```
import shelve

with shelve.open('test_shelf.db') as s:
    print(s['key1'])
    s['key1']['new_value'] = 'this was not here before'

with shelve.open('test_shelf.db', writeback=True) as s:
    print(s['key1'])
```

在这个例子中，没有再次存储 `key1` 的相应字典，所以重新打开 shelf 时，修改不会保留。

```
$ python3 shelve_create.py
$ python3 shelve_withoutwriteback.py
{'string': 'Sample data', 'int': 10, 'float': 9.5}
{'string': 'Sample data', 'int': 10, 'float': 9.5}
```

对于 shelf 中存储的可变对象，要想自动捕获对它们的修改，可以在打开 shelf 时启用写回（writeback）。`writeback` 标志会让 shelf 使用内存中缓存以记住从数据库获取的所有对象。shelf 关闭时每个缓存对象也被写回到数据库。

代码清单 7-12: shelve_writeback.py

```
import shelve
import pprint

with shelve.open('test_shelf.db', writeback=True) as s:
    print('Initial data:')
    pprint.pprint(s['key1'])

    s['key1']['new_value'] = 'this was not here before'
    print('\nModified:')
    pprint.pprint(s['key1'])

with shelve.open('test_shelf.db', writeback=True) as s:
    print('\nPreserved:')
    pprint.pprint(s['key1'])
```

尽管这会减少程序员犯错的机会，并且使对象持久存储更透明，但是并非所有情况都有必要使用写回模式。打开 shelf 时缓存会消耗额外的内存，关闭 shelf 时会暂停将各个缓存对象写回到数据库，这会减慢应用的速度。所有缓存的对象都要写回数据库，因为无法区分它们是否有修改。如果应用读取的数据多于写的数据，那么写回就会影响性能而没有太大意义。

```
$ python3 shelve_create.py
$ python3 shelve_writeback.py

Initial data:
{'float': 9.5, 'int': 10, 'string': 'Sample data'}

Modified:
{'float': 9.5,
 'int': 10,
 'new_value': 'this was not here before',
 'string': 'Sample data'}

Preserved:
{'float': 9.5,
 'int': 10,
 'new_value': 'this was not here before',
 'string': 'Sample data'}
```

7.2.3 特定 shelf 类型

之前的例子都使用了默认的 shelf 实现。可以使用 `shelve.open()` 而不是直接使用某个 shelf 实现，这是一种常见的用法，特别是使用什么类型的数据库来存储数据并不重要时。不过，有些情况下数据库格式会很重要。在这些情况下，可以直接使用 `Dbfilename-Shelf` 或 `BsdDbShelf`，或者甚至可以派生 `Shelf` 来得到一个定制解决方案。

提示：相关阅读材料

- `shelve` 的标准库文档[⊖]。
- `dbm`：`dbm` 模块查找一个可用的 DBM 库以创建新的数据库。
- `feedcache`[⊖]：`feedcache` 模块使用 `shelve` 作为默认的存储选择。
- `shove`[⊜]：`shove` 模块实现了一个类似的 API，不过提供了更多后端格式。

7.3 dbm：UNIX 键–值数据库

dbm 是面向 DBM 数据库的一个前端，DBM 数据库使用简单的字符串值作为键来访问包含字符串的记录。dbm 使用 `whichdb()` 标识数据库，然后用适当的模块打开这些数据库。dbm 还被用作 shelve 的一个后端，shelve 使用 pickle 将对象存储在一个 DBM 数据库中。

⊖ https://docs.python.org/3.5/library/shelve.html
⊖ https://bitbucket.org/dhellmann/feedcache
⊜ http://pypi.python.org/pypi/shove/

7.3.1 数据库类型

Python 提供了很多模块来访问 DBM 数据库。具体选择的默认实现取决于当前系统上可用的库以及编译 Python 时使用的选项。特定实现有单独的接口，这使得 Python 程序可以与用其他语言编写的程序（这些语言可能不会在可用格式之间自动切换）交换数据，并且可以写适用于多个平台的可移植的数据文件。

7.3.1.1 dbm.gnu

dbm.gnu 是 GNU 项目 dbm 库版本的一个接口。它的工作方式与这里介绍的其他 DBM 实现相同，只是 open() 支持的标志有些不同。

除了标准 'r'、'w'、'c' 和 'n' 标志，dbm.gnu.open() 还支持以下标志：

- 'f' 以快速（fast）模式打开数据库。在快速模式下，对数据库的写并不同步。
- 's' 以同步（synchronized）模式打开数据库。对数据库做出修改时，这些改变要直接写至文件，而不是延迟到数据库关闭或显式同步时才写至文件。
- 'u' 以不加锁（unlocked）的状态打开数据库。

7.3.1.2 dbm.ndbm

dbm.ndbm 模块为 dbm 格式的 UNIX ndbm 实现提供了一个接口，依赖于编译时这个模块如何配置。模块属性 library 指示编译扩展模块时 configure 能找到的库名。

7.3.1.3 dbm.dumb

dbm.dumb 模块是没有其他实现可用时 DBM API 的一个可移植的后备实现。使用 dbm.dumb 模块不需要依赖任何外部库，但它的速度比大多数其他实现都慢。

7.3.2 创建一个新数据库

通过按顺序查找以下各个子模块的可用版本来选择新数据库的存储格式。

- dbm.gnu
- dbm.ndbm
- dbm.dumb

open() 函数接收 flags 来控制如何管理数据库文件。如果要在必要时创建一个新的数据库，则可以使用 'c'。使用 'n' 则总会创建一个新数据库，并覆盖现有的文件。

代码清单 7-13：**dbm_new.py**

```
import dbm

with dbm.open('/tmp/example.db', 'n') as db:
    db['key'] = 'value'
    db['today'] = 'Sunday'
    db['author'] = 'Doug'
```

在这个例子中，文件总会重新初始化。

```
$ python3 dbm_new.py
```

whichdb() 会报告所创建数据库的类型。

代码清单 7-14：**dbm_whichdb.py**

```
import dbm

print(dbm.whichdb('/tmp/example.db'))
```

取决于系统上安装的模块，示例程序的输出可能有所不同。

```
$ python3 dbm_whichdb.py

dbm.ndbm
```

7.3.3 打开一个现有数据库

要打开一个现有数据库，可以使用 flags 'r'（只读）或 'w'（读写）。会自动将现有的数据库提供给 whichdb() 来识别，所以只要一个文件可以识别，便能使用一个适当的模块打开这个文件。

代码清单 7-15：**dbm_existing.py**

```
import dbm

with dbm.open('/tmp/example.db', 'r') as db:
    print('keys():', db.keys())
    for k in db.keys():
        print('iterating:', k, db[k])
    print('db["author"] =', db['author'])
```

一旦打开，则 db 是一个类字典的对象。增加到数据库时，新键总是被转换为字节串，并且作为字节串返回。

```
$ python3 dbm_existing.py

keys(): [b'key', b'today', b'author']
iterating: b'key' b'value'
iterating: b'today' b'Sunday'
iterating: b'author' b'Doug'
db["author"] = b'Doug'
```

7.3.4 错误情况

数据库的键必须是字符串。

代码清单 7-16：**dbm_intkeys.py**

```
import dbm

with dbm.open('/tmp/example.db', 'w') as db:
    try:
        db[1] = 'one'
    except TypeError as err:
        print(err)
```

如果传入其他类型则会导致一个 TypeError。

```
$ python3 dbm_intkeys.py

dbm mappings have bytes or string keys only
```

值必须是字符串或 None。

代码清单 7-17：`dbm_intvalue.py`

```
import dbm

with dbm.open('/tmp/example.db', 'w') as db:
    try:
        db['one'] = 1
    except TypeError as err:
        print(err)
```

如果值不是一个字符串，则会产生一个类似的 `TypeError`。

```
$ python3 dbm_intvalue.py

dbm mappings have byte or string elements only
```

提示：相关阅读材料
- `dbm` 的标准库文档[○]。
- `anydbm` 的 Python 2 到 Python 3 移植说明。
- `whichdb` 的 Python 2 到 Python 3 移植说明。
- `shelve`：`shelve` 模块的示例，使用 `dbm` 存储数据。

7.4 `sqlite3`：嵌入式关系数据库

`sqlite3` 模块为 SQLite 提供了一个 DB-API 2.0[○]兼容接口，SQLite 是一个进程中关系数据库。SQLite 被设计为嵌入在应用中，而不是像 MySQL、PostgreSQL 或 Oracle 那样使用一个单独的数据库服务器程序。SQLite 的速度很快，并且经过了严格的测试，很灵活，所以非常适合为一些应用建立原型和完成生产部署。

7.4.1 创建数据库

SQLite 数据库作为一个文件存储在文件系统中。`sqlite3` 会管理对这个文件的访问，包括加锁来防止多个书写器使用这个文件时造成破坏。第一次访问这个文件时会创建数据库，不过要由 应用负责管理数据库中的数据库表定义，即模式（schema）。

下面这个例子在用 `connect()` 打开数据库文件之前会先查找数据库文件，我们可以通过这个例子来了解什么时候为新数据库创建模式。

代码清单 7-18：`sqlite3_createdb.py`

```
import os
import sqlite3

db_filename = 'todo.db'
```

○ https://docs.python.org/3.5/library/dbm.html
○ www.python.org/dev/peps/pep-0249/

```python
db_is_new = not os.path.exists(db_filename)

conn = sqlite3.connect(db_filename)

if db_is_new:
    print('Need to create schema')
else:
    print('Database exists; assume schema does, too.')

conn.close()
```

将这个脚本运行两次，可以看到，如果文件还不存在，则它会创建空文件。

```
$ ls *.db

ls: *.db: No such file or directory

$ python3 sqlite3_createdb.py

Need to create schema

$ ls *.db

todo.db

$ python3 sqlite3_createdb.py

Database exists; assume schema does, too.
```

创建新的数据库文件后，下一步是创建模式来定义数据库中的表。这一节余下的例子都使用相同的数据库模式，这里包含一些管理任务的表。这个数据库模式的详细信息见表 7-1 和表 7-2。

创建这些表的数据定义语言（Data Definition Language，DDL）语句见以下代码清单。

代码清单 7-19：todo_schema.sql

```sql
-- Schema for to-do application examples

-- Projects are high-level activities made up of tasks
create table project (
    name        text primary key,
    description text,
    deadline    date
);

-- Tasks are steps that can be taken to complete a project
create table task (
    id           integer primary key autoincrement not null,
    priority     integer default 1,
    details      text,
    status       text,
    deadline     date,
    completed_on date,
    project      text not null references project(name)
);
```

表 7-1 project 表

列	类型	描述
name	text	项目名
description	text	详细的项目描述
deadline	date	整个项目的截止日期

表 7-2 task 表

列	类型	描　　述
id	number	唯一的任务标识符
priority	integer	优先级数值；值越小越重要
details	text	完整的任务详细描述
status	text	任务状态（可以是 new（新建）、pending（未完成）、done（完成）或 canceled（取消））。
deadline	date	这个任务的截止日期
completed_on	date	任务何时完成
project	text	这个任务对应的项目名

可以用 Connection 的 executescript() 方法运行创建模式的 DDL 指令。

代码清单 7-20：sqlite3_create_schema.py

```python
import os
import sqlite3

db_filename = 'todo.db'
schema_filename = 'todo_schema.sql'

db_is_new = not os.path.exists(db_filename)

with sqlite3.connect(db_filename) as conn:
    if db_is_new:
        print('Creating schema')
        with open(schema_filename, 'rt') as f:
            schema = f.read()
        conn.executescript(schema)

        print('Inserting initial data')

        conn.executescript("""
        insert into project (name, description, deadline)
        values ('pymotw', 'Python Module of the Week',
                '2016-11-01');

        insert into task (details, status, deadline, project)
        values ('write about select', 'done', '2016-04-25',
                'pymotw');

        insert into task (details, status, deadline, project)
        values ('write about random', 'waiting', '2016-08-22',
                'pymotw');

        insert into task (details, status, deadline, project)
        values ('write about sqlite3', 'active', '2017-07-31',
```

```
                'pymotw');
        """)
    else:
        print('Database exists, assume schema does, too.')
```

创建这些数据表之后，用一些 insert 语句来创建一个示例项目和相关的任务。可以用 sqlite3 命令行程序检查数据库的内容。

```
$ rm -f todo.db
$ python3 sqlite3_create_schema.py

Creating schema
Inserting initial data

$ sqlite3 todo.db 'select * from task'

1|1|write about select|done|2016-04-25||pymotw
2|1|write about random|waiting|2016-08-22||pymotw
3|1|write about sqlite3|active|2017-07-31||pymotw
```

7.4.2 获取数据

要从一个 Python 程序中获取 task 表中保存的值，可以从数据库连接创建一个 Cursor。游标（cursor）会生成一致的数据视图，这也是与类似 SQLite 的事务型数据库系统交互的主要方式。

代码清单 7-21：sqlite3_select_tasks.py

```
import sqlite3

db_filename = 'todo.db'

with sqlite3.connect(db_filename) as conn:
    cursor = conn.cursor()

    cursor.execute("""
    select id, priority, details, status, deadline from task
    where project = 'pymotw'
    """)

    for row in cursor.fetchall():
        task_id, priority, details, status, deadline = row
        print('{:2d} [{:d}] {:<25} [{:<8}] ({})'.format(
            task_id, priority, details, status, deadline))
```

查询过程包括两步。首先，用游标的 execute() 方法运行查询，告诉数据库引擎要收集哪些数据。然后，使用 fetchall() 获取结果。返回值是一个元组序列，元组中包含查询 select 子句中所包括的列的值。

```
$ python3 sqlite3_select_tasks.py

 1 [1] write about select        [done    ] (2016-04-25)
 2 [1] write about random        [waiting ] (2016-08-22)
 3 [1] write about sqlite3       [active  ] (2017-07-31)
```

可以用 fetchone() 一次获取一个结果，也可以用 fetchmany() 获取固定大小的批量结果。

代码清单 7-22：**sqlite3_select_variations.py**

```python
import sqlite3

db_filename = 'todo.db'

with sqlite3.connect(db_filename) as conn:
    cursor = conn.cursor()

    cursor.execute("""
    select name, description, deadline from project
    where name = 'pymotw'
    """)
    name, description, deadline = cursor.fetchone()

    print('Project details for {} ({})\n  due {}'.format(
        description, name, deadline))

    cursor.execute("""
    select id, priority, details, status, deadline from task
    where project = 'pymotw' order by deadline
    """)

    print('\nNext 5 tasks:')
    for row in cursor.fetchmany(5):
        task_id, priority, details, status, deadline = row
        print('{:2d} [{:d}] {:<25} [{:<8}] ({})'.format(
            task_id, priority, details, status, deadline))
```

传入 `fetchmany()` 的值是要返回的最大元素数。如果没有提供足够的元素，则返回的序列大小将小于这个最大值。

```
$ python3 sqlite3_select_variations.py

Project details for Python Module of the Week (pymotw)
  due 2016-11-01

Next 5 tasks:
 1 [1] write about select     [done    ] (2016-04-25)
 2 [1] write about random     [waiting ] (2016-08-22)
 3 [1] write about sqlite3    [active  ] (2017-07-31)
```

7.4.3 查询元数据

DB-API 2.0 规范指出：调用 `execute()` 之后，`Cursor` 应当设置其 `description` 属性来保存数据的有关信息，这些信息将由获取方法返回。API 规范指出这个描述值是一个元组序列，元组包含列名、类型、显示大小、内部大小、精度、范围和一个指示是否接受 null 值的标志。

代码清单 7-23：**sqlite3_cursor_description.py**

```python
import sqlite3

db_filename = 'todo.db'

with sqlite3.connect(db_filename) as conn:
    cursor = conn.cursor()

    cursor.execute("""
    select * from task where project = 'pymotw'
    """)

    print('Task table has these columns:')
    for colinfo in cursor.description:
        print(colinfo)
```

由于 `sqlite3` 对插入到数据库的数据没有类型或大小约束，所以只填入列名值。

```
$ python3 sqlite3_cursor_description.py

Task table has these columns:
('id', None, None, None, None, None, None)
('priority', None, None, None, None, None, None)
('details', None, None, None, None, None, None)
('status', None, None, None, None, None, None)
('deadline', None, None, None, None, None, None)
('completed_on', None, None, None, None, None, None)
('project', None, None, None, None, None, None)
```

7.4.4 行对象

默认地，获取方法从数据库返回的"行"值是元组。调用者要负责了解查询中列的顺序，并从元组中抽取单个的值。查询的值个数增加时，或者处理数据的代码分布在一个库的不同位置时，通常比较容易的做法是使用一个对象，并用它的列名来访问值。这样一来，编辑查询时，元组内容的个数和顺序会随之改变，并且依赖于查询结果的代码也不太会出问题。

Connection 对象有一个 row_factory 属性，允许调用代码控制所创建对象的类型来表示查询结果集中的各行。sqlite3 还包括一个 Row 类，这个类被用作一个行工厂。可以通过 Row 实例使用列索引或名来访问列值。

代码清单 7-24：sqlite3_row_factory.py

```python
import sqlite3

db_filename = 'todo.db'

with sqlite3.connect(db_filename) as conn:
    # Change the row factory to use Row.
    conn.row_factory = sqlite3.Row

    cursor = conn.cursor()

    cursor.execute("""
    select name, description, deadline from project
    where name = 'pymotw'
    """)
    name, description, deadline = cursor.fetchone()

    print('Project details for {} ({})\n  due {}'.format(
        description, name, deadline))

    cursor.execute("""
    select id, priority, status, deadline, details from task
    where project = 'pymotw' order by deadline
    """)

    print('\nNext 5 tasks:')
    for row in cursor.fetchmany(5):
        print('{:2d} [{:d}] {:<25} [{:<8}] ({})'.format(
            row['id'], row['priority'], row['details'],
            row['status'], row['deadline'],
        ))
```

这个版本的 sqlite3_select_variations.py 例子被重写为使用 Row 实例而不是元组。打印 project 表中的行时仍然通过位置访问列值，不过任务的 print 语句使用了关键字查找，所以查询中列顺序的改变不会有任何影响。

```
$ python3 sqlite3_row_factory.py

Project details for Python Module of the Week (pymotw)
  due 2016-11-01

Next 5 tasks:
 1 [1] write about select          [done    ] (2016-04-25)
 2 [1] write about random          [waiting ] (2016-08-22)
 3 [1] write about sqlite3         [active  ] (2017-07-31)
```

7.4.5 在查询中使用变量

如果查询被定义为字面量字符串，要嵌入到程序中，那么使用这种查询会很不灵活。例如，向数据库增加另一个项目时，显示前 5 个任务的查询就应当更新，从而能处理其中任意一个项目。要想增加灵活性，一种方法是建立一个 SQL 语句，通过在 Python 中结合相应的值来提供所需的查询。不过，以这种方式构造查询串很危险，应当尽量避免。如果未能对查询中可变部分的特殊字符正确地转义，则可能会导致 SQL 解析错误，或者更糟糕地，还有可能导致一类被称为 SQL 注入攻击（SQL-injection attack）的安全漏洞，这会使入侵者能够在数据库中执行任意的 SQL 语句。

要在查询中使用动态值，正确的方法是利用随 SQL 指令一起传入 `execute()` 的宿主变量（host variable）。执行 SQL 语句时，SQL 中的占位符值会替换为宿主变量的值。通过使用宿主变量，而不是解析之前在 SQL 语句中插入任意的值，这样可以避免注入攻击，因为不可信的值没有机会影响 SQL 语句的解析。SQLite 支持两种形式带占位符的查询，分别是位置参数和命名参数。

7.4.5.1 位置参数

问号（?）指示一个位置参数，将作为元组的一个成员被传至 `execute()`。

代码清单 7-25：**sqlite3_argument_positional.py**

```python
import sqlite3
import sys

db_filename = 'todo.db'
project_name = sys.argv[1]

with sqlite3.connect(db_filename) as conn:
    cursor = conn.cursor()

    query = """
    select id, priority, details, status, deadline from task
    where project = ?
    """
    cursor.execute(query, (project_name,))

    for row in cursor.fetchall():
        task_id, priority, details, status, deadline = row
        print('{:2d} [{:d}] {:<25} [{:<8}] ({})'.format(
            task_id, priority, details, status, deadline))
```

命令行参数会作为位置参数安全地传至查询，所以恶意数据不可能破坏数据库。

```
$ python3 sqlite3_argument_positional.py pymotw

1 [1] write about select      [done    ] (2016-04-25)
2 [1] write about random      [waiting ] (2016-08-22)
3 [1] write about sqlite3     [active  ] (2017-07-31)
```

7.4.5.2 命名参数

对于包含大量参数的更为复杂的查询，或者如果查询中某些参数会重复多次，则可以使用命名参数。命名参数前面有一个冒号前缀（例如，:param_name）。

代码清单 7-26：sqlite3_argument_named.py

```python
import sqlite3
import sys

db_filename = 'todo.db'
project_name = sys.argv[1]

with sqlite3.connect(db_filename) as conn:
    cursor = conn.cursor()

    query = """
    select id, priority, details, status, deadline from task
    where project = :project_name
    order by deadline, priority
    """

    cursor.execute(query, {'project_name': project_name})

    for row in cursor.fetchall():
        task_id, priority, details, status, deadline = row
        print('{:2d} [{:d}] {:<25} [{:<8}] ({})'.format(
            task_id, priority, details, status, deadline))
```

位置或命名参数都不需要加引号或转义，因为查询解析器会对它们做特殊处理。

```
$ python3 sqlite3_argument_positional.py pymotw

1 [1] write about select      [done    ] (2016-04-25)
2 [1] write about random      [waiting ] (2016-08-22)
3 [1] write about sqlite3     [active  ] (2017-07-31)
```

select（选择）、insert（插入）和 update（更新）语句中都可以使用查询参数。查询中能使用字面量值的地方都可以放置查询参数。

代码清单 7-27：sqlite3_argument_update.py

```python
import sqlite3
import sys

db_filename = 'todo.db'
id = int(sys.argv[1])
status = sys.argv[2]

with sqlite3.connect(db_filename) as conn:
    cursor = conn.cursor()
    query = "update task set status = :status where id = :id"
    cursor.execute(query, {'status': status, 'id': id})
```

这个 update 语句使用了两个命名参数。id 值用于查找要修改的行，status 值要写入数据表。

```
$ python3 sqlite3_argument_update.py 2 done
$ python3 sqlite3_argument_named.py pymotw

 1 [1] write about select          [done   ] (2016-04-25)
 2 [1] write about random          [done   ] (2016-08-22)
 3 [1] write about sqlite3         [active ] (2017-07-31)
```

7.4.6 批量加载

要对一个很大的数据集应用相同的 SQL 指令，可以使用 executemany()。这个方法对于加载数据很有用，因为这样可以避免在 Python 中循环处理输入，而允许底层的库应用循环优化。下面这个示例程序使用 csv 模块从一个逗号分隔值文件读取任务列表，并把它们加载到数据库。

代码清单 7-28：sqlite3_load_csv.py

```python
import csv
import sqlite3
import sys

db_filename = 'todo.db'
data_filename = sys.argv[1]

SQL = """
insert into task (details, priority, status, deadline, project)
values (:details, :priority, 'active', :deadline, :project)
"""

with open(data_filename, 'rt') as csv_file:
    csv_reader = csv.DictReader(csv_file)

    with sqlite3.connect(db_filename) as conn:
        cursor = conn.cursor()
        cursor.executemany(SQL, csv_reader)
```

文件 tasks.csv 包含以下示例数据：

```
deadline,project,priority,details
2016-11-30,pymotw,2,"finish reviewing markup"
2016-08-22,pymotw,2,"revise chapter intros"
2016-11-01,pymotw,1,"subtitle"
```

运行这个程序会生成以下结果。

```
$ python3 sqlite3_load_csv.py tasks.csv
$ python3 sqlite3_argument_named.py pymotw

 1 [1] write about select          [done   ] (2016-04-25)
 5 [2] revise chapter intros       [active ] (2016-08-20)
 2 [1] write about random          [done   ] (2016-08-22)
 6 [1] subtitle                    [active ] (2016-11-01)
 4 [2] finish reviewing markup     [active ] (2016-11-30)
 3 [1] write about sqlite3         [active ] (2017-07-31)
```

7.4.7 定义新的列类型

SQLite 对整数、浮点数和文本列提供了原生支持。sqlite3 会自动将这些类型的数据从 Python 的表示转换为可在数据库中存储的一个值，还可以根据需要从数据库中存储的值转换回 Python 表示。整数值从数据库加载为 int 或 long 变量，这取决于值的大小。文本将作为 str 保存和获取（除非改变了 Connection 的 text_factory）。

尽管 SQLite 在内部只支持几种数据类型，但 sqlite3 提供了一些便利工具，可以定义定制类型，允许 Python 应用在列中存储任意类型的数据。除了那些默认支持的类型外，还可以在数据库连接中使用 detect_types 标志来启用其他类型的转换。如果定义表时声明列使用所要求的类型，则可以使用 PARSE_DECLTYPES。

代码清单 7-29：sqlite3_date_types.py

```python
import sqlite3
import sys

db_filename = 'todo.db'

sql = "select id, details, deadline from task"

def show_deadline(conn):
    conn.row_factory = sqlite3.Row
    cursor = conn.cursor()
    cursor.execute(sql)
    row = cursor.fetchone()
    for col in ['id', 'details', 'deadline']:
        print('  {:<8}  {!r:<26} {}'.format(
            col, row[col], type(row[col])))
    return

print('Without type detection:')
with sqlite3.connect(db_filename) as conn:
    show_deadline(conn)

print('\nWith type detection:')
with sqlite3.connect(db_filename,
                     detect_types=sqlite3.PARSE_DECLTYPES,
                     ) as conn:
    show_deadline(conn)
```

sqlite3 为日期和时间戳列提供了转换器，它使用 datetime 模块的 date 和 datetime 类表示 Python 中的值。打开类型检测时，这两个与日期有关的转换器会自动启用。

```
$ python3 sqlite3_date_types.py

Without type detection:
  id        1                          <class 'int'>
  details   'write about select'       <class 'str'>
  deadline  '2016-04-25'               <class 'str'>

With type detection:
  id        1                          <class 'int'>
  details   'write about select'       <class 'str'>
  deadline  datetime.date(2016, 4, 25) <class 'datetime.date'>
```

定义一个新类型需要注册两个函数。适配器（adapter）取 Python 对象作为输入，返回一个可以存储在数据库中的字节串。转换器（converter）从数据库接收串，返回一个 Python 对象。要使用 register_adapter() 定义适配器函数，使用 register_converter() 定义转换器函数。

代码清单 7-30：sqlite3_custom_type.py

```python
import pickle
import sqlite3

db_filename = 'todo.db'

def adapter_func(obj):
    """Convert from in-memory to storage representation.
    """
    print('adapter_func({})\n'.format(obj))
    return pickle.dumps(obj)


def converter_func(data):
    """Convert from storage to in-memory representation.
    """
    print('converter_func({!r})\n'.format(data))
    return pickle.loads(data)


class MyObj:

    def __init__(self, arg):
        self.arg = arg

    def __str__(self):
        return 'MyObj({!r})'.format(self.arg)

# Register the functions for manipulating the type.
sqlite3.register_adapter(MyObj, adapter_func)
sqlite3.register_converter("MyObj", converter_func)

# Create some objects to save.  Use a list of tuples so
# the sequence can be passed directly to executemany().
to_save = [
    (MyObj('this is a value to save'),),
    (MyObj(42),),
]

with sqlite3.connect(
        db_filename,
        detect_types=sqlite3.PARSE_DECLTYPES) as conn:

    # Create a table with column of type "MyObj".
    conn.execute("""
    create table if not exists obj (
        id     integer primary key autoincrement not null,
        data   MyObj
    )
    """)
    cursor = conn.cursor()

    # Insert the objects into the database.
    cursor.executemany("insert into obj (data) values (?)",
                       to_save)

    # Query the database for the objects just saved.
    cursor.execute("select id, data from obj")
    for obj_id, obj in cursor.fetchall():
        print('Retrieved', obj_id, obj)
        print('  with type', type(obj))
        print()
```

这个例子使用 `pickle` 将一个对象保存为可以存储在数据库中的串，这对于存储任意的对象是一种很有用的技术，不过这种技术不支持按对象属性查询。真正的对象关系映射器（object-relational mapper，如 SQLAlchemy[一]）可以将属性值存储在单独的列中，这对于有大量数据时更为有用。

```
$ python3 sqlite3_custom_type.py

adapter_func(MyObj('this is a value to save'))

adapter_func(MyObj(42))

converter_func(b'\x80\x03c__main__\nMyObj\nq\x00)\x81q\x01}q\x02X\x0
3\x00\x00\x00argq\x03X\x17\x00\x00\x00this is a value to saveq\x04sb
.')

converter_func(b'\x80\x03c__main__\nMyObj\nq\x00)\x81q\x01}q\x02X\x0
3\x00\x00\x00argq\x03K*sb.')

Retrieved 1 MyObj('this is a value to save')
  with type <class '__main__.MyObj'>
Retrieved 2 MyObj(42)
  with type <class '__main__.MyObj'>
```

7.4.8 确定列类型

要得到查询数据类型的有关信息，有两个来源。可以用原表声明找出一个列的类型，这在前面已经看到。另外还可以在查询本身的 `select` 子句中包含形式为 `"name [type]"` 的类型指示符。

代码清单 7-31：sqlite3_custom_type_column.py

```python
import pickle
import sqlite3

db_filename = 'todo.db'


def adapter_func(obj):
    """Convert from in-memory to storage representation.
    """
    print('adapter_func({})\n'.format(obj))
    return pickle.dumps(obj)


def converter_func(data):
    """Convert from storage to in-memory representation.
    """
    print('converter_func({!r})\n'.format(data))
    return pickle.loads(data)


class MyObj:

    def __init__(self, arg):
        self.arg = arg
```

[一] www.sqlalchemy.org

```python
    def __str__(self):
        return 'MyObj({!r})'.format(self.arg)

# Register the functions for manipulating the type.
sqlite3.register_adapter(MyObj, adapter_func)
sqlite3.register_converter("MyObj", converter_func)

# Create some objects to save.  Use a list of tuples so we
# can pass this sequence directly to executemany().
to_save = [
    (MyObj('this is a value to save'),),
    (MyObj(42),),
]

with sqlite3.connect(
        db_filename,
        detect_types=sqlite3.PARSE_COLNAMES) as conn:
    # Create a table with column of type "text".
    conn.execute("""
    create table if not exists obj2 (
        id    integer primary key autoincrement not null,
        data  text
    )
    """)
    cursor = conn.cursor()

    # Insert the objects into the database.
    cursor.executemany("insert into obj2 (data) values (?)",
                       to_save)

    # Query the database for the objects just saved,
    # using a type specifier to convert the text
    # to objects.
    cursor.execute(
        'select id, data as "pickle [MyObj]" from obj2',
    )
    for obj_id, obj in cursor.fetchall():
        print('Retrieved', obj_id, obj)
        print('  with type', type(obj))
        print()
```

如果类型是查询的一部分而不属于原表定义，则要使用 **detect_types** 标志 **PARSE_COLNAMES**。

```
$ python3 sqlite3_custom_type_column.py

adapter_func(MyObj('this is a value to save'))

adapter_func(MyObj(42))

converter_func(b'\x80\x03c__main__\nMyObj\nq\x00)\x81q\x01}q\x02X\x0
3\x00\x00\x00argq\x03X\x17\x00\x00\x00this is a value to saveq\x04sb
.')

converter_func(b'\x80\x03c__main__\nMyObj\nq\x00)\x81q\x01}q\x02X\x0
3\x00\x00\x00argq\x03K*sb.')

Retrieved 1 MyObj('this is a value to save')
  with type <class '__main__.MyObj'>

Retrieved 2 MyObj(42)
  with type <class '__main__.MyObj'>
```

7.4.9 事务

关系数据库的关键特性之一是使用事务来维护一致的内部状态。启用事务时，在提交结果并刷新输出到真正的数据库之前，可以通过一个连接来完成多个变更而不影响任何其他用户。

7.4.9.1 保留变更

不论是通过插入（`insert`）还是更新（`update`）语句来改变数据库，都需要显式地调用 `commit()` 保存这些变更。这个要求为应用提供了一个机会，可以让多个相关的变更一同完成，使它们以一种"原子"方式保存而不是增量保存。这种方法可以避免同时连接到数据库的不同客户只看到部分更新。

可以利用一个使用了多个数据库连接的程序查看调用 `commit()` 的效果。用第一个连接插入一个新行，然后两次尝试使用不同的连接读回这个数据行。

代码清单 7-32：`sqlite3_transaction_commit.py`

```python
import sqlite3

db_filename = 'todo.db'


def show_projects(conn):
    cursor = conn.cursor()
    cursor.execute('select name, description from project')
    for name, desc in cursor.fetchall():
        print('  ', name)


with sqlite3.connect(db_filename) as conn1:
    print('Before changes:')
    show_projects(conn1)

    # Insert in one cursor.
    cursor1 = conn1.cursor()
    cursor1.execute("""
    insert into project (name, description, deadline)
    values ('virtualenvwrapper', 'Virtualenv Extensions',
            '2011-01-01')
    """)

    print('\nAfter changes in conn1:')
    show_projects(conn1)

    # Select from another connection, without committing first.
    print('\nBefore commit:')
    with sqlite3.connect(db_filename) as conn2:
        show_projects(conn2)

    # Commit, then select from another connection.
    conn1.commit()
    print('\nAfter commit:')
    with sqlite3.connect(db_filename) as conn3:
        show_projects(conn3)
```

提交 `conn1` 之前在调用 `show_projects()` 时，结果取决于使用了哪个连接。由于

这个变更通过 `conn1` 完成，所以这个连接会看到修改后的数据。不过，`conn2` 看不到这个变更。提交之后，新连接 `conn3` 会看到插入的行。

```
$ python3 sqlite3_transaction_commit.py

Before changes:
   pymotw

After changes in conn1:
   pymotw
   virtualenvwrapper

Before commit:
   pymotw

After commit:
   pymotw
   virtualenvwrapper
```

7.4.9.2 丢弃变更

还可以使用 `rollback()` 完全丢弃未提交的变更。`commit()` 和 `rollback()` 方法通常在同一个 `try:except` 块的不同部分调用，有错误时就会触发回滚。

代码清单 7-33：**sqlite3_transaction_rollback.py**

```python
import sqlite3

db_filename = 'todo.db'

def show_projects(conn):
    cursor = conn.cursor()
    cursor.execute('select name, description from project')
    for name, desc in cursor.fetchall():
        print('  ', name)

with sqlite3.connect(db_filename) as conn:

    print('Before changes:')
    show_projects(conn)

    try:

        # Insert
        cursor = conn.cursor()
        cursor.execute("""delete from project
                    where name = 'virtualenvwrapper'
                    """)

        # Show the settings.
        print('\nAfter delete:')
        show_projects(conn)

        # Pretend the processing caused an error.
        raise RuntimeError('simulated error')

    except Exception as err:
        # Discard the changes.
        print('ERROR:', err)
        conn.rollback()

    else:
```

```
        # Save the changes.
        conn.commit()

    # Show the results.
    print('\nAfter rollback:')
    show_projects(conn)
```

调用 rollback() 后，对数据库的修改不再存在。

```
$ python3 sqlite3_transaction_rollback.py

Before changes:
  pymotw
  virtualenvwrapper

After delete:
  pymotw
ERROR: simulated error

After rollback:
  pymotw
  virtualenvwrapper
```

7.4.10 隔离级别

sqlite3 支持 3 种加锁模式，也被称为隔离级别（isolation level），这会控制使用何种技术来避免连接之间不兼容的变更。打开一个连接时可以传入一个字符串作为 isolation_level 参数来设置隔离级别，所以不同的连接可以使用不同的隔离级别值。

下面这个程序展示了使用同一个数据库的不同连接时，不同的隔离级别对于线程中事件的顺序会有什么影响。这里创建了 4 个线程：两个线程更新现有的行，将变更写入数据库；另外两个线程尝试从 task 表读取所有行。

代码清单 7-34：sqlite3_isolation_levels.py

```python
import logging
import sqlite3
import sys
import threading
import time

logging.basicConfig(
    level=logging.DEBUG,
    format='%(asctime)s (%(threadName)-10s) %(message)s',
)

db_filename = 'todo.db'
isolation_level = sys.argv[1]


def writer():
    with sqlite3.connect(
            db_filename,
            isolation_level=isolation_level) as conn:
        cursor = conn.cursor()
        cursor.execute('update task set priority = priority + 1')
        logging.debug('waiting to synchronize')
        ready.wait()  # Synchronize threads
```

```python
        logging.debug('PAUSING')
        time.sleep(1)
        conn.commit()
        logging.debug('CHANGES COMMITTED')


def reader():
    with sqlite3.connect(
            db_filename,
            isolation_level=isolation_level) as conn:
        cursor = conn.cursor()
        logging.debug('waiting to synchronize')
        ready.wait()  # Synchronize threads
        logging.debug('wait over')
        cursor.execute('select * from task')
        logging.debug('SELECT EXECUTED')
        cursor.fetchall()
        logging.debug('results fetched')


if __name__ == '__main__':
    ready = threading.Event()

    threads = [
        threading.Thread(name='Reader 1', target=reader),
        threading.Thread(name='Reader 2', target=reader),
        threading.Thread(name='Writer 1', target=writer),
        threading.Thread(name='Writer 2', target=writer),
    ]

    [t.start() for t in threads]

    time.sleep(1)
    logging.debug('setting ready')
    ready.set()

    [t.join() for t in threads]
```

这些线程使用 threading 模块的一个 Event 完成同步。writer() 函数连接数据库，并完成数据库变更，不过在事件触发前并不提交。reader() 函数连接数据库，然后等待查询数据库，直到出现同步事件。

7.4.10.1 延迟

默认的隔离级别是 DEFERRED。使用延迟（deferred）模式会锁定数据库，但只是在变更真正开始时锁定一次。前面的所有例子都使用了延迟模式。

```
$ python3 sqlite3_isolation_levels.py DEFERRED

2016-08-20 17:46:26,972 (Reader 1  ) waiting to synchronize
2016-08-20 17:46:26,972 (Reader 2  ) waiting to synchronize
2016-08-20 17:46:26,973 (Writer 1  ) waiting to synchronize
2016-08-20 17:46:27,977 (MainThread) setting ready
2016-08-20 17:46:27,979 (Reader 1  ) wait over
2016-08-20 17:46:27,979 (Writer 1  ) PAUSING
2016-08-20 17:46:27,979 (Reader 2  ) wait over
2016-08-20 17:46:27,981 (Reader 1  ) SELECT EXECUTED
2016-08-20 17:46:27,982 (Reader 1  ) results fetched
2016-08-20 17:46:27,982 (Reader 2  ) SELECT EXECUTED
2016-08-20 17:46:27,982 (Reader 2  ) results fetched
```

```
2016-08-20 17:46:28,985 (Writer 1  ) CHANGES COMMITTED
2016-08-20 17:46:29,043 (Writer 2  ) waiting to synchronize
2016-08-20 17:46:29,043 (Writer 2  ) PAUSING
2016-08-20 17:46:30,044 (Writer 2  ) CHANGES COMMITTED
```

7.4.10.2 立即

采用立即(immediate)模式时,变更一开始时就会锁定数据库,在事务提交之前避免其他游标修改数据库。如果数据库有复杂的写操作,而且阅读器多于书写器,那么这种模式就很适合,因为在事务进行时不会阻塞阅读器。

```
$ python3 sqlite3_isolation_levels.py IMMEDIATE

2016-08-20 17:46:30,121 (Reader 1  ) waiting to synchronize
2016-08-20 17:46:30,121 (Reader 2  ) waiting to synchronize
2016-08-20 17:46:30,123 (Writer 1  ) waiting to synchronize
2016-08-20 17:46:31,122 (MainThread) setting ready
2016-08-20 17:46:31,122 (Reader 1  ) wait over
2016-08-20 17:46:31,122 (Reader 2  ) wait over
2016-08-20 17:46:31,122 (Writer 1  ) PAUSING
2016-08-20 17:46:31,124 (Reader 1  ) SELECT EXECUTED
2016-08-20 17:46:31,124 (Reader 2  ) SELECT EXECUTED
2016-08-20 17:46:31,125 (Reader 2  ) results fetched
2016-08-20 17:46:31,125 (Reader 1  ) results fetched
2016-08-20 17:46:32,128 (Writer 1  ) CHANGES COMMITTED
2016-08-20 17:46:32,199 (Writer 2  ) waiting to synchronize
2016-08-20 17:46:32,199 (Writer 2  ) PAUSING
2016-08-20 17:46:33,200 (Writer 2  ) CHANGES COMMITTED
```

7.4.10.3 互斥

互斥(exclusive)模式会对所有阅读器和书写器锁定数据库。如果数据库性能很重要,则应该限制使用这种模式,因为每个互斥的连接都会阻塞所有其他用户。

```
$ python3 sqlite3_isolation_levels.py EXCLUSIVE

2016-08-20 17:46:33,320 (Reader 1  ) waiting to synchronize
2016-08-20 17:46:33,320 (Reader 2  ) waiting to synchronize
2016-08-20 17:46:33,324 (Writer 1  ) waiting to synchronize
2016-08-20 17:46:34,323 (MainThread) setting ready
2016-08-20 17:46:34,323 (Reader 1  ) wait over
2016-08-20 17:46:34,323 (Writer 2  ) PAUSING
2016-08-20 17:46:34,323 (Reader 2  ) wait over
2016-08-20 17:46:35,327 (Writer 2  ) CHANGES COMMITTED
2016-08-20 17:46:35,368 (Reader 1  ) SELECT EXECUTED
2016-08-20 17:46:35,368 (Reader 2  ) results fetched
2016-08-20 17:46:35,369 (Reader 1  ) SELECT EXECUTED
2016-08-20 17:46:35,369 (Reader 2  ) results fetched
2016-08-20 17:46:35,385 (Writer 1  ) waiting to synchronize
2016-08-20 17:46:35,385 (Writer 1  ) PAUSING
2016-08-20 17:46:36,386 (Writer 1  ) CHANGES COMMITTED
```

由于第一个书写器已经开始修改,所以阅读器和第二个书写器会阻塞,直到第一个书写器提交。`sleep()`调用会在书写器线程中引入一个人为的延迟,以强调其他连接被阻塞。

7.4.10.4 自动提交

连接的`isolation_level`参数还可以被设置为None,这会启用自动提交(autocommit)模式。启用自动提交时,每个`execute()`调用会在语句完成时立即提交。自动提交模式很适合持续时间短的事务,如向一个表插入少量数据。数据库锁定时间尽可能短,所以线程间竞争资源的可能性更小。

sqlite3_autocommit.py 中删除了 commit() 的显式调用,并将隔离级别设置为 None,不过除此以外,这个方法与 sqlite3_isolation_levels.py 相同。但输出是不同的,因为两个书写器线程会在阅读器开始查询之前完成工作。

```
$ python3 sqlite3_autocommit.py

2016-08-20 17:46:36,451 (Reader 1   ) waiting to synchronize
2016-08-20 17:46:36,451 (Reader 2   ) waiting to synchronize
2016-08-20 17:46:36,455 (Writer 1   ) waiting to synchronize
2016-08-20 17:46:36,456 (Writer 2   ) waiting to synchronize
2016-08-20 17:46:37,452 (MainThread) setting ready
2016-08-20 17:46:37,452 (Reader 1   ) wait over
2016-08-20 17:46:37,452 (Writer 2   ) PAUSING
2016-08-20 17:46:37,452 (Reader 2   ) wait over
2016-08-20 17:46:37,453 (Writer 1   ) PAUSING
2016-08-20 17:46:37,453 (Reader 1   ) SELECT EXECUTED
2016-08-20 17:46:37,454 (Reader 2   ) SELECT EXECUTED
2016-08-20 17:46:37,454 (Reader 1   ) results fetched
2016-08-20 17:46:37,454 (Reader 2   ) results fetched
```

7.4.11 内存中的数据库

SQLite 支持在 RAM 中管理整个数据库,而不是依赖一个磁盘文件。内存中数据库对于自动测试很有用,在自动测试中,运行测试之间不需要保留数据库;或者想要尝试一个模式或其他数据库特性时,内存中数据库也很有用。要打开一个内存中数据库,创建 Connection 时可以使用字符串 ':memory:' 而不是一个文件名。每个 ':memory:' 连接会创建一个单独的数据库实例,所以一个连接中游标所做的变更不会影响其他连接。

7.4.12 导出数据库内容

内存中数据库的内容可以使用 Connection 的 iterdump() 方法保存。iterdump() 返回的迭代器生成一系列字符串,这些字符串将共同构造相应的 SQL 指令来重新创建数据库的状态。

代码清单 7-35:sqlite3_iterdump.py

```python
import sqlite3

schema_filename = 'todo_schema.sql'

with sqlite3.connect(':memory:') as conn:
    conn.row_factory = sqlite3.Row

    print('Creating schema')
    with open(schema_filename, 'rt') as f:
        schema = f.read()
    conn.executescript(schema)

    print('Inserting initial data')
    conn.execute("""
    insert into project (name, description, deadline)
    values ('pymotw', 'Python Module of the Week',
            '2010-11-01')
```

```python
        """)
    data = [
        ('write about select', 'done', '2010-10-03',
         'pymotw'),
        ('write about random', 'waiting', '2010-10-10',
         'pymotw'),
        ('write about sqlite3', 'active', '2010-10-17',
         'pymotw'),
    ]
    conn.executemany("""
    insert into task (details, status, deadline, project)
    values (?, ?, ?, ?)
    """, data)

    print('Dumping:')
    for text in conn.iterdump():
        print(text)
```

iterdump() 也适用于被保存到文件的数据库，不过对保留一个还没有保存的数据库最有用。这里对输出做了一些编辑，以便在这一页上显示的同时仍保证语法正确。

```
$ python3 sqlite3_iterdump.py

Creating schema
Inserting initial data
Dumping:
BEGIN TRANSACTION;
CREATE TABLE project (
    name        text primary key,
    description text,
    deadline    date
);
INSERT INTO "project" VALUES('pymotw','Python Module of the
Week','2010-11-01');
DELETE FROM "sqlite_sequence";
INSERT INTO "sqlite_sequence" VALUES('task',3);
CREATE TABLE task (
    id          integer primary key autoincrement not null,
    priority    integer default 1,
    details     text,
    status      text,
    deadline    date,
    completed_on date,
    project     text not null references project(name)
);
INSERT INTO "task" VALUES(1,1,'write about
select','done','2010-10-03',NULL,'pymotw');
INSERT INTO "task" VALUES(2,1,'write about
random','waiting','2010-10-10',NULL,'pymotw');
INSERT INTO "task" VALUES(3,1,'write about
sqlite3','active','2010-10-17',NULL,'pymotw');
COMMIT;
```

7.4.13 在 SQL 中使用 Python 函数

SQL 语法支持在查询中调用函数，可以在"列"列表中调用，也可以在 select 语句的 where 子句中调用。利用这个特性，在从查询返回数据之前可以先处理数据。这个特性可以用于在不同格式之间转换，完成一些计算（否则使用纯 SQL 会很麻烦），以及重用应用代码。

代码清单 7-36：`sqlite3_create_function.py`

```python
import codecs
import sqlite3

db_filename = 'todo.db'
def encrypt(s):
    print('Encrypting {!r}'.format(s))
    return codecs.encode(s, 'rot-13')


def decrypt(s):
    print('Decrypting {!r}'.format(s))
    return codecs.encode(s, 'rot-13')


with sqlite3.connect(db_filename) as conn:

    conn.create_function('encrypt', 1, encrypt)
    conn.create_function('decrypt', 1, decrypt)
    cursor = conn.cursor()

    # Raw values
    print('Original values:')
    query = "select id, details from task"
    cursor.execute(query)
    for row in cursor.fetchall():
        print(row)

    print('\nEncrypting...')
    query = "update task set details = encrypt(details)"
    cursor.execute(query)

    print('\nRaw encrypted values:')
    query = "select id, details from task"
    cursor.execute(query)
    for row in cursor.fetchall():
        print(row)

    print('\nDecrypting in query...')
    query = "select id, decrypt(details) from task"
    cursor.execute(query)
    for row in cursor.fetchall():
        print(row)

    print('\nDecrypting...')
    query = "update task set details = decrypt(details)"
    cursor.execute(query)
```

函数通过使用 Connection 的 `create_function()` 方法提供。参数包括函数名（即 SQL 中使用的函数名），函数的参数个数，以及要提供的 Python 函数。

```
$ python3 sqlite3_create_function.py

Original values:
(1, 'write about select')
(2, 'write about random')
(3, 'write about sqlite3')
(4, 'finish reviewing markup')
(5, 'revise chapter intros')
(6, 'subtitle')
```

```
Encrypting...
Encrypting 'write about select'
Encrypting 'write about random'
Encrypting 'write about sqlite3'
Encrypting 'finish reviewing markup'
Encrypting 'revise chapter intros'
Encrypting 'subtitle'

Raw encrypted values:
(1, 'jevgr nobhg fryrpg')
(2, 'jevgr nobhg enaqbz')
(3, 'jevgr nobhg fdyvgr3')
(4, 'svavfu erivrjvat znexhc')
(5, 'erivfr puncgre vagebf')
(6, 'fhogvgyr')

Decrypting in query...
Decrypting 'jevgr nobhg fryrpg'
Decrypting 'jevgr nobhg enaqbz'
Decrypting 'jevgr nobhg fdyvgr3'
Decrypting 'svavfu erivrjvat znexhc'
Decrypting 'erivfr puncgre vagebf'
Decrypting 'fhogvgyr'
(1, 'write about select')
(2, 'write about random')
(3, 'write about sqlite3')
(4, 'finish reviewing markup')
(5, 'revise chapter intros')
(6, 'subtitle')

Decrypting...
Decrypting 'jevgr nobhg fryrpg'
Decrypting 'jevgr nobhg enaqbz'
Decrypting 'jevgr nobhg fdyvgr3'
Decrypting 'svavfu erivrjvat znexhc'
Decrypting 'erivfr puncgre vagebf'
Decrypting 'fhogvgyr'
```

7.4.14 带正则表达式的查询

SQLite 支持很多与 SQL 语句关联的特殊用户函数。例如，可以用以下语法在查询中使用函数 regexp 检查一个列的字符串值是否与一个正则表达式匹配。

```
SELECT * FROM table
WHERE column REGEXP '.*pattern.*'
```

下面的例子将一个函数与 regexp() 关联，以使用 Python 的 re 模块来完成值的测试。

代码清单 7-37：**sqlite3_regex.py**

```
import re
import sqlite3

db_filename = 'todo.db'


def regexp(pattern, input):
    return bool(re.match(pattern, input))


with sqlite3.connect(db_filename) as conn:
```

```python
conn.row_factory = sqlite3.Row
conn.create_function('regexp', 2, regexp)
cursor = conn.cursor()

pattern = '.*[wW]rite [aA]bout.*'

cursor.execute(
    """
    select id, priority, details, status, deadline from task
    where details regexp :pattern
    order by deadline, priority
    """,
    {'pattern': pattern},
)

for row in cursor.fetchall():
    task_id, priority, details, status, deadline = row
    print('{:2d} [{:d}] {:<25} [{:<8}] ({})'.format(
        task_id, priority, details, status, deadline))
```

输出是 details 列与模式匹配的所有任务。

```
$ python3 sqlite3_regex.py

 1 [9] write about select        [done   ] (2016-04-25)
 2 [9] write about random        [done   ] (2016-08-22)
 3 [9] write about sqlite3       [active ] (2017-07-31)
```

7.4.15　定制聚集

聚集函数会收集多个单独的数据，并以某种方式汇总。`avg()`（取平均值）、`min()`、`max()` 和 `count()` 都是内置的聚集函数。

`sqlite3` 使用的聚集器 API 被定义为一个包含两个方法的类。处理查询时会对各个数据值分别调用一次 `step()` 方法。`finalize()` 方法在查询的最后调用一次，并且返回聚集值。下面这个例子为算术运算 mode（众数）实现了一个聚集器。它会返回输入中出现最频繁的值。

代码清单 7-38：**sqlite3_create_aggregate.py**

```python
import sqlite3
import collections

db_filename = 'todo.db'


class Mode:

    def __init__(self):
        self.counter = collections.Counter()

    def step(self, value):
        print('step({!r})'.format(value))
        self.counter[value] += 1

    def finalize(self):
        result, count = self.counter.most_common(1)[0]
        print('finalize() -> {!r} ({} times)'.format(
```

```
            result, count))
        return result


with sqlite3.connect(db_filename) as conn:
    conn.create_aggregate('mode', 1, Mode)

    cursor = conn.cursor()
    cursor.execute("""
    select mode(deadline) from task where project = 'pymotw'
    """)
    row = cursor.fetchone()
    print('mode(deadline) is:', row[0])
```

聚集器类用 Connection 的 create_aggregate() 方法注册。参数包括函数名（即 SQL 中使用的函数名），step() 方法的参数个数，以及要使用的类。

```
$ python3 sqlite3_create_aggregate.py

step('2016-04-25')
step('2016-08-22')
step('2017-07-31')
step('2016-11-30')
step('2016-08-20')
step('2016-11-01')
finalize() -> '2016-11-01' (1 times)
mode(deadline) is: 2016-11-01
```

7.4.16 线程和连接共享

出于历史原因，必须处理老版本的 SQLite，所以 Connection 对象不能在线程间共享。每个线程必须创建自己的数据库连接。

代码清单 7-39：**sqlite3_threading.py**

```
import sqlite3
import sys
import threading
import time

db_filename = 'todo.db'
isolation_level = None  # Autocommit mode


def reader(conn):
    print('Starting thread')
    try:
        cursor = conn.cursor()
        cursor.execute('select * from task')
        cursor.fetchall()
        print('results fetched')
    except Exception as err:
        print('ERROR:', err)


if __name__ == '__main__':
    with sqlite3.connect(db_filename,
                         isolation_level=isolation_level,
```

```
                ) as conn:
    t = threading.Thread(name='Reader 1',
                         target=reader,
                         args=(conn,),
                         )
    t.start()
    t.join()
```

如果试图在线程间共享一个连接，那么这会导致一个异常。

```
$ python3 sqlite3_threading.py

Starting thread
ERROR: SQLite objects created in a thread can only be used in that
same thread.The object was created in thread id 140735234088960
and this is thread id 123145307557888
```

7.4.17 限制对数据的访问

与其他更大的关系数据库相比，尽管 SQLite 没有用户访问控制，但其他确实提供了一种机制来限制列访问。每个连接可以安装一个授权函数（authorizer function），运行时可以根据指定的规则批准或拒绝访问列。这个授权函数会在解析 SQL 语句时调用。将传入 5 个参数：第一个参数是一个动作码，指示所完成的操作的类型（读、写、删除等）；其余的参数则取决于动作码。对于 SQLITE_READ 操作，另外 4 个参数分别是表名、列名、SQL 语句中访问出现的位置（主查询、触发器等）和 None。

代码清单 7-40：sqlite3_set_authorizer.py

```python
import sqlite3

db_filename = 'todo.db'


def authorizer_func(action, table, column, sql_location, ignore):
    print('\nauthorizer_func({}, {}, {}, {}, {})'.format(
        action, table, column, sql_location, ignore))

    response = sqlite3.SQLITE_OK  # Be permissive by default.

    if action == sqlite3.SQLITE_SELECT:
        print('requesting permission to run a select statement')
        response = sqlite3.SQLITE_OK

    elif action == sqlite3.SQLITE_READ:
        print('requesting access to column {}.{} from {}'.format(
            table, column, sql_location))
        if column == 'details':
            print('  ignoring details column')
            response = sqlite3.SQLITE_IGNORE
        elif column == 'priority':
            print('  preventing access to priority column')
            response = sqlite3.SQLITE_DENY

    return response


with sqlite3.connect(db_filename) as conn:
```

```
    conn.row_factory = sqlite3.Row
    conn.set_authorizer(authorizer_func)

    print('Using SQLITE_IGNORE to mask a column value:')
    cursor = conn.cursor()
    cursor.execute("""
    select id, details from task where project = 'pymotw'
    """)
    for row in cursor.fetchall():
        print(row['id'], row['details'])

    print('\nUsing SQLITE_DENY to deny access to a column:')
    cursor.execute("""
    select id, priority from task where project = 'pymotw'
    """)
    for row in cursor.fetchall():
        print(row['id'], row['details'])
```

下面这个例子使用了 SQLITE_IGNORE,会在查询结果中将从 task.details 列得到的字符串替换为 null 值。通过返回 SQLITE_DENY,其还会阻止所有对 task.priority 列的访问,如果尝试访问 task.priority 列,则将导致 SQLite 产生一个异常。

```
$ python3 sqlite3_set_authorizer.py

Using SQLITE_IGNORE to mask a column value:

authorizer_func(21, None, None, None, None)
requesting permission to run a select statement

authorizer_func(20, task, id, main, None)
requesting access to column task.id from main

authorizer_func(20, task, details, main, None)
requesting access to column task.details from main
  ignoring details column
authorizer_func(20, task, project, main, None)
requesting access to column task.project from main
1 None
2 None
3 None
4 None
5 None
6 None

Using SQLITE_DENY to deny access to a column:

authorizer_func(21, None, None, None, None)
requesting permission to run a select statement

authorizer_func(20, task, id, main, None)
requesting access to column task.id from main

authorizer_func(20, task, priority, main, None)
requesting access to column task.priority from main
  preventing access to priority column
Traceback (most recent call last):
  File "sqlite3_set_authorizer.py", line 53, in <module>
    """)
sqlite3.DatabaseError: access to task.priority is prohibited
```

sqlite3 中提供了一些可用的动作码,它们都作为常量提供,名字前都有前缀 SQLITE_。

可以为每一类 SQL 语句加标志，也可以控制对单个列的访问。

提示：相关阅读材料
- `sqlite3` 的标准库文档[一]。
- **PEP 249**[二]：DB API 2.0 规范（访问关系数据库的模块的一个标准接口）。
- SQLite[三]：SQLite 库的官方网站。
- `shelve`：保存任意 Python 对象的键 - 值库。
- SQLAlchemy[四]：一个流行的对象关系映射器，支持 SQLite，另外还支持很多其他关系数据库。

7.5 `xml.etree.ElementTree`：XML 操纵 API

ElementTree 库提供了一些工具，可以使用基于事件和基于文档的 API 来解析 XML，可以用 XPath 表达式搜索已解析的文档，还可以创建新文档或修改现有文档。

7.5.1 解析 XML 文档

已解析的 XML 文档在内存中由 `ElementTree` 和 `Element` 对象表示，这些对象基于 XML 文档中节点嵌套的方式按树结构相互连接。

用 `parse()` 解析一个完整的文档时，会返回一个 `ElementTree` 实例。这个树了解输入文档中的所有数据，另外可以原地搜索或操纵树中的节点。基于这种灵活性，可以更方便地处理已解析的文档，不过，与基于事件的解析方法相比，这种方法往往需要更多的内存，因为必须一次加载整个文档。

对于简单的小文档（如以下播客列表，被表示为一个 OPML 大纲），内存需求并不大。

代码清单 7-41：`podcasts.opml`

```
<?xml version="1.0" encoding="UTF-8"?>
<opml version="1.0">
<head>
    <title>My Podcasts</title>
    <dateCreated>Sat, 06 Aug 2016 15:53:26 GMT</dateCreated>
    <dateModified>Sat, 06 Aug 2016 15:53:26 GMT</dateModified>
</head>
<body>
  <outline text="Non-tech">
    <outline
        text="99% Invisible" type="rss"
        xmlUrl="http://feeds.99percentinvisible.org/99percentinvisible"
        htmlUrl="http://99percentinvisible.org" />
  </outline>
  <outline text="Python">
```

[一] https://docs.python.org/3.5/library/sqlite3.html
[二] www.python.org/dev/peps/pep-0249
[三] www.sqlite.org
[四] www.sqlalchemy.org

```
    <outline
        text="Talk Python to Me" type="rss"
        xmlUrl="https://talkpython.fm/episodes/rss"
        htmlUrl="https://talkpython.fm" />
    <outline
        text="Podcast.__init__" type="rss"
        xmlUrl="http://podcastinit.podbean.com/feed/"
        htmlUrl="http://podcastinit.com" />
  </outline>
</body>
</opml>
```

要解析这个文档，需要向parse()传递一个打开的文件句柄。

代码清单7-42：**ElementTree_parse_opml.py**

```
from xml.etree import ElementTree

with open('podcasts.opml', 'rt') as f:
    tree = ElementTree.parse(f)

print(tree)
```

这个方法会读取数据、解析XML，并返回一个ElementTree对象。

```
$ python3 ElementTree_parse_opml.py

<xml.etree.ElementTree.ElementTree object at 0x1013e5630>
```

7.5.2 遍历解析树

要按顺序访问所有子节点，可以使用iter()创建一个生成器，该生成器迭代处理这个ElementTree实例。

代码清单7-43：**ElementTree_dump_opml.py**

```
from xml.etree import ElementTree
import pprint

with open('podcasts.opml', 'rt') as f:
    tree = ElementTree.parse(f)

for node in tree.iter():
    print(node.tag)
```

这个例子会打印整个树，一次打印一个标记。

```
$ python3 ElementTree_dump_opml.py

opml
head
title
dateCreated
dateModified
body
outline
outline
outline
outline
outline
```

如果只是打印播客的名字组和提要 URL，则可以只迭代处理 outline 节点（而不考虑首部中的所有数据），并且通过查找 attrib 字典中的值来打印 text 和 xmlUrl 属性。

代码清单 7-44：**ElementTree_show_feed_urls.py**

```
from xml.etree import ElementTree

with open('podcasts.opml', 'rt') as f:
    tree = ElementTree.parse(f)

for node in tree.iter('outline'):
    name = node.attrib.get('text')
    url = node.attrib.get('xmlUrl')
    if name and url:
        print(' %s' % name)
        print('    %s' % url)
    else:
        print(name)
```

iter() 的 'outline' 参数意味着只处理标记为 'outline' 的节点。

```
$ python3 ElementTree_show_feed_urls.py

Non-tech
  99% Invisible
    http://feeds.99percentinvisible.org/99percentinvisible
Python
  Talk Python to Me
    https://talkpython.fm/episodes/rss
  Podcast.__init__
    http://podcastinit.podbean.com/feed/
```

7.5.3 查找文档中的节点

查看整个树并搜索有关的节点可能很容易出错。前面的例子必须查看每一个 outline 节点，来确定这是一个组（只有一个 text 属性的节点）还是一个播客（包含 text 和 xmlUrl 的节点）。要生成一个简单的播客提要 URL 列表而不包括名字或组，可以简化逻辑，使用 findall() 来查找有更多描述性搜索特性的节点。

对以上第一个版本做出第一次修改，用一个 XPath 参数来查找所有 outline 节点。

代码清单 7-45：**ElementTree_find_feeds_by_tag.py**

```
from xml.etree import ElementTree

with open('podcasts.opml', 'rt') as f:
    tree = ElementTree.parse(f)

for node in tree.findall('.//outline'):
    url = node.attrib.get('xmlUrl')
    if url:
        print(url)
```

这个版本中的逻辑与使用 getiterator() 的版本并没有显著区别。这里仍然必须检查是否存在 URL，只不过如果没有发现 URL，它不会打印组名。

```
$ python3 ElementTree_find_feeds_by_tag.py

http://feeds.99percentinvisible.org/99percentinvisible
https://talkpython.fm/episodes/rss
http://podcastinit.podbean.com/feed/
```

outline 节点只有两层嵌套，可以利用这一点，把搜索路径修改为 .//outline/outline，这意味着循环只处理 outline 节点的第二层。

代码清单 7-46：**ElementTree_find_feeds_by_structure.py**

```
from xml.etree import ElementTree

with open('podcasts.opml', 'rt') as f:
    tree = ElementTree.parse(f)

for node in tree.findall('.//outline/outline'):
    url = node.attrib.get('xmlUrl')
    print(url)
```

输入中所有嵌套深度为两层的 outline 节点都认为有一个 xmlURL 属性指向播客提要，所以循环在使用这个属性之前可以不做检查。

```
$ python3 ElementTree_find_feeds_by_structure.py

http://feeds.99percentinvisible.org/99percentinvisible
https://talkpython.fm/episodes/rss
http://podcastinit.podbean.com/feed/
```

不过，这个版本仅限于当前的这个结构，所以如果 outline 节点重新组织为一个更深的树，那么这个版本就无法正常工作了。

7.5.4 解析节点属性

`findall()` 和 `iter()` 返回的元素是 `Element` 对象，各个对象分别表示 XML 解析树中的一个节点。每个 `Element` 都有一些属性可以用来获取 XML 中的数据。可以用一个稍有些牵强的示例输入文件 `data.xml` 来说明这种行为。

代码清单 7-47：**data.xml**

```
1  <?xml version="1.0" encoding="UTF-8"?>
2  <top>
3    <child>Regular text.</child>
4    <child_with_tail>Regular text.</child_with_tail>"Tail" text.
5    <with_attributes name="value" foo="bar" />
6    <entity_expansion attribute="This &#38; That">
7      That &#38; This
8    </entity_expansion>
9  </top>
```

可以由 `attrib` 属性得到节点的 XML 属性，`attrib` 属性就像是一个字典。

代码清单 7-48：**ElementTree_node_attributes.py**

```
from xml.etree import ElementTree
```

```
with open('data.xml', 'rt') as f:
    tree = ElementTree.parse(f)

node = tree.find('./with_attributes')
print(node.tag)
for name, value in sorted(node.attrib.items()):
    print('  %-4s = "%s"' % (name, value))
```

输入文件第 5 行上的节点有两个属性 name 和 foo。

```
$ python3 ElementTree_node_attributes.py

with_attributes
  foo  = "bar"
  name = "value"
```

还可以得到节点的文本内容，以及结束标记后面的 tail 文本[1]。

代码清单 7-49：ElementTree_node_text.py

```
from xml.etree import ElementTree

with open('data.xml', 'rt') as f:
    tree = ElementTree.parse(f)

for path in ['./child', './child_with_tail']:
    node = tree.find(path)
    print(node.tag)
    print('  child node text:', node.text)
    print('  and tail text   :', node.tail)
```

第 3 行上的 child 节点包含嵌入文本，第 4 行的节点包含带 tail 的文本（包括空白符）。

```
$ python3 ElementTree_node_text.py

child
  child node text: Regular text.
  and tail text   :

child_with_tail
  child node text: Regular text.
  and tail text   : "Tail" text.
```

返回值之前，文档中嵌入的 XML 实体引用会被转换为适当的字符。

代码清单 7-50：ElementTree_entity_references.py

```
from xml.etree import ElementTree

with open('data.xml', 'rt') as f:
    tree = ElementTree.parse(f)

node = tree.find('entity_expansion')
print(node.tag)
print('  in attribute:', node.attrib['attribute'])
print('  in text     :', node.text.strip())
```

这个自动转换意味着可以忽略 XML 文档中表示某些字符的实现细节。

---

[1] tail 文本是指位于结束标记之后，下一元素开始或父元素结束之前所有的文本。——译者注

```
$ python3 ElementTree_entity_references.py

entity_expansion
  in attribute: This & That
  in text      : That & This
```

7.5.5 解析时监视事件

另一个处理 XML 文档的 API 是基于事件的。解析器为开始标记生成 **start** 事件,为结束标记生成 **end** 事件。解析阶段中可以通过迭代处理事件流从文档抽取数据,如果以后没有必要处理整个文档,或者没有必要将整个解析文档都保存在内存中,那么基于事件的 API 就会很方便。

有以下事件类型:

start 遇到一个新标记。会处理标记的结束尖括号,但不处理内容。

end 已经处理结束标记的结束尖括号。所有子节点都已经处理。

start-ns 开始一个命名空间声明。

end-ns 结束一个命名空间声明。

iterparse() 返回一个 iterable,它会生成元组,其中包含事件名和触发事件的节点。

代码清单 7-51:**ElementTree_show_all_events.py**

```
from xml.etree.ElementTree import iterparse

depth = 0
prefix_width = 8
prefix_dots = '.' * prefix_width
line_template = ''.join([
    '{prefix:<0.{prefix_len}}',
    '{event:<8}',
    '{suffix:<{suffix_len}} ',
    '{node.tag:<12} ',
    '{node_id}',
])

EVENT_NAMES = ['start', 'end', 'start-ns', 'end-ns']

for (event, node) in iterparse('podcasts.opml', EVENT_NAMES):
    if event == 'end':
        depth -= 1

    prefix_len = depth * 2

    print(line_template.format(
        prefix=prefix_dots,
        prefix_len=prefix_len,
        suffix='',
        suffix_len=(prefix_width - prefix_len),
        node=node,
        node_id=id(node),
        event=event,
    ))

    if event == 'start':
        depth += 1
```

默认地，只会生成 end 事件。要查看其他事件，可以将所需的事件名列表传入 iterparse()，如下例所示。

```
$ python3 ElementTree_show_all_events.py
start         opml           4312612200
..start       head           4316174520
....start     title          4316254440
....end       title          4316254440
....start     dateCreated    4316254520
....end       dateCreated    4316254520
....start     dateModified   4316254680
....end       dateModified   4316254680
..end         head           4316174520
..start       body           4316254840
....start     outline        4316254920
......start   outline        4316255080
......end     outline        4316255080
....end       outline        4316254920
....start     outline        4316255160
......start   outline        4316255240
......end     outline        4316255240
......start   outline        4316255320
......end     outline        4316255320
....end       outline        4316255160
..end         body           4316254840
end           opml           4312612200
```

以事件方式进行处理对于某些操作来说更为自然，如将 XML 输入转换为另外某种格式。可以使用这个技术将播客列表（来自前面的例子）从 XML 文件转换为一个 CSV 文件，以便把它们加载到一个电子表格或数据库应用。

代码清单 7-52：ElementTree_write_podcast_csv.py

```python
import csv
from xml.etree.ElementTree import iterparse
import sys

writer = csv.writer(sys.stdout, quoting=csv.QUOTE_NONNUMERIC)
group_name = ''

parsing = iterparse('podcasts.opml', events=['start'])

for (event, node) in parsing:
    if node.tag != 'outline':
        # Ignore anything not part of the outline.
        continue
    if not node.attrib.get('xmlUrl'):
        # Remember the current group.
        group_name = node.attrib['text']
    else:
        # Output a podcast entry.
        writer.writerow(
            (group_name, node.attrib['text'],
             node.attrib['xmlUrl'],
             node.attrib.get('htmlUrl', ''))
        )
```

这个转换程序并不需要将整个已解析的输入文件保存在内存中，其在遇到输入中的各个节点时才进行处理,这样做会更为高效。

```
$ python3 ElementTree_write_podcast_csv.py

"Non-tech","99% Invisible","http://feeds.99percentinvisible.org/\
99percentinvisible","http://99percentinvisible.org"
"Python","Talk Python to Me","https://talkpython.fm/episodes/rss\
","https://talkpython.fm"
"Python","Podcast.__init__","http://podcastinit.podbean.com/feed\
/","http://podcastinit.com"
```

说明:这里调整了 `ElementTree_write_podcast_csv.py` 的输出,以便在这一页上显示。以 \ 结尾的输出行指示这是一个人为增加的换行。

7.5.6 创建一个定制树构造器

要处理解析事件,一种可能更高效的方法是将标准的树构造器行为替换为一种定制行为。`XMLParser` 解析器使用一个 `TreeBuilder` 处理 XML,并调用目标类的方法保存结果。通常输出是由默认 `TreeBuilder` 类创建的一个 `ElementTree` 实例。可以将 `TreeBuilder` 替换为另一个类,使它在实例化 `Element` 节点之前接收事件,从而节省这部分开销。

可以将上一节的 XML-CSV 转换器重新实现为一个树构造器。

代码清单 7-53:**`ElementTree_podcast_csv_treebuilder.py`**

```python
import csv
import io
from xml.etree.ElementTree import XMLParser
import sys


class PodcastListToCSV(object):

    def __init__(self, outputFile):
        self.writer = csv.writer(
            outputFile,
            quoting=csv.QUOTE_NONNUMERIC,
        )
        self.group_name = ''

    def start(self, tag, attrib):
        if tag != 'outline':
            # Ignore anything not part of the outline.
            return
        if not attrib.get('xmlUrl'):
            # Remember the current group.
            self.group_name = attrib['text']
        else:
            # Output a podcast entry.
            self.writer.writerow(
                (self.group_name,
                 attrib['text'],
                 attrib['xmlUrl'],
                 attrib.get('htmlUrl', ''))
            )
```

```
        def end(self, tag):
            "Ignore closing tags"

        def data(self, data):
            "Ignore data inside nodes"

        def close(self):
            "Nothing special to do here"


target = PodcastListToCSV(sys.stdout)
parser = XMLParser(target=target)
with open('podcasts.opml', 'rt') as f:
    for line in f:
        parser.feed(line)
parser.close()
```

PodcastListToCSV 实现了 **TreeBuilder** 协议。每次遇到一个新的 XML 标记时，都会调用 **start()** 并提供标记名和属性。看到一个结束标记时，会根据这个标记名调用 **end()**。在这二者之间，如果一个节点有内容，则会调用 **data()**（一般认为树构造器会跟踪"当前"节点）。在所有输入都已经被处理时，将调用 **close()**。它会返回一个值，返回给 **XMLTreeBuilder** 的用户。

```
$ python3 ElementTree_podcast_csv_treebuilder.py

"Non-tech","99% Invisible","http://feeds.99percentinvisible.org/\
99percentinvisible","http://99percentinvisible.org"
"Python","Talk Python to Me","https://talkpython.fm/episodes/rss\
","https://talkpython.fm"
"Python","Podcast.__init__","http://podcastinit.podbean.com/feed\
/","http://podcastinit.com"
```

说明：这里调整了 `ElementTree_podcast_csv_treebuidler.py` 的输出，以便在这一页上显示。以 \ 结尾的输出行指示这是一个人为增加的换行。

7.5.7 解析串

如果只是处理少量的 XML 文本，特别是可能嵌入在程序源代码中的字符串字面量，则可以使用 **XML()**，将包含待解析 XML 的字符串作为它的唯一参数。

代码清单 7-54：**ElementTree_XML.py**

```
from xml.etree.ElementTree import XML


def show_node(node):
    print(node.tag)
    if node.text is not None and node.text.strip():
        print('  text: "%s"' % node.text)
    if node.tail is not None and node.tail.strip():
        print('  tail: "%s"' % node.tail)
    for name, value in sorted(node.attrib.items()):
        print('  %-4s = "%s"' % (name, value))
    for child in node:
        show_node(child)
```

```
parsed = XML('''
<root>
  <group>
    <child id="a">This is child "a".</child>
    <child id="b">This is child "b".</child>
  </group>
  <group>
    <child id="c">This is child "c".</child>
  </group>
</root>
''')

print('parsed =', parsed)

for elem in parsed:
    show_node(elem)
```

与 parse() 不同，这个函数的返回值是一个 Element 实例而不是 ElementTree。Element 直接支持迭代器协议，所以没有必要调用 getiterator()。

```
$ python3 ElementTree_XML.py

parsed = <Element 'root' at 0x10079eef8>
group
child
  text: "This is child "a"."
  id   = "a"
child
  text: "This is child "b"."
  id   = "b"
group
child
  text: "This is child "c"."
  id   = "c"
```

对于使用 id 属性来标识相应唯一节点的结构化 XML，可以使用便利方法 XMLID() 访问解析结果。

代码清单 7-55：ElementTree_XMLID.py

```
from xml.etree.ElementTree import XMLID

tree, id_map = XMLID('''
<root>
  <group>
    <child id="a">This is child "a".</child>
    <child id="b">This is child "b".</child>
  </group>
  <group>
    <child id="c">This is child "c".</child>
  </group>
</root>
''')

for key, value in sorted(id_map.items()):
    print('%s = %s' % (key, value))
```

XMLID() 将解析树作为一个 Element 对象返回，并提供一个字典，将 id 属性串映射到树中的单个节点。

```
$ python3 ElementTree_XMLID.py

a = <Element 'child' at 0x10133aea8>
b = <Element 'child' at 0x10133aef8>
c = <Element 'child' at 0x10133af98>
```

7.5.8 用元素节点构造文档

除了解析功能，`xml.etree.ElementTree` 还支持由应用中构造的 Element 对象来创建良构的 XML 文档。解析文档时使用的 `Element` 类还知道如何生成其内容的一个串行化形式，然后可以将这个串行化内容写至一个文件或其他数据流。

有 3 个辅助函数对于创建 `Element` 节点层次结构很有用。`Element()` 创建一个标准节点，`SubElement()` 将一个新节点关联到一个父节点，`Comment()` 创建一个使用 XML 注释语法串行化数据的节点。

代码清单 7-56：**ElementTree_create.py**

```python
from xml.etree.ElementTree import (
    Element, SubElement, Comment, tostring,
)

top = Element('top')

comment = Comment('Generated for PyMOTW')
top.append(comment)

child = SubElement(top, 'child')
child.text = 'This child contains text.'

child_with_tail = SubElement(top, 'child_with_tail')
child_with_tail.text = 'This child has text.'
child_with_tail.tail = 'And "tail" text.'

child_with_entity_ref = SubElement(top, 'child_with_entity_ref')
child_with_entity_ref.text = 'This & that'

print(tostring(top))
```

这个输出只包含树中的 XML 节点，而不包括含版本和编码的 XML 声明。

```
$ python3 ElementTree_create.py

b'<top><!--Generated for PyMOTW--><child>This child contains text.</
child><child_with_tail>This child has text.</child_with_tail>And "ta
il" text.<child_with_entity_ref>This &amp; that</child_with_entity_r
ef></top>'
```

`child_with_entity_ref` 文本中的 & 字符会自动转换为实体引用 &amp。

7.5.9 美观打印 XML

`ElementTree` 不会通过格式化 `tostring()` 的输出来提高可读性，因为增加额外的空白符会改变文档的内容。为了让输出更易读，后面的例子将使用 `xml.dom.minidom` 解析 XML，然后使用它的 `toprettyxml()` 方法。

代码清单 7-57：**ElementTree_pretty.py**

```python
from xml.etree import ElementTree
from xml.dom import minidom


def prettify(elem):
    """Return a pretty-printed XML string for the Element.
    """
    rough_string = ElementTree.tostring(elem, 'utf-8')
    reparsed = minidom.parseString(rough_string)
    return reparsed.toprettyxml(indent="  ")
```

更新后，这个例子如以下代码清单所示：

代码清单 7-58：**ElementTree_create_pretty.py**

```python
from xml.etree.ElementTree import Element, SubElement, Comment
from ElementTree_pretty import prettify

top = Element('top')

comment = Comment('Generated for PyMOTW')
top.append(comment)

child = SubElement(top, 'child')
child.text = 'This child contains text.'

child_with_tail = SubElement(top, 'child_with_tail')
child_with_tail.text = 'This child has text.'
child_with_tail.tail = 'And "tail" text.'

child_with_entity_ref = SubElement(top, 'child_with_entity_ref')
child_with_entity_ref.text = 'This & that'

print(prettify(top))
```

输出也会更易读。

```
$ python3 ElementTree_create_pretty.py

<?xml version="1.0" ?>
<top>
  <!--Generated for PyMOTW-->
  <child>This child contains text.</child>
  <child_with_tail>This child has text.</child_with_tail>
  And &quot;tail&quot; text.
  <child_with_entity_ref>This &amp; that</child_with_entity_ref>
</top>
```

除了增加用于格式化的额外空白符，**xml.dom.minidom** 美观打印器还会向输出增加一个 XML 声明。

7.5.10 设置元素属性

前面的例子都由标记和文本内容来创建节点，但是没有设置节点的任何属性。7.5.1 节中的很多例子都在处理一个列举播客及其提要的 OPML 文件。树中的 **outline** 节点使用了对应组名和播客特性的属性。可以用 ElementTree 由一个 CSV 输入文件构造一个类似的 XML 文件，并在构造树时设置所有元素属性。

代码清单 7-59：**ElementTree_csv_to_xml.py**

```python
import csv
from xml.etree.ElementTree import (
    Element, SubElement, Comment, tostring,
)
import datetime
from ElementTree_pretty import prettify

generated_on = str(datetime.datetime.now())

# Configure one attribute with set().
root = Element('opml')
root.set('version', '1.0')

root.append(
    Comment('Generated by ElementTree_csv_to_xml.py for PyMOTW')
)

head = SubElement(root, 'head')
title = SubElement(head, 'title')
title.text = 'My Podcasts'
dc = SubElement(head, 'dateCreated')
dc.text = generated_on
dm = SubElement(head, 'dateModified')
dm.text = generated_on

body = SubElement(root, 'body')

with open('podcasts.csv', 'rt') as f:
    current_group = None
    reader = csv.reader(f)
    for row in reader:
        group_name, podcast_name, xml_url, html_url = row
        if (current_group is None or
                group_name != current_group.text):
            # Start a new group.
            current_group = SubElement(
                body, 'outline',
                {'text': group_name},
            )
        # Add this podcast to the group,
        # setting all its attributes at
        # once.
        podcast = SubElement(
            current_group, 'outline',
            {'text': podcast_name,
             'xmlUrl': xml_url,
             'htmlUrl': html_url},
        )

print(prettify(root))
```

这个例子使用两种技术来设置新节点的属性值。根节点用 set() 配置，一次修改一个属性。另外通过向节点工厂传入一个字典，对播客节点一次性指定所有属性。

```
$ python3 ElementTree_csv_to_xml.py

<?xml version="1.0" ?>
<opml version="1.0">
  <!--Generated by ElementTree_csv_to_xml.py for PyMOTW-->
  <head>
    <title>My Podcasts</title>
    <dateCreated>2016-08-06 17:09:00.524979</dateCreated>
    <dateModified>2016-08-06 17:09:00.524979</dateModified>
  </head>
  <body>
    <outline text="Non-tech">
      <outline htmlUrl="http://99percentinvisible.org" text="99%\
 Invisible" xmlUrl="http://feeds.99percentinvisible.org/99percen\
tinvisible"/>
    </outline>
    <outline text="Python">
      <outline htmlUrl="https://talkpython.fm" text="Talk Python\
 to Me" xmlUrl="https://talkpython.fm/episodes/rss"/>
    </outline>
    <outline text="Python">
      <outline htmlUrl="http://podcastinit.com" text="Podcast.__\
init__" xmlUrl="http://podcastinit.podbean.com/feed/"/>
    </outline>
  </body>
</opml>
```

7.5.11 由节点列表构造树

利用 extend() 方法可以将多个子节点一同增加到一个 Element 实例。extend() 的参数可以是任意的 iterable，包括 list 或另一个 Element 实例。

代码清单 7-60：ElementTree_extend.py

```python
from xml.etree.ElementTree import Element, tostring
from ElementTree_pretty import prettify

top = Element('top')

children = [
    Element('child', num=str(i))
    for i in range(3)
]

top.extend(children)

print(prettify(top))
```

给定一个 list 时，列表中的节点会直接增加到新的父节点。

```
$ python3 ElementTree_extend.py

<?xml version="1.0" ?>
<top>
  <child num="0"/>
  <child num="1"/>
  <child num="2"/>
</top>
```

给定另一个 Element 实例时，该节点的子节点会增加到新的父节点。

代码清单 7-61：**ElementTree_extend_node.py**

```python
from xml.etree.ElementTree import (
    Element, SubElement, tostring, XML,
)
from ElementTree_pretty import prettify

top = Element('top')

parent = SubElement(top, 'parent')

children = XML(
    '<root><child num="0" /><child num="1" />'
    '<child num="2" /></root>'
)
parent.extend(children)

print(prettify(top))
```

在这个例子中，通过解析 XML 串创建的 root 节点有 3 个子节点，它们都增加到 parent 节点。root 节点不是输出树的一部分。

```
$ python3 ElementTree_extend_node.py

<?xml version="1.0" ?>
<top>
  <parent>
    <child num="0"/>
    <child num="1"/>
    <child num="2"/>
  </parent>
</top>
```

要了解重要的一点，extend() 并不改变节点现有的父子关系。如果传入 extend() 的值已经存在于树中的某个位置，那么它们仍在原处，并在输出中重复。

代码清单 7-62：**ElementTree_extend_node_copy.py**

```python
from xml.etree.ElementTree import (
    Element, SubElement, tostring, XML,
)
from ElementTree_pretty import prettify

top = Element('top')
parent_a = SubElement(top, 'parent', id='A')
parent_b = SubElement(top, 'parent', id='B')

# Create children.
children = XML(
    '<root><child num="0" /><child num="1" />'
    '<child num="2" /></root>'
)

# Set the id to the Python object id of the node
# to make duplicates easier to spot.
for c in children:
    c.set('id', str(id(c)))

# Add to first parent.
parent_a.extend(children)
```

```
print('A:')
print(prettify(top))
print()

# Copy nodes to second parent.
parent_b.extend(children)

print('B:')
print(prettify(top))
print()
```

这里将这些子节点的 id 属性设置为 Python 唯一对象标识符,由此强调同一个节点对象可以在输出树中出现多次。

```
$ python3 ElementTree_extend_node_copy.py

A:
<?xml version="1.0" ?>
<top>
  <parent id="A">
    <child id="4316789880" num="0"/>
    <child id="4316789960" num="1"/>
    <child id="4316790040" num="2"/>
  </parent>
  <parent id="B"/>
</top>


B:
<?xml version="1.0" ?>
<top>
  <parent id="A">
    <child id="4316789880" num="0"/>
    <child id="4316789960" num="1"/>
    <child id="4316790040" num="2"/>
  </parent>
  <parent id="B">
    <child id="4316789880" num="0"/>
    <child id="4316789960" num="1"/>
    <child id="4316790040" num="2"/>
  </parent>
</top>
```

7.5.12 将 XML 串行化至一个流

tostring() 被实现为将内容写至内存中的一个类似文件的对象,然后返回表示整个元素树的一个串。处理大量数据时,这种做法需要的内存较少,而且可以更高效地使用 I/O 库,使用 ElementTree 的 write() 方法直接写至一个文件句柄。

代码清单 7-63:**ElementTree_write.py**

```
import io
import sys
from xml.etree.ElementTree import (
    Element, SubElement, Comment, ElementTree,
)

top = Element('top')
```

```
comment = Comment('Generated for PyMOTW')
top.append(comment)

child = SubElement(top, 'child')
child.text = 'This child contains text.'

child_with_tail = SubElement(top, 'child_with_tail')
child_with_tail.text = 'This child has regular text.'
child_with_tail.tail = 'And "tail" text.'

child_with_entity_ref = SubElement(top, 'child_with_entity_ref')
child_with_entity_ref.text = 'This & that'

empty_child = SubElement(top, 'empty_child')

ElementTree(top).write(sys.stdout.buffer)
```

这个例子使用 `sys.stdout.buffer` 而不是 `sys.stdout` 写至控制台,因为 ElementTree 会生成编码字节而不是一个 Unicode 串。不过也可以写至一个以二进制模式打开的文件或者套接字。

```
$ python3 ElementTree_write.py

<top><!--Generated for PyMOTW--><child>This child contains text.</ch
ild><child_with_tail>This child has regular text.</child_with_tail>A
nd "tail" text.<child_with_entity_ref>This &amp; that</child_with_en
tity_ref><empty_child /></top>
```

树中最后一个节点不包含文本或子节点,所以它被写为一个空标记 `<empty_child />`。`write()` 有一个方法(`method`)参数,用来控制空节点的处理。

代码清单 7-64:ElementTree_write_method.py

```
import io
import sys
from xml.etree.ElementTree import (
    Element, SubElement, ElementTree,
)

top = Element('top')

child = SubElement(top, 'child')
child.text = 'Contains text.'

empty_child = SubElement(top, 'empty_child')

for method in ['xml', 'html', 'text']:
    print(method)
    sys.stdout.flush()
    ElementTree(top).write(sys.stdout.buffer, method=method)
    print('\n')
```

这里支持 3 个方法。

xml 默认方法,生成 `<empty_child />`。
html 生成标记对,HTML 文档要求必须采用这种方法(`<empty_child></empty_child>`)。
text 只打印节点的文本,完全跳过空标记。

```
$ python3 ElementTree_write_method.py

xml
<top><child>Contains text.</child><empty_child /></top>

html
<top><child>Contains text.</child><empty_child></empty_child></t
op>

text
Contains text.
```

提示：相关阅读材料
- `xml.etree.ElementTree` 的标准库文档[一]。
- `csv`：读写逗号分隔值文件。
- defusedxml[二]：这个包为实体扩展拒绝服务攻击提供了修正，这对于处理不可信的 XML 数据很有用。
- Pretty print xml with python: indenting xml[三]：Rene DudfieldR 的关于在 Python 的中美观打印 XML 的一些提示。
- ElementTree Overview[四]：Fredrick Lundh 的关于 ElementTree 的原始文档以及 ElementTree 库开发版的链接。
- Process XML in Python with ElementTree[五]：David Mertz 的 IBM DeveloperWorks 文章。
- Outline Processor Markup Language (OPML)[六]：Dave Winer 的 OPML 规范和文档。
- XML Path Language (XPath)[七]：识别 XML 文档中各部分的语法。
- XPath Support in ElementTree[八]：Fredrick Lundh 关于 ElementTree 的原始文档的一部分。

7.6 csv：逗号分隔值文件

可以用 csv 模块处理从电子表格和数据库导出的数据，并写入使用字段和记录格式的文本文件，这种格式通常被称为逗号分隔值（Comma-Separated Value，CSV）格式，因为常用逗号来分隔记录中的字段。

7.6.1 读文件

可以使用 `reader()` 创建一个对象从 CSV 文件读取数据。这个阅读器可以作为一个迭代器，按顺序处理文件中的行。

一 https://docs.python.org/3.5/library/xml.etree.elementtree.html
二 https://pypi.python.org/pypi/defusedxml
三 http://renesd.blogspot.com/2007/05/pretty-print-xml-with-python.html
四 http://effbot.org/zone/element-index.htm
五 www.ibm.com/developerworks/library/x-matters28/
六 www.opml.org
七 www.w3.org/TR/xpath
八 http://effbot.org/zone/element-xpath.htm

代码清单 7-65：**csv_reader.py**

```
import csv
import sys

with open(sys.argv[1], 'rt') as f:
    reader = csv.reader(f)
    for row in reader:
        print(row)
```

`reader()` 的第一个参数是文本行的源。在这个例子中，这是一个文件，不过也可以是任何可迭代的对象（如 `StringIO` 实例、`list` 等）。还可以指定其他可选参数来控制如何解析输入数据。

```
"Title 1","Title 2","Title 3","Title 4"
1,"a",08/18/07,"å"
2,"b∫",08/19/07,""
3,"c",08/20/07,"ç"
```

读文件时，输入数据的每一行都会被解析，并被转换为一个字符串 `list`。

```
$ python3 csv_reader.py testdata.csv

['Title 1', 'Title 2', 'Title 3', 'Title 4']
['1', 'a', '08/18/07', 'å']
['2', 'b', '08/19/07', '∫'']
['3', 'c', '08/20/07', 'ç']
```

这个解析器会处理行中嵌在字符串里的换行符，正是因为这个原因，这里的"行"（row）并不一定等同于文件的一个输入"行"（line）。

```
"Title 1","Title 2","Title 3"
1,"first line
second line",08/18/07
```

由解析器返回时，输入中带换行符的字段仍保留内部换行符。

```
$ python3 csv_reader.py testlinebreak.csv

['Title 1', 'Title 2', 'Title 3']
['1', 'first line\nsecond line', '08/18/07']
```

7.6.2 写文件

写 CSV 文件与读 CSV 文件同样容易。可以使用 `writer()` 创建一个对象来写数据，然后使用 `writerow()` 迭代处理文本行进行打印。

代码清单 7-66：**csv_writer.py**

```
import csv
import sys

unicode_chars = '∫åç'

with open(sys.argv[1], 'wt') as f:
    writer = csv.writer(f)
    writer.writerow(('Title 1', 'Title 2', 'Title 3', 'Title 4'))
    for i in range(3):
        row = (
```

```
                i + 1,
                chr(ord('a') + i),
                '08/{:02d}/07'.format(i + 1),
                unicode_chars[i],
            )
            writer.writerow(row)

print(open(sys.argv[1], 'rt').read())
```

这里的输出与阅读器示例中使用的导出数据看上去不完全相同，因为这里的值两边没有引号。

```
$ python3 csv_writer.py testout.csv

Title 1,Title 2,Title 3,Title 4
1,a,08/01/07,å
2,bʃ,08/02/07,
3,c,08/03/07,ç
```

引号

书写器默认的引号行为有所不同，所以前例中第二列和第三列没有加引号。要增加引号，需要将 `quoting` 参数设置为另外某种引号模式。

```
writer = csv.writer(f, quoting=csv.QUOTE_NONNUMERIC)
```

在这里，`QUOTE_NONNUMERIC` 会在所有包含非数值内容的列两边增加引号。

```
$ python3 csv_writer_quoted.py testout_quoted.csv

"Title 1","Title 2","Title 3","Title 4"
1,"a","08/01/07","å"
2,"bʃ","08/02/07",""
3,"c","08/03/07","ç"
```

有 4 种不同的引号选项，在 csv 模块中定义为 4 个常量。

QUOTE_ALL 不论类型是什么，对所有字段都加引号。

QUOTE_MINIMAL 对包含特殊字符的字段加引号（特殊字符是指，对于一个用相同方言和选项配置的解析器，可能会造成混淆的所有字符）。这是默认选项。

QUOTE_NONNUMERIC 对所有非整数或浮点数的字段加引号。在阅读器中使用时，不加引号的输入字段会转换为浮点数。

QUOTE_NONE 输出中所有内容都不加引号。在阅读器中使用时，引号字符包含在字段值中（正常情况下，它们会被处理为定界符并去除）。

7.6.3 方言

逗号分隔值文件没有明确定义的标准，所以解析器必须很灵活。为了提供这种灵活性，可以用很多参数来控制 csv 如何解析或写数据。并不是将各个参数单独传入阅读器和书写器，而是可以把它们组合在一起构成一个方言（dialect）对象。

方言类可以按名注册，这样 csv 模块的调用者就不需要提前知道参数设置。可以用 `list_dialects()` 获取已注册方言的完整列表。

代码清单 7-67：`csv_list_dialects.py`

```
import csv

print(csv.list_dialects())
```

标准库包括 3 个方言：`excel`、`excel-tabs` 和 `unix`。`excel` 方言用于处理采用 Microsoft Excel 默认导出格式的数据，也可以处理 LibreOffice[⊖]。`unix` 方言将所有字段都加双引号，并使用 `\n` 作为记录分隔符。

```
$ python3 csv_list_dialects.py

['excel', 'excel-tab', 'unix']
```

7.6.3.1 创建方言

可以不使用逗号来分隔字段，如果输入文件使用了竖线（`|`），则可以使用适当的定界符注册一个新的方言。

```
"Title 1"|"Title 2"|"Title 3"
1|"first line
second line"|08/18/07
```

代码清单 7-68：`csv_dialect.py`

```
import csv

csv.register_dialect('pipes', delimiter='|')

with open('testdata.pipes', 'r') as f:
    reader = csv.reader(f, dialect='pipes')
    for row in reader:
        print(row)
```

通过使用"pipes"方言，可以像读取逗号定界文件一样读取文件。

```
$ python3 csv_dialect.py

['Title 1', 'Title 2', 'Title 3']
['1', 'first line\nsecond line', '08/18/07']
```

7.6.3.2 方言参数

方言指定了解析或写一个数据文件时使用的所有 token。表 7-3 列出了可以指定的文件格式的各个方面，从如何对列定界到使用哪个字符来转义 token 都可以指定。

表 7-3　CSV 方言参数

属性	默认值	含义
`delimiter`	`,`	字段分隔符（一个字符）
`doublequote`	`True`	这个标志控制 `quotechar` 实例是否成对
`escapechar`	`None`	这个字符用来指示一个转义序列
`lineterminator`	`\r\n`	书写器使用这个字符串结束一行
`quotechar`	`"`	这个字符串用来包围包含特殊值的字段（一个字符）
`quoting`	`QUOTE_MINIMAL`	控制前面介绍的引号行为
`skipinitialspace`	`False`	忽略字段定界符后面的空白符

⊖ www.libreoffice.org

代码清单 7-69: `csv_dialect_variations.py`

```python
import csv
import sys

csv.register_dialect('escaped',
                     escapechar='\\',
                     doublequote=False,
                     quoting=csv.QUOTE_NONE,
                     )
csv.register_dialect('singlequote',
                     quotechar="'",
                     quoting=csv.QUOTE_ALL,
                     )

quoting_modes = {
    getattr(csv, n): n
    for n in dir(csv)
    if n.startswith('QUOTE_')
}

TEMPLATE = '''\
Dialect: "{name}"

  delimiter   = {dl!r:<6}    skipinitialspace = {si!r}
  doublequote = {dq!r:<6}    quoting          = {qu}
  quotechar   = {qc!r:<6}    lineterminator   = {lt!r}
  escapechar  = {ec!r:<6}
'''

for name in sorted(csv.list_dialects()):
    dialect = csv.get_dialect(name)

    print(TEMPLATE.format(
        name=name,
        dl=dialect.delimiter,
        si=dialect.skipinitialspace,
        dq=dialect.doublequote,
        qu=quoting_modes[dialect.quoting],
        qc=dialect.quotechar,
        lt=dialect.lineterminator,
        ec=dialect.escapechar,
    ))

    writer = csv.writer(sys.stdout, dialect=dialect)
    writer.writerow(
        ('col1', 1, '10/01/2010',
         'Special chars: " \' {} to parse'.format(
             dialect.delimiter))
    )
    print()
```

这个程序显示了采用多种不同的方言时同样的数据会如何显示。

```
$ python3 csv_dialect_variations.py

Dialect: "escaped"
  delimiter   = ','      skipinitialspace = 0
  doublequote = 0        quoting          = QUOTE_NONE
  quotechar   = '"'      lineterminator   = '\r\n'
  escapechar  = '\\'

col1,1,10/01/2010,Special chars: \" ' \, to parse
```

```
Dialect: "excel"
    delimiter   = ','      skipinitialspace = 0
    doublequote = 1        quoting          = QUOTE_MINIMAL
    quotechar   = '"'      lineterminator   = '\r\n'
    escapechar  = None

col1,1,10/01/2010,"Special chars: "" ' , to parse"

Dialect: "excel-tab"
    delimiter   = '\t'     skipinitialspace = 0
    doublequote = 1        quoting          = QUOTE_MINIMAL
    quotechar   = '"'      lineterminator   = '\r\n'
    escapechar  = None

col1    1    10/01/2010    "Special chars: "" '     to parse"

Dialect: "singlequote"
    delimiter   = ','      skipinitialspace = 0
    doublequote = 1        quoting          = QUOTE_ALL
    quotechar   = "'"      lineterminator   = '\r\n'
    escapechar  = None

'col1','1','10/01/2010','Special chars: " '' , to parse'

Dialect: "unix"
    delimiter   = ','      skipinitialspace = 0
    doublequote = 1        quoting          = QUOTE_ALL
    quotechar   = '"'      lineterminator   = '\n'
    escapechar  = None

"col1","1","10/01/2010","Special chars: "" ' , to parse"
```

7.6.3.3 自动检测方言

要配置方言来解析一个输入文件，最好的办法就是提前知道正确的设置。对于方言参数未知的数据，可以用 Sniffer 类来做一个有根据的猜测。sniff() 方法取一个输入数据样本和一个可选的参数（给出可能的定界字符）。

代码清单 7-70：**csv_dialect_sniffer.py**

```python
import csv
from io import StringIO
import textwrap

csv.register_dialect('escaped',
                     escapechar='\\',
                     doublequote=False,
                     quoting=csv.QUOTE_NONE)
csv.register_dialect('singlequote',
                     quotechar="'",
                     quoting=csv.QUOTE_ALL)

# Generate sample data for all known dialects.
samples = []
for name in sorted(csv.list_dialects()):
    buffer = StringIO()
    dialect = csv.get_dialect(name)
    writer = csv.writer(buffer, dialect=dialect)
    writer.writerow(
```

```
            ('col1', 1, '10/01/2010',
             'Special chars " \' {} to parse'.format(
                 dialect.delimiter))
        )
        samples.append((name, dialect, buffer.getvalue()))

# Guess the dialect for a given sample, and then use the results
# to parse the data.
sniffer = csv.Sniffer()
for name, expected, sample in samples:
    print('Dialect: "{}"'.format(name))
    print('In: {}'.format(sample.rstrip()))
    dialect = sniffer.sniff(sample, delimiters=',\t')
    reader = csv.reader(StringIO(sample), dialect=dialect)
    print('Parsed:\n   {}\n'.format(
        '\n   '.join(repr(r) for r in next(reader))))
```

sniff()会返回一个Dialect实例,其中包含用于解析数据的设置。这个结果并不总是尽善尽美,示例中的"escaped"方言可以说明这一点。

```
$ python3 csv_dialect_sniffer.py

Dialect: "escaped"
In: col1,1,10/01/2010,Special chars \" ' \, to parse
Parsed:
   'col1'
   '1'
   '10/01/2010'
   'Special chars \\" \' \\'
   ' to parse'

Dialect: "excel"
In: col1,1,10/01/2010,"Special chars "" ' , to parse"
Parsed:
   'col1'
   '1'
   '10/01/2010'
   'Special chars " \' , to parse'

Dialect: "excel-tab"
In: col1    1    10/01/2010    "Special chars "" '     to parse"
Parsed:
   'col1'
   '1'
   '10/01/2010'
   'Special chars " \' \t to parse'

Dialect: "singlequote"
In: 'col1','1','10/01/2010','Special chars " '' , to parse'
Parsed:
   'col1'
   '1'
   '10/01/2010'
   'Special chars " \' , to parse'

Dialect: "unix"
In: "col1","1","10/01/2010","Special chars "" ' , to parse"
Parsed:
   'col1'
   '1'
   '10/01/2010'
   'Special chars " \' , to parse'
```

7.6.4 使用字段名

除了处理数据序列，`csv` 模块还包括一些类可以将行作为字典来处理，从而可以对字段命名。`DictReader` 和 `DictWriter` 类将行转换为字典而不是列表。字典的键可以传入，或者可以由输入的第一行推导得出（如果行包含首部）。

代码清单 7-71：**csv_dictreader.py**

```python
import csv
import sys
with open(sys.argv[1], 'rt') as f:
    reader = csv.DictReader(f)
    for row in reader:
        print(row)
```

基于字典的阅读器和书写器会被实现为基于序列的类的包装器，它们使用相同的方法和参数。阅读器 API 中唯一的差别是：行将作为字典返回，而不是作为列表或元组。

```
$ python3 csv_dictreader.py testdata.csv

{'Title 2': 'a', 'Title 3': '08/18/07', 'Title 4': 'å', 'Title 1
': '1'}
{'Title 2': 'b', 'Title 3': '08/19/07', 'Title 4': 'ʃ', 'Title 1
': '2'}
{'Title 2': 'c', 'Title 3': '08/20/07', 'Title 4': 'ç', 'Title 1
': '3'}
```

必须为 `DictWriter` 提供一个字段名列表，使它知道如何在输出中确定列的顺序。

代码清单 7-72：**csv_dictwriter.py**

```python
import csv
import sys

fieldnames = ('Title 1', 'Title 2', 'Title 3', 'Title 4')
headers = {
    n: n
    for n in fieldnames
}
unicode_chars = 'ʃåç'

with open(sys.argv[1], 'wt') as f:

    writer = csv.DictWriter(f, fieldnames=fieldnames)
    writer.writeheader()

    for i in range(3):
        writer.writerow({
            'Title 1': i + 1,
            'Title 2': chr(ord('a') + i),
            'Title 3': '08/{:02d}/07'.format(i + 1),
            'Title 4': unicode_chars[i],
        })

print(open(sys.argv[1], 'rt').read())
```

字段名并不自动写至文件，但可以使用 `writeheader()` 方法显式地写出。

```
$ python3 csv_dictwriter.py testout.csv

Title 1,Title 2,Title 3,Title 4
1,a,08/01/07,å
2,bʃ,08/02/07,
3,c,08/03/07,ç
```

提示：相关阅读材料

- **csv** 的标准库文档[一]。
- **PEP 305**[二]：CSV 文件 API。
- **csv** 的 Python 2 到 Python 3 移植说明。

[一] https://docs.python.org/3.5/library/csv.html
[二] www.python.org/dev/peps/pep-0305

第 8 章

数据压缩与归档

尽管现代计算机系统的存储能力日益增长，但生成数据的增长是永无休止的。无损（lossless）压缩算法以压缩或解压缩数据花费的时间来换取存储数据所需要的空间，以弥补存储能力的不足。Python 为最流行的一些压缩库提供了接口，从而能使用不同压缩库读写文件。

zlib 和 gzip 提供了 GNU zip 库，另外 bz2 允许访问更新的 bzip2 格式。这些格式都处理数据流而不考虑输入格式，并且提供的接口可以透明地读写压缩文件。可以使用这些模块来压缩单个文件或数据源。

标准库还包括一些模块来管理归档（archive）格式，能够将多个文件合并到一个文件，该文件可以作为一个单元来管理。tarfile 读写 UNIX 磁带归档格式，这是一种老标准，但由于其灵活性，当前仍得到广泛使用。zipfile 根据 zip 格式来处理归档，这种格式因 PC 程序 PKZIP 得以普及，原先在 MS-DOS 和 Windows 下使用，不过由于其 API 的简单性以及这种格式的可移植性，现在也用于其他平台。

8.1 zlib：GNU zlib 压缩

zlib 模块为 GNU 项目 zlib 压缩库中的很多函数提供了底层接口。

8.1.1 处理内存中的数据

使用 zlib 最简单的方法要求把所有将要压缩或解压缩的数据存放在内存中：

代码清单 8-1：zlib_memory.py

```
import zlib
import binascii

original_data = b'This is the original text.'
print('Original      :', len(original_data), original_data)

compressed = zlib.compress(original_data)
print('Compressed    :', len(compressed),
      binascii.hexlify(compressed))
decompressed = zlib.decompress(compressed)
print('Decompressed :', len(decompressed), decompressed)
```

compress() 和 decompress() 函数都取一个字节序列参数，并且返回一个字节序列。

```
$ python3 zlib_memory.py

Original     : 26 b'This is the original text.'
Compressed   : 32 b'789c0bc9c82c5600a2928c5485fca2ccf4ccbcc41c85
92d48a123d007f2f097e'
Decompressed : 26 b'This is the original text.'
```

从前面的例子可以看到，少量数据的压缩版本可能比未压缩的版本还要大。具体的结果取决于输入数据，不过观察小数据集的压缩开销很有意思。

代码清单 8-2：`zlib_lengths.py`

```python
import zlib

original_data = b'This is the original text.'

template = '{:>15}  {:>15}'
print(template.format('len(data)', 'len(compressed)'))
print(template.format('-' * 15, '-' * 15))

for i in range(5):
    data = original_data * i
    compressed = zlib.compress(data)
    highlight = '*' if len(data) < len(compressed) else ''
    print(template.format(len(data), len(compressed)), highlight)
```

输出中的 * 突出显示了哪些行的压缩数据比未压缩版本占用的内存更多。

```
$ python3 zlib_lengths.py

      len(data)  len(compressed)
---------------  ---------------
              0                8 *
             26               32 *
             52               35
             78               35
            104               36
```

`zlib` 支持不同的压缩级别，允许在计算成本和空间缩减量之间有所平衡。默认压缩级别 `zlib.Z_DEFAULT_COMPRESSION` 为 -1，这对应着一个硬编码值，表示性能和压缩结果之间的一个折中。当前这对应级别 6。

代码清单 8-3：`zlib_compresslevel.py`

```python
import zlib

input_data = b'Some repeated text.\n' * 1024
template = '{:>5}  {:>5}'

print(template.format('Level', 'Size'))
print(template.format('-----', '----'))

for i in range(0, 10):
    data = zlib.compress(input_data, i)
    print(template.format(i, len(data)))
```

压缩级别为 0 意味着根本没有压缩。级别 9 要求的计算最多，同时会生成最小的输出。如下面的例子所示，对于一个给定的输入，可能多个压缩级别得到的空间缩减量是一样的。

```
$ python3 zlib_compresslevel.py
Level   Size
-----   ----
    0   20491
    1     172
    2     172
    3     172
    4      98
    5      98
    6      98
    7      98
    8      98
    9      98
```

8.1.2 增量压缩与解压缩

这种内存中的压缩方法有一些缺点，主要是系统需要有足够的内存，可以在内存中同时驻留未压缩和压缩版本，因此这种方法对于真实世界的用例并不实用。另一种方法是使用 `Compress` 和 `Decompress` 对象以增量方式处理数据，这样就不需要将整个数据集都放在内存中。

代码清单 8-4：zlib_incremental.py

```python
import zlib
import binascii

compressor = zlib.compressobj(1)

with open('lorem.txt', 'rb') as input:
    while True:
        block = input.read(64)
        if not block:
            break
        compressed = compressor.compress(block)
        if compressed:
            print('Compressed: {}'.format(
                binascii.hexlify(compressed)))
        else:
            print('buffering...')
    remaining = compressor.flush()
    print('Flushed: {}'.format(binascii.hexlify(remaining)))
```

这个例子从一个纯文本文件读取小数据块，并把这个数据集传至 `compress()`。压缩器维护压缩数据的一个内部缓冲区。由于压缩算法依赖于校验和以及最小块大小，所以压缩器每次接收更多输入时可能并没有准备好返回数据。如果它没有准备好一个完整的压缩块，那便会返回一个空字节串。当所有数据都已输入时，`flush()` 方法会强制压缩器结束最后一个块，并返回余下的压缩数据。

```
$ python3 zlib_incremental.py

Compressed: b'7801'
buffering...
buffering...
buffering...
buffering...
```

```
buffering...
Flushed: b'55904b6ac4400c44f73e451da0f129b20c2110c85e696b8c40dde
dd167ce1f7915025a087daa9ef4be8c07e4f21c38962e834b800647435fd3b90
747b2810eb9c4bbcc13ac123bded6e4bef1c91ee40d3c6580e3ff52aad2e8cb2
eb6062dad74a89ca904cbb0f2545e0db4b1f2e01955b8c511cb2ac08967d228a
f1447c8ec72e40c4c714116e60cdef171bb6c0feaa255dff1c507c2c4439ec96
05b7e0ba9fc54bae39355cb89fd6ebe5841d673c7b7bc68a46f575a312eebd22
0d4b32441bdc1b36ebf0aedef3d57ea4b26dd986dd39af57dfb05d32279de'
```

8.1.3 混合内容流

在压缩和未压缩数据混合在一起的情况下，还可以使用`decompressobj()`返回的`Decompress`类。

代码清单 8-5：**zlib_mixed.py**

```python
import zlib

lorem = open('lorem.txt', 'rb').read()
compressed = zlib.compress(lorem)
combined = compressed + lorem

decompressor = zlib.decompressobj()
decompressed = decompressor.decompress(combined)

decompressed_matches = decompressed == lorem
print('Decompressed matches lorem:', decompressed_matches)

unused_matches = decompressor.unused_data == lorem
print('Unused data matches lorem :', unused_matches)
```

解压缩所有数据后，`unused_data`属性会包含未用的所有数据。

```
$ python3 zlib_mixed.py

Decompressed matches lorem: True
Unused data matches lorem : True
```

8.1.4 校验和

除了压缩和解压缩函数，`zlib`还包括两个用于计算数据的校验和的函数，分别是`adler32()`和`crc32()`。这两个函数计算出的校验和都不能认为是密码安全的，它们只用于数据完整性验证。

代码清单 8-6：**zlib_checksums.py**

```python
import zlib

data = open('lorem.txt', 'rb').read()

cksum = zlib.adler32(data)
print('Adler32: {:12d}'.format(cksum))
print('       : {:12d}'.format(zlib.adler32(data, cksum)))

cksum = zlib.crc32(data)
print('CRC-32 : {:12d}'.format(cksum))
print('       : {:12d}'.format(zlib.crc32(data, cksum)))
```

这两个函数取相同的参数，包括一个包含数据的字节串和一个可选值，这个值可作为校验和的起点。这些函数会返回一个 32 位有符号整数值，这个值可以作为一个新的起点参数再传回给后续的调用，以生成一个动态变化的校验和。

```
$ python3 zlib_checksums.py

Adler32:   3542251998
     :      669447099
CRC-32:   3038370516
     :     2870078631
```

8.1.5 压缩网络数据

下一个代码清单中的服务器使用流压缩器来响应文件名请求，它将文件的一个压缩版本写至与客户通信的套接字中。

代码清单 8-7：**zlib_server.py**

```python
import zlib
import logging
import socketserver
import binascii

BLOCK_SIZE = 64


class ZlibRequestHandler(socketserver.BaseRequestHandler):

    logger = logging.getLogger('Server')

    def handle(self):
        compressor = zlib.compressobj(1)

        # Find out which file the client wants.
        filename = self.request.recv(1024).decode('utf-8')
        self.logger.debug('client asked for: %r', filename)

        # Send chunks of the file as they are compressed.
        with open(filename, 'rb') as input:
            while True:
                block = input.read(BLOCK_SIZE)
                if not block:
                    break
                self.logger.debug('RAW %r', block)
                compressed = compressor.compress(block)
                if compressed:
                    self.logger.debug(
                        'SENDING %r',
                        binascii.hexlify(compressed))
                    self.request.send(compressed)
                else:
                    self.logger.debug('BUFFERING')
        # Send any data being buffered by the compressor.
        remaining = compressor.flush()
        while remaining:
            to_send = remaining[:BLOCK_SIZE]
            remaining = remaining[BLOCK_SIZE:]
            self.logger.debug('FLUSHING %r',
                              binascii.hexlify(to_send))
```

```python
                self.request.send(to_send)
            return


if __name__ == '__main__':
    import socket
    import threading
    from io import BytesIO

    logging.basicConfig(
        level=logging.DEBUG,
        format='%(name)s: %(message)s',
    )
    logger = logging.getLogger('Client')

    # Set up a server, running in a separate thread.
    address = ('localhost', 0)  # Let the kernel assign a port.
    server = socketserver.TCPServer(address, ZlibRequestHandler)
    ip, port = server.server_address  # What port was assigned?

    t = threading.Thread(target=server.serve_forever)
    t.setDaemon(True)
    t.start()

    # Connect to the server as a client.
    logger.info('Contacting server on %s:%s', ip, port)
    s = socket.socket(socket.AF_INET, socket.SOCK_STREAM)
    s.connect((ip, port))

    # Ask for a file.
    requested_file = 'lorem.txt'
    logger.debug('sending filename: %r', requested_file)
    len_sent = s.send(requested_file.encode('utf-8'))

    # Receive a response.
    buffer = BytesIO()
    decompressor = zlib.decompressobj()
    while True:
        response = s.recv(BLOCK_SIZE)
        if not response:
            break
        logger.debug('READ %r', binascii.hexlify(response))
        # Include any unconsumed data when
        # feeding the decompressor.
        to_decompress = decompressor.unconsumed_tail + response
        while to_decompress:
            decompressed = decompressor.decompress(to_decompress)
            if decompressed:
                logger.debug('DECOMPRESSED %r', decompressed)
                buffer.write(decompressed)
                # Look for unconsumed data due to buffer overflow.
                to_decompress = decompressor.unconsumed_tail
            else:
                logger.debug('BUFFERING')
                to_decompress = None

    # Deal with data reamining inside the decompressor buffer.
    remainder = decompressor.flush()
    if remainder:
        logger.debug('FLUSHED %r', remainder)
        buffer.write(remainder)

    full_response = buffer.getvalue()
```

```
lorem = open('lorem.txt', 'rb').read()
logger.debug('response matches file contents: %s',
             full_response == lorem)

# Clean up.
s.close()
server.socket.close()
```

我们人为地将这个代码清单做了一些划分,以展示缓冲行为,如果将数据传递到 compress() 或 decompress(),但没有得到完整的压缩或未压缩输出块,此时便会进行缓冲。

客户连接到套接字,并请求一个文件。然后循环,接收压缩数据块。由于一个块可能未包含足够多的信息来完全解压缩,所以之前接收的剩余数据将与新数据结合,并且传递到解压缩器。解压缩数据时,会把它追加到一个缓冲区,处理循环结束时将与文件内容进行比较。

警告: 这个服务器存在明显的安全隐患。不要在开放的互联网或者安全问题可能产生严重影响的环境中运行这个程序。

```
$ python3 zlib_server.py

Client: Contacting server on 127.0.0.1:53658
Client: sending filename: 'lorem.txt'
Server: client asked for: 'lorem.txt'
Server: RAW b'Lorem ipsum dolor sit amet, consectetuer adipiscin
g elit. Donec\n'
Server: SENDING b'7801'
Server: RAW b'egestas, enim et consectetuer ullamcorper, lectus
ligula rutrum '
Server: BUFFERING
Server: RAW b'leo, a\nelementum elit tortor eu quam. Duis tincid
unt nisi ut ant'
Server: BUFFERING
Server: RAW b'e. Nulla\nfacilisi. Sed tristique eros eu libero.
Pellentesque ve'
Server: BUFFERING
Server: RAW b'l arcu. Vivamus\npurus orci, iaculis ac, suscipit
sit amet, pulvi'
Client: READ b'7801'
Client: BUFFERING
Server: BUFFERING
Server: RAW b'nar eu,\nlacus.\n'
Server: BUFFERING
Server: FLUSHING b'55904b6ac4400c44f73e451da0f129b20c2110c85e696
b8c40ddedd167ce1f7915025a087daa9ef4be8c07e4f21c38962e834b8006474
35fd3b90747b2810eb9'
Server: FLUSHING b'c4bbcc13ac123bded6e4bef1c91ee40d3c6580e3ff52a
ad2e8cb2eb6062dad74a89ca904cbb0f2545e0db4b1f2e01955b8c511cb2ac08
967d228af1447c8ec72'
Client: READ b'55904b6ac4400c44f73e451da0f129b20c2110c85e696b8c4
0ddedd167ce1f7915025a087daa9ef4be8c07e4f21c38962e834b800647435fd
3b90747b2810eb9'
Server: FLUSHING b'e40c4c714116e60cdef171bb6c0feaa255dff1c507c2c
4439ec9605b7e0ba9fc54bae39355cb89fd6ebe5841d673c7b7bc68a46f575a3
12eebd220d4b32441bd'
Client: DECOMPRESSED b'Lorem ipsum dolor sit amet, consectetuer
adi'
Client: READ b'c4bbcc13ac123bded6e4bef1c91ee40d3c6580e3ff52aad2e
8cb2eb6062dad74a89ca904cbb0f2545e0db4b1f2e01955b8c511cb2ac08967d
```

```
228af1447c8ec72'
Client: DECOMPRESSED b'piscing elit. Donec\negestas, enim et con
sectetuer ullamcorper, lectus ligula rutrum leo, a\nelementum el
it tortor eu quam. Duis tinci'
Client: READ b'e40c4c714116e60cdef171bb6c0feaa255dff1c507c2c4439
ec9605b7e0ba9fc54bae39355cb89fd6ebe5841d673c7b7bc68a46f575a312ee
bd220d4b32441bd'
Client: DECOMPRESSED b'dunt nisi ut ante. Nulla\nfacilisi. Sed t
ristique eros eu libero. Pellentesque vel arcu. Vivamus\npurus o
rci, iaculis ac'
Server: FLUSHING b'c1b36ebf0aedef3d57ea4b26dd986dd39af57dfb05d32
279de'
Client: READ b'c1b36ebf0aedef3d57ea4b26dd986dd39af57dfb05d32279d
e'
Client: DECOMPRESSED b', suscipit sit amet, pulvinar eu,\nlacus.
\n'
Client: response matches file contents: True
```

提示：相关阅读材料

- `zlib` 的标准库文档[1]。
- `gzip`：`gzip` 模块包含 zlib 库的一个更高层（基于文件）的接口。
- `zlib`: A Massively Spiffy Yet Delicately Unobtrusive Compression Library[2]: zlib 库的主页。
- zlib 1.2.11 Manual[3]：完整的 `zlib` 文档。
- `bz2`：`bz2` 模块为 bzip2 压缩库提供了一个类似的接口。

8.2　gzip：读写 GNU zip 文件

`gzip` 模块为 GNU zip 文件提供了一个类似文件的接口，它使用 `zlib` 来压缩和解压缩数据。

8.2.1　写压缩文件

模块级函数 `open()` 创建类似文件的 `GzipFile` 类的一个实例。它提供了读写字节的一般方法。

代码清单 8-8：gzip_write.py

```python
import gzip
import io
import os

outfilename = 'example.txt.gz'
with gzip.open(outfilename, 'wb') as output:
    with io.TextIOWrapper(output, encoding='utf-8') as enc:
        enc.write('Contents of the example file go here.\n')

print(outfilename, 'contains', os.stat(outfilename).st_size,
      'bytes')
os.system('file -b --mime {}'.format(outfilename))
```

---

[1] https://docs.python.org/3.5/library/zlib.html
[2] www.zlib.net
[3] www.zlib.net/manual.html

为了把数据写至一个压缩文件，需要用模式 `'wb'` 打开文件。这个例子用 `io` 模块的一个 `TextIOWrapper` 来包装 `GzipFile`，将 Unicode 文本编码为适合压缩的字节。

```
$ python3 gzip_write.py

application/x-gzip; charset=binary
example.txt.gz contains 75 bytes
```

通过传入一个压缩级别（`compresslevel`）参数，可以使用不同的压缩量。合法值为 0～9（包括 0 和 9）。值越小便会得到越快的处理，得到的压缩也越少。较大的值会得到较慢的处理，但压缩更多（直到某个上限）。

代码清单 8-9：**gzip_compresslevel.py**

```python
import gzip
import io
import os
import hashlib


def get_hash(data):
    return hashlib.md5(data).hexdigest()


data = open('lorem.txt', 'r').read() * 1024
cksum = get_hash(data.encode('utf-8'))

print('Level  Size       Checksum')
print('-----  ---------  --------------------------------')
print('data   {:>10}  {}'.format(len(data), cksum))

for i in range(0, 10):
    filename = 'compress-level-{}.gz'.format(i)
    with gzip.open(filename, 'wb', compresslevel=i) as output:
        with io.TextIOWrapper(output, encoding='utf-8') as enc:
            enc.write(data)
    size = os.stat(filename).st_size
    cksum = get_hash(open(filename, 'rb').read())
    print('{:>5d}  {:>10d}  {}'.format(i, size, cksum))
```

输出中，中间一列的数字显示了压缩输入所生成文件的大小（字节数）。对于这个输入数据，压缩值更高并不一定能减少存储空间。根据输入数据的不同，结果会有变化。

```
$ python3 gzip_compresslevel.py

Level  Size        Checksum
-----  ----------  --------------------------------
data       754688  e4c0f9433723971563f08a458715119c
    0      754848  7f050dafb281c7b9d30e5fccf4e0cf19
    1        9846  3b1708684b3655d136b8dca292f5bbba
    2        8267  48ceb436bf10bc6bbd60489eb285de27
    3        8227  4217663bf275f4241a8b73b1a1cfd734
    4        4167  1a5d9b968520d64ed10a4c125735d8b4
    5        4167  90d85bf6457c2eaf20307deb90d071c6
    6        4167  1798ac0cbd77d79973efd8e222bf85d8
    7        4167  7fe834b01c164a14c2d2d8e5560402e6
    8        4167  03795b47b899384cdb95f99c1b7f9f71
    9        4167  a33be56e455f8c787860f23c3b47b6f1
```

`GzipFile` 实例还包括一个 `writelines()` 方法，可以用来写字符串序列。

代码清单8-10：`gzip_writelines.py`

```python
import gzip
import io
import itertools
import os

with gzip.open('example_lines.txt.gz', 'wb') as output:
    with io.TextIOWrapper(output, encoding='utf-8') as enc:
        enc.writelines(
            itertools.repeat('The same line, over and over.\n',
                             10)
        )

os.system('gzcat example_lines.txt.gz')
```

与常规文件一样，输入行要包含一个换行符。

```
$ python3 gzip_writelines.py

The same line, over and over.
The same line, over and over.
The same line, over and over.
The same line, over and over.
The same line, over and over.
The same line, over and over.
The same line, over and over.
The same line, over and over.
The same line, over and over.
The same line, over and over.
```

8.2.2 读压缩数据

要从之前压缩的文件读回数据，可以用二进制读模式（`'rb'`）打开文件，这样就不会对行尾完成基于文本的转换或Unicode解码了。

代码清单8-11：`gzip_read.py`

```python
import gzip
import io

with gzip.open('example.txt.gz', 'rb') as input_file:
    with io.TextIOWrapper(input_file, encoding='utf-8') as dec:
        print(dec.read())
```

这个例子读取上一节`gzip_write.py`所写的文件，这里在文本解压缩后使用`TextIOWrapper`对它进行解码。

```
$ python3 gzip_read.py

Contents of the example file go here.
```

读文件时，还可以用seek定位，只读取部分数据。

代码清单8-12：`gzip_seek.py`

```python
import gzip

with gzip.open('example.txt.gz', 'rb') as input_file:
```

```python
print('Entire file:')
all_data = input_file.read()
print(all_data)

expected = all_data[5:15]

# Rewind to beginning
input_file.seek(0)

# Move ahead 5 bytes
input_file.seek(5)
print('Starting at position 5 for 10 bytes:')
partial = input_file.read(10)
print(partial)

print()
print(expected == partial)
```

seek()位置是相对未压缩数据的位置，所以调用者并不需要知道数据文件是压缩文件。

```
$ python3 gzip_seek.py

Entire file:
b'Contents of the example file go here.\n'
Starting at position 5 for 10 bytes:
b'nts of the'

True
```

8.2.3 处理流

GzipFile 类可以用来包装其他类型的数据流，使它们也能使用压缩。通过一个套接字或一个现有的（已经打开的）文件句柄传输数据时，这种方法很有用。还可以对 GzipFile 使用 BytesIO 缓冲区，以对内存中的数据完成操作。

代码清单 8-13：gzip_BytesIO.py

```python
import gzip
from io import BytesIO
import binascii

uncompressed_data = b'The same line, over and over.\n' * 10
print('UNCOMPRESSED:', len(uncompressed_data))
print(uncompressed_data)

buf = BytesIO()
with gzip.GzipFile(mode='wb', fileobj=buf) as f:
    f.write(uncompressed_data)

compressed_data = buf.getvalue()
print('COMPRESSED:', len(compressed_data))
print(binascii.hexlify(compressed_data))

inbuffer = BytesIO(compressed_data)
with gzip.GzipFile(mode='rb', fileobj=inbuffer) as f:
    reread_data = f.read(len(uncompressed_data))

print('\nREREAD:', len(reread_data))
print(reread_data)
```

使用 `GzipFile` 而不是 `zlib` 的一个好处是，`GzipFile` 支持文件 API。不过，重新读先前压缩的数据时，要向 `read()` 传递一个明确的长度。如果没有这个长度，则会导致一个 CRC 错误，这可能是因为 `BytesIO` 会在报告 EOF 之前返回一个空串。处理压缩数据流时，可以在数据前加一个整数作为前缀，表示要读取的具体数据量，也可以使用 `zlib` 中的增量解压缩 API。

```
$ python3 gzip_BytesIO.py

UNCOMPRESSED: 300
b'The same line, over and over.\nThe same line, over and over.\nT
he same line, over and over.\nThe same line, over and over.\nThe
same line, over and over.\nThe same line, over and over.\nThe sam
e line, over and over.\nThe same line, over and over.\nThe same l
ine, over and over.\nThe same line, over and over.\n'
COMPRESSED: 51
b'1f8b08006149aa5702ff0bc94855284ecc4d55c8c9cc4bd551c82f4b2d5248c
c4b0133f4b8424665916401d3e717802c010000'

REREAD: 300
b'The same line, over and over.\nThe same line, over and over.\nT
he same line, over and over.\nThe same line, over and over.\nThe
same line, over and over.\nThe same line, over and over.\nThe sam
e line, over and over.\nThe same line, over and over.\nThe same l
ine, over and over.\nThe same line, over and over.\n'
```

提示：相关阅读材料
- `gzip` 的标准库文档[⊖]。
- `zlib`：`zlib` 模块是 gzip 压缩的一个底层接口。
- `zipfile`：`zipfile` 模块提供了对 ZIP 归档文件的访问。
- `bz2`：`bz2` 模块使用 bzip2 压缩格式。
- `tarfile`：`tarfile` 模块对读取压缩 tar 归档文件提供了内置支持。
- `io`：创建输入输出管线的基本模块。

8.3 `bz2`：bzip2 压缩

`bz2` 模块是 bzip2 库的一个接口，用于压缩数据以便存储或传输。为此，它提供了 3 个 API：
- "一次性"压缩 / 解压缩函数，用以处理大数据块（blob）。
- 迭代式压缩 / 解压缩对象，用来处理数据流。
- 一个类似文件的类，支持像读写未压缩文件一样读写压缩文件。

8.3.1 内存中的一次性操作

使用 `bz2` 最简单的方法是将所有要压缩或解压缩的数据加载到内存中，然后分别使用 `compress()` 和 `decompress()` 来完成转换。

⊖ https://docs.python.org/3.5/library/gzip.html

代码清单 8-14：**bz2_memory.py**

```
import bz2
import binascii

original_data = b'This is the original text.'
print('Original     : {} bytes'.format(len(original_data)))
print(original_data)

print()
compressed = bz2.compress(original_data)
print('Compressed   : {} bytes'.format(len(compressed)))
hex_version = binascii.hexlify(compressed)
for i in range(len(hex_version) // 40 + 1):
    print(hex_version[i * 40:(i + 1) * 40])

print()
decompressed = bz2.decompress(compressed)
print('Decompressed : {} bytes'.format(len(decompressed)))
print(decompressed)
```

压缩数据包含非 ASCII 字符，所以在打印之前需要先转换为其十六进制表示。在这些例子的输出中，我们调整了十六进制表示，使每行最多有 40 个字符。

```
$ python3 bz2_memory.py

Original     : 26 bytes
b'This is the original text.'

Compressed   : 62 bytes
b'425a6839314159265359116be35a6000000002938040'
b'01040022e59c402000314c000111e93d434da223'
b'028cf9e73148cae0a0d6ed7f17724538509016be'
b'35a6'

Decompressed : 26 bytes
b'This is the original text.'
```

对于短文本，压缩版本的长度可能大大超过原来的文本。具体的结果取决于输入数据，不过观察压缩开销很有意思。

代码清单 8-15：**bz2_lengths.py**

```
import bz2

original_data = b'This is the original text.'

fmt = '{:>15}  {:>15}'
print(fmt.format('len(data)', 'len(compressed)'))
print(fmt.format('-' * 15, '-' * 15))

for i in range(5):
    data = original_data * i
    compressed = bz2.compress(data)
    print(fmt.format(len(data), len(compressed)), end='')
    print('*' if len(data) < len(compressed) else '')
```

末尾是 * 字符的输出行表示这一行压缩数据比原输入数据更长。

```
$ python3 bz2_lengths.py
```

len(data)	len(compressed)
0	14*
26	62*
52	68*
78	70
104	72

8.3.2 增量压缩和解压缩

内存中的压缩方法存在明显的缺点，对于实际用例并不实用。另一种方法是使用 **BZ2-Compressor** 和 **BZ2Decompressor** 对象以增量方式处理数据，从而不必将整个数据集都放在内存中。

代码清单 8-16：**bz2_incremental.py**

```
import bz2
import binascii
import io

compressor = bz2.BZ2Compressor()
with open('lorem.txt', 'rb') as input:
    while True:
        block = input.read(64)
        if not block:
            break
        compressed = compressor.compress(block)
        if compressed:
            print('Compressed: {}'.format(
                binascii.hexlify(compressed)))
        else:
            print('buffering...')
    remaining = compressor.flush()
    print('Flushed: {}'.format(binascii.hexlify(remaining)))
```

这个例子从一个纯文本文件读取小数据块，并将它传至 `compress()`。压缩器维护压缩数据的一个内部缓冲区。由于压缩算法取决于校验和以及最小块大小，所以压缩器每次接收更多输入时可能并没有准备好返回数据。如果它没有准备好一个完整的压缩块，则会返回一个空串。所有数据都已经输入时，`flush()` 方法会强制压缩器结束最后一个数据块，并返回余下的压缩数据。

```
$ python3 bz2_incremental.py

buffering...
buffering...
buffering...
buffering...
Flushed: b'425a6839314159265359ba83a48c000014d58000104005040 52fa
7fe003000ba9112793d4ca789068698a0d1a341901a0d53f4d1119a8d4c9e812
d755a67c10798387682c7ca7b5a3bb75da77755eb81c1cb1ca94c4b6faf209c5
2a90aaa4d16a4a1b9c167a01c8d9ef32589d831e77df7a5753a398b11660e392
126fc18a72a1088716cc8dedda5d489da410748531278043d70a8a131c2b8adc
d6a221bdb8c7ff76b88c1d5342ee48a70a12175074918'
```

8.3.3 混合内容流

在压缩和未压缩数据混合在一起的情况下，还可以使用 **BZ2Decompressor**。

代码清单 8-17：**bz2_mixed.py**

```python
import bz2

lorem = open('lorem.txt', 'rt').read().encode('utf-8')
compressed = bz2.compress(lorem)
combined = compressed + lorem

decompressor = bz2.BZ2Decompressor()
decompressed = decompressor.decompress(combined)

decompressed_matches = decompressed == lorem
print('Decompressed matches lorem:', decompressed_matches)

unused_matches = decompressor.unused_data == lorem
print('Unused data matches lorem :', unused_matches)
```

解压缩所有数据之后，`unused_data` 属性会包含所有未用的数据。

```
$ python3 bz2_mixed.py

Decompressed matches lorem: True
Unused data matches lorem : True
```

8.3.4 写压缩文件

可以用 `BZ2File` 读写 bzip2 压缩文件，并使用通常的方法读写数据。

代码清单 8-18：**bz2_file_write.py**

```python
import bz2
import io
import os

data = 'Contents of the example file go here.\n'

with bz2.BZ2File('example.bz2', 'wb') as output:
    with io.TextIOWrapper(output, encoding='utf-8') as enc:
        enc.write(data)

os.system('file example.bz2')
```

为了把数据写入一个压缩文件，需要用模式 `'wb'` 打开文件。这个例子用 `io` 模块的一个 `TextIOWrapper` 来包装 `BZ2File`，将 Unicode 文本编码为适合压缩的字节。

```
$ python3 bz2_file_write.py

example.bz2: bzip2 compressed data, block size = 900k
```

通过传入一个 `compresslevel` 参数，可以使用不同的压缩量。合法值为 1～9（包括 1 和 9）。值越小便会得到越快的处理，压缩也越少。较大的值会得到较慢的处理，但压缩更多（直到某个上限）。

代码清单 8-19：**bz2_file_compresslevel.py**

```python
import bz2
import io
import os
```

```
data = open('lorem.txt', 'r', encoding='utf-8').read() * 1024
print('Input contains {} bytes'.format(
    len(data.encode('utf-8'))))
for i in range(1, 10):
    filename = 'compress-level-{}.bz2'.format(i)
    with bz2.BZ2File(filename, 'wb', compresslevel=i) as output:
        with io.TextIOWrapper(output, encoding='utf-8') as enc:
            enc.write(data)
    os.system('cksum {}'.format(filename))
```

脚本输出中,中间一列数字显示了所生成文件的大小(字节数)。对于这个输入数据,更高的压缩值并不一定得到更少的存储空间。不过对于其他输入,结果可能有所不同。

```
$ python3 bz2_file_compresslevel.py

3018243926 8771 compress-level-1.bz2
1942389165 4949 compress-level-2.bz2
2596054176 3708 compress-level-3.bz2
1491394456 2705 compress-level-4.bz2
1425874420 2705 compress-level-5.bz2
2232840816 2574 compress-level-6.bz2
447681641  2394 compress-level-7.bz2
3699654768 1137 compress-level-8.bz2
3103658384 1137 compress-level-9.bz2
Input contains 754688 bytes
```

BZ2File 实例还包括一个 writelines() 方法,可以用来写一个字符串序列。

代码清单 8-20:**bz2_file_writelines.py**

```
import bz2
import io
import itertools
import os

data = 'The same line, over and over.\n'

with bz2.BZ2File('lines.bz2', 'wb') as output:
    with io.TextIOWrapper(output, encoding='utf-8') as enc:
        enc.writelines(itertools.repeat(data, 10))

os.system('bzcat lines.bz2')
```

与写入常规文件类似,这些行要以一个换行符结尾。

```
$ python3 bz2_file_writelines.py

The same line, over and over.
The same line, over and over.
The same line, over and over.
The same line, over and over.
The same line, over and over.
The same line, over and over.
The same line, over and over.
The same line, over and over.
The same line, over and over.
The same line, over and over.
```

8.3.5 读压缩文件

要从之前压缩的文件读回数据,需要用读模式('rb')打开文件。从 read() 返回的值是一个字节串。

代码清单 8-21: **bz2_file_read.py**

```
import bz2
import io

with bz2.BZ2File('example.bz2', 'rb') as input:
    with io.TextIOWrapper(input, encoding='utf-8') as dec:
        print(dec.read())
```

这个例子读取上一节 bz2_file_write.py 所写的文件。BZ2File 用一个 TextIOWrapper 包装，将所读取的字节解码为 Unicode 文本。

```
$ python3 bz2_file_read.py

Contents of the example file go here.
```

读文件时，还可以用 seek() 定位，然后只读取部分数据。

代码清单 8-22: **bz2_file_seek.py**

```
import bz2
import contextlib

with bz2.BZ2File('example.bz2', 'rb') as input:
    print('Entire file:')
    all_data = input.read()
    print(all_data)

    expected = all_data[5:15]

    # Rewind to beginning
    input.seek(0)

    # Move ahead 5 bytes
    input.seek(5)
    print('Starting at position 5 for 10 bytes:')
    partial = input.read(10)
    print(partial)

    print()
    print(expected == partial)
```

seek() 位置是相对未压缩数据的位置，所以调用者并不需要知道数据文件是压缩文件。这就允许将一个 BZ2File 实例传入一个原本接收常规未压缩文件的函数。

```
$ python3 bz2_file_seek.py

Entire file:
b'Contents of the example file go here.\n'
Starting at position 5 for 10 bytes:
b'nts of the'

True
```

8.3.6 读写 Unicode 数据

前面的例子直接使用 BZ2File，并在必要时用一个 io.TextIOWrapper 来管理 Unicode

文本串的编码和解码。通过使用 bz2.open() 可以避免这些额外的步骤，这会建立一个 io.TextIOWrapper 以自动处理编码或解码。

代码清单 8-23：**bz2_unicode.py**

```
import bz2
import os

data = 'Character with an åccent.'

with bz2.open('example.bz2', 'wt', encoding='utf-8') as output:
    output.write(data)

with bz2.open('example.bz2', 'rt', encoding='utf-8') as input:
    print('Full file: {}'.format(input.read()))

# Move to the beginning of the accented character.
with bz2.open('example.bz2', 'rt', encoding='utf-8') as input:
    input.seek(18)
    print('One character: {}'.format(input.read(1)))

# Move to the middle of the accented character.
with bz2.open('example.bz2', 'rt', encoding='utf-8') as input:
    input.seek(19)
    try:
        print(input.read(1))
    except UnicodeDecodeError:
        print('ERROR: failed to decode')
```

open() 返回的文件句柄支持 seek()，不过使用时要小心，因为文件指针按字节移动，而不是按字符移动，最后可能会落在一个编码字符的中间。

```
$ python3 bz2_unicode.py

Full file: Character with an åccent.
One character: å
ERROR: failed to decode
```

8.3.7 压缩网络数据

下一个例子中的代码会响应文件名请求，它将文件的一个压缩版本写至用来与客户通信的套接字。这里人为地做了一些划分，以展示当数据传递到 compress() 或 decompress() 时，如果没有得到一个完整的压缩或未压缩输出块，则会缓冲。

代码清单 8-24：**bz2_server.py**

```
import bz2
import logging
import socketserver
import binascii

BLOCK_SIZE = 32


class Bz2RequestHandler(socketserver.BaseRequestHandler):

    logger = logging.getLogger('Server')
```

```python
    def handle(self):
        compressor = bz2.BZ2Compressor()

        # Find out which file the client wants.
        filename = self.request.recv(1024).decode('utf-8')
        self.logger.debug('client asked for: "%s"', filename)

        # Send chunks of the file as they are compressed.
        with open(filename, 'rb') as input:
            while True:
                block = input.read(BLOCK_SIZE)
                if not block:
                    break
                self.logger.debug('RAW %r', block)
                compressed = compressor.compress(block)
                if compressed:
                    self.logger.debug(
                        'SENDING %r',
                        binascii.hexlify(compressed))
                    self.request.send(compressed)
                else:
                    self.logger.debug('BUFFERING')

        # Send any data being buffered by the compressor.
        remaining = compressor.flush()
        while remaining:
            to_send = remaining[:BLOCK_SIZE]
            remaining = remaining[BLOCK_SIZE:]
            self.logger.debug('FLUSHING %r',
                              binascii.hexlify(to_send))
            self.request.send(to_send)
        return
```

主程序在一个线程中结合 **SocketServer** 和 **Bz2RequestHandler** 以启动一个服务器。

```python
if __name__ == '__main__':
    import socket
    import sys
    from io import StringIO
    import threading

    logging.basicConfig(level=logging.DEBUG,
                        format='%(name)s: %(message)s',
                        )

    # Set up a server, running in a separate thread.
    address = ('localhost', 0)  # Let the kernel assign a port.
    server = socketserver.TCPServer(address, Bz2RequestHandler)
    ip, port = server.server_address  # What port was assigned?

    t = threading.Thread(target=server.serve_forever)
    t.setDaemon(True)
    t.start()

    logger = logging.getLogger('Client')

    # Connect to the server.
    logger.info('Contacting server on %s:%s', ip, port)
    s = socket.socket(socket.AF_INET, socket.SOCK_STREAM)
    s.connect((ip, port))
```

```python
        # Ask for a file.
        requested_file = (sys.argv[0]
                          if len(sys.argv) > 1
                          else 'lorem.txt')
        logger.debug('sending filename: "%s"', requested_file)
        len_sent = s.send(requested_file.encode('utf-8'))

        # Receive a response.
        buffer = StringIO()
        decompressor = bz2.BZ2Decompressor()
        while True:
            response = s.recv(BLOCK_SIZE)
            if not response:
                break
            logger.debug('READ %r', binascii.hexlify(response))

            # Include any unconsumed data when feeding the
            # decompressor.
            decompressed = decompressor.decompress(response)
            if decompressed:
                logger.debug('DECOMPRESSED %r', decompressed)
                buffer.write(decompressed.decode('utf-8'))
            else:
                logger.debug('BUFFERING')

        full_response = buffer.getvalue()
        lorem = open(requested_file, 'rt').read()
        logger.debug('response matches file contents: %s',
                     full_response == lorem)

        # Clean up.
        server.shutdown()
        server.socket.close()
        s.close()
```

然后这个程序打开一个套接字，作为客户与服务器通信，并请求文件（默认为 `lorem.txt`）。

```
Lorem ipsum dolor sit amet, consectetuer adipiscing elit. Donec
egestas, enim et consectetuer ullamcorper, lectus ligula rutrum leo,
a elementum elit tortor eu quam. Duis tincidunt nisi ut ante. Nulla
facilisi.
```

警告：这个实现存在明显的安全隐患。不要在开放的互联网或安全问题可能导致严重影响的环境中运行这个程序。

运行 `bz2_server.py` 会生成以下结果。

```
$ python3 bz2_server.py

Client: Contacting server on 127.0.0.1:57364
Client: sending filename: "lorem.txt"
Server: client asked for: "lorem.txt"
Server: RAW b'Lorem ipsum dolor sit amet, cons'
Server: BUFFERING
Server: RAW b'ectetuer adipiscing elit. Donec\n'
Server: BUFFERING
Server: RAW b'egestas, enim et consectetuer ul'
Server: BUFFERING
Server: RAW b'lamcorper, lectus ligula rutrum '
Server: BUFFERING
Server: RAW b'leo,\na elementum elit tortor eu '
```

```
Server: BUFFERING
Server: RAW b'quam. Duis tincidunt nisi ut ant'
Server: BUFFERING
Server: RAW b'e. Nulla\nfacilisi.\n'
Server: BUFFERING
Server: FLUSHING b'425a6839314159265359ba83a48c000014d5800010400
504052fa7fe003000ba'
Server: FLUSHING b'9112793d4ca789068698a0d1a341901a0d53f4d1119a8
d4c9e812d755a67c107'
Client: READ b'425a6839314159265359ba83a48c000014d58000104005040
52fa7fe003000ba'
Server: FLUSHING b'98387682c7ca7b5a3bb75da77755eb81c1cb1ca94c4b6
faf209c52a90aaa4d16'
Client: BUFFERING
Server: FLUSHING b'a4a1b9c167a01c8d9ef32589d831e77df7a5753a398b1
1660e392126fc18a72a'
Client: READ b'9112793d4ca789068698a0d1a341901a0d53f4d1119a8d4c9
e812d755a67c107'
Server: FLUSHING b'1088716cc8dedda5d489da410748531278043d70a8a13
1c2b8adcd6a221bdb8c'
Client: BUFFERING
Server: FLUSHING b'7ff76b88c1d5342ee48a70a12175074918'
Client: READ b'98387682c7ca7b5a3bb75da77755eb81c1cb1ca94c4b6faf2
09c52a90aaa4d16'
Client: BUFFERING
Client: READ b'a4a1b9c167a01c8d9ef32589d831e77df7a5753a398b11660
e392126fc18a72a'
Client: BUFFERING
Client: READ b'1088716cc8dedda5d489da410748531278043d70a8a131c2b
8adcd6a221bdb8c'
Client: BUFFERING
Client: READ b'7ff76b88c1d5342ee48a70a12175074918'
Client: DECOMPRESSED b'Lorem ipsum dolor sit amet, consectetuer
adipiscing elit. Donec\negestas, enim et consectetuer ullamcorpe
r, lectus ligula rutrum leo,\na elementum elit tortor eu quam. D
uis tincidunt nisi ut ante. Nulla\nfacilisi.\n'
Client: response matches file contents: True
```

提示：相关阅读材料

- `bz2` 的标准库文档⊖。
- bzip2⊖：bzip2 的主页。
- `zlib`：用来完成 GNU zip 压缩的 `zlib` 模块。
- `gzip`：GNU zip 压缩文件的一个类似文件的接口。
- `io`：创建输入和输出管线的基本模块。
- `bz2` 的 Python 2 到 Python 3 移植说明。

8.4 `tarfile`：tar 归档访问

`tarfile` 模块提供了对 UNIX tar 归档文件（包括压缩文件）的读写访问。除了 POSIX 标准之外，还支持多个 GNU tar 扩展。另外还能处理一些 UNIX 特殊文件类型（如硬/软链接）以及设备节点。

⊖ https://docs.python.org/3.5/library/bz2.html

⊖ www.bzip.org

说明：尽管 tarfile 实现了一种 UNIX 格式，也可以在 Microsoft Windows 中用来创建和读取 tar 归档文件。

8.4.1 测试 tar 文件

is_tarfile() 函数返回一个布尔值，指示作为参数传入的文件名是否指向一个合法的 tar 归档文件。

代码清单 8-25：**tarfile_is_tarfile.py**

```
import tarfile

for filename in ['README.txt', 'example.tar',
                 'bad_example.tar', 'notthere.tar']:
    try:
        print('{:>15}  {}'.format(filename, tarfile.is_tarfile(
            filename)))
    except IOError as err:
        print('{:>15}  {}'.format(filename, err))
```

如果文件不存在，那么 is_tarfile() 会引发一个 IOError 异常。

```
$ python3 tarfile_is_tarfile.py

    README.txt  False
   example.tar  True
bad_example.tar  False
   notthere.tar  [Errno 2] No such file or directory: 'notthere.tar'
```

8.4.2 从归档读取元数据

可以使用 TarFile 类直接处理一个 tar 归档文件。这个类支持一些方法来读取现有归档文件的有关数据，还可以通过增加更多文件来修改归档。

要读取一个现有归档文件中的文件名，可以使用 getnames()。

代码清单 8-26：**tarfile_getnames.py**

```
import tarfile

with tarfile.open('example.tar', 'r') as t:
    print(t.getnames())
```

这个函数的返回值是一个字符串列表，包含归档内容中的文件名。

```
$ python3 tarfile_getnames.py

['index.rst', 'README.txt']
```

除了文件名之外，还可以得到归档成员的元数据（作为 TarInfo 对象的实例）。

代码清单 8-27：**tarfile_getmembers.py**

```
import tarfile
import time
```

```
with tarfile.open('example.tar', 'r') as t:
    for member_info in t.getmembers():
        print(member_info.name)
        print('  Modified:', time.ctime(member_info.mtime))
        print('  Mode    :', oct(member_info.mode))
        print('  Type    :', member_info.type)
        print('  Size    :', member_info.size, 'bytes')
        print()
```

可以通过 `getmembers()` 和 `getmember()` 加载元数据。

```
$ python3 tarfile_getmembers.py

index.rst
  Modified: Fri Aug 19 16:27:54 2016
  Mode    : 0o644
  Type    : b'0'
  Size    : 9878 bytes
README.txt
  Modified: Fri Aug 19 16:27:54 2016
  Mode    : 0o644
  Type    : b'0'
  Size    : 75 bytes
```

如果提前已经知道归档成员名，则可以用 `getmember()` 获取其 `TarInfo` 对象。

代码清单 8-28：**tarfile_getmember.py**

```
import tarfile
import time

with tarfile.open('example.tar', 'r') as t:
    for filename in ['README.txt', 'notthere.txt']:
        try:
            info = t.getmember(filename)
        except KeyError:
            print('ERROR: Did not find {} in tar archive'.format(
                filename))
        else:
            print('{} is {:d} bytes'.format(
                info.name, info.size))
```

如果归档成员不存在，则 `getmember()` 会产生一个 `KeyError`。

```
$ python3 tarfile_getmember.py

README.txt is 75 bytes
ERROR: Did not find notthere.txt in tar archive
```

8.4.3 从归档抽取文件

要在程序中访问一个归档成员的数据，可以使用 `extractfile()` 方法，并且传入这个成员名。

代码清单 8-29：**tarfile_extractfile.py**

```
import tarfile

with tarfile.open('example.tar', 'r') as t:
```

```
        for filename in ['README.txt', 'notthere.txt']:
            try:
                f = t.extractfile(filename)
            except KeyError:
                print('ERROR: Did not find {} in tar archive'.format(
                    filename))
            else:
                print(filename, ':')
                print(f.read().decode('utf-8'))
```

返回值是一个类似文件的对象,可以从这个对象读取归档成员的内容。

```
$ python3 tarfile_extractfile.py

README.txt :
The examples for the tarfile module use this file and
example.tar as data.

ERROR: Did not find notthere.txt in tar archive
```

要解开归档并将文件写至文件系统,可以使用 extract() 或 extractall()。

代码清单 8-30: tarfile_extract.py

```
import tarfile
import os

os.mkdir('outdir')
with tarfile.open('example.tar', 'r') as t:
    t.extract('README.txt', 'outdir')
print(os.listdir('outdir'))
```

会从归档中读出归档成员,并写至文件系统(从参数中指定的目录开始)。

```
$ python3 tarfile_extract.py

['README.txt']
```

标准库文档中有一个说明,指出 extractall() 比 extract() 更安全,特别是处理流数据时,这是因为对于流数据来说,无法回转输入而去读之前的部分。大多数情况下都应该使用 extractall()。

代码清单 8-31: tarfile_extractall.py

```
import tarfile
import os

os.mkdir('outdir')
with tarfile.open('example.tar', 'r') as t:
    t.extractall('outdir')
print(os.listdir('outdir'))
```

使用 extractall() 时,第一个参数是一个目录名,文件将写至这个目录。

```
$ python3 tarfile_extractall.py

['README.txt', 'index.rst']
```

要从归档中抽取特定的文件,可以把这些文件名或 TarInfo 元数据容器传递到 extractall()。

代码清单8-32：**tarfile_extractall_members.py**

```python
import tarfile
import os

os.mkdir('outdir')
with tarfile.open('example.tar', 'r') as t:
    t.extractall('outdir',
                 members=[t.getmember('README.txt')],
                 )
print(os.listdir('outdir'))
```

如果提供了一个 members 列表，就只抽取指定的文件。

```
$ python3 tarfile_extractall_members.py

['README.txt']
```

8.4.4 创建新归档

要创建一个新归档，需要用模式 'w' 打开 TarFile。

代码清单8-33：**tarfile_add.py**

```python
import tarfile

print('creating archive')
with tarfile.open('tarfile_add.tar', mode='w') as out:
    print('adding README.txt')
    out.add('README.txt')

print()
print('Contents:')
with tarfile.open('tarfile_add.tar', mode='r') as t:
    for member_info in t.getmembers():
        print(member_info.name)
```

其会删除现有的文件，并重建一个新的归档。要增加文件，可以使用 add() 方法。

```
$ python3 tarfile_add.py

creating archive
adding README.txt

Contents:
README.txt
```

8.4.5 使用候选归档成员名

向归档增加一个文件时，可以不用原始文件名而用另外一个名字，利用一个候选的 arcname 构造一个 TarInfo 对象，并把它传至 addfile()。

代码清单8-34：**tarfile_addfile.py**

```python
import tarfile

print('creating archive')
```

```python
with tarfile.open('tarfile_addfile.tar', mode='w') as out:
    print('adding README.txt as RENAMED.txt')
    info = out.gettarinfo('README.txt', arcname='RENAMED.txt')
    out.addfile(info)

print()
print('Contents:')
with tarfile.open('tarfile_addfile.tar', mode='r') as t:
    for member_info in t.getmembers():
        print(member_info.name)
```

这个归档只包含修改的文件名。

```
$ python3 tarfile_addfile.py

creating archive
adding README.txt as RENAMED.txt

Contents:
RENAMED.txt
```

8.4.6 从非文件源写数据

有时可能需要将数据从内存直接写至一个归档。并不是将数据先写入一个文件，然后再把这个文件增加到归档，可以使用 addfile() 从一个返回字节的打开的类似文件句柄添加数据。

代码清单 8-35：**tarfile_addfile_string.py**

```python
import io
import tarfile

text = 'This is the data to write to the archive.'
data = text.encode('utf-8')

with tarfile.open('addfile_string.tar', mode='w') as out:
    info = tarfile.TarInfo('made_up_file.txt')
    info.size = len(data)
    out.addfile(info, io.BytesIO(data))

print('Contents:')
with tarfile.open('addfile_string.tar', mode='r') as t:
    for member_info in t.getmembers():
        print(member_info.name)
        f = t.extractfile(member_info)
        print(f.read().decode('utf-8'))
```

首先构造一个 TarInfo 对象，可以为归档成员指定所需的任何名字。设置大小之后，以一个 BytesIO 缓冲区作为数据源，使用 addfile() 把数据写至归档。

```
$ python3 tarfile_addfile_string.py

Contents:
made_up_file.txt
This is the data to write to the archive.
```

8.4.7 追加到归档

除了创建新归档，还可以使用模式 'a' 追加到一个现有的文件。

代码清单8-36：**tarfile_append.py**

```python
import tarfile

print('creating archive')
with tarfile.open('tarfile_append.tar', mode='w') as out:
    out.add('README.txt')

print('contents:',)
with tarfile.open('tarfile_append.tar', mode='r') as t:
    print([m.name for m in t.getmembers()])

print('adding index.rst')
with tarfile.open('tarfile_append.tar', mode='a') as out:
    out.add('index.rst')

print('contents:',)
with tarfile.open('tarfile_append.tar', mode='r') as t:
    print([m.name for m in t.getmembers()])
```

最后得到的归档将包含两个成员。

```
$ python3 tarfile_append.py

creating archive
contents:
['README.txt']
adding index.rst
contents:
['README.txt', 'index.rst']
```

8.4.8 处理压缩归档

除了常规的tar归档文件之外，`tarfile`模块还可以处理通过gzip或bzip2协议压缩的归档。要打开一个压缩归档，可以修改传入`open()`的模式串，根据所需压缩方法的不同，在模式串中包含":gz"或":bz2"。

代码清单8-37：**tarfile_compression.py**

```python
import tarfile
import os

fmt = '{:<30} {:<10}'
print(fmt.format('FILENAME', 'SIZE'))
print(fmt.format('README.txt', os.stat('README.txt').st_size))
FILES = [
    ('tarfile_compression.tar', 'w'),
    ('tarfile_compression.tar.gz', 'w:gz'),
    ('tarfile_compression.tar.bz2', 'w:bz2'),
]

for filename, write_mode in FILES:
    with tarfile.open(filename, mode=write_mode) as out:
        out.add('README.txt')

    print(fmt.format(filename, os.stat(filename).st_size),
          end=' ')
    print([
        m.name
        for m in tarfile.open(filename, 'r:*').getmembers()
    ])
```

在打开一个现有的归档并读取数据时,可以指定 `"r:*"` 让 `tarfile` 自动确定要使用的压缩方法。

```
$ python3 tarfile_compression.py

FILENAME                         SIZE
README.txt                       75
tarfile_compression.tar          10240   ['README.txt']
tarfile_compression.tar.gz       213     ['README.txt']
tarfile_compression.tar.bz2      199     ['README.txt']
```

提示:相关阅读材料
- `tarfile` 的标准库文档[⊖]。
- GNU tar manual[⊖]:tar 格式的文档(包括扩展)。
- `zipfile`:类似地访问 ZIP 归档。
- `gzip`:GNU zip 压缩。
- `bz2`:bzip2 压缩。

8.5 `zipfile`:ZIP 归档访问

`zipfile` 模块可以用来读写 ZIP 归档文件,这种格式因 PC 程序 PKZIP 而普及。

8.5.1 测试 ZIP 文件

`is_zipfile()` 函数返回一个布尔值,指示作为参数传入的文件名是否指向一个合法的 ZIP 归档。

代码清单 8-38:`zipfile_is_zipfile.py`

```
import zipfile

for filename in ['README.txt', 'example.zip',
                 'bad_example.zip', 'notthere.zip']:
    print('{:>15}  {}'.format(
        filename, zipfile.is_zipfile(filename)))
```

如果这个文件不存在,则 `is_zipfile()` 返回 `False`。

```
$ python3 zipfile_is_zipfile.py

     README.txt  False
    example.zip  True
bad_example.zip  False
   notthere.zip  False
```

8.5.2 从归档读取元数据

使用 `ZipFile` 类可以直接处理一个 ZIP 归档。这个类支持一些方法来读取现有归档的有关数据,还可以通过增加更多的文件来修改归档。

⊖ https://docs.python.org/3.5/library/tarfile.html

⊖ www.gnu.org/software/tar/manual/html_node/Standard.html

代码清单 8-39：`zipfile_namelist.py`

```
import zipfile

with zipfile.ZipFile('example.zip', 'r') as zf:
    print(zf.namelist())
```

`namelist()` 方法返回一个现有归档中的文件名。

```
$ python3 zipfile_namelist.py

['README.txt']
```

不过，这个文件名列表只是从归档得到的信息的一部分。要访问有关 ZIP 内容的所有元数据，可以使用 `infolist()` 或 `getinfo()` 方法。

代码清单 8-40：`zipfile_infolist.py`

```
import datetime
import zipfile
def print_info(archive_name):
    with zipfile.ZipFile(archive_name) as zf:
        for info in zf.infolist():
            print(info.filename)
            print('  Comment     :', info.comment)
            mod_date = datetime.datetime(*info.date_time)
            print('  Modified    :', mod_date)
            if info.create_system == 0:
                system = 'Windows'
            elif info.create_system == 3:
                system = 'Unix'
            else:
                system = 'UNKNOWN'
            print('  System      :', system)
            print('  ZIP version :', info.create_version)
            print('  Compressed  :', info.compress_size, 'bytes')
            print('  Uncompressed:', info.file_size, 'bytes')
            print()

if __name__ == '__main__':
    print_info('example.zip')
```

除了这里打印的字段，元数据还包括另外一些字段，但是要把这些值解释为有用的信息，需要仔细阅读 ZIP 文件规范的"PKZIP 应用说明"（PKZIP Application Note）。

```
$ python3 zipfile_infolist.py

README.txt
  Comment     : b''
  Modified    : 2010-11-15 06:48:02
  System      : Unix
  ZIP version : 30
  Compressed  : 65 bytes
  Uncompressed: 76 bytes
```

如果提前已经知道归档成员名，可以利用 `getinfo()` 直接获取其 ZipInfo 对象。

代码清单 8-41：`zipfile_getinfo.py`

```
import zipfile
```

```
with zipfile.ZipFile('example.zip') as zf:
    for filename in ['README.txt', 'notthere.txt']:
        try:
            info = zf.getinfo(filename)
        except KeyError:
            print('ERROR: Did not find {} in zip file'.format(
                filename))
        else:
            print('{} is {} bytes'.format(
                info.filename, info.file_size))
```

如果归档成员不存在,getinfo()会产生一个KeyError。

```
$ python3 zipfile_getinfo.py

README.txt is 76 bytes
ERROR: Did not find notthere.txt in zip file
```

8.5.3 从归档抽取归档文件

要从一个归档成员访问数据,可以使用read()方法,并传入该成员名。

代码清单 8-42:**zipfile_read.py**

```
import zipfile

with zipfile.ZipFile('example.zip') as zf:
    for filename in ['README.txt', 'notthere.txt']:
        try:
            data = zf.read(filename)
        except KeyError:
            print('ERROR: Did not find {} in zip file'.format(
                filename))
        else:
            print(filename, ':')
            print(data)
        print()
```

如果必要,数据会自动解压缩。

```
$ python3 zipfile_read.py

README.txt :
b'The examples for the zipfile module use \nthis file and exampl
e.zip as data.\n'

ERROR: Did not find notthere.txt in zip file
```

8.5.4 创建新归档

要创建一个新归档,需要用模式 'w' 实例化 ZipFile。其会删除所有现有的文件,并创建一个新归档。要增加文件,可以使用 write() 方法。

代码清单 8-43:**zipfile_write.py**

```
from zipfile_infolist import print_info
import zipfile
```

```
print('creating archive')
with zipfile.ZipFile('write.zip', mode='w') as zf:
    print('adding README.txt')
    zf.write('README.txt')

print()
print_info('write.zip')
```

默认地，归档的内容不会被压缩。

```
$ python3 zipfile_write.py

creating archive
adding README.txt

README.txt
  Comment       : b''
  Modified      : 2016-08-07 13:31:24
  System        : Unix
  ZIP version   : 20
  Compressed    : 76 bytes
  Uncompressed  : 76 bytes
```

要想增加压缩，需要有 `zlib` 模块。如果 `zlib` 可用，则可以使用 `zipfile.ZIP_DEFLATED` 设置单个文件的压缩模式，或者为归档整体设置压缩模式。默认的压缩模式是 `zipfile.ZIP_STORED`，它会把输入数据增加到归档而不压缩。

代码清单 8-44: **zipfile_write_compression.py**

```
from zipfile_infolist import print_info
import zipfile
try:
    import zlib
    compression = zipfile.ZIP_DEFLATED
except:
    compression = zipfile.ZIP_STORED

modes = {
    zipfile.ZIP_DEFLATED: 'deflated',
    zipfile.ZIP_STORED: 'stored',
}

print('creating archive')
with zipfile.ZipFile('write_compression.zip', mode='w') as zf:
    mode_name = modes[compression]
    print('adding README.txt with compression mode', mode_name)
    zf.write('README.txt', compress_type=compression)

print()
print_info('write_compression.zip')
```

这一次归档成员会被压缩。

```
$ python3 zipfile_write_compression.py

creating archive
adding README.txt with compression mode deflated

README.txt
```

```
    Comment       : b''
    Modified      : 2016-08-07 13:31:24
    System        : Unix
    ZIP version   : 20
    Compressed    : 65 bytes
    Uncompressed: 76 bytes
```

8.5.5 使用候选归档成员名

为文档增加文件时，可以向 `write()` 传入一个 `arcname` 值，这样可以使用原始文件名以外的另一个文件名。

代码清单 8-45：<code>zipfile_write_arcname.py</code>

```python
from zipfile_infolist import print_info
import zipfile

with zipfile.ZipFile('write_arcname.zip', mode='w') as zf:
    zf.write('README.txt', arcname='NOT_README.txt')

print_info('write_arcname.zip')
```

归档中不会再出现原来的文件名。

```
$ python3 zipfile_write_arcname.py

NOT_README.txt
    Comment       : b''
    Modified      : 2016-08-07 13:31:24
    System        : Unix
    ZIP version   : 20
    Compressed    : 76 bytes
    Uncompressed: 76 bytes
```

8.5.6 从非文件源写数据

有时，可能需要使用其他来源的数据（而非来自一个现有文件）写一个 ZIP 归档。不是将数据先写入一个文件，然后再把这个文件增加到 ZIP 归档，可以使用 `writestr()` 直接向归档写入一个字节串。

代码清单 8-46：<code>zipfile_writestr.py</code>

```python
from zipfile_infolist import print_info
import zipfile

msg = 'This data did not exist in a file.'
with zipfile.ZipFile('writestr.zip',
                     mode='w',
                     compression=zipfile.ZIP_DEFLATED,
                     ) as zf:
    zf.writestr('from_string.txt', msg)

print_info('writestr.zip')

with zipfile.ZipFile('writestr.zip', 'r') as zf:
    print(zf.read('from_string.txt'))
```

在这里，要利用 ZipFile 的 compress_type 参数压缩数据，因为 writestr() 不接收指定压缩的参数。

```
$ python3 zipfile_writestr.py

from_string.txt
  Comment      : b''
  Modified     : 2016-12-29 12:14:42
  System       : Unix
  ZIP version  : 20
  Compressed   : 36 bytes
  Uncompressed : 34 bytes

b'This data did not exist in a file.'
```

8.5.7 利用 ZipInfo 实例写数据

正常情况下，在文件或串被增加到归档时会计算修改日期。可以向 writestr() 传递一个 ZipInfo 实例来定义修改日期和其他元数据。

代码清单 8-47: zipfile_writestr_zipinfo.py

```
import time
import zipfile
from zipfile_infolist import print_info
msg = b'This data did not exist in a file.'

with zipfile.ZipFile('writestr_zipinfo.zip',
                     mode='w',
                     ) as zf:
    info = zipfile.ZipInfo('from_string.txt',
                           date_time=time.localtime(time.time()),
                           )
    info.compress_type = zipfile.ZIP_DEFLATED
    info.comment = b'Remarks go here'
    info.create_system = 0
    zf.writestr(info, msg)

print_info('writestr_zipinfo.zip')
```

在这个例子中，修改时间被设置为当前时间，并且对数据进行压缩，另外 create_system 使用了一个 false 值。还可以为这个新文件关联一个简单的注释。

```
$ python3 zipfile_writestr_zipinfo.py

from_string.txt
  Comment      : b'Remarks go here'
  Modified     : 2016-12-29 12:14:42
  System       : Windows
  ZIP version  : 20
  Compressed   : 36 bytes
  Uncompressed : 34 bytes
```

8.5.8 追加到文件

除了创建新归档，还可以追加到一个现有的归档，或者将一个归档增加到一个现有文件的末尾（如为一个自解压归档增加一个 .exe 文件）。要打开一个文件来完成追加，可以

使用模式 'a'。

代码清单 8-48：`zipfile_append.py`

```python
from zipfile_infolist import print_info
import zipfile

print('creating archive')
with zipfile.ZipFile('append.zip', mode='w') as zf:
    zf.write('README.txt')

print()
print_info('append.zip')
print('appending to the archive')
with zipfile.ZipFile('append.zip', mode='a') as zf:
    zf.write('README.txt', arcname='README2.txt')

print()
print_info('append.zip')
```

最后得到的归档将包含两个成员。

```
$ python3 zipfile_append.py

creating archive

README.txt
  Comment       : b''
  Modified      : 2016-08-07 13:31:24
  System        : Unix
  ZIP version   : 20
  Compressed    : 76 bytes
  Uncompressed: 76 bytes

appending to the archive

README.txt
  Comment       : b''
  Modified      : 2016-08-07 13:31:24
  System        : Unix
  ZIP version   : 20
  Compressed    : 76 bytes
  Uncompressed: 76 bytes

README2.txt
  Comment       : b''
  Modified      : 2016-08-07 13:31:24
  System        : Unix
  ZIP version   : 20
  Compressed    : 76 bytes
  Uncompressed: 76 bytes
```

8.5.9　Python ZIP 归档

如果归档位于 `sys.path`，Python 可以使用 `zipimport` 从这些 ZIP 归档导入模块。可以用 `PyZipFile` 类构造一个适合以这种方式使用的模块。额外的 `writepy()` 方法会告诉 `PyZipFile` 扫描一个目录来查找 `.py` 文件，并把相应的 `.pyo` 或 `.pyc` 文件添加到归档中。如果这两种编译格式都不存在，则创建并增加一个 `.pyc` 文件。

代码清单 8-49：`zipfile_pyzipfile.py`

```
import sys
import zipfile

if __name__ == '__main__':
    with zipfile.PyZipFile('pyzipfile.zip', mode='w') as zf:
        zf.debug = 3
        print('Adding python files')
        zf.writepy('.')
    for name in zf.namelist():
        print(name)

    print()
    sys.path.insert(0, 'pyzipfile.zip')
    import zipfile_pyzipfile
    print('Imported from:', zipfile_pyzipfile.__file__)
```

将 `PyZipFile` 的 `debug` 属性设置为 3，这会启用 verbose 调试，在程序编译其所找到的各个 `.py` 文件时会生成输出。

```
$ python3 zipfile_pyzipfile.py

Adding python files
Adding files from directory .
Compiling ./zipfile_append.py
Adding zipfile_append.pyc
Compiling ./zipfile_getinfo.py
Adding zipfile_getinfo.pyc
Compiling ./zipfile_infolist.py
Adding zipfile_infolist.pyc
Compiling ./zipfile_is_zipfile.py
Adding zipfile_is_zipfile.pyc
Compiling ./zipfile_namelist.py
Adding zipfile_namelist.pyc
Compiling ./zipfile_printdir.py
Adding zipfile_printdir.pyc
Compiling ./zipfile_pyzipfile.py
Adding zipfile_pyzipfile.pyc
Compiling ./zipfile_read.py
Adding zipfile_read.pyc
Compiling ./zipfile_write.py
Adding zipfile_write.pyc
Compiling ./zipfile_write_arcname.py
Adding zipfile_write_arcname.pyc
Compiling ./zipfile_write_compression.py
Adding zipfile_write_compression.pyc
Compiling ./zipfile_writestr.py
Adding zipfile_writestr.pyc
Compiling ./zipfile_writestr_zipinfo.py
Adding zipfile_writestr_zipinfo.pyc
zipfile_append.pyc
zipfile_getinfo.pyc
zipfile_infolist.pyc
zipfile_is_zipfile.pyc
zipfile_namelist.pyc
zipfile_printdir.pyc
zipfile_pyzipfile.pyc
zipfile_read.pyc
zipfile_write.pyc
zipfile_write_arcname.pyc
```

```
zipfile_write_compression.pyc
zipfile_writestr.pyc
zipfile_writestr_zipinfo.pyc

Imported from: pyzipfile.zip/zipfile_pyzipfile.pyc
```

8.5.10 限制

`zipfile`模块不支持有追加注释的 ZIP 文件或多磁盘归档,不过它支持使用 ZIP64 扩展的超过 4GB 的 ZIP 文件。

提示:相关阅读材料
- `zipfile`的标准库文档[一]。
- `zlib`:ZIP 压缩库。
- `tarfile`:读写 tar 归档。
- `zipimport`:从 ZIP 归档导入 Python 模块。
- PKZIP Application Note[二]:ZIP 归档格式的官方规范。

[一] https://docs.python.org/3.5/library/zipfile.html
[二] www.pkware.com/documents/casestudies/APPNOTE.TXT

第 9 章 加密

加密可以保护消息安全，以便验证其正确性并保护消息不被截获。Python 的加密支持包括 hashlib 和 hmac，hashlib 使用标准算法（如 MD5 和 SHA）生成消息内容的签名，hmac 则用于验证消息在传输过程中未被修改。

9.1 hashlib：密码散列

hashlib 模块定义了一个 API 来访问不同的密码散列算法。要使用一个特定的散列算法，可以用适当的构造器函数或者 new() 来创建一个散列对象。不论使用哪个具体的算法，这些对象都使用相同的 API。

9.1.1 散列算法

由于 hashlib 有 OpenSSL 提供"底层支持"，所以 OpenSSL 库提供的所有算法都可用，包括：

- MD5
- SHA1
- SHA224
- SHA256
- SHA384
- SHA512

有些算法在所有平台上都可用，而有些则依赖于底层库。这两类算法分别由 algorithms_guaranteed 和 algorithms_available 提供。

代码清单 9-1：hashlib_algorithms.py

```
import hashlib

print('Guaranteed:\n{}\n'.format(
    ', '.join(sorted(hashlib.algorithms_guaranteed))))
print('Available:\n{}'.format(
    ', '.join(sorted(hashlib.algorithms_available))))
```

```
$ python3 hashlib_algorithms.py

Guaranteed:
md5, sha1, sha224, sha256, sha384, sha512
```

```
Available:
DSA, DSA-SHA, MD4, MD5, MDC2, RIPEMD160, SHA, SHA1, SHA224,
SHA256, SHA384, SHA512, dsaEncryption, dsaWithSHA,
ecdsa-with-SHA1, md4, md5, mdc2, ripemd160, sha, sha1, sha224,
sha256, sha384, sha512
```

9.1.2 示例数据

本节中的所有例子都使用相同的示例数据,如以下代码清单所示。

代码清单 9-2: hashlib_data.py

```
import hashlib

lorem = '''Lorem ipsum dolor sit amet, consectetur adipisicing
elit, sed do eiusmod tempor incididunt ut labore et dolore magna
aliqua. Ut enim ad minim veniam, quis nostrud exercitation
ullamco laboris nisi ut aliquip ex ea commodo consequat. Duis
aute irure dolor in reprehenderit in voluptate velit esse cillum
dolore eu fugiat nulla pariatur. Excepteur sint occaecat
cupidatat non proident, sunt in culpa qui officia deserunt
mollit anim id est laborum.'''
```

9.1.3 MD5 示例

要为一个数据块(在这里就是转换为一个字节串的 Unicode 串)计算 MD5 散列或摘要,首先要创建散列对象,然后增加数据,最后调用 `digest()` 或 `hexdigest()`。

代码清单 9-3: hashlib_md5.py

```
import hashlib

from hashlib_data import lorem

h = hashlib.md5()
h.update(lorem.encode('utf-8'))
print(h.hexdigest())
```

这个例子使用了 `hexdigest()` 方法而不是 `digest()`,因为要格式化输出以便清楚地打印。如果可以接受二进制摘要值,那么可以使用 `digest()`。

```
$ python3 hashlib_md5.py

3f2fd2c9e25d60fb0fa5d593b802b7a8
```

9.1.4 SHA1 示例

SHA1 摘要也用同样的方式计算。

代码清单 9-4: hashlib_sha1.py

```
import hashlib

from hashlib_data import lorem

h = hashlib.sha1()
h.update(lorem.encode('utf-8'))
print(h.hexdigest())
```

这个例子中的摘要值有所不同，因为 MD5 和 SHA1 算法不同。

```
$ python3 hashlib_sha1.py

ea360b288b3dd178fe2625f55b2959bf1dba6eef
```

9.1.5 按名创建散列

有些情况下，利用字符串指定算法名比直接使用构造器函数更方便。例如，如果能够把散列类型存储在一个配置文件中，这会很有用。在这些情况下，可以使用 new() 来创建一个散列计算器。

代码清单 9-5：**hashlib_new.py**

```python
import argparse
import hashlib
import sys

from hashlib_data import lorem


parser = argparse.ArgumentParser('hashlib demo')
parser.add_argument(
    'hash_name',
    choices=hashlib.algorithms_available,
    help='the name of the hash algorithm to use',
)
parser.add_argument(
    'data',
    nargs='?',
    default=lorem,
    help='the input data to hash, defaults to lorem ipsum',
)
args = parser.parse_args()

h = hashlib.new(args.hash_name)
h.update(args.data.encode('utf-8'))
print(h.hexdigest())
```

运行时可以提供不同的参数，这个程序生成以下输出。

```
$ python3 hashlib_new.py sha1

ea360b288b3dd178fe2625f55b2959bf1dba6eef

$ python3 hashlib_new.py sha256

3c887cc71c67949df29568119cc646f46b9cd2c2b39d456065646bc2fc09ffd8

$ python3 hashlib_new.py sha512

a7e53384eb9bb4251a19571450465d51809e0b7046101b87c4faef96b9bc904cf7f90
035f444952dfd9f6084eeee2457433f3ade614712f42f80960b2fca43ff

$ python3 hashlib_new.py md5

3f2fd2c9e25d60fb0fa5d593b802b7a8
```

9.1.6 增量更新

散列计算器的 update() 方法可以反复调用。每次调用时，都会根据提供的附加文本更新摘要。增量更新比将整个文件读入内存更高效，而且能生成相同的结果。

代码清单 9-6：**hashlib_update.py**

```python
import hashlib

from hashlib_data import lorem

h = hashlib.md5()
h.update(lorem.encode('utf-8'))
all_at_once = h.hexdigest()


def chunkize(size, text):
    "Return parts of the text in size-based increments."
    start = 0
    while start < len(text):
        chunk = text[start:start + size]
        yield chunk
        start += size
    return

h = hashlib.md5()
for chunk in chunkize(64, lorem.encode('utf-8')):
    h.update(chunk)
line_by_line = h.hexdigest()

print('All at once :', all_at_once)
print('Line by line:', line_by_line)
print('Same        :', (all_at_once == line_by_line))
```

这个例子展示了读取或生成数据时如何以增量方式更新一个摘要。

```
$ python3 hashlib_update.py

All at once : 3f2fd2c9e25d60fb0fa5d593b802b7a8
Line by line: 3f2fd2c9e25d60fb0fa5d593b802b7a8
Same        : True
```

提示：相关阅读材料
- `hashlib` 的标准库文档[一]。
- `hmac`：hmac 模块。
- OpenSSL[二]：一个开源加密工具包。
- Cryptography[三]模块：提供加密技巧和原语的一个 Python 包。
- Voidspace: IronPython and hashlib[四]：用于 IronPython 的 `hashlib` 的一个包装器。

9.2 `hmac`：密码消息签名与验证

HMAC 算法可以用于验证信息的完整性，这些信息可能在应用之间传递，或者存储在

[一] https://docs.python.org/3.5/library/hashlib.html
[二] www.openssl.org
[三] https://pypi.python.org/pypi/cryptography
[四] www.voidspace.org.uk/python/weblog/arch_d7_2006_10_07.shtml#e497

一个可能有安全威胁的地方。基本思想是生成实际数据的一个密码散列，并提供一个共享的秘密密钥。然后使用得到的散列检查所传输或存储的消息，以确定一个信任级别，而不传输秘密密钥。

> **警告**：免责声明：我不是安全专家。要全面地了解 HMAC 的详细信息，请查看 **RFC 2104**[⊖]。

9.2.1 消息签名

`new()` 函数会创建一个新对象来计算消息签名。下面这个例子使用了默认的 MD5 散列算法。

代码清单 9-7：**hmac_simple.py**

```
import hmac

digest_maker = hmac.new(b'secret-shared-key-goes-here')

with open('lorem.txt', 'rb') as f:
    while True:
        block = f.read(1024)
        if not block:
            break
        digest_maker.update(block)

digest = digest_maker.hexdigest()
print(digest)
```

运行这段代码时，会读取一个数据文件，并为它计算一个 HMAC 签名。

```
$ python3 hmac_simple.py

4bcb287e284f8c21e87e14ba2dc40b16
```

9.2.2 候选摘要类型

尽管 `hmac` 的默认密码算法是 MD5，但这并不是最安全的方法。MD5 散列有一些缺点，如冲突（两个不同的消息生成相同的散列）。一般认为 SHA1 算法更健壮，更建议使用。

代码清单 9-8：**hmac_sha.py**

```
import hmac
import hashlib

digest_maker = hmac.new(
    b'secret-shared-key-goes-here',
    b'',
    'sha1',
)

with open('hmac_sha.py', 'rb') as f:
    while True:
        block = f.read(1024)
        if not block:
```

⊖ https://tools.ietf.org/html/rfc2104.html

```
            break
        digest_maker.update(block)

digest = digest_maker.hexdigest()
print(digest)
```

new() 函数有 3 个参数。第 1 个参数是秘密密钥，这个密钥会在通信双方之间共享，使两端都可以使用相同的值。第 2 个值是一个初始消息。如果需要认证的消息内容很小，如一个时间戳或一个 HTTP POST，则把整个消息体都传递到 new() 而不是使用 update() 方法。最后一个参数是要使用的摘要模块。默认为 `hashlib.md5`，不过这个例子传入了 `'sha1'`，其会让 hmac 使用 `hashlib.sha1`。

```
$ python3 hmac_sha.py

3c3992fa7aefb81b73a52f49713cf3faa272382a
```

9.2.3 二进制摘要

前面的例子使用 `hexdigest()` 方法来生成可打印的摘要。hexdigest 是 `digest()` 方法计算的值的一个不同表示，这是一个二进制值，可以包括不可打印的字符（包括 NUL）。有些 Web 服务（Google checkout、Amazon S3）会使用 base64 编码版本的二进制摘要而不是 hexdigest。

代码清单 9-9：**hmac_base64.py**

```python
import base64
import hmac
import hashlib

with open('lorem.txt', 'rb') as f:
    body = f.read()

hash = hmac.new(
    b'secret-shared-key-goes-here',
    body,
    hashlib.sha1,
)

digest = hash.digest()
print(base64.encodestring(digest))
```

base64 编码串以一个换行符结束，在 HTTP 首部或其他格式敏感的上下文中嵌入这个串时，通常需要去除这个换行符。

```
$ python3 hmac_base64.py

b'olW2DoXHGJEKGU0aE9f0wSVE/o4=\n'
```

9.2.4 消息签名的应用

对于所有公共网络服务，在安全性要求很高的地方存储数据，就应当使用 HMAC 认证。例如，通过一个管道或套接字发送数据时，应当对数据进行签名，然后在使用这个数据之前要检查这个签名。文件 `hmac_pickle.py` 中给出了一个扩展例子。

第一步是建立一个函数，计算一个串的摘要，另外实例化一个简单的类，并通过一个通信通道传递。

代码清单 9-10：`hmac_pickle.py`

```python
import hashlib
import hmac
import io
import pickle
import pprint

def make_digest(message):
    "Return a digest for the message."
    hash = hmac.new(
        b'secret-shared-key-goes-here',
        message,
        hashlib.sha1,
    )
    return hash.hexdigest().encode('utf-8')
class SimpleObject:
    """Demonstrate checking digests before unpickling.
    """

    def __init__(self, name):
        self.name = name

    def __str__(self):
        return self.name
```

接下来，创建一个 `BytesIO` 缓冲区表示这个套接字或管道。这个例子对数据流使用了一种易于解析的原生格式。首先写出摘要以及数据长度，后面是一个换行符。接下来是对象的串行化表示（由 pickle 生成）。实际的系统可能不希望依赖于一个长度值，毕竟如果摘要不正确，这个长度可能也是错误的。更合适的做法是使用真实数据中不太可能出现的某个终止符序列。

然后示例程序向流写两个对象。写第一个对象时使用了正确的摘要值。

```python
# Simulate a writable socket or pipe with a buffer.
out_s = io.BytesIO()

# Write a valid object to the stream:
#   digest\nlength\npickle
o = SimpleObject('digest matches')
pickled_data = pickle.dumps(o)
digest = make_digest(pickled_data)
header = b'%s %d\n' % (digest, len(pickled_data))
print('WRITING: {}'.format(header))
out_s.write(header)
out_s.write(pickled_data)
```

再用一个不正确的摘要将第二个对象写入流，这个摘要是为其他数据计算的，而并非由 pickle 生成。

```python
# Write an invalid object to the stream.
o = SimpleObject('digest does not match')
pickled_data = pickle.dumps(o)
digest = make_digest(b'not the pickled data at all')
header = b'%s %d\n' % (digest, len(pickled_data))
print('\nWRITING: {}'.format(header))
out_s.write(header)
out_s.write(pickled_data)

out_s.flush()
```

既然数据在 `BytesIO` 缓冲区中,那么可以将它再次读出。首先读取包含摘要和数据长度的数据行,然后使用得到的长度值读取其余的数据。`pickle.load()` 可以直接从流读数据,不过这种策略有一个假设,认为它是一个可信的数据流,但这个数据还不能保证足够可信到可以解除腌制。可以将 pickle 作为一个串从流读取,而不是真正将对象解除腌制,这样会更为安全。

```python
# Simulate a readable socket or pipe with a buffer.
in_s = io.BytesIO(out_s.getvalue())

# Read the data.
while True:
    first_line = in_s.readline()
    if not first_line:
        break
    incoming_digest, incoming_length = first_line.split(b' ')
    incoming_length = int(incoming_length.decode('utf-8'))
    print('\nREAD:', incoming_digest, incoming_length)
```

一旦腌制数据在内存中,那么可以重新计算摘要值,并使用 `compare_digest()` 与所读取的数据进行比较。如果摘要匹配,就可以信任这个数据,并对其解除腌制。

```python
    incoming_pickled_data = in_s.read(incoming_length)

    actual_digest = make_digest(incoming_pickled_data)
    print('ACTUAL:', actual_digest)

    if hmac.compare_digest(actual_digest, incoming_digest):
        obj = pickle.loads(incoming_pickled_data)
        print('OK:', obj)
    else:
        print('WARNING: Data corruption')
```

输出显示第一个对象通过验证,不出所料,认为第二个对象"已被破坏"。

```
$ python3 hmac_pickle.py

WRITING: b'f49cd2bf7922911129e8df37f76f95485a0b52ca 69\n'

WRITING: b'b01b209e28d7e053408ebe23b90fe5c33bc6a0ec 76\n'

READ: b'f49cd2bf7922911129e8df37f76f95485a0b52ca' 69
ACTUAL: b'f49cd2bf7922911129e8df37f76f95485a0b52ca'
OK: digest matches

READ: b'b01b209e28d7e053408ebe23b90fe5c33bc6a0ec' 76
ACTUAL: b'2ab061f9a9f749b8dd6f175bf57292e02e95c119'
WARNING: Data corruption
```

在一种计时攻击中，可能会使用简单的串或字节比较来对比两个摘要，通过传入不同长度的摘要来得到全部或部分秘密密钥。compare_digest() 实现了一种快速但时间固定的比较函数，以避免计时攻击。

提示：相关阅读材料
- hmac 的标准库文档[○]。
- **RFC 2104**[○]：HMAC：基于密钥的散列来完成消息认证。
- hashlib：hashlib 模块提供了 MD5 和 SHA1 散列生成器。
- pickle：串行化库。
- Wikipedia: MD5[○]：MD5 散列算法的描述。
- Signing and Authenticating REST Requests (Amazon AWS)[®]：使用 HMAC-SHA1 签名的凭证对 S3 完成认证的有关说明。

○ https://docs.python.org/3.5/library/hmac.html
○ https://tools.ietf.org/html/rfc2104.html
○ https://en.wikipedia.org/wiki/MD5
® http://docs.aws.amazon.com/AmazonS3/latest/dev/RESTAuthentication.html

第 10 章
使用进程、线程和协程提供并发性

Python 提供了一些复杂的工具用于管理使用进程和线程的并发操作。通过应用这些技术，使用这些模块并发地运行作业的各个部分，即使是一些相当简单的程序也可以更快地运行。

`subprocess` 提供了一个 API 可以创建子进程并与之通信。这对于运行生产或消费文本的程序尤其有好处，因为这个 API 支持通过新进程的标准输入和输出通道来回传递数据。

`signal` 模块提供了 UNIX 信号机制，可以向其他进程发送事件。信号会被异步处理，通常信号到来时要中断程序正在做的工作。信号作为一个粗粒度的消息系统很有用，不过其他进程内通信技术更可靠，而且可以传递更复杂的消息。

`threading` 包括一个面向对象的高层 API，用于处理 Python 的并发性。`Thread` 对象在同一个进程中并发地运行，并共享内存。对于 I/O 受限而不是 CPU 受限的任务来说，使用线程是这些任务实现缩放的一种简单方法。

`multiprocessing` 模块是 `threading` 的镜像，只是它提供了一个 `Process` 而非一个 `Thread` 类。每个 `Process` 都是真正的系统进程（而无共享内存），`multiprocessing` 提供了一些特性可以共享数据并传递消息，使得很多情况下从线程转换为进程很简单，只需要修改几个 `import` 语句。

`asyncio` 使用一个基于类的协议系统或协程为并发和异步 I/O 管理提供了一个框架。`asyncio` 替换了原来的 `asyncore` 和 `asynchat` 模块，这些模块仍可用，但已经废弃。

`concurrent.futures` 提供了基于线程和进程的执行器实现，用来管理资源池以运行并发的任务。

10.1 subprocess：创建附加进程

`subprocess` 模块提供了 3 个 API 来处理进程。`run()` 函数是 Python 3.5 中新增的，作为一个高层 API，其用于运行进程并收集它的输出（可选）。函数 `call()`、`check_call()` 和 `check_output()` 是从 Python 2 沿袭来的原高层 API。这些函数仍受到支持，并在现有的程序中广泛使用。类 `Popen` 是一个用于建立其他 API 的底层 API，对更复杂的进程交互很有用。`Popen` 的构造函数利用参数建立新进程，使父进程可以通过管道与之通信。它可以替换其他一些模块和函数，并能提供所替换的这些模块和函数的全部功能，甚至还更多。在所有情况下这个 API 的用法都是一致的，很多存在开销的额外步骤（如关闭额外的文件描述符，以及确保管道关闭）都已经"内置"在这个 API 中，不需要由应用代码单独处理。

`subprocess` 模块是为了替换 `os.system()`、`os.spawnv()`、`os` 和 `popen2` 模块

中不同形式的 popen() 函数,以及 commands 模块。为了更容易地比较 subprocess 和其他这些模块,这一节中的很多例子会重新创建 os 和 popen2 中使用的例子。

说明:UNIX 和 Windows 上使用的 API 大致相同,不过由于操作系统中进程模型的差异,底层实现稍有不同。这里显示的所有例子都在 Mac OS X 上测试过。在非 UNIX OS 上的行为可能会有不同。

10.1.1 运行外部命令

要运行一个外部命令,但不采用 os.system() 的方式与之交互,可以使用 run() 函数。

代码清单 10-1:subprocess_os_system.py

```
import subprocess

completed = subprocess.run(['ls', '-1'])
print('returncode:', completed.returncode)
```

命令行参数作为一个字符串列表传入,这样就无须对引号或其他可能由 shell 解释的特殊字符转义。run() 返回一个 CompletedProcess 实例,它包含进程的有关信息,如退出码和输出。

```
$ python3 subprocess_os_system.py

index.rst
interaction.py
repeater.py
signal_child.py
signal_parent.py
subprocess_check_output_error_trap_output.py
subprocess_os_system.py
subprocess_pipes.py
subprocess_popen2.py
subprocess_popen3.py
subprocess_popen4.py
subprocess_popen_read.py
subprocess_popen_write.py
subprocess_run_check.py
subprocess_run_output.py
subprocess_run_output_error.py
subprocess_run_output_error_suppress.py
subprocess_run_output_error_trap.py
subprocess_shell_variables.py
subprocess_signal_parent_shell.py
subprocess_signal_setpgrp.py
returncode: 0
```

将 shell 参数设置为 true 值会使 subprocess 创建一个中间 shell 进程,由这个进程运行命令。默认情况下会直接运行命令。

代码清单 10-2:subprocess_shell_variables.py

```
import subprocess

completed = subprocess.run('echo $HOME', shell=True)
print('returncode:', completed.returncode)
```

使用一个中间 shell 意味着运行命令之前会先处理命令串中的变量、glob 模式以及其他特殊 shell 特性。

```
$ python3 subprocess_shell_variables.py

/Users/dhellmann
returncode: 0
```

说明：如果使用 `run()` 而没有传入 `check=True`，这等价于使用 `call()`，只返回进程的退出码。

10.1.1.1 错误处理

`CompletedProcess` 的 `returncode` 属性是程序的退出码。调用者要负责解释这个返回值以检测错误。如果 `run()` 的 `check` 参数为 `True`，则会检查退出码。如果指示发生了一个错误，则会产生一个 `CalledProcessError` 异常。

代码清单 10-3：`subprocess_run_check.py`

```python
import subprocess

try:
    subprocess.run(['false'], check=True)
except subprocess.CalledProcessError as err:
    print('ERROR:', err)
```

`false` 命令退出时总有一个非 0 的状态码，`run()` 会把它解释为一个错误。

```
$ python3 subprocess_run_check.py

ERROR: Command '['false']' returned non-zero exit status 1
```

说明：向 `run()` 传入 `check=True` 就等价于使用 `check_call()`。

10.1.1.2 捕获输出

对于 `run()` 启动的进程，它的标准输入和输出通道会绑定到父进程的输入和输出。这说明调用程序无法捕获命令的输出。可以通过为 `stdout` 和 `stderr` 参数传入 `PIPE` 来捕获输出，以备以后处理。

代码清单 10-4：`subprocess_run_output.py`

```python
import subprocess

completed = subprocess.run(
    ['ls', '-1'],
    stdout=subprocess.PIPE,
)
print('returncode:', completed.returncode)
print('Have {} bytes in stdout:\n{}'.format(
    len(completed.stdout),
    completed.stdout.decode('utf-8'))
)
```

ls -1 命令成功运行，所以会捕获并返回它打印到标准输出的文本。

```
$ python3 subprocess_run_output.py

returncode: 0
Have 522 bytes in stdout:
index.rst
interaction.py
repeater.py
signal_child.py
signal_parent.py
subprocess_check_output_error_trap_output.py
subprocess_os_system.py
subprocess_pipes.py
subprocess_popen2.py
subprocess_popen3.py
subprocess_popen4.py
subprocess_popen_read.py
subprocess_popen_write.py
subprocess_run_check.py
subprocess_run_output.py
subprocess_run_output_error.py
subprocess_run_output_error_suppress.py
subprocess_run_output_error_trap.py
subprocess_shell_variables.py
subprocess_signal_parent_shell.py
subprocess_signal_setpgrp.py
```

说明：传入 check=True 并设置 stdout 为 PIPE 等价于使用 check_output()。

下一个例子在一个子 shell 中运行一系列命令。在命令返回一个错误码并退出之前，消息会发送到标准输出和标准错误输出。

代码清单 10-5：**subprocess_run_output_error.py**

```python
import subprocess

try:
    completed = subprocess.run(
        'echo to stdout; echo to stderr 1>&2; exit 1',
        check=True,
        shell=True,
        stdout=subprocess.PIPE,
    )
except subprocess.CalledProcessError as err:
    print('ERROR:', err)
else:
    print('returncode:', completed.returncode)
    print('Have {} bytes in stdout: {!r}'.format(
        len(completed.stdout),
        completed.stdout.decode('utf-8'))
    )
```

发送到标准错误输出的消息会被打印到控制台，而发送到标准输出的消息会被隐藏。

```
$ python3 subprocess_run_output_error.py

to stderr
ERROR: Command 'echo to stdout; echo to stderr 1>&2; exit 1'
returned non-zero exit status 1
```

为了避免通过 run() 运行的命令将错误消息写至控制台,可以设置 stderr 参数为常量 PIPE。

代码清单 10-6: **subprocess_run_output_error_trap.py**

```python
import subprocess

try:
    completed = subprocess.run(
        'echo to stdout; echo to stderr 1>&2; exit 1',
        shell=True,
        stdout=subprocess.PIPE,
        stderr=subprocess.PIPE,
    )
except subprocess.CalledProcessError as err:
    print('ERROR:', err)
else:
    print('returncode:', completed.returncode)
    print('Have {} bytes in stdout: {!r}'.format(
        len(completed.stdout),
        completed.stdout.decode('utf-8'))
    )
    print('Have {} bytes in stderr: {!r}'.format(
        len(completed.stderr),
        completed.stderr.decode('utf-8'))
    )
```

这个例子没有设置 check=True,所以会捕获并打印命令的输出。

```
$ python3 subprocess_run_output_error_trap.py

returncode: 1
Have 10 bytes in stdout: 'to stdout\n'
Have 10 bytes in stderr: 'to stderr\n'
```

使用 check_output() 时如果要捕获错误消息,则要把 stderr 设置为 STDOUT,错误消息将与命令的其他输出合并在一起。

代码清单 10-7: **subprocess_check_output_error_trap_output.py**

```python
import subprocess

try:
    output = subprocess.check_output(
        'echo to stdout; echo to stderr 1>&2',
        shell=True,
        stderr=subprocess.STDOUT,
    )
except subprocess.CalledProcessError as err:
    print('ERROR:', err)
else:
    print('Have {} bytes in output: {!r}'.format(
        len(output),
        output.decode('utf-8'))
    )
```

输出的顺序可能有所不同,这取决于如何对标准输出应用缓冲以及数据如何打印。

```
$ python3 subprocess_check_output_error_trap_output.py

Have 20 bytes in output: 'to stdout\nto stderr\n'
```

10.1.1.3 抑制输出

在不能显示或捕获输出的情况下，可以使用 DEVNULL 抑制输出流。下面的例子会同时抑制标准输出和错误流。

代码清单 10-8：**subprocess_run_output_error_suppress.py**

```python
import subprocess

try:
    completed = subprocess.run(
        'echo to stdout; echo to stderr 1>&2; exit 1',
        shell=True,
        stdout=subprocess.DEVNULL,
        stderr=subprocess.DEVNULL,
    )
except subprocess.CalledProcessError as err:
    print('ERROR:', err)
else:
    print('returncode:', completed.returncode)
    print('stdout is {!r}'.format(completed.stdout))
    print('stderr is {!r}'.format(completed.stderr))
```

DEVNULL 的名字来自 UNIX 特殊设备文件 /dev/null。打开文件读取时，DEVNULL 对应文件末尾，写文件时会接收并忽略所有输入。

```
$ python3 subprocess_run_output_error_suppress.py

returncode: 1
stdout is None
stderr is None
```

10.1.2 直接处理管道

函数 run()、call()、check_call() 和 check_output() 都是 Popen 类的包装器。直接使用 Popen 可以更多地控制如何运行命令以及如何处理其输入和输出流。例如，通过为 stdin、stdout 和 stderr 传递不同的参数，可以模仿不同形式的 os.popen()。

10.1.2.1 与进程的单向通信

要运行一个进程并读取它的所有输出，可以设置 stdout 值为 PIPE 并调用 communicate()。

代码清单 10-9：**subprocess_popen_read.py**

```python
import subprocess

print('read:')
proc = subprocess.Popen(
    ['echo', '"to stdout"'],
    stdout=subprocess.PIPE,
)
stdout_value = proc.communicate()[0].decode('utf-8')
print('stdout:', repr(stdout_value))
```

这与 popen() 的工作类似，只不过 Popen 实例会在内部管理数据读取。

```
$ python3 subprocess_popen_read.py

read:
stdout: '"to stdout"\n'
```

要建立一个管道,以便调用程序写数据,可以设置 `stdin` 为 `PIPE`。

代码清单 10-10: **subprocess_popen_write.py**

```python
import subprocess

print('write:')
proc = subprocess.Popen(
    ['cat', '-'],
    stdin=subprocess.PIPE,
)
proc.communicate('stdin: to stdin\n'.encode('utf-8'))
```

要将数据一次性发送到进程的标准输入通道,可以把数据传递到 `communicate()`。这类似于使用 `popen()` 并指定 `'w'` 模式。

```
$ python3 -u subprocess_popen_write.py

write:
stdin: to stdin
```

10.1.2.2 与进程的双向通信

要建立 `Popen` 实例同时完成读写,可以结合使用前面几个技术。

代码清单 10-11: **subprocess_popen2.py**

```python
import subprocess

print('popen2:')

proc = subprocess.Popen(
    ['cat', '-'],
    stdin=subprocess.PIPE,
    stdout=subprocess.PIPE,
)
msg = 'through stdin to stdout'.encode('utf-8')
stdout_value = proc.communicate(msg)[0].decode('utf-8')
print('pass through:', repr(stdout_value))
```

这会建立管道来模拟 `popen2()`。

```
$ python3 -u subprocess_popen2.py

popen2:
pass through: 'through stdin to stdout'
```

10.1.2.3 捕获错误输出

还可以监视 `stdout` 和 `stderr` 数据流,类似于 `popen3()`。

代码清单 10-12: **subprocess_popen3.py**

```python
import subprocess

print('popen3:')
```

```python
proc = subprocess.Popen(
    'cat -; echo "to stderr" 1>&2',
    shell=True,
    stdin=subprocess.PIPE,
    stdout=subprocess.PIPE,
    stderr=subprocess.PIPE,
)
msg = 'through stdin to stdout'.encode('utf-8')
stdout_value, stderr_value = proc.communicate(msg)
print('pass through:', repr(stdout_value.decode('utf-8')))
print('stderr       :', repr(stderr_value.decode('utf-8')))
```

从 stderr 读取数据与从 stdout 读取是一样的。传入 PIPE 会告诉 Popen 关联到通道，communicate() 在返回之前会从这个通道读取所有数据。

```
$ python3 -u subprocess_popen3.py

popen3:
pass through: 'through stdin to stdout'
stderr       : 'to stderr\n'
```

10.1.2.4　结合常规和错误输出

为了把错误输出从进程定向到标准输出通道，stderr 要使用 STDOUT 而不是 PIPE。

代码清单 10-13：**subprocess_popen4.py**

```python
import subprocess

print('popen4:')
proc = subprocess.Popen(
    'cat -; echo "to stderr" 1>&2',
    shell=True,
    stdin=subprocess.PIPE,
    stdout=subprocess.PIPE,
    stderr=subprocess.STDOUT,
)
msg = 'through stdin to stdout\n'.encode('utf-8')
stdout_value, stderr_value = proc.communicate(msg)
print('combined output:', repr(stdout_value.decode('utf-8')))
print('stderr value   :', repr(stderr_value))
```

以这种方式结合输出是类似于 popen4() 的做法。

```
$ python3 -u subprocess_popen4.py

popen4:
combined output: 'through stdin to stdout\nto stderr\n'
stderr value   : None
```

10.1.3　连接管道段

多个命令可以连接为一个管线（pipeline），这类似于 UNIX shell 的做法，即创建单独的 Popen 实例，把它们的输入和输出串链在一起。一个 Popen 实例的 stdout 属性被用作管线中下一个 Popen 实例的 stdin 参数，而不是常量 PIPE。输出从管线中最后一个命令的 stdout 句柄读取。

代码清单 10-14：**subprocess_pipes.py**

```
import subprocess

cat = subprocess.Popen(
    ['cat', 'index.rst'],
    stdout=subprocess.PIPE,
)

grep = subprocess.Popen(
    ['grep', '.. literalinclude::'],
    stdin=cat.stdout,
    stdout=subprocess.PIPE,
)

cut = subprocess.Popen(
    ['cut', '-f', '3', '-d:'],
    stdin=grep.stdout,
    stdout=subprocess.PIPE,
)

end_of_pipe = cut.stdout

print('Included files:')
for line in end_of_pipe:
    print(line.decode('utf-8').strip())
```

这个例子重新生成以下命令行：

```
$ cat index.rst | grep ".. literalinclude" | cut -f 3 -d:
```

这个管线读取这一节的 reStructuredText 源文件，并查找所有包含其他文件的文本行。然后打印出所包含的这些文件的文件名。

```
$ python3 -u subprocess_pipes.py

Included files:
subprocess_os_system.py
subprocess_shell_variables.py
subprocess_run_check.py
subprocess_run_output.py
subprocess_run_output_error.py
subprocess_run_output_error_trap.py
subprocess_check_output_error_trap_output.py
subprocess_run_output_error_suppress.py
subprocess_popen_read.py
subprocess_popen_write.py
subprocess_popen2.py
subprocess_popen3.py
subprocess_popen4.py
subprocess_pipes.py
repeater.py
interaction.py
signal_child.py
signal_parent.py
subprocess_signal_parent_shell.py
subprocess_signal_setpgrp.py
```

10.1.4　与其他命令交互

前面的所有例子都假设交互量是有限的。**communicate()** 方法读取所有输出，返回之前要等待子进程退出。也可以在程序运行时从 Popen 实例使用的各个管道句柄增量地进行读写。可以用一个简单的应答程序来展示这个技术，这个程序从标准输入读，并写至标准输出。

下一个例子中使用脚本 repeater.py 作为子进程。它从 stdin 读取，并将值写至 stdout，一次处理一行，直到再没有更多输入为止。开始和停止时它还会向 stderr 写一个消息，显示子进程的生命期。

代码清单 10-15：repeater.py

```python
import sys

sys.stderr.write('repeater.py: starting\n')
sys.stderr.flush()

while True:
    next_line = sys.stdin.readline()
    sys.stderr.flush()
    if not next_line:
        break
    sys.stdout.write(next_line)
    sys.stdout.flush()

sys.stderr.write('repeater.py: exiting\n')
sys.stderr.flush()
```

下一个交互例子将采用不同方式使用 Popen 实例的 stdin 和 stdout 文件句柄。在第一个例子中，将把一组 5 个数写至进程的 stdin，每写一个数就读回下一行输出。第二个例子中仍然写同样的 5 个数，但要使用 communicate() 一次读取全部输出。

代码清单 10-16：interaction.py

```python
import io
import subprocess

print('One line at a time:')
proc = subprocess.Popen(
    'python3 repeater.py',
    shell=True,
    stdin=subprocess.PIPE,
    stdout=subprocess.PIPE,
)
stdin = io.TextIOWrapper(
    proc.stdin,
    encoding='utf-8',
    line_buffering=True,  # Send data on newline
)
stdout = io.TextIOWrapper(
    proc.stdout,
    encoding='utf-8',
)
for i in range(5):
    line = '{}\n'.format(i)
    stdin.write(line)
    output = stdout.readline()
    print(output.rstrip())
remainder = proc.communicate()[0].decode('utf-8')
```

```python
    print(remainder)

print()
print('All output at once:')
proc = subprocess.Popen(
    'python3 repeater.py',
    shell=True,
    stdin=subprocess.PIPE,
    stdout=subprocess.PIPE,
)
stdin = io.TextIOWrapper(
    proc.stdin,
    encoding='utf-8',
)
for i in range(5):
    line = '{}\n'.format(i)
    stdin.write(line)
stdin.flush()

output = proc.communicate()[0].decode('utf-8')
print(output)
```

对于这两种不同的循环，"repeater.py: exiting" 行出现在输出的不同位置上。

```
$ python3 -u interaction.py

One line at a time:
repeater.py: starting
0
1
2
3
4
repeater.py: exiting

All output at once:
repeater.py: starting
repeater.py: exiting
0
1
2
3
4
```

10.1.5 进程间传递信号

os 模块的进程管理例子演示了如何使用 `os.fork()` 和 `os.kill()` 在进程间传递信号。由于每个 Popen 实例提供了一个 pid 属性，其中包含子进程的进程 ID，所以可以完成类似于 subprocess 的工作。

下一个例子结合了两个脚本。这个子进程为 USR 信号建立了一个信号处理器。

代码清单 10-17：signal_child.py

```python
import os
import signal
import time
import sys
```

```
pid = os.getpid()
received = False


def signal_usr1(signum, frame):
    "Callback invoked when a signal is received"
    global received
    received = True
    print('CHILD {:>6}: Received USR1'.format(pid))
    sys.stdout.flush()


print('CHILD {:>6}: Setting up signal handler'.format(pid))
sys.stdout.flush()
signal.signal(signal.SIGUSR1, signal_usr1)
print('CHILD {:>6}: Pausing to wait for signal'.format(pid))
sys.stdout.flush()
time.sleep(3)

if not received:
    print('CHILD {:>6}: Never received signal'.format(pid))
```

这个脚本作为父进程运行。它启动 signal_child.py，然后发送 USR1 信号。

代码清单 10-18：signal_parent.py

```
import os
import signal
import subprocess
import time
import sys

proc = subprocess.Popen(['python3', 'signal_child.py'])
print('PARENT      : Pausing before sending signal...')
sys.stdout.flush()
time.sleep(1)
print('PARENT      : Signaling child')
sys.stdout.flush()
os.kill(proc.pid, signal.SIGUSR1)
```

输出如下所示。

```
$ python3 signal_parent.py

PARENT      : Pausing before sending signal...
CHILD  26976: Setting up signal handler
CHILD  26976: Pausing to wait for signal
PARENT      : Signaling child
CHILD  26976: Received USR1
```

进程组 / 会话

如果 Popen 创建的进程创建了子进程，那么这些子进程不会接收发送给父进程的信号。这说明，使用 Popen 的 shell 参数时，很难通过发送 SIGINT 或 SIGTERM 来终止在 shell 中启动的命令。

代码清单 10-19：subprocess_signal_parent_shell.py

```
import os
import signal
```

```
import subprocess
import tempfile
import time
import sys

script = '''#!/bin/sh
echo "Shell script in process $$"
set -x
python3 signal_child.py
'''
script_file = tempfile.NamedTemporaryFile('wt')
script_file.write(script)
script_file.flush()

proc = subprocess.Popen(['sh', script_file.name])
print('PARENT      : Pausing before signaling {}...'.format(
    proc.pid))
sys.stdout.flush()
time.sleep(1)
print('PARENT      : Signaling child {}'.format(proc.pid))
sys.stdout.flush()
os.kill(proc.pid, signal.SIGUSR1)
time.sleep(3)
```

在这个例子中，发送信号所用的 pid 与等待信号的 shell 脚本子进程的 pid 不匹配，因为有 3 个不同的进程在交互。

- 程序 `subprocess_signal_parent_shell.py`
- shell 进程，其在运行主 Python 程序创建的脚本
- 程序 `signal_child.py`

```
$ python3 subprocess_signal_parent_shell.py

PARENT      : Pausing before signaling 26984...
Shell script in process 26984
+ python3 signal_child.py
CHILD  26985: Setting up signal handler
CHILD  26985: Pausing to wait for signal
PARENT      : Signaling child 26984
CHILD  26985: Never received signal
```

如果向子进程发送信号但不知道它们的进程 ID，那么可以使用一个进程组（process group）关联这些子进程，使它们能一同收到信号。进程组用 `os.setpgrp()` 创建，将进程组 ID 设置为当前进程的进程 ID。所有子进程都会从其父进程继承进程组。由于这个组只能在由 Popen 及其子进程创建的 shell 中设置，所以不能在创建 Popen 的同一进程中调用 `os.setpgrp()`。实际上，这个函数要作为 `preexec_fn` 参数传至 Popen，以便其在新进程执行 `fork()` 之后运行，之后才能使用 `exec()` 运行 shell。要向整个进程组发送信号，可以使用 `os.killpg()` 并提供 Popen 实例的 `pid` 值。

代码清单 10-20：`subprocess_signal_setpgrp.py`

```
import os
import signal
import subprocess
import tempfile
import time
import sys
```

```python
def show_setting_prgrp():
    print('Calling os.setpgrp() from {}'.format(os.getpid()))
    os.setpgrp()
    print('Process group is now {}'.format(
        os.getpid(), os.getpgrp()))
    sys.stdout.flush()


script = '''#!/bin/sh
echo "Shell script in process $$"
set -x
python3 signal_child.py
'''
script_file = tempfile.NamedTemporaryFile('wt')
script_file.write(script)
script_file.flush()

proc = subprocess.Popen(
    ['sh', script_file.name],
    preexec_fn=show_setting_prgrp,
)
print('PARENT      : Pausing before signaling {}...'.format(
    proc.pid))
sys.stdout.flush()
time.sleep(1)
print('PARENT      : Signaling process group {}'.format(
    proc.pid))
sys.stdout.flush()
os.killpg(proc.pid, signal.SIGUSR1)
time.sleep(3)
```

事件序列如下：

1. 父程序实例化 `Popen`。

2. `Popen` 实例创建一个新进程。

3. 这个新进程运行 `os.setpgrp()`。

4. 这个新进程运行 `exec()` 以启动 shell。

5. shell 运行 shell 脚本。

6. shell 脚本再次创建新进程，该进程执行 Python。

7. Python 运行 `signal_child.py`。

8. 父程序使用 shell 的 pid 向进程组传送信号。

9. shell 和 Python 进程接收到信号。

10. shell 忽略这个信号。

11. 运行 `signal_child.py` 的 Python 进程调用信号处理器。

```
$ python3 subprocess_signal_setpgrp.py

Calling os.setpgrp() from 26992
Process group is now 26992
PARENT      : Pausing before signaling 26992...
Shell script in process 26992
+ python3 signal_child.py
CHILD   26993: Setting up signal handler
CHILD   26993: Pausing to wait for signal
PARENT      : Signaling process group 26992
CHILD   26993: Received USR1
```

提示：相关阅读材料
- `subprocess` 的标准库文档[⊖]。
- `os`：尽管 `subprocess` 取代了 `os` 模块中很多处理进程的函数，但这些函数在现有代码中仍在广泛使用。
- UNIX Signals and Process Groups[⊖]：对 UNIX 信号机制以及进程组如何工作做了很好的描述。
- `signal`：关于使用 `signal` 模块的更多详细内容。
- Advanced Programming in the UNIX Environment, Third Edition[⊖]：涵盖了如何处理多个进程的内容，如处理信号和关闭重复的文件描述符。
- `pipes`：标准库中的 UNIX shell 命令管线模板。

10.2 `signal`：异步系统事件

信号是一个操作系统特性，它提供了一个途径可以通知程序这里发生了一个事件，并且异步处理这个事件。信号可以由系统本身生成，也可以从一个进程发送到另一个进程。由于信号会中断程序的正常控制流，如果在操作过程中间接收到信号，有些操作（特别是 I/O 操作）则可能会产生错误。

信号由整数标识，在操作系统 C 首部中定义。Python 在 `signal` 模块中提供了适合不同平台的多种信号（作为符号）。这一节中的例子使用了 `SIGINT` 和 `SIGUSR1`，通常会为所有 UNIX 和类 UNIX 系统定义这两个信号。

说明：使用 UNIX 信号处理器来编程不是一件容易的事情。这一节只对这个复杂主题做简要介绍，不会涵盖在每一个平台上成功使用信号所需的所有细节。不同版本的 UNIX 显然有一定程度的标准化，但也确实存在一些不同。如果遇到麻烦，可以参考操作系统文档。

10.2.1 接收信号

与其他形式基于事件的编程一样，要建立一个回调函数来接收信号，这个回调函数被称为信号处理器（signal handler），它会在出现信号时调用。信号处理器的参数包括信号编号以及程序被信号中断那一时刻的栈帧。

代码清单 10-21：`signal_signal.py`

```
import signal
import os
import time


def receive_signal(signum, stack):
    print('Received:', signum)
```

⊖ https://docs.python.org/3/5/library/subprocess.html
⊖ www.cs.ucsb.edu/~almeroth/classes/W99.276/assignment1/signals.html
⊖ https://www.amazon.com/Advanced-Programming-UNIX-Environment-3rd/dp/0321637739/

```
# Register signal handlers.
signal.signal(signal.SIGUSR1, receive_signal)
signal.signal(signal.SIGUSR2, receive_signal)

# Print the process ID so it can be used with 'kill'
# to send this program signals.
print('My PID is:', os.getpid())

while True:
    print('Waiting...')
    time.sleep(3)
```

这个示例脚本会无限循环,每次暂停几秒时间。一个信号到来时,`sleep()`调用被中断,并且信号处理器`receive_signal()`打印信号编号。信号处理器返回时,循环继续。可以使用`os.kill()`或 UNIX 命令行程序 kill 向正在运行的程序发送信号。

```
$ python3 signal_signal.py

My PID is: 71387
Waiting...
Waiting...
Waiting...
Received: 30
Waiting...
Waiting...
Received: 31
Waiting...
Waiting...
Traceback (most recent call last):
  File "signal_signal.py", line 28, in <module>
    time.sleep(3)
KeyboardInterrupt
```

在一个窗口中运行`signal_signal.py`,然后在另一个窗口中运行以下命令可以生成前面的输出。

```
$ kill -USR1 $pid
$ kill -USR2 $pid
$ kill -INT $pid
```

10.2.2 获取已注册的处理器

要查看为一个信号注册了哪些信号处理器,可以使用`getsignal()`。要将信号编号作为参数传入。返回值是已注册的处理器,或者是以下某个特殊值:`SIG_IGN`(如果信号被忽略)、`SIG_DFL`(如果使用默认行为)或 None(如果从 C 而非从 Python 注册现有信号处理器)。

代码清单 10-22: `signal_getsignal.py`

```
import signal

def alarm_received(n, stack):
    return

signal.signal(signal.SIGALRM, alarm_received)

signals_to_names = {
    getattr(signal, n): n
```

```
        for n in dir(signal)
        if n.startswith('SIG') and '_' not in n
}

for s, name in sorted(signals_to_names.items()):
    handler = signal.getsignal(s)
    if handler is signal.SIG_DFL:
        handler = 'SIG_DFL'
    elif handler is signal.SIG_IGN:
        handler = 'SIG_IGN'
    print('{:<10} ({:2d}):'.format(name, s), handler)
```

同样,由于每个操作系统可能定义了不同的信号,所以其他系统上的输出可能有所不同。以下是 OS X 的输出。

```
$ python3 signal_getsignal.py

SIGHUP     ( 1): SIG_DFL
SIGINT     ( 2): <built-in function default_int_handler>
SIGQUIT    ( 3): SIG_DFL
SIGILL     ( 4): SIG_DFL
SIGTRAP    ( 5): SIG_DFL
SIGIOT     ( 6): SIG_DFL
SIGEMT     ( 7): SIG_DFL
SIGFPE     ( 8): SIG_DFL
SIGKILL    ( 9): None
SIGBUS     (10): SIG_DFL
SIGSEGV    (11): SIG_DFL
SIGSYS     (12): SIG_DFL
SIGPIPE    (13): SIG_IGN
SIGALRM    (14): <function alarm_received at 0x100757f28>
SIGTERM    (15): SIG_DFL
SIGURG     (16): SIG_DFL
SIGSTOP    (17): None
SIGTSTP    (18): SIG_DFL
SIGCONT    (19): SIG_DFL
SIGCHLD    (20): SIG_DFL
SIGTTIN    (21): SIG_DFL
SIGTTOU    (22): SIG_DFL
SIGIO      (23): SIG_DFL
SIGXCPU    (24): SIG_DFL
SIGXFSZ    (25): SIG_IGN
SIGVTALRM  (26): SIG_DFL
SIGPROF    (27): SIG_DFL
SIGWINCH   (28): SIG_DFL
SIGINFO    (29): SIG_DFL
SIGUSR1    (30): SIG_DFL
SIGUSR2    (31): SIG_DFL
```

10.2.3 发送信号

在 Python 中发送信号的函数是 os.kill()。其用法在有关 os 模块的 17.3.10 节中介绍。

10.2.4 闹铃

闹铃(alarm)是一种特殊的信号,程序要求操作系统在过去一段时间之后再发出这个信号通知。os 的标准模块文档指出,这种方法对于避免一个 I/O 操作或其他系统调用上的无限阻塞很有用。

代码清单 10-23：**signal_alarm.py**

```python
import signal
import time

def receive_alarm(signum, stack):
    print('Alarm :', time.ctime())

# Call receive_alarm in 2 seconds.
signal.signal(signal.SIGALRM, receive_alarm)
signal.alarm(2)

print('Before:', time.ctime())
time.sleep(4)
print('After :', time.ctime())
```

在这个例子中，`sleep()` 调用会被中断，不过处理信号之后会继续。`sleep()` 返回之后打印的消息显示出程序至少在睡眠期间暂停。

```
$ python3 signal_alarm.py

Before: Sun Sep 11 11:31:18 2016
Alarm : Sun Sep 11 11:31:20 2016
After : Sun Sep 11 11:31:22 2016
```

10.2.5 忽略信号

要忽略一个信号，需要注册 `SIG_IGN` 作为处理器。下面这个脚本将 `SIGINT` 的默认处理器替换为 `SIG_IGN`，并为 SIGUSR1 注册一个处理器。然后使用 `signal.pause()` 等待接收一个信号。

代码清单 10-24：**signal_ignore.py**

```python
import signal
import os
import time

def do_exit(sig, stack):
    raise SystemExit('Exiting')

signal.signal(signal.SIGINT, signal.SIG_IGN)
signal.signal(signal.SIGUSR1, do_exit)

print('My PID:', os.getpid())

signal.pause()
```

正常情况下，`SIGINT`(用户按下 Ctrl-C 时 shell 会向程序发送这个信号) 会产生一个 `Key-boardInterrupt`。这个例子将忽略 SIGINT，并在发现 SIGUSR1 时产生一个 `System-Exit`。输出中的每个 `^C` 表示每一次尝试使用 Ctrl-C 从终端结束脚本。从另一个终端使用 `kill -USR1 72598` 才最终退出脚本。

```
$ python3 signal_ignore.py

My PID: 72598
^C^C^C^CExiting
```

10.2.6 信号和线程

信号和线程通常不能很好地合作,因为只有进程的主线程可以接收信号。下面的例子建立了一个信号处理器,它在一个线程中等待信号,而从另一个线程发送信号。

代码清单 10-25:`signal_threads.py`

```python
import signal
import threading
import os
import time


def signal_handler(num, stack):
    print('Received signal {} in {}'.format(
        num, threading.currentThread().name))


signal.signal(signal.SIGUSR1, signal_handler)


def wait_for_signal():
    print('Waiting for signal in',
          threading.currentThread().name)
    signal.pause()
    print('Done waiting')


# Start a thread that will not receive the signal.
receiver = threading.Thread(
    target=wait_for_signal,
    name='receiver',
)
receiver.start()
time.sleep(0.1)


def send_signal():
    print('Sending signal in', threading.currentThread().name)
    os.kill(os.getpid(), signal.SIGUSR1)


sender = threading.Thread(target=send_signal, name='sender')
sender.start()
sender.join()
# Wait for the thread to see the signal (not going to happen!).
print('Waiting for', receiver.name)
signal.alarm(2)
receiver.join()
```

信号处理器都在主线程中注册,因为这是 Python 的 `signal` 模块实现的一个要求,不论底层平台如何支持线程和信号的结合,都有这个要求。尽管接收者线程调用了 `signal.pause()`,但它不会接收信号。这个例子快要结束时的 `signal.alarm(2)` 调用避免了无限阻塞,因为接收者线程永远不会退出。

```
$ python3 signal_threads.py

Waiting for signal in receiver
Sending signal in sender
Received signal 30 in MainThread
Waiting for receiver
Alarm clock
```

尽管在任何线程中都能设置闹铃，但其总是由主线程接收。

代码清单 10-26：signal_threads_alarm.py

```python
import signal
import time
import threading

def signal_handler(num, stack):
    print(time.ctime(), 'Alarm in',
          threading.currentThread().name)

signal.signal(signal.SIGALRM, signal_handler)

def use_alarm():
    t_name = threading.currentThread().name
    print(time.ctime(), 'Setting alarm in', t_name)
    signal.alarm(1)
    print(time.ctime(), 'Sleeping in', t_name)
    time.sleep(3)
    print(time.ctime(), 'Done with sleep in', t_name)

# Start a thread that will not receive the signal.
alarm_thread = threading.Thread(
    target=use_alarm,
    name='alarm_thread',
)
alarm_thread.start()
time.sleep(0.1)

# Wait for the thread to see the signal (not going to happen!).
print(time.ctime(), 'Waiting for', alarm_thread.name)
alarm_thread.join()

print(time.ctime(), 'Exiting normally')
```

在这个例子中，闹铃不会中止 `use_alarm()` 中的 `sleep()` 调用。

```
$ python3 signal_threads_alarm.py

Sun Sep 11 11:31:22 2016 Setting alarm in alarm_thread
Sun Sep 11 11:31:22 2016 Sleeping in alarm_thread
Sun Sep 11 11:31:22 2016 Waiting for alarm_thread
Sun Sep 11 11:31:23 2016 Alarm in MainThread
Sun Sep 11 11:31:25 2016 Done with sleep in alarm_thread
Sun Sep 11 11:31:25 2016 Exiting normally
```

提示：相关阅读材料

- `signal` 的标准库文档[⊖]。
- **PEP 475**[⊖]：重试以 EINTR 失败的系统调用。
- `subprocess`：更多向进程发送信号的例子。
- 17.3.10 节：`kill()` 函数可以用来在进程之间发送信号。

[⊖] https://docs.python.org/3.5/library/signal.html
[⊖] www.python.org/dev/peps/pep-0475

10.3 threading：进程中管理并发操作

threading 模块提供了管理多个线程执行的 API，允许程序在同一个进程空间并发地运行多个操作。

10.3.1 Thread 对象

要使用 Thread，最简单的方法就是用一个目标函数实例化一个 Thread 对象，并调用 start() 让它开始工作。

代码清单 10-27：**threading_simple.py**

```
import threading


def worker():
    """thread worker function"""
    print('Worker')


threads = []
for i in range(5):
    t = threading.Thread(target=worker)
    threads.append(t)
    t.start()
```

输出有 5 行，每一行都是 "Worker"：

```
$ python3 threading_simple.py

Worker
Worker
Worker
Worker
Worker
```

如果能够创建一个线程，并向它传递参数告诉它要完成什么工作，那么这会很有用。任何类型的对象都可以作为参数传递到线程。下面的例子传递了一个数，线程将打印出这个数。

代码清单 10-28：**threading_simpleargs.py**

```
import threading


def worker(num):
    """thread worker function"""
    print('Worker: %s' % num)


threads = []
for i in range(5):
    t = threading.Thread(target=worker, args=(i,))
    threads.append(t)
    t.start()
```

现在这个整数参数会包含在各线程打印的消息中：

```
$ python3 threading_simpleargs.py

Worker: 0
Worker: 1
Worker: 2
Worker: 3
Worker: 4
```

10.3.2 确定当前线程

使用参数来标识或命名线程很麻烦，也没有必要。每个 Thread 实例都有一个带有默认值的名，该默认值可以在创建线程时改变。如果服务器进程中有多个服务线程处理不同的操作，那么在这样的服务器进程中，对线程命名就很有用。

代码清单 10-29：**threading_names.py**

```python
import threading
import time


def worker():
    print(threading.current_thread().getName(), 'Starting')
    time.sleep(0.2)
    print(threading.current_thread().getName(), 'Exiting')


def my_service():
    print(threading.current_thread().getName(), 'Starting')
    time.sleep(0.3)
    print(threading.current_thread().getName(), 'Exiting')


t = threading.Thread(name='my_service', target=my_service)
w = threading.Thread(name='worker', target=worker)
w2 = threading.Thread(target=worker)  # Use default name

w.start()
w2.start()
t.start()
```

调试输出的每一行中包含有当前线程的名。线程名列中有 "**Thread-1**" 的行对应未命名的线程 w2。

```
$ python3 threading_names.py

worker Starting
Thread-1 Starting
my_service Starting
worker Exiting
Thread-1 Exiting
my_service Exiting
```

大多数程序并不使用 print 来进行调试。logging 模块支持将线程名嵌入到各个日志消息中（使用格式化代码 %(threadName)s）。通过把线程名包含在日志消息中，就能跟踪这些消息的来源。

代码清单 10-30：**threading_names_log.py**

```python
import logging
import threading
import time


def worker():
    logging.debug('Starting')
    time.sleep(0.2)
    logging.debug('Exiting')


def my_service():
    logging.debug('Starting')
    time.sleep(0.3)
    logging.debug('Exiting')


logging.basicConfig(
    level=logging.DEBUG,
    format='[%(levelname)s] (%(threadName)-10s) %(message)s',
)

t = threading.Thread(name='my_service', target=my_service)
w = threading.Thread(name='worker', target=worker)
w2 = threading.Thread(target=worker)  # Use default name

w.start()
w2.start()
t.start()
```

而且 `logging` 是线程安全的，所以来自不同线程的消息在输出中会有所区分。

```
$ python3 threading_names_log.py

[DEBUG] (worker    ) Starting
[DEBUG] (Thread-1  ) Starting
[DEBUG] (my_service) Starting
[DEBUG] (worker    ) Exiting
[DEBUG] (Thread-1  ) Exiting
[DEBUG] (my_service) Exiting
```

10.3.3 守护与非守护线程

到目前为止，示例程序都在隐式地等待所有线程完成工作之后才退出。不过，程序有时会创建一个线程作为守护线程（daemon），这个线程可以一直运行而不阻塞主程序退出。如果一个服务不能很容易地中断线程，或者即使让线程工作到一半时中止也不会造成数据损失或破坏（例如，为一个服务监控工具生成"心跳"的线程），那么对于这些服务，使用守护线程就很有用。要标志一个线程为守护线程，构造线程时便要传入 daemon=True 或者要调用它的 `setDaemon()` 方法并提供参数 `True`。默认情况下线程不作为守护线程。

代码清单 10-31：**threading_daemon.py**

```python
import threading
import time
import logging


def daemon():
```

```
        logging.debug('Starting')
        time.sleep(0.2)
        logging.debug('Exiting')

    def non_daemon():
        logging.debug('Starting')
        logging.debug('Exiting')

    logging.basicConfig(
        level=logging.DEBUG,
        format='(%(threadName)-10s) %(message)s',
    )
    d = threading.Thread(name='daemon', target=daemon, daemon=True)
    t = threading.Thread(name='non-daemon', target=non_daemon)
    d.start()
    t.start()
```

这个代码的输出中不包含守护线程的"Exiting"消息,因为在从sleep()调用唤醒守护线程之前,所有非守护线程(包括主线程)已经退出。

```
$ python3 threading_daemon.py

(daemon    ) Starting
(non-daemon) Starting
(non-daemon) Exiting
```

要等待一个守护线程完成工作,需要使用join()方法。

代码清单10-32:threading_daemon_join.py

```
import threading
import time
import logging


def daemon():
    logging.debug('Starting')
    time.sleep(0.2)
    logging.debug('Exiting')


def non_daemon():
    logging.debug('Starting')
    logging.debug('Exiting')


logging.basicConfig(
    level=logging.DEBUG,
    format='(%(threadName)-10s) %(message)s',
)

d = threading.Thread(name='daemon', target=daemon, daemon=True)
t = threading.Thread(name='non-daemon', target=non_daemon)

d.start()
t.start()

d.join()
t.join()
```

使用 join() 等待守护线程退出意味着它有机会生成它的 "Exiting" 消息。

```
$ python3 threading_daemon_join.py

(daemon    ) Starting
(non-daemon) Starting
(non-daemon) Exiting
(daemon    ) Exiting
```

默认地，join() 会无限阻塞。或者，还可以传入一个浮点值，表示等待线程在多长时间（秒数）后变为不活动。即使线程在这个时间段内未完成，join() 也会返回。

代码清单 10-33：**threading_daemon_join_timeout.py**

```
import threading
import time
import logging


def daemon():
    logging.debug('Starting')
    time.sleep(0.2)
    logging.debug('Exiting')


def non_daemon():
    logging.debug('Starting')
    logging.debug('Exiting')


logging.basicConfig(
    level=logging.DEBUG,
    format='(%(threadName)-10s) %(message)s',
)

d = threading.Thread(name='daemon', target=daemon, daemon=True)

t = threading.Thread(name='non-daemon', target=non_daemon)

d.start()
t.start()

d.join(0.1)
print('d.isAlive()', d.isAlive())
t.join()
```

由于传入的超时时间小于守护线程睡眠的时间，所以 join() 返回之后这个线程仍"活着"。

```
$ python3 threading_daemon_join_timeout.py

(daemon    ) Starting
(non-daemon) Starting
(non-daemon) Exiting
d.isAlive() True
```

10.3.4 枚举所有线程

没有必要为所有守护线程维护一个显式句柄来确保它们在退出主进程之前已经完成。

enumerate() 会返回活动 Thread 实例的一个列表。这个列表也包括当前线程，由于等待当前线程终止（join）会引入一种死锁情况，所以必须跳过。

代码清单 10-34：threading_enumerate.py

```python
import random
import threading
import time
import logging


def worker():
    """thread worker function"""
    pause = random.randint(1, 5) / 10
    logging.debug('sleeping %0.2f', pause)
    time.sleep(pause)
    logging.debug('ending')


logging.basicConfig(
    level=logging.DEBUG,
    format='(%(threadName)-10s) %(message)s',
)

for i in range(3):
    t = threading.Thread(target=worker, daemon=True)
    t.start()

main_thread = threading.main_thread()
for t in threading.enumerate():
    if t is main_thread:
        continue
    logging.debug('joining %s', t.getName())
    t.join()
```

由于工作线程睡眠的时间量是随机的，所以这个程序的输出可能有变化。

```
$ python3 threading_enumerate.py

(Thread-1  ) sleeping 0.20
(Thread-2  ) sleeping 0.30
(Thread-3  ) sleeping 0.40
(MainThread) joining Thread-1
(Thread-1  ) ending
(MainThread) joining Thread-3
(Thread-2  ) ending
(Thread-3  ) ending
(MainThread) joining Thread-2
```

10.3.5 派生线程

开始时，Thread 要完成一些基本初始化，然后调用其 run() 方法，这会调用传递到构造函数的目标函数。要创建 Thread 的一个子类，需要覆盖 run() 来完成所需的工作。

代码清单 10-35：threading_subclass.py

```python
import threading
import logging
```

```python
class MyThread(threading.Thread):

    def run(self):
        logging.debug('running')


logging.basicConfig(
    level=logging.DEBUG,
    format='(%(threadName)-10s) %(message)s',
)

for i in range(5):
    t = MyThread()
    t.start()
```

`run()` 的返回值将被忽略。

```
$ python3 threading_subclass.py

(Thread-1  ) running
(Thread-2  ) running
(Thread-3  ) running
(Thread-4  ) running
(Thread-5  ) running
```

由于传递到 Thread 构造函数的 args 和 kwargs 值保存在私有变量中（这些变量名都有前缀 '__'），所以不能很容易地从子类访问这些值。要向一个定制的线程类型传递参数，需要重新定义构造函数，将这些值保存在子类可见的一个实例属性中。

代码清单 10-36：threading_subclass_args.py

```python
import threading
import logging


class MyThreadWithArgs(threading.Thread):

    def __init__(self, group=None, target=None, name=None,
                 args=(), kwargs=None, *, daemon=None):
        super().__init__(group=group, target=target, name=name,
                         daemon=daemon)
        self.args = args
        self.kwargs = kwargs

    def run(self):
        logging.debug('running with %s and %s',
                      self.args, self.kwargs)


logging.basicConfig(
    level=logging.DEBUG,
    format='(%(threadName)-10s) %(message)s',
)

for i in range(5):
    t = MyThreadWithArgs(args=(i,), kwargs={'a': 'A', 'b': 'B'})
    t.start()
```

MyThreadWithArgs 使用的 API 与 Thread 相同，不过类似于其他定制类，这个类可以轻松地修改构造函数方法，以取得更多参数或者与线程用途更直接相关的不同参数。

```
$ python3 threading_subclass_args.py

(Thread-1  ) running with (0,) and {'b': 'B', 'a': 'A'}
(Thread-2  ) running with (1,) and {'b': 'B', 'a': 'A'}
(Thread-3  ) running with (2,) and {'b': 'B', 'a': 'A'}
(Thread-4  ) running with (3,) and {'b': 'B', 'a': 'A'}
(Thread-5  ) running with (4,) and {'b': 'B', 'a': 'A'}
```

10.3.6 定时器线程

有时出于某种原因需要派生 Thread，Timer 就是这样一个例子，Timer 也包含在 threading 中。Timer 在一个延迟之后开始工作，而且可以在这个延迟期间内的任意时刻被取消。

代码清单 10-37：**threading_timer.py**

```
import threading
import time
import logging


def delayed():
    logging.debug('worker running')


logging.basicConfig(
    level=logging.DEBUG,
    format='(%(threadName)-10s) %(message)s',
)

t1 = threading.Timer(0.3, delayed)
t1.setName('t1')
t2 = threading.Timer(0.3, delayed)
t2.setName('t2')

logging.debug('starting timers')
t1.start()
t2.start()

logging.debug('waiting before canceling %s', t2.getName())
time.sleep(0.2)
logging.debug('canceling %s', t2.getName())
t2.cancel()
logging.debug('done')
```

这个例子中，第二个定时器永远不会运行，看起来第一个定时器在主程序的其余部分完成之后还会运行。由于这不是一个守护线程，所以在主线程完成时其会隐式退出。

```
$ python3 threading_timer.py

(MainThread) starting timers
(MainThread) waiting before canceling t2
(MainThread) canceling t2
(MainThread) done
(t1        ) worker running
```

10.3.7 线程间传送信号

尽管使用多线程的目的是并发地运行单独的操作，但有时也需要在两个或多个线程中同步操作。事件对象是实现线程间安全通信的一种简单方法。Event 管理一个内部标志，调用者可以用 set() 和 clear() 方法控制这个标志。其他线程可以使用 wait() 暂停，直到这个标志被设置，可有效地阻塞进程直至允许这些线程继续。

代码清单 10-38：**threading_event.py**

```python
import logging
import threading
import time


def wait_for_event(e):
    """Wait for the event to be set before doing anything"""
    logging.debug('wait_for_event starting')
    event_is_set = e.wait()
    logging.debug('event set: %s', event_is_set)


def wait_for_event_timeout(e, t):
    """Wait t seconds and then timeout"""
    while not e.is_set():
        logging.debug('wait_for_event_timeout starting')
        event_is_set = e.wait(t)
        logging.debug('event set: %s', event_is_set)
        if event_is_set:
            logging.debug('processing event')
        else:
            logging.debug('doing other work')


logging.basicConfig(
    level=logging.DEBUG,
    format='(%(threadName)-10s) %(message)s',
)

e = threading.Event()
t1 = threading.Thread(
    name='block',
    target=wait_for_event,
    args=(e,),
)
t1.start()

t2 = threading.Thread(
    name='nonblock',
    target=wait_for_event_timeout,
    args=(e, 2),
)
t2.start()

logging.debug('Waiting before calling Event.set()')
time.sleep(0.3)
e.set()
logging.debug('Event is set')
```

wait() 方法取一个参数，表示等待事件的时间（秒数），达到这个时间后就超时。它会返回一个布尔值，指示事件是否已设置，使调用者知道 wait() 为什么返回。可以对事件单独地使用 is_set() 方法而不必担心阻塞。

在这个例子中，`wait_for_event_timeout()`将检查事件状态而不会无限阻塞。`wait_for_event()`在`wait()`调用的位置阻塞，事件状态改变之前它不会返回。

```
$ python3 threading_event.py

(block     ) wait_for_event starting
(nonblock  ) wait_for_event_timeout starting
(MainThread) Waiting before calling Event.set()
(MainThread) Event is set
(nonblock  ) event set: True
(nonblock  ) processing event
(block     ) event set: True
```

10.3.8 控制资源访问

除了同步线程操作，还有一点很重要，要能够控制对共享资源的访问，从而避免破坏或丢失数据。Python的内置数据结构（列表、字典等）是线程安全的，这是Python使用原子字节码来管理这些数据结构的一个副作用（更新过程中不会释放保护Python内部数据结构的全局解释器锁GIL（Global Interpreter Lock））。Python中实现的其他数据结构或更简单的类型（如整数和浮点数）则没有这个保护。要保证同时安全地访问一个对象，可以使用一个`Lock`对象。

代码清单10-39：`threading_lock.py`

```python
import logging
import random
import threading
import time
class Counter:

    def __init__(self, start=0):
        self.lock = threading.Lock()
        self.value = start

    def increment(self):
        logging.debug('Waiting for lock')
        self.lock.acquire()
        try:
            logging.debug('Acquired lock')
            self.value = self.value + 1
        finally:
            self.lock.release()


def worker(c):
    for i in range(2):
        pause = random.random()
        logging.debug('Sleeping %0.02f', pause)
        time.sleep(pause)
        c.increment()
    logging.debug('Done')


logging.basicConfig(
    level=logging.DEBUG,
    format='(%(threadName)-10s) %(message)s',
)

counter = Counter()
```

```
    for i in range(2):
        t = threading.Thread(target=worker, args=(counter,))
        t.start()

    logging.debug('Waiting for worker threads')
    main_thread = threading.main_thread()
    for t in threading.enumerate():
        if t is not main_thread:
            t.join()
    logging.debug('Counter: %d', counter.value)
```

在这个例子中，worker() 函数使一个 Counter 实例递增，这个实例管理着一个 Lock，以避免两个线程同时改变其内部状态。如果没有使用 Lock，就有可能丢失一次对 value 属性的修改。

```
$ python3 threading_lock.py

(Thread-1  ) Sleeping 0.18
(Thread-2  ) Sleeping 0.93
(MainThread) Waiting for worker threads
(Thread-1  ) Waiting for lock
(Thread-1  ) Acquired lock
(Thread-1  ) Sleeping 0.11
(Thread-1  ) Waiting for lock
(Thread-1  ) Acquired lock
(Thread-1  ) Done
(Thread-2  ) Waiting for lock
(Thread-2  ) Acquired lock
(Thread-2  ) Sleeping 0.81
(Thread-2  ) Waiting for lock
(Thread-2  ) Acquired lock
(Thread-2  ) Done
(MainThread) Counter: 4
```

要确定是否有另一个线程请求这个锁而不影响当前线程，可以向 acquire() 的 blocking 参数传入 False。在下一个例子中，worker() 想要分别得到 3 次锁，并统计为得到锁而尝试的次数。与此同时，lock_holder() 在占有和释放锁之间循环，每个状态会短暂暂停，以模拟负载情况。

代码清单 10-40：threading_lock_noblock.py

```
import logging
import threading
import time


def lock_holder(lock):
    logging.debug('Starting')
    while True:
        lock.acquire()
        try:
            logging.debug('Holding')
            time.sleep(0.5)
        finally:
            logging.debug('Not holding')
            lock.release()
        time.sleep(0.5)
```

```python
def worker(lock):
    logging.debug('Starting')
    num_tries = 0
    num_acquires = 0
    while num_acquires < 3:
        time.sleep(0.5)
        logging.debug('Trying to acquire')
        have_it = lock.acquire(0)
        try:
            num_tries += 1
            if have_it:
                logging.debug('Iteration %d: Acquired',
                              num_tries)
                num_acquires += 1
            else:
                logging.debug('Iteration %d: Not acquired',
                              num_tries)
        finally:
            if have_it:
                lock.release()
    logging.debug('Done after %d iterations', num_tries)


logging.basicConfig(
    level=logging.DEBUG,
    format='(%(threadName)-10s) %(message)s',
)

lock = threading.Lock()

holder = threading.Thread(
    target=lock_holder,
    args=(lock,),
    name='LockHolder',
    daemon=True,
)
holder.start()

worker = threading.Thread(
    target=worker,
    args=(lock,),
    name='Worker',
)
worker.start()
```

worker() 需要超过 3 次迭代才能得到 3 次锁。

```
$ python3 threading_lock_noblock.py

(LockHolder) Starting
(LockHolder) Holding
(Worker    ) Starting
(LockHolder) Not holding
(Worker    ) Trying to acquire
(Worker    ) Iteration 1: Acquired
(LockHolder) Holding
(Worker    ) Trying to acquire
(Worker    ) Iteration 2: Not acquired
(LockHolder) Not holding
(Worker    ) Trying to acquire
(Worker    ) Iteration 3: Acquired
(LockHolder) Holding
```

```
(Worker    ) Trying to acquire
(Worker    ) Iteration 4: Not acquired
(LockHolder) Not holding
(Worker    ) Trying to acquire
(Worker    ) Iteration 5: Acquired
(Worker    ) Done after 5 iterations
```

10.3.8.1 再入锁

正常的 `Lock` 对象不能请求多次，即使是由同一个线程请求也不例外。如果同一个调用链中的多个函数访问一个锁，则可能会产生我们不希望的副作用。

代码清单 10-41：**threading_lock_reacquire.py**

```python
import threading

lock = threading.Lock()

print('First try :', lock.acquire())
print('Second try:', lock.acquire(0))
```

在这里，对第二个 `acquire()` 调用给定超时值为 0，以避免阻塞，因为锁已经被第一个调用获得。

```
$ python3 threading_lock_reacquire.py

First try : True
Second try: False
```

如果同一个线程的不同代码需要"重新获得"锁，那么在这种情况下要使用 `RLock`。

代码清单 10-42　**threading_rlock.py**

```python
import threading

lock = threading.RLock()
print('First try :', lock.acquire())
print('Second try:', lock.acquire(0))
```

与前面的例子相比，对代码唯一的修改就是用 `RLock` 替换 `Lock`。

```
$ python3 threading_rlock.py

First try : True
Second try: True
```

10.3.8.2 锁作为上下文管理器

锁实现了上下文管理器 API，并与 `with` 语句兼容。使用 `with` 则不再需要显式地获得和释放锁。

代码清单 10-43：**threading_lock_with.py**

```python
import threading
import logging


def worker_with(lock):
```

```python
    with lock:
        logging.debug('Lock acquired via with')

def worker_no_with(lock):
    lock.acquire()
    try:
        logging.debug('Lock acquired directly')
    finally:
        lock.release()

logging.basicConfig(
    level=logging.DEBUG,
    format='(%(threadName)-10s) %(message)s',
)
lock = threading.Lock()
w = threading.Thread(target=worker_with, args=(lock,))
nw = threading.Thread(target=worker_no_with, args=(lock,))
w.start()
nw.start()
```

函数 `worker_with()` 和 `worker_no_with()` 用等价的方式管理锁。

```
$ python3 threading_lock_with.py

(Thread-1  ) Lock acquired via with
(Thread-2  ) Lock acquired directly
```

10.3.9 同步线程

除了使用 Event, 还可以通过使用一个 Condition 对象来同步线程。由于 Condition 使用了一个 Lock, 所以它可以绑定到一个共享资源, 允许多个线程等待资源更新。在下一个例子中, `consumer()` 线程要等待设置了 Condition 才能继续。`producer()` 线程负责设置条件, 以及通知其他线程继续。

代码清单 10-44: `threading_condition.py`

```python
import logging
import threading
import time

def consumer(cond):
    """wait for the condition and use the resource"""
    logging.debug('Starting consumer thread')
    with cond:
        cond.wait()
        logging.debug('Resource is available to consumer')

def producer(cond):
    """set up the resource to be used by the consumer"""
    logging.debug('Starting producer thread')
    with cond:
        logging.debug('Making resource available')
```

```
            cond.notifyAll()

logging.basicConfig(
    level=logging.DEBUG,
    format='%(asctime)s (%(threadName)-2s) %(message)s',
)

condition = threading.Condition()
c1 = threading.Thread(name='c1', target=consumer,
                      args=(condition,))
c2 = threading.Thread(name='c2', target=consumer,
                      args=(condition,))
p = threading.Thread(name='p', target=producer,
                     args=(condition,))
c1.start()
time.sleep(0.2)
c2.start()
time.sleep(0.2)
p.start()
```

这些线程使用 with 来获得与 Condition 关联的锁。也可以显式地使用 acquire() 和 release() 方法。

```
$ python3 threading_condition.py

2016-07-10 10:45:28,170 (c1) Starting consumer thread
2016-07-10 10:45:28,376 (c2) Starting consumer thread
2016-07-10 10:45:28,581 (p ) Starting producer thread
2016-07-10 10:45:28,581 (p ) Making resource available
2016-07-10 10:45:28,582 (c1) Resource is available to consumer
2016-07-10 10:45:28,582 (c2) Resource is available to consumer
```

屏障（barrier）是另一种线程同步机制。Barrier 会建立一个控制点，所有参与线程会在这里阻塞，直到所有这些参与"方"都到达这一点。采用这种方法，线程可以单独启动然后暂停，直到所有线程都准备好才可以继续。

代码清单 10-45：threading_barrier.py

```
import threading
import time


def worker(barrier):
    print(threading.current_thread().name,
          'waiting for barrier with {} others'.format(
              barrier.n_waiting))
    worker_id = barrier.wait()
    print(threading.current_thread().name, 'after barrier',
          worker_id)


NUM_THREADS = 3

barrier = threading.Barrier(NUM_THREADS)

threads = [
    threading.Thread(
        name='worker-%s' % i,
```

```python
            target=worker,
            args=(barrier,),
        )
        for i in range(NUM_THREADS)
    ]
    for t in threads:
        print(t.name, 'starting')
        t.start()
        time.sleep(0.1)

    for t in threads:
        t.join()
```

在这个例子中，Barrier 被配置为会阻塞线程，直到 3 个线程都在等待。满足这个条件时，所有线程被同时释放从而越过这个控制点。wait() 的返回值指示了释放的参与线程数，可以用来限制一些线程做清理资源等动作。

```
$ python3 threading_barrier.py

worker-0 starting
worker-0 waiting for barrier with 0 others
worker-1 starting
worker-1 waiting for barrier with 1 others
worker-2 starting
worker-2 waiting for barrier with 2 others
worker-2 after barrier 2
worker-0 after barrier 0
worker-1 after barrier 1
```

Barrier 的 abort() 方法会使所有等待线程接收一个 BrokenBarrierError。如果线程在 wait() 上被阻塞而停止处理，这就允许线程完成清理工作。

代码清单 10-46：**threading_barrier_abort.py**

```python
import threading
import time


def worker(barrier):
    print(threading.current_thread().name,
          'waiting for barrier with {} others'.format(
              barrier.n_waiting))
    try:
        worker_id = barrier.wait()
    except threading.BrokenBarrierError:
        print(threading.current_thread().name, 'aborting')
    else:
        print(threading.current_thread().name, 'after barrier',
              worker_id)


NUM_THREADS = 3

barrier = threading.Barrier(NUM_THREADS + 1)

threads = [
    threading.Thread(
        name='worker-%s' % i,
        target=worker,
        args=(barrier,),
    )
```

```
        for i in range(NUM_THREADS)
]

for t in threads:
    print(t.name, 'starting')
    t.start()
    time.sleep(0.1)

barrier.abort()

for t in threads:
    t.join()
```

这个例子将 Barrier 配置为多加一个线程，即需要比实际启动的线程再多一个参与线程，所以所有线程中的处理都会阻塞。在被阻塞的各个线程中，abort() 调用会产生一个异常。

```
$ python3 threading_barrier_abort.py

worker-0 starting
worker-0 waiting for barrier with 0 others
worker-1 starting
worker-1 waiting for barrier with 1 others
worker-2 starting
worker-2 waiting for barrier with 2 others
worker-0 aborting
worker-2 aborting
worker-1 aborting
```

10.3.10　限制资源的并发访问

有时可能需要允许多个工作线程同时访问一个资源，但要限制总数。例如，连接池支持同时连接，但数目可能是固定的，或者一个网络应用可能支持固定数目的并发下载。这些连接就可以使用 Semaphore 来管理。

代码清单 10-47　**threading_semaphore.py**

```
import logging
import random
import threading
import time


class ActivePool:

    def __init__(self):
        super(ActivePool, self).__init__()
        self.active = []
        self.lock = threading.Lock()

    def makeActive(self, name):
        with self.lock:
            self.active.append(name)
            logging.debug('Running: %s', self.active)

    def makeInactive(self, name):
        with self.lock:
            self.active.remove(name)
            logging.debug('Running: %s', self.active)
```

```python
def worker(s, pool):
    logging.debug('Waiting to join the pool')
    with s:
        name = threading.current_thread().getName()
        pool.makeActive(name)
        time.sleep(0.1)
        pool.makeInactive(name)


logging.basicConfig(
    level=logging.DEBUG,
    format='%(asctime)s (%(threadName)-2s) %(message)s',
)

pool = ActivePool()
s = threading.Semaphore(2)
for i in range(4):
    t = threading.Thread(
        target=worker,
        name=str(i),
        args=(s, pool),
    )
    t.start()
```

在这个例子中，`ActivePool` 类只作为一种便利方法，用来跟踪某个给定时刻哪些线程能够运行。真正的资源池会为新的活动线程分配一个连接或另外某个值，并且当这个线程工作完成时再回收这个值。在这里，资源池只是用来保存活动线程的名，以显示至少有两个线程在并发运行。

```
$ python3 threading_semaphore.py

2016-07-10 10:45:29,398 (0 ) Waiting to join the pool
2016-07-10 10:45:29,398 (0 ) Running: ['0']
2016-07-10 10:45:29,399 (1 ) Waiting to join the pool
2016-07-10 10:45:29,399 (1 ) Running: ['0', '1']
2016-07-10 10:45:29,399 (2 ) Waiting to join the pool
2016-07-10 10:45:29,399 (3 ) Waiting to join the pool
2016-07-10 10:45:29,501 (1 ) Running: ['0']
2016-07-10 10:45:29,501 (0 ) Running: []
2016-07-10 10:45:29,502 (3 ) Running: ['3']
2016-07-10 10:45:29,502 (2 ) Running: ['3', '2']
2016-07-10 10:45:29,607 (3 ) Running: ['2']
2016-07-10 10:45:29,608 (2 ) Running: []
```

10.3.11 线程特定的数据

有些资源需要锁定以便多个线程使用，另外一些资源则需要保护，以使它们对并非是这些资源的"所有者"的线程隐藏。`local()` 函数会创建一个对象，它能够隐藏值，使其在不同线程中无法被看到。

代码清单 10-48：**threading_local.py**

```python
import random
import threading
import logging


def show_value(data):
```

```
        try:
            val = data.value
        except AttributeError:
            logging.debug('No value yet')
        else:
            logging.debug('value=%s', val)


def worker(data):
    show_value(data)
    data.value = random.randint(1, 100)
    show_value(data)
logging.basicConfig(
    level=logging.DEBUG,
    format='(%(threadName)-10s) %(message)s',
)

local_data = threading.local()
show_value(local_data)
local_data.value = 1000
show_value(local_data)

for i in range(2):
    t = threading.Thread(target=worker, args=(local_data,))
    t.start()
```

属性 local_data.value 对所有线程都不可见,除非在某个线程中设置了这个属性,这个线程才能看到它。

```
$ python3 threading_local.py

(MainThread) No value yet
(MainThread) value=1000
(Thread-1  ) No value yet
(Thread-1  ) value=33
(Thread-2  ) No value yet
(Thread-2  ) value=74
```

要初始化设置以使所有线程在开始时都有相同的值,可以使用一个子类,并在 __init__() 中设置这些属性。

代码清单 10-49: `threading_local_defaults.py`

```
import random
import threading
import logging


def show_value(data):
    try:
        val = data.value
    except AttributeError:
        logging.debug('No value yet')
    else:
        logging.debug('value=%s', val)


def worker(data):
    show_value(data)
    data.value = random.randint(1, 100)
    show_value(data)
```

```python
class MyLocal(threading.local):

    def __init__(self, value):
        super().__init__()
        logging.debug('Initializing %r', self)
        self.value = value


logging.basicConfig(
    level=logging.DEBUG,
    format='(%(threadName)-10s) %(message)s',
)

local_data = MyLocal(1000)
show_value(local_data)

for i in range(2):
    t = threading.Thread(target=worker, args=(local_data,))
    t.start()
```

这会在相同的对象上调用 `__init__()`（注意 `id()` 值），每个线程中调用一次以设置默认值。

```
$ python3 threading_local_defaults.py

(MainThread) Initializing <__main__.MyLocal object at
0x101c6c288>
(MainThread) value=1000
(Thread-1  ) Initializing <__main__.MyLocal object at
0x101c6c288>
(Thread-1  ) value=1000
(Thread-1  ) value=18
(Thread-2  ) Initializing <__main__.MyLocal object at
0x101c6c288>
(Thread-2  ) value=1000
(Thread-2  ) value=77
```

提示：相关阅读材料
- `threading` 的标准库文档[⊖]。
- `threading` 的 Python 2 到 Python 3 移植说明。
- `thread`：底层线程 API。
- `Queue`：这是一个线程安全队列，可用来在线程之间传递消息。
- `multiprocessing`：处理进程的一个 API，它是 `threading` API 的镜像。

10.4 multiprocessing：像线程一样管理进程

`multiprocessing` 模块包含一个 API，它基于 `threading` API，可以把工作划分到多个进程。有些情况下，`multiprocessing` 可以作为临时替换取代 `threading` 来利用多个 CPU 内核，相应地避免 Python 全局解释器锁所带来的计算瓶颈。

由于 `multiprocessing` 与 `threading` 模块的这种相似性，这里的前几个例子都是

[⊖] https://docs.python.org/3/library/threading.html

从 threading 例子修改得来。后面会介绍 multiprocessing 中有但 threading 未提供的特性。

10.4.1 multiprocessing 基础

要创建第二个进程，最简单的方法是用一个目标函数实例化一个 Process 对象，然后调用 start() 让它开始工作。

代码清单 10-50：multiprocessing_simple.py

```
import multiprocessing


def worker():
    """worker function"""
    print('Worker')


if __name__ == '__main__':
    jobs = []
    for i in range(5):
        p = multiprocessing.Process(target=worker)
        jobs.append(p)
        p.start()
```

输出中单词"Worker"将打印 5 次，不过取决于具体的执行顺序，无法清楚地看出孰先孰后，这是因为每个进程都在竞争访问输出流。

```
$ python3 multiprocessing_simple.py

Worker
Worker
Worker
Worker
Worker
```

大多数情况下，更有用的做法是，在创建一个进程时提供参数来告诉它要做什么。与 threading 不同，要向一个 multiprocessing Process 传递参数，这个参数必须能够用 pickle 串行化。下面这个例子向各个工作进程传递一个要打印的数。

代码清单 10-51：multiprocessing_simpleargs.py

```
import multiprocessing


def worker(num):
    """thread worker function"""
    print('Worker:', num)


if __name__ == '__main__':
    jobs = []
    for i in range(5):
        p = multiprocessing.Process(target=worker, args=(i,))
        jobs.append(p)
        p.start()
```

现在整数参数会包含在各个工作进程打印的消息中。

```
$ python3 multiprocessing_simpleargs.py

Worker: 0
Worker: 1
Worker: 2
Worker: 3
Worker: 4
```

10.4.2 可导入的目标函数

threading 与 multiprocessing 例子之间有一个区别，multiprocessing 例子中对 __main__ 使用了额外的保护。基于启动新进程的方式，要求子进程能够导入包含目标函数的脚本。可以把应用的主要部分包装在一个 __main__ 检查中，确保模块导入时不会在各个子进程中递归地运行。另一种方法是从一个单独的脚本导入目标函数。例如，multiprocessing_import_main.py 使用了第二个模块中定义的一个工作函数。

代码清单 10-52：**multiprocessing_import_main.py**

```python
import multiprocessing
import multiprocessing_import_worker

if __name__ == '__main__':
    jobs = []
    for i in range(5):
        p = multiprocessing.Process(
            target=multiprocessing_import_worker.worker,
        )
        jobs.append(p)
        p.start()
```

这个工作函数在 multiprocessing_import_worker.py 中定义。

代码清单 10-53：**multiprocessing_import_worker.py**

```python
def worker():
    """worker function"""
    print('Worker')
    return
```

调用主程序会生成与第一个例子类似的输出。

```
$ python3 multiprocessing_import_main.py

Worker
Worker
Worker
Worker
Worker
```

10.4.3 确定当前进程

通过传递参数来标识或命名进程很麻烦，也没有必要。每个 Process 实例都有一个名，可以在创建进程时改变它的默认值。对进程命名对于跟踪进程很有用，特别是如果应用中有多种类型的进程在同时运行。

代码清单 10-54：**multiprocessing_names.py**

```python
import multiprocessing
import time


def worker():
    name = multiprocessing.current_process().name
    print(name, 'Starting')
    time.sleep(2)
    print(name, 'Exiting')


def my_service():
    name = multiprocessing.current_process().name
    print(name, 'Starting')
    time.sleep(3)
    print(name, 'Exiting')


if __name__ == '__main__':
    service = multiprocessing.Process(
        name='my_service',
        target=my_service,
    )
    worker_1 = multiprocessing.Process(
        name='worker 1',
        target=worker,
    )
    worker_2 = multiprocessing.Process(  # Default name
        target=worker,
    )

    worker_1.start()
    worker_2.start()
    service.start()
```

调试输出中，每行都包含当前进程的名。进程名列为 `Process-3` 的行对应未命名的进程 `worker_1`。

```
$ python3 multiprocessing_names.py

worker 1 Starting
worker 1 Exiting
Process-3 Starting
Process-3 Exiting
my_service Starting
my_service Exiting
```

10.4.4 守护进程

默认地，在所有子进程退出之前主程序不会退出。有些情况下，可能需要启动一个后台进程，它可以一直运行而不阻塞主程序退出，如果一个服务无法用一种容易的方法中断进程，或者希望进程工作到一半时中止而不损失或破坏数据（例如为一个服务监控工具生成"心跳"的任务），那么对于这些服务，使用守护进程就很有用。

要标志一个进程为守护进程，可以将其 `daemon` 属性设置为 `True`。默认情况下进程不作为守护进程。

代码清单 10-55：`multiprocessing_daemon.py`

```python
import multiprocessing
import time
import sys


def daemon():
    p = multiprocessing.current_process()
    print('Starting:', p.name, p.pid)
    sys.stdout.flush()
    time.sleep(2)
    print('Exiting :', p.name, p.pid)
    sys.stdout.flush()


def non_daemon():
    p = multiprocessing.current_process()
    print('Starting:', p.name, p.pid)
    sys.stdout.flush()
    print('Exiting :', p.name, p.pid)
    sys.stdout.flush()


if __name__ == '__main__':
    d = multiprocessing.Process(
        name='daemon',
        target=daemon,
    )
    d.daemon = True

    n = multiprocessing.Process(
        name='non-daemon',
        target=non_daemon,
    )
    n.daemon = False

    d.start()
    time.sleep(1)
    n.start()
```

输出中没有守护进程的"Exiting"消息，因为在守护进程从其 2 秒的睡眠时间唤醒之前，所有非守护进程（包括主程序）已经退出。

```
$ python3 multiprocessing_daemon.py

Starting: daemon 70880
Starting: non-daemon 70881
Exiting : non-daemon 70881
```

守护进程会在主程序退出之前自动终止，以避免留下"孤"进程继续运行。要验证这一点，可以查找程序运行时打印的进程 ID 值，然后用一个类似 `ps` 的命令检查该进程。

10.4.5 等待进程

要等待一个进程完成工作并退出，可以使用 `join()` 方法。

代码清单 10-56：`multiprocessing_daemon_join.py`

```python
import multiprocessing
import time
import sys


def daemon():
    name = multiprocessing.current_process().name
    print('Starting:', name)
    time.sleep(2)
    print('Exiting :', name)


def non_daemon():
    name = multiprocessing.current_process().name
    print('Starting:', name)
    print('Exiting :', name)


if __name__ == '__main__':
    d = multiprocessing.Process(
        name='daemon',
        target=daemon,
    )
    d.daemon = True

    n = multiprocessing.Process(
        name='non-daemon',
        target=non_daemon,
    )
    n.daemon = False
    d.start()
    time.sleep(1)
    n.start()

    d.join()
    n.join()
```

由于主进程使用 `join()` 等待守护进程退出，所以这一次会打印"Exiting"消息。

```
$ python3 multiprocessing_daemon_join.py

Starting: non-daemon
Exiting : non-daemon
Starting: daemon
Exiting : daemon
```

默认地，`join()` 会无限阻塞。可以向这个模块传入一个超时参数（这是一个浮点数，表示在进程变为不活动之前所等待的秒数）。即使进程在这个超时期限内没有完成，`join()` 也会返回。

代码清单 10-57：`multiprocessing_daemon_join_timeout.py`

```python
import multiprocessing
import time
import sys


def daemon():
```

```python
    name = multiprocessing.current_process().name
    print('Starting:', name)
    time.sleep(2)
    print('Exiting :', name)

def non_daemon():
    name = multiprocessing.current_process().name
    print('Starting:', name)
    print('Exiting :', name)


if __name__ == '__main__':
    d = multiprocessing.Process(
        name='daemon',
        target=daemon,
    )
    d.daemon = True
    n = multiprocessing.Process(
        name='non-daemon',
        target=non_daemon,
    )
    n.daemon = False

    d.start()
    n.start()

    d.join(1)
    print('d.is_alive()', d.is_alive())
    n.join()
```

由于传入的超时值小于守护进程睡眠的时间，所以 join() 返回之后这个进程仍"活着"。

```
$ python3 multiprocessing_daemon_join_timeout.py

Starting: non-daemon
Exiting : non-daemon
d.is_alive() True
```

10.4.6 终止进程

尽管最好使用"毒药"（poison pill）方法向进程发出信号，告诉它应当退出（见 10.4.10 节），但是如果一个进程看起来已经挂起或陷入死锁，那么能够强制性地将其结束会很有用。对一个进程对象调用 terminate() 会结束子进程。

代码清单 10-58：multiprocessing_terminate.py

```python
import multiprocessing
import time


def slow_worker():
    print('Starting worker')
    time.sleep(0.1)
    print('Finished worker')


if __name__ == '__main__':
    p = multiprocessing.Process(target=slow_worker)
```

```
        print('BEFORE:', p, p.is_alive())
        p.start()
        print('DURING:', p, p.is_alive())
        p.terminate()
        print('TERMINATED:', p, p.is_alive())
        p.join()
        print('JOINED:', p, p.is_alive())
```

说明：终止进程后要使用 join() 等待进程退出，使进程管理代码有足够的时间更新对象的状态，以反映进程已经终止。

```
$ python3 multiprocessing_terminate.py

BEFORE: <Process(Process-1, initial)> False
DURING: <Process(Process-1, started)> True
TERMINATED: <Process(Process-1, started)> True
JOINED: <Process(Process-1, stopped[SIGTERM])> False
```

10.4.7 进程退出状态

进程退出时生成的状态码可以通过 exitcode 属性访问。表 10-1 列出了这个属性的可取值范围。

表 10-1 **multiprocessing** 退出码

退出码	含义
== 0	未生成任何错误
> 0	进程有一个错误，并以该错误码退出
< 0	进程以一个 -1 * exitcode 信号结束

代码清单 10-59：**multiprocessing_exitcode.py**

```python
import multiprocessing
import sys
import time


def exit_error():
    sys.exit(1)


def exit_ok():
    return


def return_value():
    return 1


def raises():
    raise RuntimeError('There was an error!')


def terminated():
    time.sleep(3)
```

```python
if __name__ == '__main__':
    jobs = []
    funcs = [
        exit_error,
        exit_ok,
        return_value,
        raises,
        terminated,
    ]
    for f in funcs:
        print('Starting process for', f.__name__)
        j = multiprocessing.Process(target=f, name=f.__name__)
        jobs.append(j)
        j.start()

    jobs[-1].terminate()

    for j in jobs:
        j.join()
        print('{:>15}.exitcode = {}'.format(j.name, j.exitcode))
```

产生异常的进程会自动得到 **exitcode** 为 1。

```
$ python3 multiprocessing_exitcode.py

Starting process for exit_error
Starting process for exit_ok
Starting process for return_value
Starting process for raises
Starting process for terminated
Process raises:
Traceback (most recent call last):
  File ".../lib/python3.5/multiprocessing/process.py", line 249,
in _bootstrap
    self.run()
  File ".../lib/python3.5/multiprocessing/process.py", line 93,
in run
    self._target(*self._args, **self._kwargs)
  File "multiprocessing_exitcode.py", line 28, in raises
    raise RuntimeError('There was an error!')
RuntimeError: There was an error!
     exit_error.exitcode = 1
        exit_ok.exitcode = 0
   return_value.exitcode = 0
         raises.exitcode = 1
     terminated.exitcode = -15
```

10.4.8 日志

调试并发问题时，如果能够访问 **multiprocessing** 所提供对象的内部状态，那么这会很有用。可以使用一个方便的模块级函数启用日志记录，名为 **log_to_stderr()**。它使用 **logging** 建立一个日志记录器对象，并增加一个处理器，使日志消息被发送到标准错误通道。

代码清单 10-60：multiprocessing_log_to_stderr.py

```
import multiprocessing
import logging
import sys
```

```python
def worker():
    print('Doing some work')
    sys.stdout.flush()


if __name__ == '__main__':
    multiprocessing.log_to_stderr(logging.DEBUG)
    p = multiprocessing.Process(target=worker)
    p.start()
    p.join()
```

默认地，日志级别被设置为 NOTSET，即不产生任何消息。通过传入一个不同的日志级别，可以初始化日志记录器并指定所需的详细程度。

```
$ python3 multiprocessing_log_to_stderr.py

[INFO/Process-1] child process calling self.run()
Doing some work
[INFO/Process-1] process shutting down
[DEBUG/Process-1] running all "atexit" finalizers with priority
>= 0
[DEBUG/Process-1] running the remaining "atexit" finalizers
[INFO/Process-1] process exiting with exitcode 0
[INFO/MainProcess] process shutting down
[DEBUG/MainProcess] running all "atexit" finalizers with
priority >= 0
[DEBUG/MainProcess] running the remaining "atexit" finalizers
```

若要直接处理日志记录器（修改其日志级别或增加处理器），可以使用 get_logger()。

代码清单 10-61：multiprocessing_get_logger.py

```python
import multiprocessing
import logging
import sys


def worker():
    print('Doing some work')
    sys.stdout.flush()


if __name__ == '__main__':
    multiprocessing.log_to_stderr()
    logger = multiprocessing.get_logger()
    logger.setLevel(logging.INFO)
    p = multiprocessing.Process(target=worker)
    p.start()
    p.join()
```

使用名 multiprocessing，还可以通过 logging 配置文件 API 来配置日志记录器。

```
$ python3 multiprocessing_get_logger.py

[INFO/Process-1] child process calling self.run()
Doing some work
[INFO/Process-1] process shutting down
[INFO/Process-1] process exiting with exitcode 0
[INFO/MainProcess] process shutting down
```

10.4.9 派生进程

要在一个单独的进程中开始工作,尽管最简单的方法是使用 `Process` 并传入一个目标函数,但也可以使用一个定制子类。

代码清单 10-62:`multiprocessing_subclass.py`

```python
import multiprocessing


class Worker(multiprocessing.Process):

    def run(self):
        print('In {}'.format(self.name))
        return


if __name__ == '__main__':
    jobs = []
    for i in range(5):
        p = Worker()
        jobs.append(p)
        p.start()
    for j in jobs:
        j.join()
```

派生类应当覆盖 `run()` 以完成工作。

```
$ python3 multiprocessing_subclass.py

In Worker-1
In Worker-2
In Worker-3
In Worker-4
In Worker-5
```

10.4.10 向进程传递消息

类似于线程,对于多个进程,一种常见的使用模式是将一个工作划分到多个工作进程中并行地运行。要想有效地使用多个进程,通常要求它们之间有某种通信,这样才能分解工作,并完成结果的聚集。利用 `multiprocessing` 完成进程间通信的一种简单方法是使用一个 `Queue` 来回传递消息。能够用 `pickle` 串行化的任何对象都可以通过 `Queue` 传递。

代码清单 10-63:`multiprocessing_queue.py`

```python
import multiprocessing


class MyFancyClass:

    def __init__(self, name):
        self.name = name

    def do_something(self):
        proc_name = multiprocessing.current_process().name
        print('Doing something fancy in {} for {}!'.format(
            proc_name, self.name))
```

```python
def worker(q):
    obj = q.get()
    obj.do_something()


if __name__ == '__main__':
    queue = multiprocessing.Queue()

    p = multiprocessing.Process(target=worker, args=(queue,))
    p.start()

    queue.put(MyFancyClass('Fancy Dan'))

    # Wait for the worker to finish.
    queue.close()
    queue.join_thread()
    p.join()
```

这个小例子只是向一个工作进程传递一个消息,然后主进程等待这个工作进程完成。

```
$ python3 multiprocessing_queue.py

Doing something fancy in Process-1 for Fancy Dan!
```

来看一个更复杂的例子,这里展示了如何管理多个工作进程,它们都消费一个 `Joinable-Queue` 的数据,并把结果传递回父进程。这里使用"毒药"技术来停止工作进程。建立具体任务后,主程序会在作业队列中为每个工作进程增加一个"stop"值。当一个工作进程遇到这个特定值时,就会退出其处理循环。主进程使用任务队列的 `join()` 方法等待所有任务都完成后才开始处理结果。

代码清单 10-64: `multiprocessing_producer_consumer.py`

```python
import multiprocessing
import time


class Consumer(multiprocessing.Process):

    def __init__(self, task_queue, result_queue):
        multiprocessing.Process.__init__(self)
        self.task_queue = task_queue
        self.result_queue = result_queue

    def run(self):
        proc_name = self.name
        while True:
            next_task = self.task_queue.get()
            if next_task is None:
                # Poison pill means shutdown.
                print('{}: Exiting'.format(proc_name))
                self.task_queue.task_done()
                break
            print('{}: {}'.format(proc_name, next_task))
            answer = next_task()
            self.task_queue.task_done()
            self.result_queue.put(answer)


class Task:
```

```python
    def __init__(self, a, b):
        self.a = a
        self.b = b

    def __call__(self):
        time.sleep(0.1)  # Pretend to take time to do the work.
        return '{self.a} * {self.b} = {product}'.format(
            self=self, product=self.a * self.b)

    def __str__(self):
        return '{self.a} * {self.b}'.format(self=self)


if __name__ == '__main__':
    # Establish communication queues.
    tasks = multiprocessing.JoinableQueue()
    results = multiprocessing.Queue()

    # Start consumers.
    num_consumers = multiprocessing.cpu_count() * 2
    print('Creating {} consumers'.format(num_consumers))
    consumers = [
        Consumer(tasks, results)
        for i in range(num_consumers)
    ]
    for w in consumers:
        w.start()
    # Enqueue jobs.
    num_jobs = 10
    for i in range(num_jobs):
        tasks.put(Task(i, i))

    # Add a poison pill for each consumer.
    for i in range(num_consumers):
        tasks.put(None)

    # Wait for all of the tasks to finish.
    tasks.join()

    # Start printing results.
    while num_jobs:
        result = results.get()
        print('Result:', result)
        num_jobs -= 1
```

尽管作业按顺序进入队列，但它们的执行却是并行的，所以不能保证它们完成的顺序。

```
$ python3 -u multiprocessing_producer_consumer.py

Creating 8 consumers
Consumer-1: 0 * 0
Consumer-2: 1 * 1
Consumer-3: 2 * 2
Consumer-4: 3 * 3
Consumer-5: 4 * 4
Consumer-6: 5 * 5
Consumer-7: 6 * 6
Consumer-8: 7 * 7
Consumer-3: 8 * 8
Consumer-7: 9 * 9
Consumer-4: Exiting
Consumer-1: Exiting
```

```
Consumer-2: Exiting
Consumer-5: Exiting
Consumer-6: Exiting
Consumer-8: Exiting
Consumer-7: Exiting
Consumer-3: Exiting
Result: 6 * 6 = 36
Result: 2 * 2 = 4
Result: 3 * 3 = 9
Result: 0 * 0 = 0
Result: 1 * 1 = 1
Result: 7 * 7 = 49
Result: 4 * 4 = 16
Result: 5 * 5 = 25
Result: 8 * 8 = 64
Result: 9 * 9 = 81
```

10.4.11 进程间信号传输

Event 类提供了一种简单的方法，可以在进程之间传递状态信息。事件可以在设置状态和未设置状态之间切换。通过使用一个可选的超时值，事件对象的用户可以等待其状态从未设置变为设置。

代码清单 10-65：multiprocessing_event.py

```python
import multiprocessing
import time


def wait_for_event(e):
    """Wait for the event to be set before doing anything"""
    print('wait_for_event: starting')
    e.wait()
    print('wait_for_event: e.is_set()->', e.is_set())


def wait_for_event_timeout(e, t):
    """Wait t seconds and then timeout"""
    print('wait_for_event_timeout: starting')
    e.wait(t)
    print('wait_for_event_timeout: e.is_set()->', e.is_set())


if __name__ == '__main__':
    e = multiprocessing.Event()
    w1 = multiprocessing.Process(
        name='block',
        target=wait_for_event,
        args=(e,),
    )
    w1.start()

    w2 = multiprocessing.Process(
        name='nonblock',
        target=wait_for_event_timeout,
        args=(e, 2),
    )
    w2.start()
    print('main: waiting before calling Event.set()')
    time.sleep(3)
    e.set()
    print('main: event is set')
```

wait()到时间时就会返回，而且没有任何错误。调用者负责使用 is_set() 检查事件的状态。

```
$ python3 -u multiprocessing_event.py

main: waiting before calling Event.set()
wait_for_event: starting
wait_for_event_timeout: starting
wait_for_event_timeout: e.is_set()-> False
main: event is set
wait_for_event: e.is_set()-> True
```

10.4.12 控制资源访问

如果需要在多个进程间共享一个资源，那么在这种情况下，可以使用一个 Lock 来避免访问冲突。

代码清单 10-66：`multiprocessing_lock.py`

```python
import multiprocessing
import sys


def worker_with(lock, stream):
    with lock:
        stream.write('Lock acquired via with\n')


def worker_no_with(lock, stream):
    lock.acquire()
    try:
        stream.write('Lock acquired directly\n')
    finally:
        lock.release()


lock = multiprocessing.Lock()
w = multiprocessing.Process(
    target=worker_with,
    args=(lock, sys.stdout),
)
nw = multiprocessing.Process(
    target=worker_no_with,
    args=(lock, sys.stdout),
)

w.start()
nw.start()

w.join()
nw.join()
```

在这个例子中，如果这两个进程没有用锁同步其输出流访问，那么打印到控制台的消息可能会纠结在一起。

```
$ python3 multiprocessing_lock.py

Lock acquired via with
Lock acquired directly
```

10.4.13 同步操作

可以用 `Condition` 对象来同步一个工作流的各个部分，使其中一些部分并行运行，而另外一些顺序运行，即使它们在不同的进程中。

代码清单 10-67：`multiprocessing_condition.py`

```python
import multiprocessing
import time


def stage_1(cond):
    """perform first stage of work,
    then notify stage_2 to continue
    """
    name = multiprocessing.current_process().name
    print('Starting', name)
    with cond:
        print('{} done and ready for stage 2'.format(name))
        cond.notify_all()


def stage_2(cond):
    """wait for the condition telling us stage_1 is done"""
    name = multiprocessing.current_process().name
    print('Starting', name)
    with cond:
        cond.wait()
        print('{} running'.format(name))


if __name__ == '__main__':
    condition = multiprocessing.Condition()
    s1 = multiprocessing.Process(name='s1',
                                 target=stage_1,
                                 args=(condition,))
    s2_clients = [
        multiprocessing.Process(
            name='stage_2[{}]'.format(i),
            target=stage_2,
            args=(condition,),
        )
        for i in range(1, 3)
    ]

    for c in s2_clients:
        c.start()
        time.sleep(1)
    s1.start()

    s1.join()
    for c in s2_clients:
        c.join()
```

在这个例子中，两个进程并行地运行一个作业的第二阶段，但前提是第一阶段已经完成。

```
$ python3 multiprocessing_condition.py

Starting s1
s1 done and ready for stage 2
Starting stage_2[2]
stage_2[2] running
Starting stage_2[1]
stage_2[1] running
```

10.4.14 控制资源的并发访问

有时可能需要允许多个工作进程同时访问一个资源，但要限制总数。例如，连接池支持同时连接，但数目可能是固定的，或者一个网络应用可能支持固定数目的并发下载。这些连接就可以使用 Semaphore 来管理。

代码清单10-68：**multiprocessing_semaphore.py**

```python
import random
import multiprocessing
import time
class ActivePool:

    def __init__(self):
        super(ActivePool, self).__init__()
        self.mgr = multiprocessing.Manager()
        self.active = self.mgr.list()
        self.lock = multiprocessing.Lock()

    def makeActive(self, name):
        with self.lock:
            self.active.append(name)

    def makeInactive(self, name):
        with self.lock:
            self.active.remove(name)

    def __str__(self):
        with self.lock:
            return str(self.active)


def worker(s, pool):
    name = multiprocessing.current_process().name
    with s:
        pool.makeActive(name)
        print('Activating {} now running {}'.format(
            name, pool))
        time.sleep(random.random())
        pool.makeInactive(name)


if __name__ == '__main__':
    pool = ActivePool()
    s = multiprocessing.Semaphore(3)
    jobs = [
        multiprocessing.Process(
            target=worker,
            name=str(i),
            args=(s, pool),
        )
        for i in range(10)
    ]

    for j in jobs:
        j.start()

    while True:
        alive = 0
        for j in jobs:
```

```
            if j.is_alive():
                alive += 1
                j.join(timeout=0.1)
                print('Now running {}'.format(pool))
        if alive == 0:
            # All done
            break
```

在这个例子中，`ActivePool` 类只作为一种便利方法，用来跟踪某个给定时刻哪些进程能够运行。真正的资源池会为新的活动进程分配一个连接或另外某个值，并且当这个进程工作完成时再回收这个值。在这里，资源池只是用来保存活动进程的名，以显示只有三个进程在并发运行。

```
$ python3 -u multiprocessing_semaphore.py

Activating 0 now running ['0', '1', '2']
Activating 1 now running ['0', '1', '2']
Activating 2 now running ['0', '1', '2']
Now running ['0', '1', '2']
Now running ['0', '1', '2']
Now running ['0', '1', '2']
Now running ['0', '1', '2']
Activating 3 now running ['0', '1', '3']
Activating 4 now running ['1', '3', '4']
Activating 6 now running ['1', '4', '6']
Now running ['1', '4', '6']
Now running ['1', '4', '6']
Activating 5 now running ['1', '4', '5']
Now running ['1', '4', '5']
Now running ['1', '4', '5']
Now running ['1', '4', '5']
Activating 8 now running ['4', '5', '8']
Now running ['4', '5', '8']
Now running ['4', '5', '8']
Now running ['4', '5', '8']
Now running ['4', '5', '8']
Now running ['4', '5', '8']
Activating 7 now running ['5', '8', '7']
Now running ['5', '8', '7']
Activating 9 now running ['8', '7', '9']
Now running ['8', '7', '9']
Now running ['8', '9']
Now running ['8', '9']
Now running ['9']
Now running ['9']
Now running ['9']
Now running []
```

10.4.15　管理共享状态

在前面的例子中，`ActivePool` 实例通过一个特殊类型列表对象（由 `Manager` 创建）集中维护活动进程列表。`Manager` 负责协调其所有用户之间共享的信息状态。

代码清单 10-69：**multiprocessing_manager_dict.py**

```
import multiprocessing
import pprint
```

```
def worker(d, key, value):
    d[key] = value


if __name__ == '__main__':
    mgr = multiprocessing.Manager()
    d = mgr.dict()
    jobs = [
        multiprocessing.Process(
            target=worker,
            args=(d, i, i * 2),
        )
        for i in range(10)
    ]
    for j in jobs:
        j.start()
    for j in jobs:
        j.join()
    print('Results:', d)
```

因为这个列表是通过管理器创建的,所以它会由所有进程共享,所有进程都能看到这个列表的更新。除了列表,管理器还支持字典。

```
$ python3 multiprocessing_manager_dict.py

Results: {0: 0, 1: 2, 2: 4, 3: 6, 4: 8, 5: 10, 6: 12, 7: 14,
8: 16, 9: 18}
```

10.4.16 共享命名空间

除了字典和列表,Manager 还可以创建一个共享 Namespace。

代码清单 10-70: **multiprocessing_namespaces.py**

```
import multiprocessing


def producer(ns, event):
    ns.value = 'This is the value'
    event.set()


def consumer(ns, event):
    try:
        print('Before event: {}'.format(ns.value))
    except Exception as err:
        print('Before event, error:', str(err))
    event.wait()
    print('After event:', ns.value)


if __name__ == '__main__':
    mgr = multiprocessing.Manager()
    namespace = mgr.Namespace()
    event = multiprocessing.Event()
    p = multiprocessing.Process(
        target=producer,
        args=(namespace, event),
    )
    c = multiprocessing.Process(
```

```
        target=consumer,
        args=(namespace, event),
    )

    c.start()
    p.start()

    c.join()
    p.join()
```

增加到 `Namespace` 的所有命名值对所有接收 `Namespace` 实例的客户都可见。

```
$ python3 multiprocessing_namespaces.py

Before event, error: 'Namespace' object has no attribute 'value'
After event: This is the value
```

对命名空间中可变值内容的更新不会自动传播，如下面的例子所示。

代码清单 10-71：**multiprocessing_namespaces_mutable.py**

```
import multiprocessing


def producer(ns, event):
    # DOES NOT UPDATE GLOBAL VALUE!
    ns.my_list.append('This is the value')
    event.set()


def consumer(ns, event):
    print('Before event:', ns.my_list)
    event.wait()
    print('After event :', ns.my_list)


if __name__ == '__main__':
    mgr = multiprocessing.Manager()
    namespace = mgr.Namespace()
    namespace.my_list = []

    event = multiprocessing.Event()
    p = multiprocessing.Process(
        target=producer,
        args=(namespace, event),
    )
    c = multiprocessing.Process(
        target=consumer,
        args=(namespace, event),
    )

    c.start()
    p.start()

    c.join()
    p.join()
```

要更新这个列表，需要将它再次关联到命名空间对象。

```
$ python3 multiprocessing_namespaces_mutable.py

Before event: []
After event : []
```

10.4.17 进程池

有些情况下，所要完成的工作可以分解并独立地分布到多个工作进程，对于这种简单的情况，可以用 Pool 类来管理固定数目的工作进程。会收集各个作业的返回值并作为一个列表返回。池（pool）参数包括进程数以及启动任务进程时要运行的函数（对每个子进程调用一次）。

代码清单 10-72：**multiprocessing_pool.py**

```python
import multiprocessing


def do_calculation(data):
    return data * 2


def start_process():
    print('Starting', multiprocessing.current_process().name)


if __name__ == '__main__':
    inputs = list(range(10))
    print('Input   :', inputs)

    builtin_outputs = list(map(do_calculation, inputs))
    print('Built-in:', builtin_outputs)

    pool_size = multiprocessing.cpu_count() * 2
    pool = multiprocessing.Pool(
        processes=pool_size,
        initializer=start_process,
    )
    pool_outputs = pool.map(do_calculation, inputs)
    pool.close()  # No more tasks
    pool.join()   # Wrap up current tasks.

    print('Pool    :', pool_outputs)
```

map() 方法的结果在功能上等价于内置 map() 的结果，只不过各个任务会并行运行。由于进程池并行地处理输入，可以用 close() 和 join() 使任务进程与主进程同步，以确保完成适当的清理。

```
$ python3 multiprocessing_pool.py

Input   : [0, 1, 2, 3, 4, 5, 6, 7, 8, 9]
Built-in: [0, 2, 4, 6, 8, 10, 12, 14, 16, 18]

Starting ForkPoolWorker-3
Starting ForkPoolWorker-4
Starting ForkPoolWorker-5
Starting ForkPoolWorker-6
Starting ForkPoolWorker-1
Starting ForkPoolWorker-7
Starting ForkPoolWorker-2
Starting ForkPoolWorker-8
Pool    : [0, 2, 4, 6, 8, 10, 12, 14, 16, 18]
```

默认地，Pool 会创建固定数目的工作进程，并向这些工作进程传递作业，直到再没有更多作业为止。设置 `maxtasksperchild` 参数可以告诉池在完成一些任务之后要重新启动一个工作进程，来避免长时间运行的工作进程消耗更多的系统资源。

代码清单 10-73：`multiprocessing_pool_maxtasksperchild.py`

```python
import multiprocessing


def do_calculation(data):
    return data * 2


def start_process():
    print('Starting', multiprocessing.current_process().name)


if __name__ == '__main__':
    inputs = list(range(10))
    print('Input   :', inputs)

    builtin_outputs = list(map(do_calculation, inputs))
    print('Built-in:', builtin_outputs)

    pool_size = multiprocessing.cpu_count() * 2
    pool = multiprocessing.Pool(
        processes=pool_size,
        initializer=start_process,
        maxtasksperchild=2,
    )
    pool_outputs = pool.map(do_calculation, inputs)
    pool.close()  # No more tasks
    pool.join()   # Wrap up current tasks.

    print('Pool    :', pool_outputs)
```

池完成其分配的任务时，即使并没有更多工作要做，也会重新启动工作进程。从下面的输出可以看到，尽管只有 10 个任务，而且每个工作进程一次可以完成两个任务，但是这里创建了 8 个工作进程。

```
$ python3 multiprocessing_pool_maxtasksperchild.py

Input   : [0, 1, 2, 3, 4, 5, 6, 7, 8, 9]
Built-in: [0, 2, 4, 6, 8, 10, 12, 14, 16, 18]
Starting ForkPoolWorker-1
Starting ForkPoolWorker-2
Starting ForkPoolWorker-4
Starting ForkPoolWorker-5
Starting ForkPoolWorker-6
Starting ForkPoolWorker-3
Starting ForkPoolWorker-7
Starting ForkPoolWorker-8
Pool    : [0, 2, 4, 6, 8, 10, 12, 14, 16, 18]
```

10.4.18 实现 MapReduce

`Pool` 类可以用于创建一个简单的单服务器 MapReduce 实现。尽管它无法充分提供分布处理的好处，但这种方法显示其能够很容易地将一些问题分解为可分布的工作单元。

在基于 MapReduce 的系统中，输入数据分解为块，由不同的工作进程实例处理。首先使用一个简单的转换将各个输入数据块映射到一个中间状态。然后将中间数据汇集在一起，基于键值分区，使所有相关的值都在一起。最后，将分区的数据归约为一个结果集。

代码清单10-74：**multiprocessing_mapreduce.py**

```
import collections
import itertools
import multiprocessing


class SimpleMapReduce:

    def __init__(self, map_func, reduce_func, num_workers=None):
        """
        map_func

          Function to map inputs to intermediate data. Takes as
          argument one input value and returns a tuple with the
          key and a value to be reduced.

        reduce_func

          Function to reduce partitioned version of intermediate
          data to final output. Takes as argument a key as
          produced by map_func and a sequence of the values
          associated with that key.

        num_workers

          The number of workers to create in the pool. Defaults
          to the number of CPUs available on the current host.
        """
        self.map_func = map_func
        self.reduce_func = reduce_func
        self.pool = multiprocessing.Pool(num_workers)

    def partition(self, mapped_values):
        """Organize the mapped values by their key.
        Returns an unsorted sequence of tuples with a key
        and a sequence of values.
        """
        partitioned_data = collections.defaultdict(list)
        for key, value in mapped_values:
            partitioned_data[key].append(value)
        return partitioned_data.items()

    def __call__(self, inputs, chunksize=1):
        """Process the inputs through the map and reduce functions
        given.

        inputs
          An iterable containing the input data to be processed.

        chunksize=1
          The portion of the input data to hand to each worker.
          This can be used to tune performance during the mapping
          phase.
        """
        map_responses = self.pool.map(
            self.map_func,
            inputs,
            chunksize=chunksize,
```

```
        )
        partitioned_data = self.partition(
            itertools.chain(*map_responses)
        )
        reduced_values = self.pool.map(
            self.reduce_func,
            partitioned_data,
        )
        return reduced_values
```

以下示例脚本使用 SimpleMapReduce 统计这篇文章 reStructuredText 源中的"单词"数，这里要忽略其中的一些标记。

代码清单 10-75：**multiprocessing_wordcount.py**

```python
import multiprocessing
import string

from multiprocessing_mapreduce import SimpleMapReduce


def file_to_words(filename):
    """Read a file and return a sequence of
    (word, occurences) values.
    """
    STOP_WORDS = set([
        'a', 'an', 'and', 'are', 'as', 'be', 'by', 'for', 'if',
        'in', 'is', 'it', 'of', 'or', 'py', 'rst', 'that', 'the',
        'to', 'with',
    ])
    TR = str.maketrans({
        p: ' '
        for p in string.punctuation
    })

    print('{} reading {}'.format(
        multiprocessing.current_process().name, filename))
    output = []

    with open(filename, 'rt') as f:
        for line in f:
            # Skip comment lines.
            if line.lstrip().startswith('..'):
                continue
            line = line.translate(TR)  # Strip punctuation.
            for word in line.split():
                word = word.lower()
                if word.isalpha() and word not in STOP_WORDS:
                    output.append((word, 1))
    return output


def count_words(item):
    """Convert the partitioned data for a word to a
    tuple containing the word and the number of occurences.
    """
    word, occurences = item
    return (word, sum(occurences))


if __name__ == '__main__':
```

```python
import operator
import glob

input_files = glob.glob('*.rst')

mapper = SimpleMapReduce(file_to_words, count_words)
word_counts = mapper(input_files)
word_counts.sort(key=operator.itemgetter(1))
word_counts.reverse()

print('\nTOP 20 WORDS BY FREQUENCY\n')
top20 = word_counts[:20]
longest = max(len(word) for word, count in top20)
for word, count in top20:
    print('{word:<{len}}: {count:5}'.format(
        len=longest + 1,
        word=word,
        count=count)
    )
```

`file_to_words()`函数将各个输入文件转换为一个元组序列，各元组包含单词和数字`1`（表示一次出现）。`partition()`使用单词作为键来划分数据，所以得到的结构包括一个键和一个`1`值序列（表示单词的每次出现）。分区数据被转换为一组元组，元组中包含一个单词和归约阶段中`count_words()`统计得出的这个单词的出现次数。

```
$ python3 -u multiprocessing_wordcount.py

ForkPoolWorker-1 reading basics.rst
ForkPoolWorker-2 reading communication.rst
ForkPoolWorker-3 reading index.rst
ForkPoolWorker-4 reading mapreduce.rst

TOP 20 WORDS BY FREQUENCY

process          :    83
running          :    45
multiprocessing  :    44
worker           :    40
starting         :    37
now              :    35
after            :    34
processes        :    31
start            :    29
header           :    27
pymotw           :    27
caption          :    27
end              :    27
daemon           :    22
can              :    22
exiting          :    21
forkpoolworker   :    21
consumer         :    20
main             :    18
event            :    16
```

提示：相关阅读材料
- `multiprocessing`的标准库文档⊖。

⊖ https://docs.python.org/3.5/library/multiprocessing.html

- threading：处理线程的高级 API。
- Wikipedia：MapReduce[⊖]维基百科上关于 MapReduce 的概述。
- MapReduce: Simplified Data Processing on Large Clusters[⊖]：Google Labs 关于 MapReduce 的演示文稿和论文。
- operator：操作符工具如 itemgetter。

10.5 asyncio：异步 I/O、事件循环和并发工具

asyncio 模块提供了使用协程构建并发应用的工具。threading 模块通过应用线程实现并发，multiprocessing 使用系统进程实现并发，asyncio 则使用一种单线程单进程方法来实现并发，应用的各个部分会彼此合作，在最优的时刻显式地切换任务。大多数情况下，会在程序阻塞等待读写数据时发生这种上下文切换，不过 asyncio 也支持调度代码在将来的某个特定时间运行，从而支持一个协程等待另一个协程完成，以处理系统信号和识别其他一些事件（这些事件可能导致应用改变其工作内容）。

10.5.1 异步并发概念

使用其他并发模型的大多数程序都采用线性方式编写，而且依赖于语言运行时系统或操作系统的底层线程或进程管理来适当地改变上下文。基于 asyncio 的应用要求应用代码显式地处理上下文切换，要正确地使用相关技术，这取决于是否能正确理解一些相关联的概念。

asyncio 提供的框架以一个事件循环（event loop）为中心，这是一个首类对象，负责高效地处理 I/O 事件、系统事件和应用上下文切换。目前已经提供了多个循环实现来高效地利用操作系统的功能。尽管通常会自动地选择一个合理的默认实现，但也完全可以在应用中选择某个特定的事件循环实现。在很多情况下这会很有用，例如，在 Windows 下，一些循环类增加了对外部进程的支持，这可能会以牺牲一些网络 I/O 效率为代价。

与事件循环交互的应用要显式地注册将运行的代码，让事件循环在资源可用时向应用代码发出必要的调用。例如，一个网络服务器打开套接字，然后注册为当这些套接字上出现输入事件时服务器要得到通知。事件循环在建立一个新的进入连接或者在数据可读取时都会提醒服务器代码。当前上下文中没有更多工作可做时，应用代码要再次短时间地交出控制。例如，如果一个套接字再没有更多的数据可以读取，那么服务器会把控制交回给事件循环。

将控制交还给事件循环的机制依赖于 Python 的协程（coroutine），这是一些特殊的函数，可以将控制交回给调用者而不丢失其状态。协程与生成器函数非常类似；实际上，在 Python 3.5 版本之前对协程未提供原生支持时，可以用生成器来实现协程。asyncio 还为协议（protocol）和传输（transport）提供了一个基于类的抽象层，可以使用回调编写代码而不是直接编写协程。在基于类的模型和协程模型中，可以通过重新进入事件循环显式地改

⊖ https://en.wikipedia.org/wiki/MapReduce
⊖ http://research.google.com/archive/mapreduce.html

变上下文，以取代 Python 多线程实现中隐式的上下文改变。

future 是一个数据结构，表示还未完成的工作结果。事件循环可以监视 Future 对象是否完成，从而允许应用的一部分等待另一部分完成一些工作。除了 future，asyncio 还包括其他并发原语，如锁和信号量。

Task 是 Future 的一个子类，它知道如何包装和管理一个协程的执行。任务所需的资源可用时，事件循环会调度任务运行，并生成一个结果，从而可以由其他协程消费。

10.5.2 利用协程合作完成多任务

协程是一个专门设计用来实现并发操作的语言构造。调用协程函数时会创建一个协程对象，然后调用者使用协程的 `send()` 方法运行这个函数的代码。协程可以使用 await 关键字（并提供另一个协程）暂停执行。暂停时，这个协程的状态会保留，使得下一次被唤醒时可以从暂停的地方恢复执行。

10.5.2.1 启动一个协程

asyncio 事件循环可以采用多种不同的方法启动一个协程。最简单的方法是使用 `run_until_complete()`，并把协程直接传入这个方法。

代码清单 10-76：**asyncio_coroutine.py**

```
import asyncio

async def coroutine():
    print('in coroutine')


event_loop = asyncio.get_event_loop()
try:
    print('starting coroutine')
    coro = coroutine()
    print('entering event loop')
    event_loop.run_until_complete(coro)
finally:
    print('closing event loop')
    event_loop.close()
```

第一步是得到事件循环的一个引用。可以使用默认的循环类型，也可以实例化一个特定的循环类。在这个例子中使用了默认循环。`run_until_complete()` 方法用这个协程启动循环；协程返回退出时这个方法会停止循环。

```
$ python3 asyncio_coroutine.py

starting coroutine
entering event loop
in coroutine
closing event loop
```

10.5.2.2 从协程返回值

协程的返回值传回给启动并等待这个协程的代码。

代码清单 10-77: **asyncio_coroutine_return.py**

```python
import asyncio

async def coroutine():
    print('in coroutine')
    return 'result'
event_loop = asyncio.get_event_loop()
try:
    return_value = event_loop.run_until_complete(
        coroutine()
    )
    print('it returned: {!r}'.format(return_value))
finally:
    event_loop.close()
```

在这里，`run_until_complete()` 还会返回它等待的协程的结果。

```
$ python3 asyncio_coroutine_return.py

in coroutine
it returned: 'result'
```

10.5.2.3　串链协程

一个协程可以启动另一个协程并等待结果，从而可以更容易地将一个任务分解为可重用的部分。下面的例子有两个阶段，它们必须按顺序执行，不过可以与其他操作并发运行。

代码清单 10-78: **asyncio_coroutine_chain.py**

```python
import asyncio

async def outer():
    print('in outer')
    print('waiting for result1')
    result1 = await phase1()
    print('waiting for result2')
    result2 = await phase2(result1)
    return (result1, result2)

async def phase1():
    print('in phase1')
    return 'result1'

async def phase2(arg):
    print('in phase2')
    return 'result2 derived from {}'.format(arg)

event_loop = asyncio.get_event_loop()
try:
    return_value = event_loop.run_until_complete(outer())
    print('return value: {!r}'.format(return_value))
finally:
    event_loop.close()
```

这里使用了 `await` 关键字而不是向循环增加新的协程。因为控制流已经在循环管理的一个协程中，所以没有必要告诉循环管理这些新协程。

```
$ python3 asyncio_coroutine_chain.py

in outer
waiting for result1
in phase1
waiting for result2
in phase2
return value: ('result1', 'result2 derived from result1')
```

10.5.2.4 生成器而不是协程

协程函数是 `asyncio` 设计中的关键部分。它们提供了一个语言构造，可以停止程序某一部分的执行，保留这个调用的状态，并在以后重新进入这个状态。所有这些动作都是并发框架很重要的功能。

Python 3.5 引入了一些新的语言特性，可以使用 `async def` 以原生方式定义这些协程，以及使用 `await` 交出控制，`asyncio` 的例子利用了这些新特性。Python 3 的早期版本可以使用由 `asyncio.coroutine()` 修饰符包装的生成器函数和 `yield from` 来达到同样的效果。

代码清单 10-79：**asyncio_generator.py**

```python
import asyncio


@asyncio.coroutine
def outer():
    print('in outer')
    print('waiting for result1')
    result1 = yield from phase1()
    print('waiting for result2')
    result2 = yield from phase2(result1)
    return (result1, result2)


@asyncio.coroutine
def phase1():
    print('in phase1')
    return 'result1'


@asyncio.coroutine
def phase2(arg):
    print('in phase2')
    return 'result2 derived from {}'.format(arg)


event_loop = asyncio.get_event_loop()
try:
    return_value = event_loop.run_until_complete(outer())
    print('return value: {!r}'.format(return_value))
finally:
    event_loop.close()
```

前面的例子使用生成器函数而不是原生协程重新实现了 `asyncio_coroutine_chain.py`。

```
$ python3 asyncio_generator.py

in outer
```

```
waiting for result1
in phase1
waiting for result2
in phase2
return value: ('result1', 'result2 derived from result1')
```

10.5.3 调度常规函数调用

除了管理协程和 I/O 回调,asyncio 事件循环还可以根据循环中保存的一个定时器值来调度常规函数调用。

10.5.3.1 "迅速"调度一个回调

如果回调的时间不重要,那么可以用 `call_soon()` 调度下一次循环迭代的调用。调用回调时,函数后面额外的位置参数会传入回调。要向回调传入关键字参数,可以使用 functools 模块的 `partial()`。

代码清单 10-80:asyncio_call_soon.py

```python
import asyncio
import functools


def callback(arg, *, kwarg='default'):
    print('callback invoked with {} and {}'.format(arg, kwarg))


async def main(loop):
    print('registering callbacks')
    loop.call_soon(callback, 1)
    wrapped = functools.partial(callback, kwarg='not default')
    loop.call_soon(wrapped, 2)

    await asyncio.sleep(0.1)


event_loop = asyncio.get_event_loop()
try:
    print('entering event loop')
    event_loop.run_until_complete(main(event_loop))
finally:
    print('closing event loop')
    event_loop.close()
```

回调会按其调度的顺序来调用。

```
$ python3 asyncio_call_soon.py

entering event loop
registering callbacks
callback invoked with 1 and default
callback invoked with 2 and not default
closing event loop
```

10.5.3.2 用 Delay 调度回调

要将一个回调推迟到将来某个时间调用,可以使用 `call_later()`。这个方法的第一个参数是延迟时间(单位为秒),第二个参数是回调。

代码清单 10-81：**asyncio_call_later.py**

```python
import asyncio

def callback(n):
    print('callback {} invoked'.format(n))

async def main(loop):
    print('registering callbacks')
    loop.call_later(0.2, callback, 1)
    loop.call_later(0.1, callback, 2)
    loop.call_soon(callback, 3)

    await asyncio.sleep(0.4)

event_loop = asyncio.get_event_loop()
try:
    print('entering event loop')
    event_loop.run_until_complete(main(event_loop))
finally:
    print('closing event loop')
    event_loop.close()
```

在这个例子中，同一个回调函数调度了多次，每次提供了不同的参数。最后一个调用使用了 `call_soon()`，这会在所有按时间调用的实例之前基于参数 3 来调用这个回调，由此可以看出"迅速"调用的延迟往往最小。

```
$ python3 asyncio_call_later.py

entering event loop
registering callbacks
callback 3 invoked
callback 2 invoked
callback 1 invoked
closing event loop
```

10.5.3.3　在指定时间内调度一个回调

还可以安排在指定时间内调度一个调用。实现这个目的的循环依赖于一个单调时钟，而不是墙上时钟时间，以确保"now"时间绝对不会逆转。要为一个调度回调选择时间，必须使用循环的 `time()` 方法从这个时钟的内部状态开始。

代码清单 10-82：**asyncio_call_at.py**

```python
import asyncio
import time

def callback(n, loop):
    print('callback {} invoked at {}'.format(n, loop.time()))

async def main(loop):
    now = loop.time()
    print('clock time: {}'.format(time.time()))
    print('loop  time: {}'.format(now))
```

```
    print('registering callbacks')
    loop.call_at(now + 0.2, callback, 1, loop)
    loop.call_at(now + 0.1, callback, 2, loop)
    loop.call_soon(callback, 3, loop)

    await asyncio.sleep(1)
event_loop = asyncio.get_event_loop()
try:
    print('entering event loop')
    event_loop.run_until_complete(main(event_loop))
finally:
    print('closing event loop')
    event_loop.close()
```

需要注意,循环的时间与 `time.time()` 返回的值并不一致。

```
$ python3 asyncio_call_at.py

entering event loop
clock time: 1479050248.66192
loop  time: 1008846.13856885
registering callbacks
callback 3 invoked at 1008846.13867956
callback 2 invoked at 1008846.239931555
callback 1 invoked at 1008846.343480996
closing event loop
```

10.5.4 异步地生成结果

Future 表示还未完成的工作的结果。事件循环可以通过监视一个 Future 对象的状态来指示它已经完成,从而允许应用的一部分等待另一部分完成一些工作。

10.5.4.1 等待 future

Future 的做法类似于协程,所以等待协程所用的技术同样可以用于等待 future 被标记为完成。下面的例子将 future 传递到事件循环的 `run_until_complete()` 方法。

代码清单 10-83:`asyncio_future_event_loop.py`

```
import asyncio


def mark_done(future, result):
    print('setting future result to {!r}'.format(result))
    future.set_result(result)


event_loop = asyncio.get_event_loop()
try:
    all_done = asyncio.Future()
    print('scheduling mark_done')
    event_loop.call_soon(mark_done, all_done, 'the result')

    print('entering event loop')
    result = event_loop.run_until_complete(all_done)
    print('returned result: {!r}'.format(result))
finally:
    print('closing event loop')
    event_loop.close()

print('future result: {!r}'.format(all_done.result()))
```

调用 set_result() 时，Future 的状态改为完成，Future 实例会保留提供给方法的结果，以备以后获取。

```
$ python3 asyncio_future_event_loop.py

scheduling mark_done
entering event loop
setting future result to 'the result'
returned result: 'the result'
closing event loop
future result: 'the result'
```

Future 还可以结合 await 关键字使用，如下例所示。

代码清单 10-84：asyncio_future_await.py

```python
import asyncio


def mark_done(future, result):
    print('setting future result to {!r}'.format(result))
    future.set_result(result)


async def main(loop):
    all_done = asyncio.Future()

    print('scheduling mark_done')
    loop.call_soon(mark_done, all_done, 'the result')

    result = await all_done
    print('returned result: {!r}'.format(result))


event_loop = asyncio.get_event_loop()
try:
    event_loop.run_until_complete(main(event_loop))
finally:
    event_loop.close()
```

Future 的结果由 await 返回，所以经常会让同样的代码处理一个常规的协程和一个 Future 实例。

```
$ python3 asyncio_future_await.py

scheduling mark_done
setting future result to 'the result'
returned result: 'the result'
```

10.5.4.2 Future 回调

除了做法与协程类似，Future 完成时也可以调用回调。回调会按其注册的顺序调用。

代码清单 10-85：asyncio_future_callback.py

```python
import asyncio
import functools
```

```
    def callback(future, n):
        print('{}: future done: {}'.format(n, future.result()))

    async def register_callbacks(all_done):
        print('registering callbacks on future')
        all_done.add_done_callback(functools.partial(callback, n=1))
        all_done.add_done_callback(functools.partial(callback, n=2))

    async def main(all_done):
        await register_callbacks(all_done)
        print('setting result of future')
        all_done.set_result('the result')

    event_loop = asyncio.get_event_loop()
    try:
        all_done = asyncio.Future()
        event_loop.run_until_complete(main(all_done))
    finally:
        event_loop.close()
```

这个回调只希望得到一个参数，即一个 Future 实例。要想为回调传递额外的参数，可以使用 functools.partial() 创建一个包装器。

```
$ python3 asyncio_future_callback.py

registering callbacks on future
setting result of future
1: future done: the result
2: future done: the result
```

10.5.5 并发地执行任务

任务是与事件循环交互的主要途径之一。任务可以包装协程，并跟踪协程何时完成。由于任务是 Future 的子类，所以其他协程可以等待任务，而且每个任务可以有一个结果，在它完成之后可以获取这个结果。

10.5.5.1 启动一个任务

要启动一个任务，可以使用 create_task() 创建一个 Task 实例。只要循环还在运行而且协程没有返回，create_task() 得到的任务便会作为事件循环管理的并发操作的一部分运行。

代码清单 10-86：asyncio_create_task.py

```
import asyncio


async def task_func():
    print('in task_func')
    return 'the result'


async def main(loop):
    print('creating task')
```

```
        task = loop.create_task(task_func())
        print('waiting for {!r}'.format(task))
        return_value = await task
        print('task completed {!r}'.format(task))
        print('return value: {!r}'.format(return_value))


event_loop = asyncio.get_event_loop()
try:
    event_loop.run_until_complete(main(event_loop))
finally:
    event_loop.close()
```

这个例子中,在 `main()` 函数退出之前,会等待任务返回一个结果。

```
$ python3 asyncio_create_task.py

creating task
waiting for <Task pending coro=<task_func() running at
asyncio_create_task.py:12>>
in task_func
task completed <Task finished coro=<task_func() done, defined at
asyncio_create_task.py:12> result='the result'>
return value: 'the result'
```

10.5.5.2 取消一个任务

通过保留 `create_task()` 返回的 `Task` 对象,可以在任务完成之前取消它的操作。

代码清单 10-87:**asyncio_cancel_task.py**

```
import asyncio


async def task_func():
    print('in task_func')
    return 'the result'


async def main(loop):
    print('creating task')
    task = loop.create_task(task_func())

    print('canceling task')
    task.cancel()

    print('canceled task {!r}'.format(task))
    try:
        await task
    except asyncio.CancelledError:
        print('caught error from canceled task')
    else:
        print('task result: {!r}'.format(task.result()))


event_loop = asyncio.get_event_loop()
try:
    event_loop.run_until_complete(main(event_loop))
finally:
    event_loop.close()
```

这个例子会在启动事件循环之前创建一个任务，然后取消这个任务。结果是 run_until_complete() 方法抛出一个 CancelledError 异常。

```
$ python3 asyncio_cancel_task.py

creating task
canceling task
canceled task <Task cancelling coro=<task_func() running at
asyncio_cancel_task.py:12>>
caught error from canceled task
```

如果一个任务正在等待另一个并发运行的操作完成，那么倘若在这个等待时刻取消任务，则其会通过此时产生的一个 CancelledError 异常通知任务将其取消。

代码清单 10-88：asyncio_cancel_task2.py

```python
import asyncio

async def task_func():
    print('in task_func, sleeping')
    try:
        await asyncio.sleep(1)
    except asyncio.CancelledError:
        print('task_func was canceled')
        raise
    return 'the result'

def task_canceller(t):
    print('in task_canceller')
    t.cancel()
    print('canceled the task')

async def main(loop):
    print('creating task')
    task = loop.create_task(task_func())
    loop.call_soon(task_canceller, task)
    try:
        await task
    except asyncio.CancelledError:
        print('main() also sees task as canceled')

event_loop = asyncio.get_event_loop()
try:
    event_loop.run_until_complete(main(event_loop))
finally:
    event_loop.close()
```

捕获异常会提供一个机会，如果必要，可以利用这个机会清理已经完成的工作。

```
$ python3 asyncio_cancel_task2.py

creating task
in task_func, sleeping
in task_canceller
canceled the task
task_func was canceled
main() also sees task as canceled
```

10.5.5.3 从协程创建任务

`ensure_future()`函数返回一个与协程执行绑定的`Task`。这个`Task`实例再传递到其他代码，这个代码可以等待这个实例，而无须知道原来的协程是如何构造或调用的。

代码清单 10-89：`asyncio_ensure_future.py`

```python
import asyncio

async def wrapped():
    print('wrapped')
    return 'result'

async def inner(task):
    print('inner: starting')
    print('inner: waiting for {!r}'.format(task))
    result = await task
    print('inner: task returned {!r}'.format(result))

async def starter():
    print('starter: creating task')
    task = asyncio.ensure_future(wrapped())
    print('starter: waiting for inner')
    await inner(task)
    print('starter: inner returned')

event_loop = asyncio.get_event_loop()
try:
    print('entering event loop')
    result = event_loop.run_until_complete(starter())
finally:
    event_loop.close()
```

需要说明，对于提供给`ensure_future()`的协程，在使用`await`之前这个协程不会启动，只有`await`才会让它执行。

```
$ python3 asyncio_ensure_future.py

entering event loop
starter: creating task
starter: waiting for inner
inner: starting
inner: waiting for <Task pending coro=<wrapped() running at
asyncio_ensure_future.py:12>>
wrapped
inner: task returned 'result'
starter: inner returned
```

10.5.6 组合协程和控制结构

一系列协程之间的线性控制流用内置关键字`await`可以很容易地管理。更复杂的结构可能允许一个协程等待多个其他协程并行完成，可以使用`asyncio`中的工具创建这些更复杂的结构。

10.5.6.1 等待多个协程

通常可以把一个操作划分为多个部分，然后分别执行，这会很有用。例如，采用这种方法，可以高效地下载多个远程资源或者查询远程 API。有些情况下，执行顺序并不重要，而且可能有任意多个操作，在这种情况下，可以使用 wait() 暂停一个协程，直到其他后台操作完成。

代码清单 10-90：asyncio_wait.py

```
import asyncio

async def phase(i):
    print('in phase {}'.format(i))
    await asyncio.sleep(0.1 * i)
    print('done with phase {}'.format(i))
    return 'phase {} result'.format(i)

async def main(num_phases):
    print('starting main')
    phases = [
        phase(i)
        for i in range(num_phases)
    ]
    print('waiting for phases to complete')
    completed, pending = await asyncio.wait(phases)
    results = [t.result() for t in completed]
    print('results: {!r}'.format(results))


event_loop = asyncio.get_event_loop()
try:
    event_loop.run_until_complete(main(3))
finally:
    event_loop.close()
```

在内部，wait() 使用一个 set 来保存它创建的 Task 实例，这说明这些实例会按一种不可预知的顺序启动和完成。wait() 的返回值是一个元组，包括两个集合，分别包括已完成和未完成的任务。

```
$ python3 asyncio_wait.py

starting main
waiting for phases to complete
in phase 0
in phase 1
in phase 2
done with phase 0
done with phase 1
done with phase 2
results: ['phase 1 result', 'phase 0 result', 'phase 2 result']
```

如果使用 wait() 时提供了一个超时值，那么达到这个超时时间后，将只保留未完成的操作。

代码清单 10-91：asyncio_wait_timeout.py

```
import asyncio

async def phase(i):
    print('in phase {}'.format(i))
    try:
        await asyncio.sleep(0.1 * i)
    except asyncio.CancelledError:
        print('phase {} canceled'.format(i))
        raise
```

```python
    else:
        print('done with phase {}'.format(i))
        return 'phase {} result'.format(i)


async def main(num_phases):
    print('starting main')
    phases = [
        phase(i)
        for i in range(num_phases)
    ]
    print('waiting 0.1 for phases to complete')
    completed, pending = await asyncio.wait(phases, timeout=0.1)
    print('{} completed and {} pending'.format(
        len(completed), len(pending),
    ))
    # Cancel remaining tasks so they do not generate errors
    # as we exit without finishing them.
    if pending:
        print('canceling tasks')
        for t in pending:
            t.cancel()
    print('exiting main')


event_loop = asyncio.get_event_loop()
try:
    event_loop.run_until_complete(main(3))
finally:
    event_loop.close()
```

其余的后台操作要显式地处理,这有多方面的原因。尽管 `wait()` 返回时未完成的任务是挂起的,但只要控制返回到事件循环它们就会恢复运行。如果没有另一个 `wait()` 调用,则将没有对象接收任务的输出;也就是说,任务会运行并消费资源,但不会带来任何好处。另外,如果程序退出时还有未完成的任务,那么 asyncio 会发出一个警告。这些警告可能打印到控制台上,应用的用户便会看到。因此,最好取消所有剩余的后台操作,或者使用 `wait()` 让它们结束运行。

```
$ python3 asyncio_wait_timeout.py

starting main
waiting 0.1 for phases to complete
in phase 1
in phase 0
in phase 2
done with phase 0
1 completed and 2 pending
cancelling tasks
exiting main
phase 1 cancelled
phase 2 cancelled
```

10.5.6.2 从协程收集结果

如果后台阶段是明确的,而且这些阶段的结果很重要,那么 `gather()` 可能对等待多个操作很有用。

代码清单 10-92：`asyncio_gather.py`

```python
import asyncio


async def phase1():
    print('in phase1')
    await asyncio.sleep(2)
    print('done with phase1')
    return 'phase1 result'


async def phase2():
    print('in phase2')
    await asyncio.sleep(1)
    print('done with phase2')
    return 'phase2 result'


async def main():
    print('starting main')
    print('waiting for phases to complete')
    results = await asyncio.gather(
        phase1(),
        phase2(),
    )
    print('results: {!r}'.format(results))


event_loop = asyncio.get_event_loop()
try:
    event_loop.run_until_complete(main())
finally:
    event_loop.close()
```

`gather()` 创建的任务不会对外提供，所以无法将其取消。返回值是一个结果列表，结果的顺序与传入 `gather()` 的参数顺序相同，而不论后台操作实际上是按什么顺序完成的。

```
$ python3 asyncio_gather.py

starting main
waiting for phases to complete
in phase2
in phase1
done with phase2
done with phase1
results: ['phase1 result', 'phase2 result']
```

10.5.6.3 后台操作完成时进行处理

`as_completed()` 是一个生成器，会管理指定的一个协程列表，并生成它们的结果，每个协程结束运行时一次生成一个结果。与 `wait()` 类似，`as_completed()` 不能保证顺序，不过执行其他动作之前没有必要等待所有后台操作完成。

代码清单 10-93：`asyncio_as_completed.py`

```python
import asyncio


async def phase(i):
```

```
        print('in phase {}'.format(i))
        await asyncio.sleep(0.5 - (0.1 * i))
        print('done with phase {}'.format(i))
        return 'phase {} result'.format(i)


async def main(num_phases):
    print('starting main')
    phases = [
        phase(i)
        for i in range(num_phases)
    ]
    print('waiting for phases to complete')
    results = []
    for next_to_complete in asyncio.as_completed(phases):
        answer = await next_to_complete
        print('received answer {!r}'.format(answer))
        results.append(answer)
    print('results: {!r}'.format(results))
    return results


event_loop = asyncio.get_event_loop()
try:
    event_loop.run_until_complete(main(3))
finally:
    event_loop.close()
```

这个例子启动了多个后台阶段,它们会按其启动顺序的逆序完成。消费生成器时,循环会使用 await 等待协程的结果。

```
$ python3 asyncio_as_completed.py

starting main
waiting for phases to complete
in phase 0
in phase 2
in phase 1
done with phase 2
received answer 'phase 2 result'
done with phase 1
received answer 'phase 1 result'
done with phase 0
received answer 'phase 0 result'
results: ['phase 2 result', 'phase 1 result', 'phase 0 result']
```

10.5.7 同步原语

尽管 asyncio 应用通常作为单线程的进程运行,不过仍被构建为并发应用。由于 I/O 以及其他外部事件的延迟和中断,每个协程或任务可能按一种不可预知的顺序执行。为了支持安全的并发执行,asyncio 包含了 threading 和 multiprocessing 模块中一些底层原语的实现。

10.5.7.1 锁

Lock 可以用来保护对一个共享资源的访问。只有锁的持有者可以使用这个资源。如果有多个请求要得到这个锁,那么其将会阻塞,以保证一次只有一个持有者。

代码清单 10-94：asyncio_lock.py

```
import asyncio
import functools


def unlock(lock):
    print('callback releasing lock')
    lock.release()


async def coro1(lock):
    print('coro1 waiting for the lock')
    with await lock:
        print('coro1 acquired lock')
    print('coro1 released lock')


async def coro2(lock):
    print('coro2 waiting for the lock')
    await lock
    try:
        print('coro2 acquired lock')
    finally:
        print('coro2 released lock')
        lock.release()


async def main(loop):
    # Create and acquire a shared lock.
    lock = asyncio.Lock()
    print('acquiring the lock before starting coroutines')
    await lock.acquire()
    print('lock acquired: {}'.format(lock.locked()))

    # Schedule a callback to unlock the lock.
    loop.call_later(0.1, functools.partial(unlock, lock))

    # Run the coroutines that want to use the lock.
    print('waiting for coroutines')
    await asyncio.wait([coro1(lock), coro2(lock)]),


event_loop = asyncio.get_event_loop()
try:
    event_loop.run_until_complete(main(event_loop))
finally:
    event_loop.close()
```

锁可以直接调用，使用 `await` 来得到，并且使用结束时可以调用 `release()` 方法释放锁，如这个例子中的 `coro2()` 所示。还可以结合 `with await` 关键字使用锁作为异步上下文管理器，如 `coro1()` 中所示。

```
$ python3 asyncio_lock.py

acquiring the lock before starting coroutines
lock acquired: True
waiting for coroutines
coro1 waiting for the lock
coro2 waiting for the lock
```

```
callback releasing lock
coro1 acquired lock
coro1 released lock
coro2 acquired lock
coro2 released lock
```

10.5.7.2 事件

asyncio.Event 基于 **threading.Event**。它允许多个消费者等待某个事件发生，而不必寻找一个特定值与通知关联。

代码清单 10-95：*asyncio_event.py*

```
import asyncio
import functools


def set_event(event):
    print('setting event in callback')
    event.set()


async def coro1(event):
    print('coro1 waiting for event')
    await event.wait()
    print('coro1 triggered')


async def coro2(event):
    print('coro2 waiting for event')
    await event.wait()
    print('coro2 triggered')


async def main(loop):
    # Create a shared event.
    event = asyncio.Event()
    print('event start state: {}'.format(event.is_set()))

    loop.call_later(
        0.1, functools.partial(set_event, event)
    )

    await asyncio.wait([coro1(event), coro2(event)])
    print('event end state: {}'.format(event.is_set()))


event_loop = asyncio.get_event_loop()
try:
    event_loop.run_until_complete(main(event_loop))
finally:
    event_loop.close()
```

与 **Lock** 一样，**coro1()** 和 **coro2()** 会等待设置事件。区别是一旦事件状态改变，它们便可以立即启动，并且它们不需要得到事件对象上的唯一的锁。

```
$ python3 asyncio_event.py

event start state: False
coro2 waiting for event
```

```
coro1 waiting for event
setting event in callback
coro2 triggered
coro1 triggered
event end state: True
```

10.5.7.3 条件

Condition 的做法与 Event 类似，只不过不是通知所有等待的协程，被唤醒的等待协程的数目由 notify() 的一个参数控制。

代码清单 10-96：asyncio_condition.py

```python
import asyncio

async def consumer(condition, n):
    with await condition:
        print('consumer {} is waiting'.format(n))
        await condition.wait()
        print('consumer {} triggered'.format(n))
    print('ending consumer {}'.format(n))

async def manipulate_condition(condition):
    print('starting manipulate_condition')

    # Pause to let consumers start
    await asyncio.sleep(0.1)

    for i in range(1, 3):
        with await condition:
            print('notifying {} consumers'.format(i))
            condition.notify(n=i)
        await asyncio.sleep(0.1)

    with await condition:
        print('notifying remaining consumers')
        condition.notify_all()

    print('ending manipulate_condition')

async def main(loop):
    # Create a condition.
    condition = asyncio.Condition()

    # Set up tasks watching the condition.
    consumers = [
        consumer(condition, i)
        for i in range(5)
    ]

    # Schedule a task to manipulate the condition variable.
    loop.create_task(manipulate_condition(condition))

    # Wait for the consumers to be done.
    await asyncio.wait(consumers)


event_loop = asyncio.get_event_loop()
try:
    result = event_loop.run_until_complete(main(event_loop))
finally:
    event_loop.close()
```

这个例子启动 Condition 的 5 个消费者。它们分别使用 wait() 方法来等待通知让它继续。manipulate_condition() 通知一个消费者，再通知两个消费者，然后通知所有其余的消费者。

```
$ python3 asyncio_condition.py

starting manipulate_condition
consumer 3 is waiting
consumer 1 is waiting
consumer 2 is waiting
consumer 0 is waiting
consumer 4 is waiting
notifying 1 consumers
consumer 3 triggered
ending consumer 3
notifying 2 consumers
consumer 1 triggered
ending consumer 1
consumer 2 triggered
ending consumer 2
notifying remaining consumers
ending manipulate_condition
consumer 0 triggered
ending consumer 0
consumer 4 triggered
ending consumer 4
```

10.5.7.4 队列

asyncio.Queue 为协程提供了一个先进先出的数据结构，这与线程的 queue.Queue 或进程的 multiprocessing.Queue 很类似。

代码清单 10-97：asyncio_queue.py

```python
import asyncio


async def consumer(n, q):
    print('consumer {}: starting'.format(n))
    while True:
        print('consumer {}: waiting for item'.format(n))
        item = await q.get()
        print('consumer {}: has item {}'.format(n, item))
        if item is None:
            # None is the signal to stop.
            q.task_done()
            break
        else:
            await asyncio.sleep(0.01 * item)
            q.task_done()
    print('consumer {}: ending'.format(n))


async def producer(q, num_workers):
    print('producer: starting')
    # Add some numbers to the queue to simulate jobs.
    for i in range(num_workers * 3):
        await q.put(i)
        print('producer: added task {} to the queue'.format(i))
    # Add None entries in the queue
```

```python
        # to signal the consumers to exit.
        print('producer: adding stop signals to the queue')
        for i in range(num_workers):
            await q.put(None)
        print('producer: waiting for queue to empty')
        await q.join()
        print('producer: ending')


    async def main(loop, num_consumers):
        # Create the queue with a fixed size so the producer
        # will block until the consumers pull some items out.
        q = asyncio.Queue(maxsize=num_consumers)

        # Schedule the consumer tasks.
        consumers = [
            loop.create_task(consumer(i, q))
            for i in range(num_consumers)
        ]

        # Schedule the producer task.
        prod = loop.create_task(producer(q, num_consumers))

        # Wait for all of the coroutines to finish.
        await asyncio.wait(consumers + [prod])


    event_loop = asyncio.get_event_loop()
    try:
        event_loop.run_until_complete(main(event_loop, 2))
    finally:
        event_loop.close()
```

用 put() 增加元素和用 get() 删除元素都是异步操作，因为队列大小可能是固定的（阻塞增加操作），或者队列可能为空（阻塞获取元素的调用）。

```
$ python3 asyncio_queue.py

consumer 0: starting
consumer 0: waiting for item
consumer 1: starting
consumer 1: waiting for item
producer: starting
producer: added task 0 to the queue
producer: added task 1 to the queue
consumer 0: has item 0
consumer 1: has item 1
producer: added task 2 to the queue
producer: added task 3 to the queue
consumer 0: waiting for item
consumer 0: has item 2
producer: added task 4 to the queue
consumer 1: waiting for item
consumer 1: has item 3
producer: added task 5 to the queue
producer: adding stop signals to the queue
consumer 0: waiting for item
consumer 0: has item 4
consumer 1: waiting for item
consumer 1: has item 5
producer: waiting for queue to empty
```

```
consumer 0: waiting for item
consumer 0: has item None
consumer 0: ending
consumer 1: waiting for item
consumer 1: has item None
consumer 1: ending
producer: ending
```

10.5.8 提供协议类抽象的异步 I/O

到目前为止，我们给出的所有例子都避免把并发和 I/O 操作混杂在一起，以确保一次只强调一个概念。不过，I/O 阻塞时切换上下文是 `asyncio` 的主要用例之一。在前面介绍的并发概念基础上，这一节将分析两个示例程序，它们实现了一个简单的回送（echo）服务器和客户端，这类似于 `socket` 和 `socketserver` 小节中使用的例子。客户端连接到服务器，发送一些数据，然后接收同样的数据作为响应。每次启动一个 I/O 操作时，执行代码都会把控制交给事件循环，允许其他任务运行，直到 I/O 完成。

10.5.8.1 回送服务器

服务器首先导入建立 `asyncio` 和 `logging` 所需的模块，然后创建一个事件循环对象。

代码清单 10-98：**asyncio_echo_server_protocol.py**

```
import asyncio
import logging
import sys

SERVER_ADDRESS = ('localhost', 10000)

logging.basicConfig(
    level=logging.DEBUG,
    format='%(name)s: %(message)s',
    stream=sys.stderr,
)
log = logging.getLogger('main')

event_loop = asyncio.get_event_loop()
```

然后服务器定义 `asyncio.Protocol` 的一个子类来处理客户端通信。可以基于与服务器套接字关联的事件来调用这个协议对象的方法。

```
class EchoServer(asyncio.Protocol):
```

每个新客户连接会触发一个 `connection_made()` 调用。`transport` 参数是 `asyncio.Transport` 的一个实例，它提供了使用套接字完成异步 I/O 的一个抽象。不同类型的通信提供了不同的传输实现，不过都有相同的 API。例如，会使用不同的传输类来处理套接字和子进程管道。可以通过 `get_extra_info()` 从 transport 得到接入客户端的地址（`get_extra_info()` 是一个特定于具体实现的方法）。

```
    def connection_made(self, transport):
        self.transport = transport
        self.address = transport.get_extra_info('peername')
        self.log = logging.getLogger(
            'EchoServer_{}_{}'.format(*self.address)
        )
        self.log.debug('connection accepted')
```

建立一个连接后，从客户端向服务器发送数据时，会调用协议的 `data_received()` 方法传入数据进行处理。数据作为一个字节串传递，要由应用采用一种适当的方式解码。在下面的代码中，结果会被记入日志，然后通过调用 `transport.write()` 立即向客户端发回一个响应。

```
def data_received(self, data):
    self.log.debug('received {!r}'.format(data))
    self.transport.write(data)
    self.log.debug('sent {!r}'.format(data))
```

有些传输类支持一个特殊的文件末尾指示符（"EOF"）。遇到一个 EOF 时会调用 `eof_received()` 方法。在这个实现中，EOF 会发回给客户端，指示已经接收到这个指示符。由于并不是所有传输类都支持一个显式的 EOF，所以这个协议会首先询问传输类发送 EOF 是否安全。

```
def eof_received(self):
    self.log.debug('received EOF')
    if self.transport.can_write_eof():
        self.transport.write_eof()
```

关闭一个连接时（可能是正常关闭，也可能是由于一个错误而关闭），会调用协议的 `connection_lost()` 方法。如果出现一个错误，那么参数会包含一个适当的异常对象；否则参数为 `None`。

```
def connection_lost(self, error):
    if error:
        self.log.error('ERROR: {}'.format(error))
    else:
        self.log.debug('closing')
    super().connection_lost(error)
```

启动服务器有两个步骤。首先，应用告诉事件循环使用协议类、主机名以及监听的套接字来创建一个新的服务器对象。`create_server()` 方法是一个协程，所以必须由事件循环处理结果来具体启动服务器。完成这个协程会生成一个绑定到事件循环的 `asyncio.Server` 实例。

```
# Create the server and let the loop finish the coroutine before
# starting the real event loop.
factory = event_loop.create_server(EchoServer, *SERVER_ADDRESS)
server = event_loop.run_until_complete(factory)
log.debug('starting up on {} port {}'.format(*SERVER_ADDRESS))
```

接下来，需要运行事件循环来处理事件和客户请求。对于一个长时间运行的服务，`run_forever()` 方法是完成这个工作最简单的方法。事件循环结束时（可能由应用代码结束，也可能通过向进程发出信号来结束），可以关闭服务器来适当地清理套接字。然后关闭事件循环，在程序退出前完成所有其他协程的处理。

```
# Enter the event loop permanently to handle all connections.
try:
    event_loop.run_forever()
finally:
    log.debug('closing server')
    server.close()
    event_loop.run_until_complete(server.wait_closed())
    log.debug('closing event loop')
    event_loop.close()
```

10.5.8.2 回送客户端

使用一个协议类来构造客户端与构造服务器非常类似。同样地,代码首先导入建立 `asyncio` 和 `logging` 所需的模块,然后创建一个事件循环对象。

代码清单 10-99: `asyncio_echo_client_protocol.py`

```
import asyncio
import functools
import logging
import sys

MESSAGES = [
    b'This is the message. ',
    b'It will be sent ',
    b'in parts.',
]
SERVER_ADDRESS = ('localhost', 10000)

logging.basicConfig(
    level=logging.DEBUG,
    format='%(name)s: %(message)s',
    stream=sys.stderr,
)
log = logging.getLogger('main')

event_loop = asyncio.get_event_loop()
```

客户端协议类定义了与服务器同样的方法,不过有不同的实现。这个类构造函数接受两个参数:一个参数是要发送的消息列表;另一个参数是一个 `Future` 实例,用来发出信号指示客户端通过接收服务器的一个响应从而完成了一个工作周期。

```
class EchoClient(asyncio.Protocol):

    def __init__(self, messages, future):
        super().__init__()
        self.messages = messages
        self.log = logging.getLogger('EchoClient')
        self.f = future
```

当客户端成功地连接到服务器时,它会立即开始通信。一次发送一个消息,不过底层网络代码可能会把多个消息结合在一起一次传输。整个消息序列发送完时,发送一个 EOF。

尽管看起来所有数据都是立即发送,但实际上传输类对象会缓冲发出的数据,并建立一个回调,当套接字的缓冲区准备好接收数据时才真正传送数据。这个处理会透明地完成,所以写应用代码时可以像立即完成 I/O 操作一样。

```
    def connection_made(self, transport):
        self.transport = transport
        self.address = transport.get_extra_info('peername')
        self.log.debug(
            'connecting to {} port {}'.format(*self.address)
        )
        # This could be transport.writelines() except that
        # would make it harder to show each part of the message
        # being sent.
        for msg in self.messages:
            transport.write(msg)
```

```
        self.log.debug('sending {!r}'.format(msg))
    if transport.can_write_eof():
        transport.write_eof()
```

从服务器接收到响应时，会记入日志。

```
def data_received(self, data):
    self.log.debug('received {!r}'.format(data))
```

最后，不论是接收到一个文件末尾标志，还是服务器端关闭了连接，本地传输对象都会关闭，并通过设置一个结果将 future 对象标记为完成。

```
def eof_received(self):
    self.log.debug('received EOF')
    self.transport.close()
    if not self.f.done():
        self.f.set_result(True)

def connection_lost(self, exc):
    self.log.debug('server closed connection')
    self.transport.close()
    if not self.f.done():
        self.f.set_result(True)
    super().connection_lost(exc)
```

正常情况下，会向事件循环传递协议类以创建连接。在这里，由于事件循环没有办法向协议构造函数传递额外的参数，所以必须创建一个 `partial` 来包装客户类，并传入要发送的消息列表和 Future 实例。然后在调用 `create_connection()` 建立客户连接时使用这个新的 callable 取代原来的类。

```
client_completed = asyncio.Future()

client_factory = functools.partial(
    EchoClient,
    messages=MESSAGES,
    future=client_completed,
)
factory_coroutine = event_loop.create_connection(
    client_factory,
    *SERVER_ADDRESS,
)
```

触发客户端运行时，要用创建客户端的协程调用一次事件循环，再用一个 Future 实例调用一次事件循环（这是客户端完成时提供给它进行通信的一个 Future 实例）。像这样使用两个调用可以避免在客户端程序中创建一个无限循环，其可能希望在与服务器的通信完成之后退出这个循环。如果只使用第一个调用来等待协程创建客户端，它可能不会处理所有响应数据并适当地清理与服务器的连接。

```
log.debug('waiting for client to complete')
try:
    event_loop.run_until_complete(factory_coroutine)
    event_loop.run_until_complete(client_completed)
finally:
    log.debug('closing event loop')
    event_loop.close()
```

10.5.8.3 输出

在一个窗口中运行服务器，在另一个窗口中运行客户端，这会生成以下输出。

```
$ python3 asyncio_echo_client_protocol.py
asyncio: Using selector: KqueueSelector
main: waiting for client to complete
EchoClient: connecting to ::1 port 10000
EchoClient: sending b'This is the message. '
EchoClient: sending b'It will be sent '
EchoClient: sending b'in parts.'
EchoClient: received b'This is the message. It will be sent in parts.'
EchoClient: received EOF
EchoClient: server closed connection
main: closing event loop

$ python3 asyncio_echo_client_protocol.py
asyncio: Using selector: KqueueSelector
main: waiting for client to complete
EchoClient: connecting to ::1 port 10000
EchoClient: sending b'This is the message. '
EchoClient: sending b'It will be sent '
EchoClient: sending b'in parts.'
EchoClient: received b'This is the message. It will be sent in parts.'
EchoClient: received EOF
EchoClient: server closed connection
main: closing event loop

$ python3 asyncio_echo_client_protocol.py
asyncio: Using selector: KqueueSelector
main: waiting for client to complete
EchoClient: connecting to ::1 port 10000
EchoClient: sending b'This is the message. '
EchoClient: sending b'It will be sent '
EchoClient: sending b'in parts.'
EchoClient: received b'This is the message. It will be sent in parts.'
EchoClient: received EOF
EchoClient: server closed connection
main: closing event loop
```

尽管客户端总是单独地发送消息,但客户端第一次运行时,服务器却会接收一个很大的消息,并把它回送给客户端。在后续的运行中,这些结果会有变化,这取决于网络有多忙,以及是否在所有数据准备好之前刷新输出网络缓冲区。

```
$ python3 asyncio_echo_server_protocol.py
asyncio: Using selector: KqueueSelector
main: starting up on localhost port 10000
EchoServer_::1_63347: connection accepted
EchoServer_::1_63347: received b'This is the message. It will
be sent in parts.'
EchoServer_::1_63347: sent b'This is the message. It will be
sent in parts.'
EchoServer_::1_63347: received EOF
EchoServer_::1_63347: closing

EchoServer_::1_63387: connection accepted
EchoServer_::1_63387: received b'This is the message. '
EchoServer_::1_63387: sent b'This is the message. '
EchoServer_::1_63387: received b'It will be sent in parts.'
EchoServer_::1_63387: sent b'It will be sent in parts.'
EchoServer_::1_63387: received EOF
EchoServer_::1_63387: closing

EchoServer_::1_63389: connection accepted
EchoServer_::1_63389: received b'This is the message. It will
```

```
be sent '
EchoServer_::1_63389: sent b'This is the message.  It will be sent '
EchoServer_::1_63389: received b'in parts.'
EchoServer_::1_63389: sent b'in parts.'
EchoServer_::1_63389: received EOF
EchoServer_::1_63389: closing
```

10.5.9 使用协程和流的异步 I/O

这一节介绍以上两个示例程序的另一个版本（分别实现一个简单的回送服务器和客户端），这里会使用协程和 asyncio 流 API，而不再使用协议和传输类抽象。与 Protocol API 相比，这些例子在更低的抽象层上操作，不过事件的处理是类似的。

10.5.9.1 回送服务器

服务器首先导入建立 asyncio 和 logging 所需的模块，然后创建一个事件循环对象。

代码清单 10-100：asyncio_echo_server_coroutine.py

```python
import asyncio
import logging
import sys

SERVER_ADDRESS = ('localhost', 10000)
logging.basicConfig(
    level=logging.DEBUG,
    format='%(name)s: %(message)s',
    stream=sys.stderr,
)
log = logging.getLogger('main')

event_loop = asyncio.get_event_loop()
```

然后服务器定义一个协程来处理通信。每次一个客户端连接时，都会调用这个协程的一个新实例；因此，在这个函数中，代码一次只与一个客户端通信。Python 的语言运行时会管理各个协程实例的状态，所以应用代码不需要管理任何额外的数据结构来跟踪各个客户端。

协程的参数是与这个新连接关联的 StreamReader 和 StreamWriter 实例。与 Transport 一样，可以通过书写器的 get_extra_info() 方法访问客户端的地址。

```python
async def echo(reader, writer):
    address = writer.get_extra_info('peername')
    log = logging.getLogger('echo_{}_{}'.format(*address))
    log.debug('connection accepted')
```

尽管建立连接时会调用协程，但此时还没有任何要读取的数据。为了避免读取时阻塞，协程对 read() 调用使用了 await，从而允许事件循环继续处理其他任务，直到有可以读取的数据。

```python
    while True:
        data = await reader.read(128)
```

如果客户端发送了数据，则其会从 await 返回，并且通过把这个数据传递到书写器从而发回给客户端。可以使用多个 write() 调用来缓冲发出的数据，然后使用 drain() 刷新输出结果。由于刷新输出网络 I/O 可能阻塞,所以再次使用 `await` 恢复对事件循环的控制,它会监视写套接字,并在可以发送更多数据时调用书写器。

```
if data:
    log.debug('received {!r}'.format(data))
    writer.write(data)
    await writer.drain()
    log.debug('sent {!r}'.format(data))
```

如果客户端还没有发送任何数据,那么 `read()` 返回一个空字节串来指示连接关闭。服务器需要关闭写客户端的套接字,然后协程返回,以指示它已经完成。

```
else:
    log.debug('closing')
    writer.close()
    return
```

启动服务器有两个步骤。首先,应用告诉事件循环使用协程、主机名以及监听的套接字来创建一个新的服务器对象。`start_server()` 方法本身是一个协程,所以必须由事件循环处理结果来具体启动服务器。完成协程时会生成一个绑定到事件循环的 `asyncio.Server` 实例。

```
# Create the server and let the loop finish the coroutine before
# starting the real event loop.
factory = asyncio.start_server(echo, *SERVER_ADDRESS)
server = event_loop.run_until_complete(factory)
log.debug('starting up on {} port {}'.format(*SERVER_ADDRESS))
```

接下来,需要运行事件循环来处理事件和客户请求。对于一个长时间运行的服务,`run_forever()` 方法是完成这个工作最简单的方法。事件循环结束时(可能由应用代码结束,也可能通过向进程发出信号来结束),可以关闭服务器以适当地清理套接字。然后可以关闭事件循环,在程序退出前完成所有其他协程的处理。

```
# Enter the event loop permanently to handle all connections.
try:
    event_loop.run_forever()
except KeyboardInterrupt:
    pass
finally:
    log.debug('closing server')
    server.close()
    event_loop.run_until_complete(server.wait_closed())
    log.debug('closing event loop')
    event_loop.close()
```

10.5.9.2 回送客户端

使用一个协程来构造客户端与构造服务器非常类似。代码同样首先导入建立 `asyncio` 和 `logging` 所需的模块,然后创建一个事件循环对象。

代码清单 10-101:`asyncio_echo_client_coroutine.py`

```
import asyncio
import logging
import sys
MESSAGES = [
    b'This is the message. ',
```

```
        b'It will be sent ',
        b'in parts.',
]
SERVER_ADDRESS = ('localhost', 10000)

logging.basicConfig(
    level=logging.DEBUG,
    format='%(name)s: %(message)s',
    stream=sys.stderr,
)
log = logging.getLogger('main')

event_loop = asyncio.get_event_loop()
```

echo_client 协程接收的参数会告诉它服务器在哪里以及要发送什么消息。

```
async def echo_client(address, messages):
```

任务启动时会调用这个协程,不过它没有可以使用的活动连接。因此,第一步是让客户端建立它自己的连接。这里使用 await 来避免在 open_connection() 协程运行时阻塞其他活动。

```
    log = logging.getLogger('echo_client')

    log.debug('connecting to {} port {}'.format(*address))
    reader, writer = await asyncio.open_connection(*address)
```

open_connection() 协程返回与新套接字关联的 StreamReader 和 StreamWriter 实例。下一步使用书写器向服务器发送数据。与服务器中一样,书写器会缓冲发出的数据,直到套接字准备好或者使用 drain() 刷新输出结果。由于缓冲输出网络 I/O 可能阻塞,所以再一次使用 await 来恢复对事件循环的控制,它会监视写套接字,并在可以发送更多数据时调用书写器。

```
    # This could be writer.writelines() except that
    # would make it harder to show each part of the message
    # being sent.
    for msg in messages:
        writer.write(msg)
        log.debug('sending {!r}'.format(msg))
    if writer.can_write_eof():
        writer.write_eof()
    await writer.drain()
```

接下来,客户端寻找服务器的响应,它会尝试读取数据,直到再没有可以读取的数据。为了避免在 read() 调用中阻塞,await 会把控制交还给事件循环。如果服务器已经发送数据,则记入日志。如果服务器还没有发送数据,那么 read() 返回一个空字节串以指示连接关闭。客户端需要首先关闭向服务器发送数据的套接字,然后返回,以指示它已经完成。

```
    log.debug('waiting for response')
    while True:
        data = await reader.read(128)
        if data:
            log.debug('received {!r}'.format(data))
        else:
            log.debug('closing')
            writer.close()
            return
```

启动客户端时，要用创建客户端的协程来调用事件循环。为此要使用 `run_until_complete()` 避免在客户端程序中创建无限循环。与协议例子中不同，这里不需要单独的 future 来指示协程何时完成，因为 `echo_client()` 包含所有客户端逻辑，除非接收到一个响应并关闭服务器连接，在此之前它不会返回。

```
try:
    event_loop.run_until_complete(
        echo_client(SERVER_ADDRESS, MESSAGES)
    )
finally:
    log.debug('closing event loop')
    event_loop.close()
```

10.5.9.3 输出

在一个窗口中运行服务器，并在另一个窗口中运行客户端，会生成以下输出。

```
$ python3 asyncio_echo_client_coroutine.py
asyncio: Using selector: KqueueSelector
echo_client: connecting to localhost port 10000
echo_client: sending b'This is the message. '
echo_client: sending b'It will be sent '
echo_client: sending b'in parts.'
echo_client: waiting for response
echo_client: received b'This is the message. It will be sent in parts.'
echo_client: closing
main: closing event loop

$ python3 asyncio_echo_client_coroutine.py
asyncio: Using selector: KqueueSelector
echo_client: connecting to localhost port 10000
echo_client: sending b'This is the message. '
echo_client: sending b'It will be sent '
echo_client: sending b'in parts.'
echo_client: waiting for response
echo_client: received b'This is the message. It will be sent in parts.'
echo_client: closing
main: closing event loop

$ python3 asyncio_echo_client_coroutine.py
asyncio: Using selector: KqueueSelector
echo_client: connecting to localhost port 10000
echo_client: sending b'This is the message. '
echo_client: sending b'It will be sent '
echo_client: sending b'in parts.'
echo_client: waiting for response
echo_client: received b'This is the message. It will be sent '
echo_client: received b'in parts.'
echo_client: closing
main: closing event loop
```

尽管客户端总是单独地发送消息，不过前两次客户端运行时，服务器会接收一个很大的消息，并回送给客户端。在后续的运行中，这些结果会有变化，这取决于网络有多忙，以及是否在所有数据准备好之前刷新输出网络缓冲区。

```
$ python3 asyncio_echo_server_coroutine.py
asyncio: Using selector: KqueueSelector
main: starting up on localhost port 10000
echo_::1_64624: connection accepted
echo_::1_64624: received b'This is the message. It will be sent
```

```
    in parts.'
    echo_::1_64624: sent b'This is the message. It will be sent in parts.'
    echo_::1_64624: closing

    echo_::1_64626: connection accepted
    echo_::1_64626: received b'This is the message. It will be sent
    in parts.'
    echo_::1_64626: sent b'This is the message. It will be sent in parts.'
    echo_::1_64626: closing

    echo_::1_64627: connection accepted
    echo_::1_64627: received b'This is the message. It will be sent '
    echo_::1_64627: sent b'This is the message. It will be sent '
    echo_::1_64627: received b'in parts.'
    echo_::1_64627: sent b'in parts.'
    echo_::1_64627: closing
```

10.5.10 使用 SSL

asyncio 提供了对 SSL 的内置支持，可以在套接字上启用 SSL 连接。向创建服务器或客户端连接的协程传递一个 SSLContext 实例就会启用这个内置支持，提供应用可以使用的套接字之前一定要先完成 SSL 协议设置。

可以进一步更新上一节中基于协程的回送服务器和客户端，再做几个小小的修改。第一步是创建证书和密钥文件。可以用如下命令创建一个自签名证书：

```
$ openssl req -newkey rsa:2048 -nodes -keyout pymotw.key \
-x509 -days 365 -out pymotw.crt
```

openssl 命令会提示输入多个值用来生成证书，然后生成所请求的输出文件。

在上一节服务器示例中建立不安全的套接字时使用了 start_server() 创建监听套接字。

```
factory = asyncio.start_server(echo, *SERVER_ADDRESS)
server = event_loop.run_until_complete(factory)
```

为了增加加密，要用刚才生成的证书和密钥创建一个 SSLContext，然后将这个上下文传递到 start_server()。

```
# The certificate is created with pymotw.com as the hostname.
# This name will not match when the example code runs elsewhere,
# so disable hostname verification.
ssl_context = ssl.create_default_context(ssl.Purpose.CLIENT_AUTH)
ssl_context.check_hostname = False
ssl_context.load_cert_chain('pymotw.crt', 'pymotw.key')

# Create the server and let the loop finish the coroutine before
# starting the real event loop.
factory = asyncio.start_server(echo, *SERVER_ADDRESS,
                               ssl=ssl_context)
```

客户端中需要做类似的修改。老版本中使用 open_connection() 来创建与服务器连接的套接字。

```
reader, writer = await asyncio.open_connection(*address)
```

同样需要一个 SSLContext 来保护套接字客户端的安全。这里对客户身份没有强制要求，所以只需要加载证书。

```python
# The certificate is created with pymotw.com as the hostname.
# This name will not match when the example code runs
# elsewhere, so disable hostname verification.
ssl_context = ssl.create_default_context(
    ssl.Purpose.SERVER_AUTH,
)
ssl_context.check_hostname = False
ssl_context.load_verify_locations('pymotw.crt')
reader, writer = await asyncio.open_connection(
    *server_address, ssl=ssl_context)
```

客户端中还需要做另一个很小的修改。由于 SSL 连接不支持发送文件末尾（EOF）通知，所以客户端使用一个 NULL 字节作为消息终止符。前面老版本的客户端使用 write_eof() 发送这个通知。

```python
# This could be writer.writelines() except that
# would make it harder to show each part of the message
# being sent.
for msg in messages:
    writer.write(msg)
    log.debug('sending {!r}'.format(msg))
if writer.can_write_eof():
    writer.write_eof()
await writer.drain()
```

新版本则发送一个 0 字节（b'\x00'）来指示消息结束。

```python
# This could be writer.writelines() except that
# would make it harder to show each part of the message
# being sent.
for msg in messages:
    writer.write(msg)
    log.debug('sending {!r}'.format(msg))
# SSL does not support EOF, so send a null byte to indicate
# the end of the message.
writer.write(b'\x00')
await writer.drain()
```

服务器中的 echo() 协程必须查找 NULL 字节，并在接收到这个字节时关闭客户连接。

```python
async def echo(reader, writer):
    address = writer.get_extra_info('peername')
    log = logging.getLogger('echo_{}_{}'.format(*address))
    log.debug('connection accepted')
    while True:
        data = await reader.read(128)
        terminate = data.endswith(b'\x00')
        data = data.rstrip(b'\x00')
        if data:
            log.debug('received {!r}'.format(data))
            writer.write(data)
            await writer.drain()
```

```
                log.debug('sent {!r}'.format(data))
            if not data or terminate:
                log.debug('message terminated, closing connection')
                writer.close()
                return
```

在一个窗口中运行服务器,在另一个窗口中运行客户端,会生成以下输出。

```
$ python3 asyncio_echo_server_ssl.py
asyncio: Using selector: KqueueSelector
main: starting up on localhost port 10000
echo_::1_53957: connection accepted
echo_::1_53957: received b'This is the message. '
echo_::1_53957: sent b'This is the message. '
echo_::1_53957: received b'It will be sent in parts.'
echo_::1_53957: sent b'It will be sent in parts.'
echo_::1_53957: message terminated, closing connection

$ python3 asyncio_echo_client_ssl.py
asyncio: Using selector: KqueueSelector
echo_client: connecting to localhost port 10000
echo_client: sending b'This is the message. '
echo_client: sending b'It will be sent '
echo_client: sending b'in parts.'
echo_client: waiting for response
echo_client: received b'This is the message. '
echo_client: received b'It will be sent in parts.'
echo_client: closing
main: closing event loop
```

10.5.11 与域名服务交互

应用使用网络与服务器通信来完成域名服务(DNS)操作,如主机名与 IP 地址的转换。`asyncio` 事件循环提供了一些便利方法,可以在后台完成这些操作,从而避免在查询中阻塞。

10.5.11.1 按名查找地址

使用协程 `getaddrinfo()` 将一个主机名和端口号转换为一个 IP 或 IPv6 地址。与 `socket` 模块中的相应函数一样,返回值是一个元组列表,元组中包含 5 部分信息:

- 地址簇
- 地址类型
- 协议
- 服务器的规范名
- 一个套接字地址元组,可用于在原先指定的端口上打开与服务器的一个连接

可以按协议过滤查询。在下面的例子中,过滤器确保只返回 TCP 响应。

代码清单 10-102:`asyncio_getaddrinfo.py`

```
import asyncio
import logging
import socket
import sys
```

```python
TARGETS = [
    ('pymotw.com', 'https'),
    ('doughellmann.com', 'https'),
    ('python.org', 'https'),
]


async def main(loop, targets):
    for target in targets:
        info = await loop.getaddrinfo(
            *target,
            proto=socket.IPPROTO_TCP,
        )

        for host in info:
            print('{:20}: {}'.format(target[0], host[4][0]))


event_loop = asyncio.get_event_loop()
try:
    event_loop.run_until_complete(main(event_loop, TARGETS))
finally:
    event_loop.close()
```

下面这个示例程序将一个主机名和协议名转换为 IP 地址和端口号。

```
$ python3 asyncio_getaddrinfo.py

pymotw.com          : 66.33.211.242
doughellmann.com    : 66.33.211.240
python.org          : 23.253.135.79
python.org          : 2001:4802:7901::e60a:1375:0:6
```

10.5.11.2 按地址查找名

协程 getnameinfo() 的工作正好反过来，在可能的情况下，可以将一个 IP 地址转换为一个主机名，以及将一个端口号转换为一个协议名。

代码清单 10-103：*asyncio_getnameinfo.py*

```python
import asyncio
import logging
import socket
import sys


TARGETS = [
    ('66.33.211.242', 443),
    ('104.130.43.121', 443),
]


async def main(loop, targets):
    for target in targets:
        info = await loop.getnameinfo(target)
        print('{:15}: {} {}'.format(target[0], *info))


event_loop = asyncio.get_event_loop()
try:
    event_loop.run_until_complete(main(event_loop, TARGETS))
finally:
    event_loop.close()
```

这个例子显示，`pymotw.com` 的 IP 地址指示 DreamHost 的一个服务器，这是运行这个网站的托管公司。检查的第二个 IP 地址指向 `python.org`，没有解析得到一个主机名。

```
$ python3 asyncio_getnameinfo.py

66.33.211.242   : apache2-echo.catalina.dreamhost.com https
104.130.43.121  : 104.130.43.121 https
```

提示：相关阅读材料

- `socket` 模块讨论包括对这些操作的一个更详细的分析。

10.5.12 使用子进程

通常需要使用其他程序和进程，从而能利用现有的代码而不是完全重写，或者利用这些程序得到 Python 未能提供的库或特性。与网络 I/O 一样，`asyncio` 包括两个抽象，其用于启动另一个程序，然后与之交互。

10.5.12.1 利用子进程使用协议抽象

下面的例子使用一个协程来启动一个进程以运行 UNIX 命令 `df`，这个命令会查找本地磁盘上的空闲空间大小。它使用 `subprocess_exec()` 启动进程，把它绑定到一个协议类，这个类知道如何读取和解析 `df` 命令输出。会根据子进程的 I/O 事件自动调用这个协议类的方法。由于 `stdin` 和 `stderr` 参数都被设置为 `None`，所以这些通信通道没有连接到这个新进程。

代码清单 10-104：`asyncio_subprocess_protocol.py`

```python
import asyncio
import functools


async def run_df(loop):
    print('in run_df')

    cmd_done = asyncio.Future(loop=loop)
    factory = functools.partial(DFProtocol, cmd_done)
    proc = loop.subprocess_exec(
        factory,
        'df', '-hl',
        stdin=None,
        stderr=None,
    )
    try:
        print('launching process')
        transport, protocol = await proc
        print('waiting for process to complete')
        await cmd_done
    finally:
        transport.close()

    return cmd_done.result()
```

类 `DFProtocol` 派生自 `SubprocessProtocol`，它定义的 API 允许一个类通过管道与另一个进程通信。`done` 参数是一个 `Future`，调用者将用它来监视进程是否完成。

```
class DFProtocol(asyncio.SubprocessProtocol):

    FD_NAMES = ['stdin', 'stdout', 'stderr']

    def __init__(self, done_future):
        self.done = done_future
        self.buffer = bytearray()
        super().__init__()
```

与套接字通信一样，在建立与新进程的输入通道时会调用 `connection_made()`。transport 参数是 `BaseSubprocessTransport` 子类的一个实例。它可以读取进程的数据输出，还可以把数据写入进程的输入流（如果进程被配置为接收输入）。

```
    def connection_made(self, transport):
        print('process started {}'.format(transport.get_pid()))
        self.transport = transport
```

进程生成输出时，会调用 `pipe_data_received()` 并提供文件描述符（数据要发送到这个文件）以及从管道读取的实际数据。协议类将从进程标准输出通道得到的输出保存在一个缓冲区中，以便以后处理。

```
    def pipe_data_received(self, fd, data):
        print('read {} bytes from {}'.format(len(data),
                                             self.FD_NAMES[fd]))
        if fd == 1:
            self.buffer.extend(data)
```

进程终止时，会调用 `process_exited()`。可以调用 `get_returncode()` 从传输类对象得到进程的退出码。在这里，如果没有报告错误，那么在通过 `Future` 实例返回输出之前会对输出解码和解析。否则，如果生成了一个错误，则认为结果为空。如果设置了 future 的结果，那么这会告诉 `run_df()` 进程已经退出，所以它首先清理，然后返回结果。

```
    def process_exited(self):
        print('process exited')
        return_code = self.transport.get_returncode()
        print('return code {}'.format(return_code))
        if not return_code:
            cmd_output = bytes(self.buffer).decode()
            results = self._parse_results(cmd_output)
        else:
            results = []
        self.done.set_result((return_code, results))
```

命令输出被解析为一个字典序列，对应每一行输出，这些字典将首部名映射到相应的值，然后返回得到的列表。

```
    def _parse_results(self, output):
        print('parsing results')
        # Output has one row of headers, all single words. The
        # remaining rows are one per file system, with columns
        # matching the headers (assuming that none of the
        # mount points has whitespace in the names).
        if not output:
            return []
        lines = output.splitlines()
        headers = lines[0].split()
        devices = lines[1:]
        results = [
```

```
            dict(zip(headers, line.split()))
            for line in devices
        ]
        return results
```

run_df() 协程用 run_until_complete() 运行。然后检查结果，并输出各个设备上的空闲空间。

```
event_loop = asyncio.get_event_loop()
try:
    return_code, results = event_loop.run_until_complete(
        run_df(event_loop)
    )
finally:
    event_loop.close()

if return_code:
    print('error exit {}'.format(return_code))
else:
    print('\nFree space:')
    for r in results:
        print('{Mounted:25}: {Avail}'.format(**r))
```

下面的输出显示了步骤序列，以及在运行这个程序的系统上3个驱动盘的空闲空间。

```
$ python3 asyncio_subprocess_protocol.py

in run_df
launching process
process started 49675
waiting for process to complete
read 332 bytes from stdout
process exited
return code 0
parsing results

Free space:
/                        : 233Gi
/Volumes/hubertinternal  : 157Gi
/Volumes/hubert-tm       : 2.3Ti
```

10.5.12.2 利用协程和流调用子进程

要使用协程直接运行一个进程，而不是通过一个 Protocol 子类来访问，可以调用 create_subprocess_exec() 并指定是否将 stdout、stderr 和 stdin 连接到管道。如果一个协程要创建子进程，那么其结果是一个 Process 实例，可以用它来管理这个子进程或者与之通信。

代码清单 10-105：asyncio_subprocess_coroutine.py

```
import asyncio
import asyncio.subprocess


async def run_df():
    print('in run_df')

    buffer = bytearray()

    create = asyncio.create_subprocess_exec(
```

```
        'df', '-hl',
        stdout=asyncio.subprocess.PIPE,
)
print('launching process')
proc = await create
print('process started {}'.format(proc.pid))
```

在这个例子中，df 除了命令行参数外不需要任何输入，所以下一步是读取所有输出。如果利用 Protocol，则无法控制一次读取多少数据。这个例子使用了 readline()，不过也可以直接调用 read() 读取非行数据（not line oriented）。与协议例子一样，这个命令的输出会缓冲，以便以后解析。

```
while True:
    line = await proc.stdout.readline()
    print('read {!r}'.format(line))
    if not line:
        print('no more output from command')
        break
    buffer.extend(line)
```

当由于程序结束而不再有更多输出时，readline() 方法会返回一个空字节串。为了确保进程得到适当的清理，下一步要等待进程完全退出。

```
print('waiting for process to complete')
await proc.wait()
```

在这里，可以检查退出状态来确定要解析输出还是处理错误（因为没有生成任何输出）。解析逻辑与上例相同，不过会放在一个单独的函数中（这里没有显示），因为没有协议类来放置这个逻辑。解析数据之后，将结果和退出码返回给调用者。

```
return_code = proc.returncode
print('return code {}'.format(return_code))
if not return_code:
    cmd_output = bytes(buffer).decode()
    results = _parse_results(cmd_output)
else:
    results = []

return (return_code, results)
```

主程序看起来与基于协议的例子很类似，因为对实现的修改被单独放在 run_df() 中。

```
event_loop = asyncio.get_event_loop()
try:
    return_code, results = event_loop.run_until_complete(
        run_df()
    )
finally:
    event_loop.close()

if return_code:
    print('error exit {}'.format(return_code))
else:
    print('\nFree space:')
    for r in results:
        print('{Mounted:25}: {Avail}'.format(**r))
```

因为可以一次读取一行 df 的输出，所以可以回送读取的每一行以展示程序的进度。否则，输出看起来与上一个例子中生成的输出很类似。

```
$ python3 asyncio_subprocess_coroutine.py

in run_df
launching process
process started 49678
read b'Filesystem      Size   Used  Avail Capacity    iused
ifree %iused  Mounted on\n'
read b'/dev/disk2s2  446Gi  213Gi  233Gi    48% 55955082
61015132   48%   /\n'
read b'/dev/disk1    465Gi  307Gi  157Gi    67% 80514922
41281172   66%   /Volumes/hubertinternal\n'
read b'/dev/disk3s2  3.6Ti  1.4Ti  2.3Ti    38% 181837749
306480579  37%   /Volumes/hubert-tm\n'
read b''
no more output from command
waiting for process to complete
return code 0
parsing results

Free space:
/                           : 233Gi
/Volumes/hubertinternal     : 157Gi
/Volumes/hubert-tm          : 2.3Ti
```

10.5.12.3 向子进程发送数据

前面的两个例子只使用了一个通信通道从第二个进程读取数据。通常有必要把数据发送给一个命令来处理。下面的例子定义了一个协程来执行 UNIX 命令 `tr`，转换其输入流中的字符。在这里，`tr` 用来把小写字母转换为大写字母。

`to_upper()` 协程接收一个事件循环和一个输入串作为参数。它会创建另一个进程运行 `"tr [:lower:] [:upper:]"`。

代码清单 10-106：**asyncio_subprocess_coroutine_write.py**

```python
import asyncio
import asyncio.subprocess


async def to_upper(input):
    print('in to_upper')

    create = asyncio.create_subprocess_exec(
        'tr', '[:lower:]', '[:upper:]',
        stdout=asyncio.subprocess.PIPE,
        stdin=asyncio.subprocess.PIPE,
    )
    print('launching process')
    proc = await create
    print('pid {}'.format(proc.pid))
```

接下来，`to_upper()` 使用 `Process` 的 `communicate()` 方法将输入串发送到这个命令，并异步地读取得到的所有输出。与 `subprocess.Popen` 中的相应方法一样，`communicate()` 会返回这个方法输出的所有字节串。如果一个命令有可能生成太多的数据（在内存中放不下），那么不能一次生成全部输入，或者必须增量地处理输出，一种更好的方法是直接使用 Process 的 stdin、stdout 和 stderr 句柄而不是调用 communicate()。

```
print('communicating with process')
stdout, stderr = await proc.communicate(input.encode())
```

I/O 完成后，要等待进程完全退出，确保它得到适当的清理。

```
print('waiting for process to complete')
await proc.wait()
```

然后可以检查返回码，并解码输出字节串以准备协程的返回值。

```
return_code = proc.returncode
print('return code {}'.format(return_code))
if not return_code:
    results = bytes(stdout).decode()
else:
    results = ''

return (return_code, results)
```

程序的主要部分建立了一个要转换的消息串，并建立事件循环来运行 to_upper()，然后输出结果。

```
MESSAGE = """
This message will be converted
to all caps.
"""

event_loop = asyncio.get_event_loop()
try:
    return_code, results = event_loop.run_until_complete(
        to_upper(MESSAGE)
    )
finally:
    event_loop.close()

if return_code:
    print('error exit {}'.format(return_code))
else:
    print('Original: {!r}'.format(MESSAGE))
    print('Changed : {!r}'.format(results))
```

输出显示了操作序列以及这个简单文本消息的转换结果。

```
$ python3 asyncio_subprocess_coroutine_write.py

in to_upper
launching process
pid 49684
communicating with process
waiting for process to complete
return code 0
Original: '\nThis message will be converted\nto all caps.\n'
Changed : '\nTHIS MESSAGE WILL BE CONVERTED\nTO ALL CAPS.\n'
```

10.5.13 接收 UNIX 信号

UNIX 系统事件通知通常会中断一个应用，触发它们的处理器。结合 asyncio 使用时，信号处理器回调会与事件循环管理的其他协程和回调交叉执行。这种集成会更少地中断函数，尽可能减少提供防护来清理未完成的操作的需要。

信号处理器必须是常规的 callable，而不是协程。

代码清单 10-107：asyncio_signal.py

```python
import asyncio
import functools
import os
import signal


def signal_handler(name):
    print('signal_handler({!r})'.format(name))
```

信号处理器用 `add_signal_handler()` 注册。第一个参数是信号；第二个参数是回调。不向回调传递任何参数，所以如果需要参数，可以用 `functools.partial()` 包装一个函数。

```python
event_loop = asyncio.get_event_loop()

event_loop.add_signal_handler(
    signal.SIGHUP,
    functools.partial(signal_handler, name='SIGHUP'),
)
event_loop.add_signal_handler(
    signal.SIGUSR1,
    functools.partial(signal_handler, name='SIGUSR1'),
)
event_loop.add_signal_handler(
    signal.SIGINT,
    functools.partial(signal_handler, name='SIGINT'),
)
```

这个示例程序使用了一个协程通过 `os.kill()` 向自身发送信号。发送各个信号之后，协程交出控制，从而允许这个处理器运行。在一个真实的应用中，可能有更多位置上的应用代码会把控制交回给事件循环，所以不需要（像这个例子中一样）人为地交出控制。

```python
async def send_signals():
    pid = os.getpid()
    print('starting send_signals for {}'.format(pid))

    for name in ['SIGHUP', 'SIGHUP', 'SIGUSR1', 'SIGINT']:
        print('sending {}'.format(name))
        os.kill(pid, getattr(signal, name))
        # Yield control to allow the signal handler to run,
        # since the signal does not interrupt the program
        # flow otherwise.
        print('yielding control')
        await asyncio.sleep(0.01)
    return
```

主程序运行 `send_signals()`，直到已经发出所有信号。

```python
try:
    event_loop.run_until_complete(send_signals())
finally:
    event_loop.close()
```

输出显示了发送一个信号后 `send_signals()` 交出控制时如何调用处理器。

```
$ python3 asyncio_signal.py

starting send_signals for 21772
sending SIGHUP
yielding control
signal_handler('SIGHUP')
sending SIGHUP
yielding control
signal_handler('SIGHUP')
sending SIGUSR1
yielding control
signal_handler('SIGUSR1')
sending SIGINT
yielding control
signal_handler('SIGINT')
```

提示：相关阅读材料
- signal：接收异步系统事件通知。

10.5.14 结合使用协程、线程与进程

很多预定义的库都不能自然地用于 asyncio。它们可能阻塞，或者可能依赖这个模块无法提供的并发特性。不过，在一个基于 asyncio 的应用中，还是可以通过使用 concurrent.futures 的一个执行器 (executor) 来使用这些库，在一个单独的线程或单独的进程中运行代码。

10.5.14.1 线程

事件循环的 run_in_executor() 方法接收 3 个参数：一个执行器实例，一个常规的要调用的 callable，以及要传入这个 callable 的所有参数。它返回一个 Future，可以用来等待函数完成工作并返回某个结果。如果没有传入执行器，则会创建一个 ThreadPoolExecutor。下面的例子会显式地创建一个执行器来限制可用的工作线程数目。

一个 ThreadPoolExecutor 启动其工作线程，然后在一个线程中调用所提供的各个函数。这个例子显示了如何结合 run_in_executor() 和 wait() 让一个协程把控制交还给事件循环（当阻塞函数在单独的线程中运行时），然后当这些函数完成时再唤醒协程。

代码清单 10-108：**asyncio_executor_thread.py**

```python
import asyncio
import concurrent.futures
import logging
import sys
import time


def blocks(n):
    log = logging.getLogger('blocks({})'.format(n))
    log.info('running')
    time.sleep(0.1)
    log.info('done')
    return n ** 2


async def run_blocking_tasks(executor):
```

```python
    log = logging.getLogger('run_blocking_tasks')
    log.info('starting')

    log.info('creating executor tasks')
    loop = asyncio.get_event_loop()
    blocking_tasks = [
        loop.run_in_executor(executor, blocks, i)
        for i in range(6)
    ]
    log.info('waiting for executor tasks')
    completed, pending = await asyncio.wait(blocking_tasks)
    results = [t.result() for t in completed]
    log.info('results: {!r}'.format(results))

    log.info('exiting')


if __name__ == '__main__':
    # Configure logging to show the name of the thread
    # where the log message originates.
    logging.basicConfig(
        level=logging.INFO,
        format='%(threadName)10s %(name)18s: %(message)s',
        stream=sys.stderr,
    )

    # Create a limited thread pool.
    executor = concurrent.futures.ThreadPoolExecutor(
        max_workers=3,
    )

    event_loop = asyncio.get_event_loop()
    try:
        event_loop.run_until_complete(
            run_blocking_tasks(executor)
        )
    finally:
        event_loop.close()
```

asyncio_executor_thread.py 使用 logging 很方便地指示正在由哪个线程和函数生成各个日志消息。由于每个 blocks() 调用中使用了一个单独的日志记录器，输出清楚地显示出这里重用了相同的线程来调用函数的多个副本，不过提供了不同的参数。

```
$ python3 asyncio_executor_thread.py

MainThread run_blocking_tasks: starting
MainThread run_blocking_tasks: creating executor tasks
  Thread-1           blocks(0): running
  Thread-2           blocks(1): running
  Thread-3           blocks(2): running
MainThread run_blocking_tasks: waiting for executor tasks
  Thread-1           blocks(0): done
  Thread-3           blocks(2): done
  Thread-1           blocks(3): running
  Thread-2           blocks(1): done
  Thread-3           blocks(4): running
  Thread-2           blocks(5): running
  Thread-1           blocks(3): done
  Thread-2           blocks(5): done
  Thread-3           blocks(4): done
MainThread run_blocking_tasks: results: [16, 4, 1, 0, 25, 9]
MainThread run_blocking_tasks: exiting
```

10.5.14.2 进程

`ProcessPoolExecutor` 的工作是类似的，但是会创建一组工作进程而不是工作线程。尽管使用单独的进程需要更多系统资源，但对于计算强度很大的操作，在各个 CPU 上运行一个单独的任务会很有意义。

代码清单 10-109：`asyncio_executor_process.py`

```
# Changes from asyncio_executor_thread.py

if __name__ == '__main__':
    # Configure logging to show the ID of the process
    # where the log message originates.
    logging.basicConfig(
        level=logging.INFO,
        format='PID %(process)5s %(name)18s: %(message)s',
        stream=sys.stderr,
    )

    # Create a limited process pool.
    executor = concurrent.futures.ProcessPoolExecutor(
        max_workers=3,
    )

    event_loop = asyncio.get_event_loop()
    try:
        event_loop.run_until_complete(
            run_blocking_tasks(executor)
        )
    finally:
        event_loop.close()
```

从线程转向进程需要做的唯一的改变是要创建一种不同类型的执行器。这个例子还修改了日志格式串，使它包含进程 ID 而不是线程名，以展示任务在单独的进程中运行。

```
$ python3 asyncio_executor_process.py

PID 16429 run_blocking_tasks: starting
PID 16429 run_blocking_tasks: creating executor tasks
PID 16429 run_blocking_tasks: waiting for executor tasks
PID 16430           blocks(0): running
PID 16431           blocks(1): running
PID 16432           blocks(2): running
PID 16430           blocks(0): done
PID 16432           blocks(2): done
PID 16431           blocks(1): done
PID 16430           blocks(3): running
PID 16432           blocks(4): running
PID 16431           blocks(5): running
PID 16431           blocks(5): done
PID 16432           blocks(4): done
PID 16430           blocks(3): done
PID 16429 run_blocking_tasks: results: [4, 0, 16, 1, 9, 25]
PID 16429 run_blocking_tasks: exiting
```

10.5.15 用 asyncio 调试

asyncio 中内置了很多有用的调试特性。例如，事件循环使用 logging 在它运行时生成状态消息。其中一些消息在应用启用日志记录时就可以得到；要得到另外一些消息，需要显式地告诉循环生成更多的调试消息。可以调用 set_debug() 并传入一个布尔值指示是否启用调试。

由于基于 asyncio 建立的应用对不能交出控制的贪婪协程非常敏感，所以事件循环中内置了相应的支持来检测慢回调。可以通过启用调试打开这个特性，将循环的 slow_callback_duration 属性设置为一个秒数（在这个时间之后要发出一个警告）以控制"慢"的定义。

最后，如果一个使用 asyncio 的应用退出时没有清理一些协程或其他资源，那么这种行为可能说明存在一个逻辑错误阻止了一些应用代码运行。启用 ResourceWarning 警告会在程序退出时报告这些情况。

代码清单 10-110：asyncio_debug.py

```python
import argparse
import asyncio
import logging
import sys
import time
import warnings

parser = argparse.ArgumentParser('debugging asyncio')
parser.add_argument(
    '-v',
    dest='verbose',
    default=False,
    action='store_true',
)
args = parser.parse_args()

logging.basicConfig(
    level=logging.DEBUG,
    format='%(levelname)7s: %(message)s',
    stream=sys.stderr,
)
LOG = logging.getLogger('')


async def inner():
    LOG.info('inner starting')
    # Use a blocking sleep to simulate
    # doing work inside the function.
    time.sleep(0.1)
    LOG.info('inner completed')


async def outer(loop):
    LOG.info('outer starting')
    await asyncio.ensure_future(loop.create_task(inner()))
    LOG.info('outer completed')


event_loop = asyncio.get_event_loop()
if args.verbose:
    LOG.info('enabling debugging')

    # Enable debugging.
    event_loop.set_debug(True)
```

```
        # Make the threshold for "slow" tasks very very small for
        # illustration. The default is 0.1, or 100 milliseconds.
        event_loop.slow_callback_duration = 0.001
        # Report all mistakes managing asynchronous resources.
        warnings.simplefilter('always', ResourceWarning)

    LOG.info('entering event loop')
    event_loop.run_until_complete(outer(event_loop))
```

运行时如果没有启用调试,那么看起来这个应用一切正常。

```
$ python3 asyncio_debug.py

  DEBUG: Using selector: KqueueSelector
   INFO: entering event loop
   INFO: outer starting
   INFO: inner starting
   INFO: inner completed
   INFO: outer completed
```

不过,打开调试就会暴露出应用中存在的一些问题。例如,尽管 inner() 能完成,但是它花费的时间要大于所设置的 slow_callback_duration。另外,程序退出时事件循环没有妥善地关闭。

```
$ python3 asyncio_debug.py -v

  DEBUG: Using selector: KqueueSelector
   INFO: enabling debugging
   INFO: entering event loop
   INFO: outer starting
   INFO: inner starting
   INFO: inner completed
WARNING: Executing <Task finished coro=<inner() done, defined at
asyncio_debug.py:34> result=None created at asyncio_debug.py:44>
took 0.102 seconds
   INFO: outer completed
.../lib/python3.5/asyncio/base_events.py:429: ResourceWarning:
unclosed event loop <_UnixSelectorEventLoop running=False
closed=False debug=True>
  DEBUG: Close <_UnixSelectorEventLoop running=False
closed=False debug=True>
```

说明:在 Python 3.5 中,`asyncio` 还是一个 provisional 模块。这个 API 在 Python 3.6 得到稳定,大部分修改回植到 Python 3.5 后来的一个补丁版本。因此,这个模块在不同版本的 Python 3.5 下的工作可能稍有不同。

提示:相关阅读材料

- `asyncio` 的标准库文档[一]。
- **PEP 3156**[二]:Asynchronous IO Support Rebooted: "asyncio" 模块。
- **PEP 380**[三]:委托到子生成器的语法。

[一] https://docs.python.org/3.5/library/asyncio.html
[二] www.python.org/dev/peps/pep-3156
[三] www.python.org/dev/peps/pep-0380

- **PEP 492**[一]：采用 async 和 await 语法的协程。
- concurrent.futures：管理并发任务池。
- socket：底层网络通信。
- select：底层异步 I/O 工具。
- socketserver：创建网络服务器的框架。
- What's New in Python 3.6: asyncio[二]：API 在 Python 3.6 中稳定后，对 asyncio 中修改的一个总结。
- trollius[三]：Tulip（asyncio 原来的版本）移植到 Python 2。
- The New asyncio moduyle in Python 3.4: Event Loops[四]：Gastón Hillar 在 "Dr. Dobb's" 上发表的文章。
- Exploring Python 3's Asyncio by Example[五]：Chad Lung 的博客文章。
- A Web Crawler with Asyncio Coroutines[六]：A. Jesse Jiryu Davis 和 Guido van Rossum 在"The Architecture of Open Source Applications"上的一篇文章。
- Playing with asyncio[七]：Nathan Hoad 的博客文章。
- Async I/O and Python[八]：Mark McLoughlin 的博客文章。
- A Curious Course on Coroutines and Concurrency[九]：David Beazley 的 PyCon 2009 教程。
- How the heck does async/await work in Python 3.5?[十]：Brett Cannon 的博客文章。
- Unix Network Programming, Volume 1: The Sockets Networking API, Third Edition, by W.Richard Stevens, Bill Fenner, and Andrew M. Rudoff; Addison-Wesley Professional, 2004. ISBN-10: 0131411551.
- Foundations of Python Network Programming, Third Edition, by Brandon Rhodes and John Goerzen; Apress, 2014. ISBN-10: 1430258543.

10.6 concurrent.futures：管理并发任务池

concurrent.futures 模块提供了使用工作线程或进程池运行任务的接口。线程和进程池的 API 是一样的，所以应用只做最小的修改就可以在线程和进程之间顺利地切换。

一 www.python.org/dev/peps/pep-0492
二 https://docs.python.org/3/whatsnew/3.6.html#asyncio
三 https://pypi.python.org/pypi/trollius
四 www.drdobbs.com/open-source/the-new-asyncio-module-in-python-34-even/240168401
五 www.giantflyingsaucer.com/blog/?p=5557
六 http://aosabook.org/en/500L/a-web-crawler-with-asyncio-coroutines.html
七 www.getoffmalawn.com/blog/playing-with-asyncio
八 https://blogs.gnome.org/markmc/2013/06/04/async-io-and-python/
九 www.dabeaz.com/coroutines/
十 www.snarky.ca/how-the-heck-does-async-await-work-in-python-3-5

这个模块提供了两种类型的类与这些池交互。执行器（executor）用来管理工作线程或进程池，future 用来管理工作线程或进程计算的结果。要使用一个工作线程或进程池，应用要创建适当的执行器类的一个实例，然后向它提交任务来运行。每个任务启动时，会返回一个 Future 实例。需要任务的结果时，应用可以使用 Future 阻塞，直到得到结果。目前已经提供了不同的 API，可以很方便地等待任务完成，所以不需要直接管理 Future 对象。

10.6.1 利用基本线程池使用 map()

ThreadPoolExecutor 管理一组工作线程，当这些线程可用于完成更多工作时，可以向它们传入任务。下面的例子使用 map() 并发地从一个输入迭代器生成一组结果。这个任务使用 time.sleep() 暂停不同的时间，从而展示不论任务的执行顺序如何，map() 总是根据输入按顺序返回值。

代码清单 10-111：`futures_thread_pool_map.py`

```
from concurrent import futures
import threading
import time


def task(n):
    print('{}: sleeping {}'.format(
        threading.current_thread().name,
        n)
    )
    time.sleep(n / 10)
    print('{}: done with {}'.format(
        threading.current_thread().name,
        n)
    )
    return n / 10


ex = futures.ThreadPoolExecutor(max_workers=2)
print('main: starting')
results = ex.map(task, range(5, 0, -1))
print('main: unprocessed results {}'.format(results))
print('main: waiting for real results')
real_results = list(results)
print('main: results: {}'.format(real_results))
```

map() 的返回值实际上是一种特殊类型的迭代器，它知道主程序迭代处理时要等待各个响应。

```
$ python3 futures_thread_pool_map.py

main: starting
Thread-1: sleeping 5
Thread-2: sleeping 4
main: unprocessed results <generator object
Executor.map.<locals>.result_iterator at 0x1013c80a0>
main: waiting for real results
Thread-2: done with 4
```

```
Thread-2: sleeping 3
Thread-1: done with 5
Thread-1: sleeping 2
Thread-1: done with 2
Thread-1: sleeping 1
Thread-2: done with 3
Thread-1: done with 1
main: results: [0.5, 0.4, 0.3, 0.2, 0.1]
```

10.6.2 调度单个任务

除了使用 map()，还可以借助 submit() 利用一个执行器调度单个任务。然后可以使用返回的 Future 实例等待这个任务的结果。

代码清单 10-112：futures_thread_pool_submit.py

```
from concurrent import futures
import threading
import time


def task(n):
    print('{}: sleeping {}'.format(
        threading.current_thread().name,
        n)
    )
    time.sleep(n / 10)
    print('{}: done with {}'.format(
        threading.current_thread().name,
        n)
    )
    return n / 10


ex = futures.ThreadPoolExecutor(max_workers=2)
print('main: starting')
f = ex.submit(task, 5)
print('main: future: {}'.format(f))
print('main: waiting for results')
result = f.result()
print('main: result: {}'.format(result))
print('main: future after result: {}'.format(f))
```

任务完成之后，Future 的状态会改变，并得到结果。

```
$ python3 futures_thread_pool_submit.py

main: starting
Thread-1: sleeping 5
main: future: <Future at 0x1010e6080 state=running>
main: waiting for results
Thread-1: done with 5
main: result: 0.5
main: future after result: <Future at 0x1010e6080 state=finished
 returned float>
```

10.6.3 按任意顺序等待任务

调用 Future 的 result() 方法会阻塞，直到任务完成（可能返回一个值，也可能抛出一个异常）或者撤销。可以使用 map() 按调度任务的顺序访问多个任务的结果。如果处理结果的顺序不重要，则可以使用 as_completed() 在每个任务完成时处理它的结果。

代码清单 10-113：**futures_as_completed.py**

```python
from concurrent import futures
import random
import time


def task(n):
    time.sleep(random.random())
    return (n, n / 10)


ex = futures.ThreadPoolExecutor(max_workers=5)
print('main: starting')

wait_for = [
    ex.submit(task, i)
    for i in range(5, 0, -1)
]

for f in futures.as_completed(wait_for):
    print('main: result: {}'.format(f.result()))
```

因为池中的工作线程与任务同样多，故而所有任务都可以启动。它们会按随机的顺序完成，所以每次运行这个示例程序时 as_completed() 生成的值都不同。

```
$ python3 futures_as_completed.py

main: starting
main: result: (3, 0.3)
main: result: (5, 0.5)
main: result: (4, 0.4)
main: result: (2, 0.2)
main: result: (1, 0.1)
```

10.6.4 Future 回调

要在任务完成时采取某个动作，不用显式地等待结果，可以使用 add_done_callback() 指示 Future 完成时要调用一个新函数。这个回调应当是有一个参数（Future 实例）的 callable 函数。

代码清单 10-114：**futures_future_callback.py**

```python
from concurrent import futures
import time


def task(n):
    print('{}: sleeping'.format(n))
    time.sleep(0.5)
    print('{}: done'.format(n))
```

```
        return n / 10


def done(fn):
    if fn.cancelled():
        print('{}: canceled'.format(fn.arg))
    elif fn.done():
        error = fn.exception()
        if error:
            print('{}: error returned: {}'.format(
                fn.arg, error))
        else:
            result = fn.result()
            print('{}: value returned: {}'.format(
                fn.arg, result))


if __name__ == '__main__':
    ex = futures.ThreadPoolExecutor(max_workers=2)
    print('main: starting')
    f = ex.submit(task, 5)
    f.arg = 5
    f.add_done_callback(done)
    result = f.result()
```

不论由于什么原因，只要认为 Future "完成"，就会调用这个回调，所以在使用它之前必须检查传入回调的对象的状态。

```
$ python3 futures_future_callback.py

main: starting
5: sleeping
5: done
5: value returned: 0.5
```

10.6.5 撤销任务

如果一个 Future 已经提交但还没有启动，那么可以调用它的 cancel() 方法将其撤销。

代码清单 10-115：futures_future_callback_cancel.py

```
from concurrent import futures
import time


def task(n):
    print('{}: sleeping'.format(n))
    time.sleep(0.5)
    print('{}: done'.format(n))
    return n / 10


def done(fn):
    if fn.cancelled():
        print('{}: canceled'.format(fn.arg))
    elif fn.done():
        print('{}: not canceled'.format(fn.arg))


if __name__ == '__main__':
```

```
    ex = futures.ThreadPoolExecutor(max_workers=2)
    print('main: starting')
    tasks = []

    for i in range(10, 0, -1):
        print('main: submitting {}'.format(i))
        f = ex.submit(task, i)
        f.arg = i
        f.add_done_callback(done)
        tasks.append((i, f))

    for i, t in reversed(tasks):
        if not t.cancel():
            print('main: did not cancel {}'.format(i))

    ex.shutdown()
```

cancel()返回一个布尔值,指示任务是否可以撤销。

```
$ python3 futures_future_callback_cancel.py

main: starting
main: submitting 10
10: sleeping
main: submitting 9
9: sleeping
main: submitting 8
main: submitting 7
main: submitting 6
main: submitting 5
main: submitting 4
main: submitting 3
main: submitting 2
main: submitting 1
1: canceled
2: canceled
3: canceled
4: canceled
5: canceled
6: canceled
7: canceled
8: canceled
main: did not cancel 9
main: did not cancel 10
10: done
10: not canceled
9: done
9: not canceled
```

10.6.6 任务中的异常

如果一个任务产生一个未处理的异常,那么它会被保存到这个任务的 Future,而且可以通过 result() 或 exception() 方法得到。

代码清单 10-116: **futures_future_exception.py**

```
from concurrent import futures


def task(n):
    print('{}: starting'.format(n))
```

```
            raise ValueError('the value {} is no good'.format(n))

ex = futures.ThreadPoolExecutor(max_workers=2)
print('main: starting')
f = ex.submit(task, 5)

error = f.exception()
print('main: error: {}'.format(error))

try:
    result = f.result()
except ValueError as e:
    print('main: saw error "{}" when accessing result'.format(e))
```

如果在一个任务函数中抛出一个未处理的异常后调用了 result()，那么会在当前上下文中再次抛出同样的异常。

```
$ python3 futures_future_exception.py

main: starting
5: starting
main: error: the value 5 is no good
main: saw error "the value 5 is no good" when accessing result
```

10.6.7　上下文管理器

执行器会与上下文管理器合作，并发地运行任务并等待它们都完成。当上下文管理器退出时，会调用执行器的 shutdown() 方法。

代码清单 10-117：futures_context_manager.py

```
from concurrent import futures


def task(n):
    print(n)


with futures.ThreadPoolExecutor(max_workers=2) as ex:
    print('main: starting')
    ex.submit(task, 1)
    ex.submit(task, 2)
    ex.submit(task, 3)
    ex.submit(task, 4)

print('main: done')
```

离开当前作用域时如果要清理线程或进程资源，那么用这种方式使用执行器就很有用。

```
$ python3 futures_context_manager.py

main: starting
1
2
3
4
main: done
```

10.6.8 进程池

`ProcessPoolExecutor` 的工作与 `ThreadPoolExecutor` 类似，不过使用进程而不是线程。这种方法允许 CPU 密集的操作使用一个单独的 CPU，而不会因为 Cpython 解释器的全局解释器锁而被阻塞。

代码清单 10-118：**futures_process_pool_map.py**

```
from concurrent import futures
import os


def task(n):
    return (n, os.getpid())


ex = futures.ProcessPoolExecutor(max_workers=2)
results = ex.map(task, range(5, 0, -1))
for n, pid in results:
    print('ran task {} in process {}'.format(n, pid))
```

与线程池一样，要重用各个工作进程以执行多个任务。

```
$ python3 futures_process_pool_map.py

ran task 5 in process 60245
ran task 4 in process 60246
ran task 3 in process 60245
ran task 2 in process 60245
ran task 1 in process 60245
```

如果某个工作进程出了问题，导致它意外退出，则认为 `ProcessPoolExecutor` "中断"，不会再调度任务。

代码清单 10-119：**futures_process_pool_broken.py**

```
from concurrent import futures
import os
import signal


with futures.ProcessPoolExecutor(max_workers=2) as ex:
    print('getting the pid for one worker')
    f1 = ex.submit(os.getpid)
    pid1 = f1.result()

    print('killing process {}'.format(pid1))
    os.kill(pid1, signal.SIGHUP)

    print('submitting another task')
    f2 = ex.submit(os.getpid)
    try:
        pid2 = f2.result()
    except futures.process.BrokenProcessPool as e:
        print('could not start new tasks: {}'.format(e))
```

`BrokenProcessPool` 异常实际上是在处理结果时抛出的，而不是在提交新任务时抛出。

```
$ python3 futures_process_pool_broken.py

getting the pid for one worker
killing process 62059
submitting another task
could not start new tasks: A process in the process pool was
terminated abruptly while the future was running or pending.
```

提示：相关阅读材料

- `concurrent.futures` 的标准库文档[一]。
- **PEP 3148**[二]：创建 `concurrent.futures` 特性集的提案。
- 10.5.14 节。
- `threading`。
- `multiprocessing`。

[一] https://docs.python.org/3.5/library/concurrent.futures.html
[二] www.python.org/dev/peps/pep-3148

第 11 章

网络通信

网络通信用于获取一个算法在本地运行所需的数据，还可以共享信息实现分布式处理，另外可以用来管理云服务。Python 的标准库提供了一些模块来创建网络服务以及远程访问现有服务。

ipaddress 模块提供了一些类来验证、比较和处理 IPv4/IPv6 网络地址。

底层 socket 库允许直接访问原生 C 套接字库，可以用于与任何网络服务通信。selectors 提供了一个高层接口，可以同时监视多个套接字，这对于支持网络服务器同时与多个客户通信很有用。select 提供了 selectors 使用的底层 API。

socketserver 中的框架抽象了创建一个新的网络服务器所需的大量重复性工作。可以结合这些类创建服务器来建立或使用线程以及支持 TCP 或 UDP。应用只需要完成实际的消息处理。

11.1 ipaddress：Internet 地址

ipaddress 模块提供了处理 IPv4 和 IPv6 网络地址的类。这些类支持验证，查找网络上的地址和主机，以及其他常见操作。

11.1.1 地址

最基本的对象表示网络地址本身。可以向 ip_address() 传入一个字符串、整数或字节序列来构造一个地址。返回值是一个 IPv4Address 或 IPv6Address 实例，这取决于使用什么类型的地址。

代码清单 11-1：ipaddress_addresses.py

```
import binascii
import ipaddress


ADDRESSES = [
    '10.9.0.6',
    'fdfd:87b5:b475:5e3e:b1bc:e121:a8eb:14aa',
]

for ip in ADDRESSES:
    addr = ipaddress.ip_address(ip)
    print('{!r}'.format(addr))
    print('   IP version:', addr.version)
```

```
    print('   is private:', addr.is_private)
    print('   packed form:', binascii.hexlify(addr.packed))
    print('       integer:', int(addr))
    print()
```

这两个类可以提供地址的不同表示以满足不同的用途,还可以回答一些基本断言,如这个地址是否为组播通信保留,或者它是否在一个专用网(private network)中。

```
$ python3 ipaddress_addresses.py

IPv4Address('10.9.0.6')
   IP version: 4
   is private: True
   packed form: b'0a090006'
       integer: 168361990

IPv6Address('fdfd:87b5:b475:5e3e:b1bc:e121:a8eb:14aa')
   IP version: 6
   is private: True
   packed form: b'fdfd87b5b4755e3eb1bce121a8eb14aa'
       integer: 337611086560236126439725644408160982186
```

11.1.2 网络

网络由一个地址范围定义。通常用一个基本地址和一个掩码表示,掩码指示地址的哪些部分表示网络,哪些部分表示该网络上的地址。可以显式表示掩码,也可以使用一个前缀长度值来表示,如下例所示。

代码清单 11-2:**ipaddress_networks.py**

```
import ipaddress

NETWORKS = [
    '10.9.0.0/24',
    'fdfd:87b5:b475:5e3e::/64',
]

for n in NETWORKS:
    net = ipaddress.ip_network(n)
    print('{!r}'.format(net))
    print('   is private:', net.is_private)
    print('    broadcast:', net.broadcast_address)
    print('   compressed:', net.compressed)
    print(' with netmask:', net.with_netmask)
    print(' with hostmask:', net.with_hostmask)
    print(' num addresses:', net.num_addresses)
    print()
```

与地址一样,对应 IPv4 和 IPv6 网络分别有两个网络类。每个类都提供了一些属性或方法来访问与网络相关的值,如广播地址和主机可以使用的网络地址。

```
$ python3 ipaddress_networks.py

IPv4Network('10.9.0.0/24')
   is private: True
    broadcast: 10.9.0.255
   compressed: 10.9.0.0/24
```

```
    with netmask: 10.9.0.0/255.255.255.0
    with hostmask: 10.9.0.0/0.0.0.255
    num addresses: 256

IPv6Network('fdfd:87b5:b475:5e3e::/64')
      is private: True
       broadcast: fdfd:87b5:b475:5e3e:ffff:ffff:ffff:ffff
      compressed: fdfd:87b5:b475:5e3e::/64
    with netmask: fdfd:87b5:b475:5e3e:::/ffff:ffff:ffff:ffff::
    with hostmask: fdfd:87b5:b475:5e3e:::/::ffff:ffff:ffff:ffff
    num addresses: 18446744073709551616
```

网络实例是可迭代的，会提供网络上的地址。

代码清单 11-3：`ipaddress_network_iterate.py`

```python
import ipaddress

NETWORKS = [
    '10.9.0.0/24',
    'fdfd:87b5:b475:5e3e::/64',
]

for n in NETWORKS:
    net = ipaddress.ip_network(n)
    print('{!r}'.format(net))
    for i, ip in zip(range(3), net):
        print(ip)
    print()
```

这个例子只打印了部分地址，因为 IPv6 网络可能包含太多的地址，无法在这里全部输出。

```
$ python3 ipaddress_network_iterate.py

IPv4Network('10.9.0.0/24')
10.9.0.0
10.9.0.1
10.9.0.2

IPv6Network('fdfd:87b5:b475:5e3e::/64')
fdfd:87b5:b475:5e3e::
fdfd:87b5:b475:5e3e::1
fdfd:87b5:b475:5e3e::2
```

迭代处理网络会提供地址，不过并不是所有这些地址都能作为主机的合法地址。例如，网络的基地址和广播地址也包含在内。要查找能够由网络上常规主机使用的地址，可以使用 `hosts()` 方法，它会生成一个生成器。

代码清单 11-4：`ipaddress_network_iterate_hosts.py`

```python
import ipaddress

NETWORKS = [
    '10.9.0.0/24',
    'fdfd:87b5:b475:5e3e::/64',
]

for n in NETWORKS:
```

```
        net = ipaddress.ip_network(n)
        print('{!r}'.format(net))
        for i, ip in zip(range(3), net.hosts()):
            print(ip)
        print()
```

将这个例子的输出与前一个例子进行比较，可以看到，主机地址中不包含迭代处理整个网络时生成的前几个值。

```
$ python3 ipaddress_network_iterate_hosts.py

IPv4Network('10.9.0.0/24')
10.9.0.1
10.9.0.2
10.9.0.3

IPv6Network('fdfd:87b5:b475:5e3e::/64')
fdfd:87b5:b475:5e3e::1
fdfd:87b5:b475:5e3e::2
fdfd:87b5:b475:5e3e::3
```

除了迭代器协议，网络还支持 in 操作符，可以用来确定一个地址是否是一个网络的一部分。

代码清单 11-5：**ipaddress_network_membership.py**

```
import ipaddress


NETWORKS = [
    ipaddress.ip_network('10.9.0.0/24'),
    ipaddress.ip_network('fdfd:87b5:b475:5e3e::/64'),
]

ADDRESSES = [
    ipaddress.ip_address('10.9.0.6'),
    ipaddress.ip_address('10.7.0.31'),
    ipaddress.ip_address(
        'fdfd:87b5:b475:5e3e:b1bc:e121:a8eb:14aa'
    ),
    ipaddress.ip_address('fe80::3840:c439:b25e:63b0'),
]


for ip in ADDRESSES:
    for net in NETWORKS:
        if ip in net:
            print('{}\nis on {}'.format(ip, net))
            break
    else:
        print('{}\nis not on a known network'.format(ip))
    print()
```

in 的实现使用网络掩码来测试地址，所以这比展开网络上的完整地址列表要高效得多。

```
$ python3 ipaddress_network_membership.py

10.9.0.6
```

```
is on 10.9.0.0/24

10.7.0.31
is not on a known network

fdfd:87b5:b475:5e3e:b1bc:e121:a8eb:14aa
is on fdfd:87b5:b475:5e3e::/64
fe80::3840:c439:b25e:63b0
is not on a known network
```

11.1.3 接口

网络接口表示网络上的一个特定地址，可以表示为一个主机地址和一个网络前缀或网络掩码。

代码清单 11-6：**ipaddress_interfaces.py**

```
import ipaddress

ADDRESSES = [
    '10.9.0.6/24',
    'fdfd:87b5:b475:5e3e:b1bc:e121:a8eb:14aa/64',
]

for ip in ADDRESSES:
    iface = ipaddress.ip_interface(ip)
    print('{!r}'.format(iface))
    print('network:\n  ', iface.network)
    print('ip:\n  ', iface.ip)
    print('IP with prefixlen:\n  ', iface.with_prefixlen)
    print('netmask:\n  ', iface.with_netmask)
    print('hostmask:\n  ', iface.with_hostmask)
    print()
```

接口对象包含一些属性，可以分别访问完整的网络和地址，另外提供了多种不同方法来表示接口和网络掩码。

```
$ python3 ipaddress_interfaces.py

IPv4Interface('10.9.0.6/24')
network:
   10.9.0.0/24
ip:
   10.9.0.6
IP with prefixlen:
   10.9.0.6/24
netmask:
   10.9.0.6/255.255.255.0
hostmask:
   10.9.0.6/0.0.0.255

IPv6Interface('fdfd:87b5:b475:5e3e:b1bc:e121:a8eb:14aa/64')
network:
   fdfd:87b5:b475:5e3e::/64
ip:
   fdfd:87b5:b475:5e3e:b1bc:e121:a8eb:14aa
IP with prefixlen:
```

```
    fdfd:87b5:b475:5e3e:b1bc:e121:a8eb:14aa/64
netmask:
    fdfd:87b5:b475:5e3e:b1bc:e121:a8eb:14aa/ffff:ffff:ffff:ffff::
hostmask:
    fdfd:87b5:b475:5e3e:b1bc:e121:a8eb:14aa/::ffff:ffff:ffff:ffff
```

提示：相关阅读材料
- **ipaddress** 的标准库文档[一]。
- **PEP 3144**[二]：Python 标准库的 IP 地址管理库。
- **ipaddress** 模块的一个介绍[三]。
- Wikipedia：IP address[四]：IP 地址和网络的一个介绍。
- *Computer Networks*，第 5 版，作者 Andrew S. Tanenbaum 和 David J. Wetherall. Pearson, 2010. ISBN-10：0132126958。[五]

11.2 socket：网络通信

socket 模块提供了一个底层 C API，可以使用 BSD 套接字接口实现网络通信。它包括 socket 类，用于处理具体的数据通道，还包括用来完成网络相关任务的函数，如将一个服务器名转换为一个地址以及格式化数据以便在网络上发送。

11.2.1 寻址、协议簇和套接字类型

套接字（socket）是程序在本地或者通过互联网来回传递数据时所用通信通道的一个端点。套接字有两个主要属性用于控制如何发送数据：地址簇（address family）控制所用的 OSI 网络层协议；套接字类型（socket type）控制传输层协议。

Python 支持 3 个地址簇。最常用的是 AF_INET，用于 IPv4 Internet 寻址。IPv4 地址长度为 4 个字节，通常表示为 4 个数的序列，每个字节对应一个数，用点号分隔（如 10.1.1.5 和 127.0.0.1）。这些值通常被称为 "IP 地址"。目前几乎所有互联网网络通信都使用 IPv4。

AF_INET6 用于 IPv6 Internet 寻址。IPv6 是 "下一代" Internet 协议，它支持 128 位地址和通信流调整，还支持 IPv4 不支持的一些路由特性。采用 IPv6 的应用在不断增多，特别是随着云计算的大量普及以及由于物联网项目而为网络增加了很多额外的设备，都促使 IPv6 得到更广泛的应用。

AF_UNIX 是 UNIX 域套接字（UNIX Domain Socket，UDS）的地址簇，这是一种 POSIX 兼容系统上的进程间通信协议。UDS 的实现通常允许操作系统直接从进程向进程传递数据，而不用通过网络栈。这比使用 AF_INET 更高效，但是由于要用文件系统作为寻址

[一] https://docs.python.org/3.5/library/ipaddress.html
[二] www.python.org/dev/peps/pep-3144
[三] https://docs.python.org/3.5/howto/ipaddress.html#ipaddress-howto
[四] https://en.wikipedia.org/wiki/IP_address
[五] 英文影印版已由机械工业出版社引进出版，书号是 978-7-111-35925-8。——编辑注

的命名空间，所以 UDS 仅限于同一个系统上的进程。相比其他 IPC 机制（如命名管道或共享内存），使用 UDS 的优势在于它与 IP 网络应用的编程接口是一样的。这说明，应用在单个主机上运行时可以利用高效的通信，在网络上发送数据时仍然可以使用同样的代码。

说明：AF_UNIX 常量仅在支持 UDS 的系统上定义。

套接字类型往往是 SOCK_DGRAM 或 SOCK_STREAM，其中 SOCK_DGRAM 对应面向消息的数据报传输，而 SOCK_STREAM 对应面向流的传输。数据报套接字通常与 UDP 关联，即用户数据报协议（user datagram protocol）。这些套接字能提供不可靠的消息传送。面向流的套接字与 TCP 相关，即传输控制协议（transmission control protocol）。它们可以在客户和服务器之间提供字节流，通过超时管理、重传和其他特性确保提供消息传送或失败通知。

大多数传送大量数据的应用协议（如 HTTP）都建立在 TCP 基础上，因为这样可以更容易地创建自动处理消息排序和传送的复杂应用。UDP 通常用于顺序不太重要的协议（因为消息是自包含的，而且通常很小，如通过 DNS 的名字查找），或者用于组播（向多个主机发送相同的数据）。UDP 和 TCP 都可以用于 IPv4 或 IPv6 寻址。

说明：Python 的 socket 模块还支持其他套接字类型，不过它们不太常用，所以这里不做介绍。相关的更多详细信息可以参考标准库文档。

11.2.1.1 在网络上查找主机

socket 包含一些与网络上的域名服务交互的函数，这使得程序可以将服务器的主机名转换为其数字网络地址。应用使用地址连接服务器之前并不需要显式地转换地址，不过报告错误时除了报告所用的名字之外，如果还能包含这个数字地址，那么便会很有用。

要查找当前主机的正式名，可以使用 gethostname()。

代码清单 11-7：socket_gethostname.py

```
import socket

print(socket.gethostname())
```

所返回的名字取决于当前系统的网络设置，在不同的网络上返回的名字可能有变化（如连接到无线 LAN 的一个笔记本电脑）。

```
$ python3 socket_gethostname.py

apu.hellfly.net
```

这里使用 gethostbyname() 访问操作系统主机名解析 API，并且将服务器名转换为其数字地址。

代码清单 11-8：socket_gethostbyname.py

```
import socket

HOSTS = [
    'apu',
```

```
        'pymotw.com',
        'www.python.org',
        'nosuchname',
    ]

    for host in HOSTS:
        try:
            print('{} : {}'.format(host, socket.gethostbyname(host)))
        except socket.error as msg:
            print('{} : {}'.format(host, msg))
```

如果当前系统的 DNS 配置在搜索中包括一个或多个域,那么不要求名字(name)参数是完全限定名(也就是说,不需要包含域名以及基主机名)。如果无法找到一个名字,则会产生一个 socket.error 类型的异常。

```
$ python3 socket_gethostbyname.py

apu : 10.9.0.10
pymotw.com : 66.33.211.242
www.python.org : 151.101.32.223
nosuchname : [Errno 8] nodename nor servname provided, or not
known
```

要访问有关服务器的更多命名信息,可以使用函数 gethostbyname_ex()。它会返回服务器的标准主机名、所有别名,以及可以到达这个主机的所有可用 IP 地址。

代码清单 11-9:socket_gethostbyname_ex.py

```
import socket

HOSTS = [
    'apu',
    'pymotw.com',
    'www.python.org',
    'nosuchname',
]

for host in HOSTS:
    print(host)
    try:
        name, aliases, addresses = socket.gethostbyname_ex(host)
        print('  Hostname:', name)
        print('  Aliases :', aliases)
        print(' Addresses:', addresses)
    except socket.error as msg:
        print('ERROR:', msg)
    print()
```

如果能得到一个服务器的所有已知 IP 地址,客户就可以实现自己的负载平衡或故障恢复算法。

```
$ python3 socket_gethostbyname_ex.py

apu
  Hostname: apu.hellfly.net
  Aliases : ['apu']
 Addresses: ['10.9.0.10']

pymotw.com
```

```
Hostname: pymotw.com
 Aliases : []
 Addresses: ['66.33.211.242']

www.python.org
 Hostname: prod.python.map.fastlylb.net
 Aliases : ['www.python.org', 'python.map.fastly.net']
 Addresses: ['151.101.32.223']

nosuchname
ERROR: [Errno 8] nodename nor servname provided, or not known
```

使用 `getfqdn()` 可以将一个部分名转换为完全限定域名。

代码清单 11-10：**socket_getfqdn.py**

```
import socket

for host in ['apu', 'pymotw.com']:
    print('{:>10} : {}'.format(host, socket.getfqdn(host)))
```

如果输入是一个别名（如这里的 www），那么返回的名字不一定与输入参数一致。

```
$ python3 socket_getfqdn.py

       apu : apu.hellfly.net
pymotw.com : apache2-echo.catalina.dreamhost.com
```

如果得到一个服务器的地址，那么可以使用 `gethostbyaddr()` 完成一个"逆向"查找以得到主机名。

代码清单 11-11：**socket_gethostbyaddr.py**

```
import socket

hostname, aliases, addresses = socket.gethostbyaddr('10.9.0.10')

print('Hostname :', hostname)
print('Aliases  :', aliases)
print('Addresses:', addresses)
```

返回值是一个元组，其中包含完全主机名、所有别名，以及与这个名关联的所有 IP 地址。

```
$ python3 socket_gethostbyaddr.py

Hostname : apu.hellfly.net
Aliases  : ['apu']
Addresses: ['10.9.0.10']
```

11.2.1.2　查找服务信息

除了 IP 地址之外，每个套接字地址还包括一个整数端口号（port number）。很多应用可以在同一个主机上运行并监听一个 IP 地址，不过一次只有一个套接字可以使用该地址的端口。通过结合 IP 地址、协议和端口号，可以唯一地标识一个通信通道，并确保通过一个套接字发送的消息能到达正确的目标。

有些端口号已经预先分配给特定的协议。例如，使用 SMTP 的 email 服务器使用 TCP 在端口 25 完成通信，Web 客户和服务器使用端口 80 完成 HTTP 通信。可以使用

getservbyname()查找网络服务的端口号和标准名。

代码清单 11-12：**socket_getservbyname.py**

```
import socket
from urllib.parse import urlparse

URLS = [
    'http://www.python.org',
    'https://www.mybank.com',
    'ftp://prep.ai.mit.edu',
    'gopher://gopher.micro.umn.edu',
    'smtp://mail.example.com',
    'imap://mail.example.com',
    'imaps://mail.example.com',
    'pop3://pop.example.com',
    'pop3s://pop.example.com',
]

for url in URLS:
    parsed_url = urlparse(url)
    port = socket.getservbyname(parsed_url.scheme)
    print('{:>6} : {}'.format(parsed_url.scheme, port))
```

尽管标准化服务不太可能改变端口，但最好还是用一个系统调用查找端口值，而不是在程序中硬编码写出端口号，这样将来增加新服务时会更灵活。

```
$ python3 socket_getservbyname.py

  http : 80
 https : 443
   ftp : 21
gopher : 70
  smtp : 25
  imap : 143
 imaps : 993
  pop3 : 110
 pop3s : 995
```

要逆向完成服务端口查找，可以使用getservbyport()。

代码清单 11-13：**socket_getservbyport.py**

```
import socket
from urllib.parse import urlunparse
for port in [80, 443, 21, 70, 25, 143, 993, 110, 995]:
    url = '{}://example.com/'.format(socket.getservbyport(port))
    print(url)
```

要从任意的地址构造服务 URL，这个逆向查找就很有用。

```
$ python3 socket_getservbyport.py

http://example.com/
https://example.com/
ftp://example.com/
gopher://example.com/
smtp://example.com/
imap://example.com/
imaps://example.com/
```

```
pop3://example.com/
pop3s://example.com/
```

可以使用`getprotobyname()`获取分配给一个传输协议的端口号。

代码清单 11-14：socket_getprotobyname.py

```python
import socket


def get_constants(prefix):
    """Create a dictionary mapping socket module
    constants to their names.
    """
    return {
        getattr(socket, n): n
        for n in dir(socket)
        if n.startswith(prefix)
    }


protocols = get_constants('IPPROTO_')

for name in ['icmp', 'udp', 'tcp']:
    proto_num = socket.getprotobyname(name)
    const_name = protocols[proto_num]
    print('{:>4} -> {:2d} (socket.{:<12} = {:2d})'.format(
        name, proto_num, const_name,
        getattr(socket, const_name)))
```

协议码值是标准化的，在`socket`中被定义为常量，这些协议码都有前缀`IPPROTO_`。

```
$ python3 socket_getprotobyname.py

icmp ->  1 (socket.IPPROTO_ICMP =  1)
 udp -> 17 (socket.IPPROTO_UDP = 17)
 tcp ->  6 (socket.IPPROTO_TCP =  6)
```

11.2.1.3 查找服务器地址

`getaddrinfo()`将一个服务的基本地址转换为一个元组列表，其中包含建立一个连接所需的全部信息。每个元组可能包含不同的网络簇或协议。

代码清单 11-15：socket_getaddrinfo.py

```python
import socket


def get_constants(prefix):
    """Create a dictionary mapping socket module
    constants to their names.
    """
    return {
        getattr(socket, n): n
        for n in dir(socket)
        if n.startswith(prefix)
    }


families = get_constants('AF_')
types = get_constants('SOCK_')
```

```
protocols = get_constants('IPPROTO_')

for response in socket.getaddrinfo('www.python.org', 'http'):

    # Unpack the response tuple.
    family, socktype, proto, canonname, sockaddr = response

    print('Family          :', families[family])
    print('Type            :', types[socktype])
    print('Protocol        :', protocols[proto])
    print('Canonical name:', canonname)
    print('Socket address:', sockaddr)
    print()
```

这个程序展示了如何查找 www.python.org 的连接信息。

```
$ python3 socket_getaddrinfo.py

Family          : AF_INET
Type            : SOCK_DGRAM
Protocol        : IPPROTO_UDP
Canonical name:
Socket address: ('151.101.32.223', 80)

Family          : AF_INET
Type            : SOCK_STREAM
Protocol        : IPPROTO_TCP
Canonical name:
Socket address: ('151.101.32.223', 80)

Family          : AF_INET6
Type            : SOCK_DGRAM
Protocol        : IPPROTO_UDP
Canonical name:
Socket address: ('2a04:4e42:8::223', 80, 0, 0)

Family          : AF_INET6
Type            : SOCK_STREAM
Protocol        : IPPROTO_TCP
Canonical name:
Socket address: ('2a04:4e42:8::223', 80, 0, 0)
```

getaddrinfo() 有多个参数用于过滤结果列表。例子中给出的主机（host）和端口（port）值是必要参数。可选参数包括 family、socktype、proto 和 flags。这些可选值可以取 0 或由 socket 定义的某个常量。

代码清单 11-16：**socket_getaddrinfo_extra_args.py**

```python
import socket


def get_constants(prefix):
    """Create a dictionary mapping socket module
    constants to their names.
    """
    return {
        getattr(socket, n): n
        for n in dir(socket)
        if n.startswith(prefix)
    }
```

```python
families = get_constants('AF_')
types = get_constants('SOCK_')
protocols = get_constants('IPPROTO_')

responses = socket.getaddrinfo(
    host='www.python.org',
    port='http',
    family=socket.AF_INET,
    type=socket.SOCK_STREAM,
    proto=socket.IPPROTO_TCP,
    flags=socket.AI_CANONNAME,
)

for response in responses:
    # Unpack the response tuple.
    family, socktype, proto, canonname, sockaddr = response

    print('Family        :', families[family])
    print('Type          :', types[socktype])
    print('Protocol      :', protocols[proto])
    print('Canonical name:', canonname)
    print('Socket address:', sockaddr)
    print()
```

这一次由于标志（flags）包括 AI_CANONNAME，如果这个主机有别名，那么结果中会包含服务器的标准名（可能与查找所用的值不同）。如果没有这个标志，标准名值则仍为空。

```
$ python3 socket_getaddrinfo_extra_args.py

Family        : AF_INET
Type          : SOCK_STREAM
Protocol      : IPPROTO_TCP
Canonical name: prod.python.map.fastlylb.net
Socket address: ('151.101.32.223', 80)
```

11.2.1.4　IP 地址表示

用 C 编写的网络程序使用数据类型 struct sockaddr 将 IP 地址表示为二进制值（而不是 Python 程序中常见的字符串地址）。要完成 IPv4 地址的 Python 表示和 C 表示之间的转换，可以使用 inet_aton() 和 inet_ntoa()。

代码清单 11-17：socket_address_packing.py

```python
import binascii
import socket
import struct
import sys

for string_address in ['192.168.1.1', '127.0.0.1']:
    packed = socket.inet_aton(string_address)
    print('Original:', string_address)
    print('Packed  :', binascii.hexlify(packed))
    print('Unpacked:', socket.inet_ntoa(packed))
    print()
```

数据包格式中的 4 个字节可以被传递到 C 库，通过网络安全地传输，或者紧凑地保存在数据库中。

```
$ python3 socket_address_packing.py

Original: 192.168.1.1
Packed   : b'c0a80101'
Unpacked: 192.168.1.1

Original: 127.0.0.1
Packed   : b'7f000001'
Unpacked: 127.0.0.1
```

相关函数 inet_pton() 和 inet_ntop() 能处理 IPv4 和 IPv6 地址，根据传入的地址簇参数生成适当的格式。

代码清单 11-18：socket_ipv6_address_packing.py

```python
import binascii
import socket
import struct
import sys

string_address = '2002:ac10:10a:1234:21e:52ff:fe74:40e'
packed = socket.inet_pton(socket.AF_INET6, string_address)

print('Original:', string_address)
print('Packed  :', binascii.hexlify(packed))
print('Unpacked:', socket.inet_ntop(socket.AF_INET6, packed))
```

IPv6 地址已经是十六进制值，所以将打包版本转换为一系列十六进制数位时会生成一个与原值类似的串。

```
$ python3 socket_ipv6_address_packing.py

Original: 2002:ac10:10a:1234:21e:52ff:fe74:40e
Packed   : b'2002ac10010a1234021e52fffe74040e'
Unpacked: 2002:ac10:10a:1234:21e:52ff:fe74:40e
```

提示：相关阅读材料

- Wikipedia：IPv6⊖：讨论 Internet Protocol 6 (IPv6) 的维基百科文章。
- Wikipedia：OSI model⊜：维基百科文章，介绍网络实现的 7 层模型。
- Assigned Internet Protocol Numbers⊜：标准协议名和协议码列表。

11.2.2 TCP/IP 客户和服务器

套接字可以被配置为一个服务器，监听到来的消息，或者也可以被配置为连接到其他应用的客户。TCP/IP 套接字的两端连接后，可以完成双向通信。

11.2.2.1 回送服务器

下面这个示例程序以标准库文档中的一个例子为基础，它接收到来的消息，再回送给

⊖ https://en.wikipedia.org/wiki/IPv6
⊜ https://en.wikipedia.org/wiki/OSI_model
⊜ www.iana.org/assignments/protocol-numbers/protocol-numbers.xml

发送者。首先创建一个TCP/IP套接字，然后使用`bind()`将这个套接字与服务器地址关联。在这里，地址是`localhost`（指示当前服务器），端口号为10000。

代码清单11-19：`socket_echo_server.py`

```python
import socket
import sys

# Create a TCP/IP socket.
sock = socket.socket(socket.AF_INET, socket.SOCK_STREAM)

# Bind the socket to the port.
server_address = ('localhost', 10000)
print('starting up on {} port {}'.format(*server_address))
sock.bind(server_address)

# Listen for incoming connections.
sock.listen(1)

while True:
    # Wait for a connection.
    print('waiting for a connection')
    connection, client_address = sock.accept()
    try:
        print('connection from', client_address)

        # Receive the data in small chunks and retransmit it.
        while True:
            data = connection.recv(16)
            print('received {!r}'.format(data))
            if data:
                print('sending data back to the client')
                connection.sendall(data)
            else:
                print('no data from', client_address)
                break

    finally:
        # Clean up the connection.
        connection.close()
```

调用`listen()`将这个套接字设置为服务器模式，调用`accept()`等待入站连接。整数参数是在后台排队的连接数，达到这个连接数后，系统会拒绝连接新客户。这个例子希望一次只处理一个连接。

`accept()`返回服务器和客户之间的一个打开的连接，并返回客户地址。这个连接实际上是另一个端口上的一个不同的套接字（由内核分配）。使用`recv()`从连接读取数据，并用`sendall()`传输数据。

与一个客户的通信完成时，需要用`close()`清理这个连接。这个例子使用了一个`try:finally`块来确保总会调用`close()`，即使出现了错误也不例外。

11.2.2.2 回送客户

与服务器不同，客户程序采用另外一种方式建立`socket`。它不是绑定到一个端口并监听，而是使用`connect()`将套接字直接关联到远程地址。

代码清单 11-20：`socket_echo_client.py`

```python
import socket
import sys

# Create a TCP/IP socket.
sock = socket.socket(socket.AF_INET, socket.SOCK_STREAM)

# Connect the socket to the port where the server is listening.
server_address = ('localhost', 10000)
print('connecting to {} port {}'.format(*server_address))
sock.connect(server_address)

try:
    # Send data.
    message = b'This is the message.  It will be repeated.'
    print('sending {!r}'.format(message))
    sock.sendall(message)

    # Look for the response.
    amount_received = 0
    amount_expected = len(message)

    while amount_received < amount_expected:
        data = sock.recv(16)
        amount_received += len(data)
        print('received {!r}'.format(data))

finally:
    print('closing socket')
    sock.close()
```

建立连接之后，可以通过 `socket` 利用 `sendall()` 发送数据，并用 `recv()` 接收数据，这与服务器中是一样的。发送整个消息并接收到一个副本时，套接字会被关闭以释放端口。

11.2.2.3 客户与服务器

要在不同的终端窗口运行客户和服务器，使它们能够相互通信。服务器输出显示了入站连接和数据，以及发回给客户的响应。

```
$ python3 socket_echo_server.py
starting up on localhost port 10000
waiting for a connection
connection from ('127.0.0.1', 65141)
received b'This is the mess'
sending data back to the client
received b'age.  It will be'
sending data back to the client
received b' repeated.'
sending data back to the client
received b''
no data from ('127.0.0.1', 65141)
waiting for a connection
```

客户输出显示了发出的消息和来自服务器的响应。

```
$ python3 socket_echo_client.py
connecting to localhost port 10000
sending b'This is the message.  It will be repeated.'
received b'This is the mess'
```

```
received b'age.  It will be'
received b' repeated.'
closing socket
```

11.2.2.4 简易客户连接

如果使用便利函数 `create_connection()` 来连接服务器，那么 TCP/IP 客户可以省去几步。这个函数只有一个参数，这是一个包含服务器地址的二值元组，函数将由这个参数推导出用于连接的最佳地址。

代码清单 11-21: `socket_echo_client_easy.py`

```python
import socket
import sys


def get_constants(prefix):
    """Create a dictionary mapping socket module
    constants to their names.
    """
    return {
        getattr(socket, n): n
        for n in dir(socket)
        if n.startswith(prefix)
    }


families = get_constants('AF_')
types = get_constants('SOCK_')
protocols = get_constants('IPPROTO_')

# Create a TCP/IP socket.
sock = socket.create_connection(('localhost', 10000))

print('Family  :', families[sock.family])
print('Type    :', types[sock.type])
print('Protocol:', protocols[sock.proto])
print()

try:

    # Send data.
    message = b'This is the message.  It will be repeated.'
    print('sending {!r}'.format(message))
    sock.sendall(message)

    amount_received = 0
    amount_expected = len(message)
    while amount_received < amount_expected:
        data = sock.recv(16)
        amount_received += len(data)
        print('received {!r}'.format(data))

finally:
    print('closing socket')
    sock.close()
```

`create_connection()` 使用 `getaddrinfo()` 查找候选连接参数，并返回一个打开的 `socket`，这是能成功创建一个连接的第一个配置。可以检查 `family`、`type` 和 `proto` 属性来确定返回的 `socket` 的类型。

```
$ python3 socket_echo_client_easy.py
Family  : AF_INET
Type    : SOCK_STREAM
Protocol: IPPROTO_TCP

sending b'This is the message.  It will be repeated.'
received b'This is the mess'
received b'age.  It will be'
received b' repeated.'
closing socket
```

11.2.2.5 选择监听地址

将服务器绑定到正确的地址很重要,这样客户才能与之通信。前面的例子都使用 `'localhost'` 作为 IP 地址,这会限制其与在同一服务器上运行的客户的连接。可以使用服务器的一个公共地址,如 `gethostname()` 返回的值,来允许其他主机连接。下面这个例子修改了回送服务器,使它监听通过一个命令行参数指定的地址。

代码清单 11-22: `socket_echo_server_explicit.py`

```python
import socket
import sys

# Create a TCP/IP socket.
sock = socket.socket(socket.AF_INET, socket.SOCK_STREAM)

# Bind the socket to the address given on the command line.
server_name = sys.argv[1]
server_address = (server_name, 10000)
print('starting up on {} port {}'.format(*server_address))
sock.bind(server_address)
sock.listen(1)

while True:
    print('waiting for a connection')
    connection, client_address = sock.accept()
    try:
        print('client connected:', client_address)
        while True:
            data = connection.recv(16)
            print('received {!r}'.format(data))
            if data:
                connection.sendall(data)
            else:
                break
    finally:
        connection.close()
```

测试这个服务器之前,需要对客户程序做类似的修改。

代码清单 11-23: `socket_echo_client_explicit.py`

```python
import socket
import sys

# Create a TCP/IP socket.
sock = socket.socket(socket.AF_INET, socket.SOCK_STREAM)

# Connect the socket to the port on the server
```

```python
    # given by the caller.
    server_address = (sys.argv[1], 10000)
    print('connecting to {} port {}'.format(*server_address))
    sock.connect(server_address)

    try:

        message = b'This is the message.  It will be repeated.'
        print('sending {!r}'.format(message))
        sock.sendall(message)

        amount_received = 0
        amount_expected = len(message)
        while amount_received < amount_expected:
            data = sock.recv(16)
            amount_received += len(data)
            print('received {!r}'.format(data))

    finally:
        sock.close()
```

启动服务器并提供参数 hubert 之后，netstat 命令显示出它在监听指定主机的地址。

```
$ host hubert.hellfly.net

hubert.hellfly.net has address 10.9.0.6

$ netstat -an | grep 10000

Active Internet connections (including servers)
Proto Recv-Q Send-Q  Local Address      Foreign Address     (state)
...
tcp4       0      0  10.9.0.6.10000     *.*                 LISTEN
...
```

在另一个主机上运行这个客户时，传入 hubert.hellfly.net 作为运行服务器的主机，会生成以下输出：

```
$ hostname

apu

$ python3 ./socket_echo_client_explicit.py hubert.hellfly.net
connecting to hubert.hellfly.net port 10000
sending b'This is the message.  It will be repeated.'
received b'This is the mess'
received b'age.  It will be'
received b' repeated.'
```

服务器输出如下：

```
$ python3 socket_echo_server_explicit.py hubert.hellfly.net
starting up on hubert.hellfly.net port 10000
waiting for a connection
client connected: ('10.9.0.10', 33139)
received b''
waiting for a connection
client connected: ('10.9.0.10', 33140)
received b'This is the mess'
received b'age.  It will be'
received b' repeated.'
received b''
waiting for a connection
```

很多服务器有不止一个网络接口,相应地也会有不止一个 IP 地址。不需要运行服务的不同副本分别被绑定到各个 IP 地址,可以使用一个特殊的地址 `INADDR_ANY` 同时监听所有地址。尽管 `socket` 为 `INADDR_ANY` 定义了一个常量,这是一个整数值,但在传递到 `bind()` 之前必须将它转换为采用点记法的地址字符串。作为一种快捷方式,可以使用 `0.0.0.0` 或空串(`''`)而不是完成这个转换。

代码清单 11-24:<code>socket_echo_server_any.py</code>

```python
import socket
import sys

# Create a TCP/IP socket.
sock = socket.socket(socket.AF_INET, socket.SOCK_STREAM)

# Bind the socket to the address given on the command line.
server_address = ('', 10000)
sock.bind(server_address)
print('starting up on {} port {}'.format(*sock.getsockname()))
sock.listen(1)

while True:
    print('waiting for a connection')
    connection, client_address = sock.accept()
    try:
        print('client connected:', client_address)
        while True:
            data = connection.recv(16)
            print('received {!r}'.format(data))
            if data:
                connection.sendall(data)
            else:
                break
    finally:
        connection.close()
```

要看一个套接字使用的具体地址,可以调用它的 `getsockname()` 方法。启动服务后,再次运行 `netstat`,可以显示出它在监听所有地址的入站连接。

```
$ netstat -an

Active Internet connections (including servers)
Proto Recv-Q Send-Q  Local Address          Foreign Address        (state)
...
tcp4       0      0  *.10000                *.*                    LISTEN
...
```

11.2.3 用户数据报客户和服务器

用户数据报协议(User Datagram Protocol,UDP)的工作方式与 TCP/IP 不同。TCP 是一个面向流(stream-oriented)的协议,确保所有数据以正确的顺序传输,而 UDP 是一个面向消息(message-oriented)的协议。一方面,UDP 不需要一个长期活动的连接,所以建立 UDP 套接字稍简单一些。另一方面,UDP 消息必须放在一个数据报中(对于 IPv4,这意味着它们可以包含 65 507 个字节,因为 65 535 个字节大小的数据包还包括首部信息),而且无法得到 TCP 所能提供的传输保障。

11.2.3.1 回送服务器

由于实际上并没有连接,所以服务器并不需要监听和接收连接。它只需要使用 `bind()` 将其套接字与一个端口关联,然后等待各个消息。

代码清单 11-25:**socket_echo_server_dgram.py**

```python
import socket
import sys

# Create a UDP socket.
sock = socket.socket(socket.AF_INET, socket.SOCK_DGRAM)

# Bind the socket to the port.
server_address = ('localhost', 10000)
print('starting up on {} port {}'.format(*server_address))
sock.bind(server_address)

while True:
    print('\nwaiting to receive message')
    data, address = sock.recvfrom(4096)

    print('received {} bytes from {}'.format(
        len(data), address))
    print(data)

    if data:
        sent = sock.sendto(data, address)
        print('sent {} bytes back to {}'.format(
            sent, address))
```

使用 `recvfrom()` 从套接字读取消息,这个函数会返回数据,还会返回发出这个数据的客户的地址。

11.2.3.2 回送客户

UDP 回送客户与服务器类似,但是不使用 `bind()` 将套接字关联到一个地址。它使用 `sendto()` 将消息直接传送到服务器,并使用 `recvfrom()` 接收响应。

代码清单 11-26:**socket_echo_client_dgram.py**

```python
import socket
import sys

# Create a UDP socket.
sock = socket.socket(socket.AF_INET, socket.SOCK_DGRAM)

server_address = ('localhost', 10000)
message = b'This is the message.  It will be repeated.'

try:
    # Send data.
    print('sending {!r}'.format(message))
    sent = sock.sendto(message, server_address)

    # Receive response.
    print('waiting to receive')
    data, server = sock.recvfrom(4096)
    print('received {!r}'.format(data))
```

```
finally:
    print('closing socket')
    sock.close()
```

11.2.3.3 客户与服务器

运行这个服务器会生成以下输出：

```
$ python3 socket_echo_server_dgram.py
starting up on localhost port 10000

waiting to receive message
received 42 bytes from ('127.0.0.1', 57870)
b'This is the message.  It will be repeated.'
sent 42 bytes back to ('127.0.0.1', 57870)

waiting to receive message
```

客户输出如下：

```
$ python3 socket_echo_client_dgram.py
sending b'This is the message.  It will be repeated.'
waiting to receive
received b'This is the message.  It will be repeated.'
closing socket
```

11.2.4 UNIX 域套接字

从程序员的角度来看，使用 UNIX 域套接字和 TCP/IP 套接字存在两个根本区别。首先，套接字的地址是文件系统上的一个路径，而不是一个包含服务器名和端口的元组。其次，文件系统中创建的表示套接字的节点会持久保存，即使套接字关闭也仍然存在，所以每次服务器启动时都需要将其删除。只需在设置部分做一些修改，就可以把前面的回送服务器例子更新为使用 UDS。

需要基于地址簇 AF_Unix 创建 socket。套接字的绑定和入站连接的管理与对 TCP/IP 套接字的做法相同。

代码清单 11-27：socket_echo_server_uds.py

```
import socket
import sys
import os

server_address = './uds_socket'

# Make sure the socket does not already exist.
try:
    os.unlink(server_address)
except OSError:
    if os.path.exists(server_address):
        raise

# Create a UDS socket.
sock = socket.socket(socket.AF_UNIX, socket.SOCK_STREAM)

# Bind the socket to the address.
print('starting up on {}'.format(server_address))
sock.bind(server_address)
```

```python
# Listen for incoming connections.
sock.listen(1)

while True:
    # Wait for a connection.
    print('waiting for a connection')
    connection, client_address = sock.accept()
    try:
        print('connection from', client_address)

        # Receive the data in small chunks and retransmit it.
        while True:
            data = connection.recv(16)
            print('received {!r}'.format(data))
            if data:
                print('sending data back to the client')
                connection.sendall(data)
            else:
                print('no data from', client_address)
                break

    finally:
        # Clean up the connection.
        connection.close()
```

还需要修改客户设置以使用 UDS。要假设套接字的相应文件系统节点存在,因为服务器要通过绑定这个地址创建套接字。UDS 客户中发送和接收数据的做法与前面的 TCP/IP 客户是一样的。

代码清单 11-28:socket_echo_client_uds.py

```python
import socket
import sys

# Create a UDS socket.
sock = socket.socket(socket.AF_UNIX, socket.SOCK_STREAM)

# Connect the socket to the port where the server is listening.
server_address = './uds_socket'
print('connecting to {}'.format(server_address))
try:
    sock.connect(server_address)
except socket.error as msg:
    print(msg)
    sys.exit(1)

try:

    # Send data.
    message = b'This is the message.  It will be repeated.'
    print('sending {!r}'.format(message))
    sock.sendall(message)

    amount_received = 0
    amount_expected = len(message)

    while amount_received < amount_expected:
        data = sock.recv(16)
        amount_received += len(data)
        print('received {!r}'.format(data))
```

```
finally:
    print('closing socket')
    sock.close()
```

程序输出基本上相同,但对地址信息有适当的更新。服务器显示接收的消息和发回给客户的消息。

```
$ python3 socket_echo_server_uds.py
starting up on ./uds_socket
waiting for a connection
connection from
received b'This is the mess'
sending data back to the client
received b'age.  It will be'
sending data back to the client
received b' repeated.'
sending data back to the client
received b''
no data from
waiting for a connection
```

客户立即发送所有消息,并采用增量方式逐个部分地接收消息。

```
$ python3 socket_echo_client_uds.py
connecting to ./uds_socket
sending b'This is the message.  It will be repeated.'
received b'This is the mess'
received b'age.  It will be'
received b' repeated.'
closing socket
```

11.2.4.1 权限

由于 UDS 套接字由文件系统上的一个节点表示,所以可以使用标准文件系统权限来控制对服务器的访问。

```
$ ls -l ./uds_socket

srwxr-xr-x  1 dhellmann   dhellmann   0 Aug 21 11:19 uds_socket

$ sudo chown root ./uds_socket

$ ls -l ./uds_socket

srwxr-xr-x  1 root    dhellmann   0 Aug 21 11:19 uds_socket
```

如果现在客户作为一个用户运行而不是作为 root 运行,那么这会导致一个错误,因为这个进程没有打开套接字的权限。

```
$ python3 socket_echo_client_uds.py

connecting to ./uds_socket
[Errno 13] Permission denied
```

11.2.4.2 父进程与子进程间通信

在 UNIX 下,`socketpair()` 函数对于建立 UDS 套接字完成进程间通信很有用。它会创建一对连接的套接字,创建子进程之后,可以用来在父进程和子进程之间通信。

代码清单 11-29: **socket_socketpair.py**

```python
import socket
import os

parent, child = socket.socketpair()

pid = os.fork()

if pid:
    print('in parent, sending message')
    child.close()
    parent.sendall(b'ping')
    response = parent.recv(1024)
    print('response from child:', response)
    parent.close()

else:
    print('in child, waiting for message')
    parent.close()
    message = child.recv(1024)
    print('message from parent:', message)
    child.sendall(b'pong')
    child.close()
```

默认地会创建一个 UDS 套接字，不过调用者还可以通过传递地址簇，套接字类型，甚至协议选项来指定如何创建套接字。

```
$ python3 -u socket_socketpair.py

in parent, sending message
in child, waiting for message
message from parent: b'ping'
response from child: b'pong'
```

11.2.5 组播

点对点连接可以处理很多通信需求，不过随着直接连接数的增加，在多对通信方之间传递相同的消息会变得越来越困难。单独地向各个接收方发送消息会耗费额外的处理时间和带宽，这对于诸如完成流视频或音频操作的应用来说会带来问题。使用组播（multicast）向多个端点同时发送消息可以得到更好的效率，因为网络基础设施可以确保数据包会被传送到所有接收方。

组播消息总是使用 UDP 发送，因为 TCP 需要提供一对通信系统。组播的地址被称为组播组（multicast group），这是常规的 IPv4 地址范围的一个子集（224.0.0.0 ～ 230.255.255.255），专门为组播通信预留。这些地址会由网络路由器和交换机进行特殊处理，所以发送到组的消息可以在互联网上被分发到加入这个组的所有接收方。

说明： 一些托管交换机和路由器默认地会禁用组播通信。如果运行这些示例程序有问题，那么可以检查你的网络设置。

11.2.5.1 发送组播消息

下一个例子中修改后的回送客户会向一个组播组发送一个消息，然后报告它收到的所有响应。由于无法知道会收到多少响应,所以它对套接字使用了一个超时值,以避免等待回答时无限阻塞。

配置这个套接字时还需要提供消息的一个生存时间值(Time-To-Live value,TTL)。TTL会控制多少网络接收这个数据包。要使用 `IP_MULTICAST_TTL` 选项和 `setsockopt()` 来设置 TTL。默认值 1 表示路由器不会把数据包转发到当前网段之外。TTL 的取值范围最大为 255,应包装为一个字节。

代码清单 11-30:`socket_multicast_sender.py`

```python
import socket
import struct
import sys

message = b'very important data'
multicast_group = ('224.3.29.71', 10000)

# Create the datagram socket.
sock = socket.socket(socket.AF_INET, socket.SOCK_DGRAM)

# Set a timeout so the socket does not block
# indefinitely when trying to receive data.
sock.settimeout(0.2)

# Set the time-to-live for messages to 1 so they do not
# go past the local network segment.
ttl = struct.pack('b', 1)
sock.setsockopt(socket.IPPROTO_IP, socket.IP_MULTICAST_TTL, ttl)

try:

    # Send data to the multicast group.
    print('sending {!r}'.format(message))
    sent = sock.sendto(message, multicast_group)

    # Look for responses from all recipients.
    while True:
        print('waiting to receive')
        try:
            data, server = sock.recvfrom(16)
        except socket.timeout:
            print('timed out, no more responses')
            break
        else:
            print('received {!r} from {}'.format(
                data, server))

finally:
    print('closing socket')
    sock.close()
```

发送者程序的其余代码类似于 UDP 回送客户,只是它可能会接收多个响应,所以这里使用一个循环来调用 `recvfrom()`,直到超时。

11.2.5.2 接收组播消息

建立组播接收者的第一步是创建 UDP 套接字。创建常规的套接字并绑定到一个端口后,可以使用 `setsockopt()` 改变 `IP_ADD_MEMBERSHIP` 选项,把它增加到组播组。这个选项值是组播组地址的一个 8 字节的打包表示，后面是服务器监听通信流的网络接口（由其 IP 地址标识）。在这里，接收者使用 INADDR_ANY 监听所有接口。

代码清单 11-31：**socket_multicast_receiver.py**

```python
import socket
import struct
import sys

multicast_group = '224.3.29.71'
server_address = ('', 10000)

# Create the socket.
sock = socket.socket(socket.AF_INET, socket.SOCK_DGRAM)

# Bind to the server address.
sock.bind(server_address)

# Tell the operating system to add the socket to
# the multicast group on all interfaces.
group = socket.inet_aton(multicast_group)
mreq = struct.pack('4sL', group, socket.INADDR_ANY)
sock.setsockopt(
    socket.IPPROTO_IP,
    socket.IP_ADD_MEMBERSHIP,
    mreq)

# Receive/respond loop
while True:
    print('\nwaiting to receive message')
    data, address = sock.recvfrom(1024)

    print('received {} bytes from {}'.format(
        len(data), address))
    print(data)

    print('sending acknowledgement to', address)
    sock.sendto(b'ack', address)
```

接收者的主循环与常规 UDP 回送服务器类似。

11.2.5.3 示例输出

这个例子显示了组播接收者在两个不同的主机上运行。A 地址为 `192.168.1.13`，B 地址为 `192.168.1.14`。

```
[A]$ python3 socket_multicast_receiver.py

waiting to receive message
received 19 bytes from ('192.168.1.14', 62650)
b'very important data'
sending acknowledgement to ('192.168.1.14', 62650)

waiting to receive message

[B]$ python3 source/socket/socket_multicast_receiver.py

waiting to receive message
received 19 bytes from ('192.168.1.14', 64288)
b'very important data'
```

```
sending acknowledgement to ('192.168.1.14', 64288)

waiting to receive message
```

发送者在主机 B 上运行。

```
[B]$ python3 socket_multicast_sender.py
sending b'very important data'
waiting to receive
received b'ack' from ('192.168.1.14', 10000)
waiting to receive
received b'ack' from ('192.168.1.13', 10000)
waiting to receive

timed out, no more responses
closing socket
```

消息只发送一次,但是会接收到相对发出消息的两个应答,分别来自主机 A 和主机 B。

提示:相关阅读材料
- Wikipedia:Multicast[⊖]:维基百科文章,介绍组播的技术细节。
- Wikipedia:IP multicast[⊖]:有关 IP 组播的维基百科文章,提供了寻址的有关信息。

11.2.6 发送二进制数据

套接字可以传输字节流。这些字节可能包含编码为字节的文本消息(如前面的例子所示),或者它们也可能由二进制数据构成,这些二进制数据已经用 `struct` 打包到一个缓冲区以便传输。

下面这个客户程序将一个整数、一个包含两字符的字符串和一个浮点值编码为一个字节序列,从而能传递到套接字完成传输。

代码清单 11-32:`socket_binary_client.py`

```python
import binascii
import socket
import struct
import sys

# Create a TCP/IP socket.
sock = socket.socket(socket.AF_INET, socket.SOCK_STREAM)
server_address = ('localhost', 10000)
sock.connect(server_address)

values = (1, b'ab', 2.7)
packer = struct.Struct('I 2s f')
packed_data = packer.pack(*values)

print('values =', values)

try:
    # Send data.
    print('sending {!r}'.format(binascii.hexlify(packed_data)))
```

[⊖] https://en.wikipedia.org/wiki/Multicast
[⊖] https://en.wikipedia.org/wiki/IP_multicast

```
        sock.sendall(packed_data)
finally:
    print('closing socket')
    sock.close()
```

在两个系统之间发送多字节的二进制数据时,有一点很重要,要确保连接的两端都知道采用怎样的字节顺序,以及如何把它们重新组装为适合本地体系结构的正确顺序。服务器程序使用相同的 Struct 指示符(specifier)来解开接收到的字节,以便使用正确的顺序解释。

代码清单 11-33:`socket_binary_server.py`

```
import binascii
import socket
import struct
import sys

# Create a TCP/IP socket.
sock = socket.socket(socket.AF_INET, socket.SOCK_STREAM)
server_address = ('localhost', 10000)
sock.bind(server_address)
sock.listen(1)

unpacker = struct.Struct('I 2s f')

while True:
    print('\nwaiting for a connection')
    connection, client_address = sock.accept()
    try:
        data = connection.recv(unpacker.size)
        print('received {!r}'.format(binascii.hexlify(data)))

        unpacked_data = unpacker.unpack(data)
        print('unpacked:', unpacked_data)

    finally:
        connection.close()
```

运行客户会生成以下输出:

```
$ python3 source/socket/socket_binary_client.py
values = (1, b'ab', 2.7)
sending b'0100000061620000cdcc2c40'
closing socket
```

服务器显示了它接收的值:

```
$ python3 socket_binary_server.py

waiting for a connection
received b'0100000061620000cdcc2c40'
unpacked: (1, b'ab', 2.700000047683716)

waiting for a connection
```

浮点值在打包和解包时会损失一些精度,不过数据确实能按期望的方式传输。有一点要记住,根据整数的值的不同,有时将它转换为文本然后再传输可能比使用 struct 更高效。整数 1 表示为字符串时只占用一个字节,而打包到结构中时会占用 4 个字节。

提示：相关阅读材料

- `struct`：在字符串和其他数据类型之间转换。

11.2.7 非阻塞通信和超时

默认地，`socket` 被配置为发送或接收数据时会阻塞，在套接字准备就绪之前将停止程序的执行。`send()` 调用等待有缓冲区空间来存放发出的数据，`recv()` 则等待其他程序发出数据来读取。这种形式的 I/O 操作很容易理解，不过可能导致操作很低效，如果两个程序最后都在等待对方发送或接收数据，那么可能会导致死锁。

有很多种方法来绕开这种情况。一种做法是对各个套接字分别使用单独的线程完成通信。不过，这可能引入线程间通信的其他复杂性。另一种选择是将套接字改为根本不阻塞，即使没有准备好来处理操作，也会立即返回。可以使用 `setblocking()` 方法改变一个套接字的阻塞标志。默认值为 1，这表示会阻塞。传入值 0 则会关闭阻塞。如果套接字将阻塞关闭，而且没有为处理操作做好准备，则会产生一个 `socket.error`。

一种折中的解决方案是为套接字操作设置一个超时值。可以使用 `settimeout()` 将 `socket` 的超时值改为一个浮点值，表示确定这个套接字未做好操作准备之前所阻塞的时间（秒数）。超过这个超时期限时，会产生一个 `timeout` 异常。

提示：相关阅读材料

- `socket` 的标准库文档[⊖]。
- `socket` 的 Python 2 到 Python 3 移植说明。
- `select`：测试一个套接字，查看其是否准备好完成非阻塞 I/O 读写。
- `SocketServer`：创建网络服务器的框架。
- `asyncio`：异步 I/O 和并发工具。
- `urllib` 和 `urllib2`：大多数网络客户都应当使用更方便的库通过 URL 来访问远程资源。
- Socket Programming HOWTO[⊖]：Gordon McMillan 提供的一个指南，包含在标准库文档中。
- *Foundations of Python Network Programming, Third Edition*, by Brandon Rhodes and John Goerzen. Apress, 2014. ISBN-10: 1430258543.
- *Unix Network Programming, Volume 1: The Sockets Networking API, Third Edition*, by W. Richard Stevens, Bill Fenner and Andrew M. Rudoff. Addison-Wesley, 2004. ISBN-10:0131411551.[⊜]

11.3 selectors：I/O 多路复用抽象

selectors 模块在 select 中平台特定的 I/O 监视函数之上提供了一个平台独立的抽象层。

⊖ https://docs.python.org/3.5/library/socket.html

⊖ https://docs.python.org/3/howto/sockets.html

⊜ 英文影印版已由机械工业出版社引进出版，书号为 7-111-14685-9。——编辑注

11.3.1 操作模型

`selectors` 中的 API 是基于事件的,与 `select` 中的 `poll()` 类似。它有多个实现,并且这个模块会自动设置别名 `DefaultSelector` 来指示对当前系统配置最为高效的一个实现。

选择器对象提供了一些方法,可以指定在一个套接字上查找哪些事件,然后以一种平台独立的方式让调用者等待事件。注册对事件的兴趣会创建一个 `SelectorKey`,其中包含套接字、所注册事件的有关信息,可能还有可选的应用数据。选择器的所有者调用它的 `select()` 方法来了解事件。返回值是一个键对象序列和一个指示发生了哪些事件的位掩码。使用选择器的程序要反复调用 `select()`,然后适当地处理事件。

11.3.2 回送服务器

这里给出的回送服务器例子使用了 `SelectorKey` 中的应用数据来注册发生新事件时要调用的一个回调函数。主循环从键得到这个回调,并把套接字和事件掩码传递给该回调。服务器启动时,其会注册当主服务器套接字上发生读事件时要调用的 `accept()` 函数。接受连接会产生一个新的套接字,然后注册 `read()` 函数作为读事件的一个回调。

代码清单 11-34:`selectors_echo_server.py`

```
import selectors
import socket

mysel = selectors.DefaultSelector()
keep_running = True


def read(connection, mask):
    "Callback for read events"
    global keep_running

    client_address = connection.getpeername()
    print('read({})'.format(client_address))
    data = connection.recv(1024)
    if data:
        # A readable client socket has data.
        print('  received {!r}'.format(data))
        connection.sendall(data)
    else:
        # Interpret empty result as closed connection.
        print('  closing')
        mysel.unregister(connection)
        connection.close()
        # Tell the main loop to stop.
        keep_running = False


def accept(sock, mask):
    "Callback for new connections"
    new_connection, addr = sock.accept()
    print('accept({})'.format(addr))
    new_connection.setblocking(False)
    mysel.register(new_connection, selectors.EVENT_READ, read)
```

```python
server_address = ('localhost', 10000)
print('starting up on {} port {}'.format(*server_address))
server = socket.socket(socket.AF_INET, socket.SOCK_STREAM)
server.setblocking(False)
server.bind(server_address)
server.listen(5)

mysel.register(server, selectors.EVENT_READ, accept)

while keep_running:
    print('waiting for I/O')
    for key, mask in mysel.select(timeout=1):
        callback = key.data
        callback(key.fileobj, mask)

print('shutting down')
mysel.close()
```

如果 read() 没有从套接字接收到任何数据,那么当连接的另一端关闭时,它会中断读事件而不是发送数据。之后,会从选择器删除这个套接字,并将其关闭。由于这只是一个示例程序,所以这个服务器与唯一的客户结束通信后还会关闭服务器自身。

11.3.3 回送客户

下面的回送客户例子会处理主循环中的所有 I/O 事件,而不是使用回调。它会建立选择器来报告套接字上的读事件,并报告套接字什么时候准备好可以发送数据。由于它查看两种类型的事件,所以客户必须通过查看掩码值来检查发生了哪个事件。所有数据都发出后,它会修改选择器配置,只在有可读取的数据时才会报告。

代码清单 11-35:**selectors_echo_client.py**

```python
import selectors
import socket

mysel = selectors.DefaultSelector()
keep_running = True
outgoing = [
    b'It will be repeated.',
    b'This is the message.  ',
]
bytes_sent = 0
bytes_received = 0

# Connecting is a blocking operation, so call setblocking()
# after it returns.
server_address = ('localhost', 10000)
print('connecting to {} port {}'.format(*server_address))
sock = socket.socket(socket.AF_INET, socket.SOCK_STREAM)
sock.connect(server_address)
sock.setblocking(False)

# Set up the selector to watch for when the socket is ready
# to send data as well as when there is data to read.
mysel.register(
    sock,
    selectors.EVENT_READ | selectors.EVENT_WRITE,
)
```

```
    while keep_running:
        print('waiting for I/O')
        for key, mask in mysel.select(timeout=1):
            connection = key.fileobj
            client_address = connection.getpeername()
            print('client({})'.format(client_address))

            if mask & selectors.EVENT_READ:
                print('  ready to read')
                data = connection.recv(1024)
                if data:
                    # A readable client socket has data.
                    print('  received {!r}'.format(data))
                    bytes_received += len(data)

                # Interpret empty result as closed connection,
                # and also close when we have received a copy
                # of all of the data sent.
                keep_running = not (
                    data or
                    (bytes_received and
                     (bytes_received == bytes_sent))
                )

            if mask & selectors.EVENT_WRITE:
                print('  ready to write')
                if not outgoing:
                    # We are out of messages, so we no longer need to
                    # write anything. Change our registration to let
                    # us keep reading responses from the server.
                    print('  switching to read-only')
                    mysel.modify(sock, selectors.EVENT_READ)
                else:
                    # Send the next message.
                    next_msg = outgoing.pop()
                    print('  sending {!r}'.format(next_msg))
                    sock.sendall(next_msg)
                    bytes_sent += len(next_msg)

    print('shutting down')
    mysel.unregister(connection)
    connection.close()
    mysel.close()
```

这个客户不仅跟踪它发出的数据量,还会跟踪接收的数据量。当这些值一致而且非 0 时,客户退出处理循环,并妥善地关闭,它将从选择器删除套接字,并关闭套接字和选择器。

11.3.4 服务器和客户

要在不同的终端窗口运行客户和服务器,使它们能够相互通信。服务器输出显示了入站连接和数据,以及发回给客户的响应。

```
$ python3 source/selectors/selectors_echo_server.py
starting up on localhost port 10000
waiting for I/O
waiting for I/O
accept(('127.0.0.1', 59850))
waiting for I/O
read(('127.0.0.1', 59850))
```

```
  received b'This is the message.  It will be repeated.'
waiting for I/O
read(('127.0.0.1', 59850))
  closing
shutting down
```

客户输出显示了发出的消息和从服务器得到的响应。

```
$ python3 source/selectors/selectors_echo_client.py
connecting to localhost port 10000
waiting for I/O
client(('127.0.0.1', 10000))
  ready to write
  sending b'This is the message.  '
waiting for I/O
client(('127.0.0.1', 10000))
  ready to write
  sending b'It will be repeated.'
waiting for I/O
client(('127.0.0.1', 10000))
  ready to write
  switching to read-only
waiting for I/O
client(('127.0.0.1', 10000))
  ready to read
  received b'This is the message.  It will be repeated.'
shutting down
```

提示：相关阅读材料
- selectors 的标准库文档[⊖]。
- select：高效处理 I/O 的底层 API。

11.4 select：高效等待 I/O

select 模块允许访问特定于平台的 I/O 监视函数。最可移植的接口是 POSIX 函数 select()，UNIX 和 Windows 都提供了这个函数。这个模块还包括函数 poll()（这个 API 只适用于 UNIX），另外还提供了很多只适用于一些 UNIX 特定版本的选项。

说明：新的 selectors 模块提供了建立在 select API 之上的一个更高层的接口。使用 selectors 可以更容易地构建可移植的代码，所以除非出于某种原因必须使用 select 提供的底层 API，否则最好使用 selectors 模块。

11.4.1 使用 select()

Python 的 select() 函数是底层操作系统实现的一个直接接口。它会监视套接字、打开的文件和管道（可以是任何有 fileno() 方法的对象，这个方法会返回一个合法的文件描述符），直到这个对象可读或可写，或者出现一个通信错误。利用 select() 可以很容易地同时监视多个连接，这比在 Python 中使用套接字超时编写一个轮询循环更为高效，因为

⊖ https://docs.python.org/3.5/library/selectors.html

监视发生在操作系统网络层而不是在解释器中完成。

说明: 对 Python 文件对象使用 `select()` 只适用于 UNIX,在 Windows 下并不支持。

可以扩展 `socket` 一节的回送服务器例子,通过使用 `select()` 同时监视多个连接。这个新版本首先创建一个非阻塞的 TCP/IP 套接字,将它配置为监听一个地址。

代码清单 11-36: **select_echo_server.py**

```python
import select
import socket
import sys
import queue

# Create a TCP/IP socket.
server = socket.socket(socket.AF_INET, socket.SOCK_STREAM)
server.setblocking(0)

# Bind the socket to the port.
server_address = ('localhost', 10000)
print('starting up on {} port {}'.format(*server_address),
      file=sys.stderr)
server.bind(server_address)

# Listen for incoming connections.
server.listen(5)
```

`select()` 的参数是 3 个列表,包含要监视的通信通道。首先检查第一个对象列表中的对象以得到要读取的数据,第二个列表中包含的对象将接收发出的数据(如果缓冲区有空间),第三个列表包含那些可能有错误的对象(通常是输入和输出通道对象的组合)。下一步是建立这些列表,其中包含要传至 `select()` 的输入源和输出目标。

```python
# Sockets from which we expect to read
inputs = [server]

# Sockets to which we expect to write
outputs = []
```

由服务器主循环向这些列表增加或删除连接。由于这个服务器版本在发送数据之前要等待套接字变为可写(而不是立即发送应答),所以每个输出连接都需要一个队列作为通过这个套接字发送的数据的缓冲区。

```python
# Outgoing message queues (socket:Queue)
message_queues = {}
```

服务器程序的主要部分会循环调用 `select()` 来阻塞并等待网络活动。

```python
while inputs:

    # Wait for at least one of the sockets to be
    # ready for processing.
    print('waiting for the next event', file=sys.stderr)
    readable, writable, exceptional = select.select(inputs,
                                                    outputs,
                                                    inputs)
```

`select()` 返回 3 个新列表,包含所传入列表内容的子集。`readable` 列表中的所有套接字会缓存到来的数据,可供读取。`writable`列表中所有套接字的缓冲区中有自由空间,可以写入数据。`exceptional`中返回的套接字都有一个错误("异常条件"的具体定义取决于平台)。

"可读"套接字表示3种可能的情况。如果套接字是主"服务器"套接字,即用来监听连接的套接字,那么"可读"条件就意味着它已经准备就绪,可以接受另一个入站连接。除了将新连接增加到要监视的输入列表,这里还会将客户套接字设置为非阻塞。

```
# Handle inputs.
for s in readable:

    if s is server:
        # A "readable" socket is ready to accept a connection.
        connection, client_address = s.accept()
        print('  connection from', client_address,
              file=sys.stderr)
        connection.setblocking(0)
        inputs.append(connection)

        # Give the connection a queue for data
        # we want to send.
        message_queues[connection] = queue.Queue()
```

下一种情况是与一个已发送数据的客户建立连接。数据用`recv()`读取,然后放置在队列中,以便通过套接字发送并返回给客户。

```
    else:
        data = s.recv(1024)
        if data:
            # A readable client socket has data.
            print('  received {!r} from {}'.format(
                data, s.getpeername()), file=sys.stderr,
            )
            message_queues[s].put(data)
            # Add output channel for response.
            if s not in outputs:
                outputs.append(s)
```

如果一个可读套接字未从`recv()`返回数据,那么说明这个套接字来自已经断开连接的客户,此时可以关闭流。

```
        else:
            # Interpret empty result as closed connection.
            print('  closing', client_address,
                  file=sys.stderr)
            # Stop listening for input on the connection.
            if s in outputs:
                outputs.remove(s)
            inputs.remove(s)
            s.close()

            # Remove message queue.
            del message_queues[s]
```

对于可写连接,情况要少一些。如果一个连接的相应队列中有数据,那么就发送下一个消息。否则,将这个连接从输出连接列表中删除,这样下一次循环时,`select()`不再指示这个套接字已准备好发送数据。

```python
# Handle outputs.
for s in writable:
    try:
        next_msg = message_queues[s].get_nowait()
    except queue.Empty:
        # No messages waiting, so stop checking
        # for writability.
        print('  ', s.getpeername(), 'queue empty',
              file=sys.stderr)
        outputs.remove(s)
    else:
        print('  sending {!r} to {}'.format(next_msg,
                                            s.getpeername()),
              file=sys.stderr)
        s.send(next_msg)
```

最后一种情况，exceptional 列表中的套接字会关闭。

```python
# Handle "exceptional conditions."
for s in exceptional:
    print('exception condition on', s.getpeername(),
          file=sys.stderr)
    # Stop listening for input on the connection.
    inputs.remove(s)
    if s in outputs:
        outputs.remove(s)
    s.close()

    # Remove message queue.
    del message_queues[s]
```

这个示例客户程序使用了两个套接字来展示服务器如何利用 select() 同时管理多个连接。客户首先将各个 TCP/IP 套接字连接到服务器。

代码清单 11-37：`select_echo_multiclient.py`

```python
import socket
import sys

messages = [
    'This is the message. ',
    'It will be sent ',
    'in parts.',
]
server_address = ('localhost', 10000)

# Create a TCP/IP socket.
socks = [
    socket.socket(socket.AF_INET, socket.SOCK_STREAM),
    socket.socket(socket.AF_INET, socket.SOCK_STREAM),
]
# Connect the socket to the port where the server is listening.
print('connecting to {} port {}'.format(*server_address),
      file=sys.stderr)
for s in socks:
    s.connect(server_address)
```

然后通过各个套接字一次发送一个消息，写新数据之后再读取得到的所有响应。

```python
for message in messages:
    outgoing_data = message.encode()
```

```
    # Send messages on both sockets.
    for s in socks:
        print('{}: sending {!r}'.format(s.getsockname(),
                                        outgoing_data),
              file=sys.stderr)
        s.send(outgoing_data)

    # Read responses on both sockets.
    for s in socks:
        data = s.recv(1024)
        print('{}: received {!r}'.format(s.getsockname(),
                                         data),
              file=sys.stderr)
        if not data:
            print('closing socket', s.getsockname(),
                  file=sys.stderr)
            s.close()
```

在一个窗口中运行服务器,在另一个窗口中运行客户程序。输出如下所示(不过端口号可能不同)。

```
$ python3 select_echo_server.py
starting up on localhost port 10000
waiting for the next event
  connection from ('127.0.0.1', 61003)
waiting for the next event
  connection from ('127.0.0.1', 61004)
waiting for the next event
  received b'This is the message. ' from ('127.0.0.1', 61003)
  received b'This is the message. ' from ('127.0.0.1', 61004)
waiting for the next event
  sending b'This is the message. ' to ('127.0.0.1', 61003)
  sending b'This is the message. ' to ('127.0.0.1', 61004)
waiting for the next event
   ('127.0.0.1', 61003) queue empty
   ('127.0.0.1', 61004) queue empty
waiting for the next event
  received b'It will be sent ' from ('127.0.0.1', 61003)
  received b'It will be sent ' from ('127.0.0.1', 61004)
waiting for the next event
  sending b'It will be sent ' to ('127.0.0.1', 61003)
  sending b'It will be sent ' to ('127.0.0.1', 61004)
waiting for the next event
   ('127.0.0.1', 61003) queue empty
   ('127.0.0.1', 61004) queue empty
waiting for the next event
  received b'in parts.' from ('127.0.0.1', 61003)
waiting for the next event
  received b'in parts.' from ('127.0.0.1', 61004)
  sending b'in parts.' to ('127.0.0.1', 61003)
waiting for the next event
   ('127.0.0.1', 61003) queue empty
  sending b'in parts.' to ('127.0.0.1', 61004)
waiting for the next event
   ('127.0.0.1', 61004) queue empty
waiting for the next event
  closing ('127.0.0.1', 61004)
  closing ('127.0.0.1', 61004)
waiting for the next event
```

客户程序的输出显示了使用这两个套接字发送和接收到的数据。

```
$ python3 select_echo_multiclient.py
connecting to localhost port 10000
('127.0.0.1', 61003): sending b'This is the message. '
('127.0.0.1', 61004): sending b'This is the message. '
('127.0.0.1', 61003): received b'This is the message. '
('127.0.0.1', 61004): received b'This is the message. '
('127.0.0.1', 61003): sending b'It will be sent '
('127.0.0.1', 61004): sending b'It will be sent '
('127.0.0.1', 61003): received b'It will be sent '
('127.0.0.1', 61004): received b'It will be sent '
('127.0.0.1', 61003): sending b'in parts.'
('127.0.0.1', 61004): sending b'in parts.'
('127.0.0.1', 61003): received b'in parts.'
('127.0.0.1', 61004): received b'in parts.'
```

11.4.2 带超时的非阻塞 I/O

select() 还可以有可选的第 4 个参数——如果没有活动通道，它在停止监视之前会等待一定时间，这个参数就是这里等待的秒数。通过使用一个超时值，可以让主程序在一个更大的处理循环中调用 select()，在检查网络输入的间隙完成其他动作。

超过这个超时期限时，select() 会返回 3 个空列表。如果要更新前面的服务器示例来使用一个超时值，则需要向 select() 调用增加一个额外的参数，并在 select() 返回后处理空列表。

代码清单 11-38：**select_echo_server_timeout.py**

```
readable, writable, exceptional = select.select(inputs,
                                                outputs,
                                                inputs,
                                                timeout)

if not (readable or writable or exceptional):
    print('  timed out, do some other work here',
          file=sys.stderr)
    continue
```

客户程序的这个"慢"版本会在发送各个消息之后暂停，模拟传输中的时延或其他延迟。

代码清单 11-39：**select_echo_slow_client.py**

```
import socket
import sys
import time

# Create a TCP/IP socket.
sock = socket.socket(socket.AF_INET, socket.SOCK_STREAM)

# Connect the socket to the port where the server is listening.
server_address = ('localhost', 10000)
print('connecting to {} port {}'.format(*server_address),
      file=sys.stderr)
sock.connect(server_address)

time.sleep(1)

messages = [
    'Part one of the message.',
```

```python
        'Part two of the message.',
    ]
    amount_expected = len(''.join(messages))

    try:
        # Send data.
        for message in messages:
            data = message.encode()
            print('sending {!r}'.format(data), file=sys.stderr)
            sock.sendall(data)
            time.sleep(1.5)

        # Look for the response.
        amount_received = 0

        while amount_received < amount_expected:
            data = sock.recv(16)
            amount_received += len(data)
            print('received {!r}'.format(data), file=sys.stderr)

    finally:
        print('closing socket', file=sys.stderr)
        sock.close()
```

运行这个新服务器和慢客户,服务器会生成以下结果:

```
$ python3 select_echo_server_timeout.py
starting up on localhost port 10000
waiting for the next event
  timed out, do some other work here
waiting for the next event
  connection from ('127.0.0.1', 61144)
waiting for the next event
  timed out, do some other work here
waiting for the next event
  received b'Part one of the message.' from ('127.0.0.1', 61144)
waiting for the next event
  sending b'Part one of the message.' to ('127.0.0.1', 61144)
waiting for the next event
('127.0.0.1', 61144) queue empty
waiting for the next event
  timed out, do some other work here
waiting for the next event
  received b'Part two of the message.' from ('127.0.0.1', 61144)
waiting for the next event
  sending b'Part two of the message.' to ('127.0.0.1', 61144)
waiting for the next event
('127.0.0.1', 61144) queue empty
waiting for the next event
  timed out, do some other work here
waiting for the next event
closing ('127.0.0.1', 61144)
waiting for the next event
  timed out, do some other work here
```

以下是客户输出:

```
$ python3 select_echo_slow_client.py
connecting to localhost port 10000
sending b'Part one of the message.'
sending b'Part two of the message.'
```

```
received b'Part one of the '
received b'message.Part two'
received b' of the message.'
closing socket
```

11.4.3 使用poll()

poll()函数提供的特性与select()类似,不过底层实现更为高效。其缺点是Windows不支持poll(),所以使用poll()的程序可移植性较差。

利用poll()建立的回送服务器最前面与其他例子一样,也使用了相同的套接字配置代码。

代码清单11-40:select_poll_echo_server.py

```python
import select
import socket
import sys
import queue

# Create a TCP/IP socket.
server = socket.socket(socket.AF_INET, socket.SOCK_STREAM)
server.setblocking(0)

# Bind the socket to the port.
server_address = ('localhost', 10000)
print('starting up on {} port {}'.format(*server_address),
      file=sys.stderr)
server.bind(server_address)

# Listen for incoming connections.
server.listen(5)

# Keep up with the queues of outgoing messages.
message_queues = {}
```

传入poll()的超时值用毫秒表示,而不是秒,所以如果要暂停1秒,则超时值必须设置为1000。

```python
# Do not block forever (milliseconds).
TIMEOUT = 1000
```

Python用一个类实现poll(),由这个类管理所监视的注册的数据通道。通过调用register()增加通道,同时利用标志指示该通道对哪些事件感兴趣。表11-1列出了所有标志。

表11-1 poll()的事件标志

事　　件	描　　述
POLLIN	输入准备就绪
POLLPRI	优先级输入准备就绪
POLLOUT	能够接收输出
POLLERR	错误
POLLHUP	通道关闭
POLLNVAL	通道未打开

回送服务器将建立一些只读的套接字，以及另外一些可以读写的套接字。局部变量
READ_ONLY 和 READ_WRITE 中保存了相应的标志组合。

```python
# Commonly used flag sets
READ_ONLY = (
    select.POLLIN |
    select.POLLPRI |
    select.POLLHUP |
    select.POLLERR
)
READ_WRITE = READ_ONLY | select.POLLOUT
```

这里注册了 `server` 套接字，入站连接或数据会触发一个事件。

```python
# Set up the poller.
poller = select.poll()
poller.register(server, READ_ONLY)
```

由于 `poll()` 返回一个元组列表，元组中包含套接字的文件描述符和事件标志，因此需要一个文件描述符编号到对象的映射来获取 `socket`，以便读取或写入。

```python
# Map file descriptors to socket objects.
fd_to_socket = {
    server.fileno(): server,
}
```

服务器的循环调用 `poll()`，然后处理由查找套接字返回的"事件"，并根据事件中的标志采取行动。

```python
while True:

    # Wait for at least one of the sockets to be
    # ready for processing.
    print('waiting for the next event', file=sys.stderr)
    events = poller.poll(TIMEOUT)

    for fd, flag in events:

        # Retrieve the actual socket from its file descriptor.
        s = fd_to_socket[fd]
```

与 `select()` 类似，如果主服务器套接字是"可读的"，那么实际上这表示有来自客户的一个连接。用 `READ_ONLY` 标志注册这个新连接，以便监视通过它的新数据。

```python
        # Handle inputs.
        if flag & (select.POLLIN | select.POLLPRI):

            if s is server:
                # A readable socket is ready
                # to accept a connection.
                connection, client_address = s.accept()
                print('  connection', client_address,
                      file=sys.stderr)
                connection.setblocking(0)
                fd_to_socket[connection.fileno()] = connection
                poller.register(connection, READ_ONLY)

                # Give the connection a queue for data to send.
                message_queues[connection] = queue.Queue()
```

除了服务器以外，其他套接字都是现有的客户，可以使用 `recv()` 访问等待读取的数据。

```
else:
    data = s.recv(1024)
```

如果recv()返回了数据,那么其会被放置在这个套接字相应的发出队列中。使用modify()改变套接字的标志,使poll()监视这个套接字是否准备好接收数据。

```
if data:
    # A readable client socket has data.
    print('  received {!r} from {}'.format(
        data, s.getpeername()), file=sys.stderr,
    )
    message_queues[s].put(data)
    # Add output channel for response.
    poller.modify(s, READ_WRITE)
```

recv()返回的空串表示客户已经断开连接,所以使用unregister()告诉poll对象忽略这个套接字。

```
else:
    # Interpret empty result as closed connection.
    print('  closing', client_address,
        file=sys.stderr)
    # Stop listening for input on the connection.
    poller.unregister(s)
    s.close()

    # Remove message queue.
    del message_queues[s]
```

POLLHUP标志指示一个客户"挂起"连接而没有将其妥善地关闭。服务器停止轮询消失的客户。

```
elif flag & select.POLLHUP:
    # Client hung up
    print('  closing', client_address, '(HUP)',
        file=sys.stderr)
    # Stop listening for input on the connection.
    poller.unregister(s)
    s.close()
```

可写套接字的处理看起来与select()例子中使用的版本类似,但却使用了modify()来改变轮询服务器中套接字的标志,而不是将它从输出列表中删除。

```
elif flag & select.POLLOUT:
    # Socket is ready to send data,
    # if there is any to send.
    try:
        next_msg = message_queues[s].get_nowait()
    except queue.Empty:
        # No messages waiting, so stop checking.
        print(s.getpeername(), 'queue empty',
            file=sys.stderr)
        poller.modify(s, READ_ONLY)
    else:
        print('  sending {!r} to {}'.format(
            next_msg, s.getpeername()), file=sys.stderr,
        )
        s.send(next_msg)
```

最后一点,任何有POLLERR错误的事件都会导致服务器关闭套接字。

```
    elif flag & select.POLLERR:
        print('  exception on', s.getpeername(),
              file=sys.stderr)
        # Stop listening for input on the connection.
        poller.unregister(s)
        s.close()

        # Remove message queue.
        del message_queues[s]
```

基于轮询的服务器与 `select_echo_multiclient.py`（使用多个套接字的客户程序）一起运行时，会生成以下输出。

```
$ python3 select_poll_echo_server.py
starting up on localhost port 10000
waiting for the next event
waiting for the next event
waiting for the next event
waiting for the next event
  connection ('127.0.0.1', 61253)
waiting for the next event
  connection ('127.0.0.1', 61254)
waiting for the next event
  received b'This is the message. ' from ('127.0.0.1', 61253)
  received b'This is the message. ' from ('127.0.0.1', 61254)
waiting for the next event
  sending b'This is the message. ' to ('127.0.0.1', 61253)
  sending b'This is the message. ' to ('127.0.0.1', 61254)
waiting for the next event
('127.0.0.1', 61253) queue empty
('127.0.0.1', 61254) queue empty
waiting for the next event
  received b'It will be sent ' from ('127.0.0.1', 61253)
  received b'It will be sent ' from ('127.0.0.1', 61254)
waiting for the next event
  sending b'It will be sent ' to ('127.0.0.1', 61253)
  sending b'It will be sent ' to ('127.0.0.1', 61254)
waiting for the next event
('127.0.0.1', 61253) queue empty
('127.0.0.1', 61254) queue empty
waiting for the next event
  received b'in parts.' from ('127.0.0.1', 61253)
  received b'in parts.' from ('127.0.0.1', 61254)
waiting for the next event
  sending b'in parts.' to ('127.0.0.1', 61253)
  sending b'in parts.' to ('127.0.0.1', 61254)
waiting for the next event
('127.0.0.1', 61253) queue empty
('127.0.0.1', 61254) queue empty
waiting for the next event
  closing ('127.0.0.1', 61254)
waiting for the next event
  closing ('127.0.0.1', 61254)
waiting for the next event
```

11.4.4 平台特定的选项

`select` 还提供了一些可移植性较差的选项，包括 epoll（Linux 支持的边界轮询 API）、kqueue（使用了 BSD 的内核队列）和 kevent（BSD 的内核事件接口）。有关这些选项如何工作的更多详细内容请参考操作系统库文档。

提示：相关阅读材料
- `select` 的标准库文档[一]。
- `selectors`：`select` 之上的更高层抽象。
- Socket Programming HOWTO[二]：Gordon McMillan 提供的指南，包含在标准库文档中。
- `socket`：底层网络通信。
- `SocketServer`：创建网络服务器应用的框架。
- `asyncio`：异步 I/O 框架。
- *Unix Network Programming, Volume 1*：*The Sockets Networking API, Third Edition*, by W. Richard Stevens, Bill Fenner, and Andrew M. Rudoff. Addison-Wesley, 2004. ISBN-10:0131411551.
- *Foundations of Python Network Programming, Third Edition*, by Brandon Rhodes and John Goerzen. Apress, 2014. ISBN-10：1430258543.

11.5 `socketserver`：创建网络服务器

　　`socketserver` 模块是创建网络服务器的一个框架。它定义了一些类来处理 TCP、UDP、UNIX 流和 UNIX 数据报之上的同步网络请求（服务器请求处理器会阻塞，直至请求完成）。它还提供了一些"混入"类（mix-in[三]），可以很容易地转换服务器，为每个请求使用一个单独的线程或进程。

　　处理请求的责任被划分到一个服务器类和一个请求处理器类之间。服务器处理通信问题，如监听一个套接字并接受连接，请求处理器处理"协议"问题，如解释到来的数据、处理数据并把数据发回给客户。这种责任划分意味着很多应用可以使用某个现有的服务器类而不需要任何修改，另外可以为它提供一个请求处理器类以处理定制协议。

11.5.1 服务器类型

　　`socketserver` 中定义了 5 个服务器类。`BaseServer` 定义了 API，这个类不能实例化，也不能直接使用。`TCPServer` 使用 TCP/IP 套接字来通信，`UDPServer` 使用数据报套接字。`UnixStreamServer` 和 `UnixDatagramServer` 使用 UNIX 域套接字，只适用于 UNIX 平台。

11.5.2 服务器对象

　　要构造一个服务器，需要向它传递一个地址（服务器将在这个地址监听请求），以及一个请求处理器类（而非实例）。地址格式取决于所使用的服务器类型和套接字簇。可以参考

[一] https://docs.python.org/3/5/library/select.html
[二] https://docs.python.org/3/howto/sockets.html
[三] mix-in 的作用是在运行期间动态改变类的基类或类的方法，使类的表现可以发生变化。——译者注

`socket`模块文档来了解有关的详细内容。

一旦实例化服务器对象,则可以使用`handle_request()`或`serve_forever()`来处理请求。`serve_forever()`方法在一个无限循环中调用`handle_request()`,不过如果应用需要将服务器与另一个事件循环集成,或者要使用`select()`监视对应不同服务器的多个套接字,那么也可以直接调用`handle_request()`。

11.5.3 实现服务器

创建一个服务器时,通常重用某个现有的类并提供一个定制请求处理器类就足够了。但对于另外一些情况,`BaseServer`还包含一些方法,可以在子类中覆盖这些方法。

- `verify_request(request, client_address)`:返回`True`表示要处理请求,返回`False`则忽略请求。例如,一个服务器可能拒绝来自某个IP段的请求,或者如果服务器负载过重则会拒绝请求。
- `process_request(request, client_address)`:调用`finish_request()`来具体完成处理请求的工作。这个方法还可以创建一个单独的线程或进程,就像mix-in类一样。
- `finish_request(request, client_address)`:使用提供给服务器构造函数的类创建一个请求处理器实例。在这个请求处理器上调用`handle()`处理请求。

11.5.4 请求处理器

要接收到来的请求以及确定采取什么动作,大部分工作都由请求处理器完成。处理器负责在套接字层之上实现协议(即HTTP、XML-RPC或AMQP)。请求处理器从入站数据通道读取请求,处理这个请求,并写回一个响应。可以覆盖3个方法。

- `setup()`:为请求准备请求处理器。在`StreamRequestHandler`中,`setup()`方法会创建类似文件的对象来读写套接字。
- `handle()`:对请求完成具体工作。解析到来的请求,处理数据,并发出一个响应。
- `finish()`:清理`setup()`期间创建的所有数据。

很多处理器实现时可以只有一个`handle()`方法。

11.5.5 回送示例

下面这个例子实现了一对简单的服务器/请求处理器,将接受TCP连接,并回送客户发送的所有数据。首先来看请求处理器。

代码清单11-41:`socketserver_echo.py`

```
import logging
import sys
import socketserver

logging.basicConfig(level=logging.DEBUG,
                    format='%(name)s: %(message)s',
                    )
```

```python
class EchoRequestHandler(socketserver.BaseRequestHandler):

    def __init__(self, request, client_address, server):
        self.logger = logging.getLogger('EchoRequestHandler')
        self.logger.debug('__init__')
        socketserver.BaseRequestHandler.__init__(self, request,
                                                 client_address,
                                                 server)
        return

    def setup(self):
        self.logger.debug('setup')
        return socketserver.BaseRequestHandler.setup(self)

    def handle(self):
        self.logger.debug('handle')

        # Echo the data back to the client.
        data = self.request.recv(1024)
        self.logger.debug('recv()->"%s"', data)
        self.request.send(data)
        return

    def finish(self):
        self.logger.debug('finish')
        return socketserver.BaseRequestHandler.finish(self)
```

真正需要实现的只有一个方法，即 EchoRequestHandler.handle()，不过这里包含了前面提到的所有方法以便展示调用顺序。EchoServer 类的工作与 TCPServer 相同，只不过在调用各个方法时会记录日志。

```python
class EchoServer(socketserver.TCPServer):

    def __init__(self, server_address,
                 handler_class=EchoRequestHandler,
                 ):
        self.logger = logging.getLogger('EchoServer')
        self.logger.debug('__init__')
        socketserver.TCPServer.__init__(self, server_address,
                                         handler_class)
        return

    def server_activate(self):
        self.logger.debug('server_activate')
        socketserver.TCPServer.server_activate(self)
        return

    def serve_forever(self, poll_interval=0.5):
        self.logger.debug('waiting for request')
        self.logger.info(
            'Handling requests, press <Ctrl-C> to quit'
        )
        socketserver.TCPServer.serve_forever(self, poll_interval)
        return

    def handle_request(self):
        self.logger.debug('handle_request')
        return socketserver.TCPServer.handle_request(self)

    def verify_request(self, request, client_address):
        self.logger.debug('verify_request(%s, %s)',
```

```python
                request, client_address)
        return socketserver.TCPServer.verify_request(
            self, request, client_address,
        )

    def process_request(self, request, client_address):
        self.logger.debug('process_request(%s, %s)',
                          request, client_address)
        return socketserver.TCPServer.process_request(
            self, request, client_address,
        )

    def server_close(self):
        self.logger.debug('server_close')
        return socketserver.TCPServer.server_close(self)

    def finish_request(self, request, client_address):
        self.logger.debug('finish_request(%s, %s)',
                          request, client_address)
        return socketserver.TCPServer.finish_request(
            self, request, client_address,
        )

    def close_request(self, request_address):
        self.logger.debug('close_request(%s)', request_address)
        return socketserver.TCPServer.close_request(
            self, request_address,
        )

    def shutdown(self):
        self.logger.debug('shutdown()')
        return socketserver.TCPServer.shutdown(self)
```

最后一步是增加一个主程序，这会建立服务器，使它在一个线程中运行，并且向这个服务器发送数据以展示回送数据时会调用哪些方法。

```python
if __name__ == '__main__':
    import socket
    import threading

    address = ('localhost', 0)  # Let the kernel assign a port.
    server = EchoServer(address, EchoRequestHandler)
    ip, port = server.server_address  # What port was assigned?

    # Start the server in a thread.
    t = threading.Thread(target=server.serve_forever)
    t.setDaemon(True)  # Don't hang on exit.
    t.start()

    logger = logging.getLogger('client')
    logger.info('Server on %s:%s', ip, port)

    # Connect to the server.
    logger.debug('creating socket')
    s = socket.socket(socket.AF_INET, socket.SOCK_STREAM)
    logger.debug('connecting to server')
    s.connect((ip, port))

    # Send the data.
    message = 'Hello, world'.encode()
    logger.debug('sending data: %r', message)
    len_sent = s.send(message)
```

```
    # Receive a response.
    logger.debug('waiting for response')
    response = s.recv(len_sent)
    logger.debug('response from server: %r', response)

    # Clean up.
    server.shutdown()
    logger.debug('closing socket')
    s.close()
    logger.debug('done')
    server.socket.close()
```

运行这个程序会生成以下输出。

```
$ python3 socketserver_echo.py

EchoServer: __init__
EchoServer: server_activate
EchoServer: waiting for request
EchoServer: Handling requests, press <Ctrl-C> to quit
client: Server on 127.0.0.1:55484
client: creating socket
client: connecting to server
client: sending data: b'Hello, world'
EchoServer: verify_request(<socket.socket fd=7, family=AddressFamily
.AF_INET, type=SocketKind.SOCK_STREAM, proto=0, laddr=('127.0.0.1',
 55484), raddr=('127.0.0.1', 55485)>, ('127.0.0.1', 55485))
EchoServer: process_request(<socket.socket fd=7, family=AddressFamil
y.AF_INET, type=SocketKind.SOCK_STREAM, proto=0, laddr=('127.0.0.1',
 55484), raddr=('127.0.0.1', 55485)>, ('127.0.0.1', 55485))
EchoServer: finish_request(<socket.socket fd=7, family=AddressFamily
.AF_INET, type=SocketKind.SOCK_STREAM, proto=0, laddr=('127.0.0.1',
 55484), raddr=('127.0.0.1', 55485)>, ('127.0.0.1', 55485))
EchoRequestHandler: __init__
EchoRequestHandler: setup
EchoRequestHandler: handle
client: waiting for response
EchoRequestHandler: recv()->"b'Hello, world'"
EchoRequestHandler: finish
client: response from server: b'Hello, world'
EchoServer: shutdown()
EchoServer: close_request(<socket.socket fd=7, family=AddressFamily.
AF_INET, type=SocketKind.SOCK_STREAM, proto=0, laddr=('127.0.0.1', 5
5484), raddr=('127.0.0.1', 55485)>)
client: closing socket
client: done
```

说明：每次程序运行时使用的端口号都会改变，因为内核会自动分配可用的端口。要让服务器每次监听一个特定的端口，需要在地址元组中提供该端口号而不是 0。

以下是这个服务器的 "压缩" 版本，这里没有日志记录调用。请求处理器类中只需要提供 handle() 方法。

代码清单 11-42：**socketserver_echo_simple.py**

```
import socketserver


class EchoRequestHandler(socketserver.BaseRequestHandler):

    def handle(self):
```

```python
        # Echo the data back to the client.
        data = self.request.recv(1024)
        self.request.send(data)
        return


if __name__ == '__main__':
    import socket
    import threading

    address = ('localhost', 0)  # Let the kernel assign a port.
    server = socketserver.TCPServer(address, EchoRequestHandler)
    ip, port = server.server_address  # What port was assigned?

    t = threading.Thread(target=server.serve_forever)
    t.setDaemon(True)  # Don't hang on exit.
    t.start()

    # Connect to the server.
    s = socket.socket(socket.AF_INET, socket.SOCK_STREAM)
    s.connect((ip, port))

    # Send the data.
    message = 'Hello, world'.encode()
    print('Sending : {!r}'.format(message))
    len_sent = s.send(message)

    # Receive a response.
    response = s.recv(len_sent)
    print('Received: {!r}'.format(response))

    # Clean up.
    server.shutdown()
    s.close()
    server.socket.close()
```

这里并不需要特殊的服务器类，因为 `TCPServer` 会处理所有服务器需求。

```
$ python3 socketserver_echo_simple.py

Sending : b'Hello, world'
Received: b'Hello, world'
```

11.5.6 线程和进程

要为一个服务器增加线程或进程支持，需要在服务器的类层次结构中包括适当的 mix-in 类。这些 mix-in 类要覆盖 `process_request()`，准备好处理一个请求时会开始一个新的线程或进程，具体工作就在这个新的子线程或进程中完成。

对于线程，要使用 `ThreadingMixIn`。

代码清单 11-43: socketserver_threaded.py

```python
import threading
import socketserver


class ThreadedEchoRequestHandler(
        socketserver.BaseRequestHandler,
):
```

```python
    def handle(self):
        # Echo the data back to the client.
        data = self.request.recv(1024)
        cur_thread = threading.currentThread()
        response = b'%s: %s' % (cur_thread.getName().encode(),
                                data)
        self.request.send(response)
        return


class ThreadedEchoServer(socketserver.ThreadingMixIn,
                         socketserver.TCPServer,
                         ):
    pass


if __name__ == '__main__':
    import socket

    address = ('localhost', 0)  # Let the kernel assign a port.
    server = ThreadedEchoServer(address,
                                ThreadedEchoRequestHandler)
    ip, port = server.server_address  # What port was assigned?

    t = threading.Thread(target=server.serve_forever)
    t.setDaemon(True)  # Don't hang on exit.
    t.start()
    print('Server loop running in thread:', t.getName())

    # Connect to the server.
    s = socket.socket(socket.AF_INET, socket.SOCK_STREAM)
    s.connect((ip, port))

    # Send the data.
    message = b'Hello, world'
    print('Sending : {!r}'.format(message))
    len_sent = s.send(message)

    # Receive a response.
    response = s.recv(1024)
    print('Received: {!r}'.format(response))

    # Clean up.
    server.shutdown()
    s.close()
    server.socket.close()
```

对于这个使用线程的服务器，其响应包含了处理请求的那个线程的标识符。

```
$ python3 socketserver_threaded.py

Server loop running in thread: Thread-1
Sending : b'Hello, world'
Received: b'Thread-2: Hello, world'
```

对于不同的进程，要使用 ForkingMixIn。

代码清单 11-44：socketserver_forking.py

```
import os
import socketserver
```

```python
class ForkingEchoRequestHandler(socketserver.BaseRequestHandler):

    def handle(self):
        # Echo the data back to the client.
        data = self.request.recv(1024)
        cur_pid = os.getpid()
        response = b'%d: %s' % (cur_pid, data)
        self.request.send(response)
        return
class ForkingEchoServer(socketserver.ForkingMixIn,
                       socketserver.TCPServer,
                       ):
    pass


if __name__ == '__main__':
    import socket
    import threading

    address = ('localhost', 0)  # Let the kernel assign a port.
    server = ForkingEchoServer(address,
                               ForkingEchoRequestHandler)
    ip, port = server.server_address  # What port was assigned?

    t = threading.Thread(target=server.serve_forever)
    t.setDaemon(True)  # Don't hang on exit.
    t.start()
    print('Server loop running in process:', os.getpid())

    # Connect to the server.
    s = socket.socket(socket.AF_INET, socket.SOCK_STREAM)
    s.connect((ip, port))

    # Send the data.
    message = 'Hello, world'.encode()
    print('Sending : {!r}'.format(message))
    len_sent = s.send(message)

    # Receive a response.
    response = s.recv(1024)
    print('Received: {!r}'.format(response))

    # Clean up.
    server.shutdown()
    s.close()
    server.socket.close()
```

在这里，服务器的响应中包含了子进程的进程 ID。

```
$ python3 socketserver_forking.py

Server loop running in process: 22599
Sending : b'Hello, world'
Received: b'22600: Hello, world'
```

提示：相关阅读材料
- **socketserver** 的标准库文档[⊖]。
- **socket**：底层网络通信。

⊖ https://docs.python.org/3.5/library/socketserver.html

- `select`：底层异步 I/O 工具。
- `asyncio`：异步 I/O、事件循环和并发工具。
- `SimpleXMLRPCServer`：使用 `socketserver` 建立的 XML-RPC 服务器。
- *Unix Network Programming, Volume 1：The Sockets Networking API, Third Edition,* by W. Richard Stevens, Bill Fenner, and Andrew M. Rudoff. Addison-Wesley, 2004. ISBN-10:0131411551.
- *Foundations of Python Network Programming, Third Edition,* by Brandon Rhodes and John Goerzen. Apress, 2014. ISBN-10：1430258543.

第 12 章

互联网

互联网是现代计算很普及的一个领域。甚至很小的、单一用途的脚本也会频繁与远程服务交互，以便发送或接收数据。Python 提供了一组丰富的工具来处理 Web 协议，因此非常适合编写基于 Web 的应用（不论作为客户还是服务器）。

urllib.parse 模块管理 URL 串，可以分解 URL 串或组合 URL 串的组成部分，在客户和服务器中很有用。

urllib.request 模块实现了一个 API 来远程获取内容。

HTTP POST 请求通常用 urllib 完成"表单编码"。通过 POST 发送的二进制数据应当首先用 base64 编码，以符合消息格式标准。

合法客户作为蜘蛛或爬虫程序访问大量网站时[○]，应当使用 urllib.robotparser，从而在对远程服务器带来过重负载之前能确保确实有权限。

要用 Python 创建一个定制 Web 服务器，而不需要任何外部框架，可以使用 http.server 作为起点。它会处理 HTTP 协议，所以唯一需要的定制就是响应到来请求的应用代码。

服务器中的会话状态可以通过 cookie 来管理，cookie 由 http.cookies 模块创建和解析。由于这个模块提供了对 cookie 到期、路径、域以及其他设置的完全支持，因此很容易配置会话。

uuid 模块用于为需要唯一值的资源生成标识符。UUID 很适合自动生成统一资源名（Uniform Resource Name，URN）值，其中资源名必须唯一，但并不需要传达任何具体含义。

Python 的标准库支持两个基于 Web 的远程过程调用机制。AJAX 通信中使用的 JavaScript 对象记法（JavaScript Object Notation，JSON）编码机制和 REST API 在 json 中实现。它在客户或服务器中同样适用。另外 xmlrpc.client 和 xmlrpc.server 中分别包含了完整的 XML-RPC 客户和服务器库。

12.1 urllib.parse：分解 URL

urllib.parse 模块提供了一些函数，可以管理 URL 及其组成部分，这包括将 URL 分解为组成部分以及由组成部分构成 URL。

○ 蜘蛛（spider）或爬虫（crawler）是一段计算机程序，它从互联网上按照一定的逻辑和算法抓取和下载互联网的网页，是搜索引擎的一个重要组成部分。——译者注

12.1.1 解析

urlparse()函数的返回值是一个 ParseResult 对象,其相当于一个包含 6 个元素的 tuple。

代码清单 12-1: **urllib_parse_urlparse.py**

```
from urllib.parse import urlparse

url = 'http://netloc/path;param?query=arg#frag'
parsed = urlparse(url)
print(parsed)
```

通过元组接口得到的 URL 各部分分别是机制、网络位置、路径、路径段参数(由一个分号将路径分开)、查询以及片段。

```
$ python3 urllib_parse_urlparse.py

ParseResult(scheme='http', netloc='netloc', path='/path',
params='param', query='query=arg', fragment='frag')
```

尽管返回值相当于一个元组,但实际上它基于一个 namedtuple,这是 tuple 的一个子类,除了可以通过索引访问,它还支持通过命名属性访问 URL 的各部分。属性 API 不仅更易于程序员使用,还允许访问 tuple API 中未提供的很多值。

代码清单 12-2: **urllib_parse_urlparseattrs.py**

```
from urllib.parse import urlparse

url = 'http://user:pwd@NetLoc:80/path;param?query=arg#frag'
parsed = urlparse(url)
print('scheme  :', parsed.scheme)
print('netloc  :', parsed.netloc)
print('path    :', parsed.path)
print('params  :', parsed.params)
print('query   :', parsed.query)
print('fragment:', parsed.fragment)
print('username:', parsed.username)
print('password:', parsed.password)
print('hostname:', parsed.hostname)
print('port    :', parsed.port)
```

输入 URL 中可能提供用户名(username)和密码(password),如果没有提供就设置为 None。主机名(hostname)与 netloc 值相同,全为小写并且去除端口值。如果有端口(port),则转换为一个整数,如果没有则设置为 None。

```
$ python3 urllib_parse_urlparseattrs.py

scheme  : http
netloc  : user:pwd@NetLoc:80
path    : /path
params  : param
query   : query=arg
fragment: frag
username: user
password: pwd
```

```
hostname: netloc
port    : 80
```

`urlsplit()` 函数可以替换 `urlparse()`，但行为稍有不同，因为它不会从 URL 分解参数。这对于遵循 **RFC 2396**⊖的 URL 很有用（支持对应路径每一段的参数）。

代码清单 12-3：urllib_parse_urlsplit.py

```
from urllib.parse import urlsplit

url = 'http://user:pwd@NetLoc:80/p1;para/p2;para?query=arg#frag'
parsed = urlsplit(url)
print(parsed)
print('scheme  :', parsed.scheme)
print('netloc  :', parsed.netloc)
print('path    :', parsed.path)
print('query   :', parsed.query)
print('fragment:', parsed.fragment)
print('username:', parsed.username)
print('password:', parsed.password)
print('hostname:', parsed.hostname)
print('port    :', parsed.port)
```

由于没有分解参数，tuple API 会显示 5 个元素而不是 6 个，并且这里没有 `params` 属性。

```
$ python3 urllib_parse_urlsplit.py

SplitResult(scheme='http', netloc='user:pwd@NetLoc:80',
path='/p1;para/p2;para', query='query=arg', fragment='frag')
scheme   : http
netloc   : user:pwd@NetLoc:80
path     : /p1;para/p2;para
query    : query=arg
fragment: frag
username: user
password: pwd
hostname: netloc
port     : 80
```

要想从一个 URL 剥离出片段标识符，如从一个 URL 查找基页面名，可以使用 `urldefrag()`。

代码清单 12-4：urllib_parse_urldefrag.py

```
from urllib.parse import urldefrag

original = 'http://netloc/path;param?query=arg#frag'
print('original:', original)
d = urldefrag(original)
print('url     :', d.url)
print('fragment:', d.fragment)
```

返回值是一个基于 `namedtuple` 的 `DefragResult`，其中包含基 URL 和片段。

```
$ python3 urllib_parse_urldefrag.py

original: http://netloc/path;param?query=arg#frag
```

⊖ https://tools.ietf.org/html/rfc2396.html

```
url     : http://netloc/path;param?query=arg
fragment: frag
```

12.1.2 反解析

还可以利用一些方法把分解的 URL 的各个部分重新组装在一起，形成一个串。解析的 URL 对象有一个 `geturl()` 方法。

代码清单 12-5：**urllib_parse_geturl.py**

```
from urllib.parse import urlparse

original = 'http://netloc/path;param?query=arg#frag'
print('ORIG  :', original)
parsed = urlparse(original)
print('PARSED:', parsed.geturl())
```

`geturl()` 只适用于 `urlparse()` 或 `urlsplit()` 返回的对象。

```
$ python3 urllib_parse_geturl.py

ORIG  : http://netloc/path;param?query=arg#frag
PARSED: http://netloc/path;param?query=arg#frag
```

利用 `urlunparse()` 可以将包含串的普通元组重新组合为一个 URL。

代码清单 12-6：**urllib_parse_urlunparse.py**

```
from urllib.parse import urlparse, urlunparse

original = 'http://netloc/path;param?query=arg#frag'
print('ORIG  :', original)
parsed = urlparse(original)
print('PARSED:', type(parsed), parsed)
t = parsed[:]
print('TUPLE :', type(t), t)
print('NEW   :', urlunparse(t))
```

尽管 `urlparse()` 返回的 `ParseResult` 可以作为一个元组，但这个例子却显式地创建了一个新元组，来展示 `urlunparse()` 也适用于普通元组。

```
$ python3 urllib_parse_urlunparse.py

ORIG  : http://netloc/path;param?query=arg#frag
PARSED: <class 'urllib.parse.ParseResult'>
ParseResult(scheme='http', netloc='netloc', path='/path',
params='param', query='query=arg', fragment='frag')
TUPLE : <class 'tuple'> ('http', 'netloc', '/path', 'param',
'query=arg', 'frag')
NEW   : http://netloc/path;param?query=arg#frag
```

如果输入 URL 包含多余的部分，那么重新构造的 URL 可能会将其去除。

代码清单 12-7：**urllib_parse_urlunparseextra.py**

```
from urllib.parse import urlparse, urlunparse

original = 'http://netloc/path;?#'
```

```
print('ORIG   :', original)
parsed = urlparse(original)
print('PARSED:', type(parsed), parsed)
t = parsed[:]
print('TUPLE :', type(t), t)
print('NEW    :', urlunparse(t))
```

在这里,原 URL 中没有参数、查询和片段。新 URL 看起来与原 URL 并不相同,不过按照标准它们是等价的。

```
$ python3 urllib_parse_urlunparseextra.py

ORIG   : http://netloc/path;?#
PARSED: <class 'urllib.parse.ParseResult'>
ParseResult(scheme='http', netloc='netloc', path='/path',
params='', query='', fragment='')
TUPLE  : <class 'tuple'> ('http', 'netloc', '/path', '', '', '')
NEW    : http://netloc/path
```

12.1.3 连接

除了解析 URL,urlparse 还包括一个 urljoin() 方法,可以由相对片段构造绝对 URL。

代码清单 12-8:**urllib_parse_urljoin.py**

```
from urllib.parse import urljoin

print(urljoin('http://www.example.com/path/file.html',
              'anotherfile.html'))
print(urljoin('http://www.example.com/path/file.html',
              '../anotherfile.html'))
```

在这个例子中,计算第二个 URL 时要考虑路径的相对部分("../")。

```
$ python3 urllib_parse_urljoin.py

http://www.example.com/path/anotherfile.html
http://www.example.com/anotherfile.html
```

非相对路径的处理与 os.path.join() 的处理方式相同。

代码清单 12-9:**urllib_parse_urljoin_with_path.py**

```
from urllib.parse import urljoin

print(urljoin('http://www.example.com/path/',
              '/subpath/file.html'))
print(urljoin('http://www.example.com/path/',
              'subpath/file.html'))
```

如果连接到 URL 的路径以一个斜线开头(/),那么 urljoin() 会把 URL 的路径重置为顶级路径。如果不是以一个斜线开头,那么新路径值则追加到 URL 当前路径的末尾。

```
$ python3 urllib_parse_urljoin_with_path.py

http://www.example.com/subpath/file.html
http://www.example.com/path/subpath/file.html
```

12.1.4　解码查询参数

参数在被增加到一个 URL 之前，需要先编码。

代码清单 12-10：urllib_parse_urlencode.py

```
from urllib.parse import urlencode

query_args = {
    'q': 'query string',
    'foo': 'bar',
}
encoded_args = urlencode(query_args)
print('Encoded:', encoded_args)
```

编码会替换诸如空格之类的特殊字符，以确保采用一种符合标准的格式将它们传递到服务器。

```
$ python3 urllib_parse_urlencode.py

Encoded: q=query+string&foo=bar
```

如果要利用查询串中的变量传递一个值序列，那么需要在调用 `urlencode()` 时将 `doseq` 设置为 `True`。

代码清单 12-11：urllib_parse_urlencode_doseq.py

```
from urllib.parse import urlencode

query_args = {
    'foo': ['foo1', 'foo2'],
}
print('Single   :', urlencode(query_args))
print('Sequence:', urlencode(query_args, doseq=True))
```

结果是一个查询串，包含与同一个名关联的多个值。

```
$ python3 urllib_parse_urlencode_doseq.py

Single   : foo=%5B%27foo1%27%2C+%27foo2%27%5D
Sequence: foo=foo1&foo=foo2
```

要解码这个查询串，可以使用 `parse_qs()` 或 `parse_qsl()`。

代码清单 12-12：urllib_parse_parse_qs.py

```
from urllib.parse import parse_qs, parse_qsl

encoded = 'foo=foo1&foo=foo2'
print('parse_qs :', parse_qs(encoded))
print('parse_qsl:', parse_qsl(encoded))
```

`parse_qs()` 的返回值是一个将名映射到值的字典，而 `parse_qsl()` 返回一个元组列表，每个元组包含一个名和一个值。

```
$ python3 urllib_parse_parse_qs.py
```

```
parse_qs  : {'foo': ['foo1', 'foo2']}
parse_qsl : [('foo', 'foo1'), ('foo', 'foo2')]
```

查询参数中可能有一些特殊字符，会导致服务器端在解析 URL 时出问题，所以在传递到 urlencode() 时要对这些特殊字符"加引号"。要在本地对它们加引号以建立这些串的安全版本，可以直接使用 quote() 或 quote_plus() 函数。

代码清单 12-13：urllib_parse_quote.py

```
from urllib.parse import quote, quote_plus, urlencode

url = 'http://localhost:8080/~hellmann/'
print('urlencode() :', urlencode({'url': url}))
print('quote()     :', quote(url))
print('quote_plus():', quote_plus(url))
```

quote_plus() 中的加引号实现会更大程度地替换字符。

```
$ python3 urllib_parse_quote.py

urlencode() : url=http%3A%2F%2Flocalhost%3A8080%2F%7Ehellmann%2F
quote()     : http%3A//localhost%3A8080/%7Ehellmann/
quote_plus(): http%3A%2F%2Flocalhost%3A8080%2F%7Ehellmann%2F
```

要完成加引号操作的逆过程，可以在适当的时候使用 unquote() 或 unquote_plus()。

代码清单 12-14：urllib_parse_unquote.py

```
from urllib.parse import unquote, unquote_plus

print(unquote('http%3A//localhost%3A8080/%7Ehellmann/'))
print(unquote_plus(
    'http%3A%2F%2Flocalhost%3A8080%2F%7Ehellmann%2F'
))
```

编码的值会转换回一个普通的 URL 串。

```
$ python3 urllib_parse_unquote.py

http://localhost:8080/~hellmann/
http://localhost:8080/~hellmann/
```

提示：相关阅读材料

- urllib.parse 的标准库文档[一]。
- urllib.request：获取一个 URL 标识的资源的内容。
- **RFC 1738**[二]：统一资源定位符（Uniform Resource Locator，URL）语法。
- **RFC 1808**[三]：相对 URL。
- **RFC 2396**[四]：统一资源标识符（Uniform Resource Identifier，URI）通用语法。
- **RFC 3986**[五]：统一资源标识符语法。

[一] https://docs.python.org/3.5/library/urllib.parse.html
[二] https://tools.ietf.org/html/rfc1738.html
[三] https://tools.ietf.org/html/rfc1808.html
[四] https://tools.ietf.org/html/rfc2396.html
[五] https://tools.ietf.org/html/rfc3986.html

12.2 `urllib.request`：网络资源访问

`urllib.request`模块提供了一个API来使用URL标识的Internet资源。各个应用可以扩展这个模块来支持新协议或者增加现有协议的变种（如处理HTTP基本认证）。

12.2.1 HTTP GET

说明：这些例子的测试服务器见`http_server_GET.py`，取自`http.server`模块的例子。在一个终端窗口启动这个服务器，再在另一个终端窗口中运行这些例子。

HTTP GET操作是`urllib.request`最简单的用法。通过将URL传递到`urlopen()`来得到远程数据的一个"类似文件"的句柄。

代码清单12-15：`urllib_request_urlopen.py`

```
from urllib import request

response = request.urlopen('http://localhost:8080/')
print('RESPONSE:', response)
print('URL      :', response.geturl())

headers = response.info()
print('DATE     :', headers['date'])
print('HEADERS :')
print('---------')
print(headers)

data = response.read().decode('utf-8')
print('LENGTH   :', len(data))
print('DATA     :')
print('---------')
print(data)
```

这个示例服务器接收到来的值，格式化一个纯文本响应并发回客户。利用`urlopen()`的返回值，可以通过`info()`方法从HTTP服务器访问首部，还可以通过类似`read()`和`readlines()`等方法访问远程资源的相应数据。

```
$ python3 urllib_request_urlopen.py

RESPONSE: <http.client.HTTPResponse object at 0x101744d68>
URL     : http://localhost:8080/
DATE    : Sat, 08 Oct 2016 18:08:54 GMT
HEADERS :
---------
Server: BaseHTTP/0.6 Python/3.5.2
Date: Sat, 08 Oct 2016 18:08:54 GMT
Content-Type: text/plain; charset=utf-8


LENGTH  : 349
DATA    :
---------
CLIENT VALUES:
client_address=('127.0.0.1', 58420) (127.0.0.1)
```

```
command=GET
path=/
real path=/
query=
request_version=HTTP/1.1

SERVER VALUES:
server_version=BaseHTTP/0.6
sys_version=Python/3.5.2
protocol_version=HTTP/1.0

HEADERS RECEIVED:
Accept-Encoding=identity
Connection=close
Host=localhost:8080
User-Agent=Python-urllib/3.5
```

urlopen()返回的类似文件对象是可迭代的(iterable)。

代码清单12-16：**urllib_request_urlopen_iterator.py**

```
from urllib import request

response = request.urlopen('http://localhost:8080/')
for line in response:
    print(line.decode('utf-8').rstrip())
```

这个例子在打印输出之前去除了末尾的换行符和回车。

```
$ python3 urllib_request_urlopen_iterator.py

CLIENT VALUES:
client_address=('127.0.0.1', 58444) (127.0.0.1)
command=GET
path=/
real path=/
query=
request_version=HTTP/1.1

SERVER VALUES:
server_version=BaseHTTP/0.6
sys_version=Python/3.5.2
protocol_version=HTTP/1.0

HEADERS RECEIVED:
Accept-Encoding=identity
Connection=close
Host=localhost:8080
User-Agent=Python-urllib/3.5
```

12.2.2 编码参数

可以利用urllib.parse.urlencode()对参数编码，并追加到URL，从而将参数传递到服务器。

代码清单12-17：**urllib_request_http_get_args.py**

```
from urllib import parse
from urllib import request
```

```
query_args = {'q': 'query string', 'foo': 'bar'}
encoded_args = parse.urlencode(query_args)
print('Encoded:', encoded_args)

url = 'http://localhost:8080/?' + encoded_args
print(request.urlopen(url).read().decode('utf-8'))
```

示例输出中返回的客户值列表包含了已编码的查询参数。

```
$ python urllib_request_http_get_args.py
Encoded: q=query+string&foo=bar
CLIENT VALUES:
client_address=('127.0.0.1', 58455) (127.0.0.1)
command=GET
path=/?q=query+string&foo=bar
real path=/
query=q=query+string&foo=bar
request_version=HTTP/1.1

SERVER VALUES:
server_version=BaseHTTP/0.6
sys_version=Python/3.5.2
protocol_version=HTTP/1.0

HEADERS RECEIVED:
Accept-Encoding=identity
Connection=close
Host=localhost:8080
User-Agent=Python-urllib/3.5
```

12.2.3　HTTP POST

说明：这些例子的测试服务器见 `http_server_POST.py`，取自 `http.server` 模块的例子。在一个终端窗口启动这个服务器，再在另一个终端窗口中运行这些例子。

要使用 POST 而不是 GET 向远程服务器发送表单编码数据，可以把编码的查询参数作为数据传递到 `urlopen()`。

代码清单 12-18：**urllib_request_urlopen_post.py**

```
from urllib import parse
from urllib import request

query_args = {'q': 'query string', 'foo': 'bar'}
encoded_args = parse.urlencode(query_args).encode('utf-8')
url = 'http://localhost:8080/'
print(request.urlopen(url, encoded_args).read().decode('utf-8'))
```

这个服务器可以对表单数据解码，并按名访问各个值。

```
$ python3 urllib_request_urlopen_post.py

Client: ('127.0.0.1', 58568)
User-agent: Python-urllib/3.5
Path: /
Form data:
    foo=bar
    q=query string
```

12.2.4 添加发出首部

`urlopen()` 是一个便利函数,隐藏了如何建立和处理请求的一些细节。可以直接使用一个 `Request` 实例提供更精确的控制。例如,可以向发出的请求增加定制首部,以便控制所返回数据的格式,指定本地缓存的文档的版本,以及告诉远程服务器与它通信的软件客户的名字。

如前例的输出所示,默认的 User-agent 首部值包括常量 `Python-urllib`,后面是 Python 解释器版本。如果要创建一个应用访问其他人拥有的 Web 资源,那么最好在请求中包含真实的用户代理信息,这样比较礼貌,可以更容易地标识目标来源。使用一个定制代理,还允许它们用一个 `robots.txt` 文件控制爬虫(参见 `http.robotparser` 模块)。

代码清单 12-19:`urllib_request_request_header.py`

```
from urllib import request

r = request.Request('http://localhost:8080/')
r.add_header(
    'User-agent',
    'PyMOTW (https://pymotw.com/)',
)

response = request.urlopen(r)
data = response.read().decode('utf-8')
print(data)
```

创建一个 `Request` 对象后,打开请求之前使用 `add_header()` 设置用户代理值。输出的最后一行显示了这个定制值。

```
$ python3 urllib_request_request_header.py

CLIENT VALUES:
client_address=('127.0.0.1', 58585) (127.0.0.1)
command=GET
path=/
real path=/
query=
request_version=HTTP/1.1

SERVER VALUES:
server_version=BaseHTTP/0.6
sys_version=Python/3.5.2
protocol_version=HTTP/1.0

HEADERS RECEIVED:
Accept-Encoding=identity
Connection=close
Host=localhost:8080
User-Agent=PyMOTW (https://pymotw.com/)
```

12.2.5 从请求提交表单数据

建立 `Request` 时,可以指定发出的数据,从而将其提交给服务器。

代码清单 12-20：`urllib_request_request_post.py`

```python
from urllib import parse
from urllib import request

query_args = {'q': 'query string', 'foo': 'bar'}

r = request.Request(
    url='http://localhost:8080/',
    data=parse.urlencode(query_args).encode('utf-8'),
)
print('Request method :', r.get_method())
r.add_header(
    'User-agent',
    'PyMOTW (https://pymotw.com/)',
)

print()
print('OUTGOING DATA:')
print(r.data)

print()
print('SERVER RESPONSE:')
print(request.urlopen(r).read().decode('utf-8'))
```

增加了数据之后，Request 使用的 HTTP 方法会从 GET 自动变成 POST。

```
$ python3 urllib_request_request_post.py

Request method : POST

OUTGOING DATA:
b'q=query+string&foo=bar'

SERVER RESPONSE:
Client: ('127.0.0.1', 58613)
User-agent: PyMOTW (https://pymotw.com/)
Path: /
Form data:
    foo=bar
    q=query string
```

12.2.6 上传文件

要对文件编码以完成上传，与使用简单表单相比，这需要多做一些工作。要在请求体中构造一个完整的 MIME 消息，使得服务器可以把收到的表单域与上传的文件区分开。

代码清单 12-21：`urllib_request_upload_files.py`

```python
import io
import mimetypes
from urllib import request
import uuid


class MultiPartForm:
    """Accumulate the data to be used when posting a form."""

    def __init__(self):
        self.form_fields = []
```

```python
        self.files = []
        # Use a large random byte string to separate
        # parts of the MIME data.
        self.boundary = uuid.uuid4().hex.encode('utf-8')
        return

    def get_content_type(self):
        return 'multipart/form-data; boundary={}'.format(
            self.boundary.decode('utf-8'))

    def add_field(self, name, value):
        """Add a simple field to the form data."""
        self.form_fields.append((name, value))

    def add_file(self, fieldname, filename, fileHandle,
                 mimetype=None):
        """Add a file to be uploaded."""
        body = fileHandle.read()
        if mimetype is None:
            mimetype = (
                mimetypes.guess_type(filename)[0] or
                'application/octet-stream'
            )
        self.files.append((fieldname, filename, mimetype, body))
        return

    @staticmethod
    def _form_data(name):
        return ('Content-Disposition: form-data; '
                'name="{}"\r\n').format(name).encode('utf-8')

    @staticmethod
    def _attached_file(name, filename):
        return ('Content-Disposition: file; '
                'name="{}"; filename="{}"\r\n').format(
                    name, filename).encode('utf-8')

    @staticmethod
    def _content_type(ct):
        return 'Content-Type: {}\r\n'.format(ct).encode('utf-8')

    def __bytes__(self):
        """Return a byte-string representing the form data,
        including attached files.
        """
        buffer = io.BytesIO()
        boundary = b'--' + self.boundary + b'\r\n'

        # Add the form fields.
        for name, value in self.form_fields:
            buffer.write(boundary)
            buffer.write(self._form_data(name))
            buffer.write(b'\r\n')
            buffer.write(value.encode('utf-8'))
            buffer.write(b'\r\n')

        # Add the files to upload.
        for f_name, filename, f_content_type, body in self.files:
            buffer.write(boundary)
            buffer.write(self._attached_file(f_name, filename))
            buffer.write(self._content_type(f_content_type))
            buffer.write(b'\r\n')
            buffer.write(body)
```

```python
            buffer.write(b'\r\n')

        buffer.write(b'--' + self.boundary + b'--\r\n')
        return buffer.getvalue()


if __name__ == '__main__':
    # Create the form with simple fields.
    form = MultiPartForm()
    form.add_field('firstname', 'Doug')
    form.add_field('lastname', 'Hellmann')

    # Add a fake file.
    form.add_file(
        'biography', 'bio.txt',
        fileHandle=io.BytesIO(b'Python developer and blogger.'))

    # Build the request, including the byte-string
    # for the data to be posted.
    data = bytes(form)
    r = request.Request('http://localhost:8080/', data=data)
    r.add_header(
        'User-agent',
        'PyMOTW (https://pymotw.com/)',
    )
    r.add_header('Content-type', form.get_content_type())
    r.add_header('Content-length', len(data))

    print()
    print('OUTGOING DATA:')
    for name, value in r.header_items():
        print('{}: {}'.format(name, value))
    print()
    print(r.data.decode('utf-8'))

    print()
    print('SERVER RESPONSE:')
    print(request.urlopen(r).read().decode('utf-8'))
```

MultiPartForm 类可以把一个任意的表单表示为一个带附加文件的多部分 MIME 消息。

```
$ python3 urllib_request_upload_files.py

OUTGOING DATA:
User-agent: PyMOTW (https://pymotw.com/)
Content-type: multipart/form-data;
    boundary=d99b5dc60871491b9d63352eb24972b4
Content-length: 389

--d99b5dc60871491b9d63352eb24972b4
Content-Disposition: form-data; name="firstname"

Doug
--d99b5dc60871491b9d63352eb24972b4
Content-Disposition: form-data; name="lastname"

Hellmann
--d99b5dc60871491b9d63352eb24972b4
Content-Disposition: file; name="biography";
    filename="bio.txt"
Content-Type: text/plain
```

```
Python developer and blogger.
--d99b5dc60871491b9d63352eb24972b4--

SERVER RESPONSE:
Client: ('127.0.0.1', 59310)
User-agent: PyMOTW (https://pymotw.com/)
Path: /
Form data:
    Uploaded biography as 'bio.txt' (29 bytes)
    firstname=Doug
    lastname=Hellmann
```

12.2.7 创建定制协议处理器

`urllib.request` 提供了对 HTTP (S)、FTP 和本地文件访问的内置支持。为了增加对其他 URL 类型的支持，可以注册另外的协议处理器。例如，为了支持指向远程 NFS 服务器上任意文件的 URL，而不需要用户在访问文件之前先装载路径，可以创建一个派生 `BaseHandler` 的类，并包含一个 `nfs_open()` 方法。

协议特定的 `open()` 方法有一个参数，即 `Request` 实例，它会返回一个对象，这个对象有一个 `read()` 方法来读取数据，一个 `info()` 方法来返回响应首部，还有一个 `geturl()` 方法返回所读取文件的具体 URL。要满足这些需求，一种简单的办法是创建 `urllib.response.addinfourl` 的一个实例，然后把首部、URL 和打开的文件句柄传入它的构造函数。

代码清单 12-22：**urllib_request_nfs_handler.py**

```python
import io
import mimetypes
import os
import tempfile
from urllib import request
from urllib import response


class NFSFile:

    def __init__(self, tempdir, filename):
        self.tempdir = tempdir
        self.filename = filename
        with open(os.path.join(tempdir, filename), 'rb') as f:
            self.buffer = io.BytesIO(f.read())

    def read(self, *args):
        return self.buffer.read(*args)

    def readline(self, *args):
        return self.buffer.readline(*args)

    def close(self):
        print('\nNFSFile:')
        print('  unmounting {}'.format(
            os.path.basename(self.tempdir)))
        print('  when {} is closed'.format(
            os.path.basename(self.filename)))
```

```python
class FauxNFSHandler(request.BaseHandler):
    def __init__(self, tempdir):
        self.tempdir = tempdir
        super().__init__()

    def nfs_open(self, req):
        url = req.full_url
        directory_name, file_name = os.path.split(url)
        server_name = req.host
        print('FauxNFSHandler simulating mount:')
        print('  Remote path: {}'.format(directory_name))
        print('  Server     : {}'.format(server_name))
        print('  Local path : {}'.format(
            os.path.basename(tempdir)))
        print('  Filename   : {}'.format(file_name))
        local_file = os.path.join(tempdir, file_name)
        fp = NFSFile(tempdir, file_name)
        content_type = (
            mimetypes.guess_type(file_name)[0] or
            'application/octet-stream'
        )
        stats = os.stat(local_file)
        size = stats.st_size
        headers = {
            'Content-type': content_type,
            'Content-length': size,
        }
        return response.addinfourl(fp, headers,
                                   req.get_full_url())


if __name__ == '__main__':
    with tempfile.TemporaryDirectory() as tempdir:
        # Populate the temporary file for the simulation.
        filename = os.path.join(tempdir, 'file.txt')
        with open(filename, 'w', encoding='utf-8') as f:
            f.write('Contents of file.txt')

        # Construct an opener with our NFS handler
        # and register it as the default opener.
        opener = request.build_opener(FauxNFSHandler(tempdir))
        request.install_opener(opener)

        # Open the file through a URL.
        resp = request.urlopen(
            'nfs://remote_server/path/to/the/file.txt'
        )
        print()
        print('READ CONTENTS:', resp.read())
        print('URL          :', resp.geturl())
        print('HEADERS:')
        for name, value in sorted(resp.info().items()):
            print('  {:<15} = {}'.format(name, value))
        resp.close()
```

FauxNFSHandler 和 NFSFile 类可以打印消息,以展示一个真实实现会在哪里增加装载 (mount) 和卸载 (unmount) 调用。由于这只是一个模拟,所以只向 FauxNFSHandler 提供了一个临时目录名,它会在这个目录中查找所有文件。

```
$ python3 urllib_request_nfs_handler.py

FauxNFSHandler simulating mount:
  Remote path: nfs://remote_server/path/to/the
  Server     : remote_server
  Local path : tmprucom5sb
  Filename   : file.txt

READ CONTENTS: b'Contents of file.txt'
URL          : nfs://remote_server/path/to/the/file.txt
HEADERS:
  Content-length = 20
  Content-type   = text/plain

NFSFile:
  unmounting tmprucom5sb
  when file.txt is closed
```

提示：相关阅读材料

- urllib.request 的标准库文档[⊖]。
- urllib.parse：处理 URL 串本身。
- Form content types[⊜]：通过 HTTP 表单提交文件或大量数据的 W3C 规范。
- mimetypes：将文件名映射到 mimetype。
- Requests[⊗]：一个第三方 HTTP 库，可以更好地支持安全连接，而且有一个更易于使用的 API。Python 核心开发团队推荐大部分开发人员都应该使用 requests，部分原因在于这个模块比标准库的安全更新更频繁。

12.3 urllib.robotparser：Internet 蜘蛛访问控制

robotparser 为 robots.txt 文件格式实现一个解析器，提供了一个函数来检查给定的用户代理是否可以访问一个资源。这个模块可以用于合法蜘蛛或者需要抑制或限制的其他爬虫应用中。

12.3.1 robots.txt

robots.txt 文件格式是一个基于文本的简单访问控制系统，用于自动访问 Web 资源的计算机程序（如"蜘蛛""爬虫"等）。这个文件由记录构成，各记录会指定程序的用户代理标识符，后面是该代理不能访问的一个 URL（或 URL 前缀）列表。

以下代码清单显示了 https://pymotw.com/ 的 robots.txt 文件。

代码清单 12-23：robots.txt

```
Sitemap: https://pymotw.com/sitemap.xml
User-agent: *
Disallow: /admin/
```

⊖ https://docs.python.org/3.5/library/urllib.request.html
⊜ www.w3.org/TR/REC-html40/interact/forms.html#h-17.13.4
⊗ https://pypi.python.org/pypi/requests

```
Disallow: /downloads/
Disallow: /media/
Disallow: /static/
Disallow: /codehosting/
```

这个文件会阻止访问网站中某些计算资源代价昂贵的部分，如果搜索引擎试图索引这些部分，那么可能会让服务器负载过重。要得到更完整的 robots.txt 示例集，可以参考 Web Robots 页面[⊖]。

12.3.2 测试访问权限

基于之前提供的数据，一个简单的爬虫应用可以使用 RobotFileParser.can_fetch() 测试是否允许下载一个页面。

代码清单 12-24：urllib_robotparser_simple.py

```python
from urllib import parse
from urllib import robotparser

AGENT_NAME = 'PyMOTW'
URL_BASE = 'https://pymotw.com/'
parser = robotparser.RobotFileParser()
parser.set_url(parse.urljoin(URL_BASE, 'robots.txt'))
parser.read()

PATHS = [
    '/',
    '/PyMOTW/',
    '/admin/',
    '/downloads/PyMOTW-1.92.tar.gz',
]

for path in PATHS:
    print('{!r:>6} : {}'.format(
        parser.can_fetch(AGENT_NAME, path), path))
    url = parse.urljoin(URL_BASE, path)
    print('{!r:>6} : {}'.format(
        parser.can_fetch(AGENT_NAME, url), url))
    print()
```

can_fetch() 的 URL 参数可以是一个相对于网站根目录的相对路径，也可以是一个完全 URL。

```
$ python3 urllib_robotparser_simple.py

  True : /
  True : https://pymotw.com/

  True : /PyMOTW/
  True : https://pymotw.com/PyMOTW/

 False : /admin/
 False : https://pymotw.com/admin/

 False : /downloads/PyMOTW-1.92.tar.gz
 False : https://pymotw.com/downloads/PyMOTW-1.92.tar.gz
```

⊖ www.robotstxt.org/orig.html

12.3.3 长寿命蜘蛛

如果一个应用需要花很长时间来处理它下载的资源，或者受到抑制，需要在下载之间暂停，那么这样的应用应当以它已下载内容的寿命为根据，定期检查新的 robots.txt 文件。这个寿命并不是自动管理的，不过有一些简便方法可以方便地跟踪其寿命。

代码清单 12-25：urllib_robotparser_longlived.py

```
from urllib import robotparser
import time

AGENT_NAME = 'PyMOTW'
parser = robotparser.RobotFileParser()
# Use the local copy.
parser.set_url('file:robots.txt')
parser.read()
parser.modified()

PATHS = [
    '/',
    '/PyMOTW/',
    '/admin/',
    '/downloads/PyMOTW-1.92.tar.gz',
]

for path in PATHS:
    age = int(time.time() - parser.mtime())
    print('age:', age, end=' ')
    if age > 1:
        print('rereading robots.txt')
        parser.read()
        parser.modified()
    else:
        print()
    print('{!r:>6} : {}'.format(
        parser.can_fetch(AGENT_NAME, path), path))
    # Simulate a delay in processing.
    time.sleep(1)
    print()
```

这个例子有些极端，如果已下载的文件寿命超过了 1 秒，那么它就会下载一个新的 robots.txt 文件。

```
$ python3 urllib_robotparser_longlived.py

age: 0
  True : /

age: 1
  True : /PyMOTW/

age: 2 rereading robots.txt
 False : /admin/

age: 1
 False : /downloads/PyMOTW-1.92.tar.gz
```

作为一个更好的长寿命应用，在下载整个文件之前可能会请求文件的修改时间。另一方面，robots.txt 文件通常很小，所以再次获取整个文档的开销并不昂贵。

提示：相关阅读材料
- urllib.robotparser 的标准库文档[⊖]。
- The Web Robots Page[⊖]：robots.txt 格式的描述。

12.4 base64：用 ASCII 编码二进制数据

base64 模块包含一些函数可以将二进制数据转换为适合使用纯文本协议传输的 ASCII 的一个子集。Base64、Base32、Base16 和 Base85 编码将 8 位字节转换为 ASCII 可打印字符范围内的字符，留出更多的位来表示数据，保证与只支持 ASCII 数据的系统兼容，如 SMTP。base（进制）值对应各编码中使用的字母表长度。这些原始编码还有一些"URL 安全"(URL-safe) 的变形，其使用的字母表稍有不同。

12.4.1 Base64 编码

以下代码清单给出了一个简单的例子，其中对一些文本进行了编码。

代码清单 12-26：base64_b64encode.py

```
import base64
import textwrap

# Load this source file and strip the header.
with open(__file__, 'r', encoding='utf-8') as input:
    raw = input.read()
    initial_data = raw.split('#end_pymotw_header')[1]

byte_string = initial_data.encode('utf-8')
encoded_data = base64.b64encode(byte_string)

num_initial = len(byte_string)

# There will never be more than 2 padding bytes.
padding = 3 - (num_initial % 3)

print('{} bytes before encoding'.format(num_initial))
print('Expect {} padding bytes'.format(padding))
print('{} bytes after encoding\n'.format(len(encoded_data)))
print(encoded_data)
```

输入必须是一个字节串，所以首先将 Unicode 字符串编码为 UTF-8。输出显示了 UTF-8 源文件的 185 个字节在编码后扩展为 248 字节。

说明：由库生成的编码数据中并没有回车，不过这里的输出中人工增加了换行符，以便能更美观地在页面上显示。

⊖ https://docs.python.org/3.5/library/urllib.robotparser.html
⊖ www.robotstxt.org/orig.html

```
$ python3 base64_b64encode.py

185 bytes before encoding
Expect 1 padding bytes
248 bytes after encoding

b'CgppbXBvcnQgYmFzZTY0CmltcG9ydCB0ZXh0d3JhcAoKIyBMb2FkIHRoaXMgc2
91cmNlIGZpbGUgYW5kIHN0cmlwIHRoZSBoZWFkZXIuCndpdGggb3BlbihfX2ZpbG
VfXywgJ3InLCBlbmNvZGluZz0ndXRmLTgnKSBhcyBpbnB1dDoKICAgIHJhdyA9IG
lucHV0LnJlYWQoKQogICAgaW5pdGlhbF9kYXRhID0gcmF3LnNwbGl0KCc='
```

12.4.2 Base64 解码

`b64decode()` 将编码的串转换回原来的形式,它取 4 个字节,利用一个查找表将这 4 个字节转换回原来的 3 个字节。

代码清单 12-27:`base64_b64decode.py`

```
import base64

encoded_data = b'VGhpcyBpcyB0aGUgZGF0YSwgaW4gdGhlIGNsZWFyLg=='
decoded_data = base64.b64decode(encoded_data)
print('Encoded :', encoded_data)
print('Decoded :', decoded_data)
```

编码过程中,会查看输入中的各个 24 位序列(3 个字节),然后将这 24 位编码为输出中的 4 个字节。输出末尾插入了等号作为填充,因为在这个例子中,原始串中的位数不能被 24 整除。

```
$ python3 base64_b64decode.py

Encoded : b'VGhpcyBpcyB0aGUgZGF0YSwgaW4gdGhlIGNsZWFyLg=='
Decoded : b'This is the data, in the clear.'
```

`b64decode()` 的返回值是一个字节串。如果已知内容是文本,那么这个字节串可以转换为一个 Unicode 对象。不过,由于使用 Base64 编码的意义在于能够传输二进制数据,所以假设解码值是文本的做法并不一定安全。

12.4.3 URL 安全的变种

因为默认的 Base64 字母表可能使用 `+` 和 `/`,这两个字符在 URL 中会用到,所以通常很有必要使用一个候选编码替换这些字符。

代码清单 12-28:`base64_urlsafe.py`

```
import base64

encodes_with_pluses = b'\xfb\xef'
encodes_with_slashes = b'\xff\xff'

for original in [encodes_with_pluses, encodes_with_slashes]:
    print('Original         :', repr(original))
    print('Standard encoding:',
          base64.standard_b64encode(original))
    print('URL-safe encoding:',
```

```
        base64.urlsafe_b64encode(original))
print()
```

+替换为 -，/ 替换为下划线（_）。除此之外，字母表是一样的。

```
$ python3 base64_urlsafe.py

Original        : b'\xfb\xef'
Standard encoding: b'++8='
URL-safe encoding: b'--8='

Original        : b'\xff\xff'
Standard encoding: b'//8='
URL-safe encoding: b'__8='
```

12.4.4 其他编码

除了 Base64，这个模块还提供了一些函数来处理 Base85、Base32 和 Base16（十六进制）编码数据。

代码清单 12-29：base64_base32.py

```python
import base64

original_data = b'This is the data, in the clear.'
print('Original:', original_data)

encoded_data = base64.b32encode(original_data)
print('Encoded :', encoded_data)

decoded_data = base64.b32decode(encoded_data)
print('Decoded :', decoded_data)
```

Base32 字母表包括 ASCII 集中的 26 个大写字母以及数字 2 到 7。

```
$ python3 base64_base32.py

Original: b'This is the data, in the clear.'
Encoded : b'KRUGS4ZANFZSA5DIMUQGIYLUMEWCA2LOEB2GQZJAMNWGKYLSFY==
===='
Decoded : b'This is the data, in the clear.'
```

Base16 函数处理十六进制字母表。

代码清单 12-30：base64_base16.py

```python
import base64

original_data = b'This is the data, in the clear.'
print('Original:', original_data)

encoded_data = base64.b16encode(original_data)
print('Encoded :', encoded_data)

decoded_data = base64.b16decode(encoded_data)
print('Decoded :', decoded_data)
```

每次编码位数下降时，采用编码格式的输出就会占用更多空间。

```
$ python3 base64_base16.py

Original: b'This is the data, in the clear.'
Encoded : b'546869732069732074686520646174612C20696E207468652063
6C6561722E'
Decoded : b'This is the data, in the clear.'
```

Base85 函数使用一个扩展的字母表，与 Base64 编码使用的字母表相比，在空间上更节省。

代码清单 12-31：base64_base85.py

```
import base64

original_data = b'This is the data, in the clear.'
print('Original      : {} bytes {!r}'.format(
    len(original_data), original_data))

b64_data = base64.b64encode(original_data)
print('b64 Encoded : {} bytes {!r}'.format(
    len(b64_data), b64_data))

b85_data = base64.b85encode(original_data)
print('b85 Encoded : {} bytes {!r}'.format(
    len(b85_data), b85_data))

a85_data = base64.a85encode(original_data)
print('a85 Encoded : {} bytes {!r}'.format(
    len(a85_data), a85_data))
```

Mercurial、git 和 PDF 文件格式中就使用了很多 Base85 编码和变种。Python 包含两个实现，`b85encode()` 实现了 Git Mercurial 中使用的版本，`a85encode()` 实现了 PDF 文件中使用的 Ascii85 变种版本。

```
$ python3 base64_base85.py

Original    : 31 bytes b'This is the data, in the clear.'
b64 Encoded : 44 bytes b'VGhpcyBpcyB0aGUgZGF0YSwgaW4gdGhlIGNsZWFF
yLg=='
b85 Encoded : 39 bytes b'RA^~)AZc?TbZBKDWMOn+EFfuaAarPDAY*K0VR9}
'
a85 Encoded : 39 bytes b'<+oue+DGm>FD,5.A79Rg/0JYE+EV:.+Cf5!@<*t
'
```

提示：相关阅读材料

- `base64` 的标准库文档[一]。
- **RFC 3548**[二]：Base16、Base32 和 Base64 数据编码。
- **RFC 1924**[三]：IPv6 地址的一个紧凑表示（建议 IPv6 网络地址采用 Base85 编码）。
- Wikipedia：Ascii85[四]。
- `base64` 的 Python 2 到 Python 3 移植说明。

[一] https://docs.python.org/3.5/library/base64.html
[二] https://tools.ietf.org/html/rfc3548.html
[三] https://tools.ietf.org/html/rfc1924.html
[四] https://en.wikipedia.org/wiki/Ascii85

12.5 `http.server`：实现 Web 服务器的基类

`http.server` 使用 `socketserver` 的类创建基类，用来建立 HTTP 服务器。`HTTPServer` 可以直接使用，不过需要扩展 `BaseHTTPRequestHandler` 来处理各个协议方法（如 GET、POST 等）。

12.5.1 HTTP GET

要在一个请求处理器类中增加一个 HTTP 方法支持，需要实现方法 `do_METHOD()`，这里的 `METHOD` 要替换为具体的 HTTP 方法名（例如，`do_GET()`、`do_POST()` 等）。为保持一致，请求处理器方法不带任何参数。请求的所有参数都由 `BaseHTTPRequestHandler` 解析，并存储为请求实例的实例属性。

下面这个示例请求处理器展示了如何向客户返回一个响应，以及对构建响应可能有用的一些本地属性。

代码清单 12-32：`http_server_GET.py`

```
from http.server import BaseHTTPRequestHandler
from urllib import parse
class GetHandler(BaseHTTPRequestHandler):

    def do_GET(self):
        parsed_path = parse.urlparse(self.path)
        message_parts = [
            'CLIENT VALUES:',
            'client_address={} ({})'.format(
                self.client_address,
                self.address_string()),
            'command={}'.format(self.command),
            'path={}'.format(self.path),
            'real path={}'.format(parsed_path.path),
            'query={}'.format(parsed_path.query),
            'request_version={}'.format(self.request_version),
            '',
            'SERVER VALUES:',
            'server_version={}'.format(self.server_version),
            'sys_version={}'.format(self.sys_version),
            'protocol_version={}'.format(self.protocol_version),
            '',
            'HEADERS RECEIVED:',
        ]
        for name, value in sorted(self.headers.items()):
            message_parts.append(
                '{}={}'.format(name, value.rstrip())
            )
        message_parts.append('')
        message = '\r\n'.join(message_parts)
        self.send_response(200)
        self.send_header('Content-Type',
                         'text/plain; charset=utf-8')
        self.end_headers()
        self.wfile.write(message.encode('utf-8'))


if __name__ == '__main__':
```

```
from http.server import HTTPServer
server = HTTPServer(('localhost', 8080), GetHandler)
print('Starting server, use <Ctrl-C> to stop')
server.serve_forever()
```

先组装消息文本，然后写至 `wfile`，这是包装了响应套接字的文件句柄。每个响应需要一个响应码，通过 `send_response()` 设置。如果使用了一个错误码（如 404, 501 等），那么首部会包含一个适当的默认错误消息，或者可能会随这个错误码传递一个消息。

要在服务器中运行请求处理器，需要将它传递到 `HTTPServer` 的构造函数，如示例脚本中 `__main__` 处理部分所示。然后启动服务器。

```
$ python3 http_server_GET.py

Starting server, use <Ctrl-C> to stop
```

在另外一个终端使用 `curl` 来访问这个服务器。

```
$ curl -v -i http://127.0.0.1:8080/?foo=bar

*   Trying 127.0.0.1...
* Connected to 127.0.0.1 (127.0.0.1) port 8080 (#0)
> GET /?foo=bar HTTP/1.1
> Host: 127.0.0.1:8080
> User-Agent: curl/7.43.0
> Accept: */*
>
HTTP/1.0 200 OK
Content-Type: text/plain; charset=utf-8
Server: BaseHTTP/0.6 Python/3.5.2
Date: Thu, 06 Oct 2016 20:44:11 GMT

CLIENT VALUES:
client_address=('127.0.0.1', 52934) (127.0.0.1)
command=GET
path=/?foo=bar
real path=/
query=foo=bar
request_version=HTTP/1.1

SERVER VALUES:
server_version=BaseHTTP/0.6
sys_version=Python/3.5.2
protocol_version=HTTP/1.0

HEADERS RECEIVED:
Accept=*/*
Host=127.0.0.1:8080
User-Agent=curl/7.43.0
* Connection #0 to host 127.0.0.1 left intact
```

说明： 不同的 `curl` 版本生成的输出可能会有变化。如果运行这些例子生成不同的输出，则可以检查 `curl` 报告的版本号。

12.5.2 HTTP POST

支持 POST 请求需要多做一些工作，因为基类不会自动解析表单数据。`cgi` 模块提供了 FieldStorage 类，如果给定了正确的输入，它便知道如何解析表单。

代码清单 12-33：`http_server_POST.py`

```python
import cgi
from http.server import BaseHTTPRequestHandler
import io

class PostHandler(BaseHTTPRequestHandler):

    def do_POST(self):
        # Parse the form data posted.
        form = cgi.FieldStorage(
            fp=self.rfile,
            headers=self.headers,
            environ={
                'REQUEST_METHOD': 'POST',
                'CONTENT_TYPE': self.headers['Content-Type'],
            }
        )

        # Begin the response.
        self.send_response(200)
        self.send_header('Content-Type',
                         'text/plain; charset=utf-8')
        self.end_headers()

        out = io.TextIOWrapper(
            self.wfile,
            encoding='utf-8',
            line_buffering=False,
            write_through=True,
        )

        out.write('Client: {}\n'.format(self.client_address))
        out.write('User-agent: {}\n'.format(
            self.headers['user-agent']))
        out.write('Path: {}\n'.format(self.path))
        out.write('Form data:\n')

        # Echo back information about what was posted in the form.
        for field in form.keys():
            field_item = form[field]
            if field_item.filename:
                # The field contains an uploaded file.
                file_data = field_item.file.read()
                file_len = len(file_data)
                del file_data
                out.write(
                    '\tUploaded {} as {!r} ({} bytes)\n'.format(
                        field, field_item.filename, file_len)
                )
            else:
                # Regular form value
                out.write('\t{}={}\n'.format(
                    field, form[field].value))

        # Disconnect the encoding wrapper from the underlying
        # buffer so that deleting the wrapper doesn't close
        # the socket, which is still being used by the server.
        out.detach()
```

```python
if __name__ == '__main__':
    from http.server import HTTPServer
    server = HTTPServer(('localhost', 8080), PostHandler)
    print('Starting server, use <Ctrl-C> to stop')
    server.serve_forever()
```

在一个窗口中运行这个服务器。

```
$ python3 http_server_POST.py

Starting server, use <Ctrl-C> to stop
```

通过使用 `-F` 选项，`curl` 的参数会包括提交给服务器的表单数据。最后一个参数（`-F datafile=@http_server_GET.py`）将提交文件 `http_server_GET.py` 的内容，以展示如何从表单读取文件数据。

```
$ curl -v http://127.0.0.1:8080/ -F name=dhellmann -F foo=bar \
-F datafile=@http_server_GET.py

*   Trying 127.0.0.1...
* Connected to 127.0.0.1 (127.0.0.1) port 8080 (#0)
> POST / HTTP/1.1
> Host: 127.0.0.1:8080
> User-Agent: curl/7.43.0
> Accept: */*
> Content-Length: 1974
> Expect: 100-continue
> Content-Type: multipart/form-data;
boundary=------------------------a2b3c7485cf8def2
>
* Done waiting for 100-continue
HTTP/1.0 200 OK
Content-Type: text/plain; charset=utf-8
Server: BaseHTTP/0.6 Python/3.5.2
Date: Thu, 06 Oct 2016 20:53:48 GMT

Client: ('127.0.0.1', 53121)
User-agent: curl/7.43.0
Path: /
Form data:
    name=dhellmann
    Uploaded datafile as 'http_server_GET.py' (1612 bytes)
    foo=bar
* Connection #0 to host 127.0.0.1 left intact
```

12.5.3 线程和进程

`HTTPServer` 是 `socketserver.TCPServer` 的一个简单子类，并不使用多线程或进程来处理请求。要增加线程或进程，需要使用适当的 mix-in 技术从 `socketserver` 创建一个新类。

代码清单 12-34：`http_server_threads.py`

```python
from http.server import HTTPServer, BaseHTTPRequestHandler
from socketserver import ThreadingMixIn
import threading


class Handler(BaseHTTPRequestHandler):
```

```python
    def do_GET(self):
        self.send_response(200)
        self.send_header('Content-Type',
                         'text/plain; charset=utf-8')
        self.end_headers()
        message = threading.currentThread().getName()
        self.wfile.write(message.encode('utf-8'))
        self.wfile.write(b'\n')

class ThreadedHTTPServer(ThreadingMixIn, HTTPServer):
    """Handle requests in a separate thread."""

if __name__ == '__main__':
    server = ThreadedHTTPServer(('localhost', 8080), Handler)
    print('Starting server, use <Ctrl-C> to stop')
    server.serve_forever()
```

与其他例子类似，采用同样的方式运行服务器。

```
$ python3 http_server_threads.py

Starting server, use <Ctrl-C> to stop
```

每次服务器接收到一个请求时，它都会开始一个新线程或进程以处理这个请求。

```
$ curl http://127.0.0.1:8080/

Thread-1

$ curl http://127.0.0.1:8080/

Thread-2

$ curl http://127.0.0.1:8080/

Thread-3
```

用 `ForkingMixIn` 替换 `ThreadingMixIn` 会得到类似的结果，不过要使用单独的进程而不是线程。

12.5.4 处理错误

处理错误时要调用 `send_error()`，并传入适当的错误码和一个可选的错误消息。整个响应（包括首部、状态码和响应体）会自动生成。

代码清单 12-35：`http_server_errors.py`

```python
from http.server import BaseHTTPRequestHandler

class ErrorHandler(BaseHTTPRequestHandler):

    def do_GET(self):
        self.send_error(404)


if __name__ == '__main__':
```

```
from http.server import HTTPServer
server = HTTPServer(('localhost', 8080), ErrorHandler)
print('Starting server, use <Ctrl-C> to stop')
server.serve_forever()
```

在这里，总是返回一个 404 错误。

```
$ python3 http_server_errors.py

Starting server, use <Ctrl-C> to stop
```

这里使用一个 HTML 文档将错误消息报告给客户，并提供一个首部来指示错误码。

```
$ curl -i http://127.0.0.1:8080/

HTTP/1.0 404 Not Found
Server: BaseHTTP/0.6 Python/3.5.2
Date: Thu, 06 Oct 2016 20:58:08 GMT
Connection: close
Content-Type: text/html;charset=utf-8
Content-Length: 447

<!DOCTYPE HTML PUBLIC "-//W3C//DTD HTML 4.01//EN"
        "http://www.w3.org/TR/html4/strict.dtd">
<html>
    <head>
        <meta http-equiv="Content-Type"
        content="text/html;charset=utf-8">
        <title>Error response</title>
    </head>
    <body>
        <h1>Error response</h1>
        <p>Error code: 404</p>
        <p>Message: Not Found.</p>
        <p>Error code explanation: 404 - Nothing matches the
        given URI.</p>
    </body>
</html>
```

12.5.5 设置首部

send_header 方法向 HTTP 响应增加首部数据。这个方法有两个参数：首部名和值。

代码清单 12-36：http_server_send_header.py

```
from http.server import BaseHTTPRequestHandler
import time


class GetHandler(BaseHTTPRequestHandler):

    def do_GET(self):
        self.send_response(200)
        self.send_header(
            'Content-Type',
            'text/plain; charset=utf-8',
        )
        self.send_header(
            'Last-Modified',
            self.date_time_string(time.time())
        )
```

```python
        self.end_headers()
        self.wfile.write('Response body\n'.encode('utf-8'))

if __name__ == '__main__':
    from http.server import HTTPServer
    server = HTTPServer(('localhost', 8080), GetHandler)
    print('Starting server, use <Ctrl-C> to stop')
    server.serve_forever()
```

这个例子将 Last-Modified 首部设置为当前时间戳（按照 RFC 7231 格式化）。

```
$ curl -i http://127.0.0.1:8080/

HTTP/1.0 200 OK
Server: BaseHTTP/0.6 Python/3.5.2
Date: Thu, 06 Oct 2016 21:00:54 GMT
Content-Type: text/plain; charset=utf-8
Last-Modified: Thu, 06 Oct 2016 21:00:54 GMT

Response body
```

与其他例子类似，服务器将请求记录到终端。

```
$ python3 http_server_send_header.py

Starting server, use <Ctrl-C> to stop
127.0.0.1 - - [06/Oct/2016 17:00:54] "GET / HTTP/1.1" 200 -
```

12.5.6　命令行用法

http.server 包含一个内置服务器，可以从本地文件系统提供文件。在命令行上为 Python 解释器使用 -m 选项来启动这个服务器。

```
$ python3 -m http.server 8080

Serving HTTP on 0.0.0.0 port 8080 ...
127.0.0.1 - - [06/Oct/2016 17:12:48] "HEAD /index.rst HTTP/1.1" 200 -
```

这个服务器的根目录就是启动服务器的工作目录。

```
$ curl -I http://127.0.0.1:8080/index.rst

HTTP/1.0 200 OK
Server: SimpleHTTP/0.6 Python/3.5.2
Date: Thu, 06 Oct 2016 21:12:48 GMT
Content-type: application/octet-stream
Content-Length: 8285
Last-Modified: Thu, 06 Oct 2016 21:12:10 GMT
```

提示：相关阅读材料

- http.server 的标准库文档[一]。
- socketserver：socketserver 模块提供了处理原始套接字连接的基类。
- **RFC 7231**[二]：Hypertext Transfer Protocol (HTTP/1.1)：Semantics and Content，这个 RFC 包含 HTTP 首部和日期格式的一个规范。

[一] https://docs.python.org/3.5/library/http.server.html
[二] https://tools.ietf.org/html/rfc7231.html

12.6 `http.cookies`：HTTP cookie

`http.cookies` 模块为大多数符合 RFC 2109[①]的 cookie 实现一个解析器。这个实现没有标准那么严格，因为 MSIE 3.0x 不支持完整的标准。

12.6.1 创建和设置 cookie

可以用 cookie 为基于浏览器的应用实现状态管理，因此，cookie 通常由服务器设置，并由客户存储和返回。下面给出一个最简单的例子，创建一个 cookie 设置一个名 – 值对。

代码清单 12-37：`http_cookies_setheaders.py`

```
from http import cookies

c = cookies.SimpleCookie()
c['mycookie'] = 'cookie_value'
print(c)
```

输出是一个合法的 `Set-Cookie` 首部，可以作为 HTTP 响应的一部分传递到客户。

```
$ python3 http_cookies_setheaders.py

Set-Cookie: mycookie=cookie_value
```

12.6.2 Morsel

还可以控制 cookie 的其他方面，如到期时间、路径和域。实际上，cookie 的所有 RFC 属性都可以通过表示 cookie 值的 `Morsel` 对象来管理。

代码清单 12-38：`http_cookies_Morsel.py`

```
from http import cookies
import datetime


def show_cookie(c):
    print(c)
    for key, morsel in c.items():
        print()
        print('key =', morsel.key)
        print('  value =', morsel.value)
        print('  coded_value =', morsel.coded_value)
        for name in morsel.keys():
            if morsel[name]:
                print('  {} = {}'.format(name, morsel[name]))


c = cookies.SimpleCookie()

# A cookie with a value that has to be encoded
# to fit into the header
c['encoded_value_cookie'] = '"cookie,value;"'
c['encoded_value_cookie']['comment'] = 'Has escaped punctuation'
```

[①] https://tools.ietf.org/html/rfc2109.html

```
# A cookie that applies to only part of a site
c['restricted_cookie'] = 'cookie_value'
c['restricted_cookie']['path'] = '/sub/path'
c['restricted_cookie']['domain'] = 'PyMOTW'
c['restricted_cookie']['secure'] = True

# A cookie that expires in 5 minutes
c['with_max_age'] = 'expires in 5 minutes'
c['with_max_age']['max-age'] = 300  # Seconds

# A cookie that expires at a specific time
c['expires_at_time'] = 'cookie_value'
time_to_live = datetime.timedelta(hours=1)
expires = (datetime.datetime(2009, 2, 14, 18, 30, 14) +
           time_to_live)

# Date format: Wdy, DD-Mon-YY HH:MM:SS GMT
expires_at_time = expires.strftime('%a, %d %b %Y %H:%M:%S')
c['expires_at_time']['expires'] = expires_at_time

show_cookie(c)
```

这个例子使用两个不同的方法设置到期的 cookie。其中一个方法将 max-age 设置为一个秒数，另一个方法将 expires 设置为一个日期时间，达到这个日期时间就会丢弃这个 cookie。

```
$ python3 http_cookies_Morsel.py

Set-Cookie: encoded_value_cookie="\"cookie\054value\073\"";
Comment=Has escaped punctuation
Set-Cookie: expires_at_time=cookie_value; expires=Sat, 14 Feb
2009 19:30:14
Set-Cookie: restricted_cookie=cookie_value; Domain=PyMOTW;
Path=/sub/path; Secure
Set-Cookie: with_max_age="expires in 5 minutes"; Max-Age=300

key = with_max_age
  value = expires in 5 minutes
  coded_value = "expires in 5 minutes"
  max-age = 300

key = expires_at_time
  value = cookie_value
  coded_value = cookie_value
  expires = Sat, 14 Feb 2009 19:30:14

key = restricted_cookie
  value = cookie_value
  coded_value = cookie_value
  domain = PyMOTW
  path = /sub/path
  secure = True

key = encoded_value_cookie
  value = "cookie,value;"
  coded_value = "\"cookie\054value\073\""
  comment = Has escaped punctuation
```

Cookie 和 Morsel 对象都相当于字典。Morsel 响应一个固定的键集：

- expires

- path
- comment
- domain
- max-age
- secure
- version

Cookie 实例的键是所存储的各个 cookie 的名。这个信息也可以从 Morsel 的键属性得到。

12.6.3 编码的值

cookie 首部值必须经过编码才能被正确地解析。

代码清单 12-39：http_cookies_coded_value.py

```
from http import cookies

c = cookies.SimpleCookie()
c['integer'] = 5
c['with_quotes'] = 'He said, "Hello, World!"'

for name in ['integer', 'with_quotes']:
    print(c[name].key)
    print('  {}'.format(c[name]))
    print('  value={!r}'.format(c[name].value))
    print('  coded_value={!r}'.format(c[name].coded_value))
    print()
```

Morsel.value 是 cookie 的解码值，而 Morsel.coded_value 表示总是用来将值传输到客户。这两个值都是串。如果保存到一个 cookie 的值不是串，那么其将会自动转换为串。

```
$ python3 http_cookies_coded_value.py

integer
  Set-Cookie: integer=5
  value='5'
  coded_value='5'

with_quotes
  Set-Cookie: with_quotes="He said\054 \"Hello\054 World!\""
  value='He said, "Hello, World!"'
  coded_value='"He said\\054 \\"Hello\\054 World!\\""'
```

12.6.4 接收和解析 Cookie 首部

一旦客户接收到 Set-Cookie 首部，在后续请求中它会使用一个 Cookie 首部把这些 cookie 返回到服务器。到来的 Cookie 首部串可能包含多个 cookie 值，由分号分隔（;）。

Cookie: integer=5; with_quotes="He said, \"Hello, World!\""

取决于 Web 服务器和框架，可以直接从首部或 HTTP_COOKIE 环境变量得到 cookie。

代码清单 12-40：`http_cookies_parse.py`

```
from http import cookies

HTTP_COOKIE = '; '.join([
    r'integer=5',
    r'with_quotes="He said, \"Hello, World!\""',
])

print('From constructor:')
c = cookies.SimpleCookie(HTTP_COOKIE)
print(c)

print()
print('From load():')
c = cookies.SimpleCookie()
c.load(HTTP_COOKIE)
print(c)
```

要对它们解码，实例化时可以将串（但不包括首部前缀）传递到 `SimpleCookie`，或者使用 `load()` 方法。

```
$ python3 http_cookies_parse.py

From constructor:
Set-Cookie: integer=5
Set-Cookie: with_quotes="He said, \"Hello, World!\""

From load():
Set-Cookie: integer=5
Set-Cookie: with_quotes="He said, \"Hello, World!\""
```

12.6.5 候选输出格式

除了使用 `Set-Cookie` 首部，服务器还可以提供 JavaScript 向客户增加 cookie。`SimpleCookie` 和 `Morsel` 通过 `js_output()` 方法来提供 JavaScript 输出。

代码清单 12-41：`http_cookies_js_output.py`

```
from http import cookies
import textwrap

c = cookies.SimpleCookie()
c['mycookie'] = 'cookie_value'
c['another_cookie'] = 'second value'
js_text = c.js_output()
print(textwrap.dedent(js_text).lstrip())
```

结果是一个完整的 `script` 标记，其中包含设置 cookie 的语句。

```
$ python3 http_cookies_js_output.py

<script type="text/javascript">
<!-- begin hiding
document.cookie = "another_cookie=\"second value\"";
// end hiding -->
</script>
```

```
<script type="text/javascript">
<!-- begin hiding
document.cookie = "mycookie=cookie_value";
// end hiding -->
</script>
```

提示：相关阅读材料
- `http.cookies` 的标准库文档[¤]。
- `http.cookiejar`：`cookielib` 模块，在客户端处理 cookie。
- **RFC 2109**[¤]：HTTP 状态管理机制。

12.7 webbrowser：显示 Web 页面

webbrowser 模块包含一些函数，可以在交互式的浏览器应用中打开 URL。它提供了一个可用浏览器的注册表，因为系统上可能有多个可用的浏览器。还可以用 BROWSER 环境变量控制浏览器。

12.7.1 简单示例

要在一个浏览器中打开一个页面，可以使用 `open()` 函数。

代码清单 12-42：webbrowser_open.py

```
import webbrowser

webbrowser.open(
    'https://docs.python.org/3/library/webbrowser.html'
)
```

会在一个浏览器窗口中打开这个 URL，这个窗口将被增加到窗口栈的栈顶。文档中指出，如果可能的话，会重用一个现有的窗口，不过具体行为可能取决于你的浏览器的设置。如果在 Mac OS X 上使用 Firefox，便总会创建一个新窗口。

12.7.2 窗口与标签页

如果总是想使用一个新窗口，则可以使用 `open_new()`。

代码清单 12-43：webbrowser_open_new.py

```
import webbrowser

webbrowser.open_new(
    'https://docs.python.org/3/library/webbrowser.html'
)
```

如果想创建一个新的标签页，则可以使用 `open_new_tab()`。

¤ https://docs.python.org/3.5/library/http.cookies.html
¤ https://tools.ietf.org/html/rfc2109.html

12.7.3 使用特定浏览器

如果出于某种原因你的应用需要使用一个特定的浏览器，那么可以使用 `get()` 函数访问已注册的一组浏览器控制器。浏览器控制器包含 `open()`、`open_new()` 和 `open_new_tab()` 方法。下面的例子会强制使用 lynx 浏览器。

代码清单 12-44：**webbrowser_get.py**

```
import webbrowser

b = webbrowser.get('lynx')
b.open('https://docs.python.org/3/library/webbrowser.html')
```

关于可用浏览器类型的完整列表，请参考模块的文档。

12.7.4 BROWSER 变量

用户可以把环境变量 `BROWSER` 设置为想要尝试的浏览器名或命令，从应用外部控制 `webbrowser` 模块。所用的值包含一系列浏览器名，各个名字之间用 `os.pathsep` 分隔。如果名字中包含 `%s`，这个名字会被解释为一个字面量命令，将直接执行，并把 `%s` 替换为 URL。否则，这个名字会被传递到 `get()`，以便从注册表得到一个控制器对象。

例如，下面的命令会在 lynx 中打开 Web 页面（假设存在这个页面），而不论另外注册了哪些浏览器：

```
$ BROWSER=lynx python3 webbrowser_open.py
```

如果 `BROWSER` 中的所有浏览器名都不适用，那么 `webbrowser` 就会退回为其默认行为。

12.7.5 命令行接口

不仅可以从 Python 程序中访问 `webbrowser` 模块的所有特性，也可以通过命令行得到。

```
$ python3 -m webbrowser

Usage: .../lib/python3.5/webbrowser.py [-n | -t] url
    -n: open new window
    -t: open new tab
```

提示：相关阅读材料

- `webbrowser` 的标准库文档[⊖]。
- What the What?[⊖]：运行你的 Python 程序，然后利用 Google 搜索得到所生成的异常消息。

⊖ https://docs.python.org/3.5/library/webbrowser.html
⊖ https://github.com/dhellmann/whatthewhat

12.8 uuid：全局唯一标识符

uuid 模块实现了 **RFC 4122**[①]中描述的全局唯一标识符（Universally Unique Identifier）；这个 RFC 定义了一个系统，可以为资源创建唯一的标识符，这里采用一种不需要集中注册机的方式。UUID 值为 128 位，正如参考指南所述，"UUID 可以保证跨空间和时间的唯一性"。对于文档、主机、应用客户以及其他需要唯一值的情况，UUID 可以用来生成标识符。这个 RFC 特别强调创建一个统一资源名（Uniform Resource Name）命名空间，并且涵盖了 3 个主要算法。

- 使用 IEEE 802 MAC 地址作为唯一性来源
- 使用伪随机数
- 使用公开的串并结合密码散列

在上述所有情况下，种子值都要与系统时钟结合，如果向后设置时钟，则要用一个时钟序列值维护唯一性。

12.8.1 UUID 1：IEEE 802 MAC 地址

UUID 1 值使用主机的 MAC 地址计算。uuid 模块使用 `getnode()` 来获取当前系统的 MAC 值。

代码清单 12-45：`uuid_getnode.py`

```
import uuid

print(hex(uuid.getnode()))
```

如果一个系统有多个网卡，那么相应地便会有多个 MAC 地址，并且可能返回其中任意一个值。

```
$ python3 uuid_getnode.py

0xc82a14598875
```

要为一个主机（由其 MAC 地址标识）生成一个 UUID，需要使用 `uuid1()` 函数。节点标识符参数是可选的；如果没有设置这个域，那么便会使用 `getnode()` 返回的值。

代码清单 12-46：`uuid_uuid1.py`

```
import uuid

u = uuid.uuid1()

print(u)
print(type(u))
print('bytes   :', repr(u.bytes))
print('hex     :', u.hex)
print('int     :', u.int)
print('urn     :', u.urn)
```

[①] https://tools.ietf.org/html/rfc4122.html

```
print('variant :', u.variant)
print('version :', u.version)
print('fields  :', u.fields)
print('   time_low              : ', u.time_low)
print('   time_mid              : ', u.time_mid)
print('   time_hi_version       : ', u.time_hi_version)
print('   clock_seq_hi_variant: ', u.clock_seq_hi_variant)
print('   clock_seq_low         : ', u.clock_seq_low)
print('   node                  : ', u.node)
print('   time                  : ', u.time)
print('   clock_seq             : ', u.clock_seq)
```

对于返回的 UUID 对象，可以通过只读的实例属性访问它的各个部分。有些属性是 UUID 值的不同表示，如 `hex`、`int` 和 `urn`。

```
$ python3 uuid_uuid1.py

335ea282-cded-11e6-9ede-c82a14598875
<class 'uuid.UUID'>
bytes   : b'3^\xa2\x82\xcd\xed\x11\xe6\x9e\xde\xc8*\x14Y\x88u'
hex     : 335ea282cded11e69edec82a14598875
int     : 68281999803480928707202152670695098485
urn     : urn:uuid:335ea282-cded-11e6-9ede-c82a14598875
variant : specified in RFC 4122
version : 1
fields  : (861840002, 52717, 4582, 158, 222, 220083055593589)
   time_low              : 861840002
   time_mid              : 52717
   time_hi_version       : 4582
   clock_seq_hi_variant: 158
   clock_seq_low         : 222
   node                  : 220083055593589
   time                  : 137023257334162050
   clock_seq             : 7902
```

由于有时间分量（time），所以每次调用 `uuid1()` 都会返回一个新值。

代码清单 12-47：`uuid_uuid1_repeat.py`

```
import uuid

for i in range(3):
    print(uuid.uuid1())
```

在这个输出中，只有时间分量（串的开始部分）有变化。

```
$ python3 uuid_uuid1_repeat.py

3369ab5c-cded-11e6-8d5e-c82a14598875
336eea22-cded-11e6-9943-c82a14598875
336eeb5e-cded-11e6-9e22-c82a14598875
```

由于每个计算机有不同的 MAC 地址，所以在不同系统上运行这个示例程序会生成完全不同的值。下一个例子传递不同的节点 ID 来模拟在不同主机上运行。

代码清单 12-48：`uuid_uuid1_othermac.py`

```
import uuid
```

```
for node in [0x1ec200d9e0, 0x1e5274040e]:
    print(uuid.uuid1(node), hex(node))
```

除了返回不同的时间值，UUID 末尾的节点标识符也有变化。

```
$ python3 uuid_uuid1_othermac.py

337969be-cded-11e6-97fa-001ec200d9e0 0x1ec200d9e0
3379b7e6-cded-11e6-9d72-001e5274040e 0x1e5274040e
```

12.8.2　UUID 3 和 5：基于名字的值

有些情况下可能需要根据名字创建 UUID 值，而不是根据随机值或基于时间的值来创建。UUID 3 和 5 规范使用密码散列值（分别使用 MD5 或 SHA-1），将特定于命名空间的种子值与名字相结合。有一些由预定义 UUID 值标识的公开的命名空间，分别用于处理 DNS、URL、ISO OID 和 X.500 识别名（Distinguished Name）。通过生成和保存 UUID 值，还可以定义新的特定于应用的命名空间。

代码清单 12-49：**uuid_uuid3_uuid5.py**

```
import uuid

hostnames = ['www.doughellmann.com', 'blog.doughellmann.com']

for name in hostnames:
    print(name)
    print('  MD5   :', uuid.uuid3(uuid.NAMESPACE_DNS, name))
    print('  SHA-1 :', uuid.uuid5(uuid.NAMESPACE_DNS, name))
    print()
```

要从一个 DNS 名创建 UUID，可以把 uuid.NAMESPACE_DNS 作为命名空间参数传入 uuid3() 或 uuid5()。

```
$ python3 uuid_uuid3_uuid5.py

www.doughellmann.com
  MD5   : bcd02e22-68f0-3046-a512-327cca9def8f
  SHA-1 : e3329b12-30b7-57c4-8117-c2cd34a87ce9

blog.doughellmann.com
  MD5   : 9bdabfce-dfd6-37ab-8a3f-7f7293bcf111
  SHA-1 : fa829736-7ef8-5239-9906-b4775a5abacb
```

不论什么时间计算或者在哪里计算，一个命名空间中给定名的 UUID 值总是相同的。

代码清单 12-50：**uuid_uuid3_repeat.py**

```
import uuid

namespace_types = sorted(
    n
    for n in dir(uuid)
    if n.startswith('NAMESPACE_')
)
name = 'www.doughellmann.com'

for namespace_type in namespace_types:
```

```
    print(namespace_type)
    namespace_uuid = getattr(uuid, namespace_type)
    print(' ', uuid.uuid3(namespace_uuid, name))
    print(' ', uuid.uuid3(namespace_uuid, name))
    print()
```

但是命名空间中相同名字的 UUID 值则是不同的。

```
$ python3 uuid_uuid3_repeat.py

NAMESPACE_DNS
  bcd02e22-68f0-3046-a512-327cca9def8f
  bcd02e22-68f0-3046-a512-327cca9def8f

NAMESPACE_OID
  e7043ac1-4382-3c45-8271-d5c083e41723
  e7043ac1-4382-3c45-8271-d5c083e41723

NAMESPACE_URL
  5d0fdaa9-eafd-365e-b4d7-652500dd1208
  5d0fdaa9-eafd-365e-b4d7-652500dd1208

NAMESPACE_X500
  4a54d6e7-ce68-37fb-b0ba-09acc87cabb7
  4a54d6e7-ce68-37fb-b0ba-09acc87cabb7
```

12.8.3 UUID 4：随机值

有时，基于主机和基于命名空间的 UUID 值"差别还不够大"。例如，如果 UUID 要作为散列键，则需要有区分度更大、更随机的值序列来避免散列表中出现冲突。让值有更少的共同数字也能更容易地在日志文件中查找这些值。为了增加 UUID 的区分度，可以使用 `uuid4()` 利用随机的输入值生成 UUID。

代码清单 12-51：**uuid_uuid4.py**

```
import uuid

for i in range(3):
    print(uuid.uuid4())
```

随机性的来源取决于导入 `uuid` 时哪些 C 库可用。如果可以加载 libuuid（或 uuid.dll），而且其中包含一个生成随机值的函数，那么便使用这个函数。否则，使用 `os.urandom()` 或 `random` 模块。

```
$ python3 uuid_uuid4.py

7821863a-06f0-4109-9b88-59ba1ca5cc04
44846e16-4a59-4a21-8c8e-008f169c2dd5
1f3cef3c-e2bc-4877-96c8-eba43bf15bb6
```

12.8.4 处理 UUID 对象

除了生成新的 UUID 值，还可以解析标准格式的串以创建 UUID 对象，使比较和排序操作的处理更为容易。

代码清单 12-52：**uuid_uuid_objects.py**

```
import uuid


def show(msg, l):
    print(msg)
    for v in l:
        print(' ', v)
    print()

input_values = [
    'urn:uuid:f2f84497-b3bf-493a-bba9-7c68e6def80b',
    '{417a5ebb-01f7-4ed5-aeac-3d56cd5037b0}',
    '2115773a-5bf1-11dd-ab48-001ec200d9e0',
]
show('input_values', input_values)
uuids = [uuid.UUID(s) for s in input_values]
show('converted to uuids', uuids)

uuids.sort()
show('sorted', uuids)
```

从输入中去除外围大括号，另外将短横线（-）也去除。如果串有一个包含 `urn:` 或 `uuid:` 的前缀，则这个前缀也会被删除。剩下的文本必然是由十六进制数构成的串，然后再将它解释为一个 UUID 值。

```
$ python3 uuid_uuid_objects.py
input_values
  urn:uuid:f2f84497-b3bf-493a-bba9-7c68e6def80b
  {417a5ebb-01f7-4ed5-aeac-3d56cd5037b0}
  2115773a-5bf1-11dd-ab48-001ec200d9e0

converted to uuids
  f2f84497-b3bf-493a-bba9-7c68e6def80b
  417a5ebb-01f7-4ed5-aeac-3d56cd5037b0
  2115773a-5bf1-11dd-ab48-001ec200d9e0

sorted
  2115773a-5bf1-11dd-ab48-001ec200d9e0
  417a5ebb-01f7-4ed5-aeac-3d56cd5037b0
  f2f84497-b3bf-493a-bba9-7c68e6def80b
```

提示：相关阅读材料

- **uuid** 的标准库文档[⊖]。
- **uuid** 的 Python 2 到 Python 3 移植说明。
- **RFC 4122**[⊖]：全局唯一标识符（UUID）URN 命名空间。

12.9 json：JavaScript 对象记法

json 模块提供了一个与 **pickle** 类似的 API，可以将内存中 Python 对象转换为一个串

⊖ https://docs.python.org/3.5/library/uuid.html
⊖ https://tools.ietf.org/html/rfc4122.html

行化表示，被称为 JavaScript 对象记法（JavaScript Object Notation，JSON）。不同于 `pickle`，JSON 有一个优点，它有多种语言的实现（特别是 JavaScript）。JSON 对于 REST API 中 Web 服务器和客户之间的通信使用最广泛，不过也可以用于满足其他应用间的通信需求。

12.9.1　编码和解码简单数据类型

默认地，编码器理解 Python 的一些内置类型（即 `str`、`int`、`float`、`list`、`tuple` 和 `dict`）。

代码清单 12-53：**json_simple_types.py**

```
import json

data = [{'a': 'A', 'b': (2, 4), 'c': 3.0}]
print('DATA:', repr(data))

data_string = json.dumps(data)
print('JSON:', data_string)
```

对值编码时，表面上类似于 Python 的 `repr()` 输出。

```
$ python3 json_simple_types.py

DATA: [{'c': 3.0, 'b': (2, 4), 'a': 'A'}]
JSON: [{"c": 3.0, "b": [2, 4], "a": "A"}]
```

编码然后再重新解码时，可能不会得到完全相同的对象类型。

代码清单 12-54：**json_simple_types_decode.py**

```
import json

data = [{'a': 'A', 'b': (2, 4), 'c': 3.0}]
print('DATA   :', data)

data_string = json.dumps(data)
print('ENCODED:', data_string)

decoded = json.loads(data_string)
print('DECODED:', decoded)

print('ORIGINAL:', type(data[0]['b']))
print('DECODED :', type(decoded[0]['b']))
```

具体地，元组会变成列表。

```
$ python3 json_simple_types_decode.py

DATA    : [{'c': 3.0, 'b': (2, 4), 'a': 'A'}]
ENCODED: [{"c": 3.0, "b": [2, 4], "a": "A"}]
DECODED: [{'c': 3.0, 'b': [2, 4], 'a': 'A'}]
ORIGINAL: <class 'tuple'>
DECODED : <class 'list'>
```

12.9.2　人类可读和紧凑输出

JSON 优于 `pickle` 的另一个好处是，JSON 会生成人类可读的结果。`dumps()` 函数接受多个参数从而使输出更容易理解。例如，sort_keys 标志会告诉编码器按有序顺序而不是随机顺序输出字典的键。

代码清单 12-55：**json_sort_keys.py**

```
import json

data = [{'a': 'A', 'b': (2, 4), 'c': 3.0}]
print('DATA:', repr(data))

unsorted = json.dumps(data)
print('JSON:', json.dumps(data))
print('SORT:', json.dumps(data, sort_keys=True))

first = json.dumps(data, sort_keys=True)
second = json.dumps(data, sort_keys=True)

print('UNSORTED MATCH:', unsorted == first)
print('SORTED MATCH  :', first == second)
```

排序后，会让人更容易地查看结果，而且还可以在测试中比较 JSON 输出。

```
$ python3 json_sort_keys.py

DATA: [{'c': 3.0, 'b': (2, 4), 'a': 'A'}]
JSON: [{"c": 3.0, "b": [2, 4], "a": "A"}]
SORT: [{"a": "A", "b": [2, 4], "c": 3.0}]
UNSORTED MATCH: False
SORTED MATCH  : True
```

对于高度嵌套的数据结构，还可以指定一个缩进（indent）值来得到格式美观的输出。

代码清单 12-56：**json_indent.py**

```
import json

data = [{'a': 'A', 'b': (2, 4), 'c': 3.0}]
print('DATA:', repr(data))
print('NORMAL:', json.dumps(data, sort_keys=True))
print('INDENT:', json.dumps(data, sort_keys=True, indent=2))
```

当缩进是一个非负整数时，输出更类似于 pprint 的输出，数据结构中每一级的前导空格与缩进级别匹配。

```
$ python3 json_indent.py

DATA: [{'c': 3.0, 'b': (2, 4), 'a': 'A'}]
NORMAL: [{"a": "A", "b": [2, 4], "c": 3.0}]
INDENT: [
  {
    "a": "A",
    "b": [
      2,
      4
    ],
    "c": 3.0
  }
]
```

这种详细输出会增加传输等量数据所需的字节数，所以生产环境中往往不使用这种输出。实际上，可以调整编码输出中分隔数据的设置，从而使其比默认格式更紧凑。

代码清单 12-57：`json_compact_encoding.py`

```python
import json

data = [{'a': 'A', 'b': (2, 4), 'c': 3.0}]
print('DATA:', repr(data))

print('repr(data)             :', len(repr(data)))

plain_dump = json.dumps(data)
print('dumps(data)            :', len(plain_dump))

small_indent = json.dumps(data, indent=2)
print('dumps(data, indent=2)  :', len(small_indent))

with_separators = json.dumps(data, separators=(',', ':'))
print('dumps(data, separators):', len(with_separators))
```

`dumps()` 的 `separators` 参数应当是一个元组，其中包含用来分隔列表中各项的字符串，以及分隔字典中键和值的字符串。默认为（'，'，'：'）。通过去除空白符，可以生成一个更为紧凑的输出。

```
$ python3 json_compact_encoding.py

DATA: [{'c': 3.0, 'b': (2, 4), 'a': 'A'}]
repr(data)             : 35
dumps(data)            : 35
dumps(data, indent=2)  : 73
dumps(data, separators): 29
```

12.9.3 编码字典

JSON 格式要求字典的键是字符串。如果一个字典以非字符串类型作为键，那么对这个字典编码时，便会生成一个 `TypeError`。要想绕开这个限制，一种办法是使用 `skipkeys` 参数告诉编码器跳过非串的键。

代码清单 12-58：`json_skipkeys.py`

```python
import json

data = [{'a': 'A', 'b': (2, 4), 'c': 3.0, ('d',): 'D tuple'}]

print('First attempt')
try:
    print(json.dumps(data))
except TypeError as err:
    print('ERROR:', err)

print()
print('Second attempt')
print(json.dumps(data, skipkeys=True))
```

这里不会产生一个异常，而是会忽略非串的键。

```
$ python3 json_skipkeys.py

First attempt
ERROR: keys must be a string

Second attempt
[{"c": 3.0, "b": [2, 4], "a": "A"}]
```

12.9.4　处理定制类型

目前为止，所有例子都使用 Python 的内置类型，因为这些类型得到了 `json` 的内置支持。通常还需要对定制类编码，有两种办法可以做到。

假设以下代码清单中的类需要进行编码。

代码清单 12-59：json_myobj.py

```python
class MyObj:

    def __init__(self, s):
        self.s = s

    def __repr__(self):
        return '<MyObj({})>'.format(self.s)
```

要对 `MyObj` 实例编码，一个简单的方法是定义一个函数，将未知类型转换为已知类型。这个函数并不需要具体完成编码，它只是将一个类型的对象转换为另一个类型。

代码清单 12-60：json_dump_default.py

```python
import json
import json_myobj

obj = json_myobj.MyObj('instance value goes here')

print('First attempt')
try:
    print(json.dumps(obj))
except TypeError as err:
    print('ERROR:', err)


def convert_to_builtin_type(obj):
    print('default(', repr(obj), ')')
    # Convert objects to a dictionary of their representation.
    d = {
        '__class__': obj.__class__.__name__,
        '__module__': obj.__module__,
    }
    d.update(obj.__dict__)
    return d

print()
print('With default')
print(json.dumps(obj, default=convert_to_builtin_type))
```

在 `convert_to_builtin_type()` 中，`json` 无法识别的类实例会被转换为字典，其中包含足够多的信息，如果程序能访问这个处理所需的 Python 模块，就能利用这些信息重新创建对象。

```
$ python3 json_dump_default.py

First attempt
ERROR: <MyObj(instance value goes here)> is not JSON serializable

With default
default( <MyObj(instance value goes here)> )
{"s": "instance value goes here", "__module__": "json_myobj",
"__class__": "MyObj"}
```

要对结果解码并创建一个 `MyObj()` 实例，可以使用 `loads()` 的 `object_hook` 参数关联解码器，从而可以从模块导入这个类，并将该类用来创建实例。对于从到来数据流解码的各个字典，都会调用 `object_hook`，这就提供了一个机会，可以把字典转换为另外一种类型的对象。hook 函数要返回调用应用要接收的对象而不是字典。

代码清单 12-61：`json_load_object_hook.py`

```python
import json

def dict_to_object(d):
    if '__class__' in d:
        class_name = d.pop('__class__')
        module_name = d.pop('__module__')
        module = __import__(module_name)
        print('MODULE:', module.__name__)
        class_ = getattr(module, class_name)
        print('CLASS:', class_)
        args = {
            key: value
            for key, value in d.items()
        }
        print('INSTANCE ARGS:', args)
        inst = class_(**args)
    else:
        inst = d
    return inst


encoded_object = '''
    [{"s": "instance value goes here",
      "__module__": "json_myobj", "__class__": "MyObj"}]
    '''

myobj_instance = json.loads(
    encoded_object,
    object_hook=dict_to_object,
)
print(myobj_instance)
```

由于 `json` 将串值转换为 Unicode 对象，因此，在其被用作类构造函数的关键字参数之前，需要将它们重新编码为 ASCII 串。

```
$ python3 json_load_object_hook.py

MODULE: json_myobj
CLASS: <class 'json_myobj.MyObj'>
INSTANCE ARGS: {'s': 'instance value goes here'}
[<MyObj(instance value goes here)>]
```

内置类型也有类似的 hook，如整数（`parse_int`）、浮点数（`parse_float`）和常量（`parse_constant`）。

12.9.5　编码器和解码器类

除了之前介绍的便利函数，`json` 模块还提供了一些类来完成编码和解码。直接使用这些类可以访问另外的 API 来定制其行为。

`JSONEncoder` 使用一个 `iterable` 接口生成编码数据"块"，从而更容易将其写至文件或网络套接字，而不必在内存中表示完整的数据结构。

代码清单 12-62：**json_encoder_iterable.py**

```python
import json

encoder = json.JSONEncoder()
data = [{'a': 'A', 'b': (2, 4), 'c': 3.0}]

for part in encoder.iterencode(data):
    print('PART:', part)
```

输出按逻辑单元生成，而不是根据某个大小值。

```
$ python3 json_encoder_iterable.py

PART: [
PART: {
PART: "c"
PART: :
PART: 3.0
PART: ,
PART: "b"
PART: :
PART: [2
PART: , 4
PART: ]
PART: ,
PART: "a"
PART: :
PART: "A"
PART: }
PART: ]
```

`encode()` 方法基本上等价于 `''.join(encoder.iterencode())`，只不过之前会做一些额外的错误检查。

要对任意的对象编码，需要用一个实现覆盖 `default()` 方法，这个实现类似于 `convert_to_builtin_type()` 中的实现。

代码清单 12-63：**json_encoder_default.py**

```python
import json
import json_myobj


class MyEncoder(json.JSONEncoder):

    def default(self, obj):
        print('default(', repr(obj), ')')
        # Convert objects to a dictionary of their representation.
        d = {
            '__class__': obj.__class__.__name__,
            '__module__': obj.__module__,
        }
        d.update(obj.__dict__)
        return d


obj = json_myobj.MyObj('internal data')
print(obj)
print(MyEncoder().encode(obj))
```

输出与前一个实现的输出相同。

```
$ python3 json_encoder_default.py

<MyObj(internal data)>
default( <MyObj(internal data)> )
{"s": "internal data", "__module__": "json_myobj", "__class__": "MyObj"}
```

这里要解码文本，然后将字典转换为一个对象，与前面的实现相比，这需要多做一些工作，不过不算太多。

代码清单 12-64：**json_decoder_object_hook.py**

```python
import json


class MyDecoder(json.JSONDecoder):

    def __init__(self):
        json.JSONDecoder.__init__(
            self,
            object_hook=self.dict_to_object,
        )

    def dict_to_object(self, d):
        if '__class__' in d:
            class_name = d.pop('__class__')
            module_name = d.pop('__module__')
            module = __import__(module_name)
            print('MODULE:', module.__name__)
            class_ = getattr(module, class_name)
            print('CLASS:', class_)
            args = {
                key: value
                for key, value in d.items()
            }
            print('INSTANCE ARGS:', args)
```

```
                inst = class_(**args)
            else:
                inst = d
        return inst


encoded_object = '''
[{"s": "instance value goes here",
  "__module__": "json_myobj", "__class__": "MyObj"}]
'''

myobj_instance = MyDecoder().decode(encoded_object)
print(myobj_instance)
```

输出与前面的例子相同。

```
$ python3 json_decoder_object_hook.py

MODULE: json_myobj
CLASS: <class 'json_myobj.MyObj'>
INSTANCE ARGS: {'s': 'instance value goes here'}
[<MyObj(instance value goes here)>]
```

12.9.6 处理流和文件

目前为止，所有例子都假设整个数据结构的编码版本可以一次完全放在内存中。对于很大的数据结构，更合适的做法可能是将编码直接写至一个类似文件的对象。便利函数 `load()` 和 `dump()` 会接收一个类似文件对象的引用用于读写。

代码清单 12-65：**json_dump_file.py**

```
import io
import json

data = [{'a': 'A', 'b': (2, 4), 'c': 3.0}]

f = io.StringIO()
json.dump(data, f)

print(f.getvalue())
```

类似于这个例子中使用的 `StringIO` 缓冲区，也可以使用套接字或常规的文件句柄。

```
$ python3 json_dump_file.py

[{"c": 3.0, "b": [2, 4], "a": "A"}]
```

尽管没有优化，即一次只读取数据的一部分，但 `load()` 函数还提供了一个好处，它封装了从流输入生成对象的逻辑。

代码清单 12-66：**json_load_file.py**

```
import io
import json

f = io.StringIO('[{"a": "A", "c": 3.0, "b": [2, 4]}]')
print(json.load(f))
```

类似于 dump()，任何类似文件对象都可以被传递到 load()。

```
$ python3 json_load_file.py

[{'c': 3.0, 'b': [2, 4], 'a': 'A'}]
```

12.9.7 混合数据流

JSONDecoder 包含一个 raw_decode() 方法，如果一个数据结构后面跟有更多数据，如带尾部文本的 JSON 数据，则可以用这个方法完成解码。返回值是对输入数据解码创建的对象，以及该数据的一个索引（指示在哪里结束解码）。

代码清单 12-67：**json_mixed_data.py**

```python
import json

decoder = json.JSONDecoder()


def get_decoded_and_remainder(input_data):
    obj, end = decoder.raw_decode(input_data)
    remaining = input_data[end:]
    return (obj, end, remaining)


encoded_object = '[{"a": "A", "c": 3.0, "b": [2, 4]}]'
extra_text = 'This text is not JSON.'

print('JSON first:')
data = ' '.join([encoded_object, extra_text])
obj, end, remaining = get_decoded_and_remainder(data)

print('Object            :', obj)
print('End of parsed input :', end)
print('Remaining text    :', repr(remaining))

print()
print('JSON embedded:')
try:
    data = ' '.join([extra_text, encoded_object, extra_text])
    obj, end, remaining = get_decoded_and_remainder(data)
except ValueError as err:
    print('ERROR:', err)
```

遗憾的是，这种做法只适用于对象出现在输入起始位置的情况。

```
$ python3 json_mixed_data.py

JSON first:
Object              : [{'c': 3.0, 'b': [2, 4], 'a': 'A'}]
End of parsed input : 35
Remaining text      : ' This text is not JSON.'

JSON embedded:
ERROR: Expecting value: line 1 column 1 (char 0)
```

12.9.8 命令行上处理 JSON

json.tool 模块实现了一个命令行程序来重新格式化 JSON 数据，使数据更易读。

```
[{"a": "A", "c": 3.0, "b": [2, 4]}]
```

输入文件 example.json 包含一个映射,其中键采用字母表顺序。第一个例子显示了按顺序重新格式化的数据,第二个例子使用了 --sort-keys 在打印输出之前先对映射键排序。

```
$ python3 -m json.tool example.json

[
    {
        "a": "A",
        "c": 3.0,
        "b": [
            2,
            4
        ]
    }
]

$ python3 -m json.tool --sort-keys example.json

[
    {
        "a": "A",
        "b": [
            2,
            4
        ],
        "c": 3.0
    }
]
```

提示:相关阅读材料

- json 的标准库文档[⊖]。
- json 的 Python 2 到 Python 3 移植说明。
- JavaScript Object Notation[⊖]:JSON 主页,提供了文档以及其他语言的实现。
- jsonpickle[⊖]:jsonpickle 可以将任意 Python 对象串行化为 JSON。

12.10 xmlrpc.client:XML-RPC 的客户库

XML-RPC 是一个轻量级远程过程调用协议,建立在 HTTP 和 XML 之上。xmlrpclib 模块允许 Python 程序与使用任何语言编写的 XML-RPC 服务器通信。

这一节中的所有例子都使用了 xmlrpc-server.py 中定义的服务器,可以在源发布包中找到,这里给出这个服务器以供参考。

代码清单 12-68:xmlrpc_server.py

```
from xmlrpc.server import SimpleXMLRPCServer
from xmlrpc.client import Binary
```

⊖ https://docs.python.org/3.5/library/json.html
⊖ http://json.org/
⊖ https://jsonpickle.github.io

```python
import datetime


class ExampleService:

    def ping(self):
        """Simple function to respond when called
        to demonstrate connectivity.
        """
        return True

    def now(self):
        """Returns the server current date and time."""
        return datetime.datetime.now()

    def show_type(self, arg):
        """Illustrates how types are passed in and out of
        server methods.

        Accepts one argument of any type.

        Returns a tuple with string representation of the value,
        the name of the type, and the value itself.

        """
        return (str(arg), str(type(arg)), arg)
    def raises_exception(self, msg):
        "Always raises a RuntimeError with the message passed in."
        raise RuntimeError(msg)

    def send_back_binary(self, bin):
        """Accepts a single Binary argument, and unpacks and
        repacks it to return it."""
        data = bin.data
        print('send_back_binary({!r})'.format(data))
        response = Binary(data)
        return response


if __name__ == '__main__':
    server = SimpleXMLRPCServer(('localhost', 9000),
                                logRequests=True,
                                allow_none=True)
    server.register_introspection_functions()
    server.register_multicall_functions()

    server.register_instance(ExampleService())

    try:
        print('Use Control-C to exit')
        server.serve_forever()
    except KeyboardInterrupt:
        print('Exiting')
```

12.10.1 连接服务器

要将一个客户连接到服务器，最简单的方法是实例化一个 **ServerProxy** 对象，为它指定服务器的 URI。例如，演示服务器在 localhost 的端口 9000 上运行。

代码清单 12-69：**xmlrpc_ServerProxy.py**

```python
import xmlrpc.client

server = xmlrpc.client.ServerProxy('http://localhost:9000')
print('Ping:', server.ping())
```

在这种情况下，服务的 `ping()` 方法没有任何参数，它会返回一个布尔值。

```
$ python3 xmlrpc_ServerProxy.py

Ping: True
```

还可以有其他选项支持其他类型的传输以便连接服务器。HTTP 和 HTTPS 已经明确得到支持，二者都提供基本认证。要实现一个新的通信通道，只需要一个新的传输类。例如，可以在 SMTP 之上实现 XML-RPC，这是一个很有意思的练习。

代码清单 12-70：**xmlrpc_ServerProxy_verbose.py**

```python
import xmlrpc.client

server = xmlrpc.client.ServerProxy('http://localhost:9000',
                                   verbose=True)
print('Ping:', server.ping())
```

指定 `verbose` 选项会提供调试信息，这对于解决通信错误很有用。

```
$ python3 xmlrpc_ServerProxy_verbose.py

send: b'POST /RPC2 HTTP/1.1\r\nHost: localhost:9000\r\n
Accept-Encoding: gzip\r\nContent-Type: text/xml\r\n
User-Agent: Python-xmlrpc/3.5\r\nContent-Length: 98\r\n\r\n'
send: b"<?xml version='1.0'?>\n<methodCall>\n<methodName>
ping</methodName>\n<params>\n</params>\n</methodCall>\n"
reply: 'HTTP/1.0 200 OK\r\n'
header: Server header: Date header: Content-type header:
Content-length body: b"<?xml version='1.0'?>\n<methodResponse>\n
<params>\n<param>\n<value><boolean>1</boolean></value></param>
\n</params>\n</methodResponse>\n"
Ping: True
```

如果需要其他系统，则可以将默认编码由 UTF-8 改为其他编码。

代码清单 12-71：**xmlrpc_ServerProxy_encoding.py**

```python
import xmlrpc.client

server = xmlrpc.client.ServerProxy('http://localhost:9000',
                                   encoding='ISO-8859-1')
print('Ping:', server.ping())
```

服务器会自动检测正确的编码。

```
$ python3 xmlrpc_ServerProxy_encoding.py

Ping: True
```

`allow_none` 选项会控制 Python 的 `None` 值是自动转换为 nil 值还是会导致一个错误。

代码清单12-72：**xmlrpc_ServerProxy_allow_none.py**

```python
import xmlrpc.client

server = xmlrpc.client.ServerProxy('http://localhost:9000',
                                   allow_none=False)
try:
    server.show_type(None)
except TypeError as err:
    print('ERROR:', err)

server = xmlrpc.client.ServerProxy('http://localhost:9000',
                                   allow_none=True)
print('Allowed:', server.show_type(None))
```

如果客户不允许None，则会在本地产生一个错误，不过如果未配置允许None，那么也有可能从服务器产生错误。

```
$ python3 xmlrpc_ServerProxy_allow_none.py

ERROR: cannot marshal None unless allow_none is enabled
Allowed: ['None', "<class 'NoneType'>", None]
```

12.10.2 数据类型

XML-RPC协议能够识别一组有限的常用数据类型。这些类型可以作为参数或返回值传递，还可以结合使用来创建更复杂的数据结构。

代码清单12-73：**xmlrpc_types.py**

```python
import xmlrpc.client
import datetime

server = xmlrpc.client.ServerProxy('http://localhost:9000')

data = [
    ('boolean', True),
    ('integer', 1),
    ('float', 2.5),
    ('string', 'some text'),
    ('datetime', datetime.datetime.now()),
    ('array', ['a', 'list']),
    ('array', ('a', 'tuple')),
    ('structure', {'a': 'dictionary'}),
]

for t, v in data:
    as_string, type_name, value = server.show_type(v)
    print('{:<12}: {}'.format(t, as_string))
    print('{:12}  {}'.format('', type_name))
    print('{:12}  {}'.format('', value))
```

下面给出简单类型：

```
$ python3 xmlrpc_types.py

boolean     : True
              <class 'bool'>
              True
```

```
integer    : 1
             <class 'int'>
             1
float      : 2.5
             <class 'float'>
             2.5
string     : some text
             <class 'str'>
             some text
datetime   : 20160618T19:31:47
             <class 'xmlrpc.client.DateTime'>
             20160618T19:31:47
array      : ['a', 'list']
             <class 'list'>
             ['a', 'list']
array      : ['a', 'tuple']
             <class 'list'>
             ['a', 'tuple']
structure  : {'a': 'dictionary'}
             <class 'dict'>
             {'a': 'dictionary'}
```

可以嵌套支持的类型来创建任意复杂的值。

代码清单12-74：xmlrpc_types_nested.py

```python
import xmlrpc.client
import datetime
import pprint

server = xmlrpc.client.ServerProxy('http://localhost:9000')

data = {
    'boolean': True,
    'integer': 1,
    'floating-point number': 2.5,
    'string': 'some text',
    'datetime': datetime.datetime.now(),
    'array': ['a', 'list'],
    'array': ('a', 'tuple'),
    'structure': {'a': 'dictionary'},
}
arg = []
for i in range(3):
    d = {}
    d.update(data)
    d['integer'] = i
    arg.append(d)

print('Before:')
pprint.pprint(arg, width=40)

print('\nAfter:')
pprint.pprint(server.show_type(arg)[-1], width=40)
```

这个程序向示例服务器传递一个字典列表，其中包含支持的所有类型，由示例服务器返回数据。元组会转换为列表，datetime实例转换为DateTime对象。否则，其他数据不变。

```
$ python3 xmlrpc_types_nested.py

Before:
```

```
[{'array': ('a', 'tuple'),
  'boolean': True,
  'datetime': datetime.datetime(2016, 6, 18, 19, 27, 30, 45333),
  'floating-point number': 2.5,
  'integer': 0,
  'string': 'some text',
  'structure': {'a': 'dictionary'}},
 {'array': ('a', 'tuple'),
  'boolean': True,
  'datetime': datetime.datetime(2016, 6, 18, 19, 27, 30, 45333),
  'floating-point number': 2.5,
  'integer': 1,
  'string': 'some text',
  'structure': {'a': 'dictionary'}},
 {'array': ('a', 'tuple'),
  'boolean': True,
  'datetime': datetime.datetime(2016, 6, 18, 19, 27, 30, 45333),
  'floating-point number': 2.5,
  'integer': 2,
  'string': 'some text',
  'structure': {'a': 'dictionary'}}]

After:
[{'array': ['a', 'tuple'],
  'boolean': True,
  'datetime': <DateTime '20160618T19:27:30' at 0x101ecfac8>,
  'floating-point number': 2.5,
  'integer': 0,
  'string': 'some text',
  'structure': {'a': 'dictionary'}},
 {'array': ['a', 'tuple'],
  'boolean': True,
  'datetime': <DateTime '20160618T19:27:30' at 0x101ecfcc0>,
  'floating-point number': 2.5,
  'integer': 1,
  'string': 'some text',
  'structure': {'a': 'dictionary'}},
 {'array': ['a', 'tuple'],
  'boolean': True,
  'datetime': <DateTime '20160618T19:27:30' at 0x101ecfe10>,
  'floating-point number': 2.5,
  'integer': 2,
  'string': 'some text',
  'structure': {'a': 'dictionary'}}]
```

XML-RPC 支持日期作为一个内置类型，xmlrpclib 可以使用两个类在发出代理中或者从服务器接收日期时表示日期值。

代码清单 12-75：xmlrpc_ServerProxy_use_datetime.py

```
import xmlrpc.client

server = xmlrpc.client.ServerProxy('http://localhost:9000',
                                   use_datetime=True)
now = server.now()
print('With:', now, type(now), now.__class__.__name__)

server = xmlrpc.client.ServerProxy('http://localhost:9000',
                                   use_datetime=False)
now = server.now()
print('Without:', now, type(now), now.__class__.__name__)
```

默认地，会使用 DateTime 的一个内部版本，不过如果设置了 use_datetime 选项，则会打开 datetime 支持，使用 datetime 模块中的类。

```
$ python3 source/xmlrpc.client/xmlrpc_ServerProxy_use_datetime.py

With: 2016-06-18 19:18:31 <class 'datetime.datetime'> datetime
Without: 20160618T19:18:31 <class 'xmlrpc.client.DateTime'> DateTime
```

12.10.3 传递对象

Python 类的实例被处理为结构，并作为字典传递，对象的属性将作为字典中的值。

代码清单 12-76：**xmlrpc_types_object.py**

```python
import xmlrpc.client
import pprint


class MyObj:

    def __init__(self, a, b):
        self.a = a
        self.b = b

    def __repr__(self):
        return 'MyObj({!r}, {!r})'.format(self.a, self.b)


server = xmlrpc.client.ServerProxy('http://localhost:9000')

o = MyObj(1, 'b goes here')
print('o  :', o)
pprint.pprint(server.show_type(o))

o2 = MyObj(2, o)
print('\no2 :', o2)
pprint.pprint(server.show_type(o2))
```

值从服务器发送回客户时，结果将是客户中的一个字典。这个结果反映了一个事实：这些值中没有编码相应的信息来告诉服务器（或客户）要把它实例化为一个类的一部分。

```
$ python3 xmlrpc_types_object.py

o  : MyObj(1, 'b goes here')
["{'b': 'b goes here', 'a': 1}", "<class 'dict'>",
 {'a': 1, 'b': 'b goes here'}]

o2 : MyObj(2, MyObj(1, 'b goes here'))
["{'b': {'b': 'b goes here', 'a': 1}, 'a': 2}",
 "<class 'dict'>",
 {'a': 2, 'b': {'a': 1, 'b': 'b goes here'}}]
```

12.10.4 二进制数据

所有传递到服务器的值都会被编码，并自动转义。不过，有些数据类型包含的字符可能不是合法的 XML。例如，二进制图像数据可能包括 ASCII 控制范围 0 到 31 中的字节值。要传递二进制数据，最好使用 Binary 类对其编码来传输。

代码清单 12-77：xmlrpc_Binary.py

```python
import xmlrpc.client
import xml.parsers.expat

server = xmlrpc.client.ServerProxy('http://localhost:9000')

s = b'This is a string with control characters\x00'
print('Local string:', s)

data = xmlrpc.client.Binary(s)
response = server.send_back_binary(data)
print('As binary:', response.data)

try:
    print('As string:', server.show_type(s))
except xml.parsers.expat.ExpatError as err:
    print('\nERROR:', err)
```

如果将一个包含 NULL 字节的串传递到 show_type()，那么 XML 解析器处理响应时会产生一个异常。

```
$ python3 xmlrpc_Binary.py

Local string: b'This is a string with control characters\x00'
As binary: b'This is a string with control characters\x00'

ERROR: not well-formed (invalid token): line 6, column 55
```

还可以利用 pickle 使用 Binary 对象发送对象。通过网络向可执行代码发送大量数据时，通常存在一些安全问题，这里也不例外（也就是说，除非通信通道是安全的，否则不要这么做）。

```python
import xmlrpc.client
import pickle
import pprint


class MyObj:

    def __init__(self, a, b):
        self.a = a
        self.b = b

    def __repr__(self):
        return 'MyObj({!r}, {!r})'.format(self.a, self.b)

server = xmlrpc.client.ServerProxy('http://localhost:9000')
o = MyObj(1, 'b goes here')
print('Local:', id(o))
print(o)

print('\nAs object:')
pprint.pprint(server.show_type(o))

p = pickle.dumps(o)
b = xmlrpc.client.Binary(p)
r = server.send_back_binary(b)

o2 = pickle.loads(r.data)
```

```
print('\nFrom pickle:', id(o2))
pprint.pprint(o2)
```

`Binary`实例的数据属性包含对象的腌制版本，所以在使用之前必须解除腌制。这一步会得到一个不同的对象（有一个新的 ID 值）。

```
$ python3 xmlrpc_Binary_pickle.py

Local: 4327262304
MyObj(1, 'b goes here')

As object:
["{'a': 1, 'b': 'b goes here'}", "<class 'dict'>",
 {'a': 1, 'b': 'b goes here'}]

From pickle: 4327262472
MyObj(1, 'b goes here')
```

12.10.5 异常处理

由于 XML-RPC 服务器有可能用任何语言编写，所以不能直接传输异常类。实际上，服务器中产生的异常会被转换为 `Fault` 对象，并在客户端本地作为异常产生。

代码清单 12-78：`xmlrpc_exception.py`

```python
import xmlrpc.client

server = xmlrpc.client.ServerProxy('http://localhost:9000')
try:
    server.raises_exception('A message')
except Exception as err:
    print('Fault code:', err.faultCode)
    print('Message   :', err.faultString)
```

原来的错误消息被保存在 `faultString` 属性中，并且 `faultCode` 被设置为一个 XML-RPC 错误码。

```
$ python3 xmlrpc_exception.py

Fault code: 1
Message   : <class 'RuntimeError'>:A message
```

12.10.6 将调用组合在一个消息中

多调用（multicall）是对 XML-RPC 协议的一个扩展，允许同时发送多个调用，并收集响应，返回给调用者。

代码清单 12-79：`xmlrpc_MultiCall.py`

```python
import xmlrpc.client

server = xmlrpc.client.ServerProxy('http://localhost:9000')

multicall = xmlrpc.client.MultiCall(server)
multicall.ping()
multicall.show_type(1)
multicall.show_type('string')
```

```
for i, r in enumerate(multicall()):
    print(i, r)
```

要使用一个 `MultiCall` 实例,可以像 `ServerProxy` 一样调用方法,然后不带参数地调用这个对象来具体运行远程函数。返回值是一个迭代器,可以得到所有调用的结果。

```
$ python3 xmlrpc_MultiCall.py

0 True
1 ['1', "<class 'int'>", 1]
2 ['string', "<class 'str'>", 'string']
```

如果某个调用导致一个 `Fault`,那么从迭代器生成结果时会产生异常,所以不再生成更多结果。

代码清单 12-80:**xmlrpc_MultiCall_exception.py**

```python
import xmlrpc.client

server = xmlrpc.client.ServerProxy('http://localhost:9000')

multicall = xmlrpc.client.MultiCall(server)
multicall.ping()
multicall.show_type(1)
multicall.raises_exception('Next-to-last call stops execution')
multicall.show_type('string')

try:
    for i, r in enumerate(multicall()):
        print(i, r)
except xmlrpc.client.Fault as err:
    print('ERROR:', err)
```

由于第 3 个响应(来自 `raises_exception()`)生成了一个异常,所以无法再访问 `show_type()` 的响应。

```
$ python3 xmlrpc_MultiCall_exception.py

0 True
1 ['1', "<class 'int'>", 1]
ERROR: <Fault 1: "<class 'RuntimeError'>:Next-to-last call stops execution">
```

提示:相关阅读材料

- `xmlrpc.client` 的标准库文档[⊖]。
- `xmlrpc.server`:XML-RPC 服务器实现。
- `http.server`:HTTP 服务器实现。
- XML-RPC HOWTO[⊖]:描述如何使用 XML-RPC 来以多种语言实现客户和服务器。

⊖ https://docs.python.org/3.5/library/xmlrpc.client.html
⊖ www.tldp.org/HOWTO/XML-RPC-HOWTO/index.html

12.11 `xmlrpc.server`：一个 XML-RPC 服务器

`xmlrpc.server` 模块包含一些类，可以使用 XML-RPC 协议创建跨平台、语言独立的服务器。除了 Python 之外，还有很多其他语言的客户库，这使得 XML-RPC 成为构建 RPC 式服务的一个很好的选择。

说明：这里提供的所有例子还包含一个与演示服务器交互的客户模块。要运行这些例子，需要两个单独的 shell 窗口，一个运行服务器，以及另一个运行客户。

12.11.1 一个简单的服务器

这个简单的服务器示例提供了一个函数，该函数取一个字典的名，返回这个字典的内容。第一步是创建 `SimpleXMLRPCServer` 实例，告诉它在哪里监听到来的请求（这里要在 localhost 的端口 9000 监听）。然后定义一个函数作为服务的一部分，注册这个函数，使服务器知道如何调用该函数。最后一步是将这个服务器放在一个接收和响应请求的无限循环中。

警告：这个实现存在明显的安全隐患。如果服务器位于开放的互联网中或在安全问题可能导致严重后果的环境中，那么不要在这样的服务器上运行这个实现。

代码清单 12-81：`xmlrpc_function.py`

```
from xmlrpc.server import SimpleXMLRPCServer
import logging
import os

# Set up logging.
logging.basicConfig(level=logging.INFO)

server = SimpleXMLRPCServer(
    ('localhost', 9000),
    logRequests=True,
)

# Expose a function.
def list_contents(dir_name):
    logging.info('list_contents(%s)', dir_name)
    return os.listdir(dir_name)
server.register_function(list_contents)

# Start the server.
try:
    print('Use Control-C to exit')
    server.serve_forever()
except KeyboardInterrupt:
    print('Exiting')
```

通过使用 `xmlrpc.client`，可以在 URL `http://localhost:9000` 处访问这个服务器。以下代码清单中的客户代码展示了如何从 Python 调用 `list_contents()` 服务。

代码清单 12-82：**xmlrpc_function_client.py**

```
import xmlrpc.client

proxy = xmlrpc.client.ServerProxy('http://localhost:9000')
print(proxy.list_contents('/tmp'))
```

ServerProxy 使用基 URL 连接到服务器，然后在代理上直接调用方法。代理上调用的各个方法会被转换为对服务器的请求。参数使用 XML 格式化，然后通过一个 POST 消息发送到服务器。服务器解包 XML，根据从客户调用的方法名来确定调用哪个函数。参数将被传递到这个函数，返回值转换回 XML 以便返回给客户。

启动服务器会得到以下输出。

```
$ python3 xmlrpc_function.py

Use Control-C to exit
```

在第二个窗口运行客户，会显示 /tmp 目录的内容。

```
$ python3 xmlrpc_function_client.py

['com.apple.launchd.aoGXonn8nV', 'com.apple.launchd.ilryIaQugf',
'example.db.db',
'KSOutOfProcessFetcher.501.ppfIhqX0vjaTSb8AJYobDV7Cu68=',
'pymotw_import_example.shelve.db']
```

完成这个请求之后，日志输出会出现在服务器窗口中。

```
$ python3 xmlrpc_function.py

Use Control-C to exit
INFO:root:list_contents(/tmp)
127.0.0.1 - - [18/Jun/2016 19:54:54] "POST /RPC2 HTTP/1.1" 200 -
```

输出的第一行来自 `list_contents()` 中的 `logging.info()` 调用。第二行来自记录请求的服务器，因为 `logRequests` 为 `True`。

12.11.2 候选 API 名

有时，模块或库中使用的函数名并不是外部 API 中要使用的名。函数名之所以有变化，可能是因为加载了一个平台特定的实现，或者要根据一个配置文件动态地构建服务 API，也可能实际函数要用桩函数替换来完成测试。要注册一个有候选名的函数，需要将这个名作为第二个参数传递到 `register_function()`。

代码清单 12-83：**xmlrpc_alternate_name.py**

```
from xmlrpc.server import SimpleXMLRPCServer
import os

server = SimpleXMLRPCServer(('localhost', 9000))


def list_contents(dir_name):
    "Expose a function with an alternate name"
    return os.listdir(dir_name)
```

```
server.register_function(list_contents, 'dir')

try:
    print('Use Control-C to exit')
    server.serve_forever()
except KeyboardInterrupt:
    print('Exiting')
```

客户现在应当使用 `dir()` 而不是 `list_contents()`。

代码清单 12-84: **xmlrpc_alternate_name_client.py**

```
import xmlrpc.client

proxy = xmlrpc.client.ServerProxy('http://localhost:9000')
print('dir():', proxy.dir('/tmp'))
try:
    print('\nlist_contents():', proxy.list_contents('/tmp'))
except xmlrpc.client.Fault as err:
    print('\nERROR:', err)
```

调用 `list_contents()` 会得到一个错误,因为服务器上不再有以这个名字注册的处理器。

```
$ python3 xmlrpc_alternate_name_client.py

dir(): ['com.apple.launchd.aoGXonn8nV',
 'com.apple.launchd.ilryIaQugf', 'example.db.db',
 'KSOutOfProcessFetcher.501.ppfIhqX0vjaTSb8AJYobDV7Cu68=',
 'pymotw_import_example.shelve.db']

ERROR: <Fault 1: '<class \'Exception\'>:method "list_contents"
is not supported'>
```

12.11.3 加点的 API 名

还可以用通常情况下不能作为合法 Python 标识符的名字来注册各个函数。例如,可以在名字中包含一个点号(.)来分隔服务中的命名空间。下面的例子扩展了"目录"服务,增加了"创建"和"删除"调用。所有函数注册时都使用了前缀 `dir.`,这样同一个服务器就可以通过使用不同的前缀来提供其他服务。这个例子中还有一点不同,有些函数会返回 `None`,所以必须告诉服务器将 `None` 值转换为一个 nil 值。

代码清单 12-85: **xmlrpc_dotted_name.py**

```
from xmlrpc.server import SimpleXMLRPCServer
import os

server = SimpleXMLRPCServer(('localhost', 9000), allow_none=True)

server.register_function(os.listdir, 'dir.list')
server.register_function(os.mkdir, 'dir.create')
server.register_function(os.rmdir, 'dir.remove')

try:
    print('Use Control-C to exit')
    server.serve_forever()
```

```
except KeyboardInterrupt:
    print('Exiting')
```

要在客户中调用服务函数,只需用加点的名来指示函数。

代码清单 12-86: `xmlrpc_dotted_name_client.py`

```
import xmlrpc.client

proxy = xmlrpc.client.ServerProxy('http://localhost:9000')
print('BEFORE       :', 'EXAMPLE' in proxy.dir.list('/tmp'))
print('CREATE       :', proxy.dir.create('/tmp/EXAMPLE'))
print('SHOULD EXIST :', 'EXAMPLE' in proxy.dir.list('/tmp'))
print('REMOVE       :', proxy.dir.remove('/tmp/EXAMPLE'))
print('AFTER        :', 'EXAMPLE' in proxy.dir.list('/tmp'))
```

假设当前系统上没有 **/tmp/EXAMPLE** 文件,那么示例客户脚本会生成以下输出。

```
$ python3 xmlrpc_dotted_name_client.py

BEFORE        : False
CREATE        : None
SHOULD EXIST  : True
REMOVE        : None
AFTER         : False
```

12.11.4 任意 API 名

还有一个有趣的特性,可以用一些非法的 Python 对象属性名来注册函数。下面的示例服务用名字 **multiply args** 注册了一个函数。

代码清单 12-87: `xmlrpc_arbitrary_name.py`

```
from xmlrpc.server import SimpleXMLRPCServer

server = SimpleXMLRPCServer(('localhost', 9000))

def my_function(a, b):
    return a * b
server.register_function(my_function, 'multiply args')

try:
    print('Use Control-C to exit')
    server.serve_forever()
except KeyboardInterrupt:
    print('Exiting')
```

由于所注册的名字包含一个空格,因此不能使用点记法直接从代理访问。不过,使用 **getattr()** 是可以的。

代码清单 12-88: `xmlrpc_arbitrary_name_client.py`

```
import xmlrpc.client

proxy = xmlrpc.client.ServerProxy('http://localhost:9000')
print(getattr(proxy, 'multiply args')(5, 5))
```

但是并不建议用类似这样的名字创建服务。给出这个例子并不是因为这是一个好的想法，而是因为确实存在这种有任意名字的服务，新程序可能需要调用这些服务。

```
$ python3 xmlrpc_arbitrary_name_client.py

25
```

12.11.5 公布对象的方法

前面几节讨论了使用好的命名约定和命名空间建立 API 的技术。要在 API 中结合命名空间，另一种方法是使用类的实例并公布其方法。可以使用只有一个方法的实例重新创建第一个例子。

代码清单 12-89：**xmlrpc_instance.py**

```python
from xmlrpc.server import SimpleXMLRPCServer
import os
import inspect
server = SimpleXMLRPCServer(
    ('localhost', 9000),
    logRequests=True,
)


class DirectoryService:
    def list(self, dir_name):
        return os.listdir(dir_name)

server.register_instance(DirectoryService())

try:
    print('Use Control-C to exit')
    server.serve_forever()
except KeyboardInterrupt:
    print('Exiting')
```

客户可以直接调用这个方法。

代码清单 12-90：**xmlrpc_instance_client.py**

```python
import xmlrpc.client

proxy = xmlrpc.client.ServerProxy('http://localhost:9000')
print(proxy.list('/tmp'))
```

输出显示了这个目录的内容。

```
$ python3 xmlrpc_instance_client.py

['com.apple.launchd.aoGXonn8nV', 'com.apple.launchd.ilryIaQugf',
'example.db.db',
'KSOutOfProcessFetcher.501.ppfIhqX0vjaTSb8AJYobDV7Cu68=',
'pymotw_import_example.shelve.db']
```

不过，服务的 `dir.` 前缀已丢失。可以定义一个类以便建立一个服务树（可以从客户调用）来恢复。

代码清单12-91：`xmlrpc_instance_dotted_names.py`

```python
from xmlrpc.server import SimpleXMLRPCServer
import os
import inspect

server = SimpleXMLRPCServer(
    ('localhost', 9000),
    logRequests=True,
)
class ServiceRoot:
    pass

class DirectoryService:

    def list(self, dir_name):
        return os.listdir(dir_name)

root = ServiceRoot()
root.dir = DirectoryService()

server.register_instance(root, allow_dotted_names=True)

try:
    print('Use Control-C to exit')
    server.serve_forever()
except KeyboardInterrupt:
    print('Exiting')
```

由于注册了 `ServiceRoot` 实例，并启用了 `allow_dotted_names`，请求到来时，服务器有权限遍历这个对象树从而使用 `getattr()` 查找指定的方法。

代码清单12-92：`xmlrpc_instance_dotted_names_client.py`

```python
import xmlrpc.client

proxy = xmlrpc.client.ServerProxy('http://localhost:9000')
print(proxy.dir.list('/tmp'))
```

`dir.list()` 的输出与之前实现的输出相同。

```
$ python3 xmlrpc_instance_dotted_names_client.py

['com.apple.launchd.aoGXonn8nV', 'com.apple.launchd.ilryIaQugf',
'example.db.db',
'KSOutOfProcessFetcher.501.ppfIhqX0vjaTSb8AJYobDV7Cu68=',
'pymotw_import_example.shelve.db']
```

12.11.6 分派调用

默认地，`register_instance()` 会查找实例的所有可调用属性（属性名以一个下划线（`_`）开头的除外），并用它们的名字注册。为了更谨慎地处理公布的方法，可以使用定制的分派逻辑。

代码清单12-93：`xmlrpc_instance_with_prefix.py`

```python
from xmlrpc.server import SimpleXMLRPCServer
import os
```

```python
import inspect

server = SimpleXMLRPCServer(
    ('localhost', 9000),
    logRequests=True,
)


def expose(f):
    "Decorator to set exposed flag on a function."
    f.exposed = True
    return f


def is_exposed(f):
    "Test whether another function should be publicly exposed."
    return getattr(f, 'exposed', False)


class MyService:
    PREFIX = 'prefix'

    def _dispatch(self, method, params):
        # Remove our prefix from the method name.
        if not method.startswith(self.PREFIX + '.'):
            raise Exception(
                'method "{}" is not supported'.format(method)
            )

        method_name = method.partition('.')[2]
        func = getattr(self, method_name)
        if not is_exposed(func):
            raise Exception(
                'method "{}" is not supported'.format(method)
            )

        return func(*params)

    @expose
    def public(self):
        return 'This is public'

    def private(self):
        return 'This is private'
server.register_instance(MyService())

try:
    print('Use Control-C to exit')
    server.serve_forever()
except KeyboardInterrupt:
    print('Exiting')
```

MyService的public()方法被标记为要公布到XML-RPC服务,而private()方法不公布。在客户试图访问MyService中的一个函数时会调用_dispatch()方法。它首先强制使用一个前缀(在这里是prefix.,不过也可以使用任意的串)。然后要求这个函数有一个名为exposed的属性,而且值为true。为方便起见,这里使用一个修饰符为函数设置这个exposed标志。下面的例子包含了一些示例客户调用。

代码清单 12-94：`xmlrpc_instance_with_prefix_client.py`

```
import xmlrpc.client

proxy = xmlrpc.client.ServerProxy('http://localhost:9000')
print('public():', proxy.prefix.public())
try:
    print('private():', proxy.prefix.private())
except Exception as err:
    print('\nERROR:', err)
try:
    print('public() without prefix:', proxy.public())
except Exception as err:
    print('\nERROR:', err)
```

下面给出得到的输出，这里捕获并报告了预料之中的错误消息。

```
$ python3 xmlrpc_instance_with_prefix_client.py

public(): This is public

ERROR: <Fault 1: '<class \'Exception\'>:method "prefix.private" is
not supported'>

ERROR: <Fault 1: '<class \'Exception\'>:method "public" is not
supported'>
```

还有另外一些方法可以覆盖分派机制，包括直接从 `SimpleXMLRPCServer` 派生子类。参考这个模块中的 `docstring` 来了解更多详细内容。

12.11.7 自省 API

与很多网络服务一样，可以查询一个 XML-RPC 服务器来确定它支持哪些方法，并了解如何使用这些方法。`SimpleXMLRPCServer` 包括一组用于完成这个自省的公共方法。默认地，这些方法是关闭的，不过可以用 `register_introspection_functions()` 启用。通过在服务类上定义 `_listMethods()` 和 `_methodHelp()`，可以在服务中增加对 `system.listMethods()` 和 `system.methodHelp()` 的支持。

代码清单 12-95：`xmlrpc_introspection.py`

```
from xmlrpc.server import (SimpleXMLRPCServer,
                           list_public_methods)
import os
import inspect

server = SimpleXMLRPCServer(
    ('localhost', 9000),
    logRequests=True,
)
server.register_introspection_functions()

class DirectoryService:

    def _listMethods(self):
        return list_public_methods(self)
```

```python
    def _methodHelp(self, method):
        f = getattr(self, method)
        return inspect.getdoc(f)

    def list(self, dir_name):
        """list(dir_name) => [<filenames>]

        Returns a list containing the contents of
        the named directory.

        """
        return os.listdir(dir_name)

server.register_instance(DirectoryService())

try:
    print('Use Control-C to exit')
    server.serve_forever()
except KeyboardInterrupt:
    print('Exiting')
```

在这里，便利函数 `list_public_methods()` 扫描一个实例，返回不是以下划线（_）开头的可调用属性的名字。重新定义 `_listMethods()` 来应用所需的规则。类似地，对于这个基本例子，`_methodHelp()` 返回了函数的 docstring，不过也可以写为从其他来源构建一个帮助文本串。

这个客户会查询服务器，并报告所有可公开调用的方法。

代码清单 12-96：xmlrpc_introspection_client.py

```python
import xmlrpc.client

proxy = xmlrpc.client.ServerProxy('http://localhost:9000')
for method_name in proxy.system.listMethods():
    print('=' * 60)
    print(method_name)
    print('-' * 60)
    print(proxy.system.methodHelp(method_name))
    print()
```

结果中还包含了系统方法。

```
$ python3 xmlrpc_introspection_client.py

============================================================
list
------------------------------------------------------------
list(dir_name) => [<filenames>]

Returns a list containing the contents of
the named directory.

============================================================
system.listMethods
------------------------------------------------------------
system.listMethods() => ['add', 'subtract', 'multiple']

Returns a list of the methods supported by the server.
```

```
============================================================
system.methodHelp
------------------------------------------------------------
system.methodHelp('add') => "Adds two integers together"

Returns a string containing documentation for the specified method.

============================================================
system.methodSignature
------------------------------------------------------------
system.methodSignature('add') => [double, int, int]

Returns a list describing the signature of the method. In the
above example, the add method takes two integers as arguments
and returns a double result.

This server does NOT support system.methodSignature.
```

提示：相关阅读材料

- `xmlrpc.server` 的标准库文档[⊖]。
- `xmlrpc.client`：XML-RPC 客户。
- XML-RPC HOWTO[⊖]：描述如何使用 XML-RPC 来以多种语言实现客户和服务器。

[⊖] https://docs.python.org/3.5/library/xmlrpc.server.html
[⊖] www.tldp.org/HOWTO/XML-RPC-HOWTO/index.html

第13章 email

email是最古老的数字通信方式之一,也是最流行的方式之一。Python的标准库提供了发送、接收和存储email消息的模块。

smtplib与邮件服务器通信以传送消息。smtpd可以用于创建定制的邮件服务器,并提供一些很有用的类,可以调试其他应用中的email传输。

imaplib使用IMAP协议来管理存储在一个服务器上的消息。它为IMAP客户提供了一个底层API,可以查询、获取、移动和删除消息。

利用mailbox,可以使用多种标准格式来创建和修改本地消息归档,包括流行的mbox和Maildir格式,这是很多email客户程序使用的格式。

13.1 smtplib:简单邮件传输协议客户

smtplib包括一个SMTP类,可以用来与邮件服务器通信发送邮件。

说明:后面的例子中,email地址、主机名和IP地址都被故意修改为没有实际意义。除此以外,这些脚本展示的命令和响应序列都是准确的。

13.1.1 发送email消息

SMTP最常用的一种用法就是连接到一个邮件服务器并发送一个消息。可以把邮件服务器主机名和端口传递到构造函数,也可以显式调用connect()。一旦连接,可以调用sendmail()并提供信封参数和消息体。消息文本要完整并遵循RFC 5322[⊖],因为smtplib根本不会修改内容或首部。这说明,调用者需要增加From和To首部。

代码清单13-1:smtplib_sendmail.py

```
import smtplib
import email.utils
from email.mime.text import MIMEText
# Create the message.
msg = MIMEText('This is the body of the message.')
msg['To'] = email.utils.formataddr(('Recipient',
                                    'recipient@example.com'))
msg['From'] = email.utils.formataddr(('Author',
                                      'author@example.com'))
```

⊖ https://tools.ietf.org/html/rfc5322

```
msg['Subject'] = 'Simple test message'
server = smtplib.SMTP('localhost', 1025)
server.set_debuglevel(True)  # Show communication with the server.
try:
    server.sendmail('author@example.com',
                   ['recipient@example.com'],
                   msg.as_string())
finally:
    server.quit()
```

这个例子中还打开了调试,以便显示客户与服务器之间的通信。否则,这个示例根本不会产生任何输出。

```
$ python3 smtplib_sendmail.py

send: 'ehlo 1.0.0.0.0.0.0.0.0.0.0.0.0.0.0.0.0.0.0.0.0.0.0.0.0.
0.0.0.0.0.ip6.arpa\r\n'
reply: b'250-1.0.0.0.0.0.0.0.0.0.0.0.0.0.0.0.0.0.0.0.0.0.0.0.0
.0.0.0.0.0.ip6.arpa\r\n'
reply: b'250-SIZE 33554432\r\n'
reply: b'250 HELP\r\n'
reply: retcode (250); Msg: b'1.0.0.0.0.0.0.0.0.0.0.0.0.0.0.0.0
.0.0.0.0.0.0.0.0.0.0.0.0.0.ip6.arpa\nSIZE 33554432\nHELP'
send: 'mail FROM:<author@example.com> size=236\r\n'
reply: b'250 OK\r\n'
reply: retcode (250); Msg: b'OK'
send: 'rcpt TO:<recipient@example.com>\r\n'
reply: b'250 OK\r\n'
reply: retcode (250); Msg: b'OK'
send: 'data\r\n'
reply: b'354 End data with <CR><LF>.<CR><LF>\r\n'
reply: retcode (354); Msg: b'End data with <CR><LF>.<CR><LF>'
data: (354, b'End data with <CR><LF>.<CR><LF>')
send: b'Content-Type: text/plain; charset="us-ascii"\r\nMIME-Ver
sion: 1.0\r\nContent-Transfer-Encoding: 7bit\r\nTo: Recipient <r
ecipient@example.com>\r\nFrom: Author <author@example.com>\r\nSu
bject: Simple test message\r\n\r\nThis is the body of the messag
e.\r\n.\r\n'
reply: b'250 OK\r\n'
reply: retcode (250); Msg: b'OK'
data: (250, b'OK')
send: 'quit\r\n'
reply: b'221 Bye\r\n'
reply: retcode (221); Msg: b'Bye'
```

`sendmail()`的第二个参数(即接收者)会作为一个列表传递。这个列表中可以包括任意多个地址,消息将被逐一传送到各个地址。由于信封信息与消息首部是分开的,所以将地址包含在方法参数中而不是置于消息首部中,这样可以实现暗送(Blind Carbon-Copy,BCC)。

13.1.2 认证和加密

SMTP类还会处理认证和传输层安全(Transport Layer Security,TLS)加密(如果服务器提供了支持)。要确定服务器是否支持TLS,可以直接调用`ehlo()`为服务器标识客户,询问可以得到哪些扩展。然后调用`has_extn()`来检查结果。启动TLS之后,在认证用户之前必须再次调用 ehlo()。很多邮件托管提供商现在只支持基于 TLS 的连接。要与这些服务器通信，可以使用 SMTP_SSL 来启动一个加密连接。

代码清单 13-2：**smtplib_authenticated.py**

```python
import smtplib
import email.utils
from email.mime.text import MIMEText
import getpass

# Prompt the user for connection info.
to_email = input('Recipient: ')
servername = input('Mail server name: ')
serverport = input('Server port: ')
if serverport:
    serverport = int(serverport)
else:
    serverport = 25
use_tls = input('Use TLS? (yes/no): ').lower()
username = input('Mail username: ')
password = getpass.getpass("%s's password: " % username)

# Create the message.
msg = MIMEText('Test message from PyMOTW.')
msg.set_unixfrom('author')
msg['To'] = email.utils.formataddr(('Recipient', to_email))
msg['From'] = email.utils.formataddr(('Author',
                                      'author@example.com'))
msg['Subject'] = 'Test from PyMOTW'

if use_tls == 'yes':
    print('starting with a secure connection')
    server = smtplib.SMTP_SSL(servername, serverport)
else:
    print('starting with an insecure connection')
    server = smtplib.SMTP(servername, serverport)
try:
    server.set_debuglevel(True)

    # Identify ourselves, prompting server for supported features.
    server.ehlo()

    # If we can encrypt this session, do it.
    if server.has_extn('STARTTLS'):
        print('(starting TLS)')
        server.starttls()
        server.ehlo()  # Reidentify ourselves over TLS connection.
    else:
        print('(no STARTTLS)')

    if server.has_extn('AUTH'):
        print('(logging in)')
        server.login(username, password)
    else:
        print('(no AUTH)')

    server.sendmail('author@example.com',
                    [to_email],
                    msg.as_string())
finally:
    server.quit()
```

启用 TLS 之后，STARTTLS 扩展不会出现在对 EHLO 的应答中。

```
$ python3 source/smtplib/smtplib_authenticated.py
Recipient: doug@pymotw.com
Mail server name: localhost
Server port: 1025
Use TLS? (yes/no): no
Mail username: test
test's password:
starting with an insecure connection
send: 'ehlo 1.0.0.0.0.0.0.0.0.0.0.0.0.0.0.0.0.0.0.0.0.0.0.0.0
.0.0.0.0.0.ip6.arpa\r\n'
reply: b'250-1.0.0.0.0.0.0.0.0.0.0.0.0.0.0.0.0.0.0.0.0.0.0.0.
0.0.0.0.0.ip6.arpa\r\n'
reply: b'250-SIZE 33554432\r\n'
reply: b'250 HELP\r\n'
reply: retcode (250); Msg: b'1.0.0.0.0.0.0.0.0.0.0.0.0.0.0.0.
0.0.0.0.0.0.0.0.0.0.0.0.0.ip6.arpa\nSIZE 33554432\nHELP'
(no STARTTLS)
(no AUTH)
send: 'mail FROM:<author@example.com> size=220\r\n'
reply: b'250 OK\r\n'
reply: retcode (250); Msg: b'OK'
send: 'rcpt TO:<doug@pymotw.com>\r\n'
reply: b'250 OK\r\n'
reply: retcode (250); Msg: b'OK'
send: 'data\r\n'
reply: b'354 End data with <CR><LF>.<CR><LF>\r\n'
reply: retcode (354); Msg: b'End data with <CR><LF>.<CR><LF>'
data: (354, b'End data with <CR><LF>.<CR><LF>')
send: b'Content-Type: text/plain; charset="us-ascii"\r\n
MIME-Version: 1.0\r\nContent-Transfer-Encoding: 7bit\r\nTo:
Recipient <doug@pymotw.com>\r\nFrom: Author <author@example.com>
\r\nSubject: Test from PyMOTW\r\n\r\nTest message from PyMOTW.
\r\n.\r\n'
reply: b'250 OK\r\n'
reply: retcode (250); Msg: b'OK'
data: (250, b'OK')
send: 'quit\r\n'
reply: b'221 Bye\r\n'
reply: retcode (221); Msg: b'Bye'

$ python3 source/smtplib/smtplib_authenticated.py
Recipient: doug@pymotw.com
Mail server name: mail.isp.net
Server port: 465
Use TLS? (yes/no): yes
Mail username: doughellmann@isp.net
doughellmann@isp.net's password:
starting with a secure connection
send: 'ehlo 1.0.0.0.0.0.0.0.0.0.0.0.0.0.0.0.0.0.0.0.0.0.0.0.0
.0.0.0.0.0.ip6.arpa\r\n'
reply: b'250-mail.isp.net\r\n'
reply: b'250-PIPELINING\r\n'
reply: b'250-SIZE 71000000\r\n'
reply: b'250-ENHANCEDSTATUSCODES\r\n'
reply: b'250-8BITMIME\r\n'
reply: b'250-AUTH PLAIN LOGIN\r\n'
reply: b'250 AUTH=PLAIN LOGIN\r\n'
reply: retcode (250); Msg: b'mail.isp.net\nPIPELINING\nSIZE
71000000\nENHANCEDSTATUSCODES\n8BITMIME\nAUTH PLAIN LOGIN\n
AUTH=PLAIN LOGIN'
(no STARTTLS)
```

```
(logging in)
send: 'AUTH PLAIN AGRvdWdoZWxsbWFubkBmYXN0bWFpbC5mbQBUTUZ3MDBmZmF
zdG1haWw=\r\n'
reply: b'235 2.0.0 OK\r\n'
reply: retcode (235); Msg: b'2.0.0 OK'
send: 'mail FROM:<author@example.com> size=220\r\n'
reply: b'250 2.1.0 Ok\r\n'
reply: retcode (250); Msg: b'2.1.0 Ok'
send: 'rcpt TO:<doug@pymotw.com>\r\n'
reply: b'250 2.1.5 Ok\r\n'
reply: retcode (250); Msg: b'2.1.5 Ok'
send: 'data\r\n'
reply: b'354 End data with <CR><LF>.<CR><LF>\r\n'
reply: retcode (354); Msg: b'End data with <CR><LF>.<CR><LF>'
data: (354, b'End data with <CR><LF>.<CR><LF>')
send: b'Content-Type: text/plain; charset="us-ascii"\r\n
MIME-Version: 1.0\r\nContent-Transfer-Encoding: 7bit\r\nTo:
Recipient <doug@pymotw.com>\r\nFrom: Author <author@example.com>
\r\nSubject: Test from PyMOTW\r\n\r\nTest message from PyMOTW.
\r\n.\r\n'
reply: b'250 2.0.0 Ok: queued as A0EF7F2983\r\n'
reply: retcode (250); Msg: b'2.0.0 Ok: queued as A0EF7F2983'
data: (250, b'2.0.0 Ok: queued as A0EF7F2983')
send: 'quit\r\n'
reply: b'221 2.0.0 Bye\r\n'
reply: retcode (221); Msg: b'2.0.0 Bye'
```

13.1.3 验证 email 地址

SMTP 协议包括一个命令来询问服务器一个地址是否合法。通常，VRFY 是禁用的，以避免垃圾邮件工具（spammer）查找到合法的 email 地址。不过，如果启用这个命令，那么客户可以询问服务器一个地址是否合法，并接收一个状态码指示其合法性，同时提供用户的全名（如果有）。

代码清单 13-3：**smtplib_verify.py**

```
import smtplib

server = smtplib.SMTP('mail')
server.set_debuglevel(True)  # Show communication with the server.
try:
    dhellmann_result = server.verify('dhellmann')
    notthere_result = server.verify('notthere')
finally:
    server.quit()

print('dhellmann:', dhellmann_result)
print('notthere :', notthere_result)
```

如输出中最后两行所示，地址 dhellmann 是合法的，而 notthere 不合法。

```
$ python3 smtplib_verify.py

send: 'vrfy <dhellmann>\r\n'
reply: '250 2.1.5 Doug Hellmann <dhellmann@mail>\r\n'
reply: retcode (250); Msg: 2.1.5 Doug Hellmann <dhellmann@mail>
send: 'vrfy <notthere>\r\n'
reply: '550 5.1.1 <notthere>... User unknown\r\n'
reply: retcode (550); Msg: 5.1.1 <notthere>... User unknown
```

```
send: 'quit\r\n'
reply: '221 2.0.0 mail closing connection\r\n'
reply: retcode (221); Msg: 2.0.0 mail closing connection
dhellmann: (250, '2.1.5 Doug Hellmann <dhellmann@mail>')
notthere : (550, '5.1.1 <notthere>... User unknown')
```

提示：相关阅读材料
- **smtplib** 的标准库文档[一]。
- **RFC 821**[二]：简单邮件传输协议（Simple Mail Transfer Protocol，SMTP）规范。
- **RFC 1869**[三]：基本协议的 SMTP 服务扩展。
- **RFC 822**[四]：ARPA Internet 文本消息格式标准，原来的 email 消息格式规范。
- **RFC 5322**[五]：Internet 消息格式；更新为 email 消息格式。
- **email**：构建和解析 email 消息的标准库模块。
- **smtpd**：实现一个简单的 SMTP 服务器。

13.2 smtpd：示例邮件服务器

smtpd 模块包括一些用于构建简单邮件传输协议（SMTP）服务器的类。这是 smtplib 所使用协议的服务器端。

13.2.1 邮件服务器基类

已经提供的所有示例服务器的基类都是 SMTPServer。它会处理与客户的通信以及接收到来的数据，还提供了一个方便的 hook，可以覆盖这个 hook，一旦得到完整的消息就可以进行处理。

构造函数参数包括监听连接的本地地址和要发送代理消息的远程地址。process_message() 方法作为一个 hook 提供，要由派生类覆盖。接收到完整的消息时会调用这个方法，并指定以下参数：

peer：客户的地址，这是一个包含 IP 和入站端口的元组。

mailfrom：消息信封中的"from"信息，传送消息时由客户提供给服务器。这个消息不一定总与 From 首部匹配。

rcpttos：消息信封中的接收者列表。同样，这个列表不一定总与 To 首部匹配，特别是暗送给接收者时。

data：完整的 RFC 5322 消息体。

process_message() 的默认实现会产生 NotImplementedError。下面的例子定义了一个子类，它覆盖了这个方法，会打印接收到的消息的有关信息。

[一] https://docs.python.org/3.5/library/smtplib.html
[二] https://tools.ietf.org/html/rfc821.html
[三] https://tools.ietf.org/html/rfc1869.html
[四] https://tools.ietf.org/html/rfc822.html
[五] https://tools.ietf.org/html/rfc5322.html

代码清单 13-4：**smtpd_custom.py**

```
import smtpd
import asyncore

class CustomSMTPServer(smtpd.SMTPServer):

    def process_message(self, peer, mailfrom, rcpttos, data):
        print('Receiving message from:', peer)
        print('Message addressed from:', mailfrom)
        print('Message addressed to  :', rcpttos)
        print('Message length         :', len(data))

server = CustomSMTPServer(('127.0.0.1', 1025), None)

asyncore.loop()
```

SMTPServer 使用了 asyncore，所以如果要运行服务器，则需要调用 asyncore.loop()。

还需要一个客户来展示服务器。可以修改 smtplib 一节中的某个例子来创建一个客户，向在端口 1025 上本地运行的测试服务器发送数据。

代码清单 13-5：**smtpd_senddata.py**

```
import smtplib
import email.utils
from email.mime.text import MIMEText

# Create the message.
msg = MIMEText('This is the body of the message.')
msg['To'] = email.utils.formataddr(('Recipient',
                                    'recipient@example.com'))
msg['From'] = email.utils.formataddr(('Author',
                                      'author@example.com'))
msg['Subject'] = 'Simple test message'

server = smtplib.SMTP('127.0.0.1', 1025)
server.set_debuglevel(True)  # Show communication with the server.
try:
    server.sendmail('author@example.com',
                    ['recipient@example.com'],
                    msg.as_string())
finally:
    server.quit()
```

要测试这些程序，可以在一个终端窗口运行 smtpd_custom.py，在另一个终端窗口运行 smtpd_senddata.py。

```
$ python3 smtpd_custom.py

Receiving message from: ('127.0.0.1', 58541)
Message addressed from: author@example.com
Message addressed to  : ['recipient@example.com']
Message length         : 229
```

smtpd_senddata.py 的调试输出显示了与服务器的所有通信。

```
$ python3 smtpd_senddata.py
send: 'ehlo 1.0.0.0.0.0.0.0.0.0.0.0.0.0.0.0.0.0.0.0.0.0.0.0.0.0.0.
0.0.0.0.0.0.ip6.arpa\r\n'
reply: b'250-1.0.0.0.0.0.0.0.0.0.0.0.0.0.0.0.0.0.0.0.0.0.0.0.0.0.0
.0.0.0.0.0.0.ip6.arpa\r\n'
reply: b'250-SIZE 33554432\r\n'
reply: b'250 HELP\r\n'
reply: retcode (250); Msg: b'1.0.0.0.0.0.0.0.0.0.0.0.0.0.0.0.0.0.0
.0.0.0.0.0.0.0.0.0.0.0.0.0.0.ip6.arpa\nSIZE 33554432\nHELP'
send: 'mail FROM:<author@example.com> size=236\r\n'
reply: b'250 OK\r\n'
reply: retcode (250); Msg: b'OK'
send: 'rcpt TO:<recipient@example.com>\r\n'
reply: b'250 OK\r\n'
reply: retcode (250); Msg: b'OK'
send: 'data\r\n'
reply: b'354 End data with <CR><LF>.<CR><LF>\r\n'
reply: retcode (354); Msg: b'End data with <CR><LF>.<CR><LF>'
data: (354, b'End data with <CR><LF>.<CR><LF>')
send: b'Content-Type: text/plain; charset="us-ascii"\r\nMIME-Ver
sion: 1.0\r\nContent-Transfer-Encoding: 7bit\r\nTo: Recipient <r
ecipient@example.com>\r\nFrom: Author <author@example.com>\r\nSu
bject: Simple test message\r\n\r\nThis is the body of the messag
e.\r\n.\r\n'
reply: b'250 OK\r\n'
reply: retcode (250); Msg: b'OK'
data: (250, b'OK')
send: 'quit\r\n'
reply: b'221 Bye\r\n'
reply: retcode (221); Msg: b'Bye'
```

要停止服务器，只需按下 **Ctrl-C**。

13.2.2 调试服务器

前面的例子显示了 `process_message()` 的参数，不过 `smtpd` 还包括一个专门设计的服务器，名为 `DebuggingServer`，用来完成更完备的调试。它会把到来的消息完整地打印到控制台，然后停止处理（它不会把消息转发给一个真正的邮件服务器）。

代码清单 13-6：**smtpd_debug.py**

```python
import smtpd
import asyncore

server = smtpd.DebuggingServer(('127.0.0.1', 1025), None)

asyncore.loop()
```

使用前面的 `smtpd_senddata.py` 客户程序，将从 `DebuggingServer` 生成以下输出。

```
---------- MESSAGE FOLLOWS ----------
Content-Type: text/plain; charset="us-ascii"
MIME-Version: 1.0
Content-Transfer-Encoding: 7bit
To: Recipient <recipient@example.com>
From: Author <author@example.com>
Subject: Simple test message
X-Peer: 127.0.0.1
```

```
This is the body of the message.
------------ END MESSAGE ------------
```

13.2.3 代理服务器

`PureProxy` 类实现了一个简单的代理服务器。到来的消息将作为构造函数的一个参数，向上转发给服务器。

> **警告**：`smtpd` 的标准库文档指出，"运行这个模块时，很有可能进入一种开放转发（open relay，又称匿名转发），所以请务必谨慎"。

建立代理服务器的步骤与建立调试服务器的步骤类似。

代码清单 13-7：`smtpd_proxy.py`

```python
import smtpd
import asyncore

server = smtpd.PureProxy(('127.0.0.1', 1025), ('mail', 25))

asyncore.loop()
```

这个程序不打印任何输出，所以要验证它是否正常工作，需要查看邮件服务器日志。

```
Aug 20 19:16:34 homer sendmail[6785]: m9JNGXJb006785:
from=<author@example.com>, size=248, class=0, nrcpts=1,
msgid=<200810192316.m9JNGXJb006785@homer.example.com>,
proto=ESMTP, daemon=MTA, relay=[192.168.1.17]
```

> **提示**：相关阅读材料
> - `smtpd` 的标准库文档[⊖]。
> - `smtplib`：提供一个客户接口。
> - `email`：解析 email 消息。
> - `asyncore`：编写异步服务器的基本模块。
> - **RFC 2822**[⊖]：Internet 消息格式；定义 email 消息格式。
> - **RFC 5322**[⊖]：取代 RFC 2822。

13.3 `mailbox`：管理 email 归档

`mailbox` 模块定义了一个通用 API，用来访问采用本地磁盘格式存储的邮件消息，包括：
- Maildir
- mbox
- MH

⊖ https://docs.python.org/3.5/library/smtpd.html
⊖ https://tools.ietf.org/html/rfc2822.html
⊖ https://tools.ietf.org/html/rfc5322.html

- Babyl
- MMDF

`mailbox`模块提供了`Mailbox`和`Message`基类,每个邮箱格式分别包含相应的一对子类,以实现相应格式的有关细节。

13.3.1 mbox

mbox格式是文档中展示的最简单的格式,因为它完全是纯文本的。每个邮箱都被存储为一个文件,并且所有消息都联接在一起。每次遇到一个以"From"开头的行时("From"后面有一个空格),就会将其处理为一个新消息的开始。只要这些字符出现在消息体中某一行的开头,就会将其转义,在这一行前面增加">"前缀。

13.3.1.1 创建mbox邮箱

将文件名传递到构造函数来实例化mbox类。如果该文件不存在,那么在使用add()追加消息时会创建这个文件。

代码清单13-8:**mailbox_mbox_create.py**

```
import mailbox
import email.utils

from_addr = email.utils.formataddr(('Author',
                                    'author@example.com'))
to_addr = email.utils.formataddr(('Recipient',
                                  'recipient@example.com'))

payload = '''This is the body.
From (will not be escaped).
There are 3 lines.
'''

mbox = mailbox.mbox('example.mbox')
mbox.lock()
try:
    msg = mailbox.mboxMessage()
    msg.set_unixfrom('author Sat Feb  7 01:05:34 2009')
    msg['From'] = from_addr
    msg['To'] = to_addr
    msg['Subject'] = 'Sample message 1'
    msg.set_payload(payload)
    mbox.add(msg)
    mbox.flush()

    msg = mailbox.mboxMessage()
    msg.set_unixfrom('author')
    msg['From'] = from_addr
    msg['To'] = to_addr
    msg['Subject'] = 'Sample message 2'
    msg.set_payload('This is the second body.\n')
    mbox.add(msg)
    mbox.flush()
finally:
    mbox.unlock()

print(open('example.mbox', 'r').read())
```

这个脚本的结果是一个新的邮箱文件，其中包含两个邮件消息。

```
$ python3 mailbox_mbox_create.py

From MAILER-DAEMON Thu Dec 29 17:23:56 2016
From: Author <author@example.com>
To: Recipient <recipient@example.com>
Subject: Sample message 1

This is the body.
>From (will not be escaped).
There are 3 lines.

From MAILER-DAEMON Thu Dec 29 17:23:56 2016
From: Author <author@example.com>
To: Recipient <recipient@example.com>
Subject: Sample message 2

This is the second body.
```

13.3.1.2 读取 mbox 邮箱

要读取一个已有的邮箱，需要打开这个邮箱，像字典一样处理这个 mbox 对象。键是邮箱实例定义的任意值，它们只作为消息对象的内部标识符，并不一定有实际意义。

代码清单 13-9：**mailbox_mbox_read.py**

```python
import mailbox

mbox = mailbox.mbox('example.mbox')
for message in mbox:
    print(message['subject'])
```

打开的邮箱支持迭代器协议，不过与真正的字典对象不同，邮箱的默认迭代器会处理值（values）而不是键（keys）。

```
$ python3 mailbox_mbox_read.py

Sample message 1
Sample message 2
```

13.3.1.3 从 mbox 邮箱删除消息

要从一个 mbox 文件删除已有的消息，可以使用 remove() 并提供这个消息的键，也可以使用 del。

代码清单 13-10：**mailbox_mbox_remove.py**

```python
import mailbox

mbox = mailbox.mbox('example.mbox')
mbox.lock()
try:
    to_remove = []
    for key, msg in mbox.iteritems():
        if '2' in msg['subject']:
            print('Removing:', key)
            to_remove.append(key)
    for key in to_remove:
        mbox.remove(key)
```

```
finally:
    mbox.flush()
    mbox.close()

print(open('example.mbox', 'r').read())
```

可以使用 lock() 和 unlock() 方法来避免同时访问文件可能导致的问题，flush() 强制将修改写入磁盘。

```
$ python3 mailbox_mbox_remove.py

Removing: 1
From MAILER-DAEMON Thu Dec 29 17:23:56 2016
From: Author <author@example.com>
To: Recipient <recipient@example.com>
Subject: Sample message 1

This is the body.
>From (will not be escaped).
There are 3 lines.
```

13.3.2 Maildir

创建 Maildir 格式是为了消除 mbox 文件并发修改存在的问题。这里不再使用单个文件，Maildir 邮箱会被组织为一个目录，其中各个消息分别包含在自己单独的文件中。这种机制还允许邮箱嵌套，所以可以扩展 Maildir 邮箱的 API，增加了一些方法来处理子文件夹。

13.3.2.1 创建 Maildir 邮箱

创建 Maildir 和创建 mbox 很类似，唯一的区别是 Maildir 构造函数的参数是一个目录名而不是文件名。与前面一样，如果邮箱不存在，那么会在具体增加消息时创建邮箱。

代码清单 13-11：mailbox_maildir_create.py

```
import mailbox
import email.utils
import os

from_addr = email.utils.formataddr(('Author',
                                    'author@example.com'))
to_addr = email.utils.formataddr(('Recipient',
                                  'recipient@example.com'))

payload = '''This is the body.
From (will not be escaped).
There are 3 lines.
'''

mbox = mailbox.Maildir('Example')
mbox.lock()
try:
    msg = mailbox.mboxMessage()
    msg.set_unixfrom('author Sat Feb  7 01:05:34 2009')
    msg['From'] = from_addr
    msg['To'] = to_addr
    msg['Subject'] = 'Sample message 1'
    msg.set_payload(payload)
    mbox.add(msg)
    mbox.flush()
```

```
        msg = mailbox.mboxMessage()
        msg.set_unixfrom('author Sat Feb  7 01:05:34 2009')
        msg['From'] = from_addr
        msg['To'] = to_addr
        msg['Subject'] = 'Sample message 2'
        msg.set_payload('This is the second body.\n')
        mbox.add(msg)
        mbox.flush()
finally:
    mbox.unlock()

for dirname, subdirs, files in os.walk('Example'):
    print(dirname)
    print('  Directories:', subdirs)
    for name in files:
        fullname = os.path.join(dirname, name)
        print('\n***', fullname)
        print(open(fullname).read())
        print('*' * 20)
```

消息增加到邮箱时,它们会放在 new 子目录中。

警告:尽管从多个进程写同一个 Maildir 是安全的,但 add() 不是线程安全的。需要使用信号量或其他锁定机制,避免同一个进程的多个线程同时修改邮箱。

```
$ python3 mailbox_maildir_create.py

Example
  Directories: ['cur', 'new', 'tmp']
Example/cur
  Directories: []
Example/new
  Directories: []

*** Example/new/1483032236.M378880P24253Q1.hubert.local
From: Author <author@example.com>
To: Recipient <recipient@example.com>
Subject: Sample message 1

This is the body.
From (will not be escaped).
There are 3 lines.

********************

*** Example/new/1483032236.M381366P24253Q2.hubert.local
From: Author <author@example.com>
To: Recipient <recipient@example.com>
Subject: Sample message 2

This is the second body.

********************
Example/tmp
  Directories: []
```

读取之后,客户可能使用 MaildirMessage 的 set_subdir() 方法把消息移动到 cur 子目录。

代码清单13-12：**mailbox_maildir_set_subdir.py**

```
import mailbox
import os

print('Before:')
mbox = mailbox.Maildir('Example')
mbox.lock()
try:
    for message_id, message in mbox.iteritems():
        print('{:6} "{}"'.format(message.get_subdir(),
                                 message['subject']))
        message.set_subdir('cur')
        # Tell the mailbox to update the message.
        mbox[message_id] = message
finally:
    mbox.flush()
    mbox.close()

print('\nAfter:')
mbox = mailbox.Maildir('Example')
for message in mbox:
    print('{:6} "{}"'.format(message.get_subdir(),
                             message['subject']))

print()
for dirname, subdirs, files in os.walk('Example'):
    print(dirname)
    print('  Directories:', subdirs)
    for name in files:
        fullname = os.path.join(dirname, name)
        print(fullname)
```

尽管 Maildir 包含一个 tmp 目录，但 set_subdir() 的合法参数只有 cur 和 new。

```
$ python3 mailbox_maildir_set_subdir.py

Before:
new    "Sample message 2"
new    "Sample message 1"

After:
cur    "Sample message 2"
cur    "Sample message 1"

Example
  Directories: ['cur', 'new', 'tmp']
Example/cur
  Directories: []
Example/cur/1483032236.M378880P24253Q1.hubert.local
Example/cur/1483032236.M381366P24253Q2.hubert.local
Example/new
  Directories: []
Example/tmp
  Directories: []
```

13.3.2.2　读取 Maildir 邮箱

读取一个已有的 Maildir 邮箱与读取 mbox 邮箱很类似。

代码清单 13-13：mailbox_maildir_read.py

```
import mailbox

mbox = mailbox.Maildir('Example')
for message in mbox:
    print(message['subject'])
```

不能保证以某种特定的顺序读取消息。

```
$ python3 mailbox_maildir_read.py

Sample message 2
Sample message 1
```

13.3.2.3 从 Maildir 邮箱删除消息

要从一个 Maildir 邮箱删除已有的消息，可以将消息的键传递到 remove() 或者使用 del。

代码清单 13-14：mailbox_maildir_remove.py

```
import mailbox
import os

mbox = mailbox.Maildir('Example')
mbox.lock()
try:
    to_remove = []
    for key, msg in mbox.iteritems():
        if '2' in msg['subject']:
            print('Removing:', key)
            to_remove.append(key)
    for key in to_remove:
        mbox.remove(key)
finally:
    mbox.flush()
    mbox.close()

for dirname, subdirs, files in os.walk('Example'):
    print(dirname)
    print('  Directories:', subdirs)
    for name in files:
        fullname = os.path.join(dirname, name)
        print('\n***', fullname)
        print(open(fullname).read())
        print('*' * 20)
```

没有办法计算一个消息的键，所以应当使用 items() 或 iteritems() 从邮箱同时获取键和消息对象。

```
$ python3 mailbox_maildir_remove.py

Removing: 1483032236.M381366P24253Q2.hubert.local
Example
  Directories: ['cur', 'new', 'tmp']
Example/cur
  Directories: []

*** Example/cur/1483032236.M378880P24253Q1.hubert.local
From: Author <author@example.com>
To: Recipient <recipient@example.com>
```

```
Subject: Sample message 1

This is the body.
From (will not be escaped).
There are 3 lines.
********************
Example/new
  Directories: []
Example/tmp
  Directories: []
```

13.3.2.4 Maildir 文件夹

Maildir 邮箱的子目录或文件夹（folder）可以通过 `Maildir` 类的方法直接管理。调用者可以列出、获取、创建和删除一个给定邮箱的子目录。

代码清单 13-15：**mailbox_maildir_folders.py**

```python
import mailbox
import os


def show_maildir(name):
    os.system('find {} -print'.format(name))


mbox = mailbox.Maildir('Example')
print('Before:', mbox.list_folders())
show_maildir('Example')

print('\n{:#^30}\n'.format(''))

mbox.add_folder('subfolder')
print('subfolder created:', mbox.list_folders())
show_maildir('Example')

subfolder = mbox.get_folder('subfolder')
print('subfolder contents:', subfolder.list_folders())

print('\n{:#^30}\n'.format(''))

subfolder.add_folder('second_level')
print('second_level created:', subfolder.list_folders())
show_maildir('Example')

print('\n{:#^30}\n'.format(''))

subfolder.remove_folder('second_level')
print('second_level removed:', subfolder.list_folders())
show_maildir('Example')
```

构造文件夹的目录名时，要在文件夹名前面加一个点号（.）作为前缀。

```
$ python3 mailbox_maildir_folders.py

Example
Example/cur
Example/cur/1483032236.M378880P24253Q1.hubert.local
Example/new
Example/tmp
Example
```

```
Example/.subfolder
Example/.subfolder/cur
Example/.subfolder/maildirfolder
Example/.subfolder/new
Example/.subfolder/tmp
Example/cur
Example/cur/1483032236.M378880P24253Q1.hubert.local
Example/new
Example/tmp
Example
Example/.subfolder
Example/.subfolder/.second_level
Example/.subfolder/.second_level/cur
Example/.subfolder/.second_level/maildirfolder
Example/.subfolder/.second_level/new
Example/.subfolder/.second_level/tmp
Example/.subfolder/cur
Example/.subfolder/maildirfolder
Example/.subfolder/new
Example/.subfolder/tmp
Example/cur
Example/cur/1483032236.M378880P24253Q1.hubert.local
Example/new
Example/tmp
Example
Example/.subfolder
Example/.subfolder/cur
Example/.subfolder/maildirfolder
Example/.subfolder/new
Example/.subfolder/tmp
Example/cur
Example/cur/1483032236.M378880P24253Q1.hubert.local
Example/new
Example/tmp
Before: []

#############################

subfolder created: ['subfolder']
subfolder contents: []

#############################

second_level created: ['second_level']

#############################

second_level removed: []
```

13.3.3 消息标志

邮箱中的消息有一些标志，可以用来跟踪消息是否已读、是否被标记为重要消息或者是否被标记为以后要删除等方面。这些标志被存储为一个特定于格式的字母编码序列，Message 类提供了一些方法来获取和改变这些标志的值。下面这个例子在增加标志来指示消息很重要之前，首先显示 Example Maildir 中消息的标志。

代码清单 13-16：mailbox_maildir_add_flag.py

```
import mailbox
```

```
print('Before:')
mbox = mailbox.Maildir('Example')
mbox.lock()
try:
    for message_id, message in mbox.iteritems():
        print('{:6} "{}"'.format(message.get_flags(),
                                 message['subject']))
        message.add_flag('F')
        # Tell the mailbox to update the message.
        mbox[message_id] = message
finally:
    mbox.flush()
    mbox.close()

print('\nAfter:')
mbox = mailbox.Maildir('Example')
for message in mbox:
    print('{:6} "{}"'.format(message.get_flags(),
                             message['subject']))
```

默认地,消息没有标志。增加一个标志会改变内存中的消息,不过不会更新磁盘上的消息。要更新磁盘上的消息,可以使用现有的标识符把消息对象存储在邮箱里。

```
$ python3 mailbox_maildir_add_flag.py

Before:
       "Sample message 1"

After:
F      "Sample message 1"
```

用 `add_flag()` 增加标志会保留所有现有的标志。使用 `set_flags()` 则会覆盖现有的标志集,把它替换为传递给这个方法的新值。

代码清单 13-17:`mailbox_maildir_set_flags.py`

```
import mailbox

print('Before:')
mbox = mailbox.Maildir('Example')
mbox.lock()
try:
    for message_id, message in mbox.iteritems():
        print('{:6} "{}"'.format(message.get_flags(),
                                 message['subject']))
        message.set_flags('S')
        # Tell the mailbox to update the message.
        mbox[message_id] = message
finally:
    mbox.flush()
    mbox.close()

print('\nAfter:')
mbox = mailbox.Maildir('Example')
for message in mbox:
    print('{:6} "{}"'.format(message.get_flags(),
                             message['subject']))
```

在这个例子中,当 `set_flags()` 把标志替换为 S 时,上例增加的 F 标志就丢失了。

```
$ python3 mailbox_maildir_set_flags.py

Before:
F       "Sample message 1"

After:
S       "Sample message 1"
```

13.3.4 其他格式

`mailbox`还支持另外一些格式，不过这些格式都没有mbox或Maildir流行。MH也是一种多文件邮箱格式，一些邮箱处理器就使用了这种格式。Babyl和MMDF是单文件格式，使用了不同于mbox的消息分隔符。单文件格式支持的API与mbox相同，MH则包括Maildir类中与文件夹相关的方法。

提示：相关阅读材料
- `mailbox`的标准库文档[一]。
- `mailbox`的Python 2到Python 3移植说明。
- mbox manpage from qmail[二]：mbox格式的文档。
- Maildir manpage from qmail[三]：Maildir格式的文档。
- email：`email`模块。
- `imaplib`：`imaplib`模块可以处理IMAP服务器上保存的邮件消息。

13.4 `imaplib`：IMAP4 客户库

`imaplib`实现了一个可以与IMAP 4服务器通信的客户，IMAP 表示 Internet 消息访问协议（Internet Message Access Protocol）。IMAP协议定义了一组发送到服务器的命令，以及发送回客户的响应。大多数命令都可以作为`IMAP4`对象（用来与服务器通信）的方法来提供。

下面的例子将讨论IMAP协议的一部分，不过并不完备。要想全面地了解有关的详细信息，可以参考 **RFC 3501**[四]。

13.4.1 变种

提供了3个客户类用于借助不同的机制与服务器通信。第一个是`IMAP4`，使用明文套接字；第二个是`IMAP4_SSL`，使用基于SSL套接字的加密通信；最后一个是`IMAP4_stream`，使用一个外部命令的标准输入和标准输出。这里的所有例子都使用`IMAP4_`

[一] https://docs.python.org/3.5/library/mailbox.html
[二] www.qmail.org/man/man5/mbox.html
[三] www.qmail.org/man/man5/maildir.html
[四] https://tools.ietf.org/html/rfc3501

SSL，不过其他类的 API 也是类似的。

13.4.2 连接服务器

要建立与一个 IMAP 服务器的连接，有 2 个步骤。首先，建立套接字连接本身。其次，用服务器上的一个账户认证为用户。下面的示例代码会从一个配置文件读取服务器和用户信息。

代码清单 13-18：imaplib_connect.py

```python
import imaplib
import configparser
import os


def open_connection(verbose=False):
    # Read the config file.
    config = configparser.ConfigParser()
    config.read([os.path.expanduser('~/.pymotw')])

    # Connect to the server.
    hostname = config.get('server', 'hostname')
    if verbose:
        print('Connecting to', hostname)
    connection = imaplib.IMAP4_SSL(hostname)

    # Log in to our account.
    username = config.get('account', 'username')
    password = config.get('account', 'password')
    if verbose:
        print('Logging in as', username)
    connection.login(username, password)
    return connection


if __name__ == '__main__':
    with open_connection(verbose=True) as c:
        print(c)
```

运行时，`open_connection()` 从用户主目录中的一个文件读取配置信息，然后打开 `IMAP4_SSL` 连接并认证用户。

```
$ python3 imaplib_connect.py

Connecting to pymotw.hellfly.net
Logging in as example
<imaplib.IMAP4_SSL object at 0x10421e320>
```

这一节的其他例子还会重用这个模块，以避免重复代码。

13.4.2.1 认证失败

如果建立了连接，但是认证失败，那么便会产生一个异常。

代码清单 13-19：imaplib_connect_fail.py

```python
import imaplib
import configparser
```

```
import os

# Read the config file.
config = configparser.ConfigParser()
config.read([os.path.expanduser('~/.pymotw')])

# Connect to the server.
hostname = config.get('server', 'hostname')
print('Connecting to', hostname)
connection = imaplib.IMAP4_SSL(hostname)

# Log in to our account.
username = config.get('account', 'username')
password = 'this_is_the_wrong_password'
print('Logging in as', username)
try:
    connection.login(username, password)
except Exception as err:
    print('ERROR:', err)
```

这个例子故意用错误的密码来触发这个异常。

```
$ python3 imaplib_connect_fail.py

Connecting to pymotw.hellfly.net
Logging in as example
ERROR: b'[AUTHENTICATIONFAILED] Authentication failed.'
```

13.4.3 示例配置

示例账户有多个邮箱，邮箱的层次结构如下：

- INBOX
- Deleted Messages
- Archive
- Example
    - 2016

INBOX 文件夹下有一个未读的消息，Example/2016 中有一个已读的消息。

13.4.4 列出邮箱

要获取一个账户的可用邮箱，可以使用 `list()` 方法。

代码清单 13-20：**imaplib_list.py**

```
import imaplib
from pprint import pprint
from imaplib_connect import open_connection

with open_connection() as c:
    typ, data = c.list()
    print('Response code:', typ)
    print('Response:')
    pprint(data)
```

返回值是一个 `tuple`，其中包含一个响应码，以及由服务器返回的数据。响应码为

OK,除非出现错误。`list()` 的数据是一个字符串序列,其中包含标志、层次结构定界符和每个邮箱的邮箱名。

```
$ python3 imaplib_list.py

Response code: OK
Response:
[b'(\\HasChildren) "." Example',
 b'(\\HasNoChildren) "." Example.2016',
 b'(\\HasNoChildren) "." Archive',
 b'(\\HasNoChildren) "." "Deleted Messages"',
 b'(\\HasNoChildren) "." INBOX']
```

可以使用 `re` 或 `csv` 将各个响应串划分为 3 个部分(参见本节最后参考资料中的"IMAP Backup Script",其中给出了一个使用 `csv` 的例子)。

代码清单 13-21:`imaplib_list_parse.py`

```python
import imaplib
import re

from imaplib_connect import open_connection

list_response_pattern = re.compile(
    r'\((?P<flags>.*?)\) "(?P<delimiter>.*)" (?P<name>.*)'
)


def parse_list_response(line):
    match = list_response_pattern.match(line.decode('utf-8'))
    flags, delimiter, mailbox_name = match.groups()
    mailbox_name = mailbox_name.strip('"')
    return (flags, delimiter, mailbox_name)

with open_connection() as c:
    typ, data = c.list()
print('Response code:', typ)

for line in data:
    print('Server response:', line)
    flags, delimiter, mailbox_name = parse_list_response(line)
    print('Parsed response:', (flags, delimiter, mailbox_name))
```

如果邮箱名包含空格,那么服务器会对邮箱名加引号,不过以后在对服务器的其他调用中使用邮箱名时需要将这些引号去掉。

```
$ python3 imaplib_list_parse.py

Response code: OK
Server response: b'(\\HasChildren) "." Example'
Parsed response: ('\\HasChildren', '.', 'Example')
Server response: b'(\\HasNoChildren) "." Example.2016'
Parsed response: ('\\HasNoChildren', '.', 'Example.2016')
Server response: b'(\\HasNoChildren) "." Archive'
Parsed response: ('\\HasNoChildren', '.', 'Archive')
Server response: b'(\\HasNoChildren) "." "Deleted Messages"'
Parsed response: ('\\HasNoChildren', '.', 'Deleted Messages')
Server response: b'(\\HasNoChildren) "." INBOX'
Parsed response: ('\\HasNoChildren', '.', 'INBOX')
```

`list()` 有一些参数可以指定层次结构中的邮箱。例如,要列出 `Example` 中的子文件夹，需要传入 "Example" 作为 directory 参数。

代码清单 13-22：imaplib_list_subfolders.py

```
import imaplib

from imaplib_connect import open_connection

with open_connection() as c:
    typ, data = c.list(directory='Example')

print('Response code:', typ)

for line in data:
    print('Server response:', line)
```

这会返回父文件夹和子文件夹。

```
$ python3 imaplib_list_subfolders.py

Response code: OK
Server response: b'(\\HasChildren) "." Example'
Server response: b'(\\HasNoChildren) "." Example.2016'
```

或者，要列出与一个模式匹配的文件夹，需要传入 pattern 参数。

代码清单 13-23：imaplib_list_pattern.py

```
import imaplib

from imaplib_connect import open_connection

with open_connection() as c:
    typ, data = c.list(pattern='*Example*')

print('Response code:', typ)

for line in data:
    print('Server response:', line)
```

在这种情况下，Example 和 Example.2016 都会包含在响应中。

```
$ python3 imaplib_list_pattern.py

Response code: OK
Server response: b'(\\HasChildren) "." Example'
Server response: b'(\\HasNoChildren) "." Example.2016'
```

13.4.5 邮箱状态

可以使用 status() 询问内容的有关合计信息。表 13-1 列出了标准中定义的状态条件。

表 13-1　IMAP 4 邮箱状态条件

条件	含义
MESSAGES	邮箱中的消息数
RECENT	设置了 \Recent 标志的消息数

（续）

条　件	含　义
UIDNEXT	邮箱的下一个唯一标识符值
UIDVALIDITY	邮箱的唯一标识符合法性值
UNSEEN	未设置 \Seen 标志的消息数

状态条件必须被格式化为用空格分隔的字符串，并被包围在括号中——换句话说，使用 IMAP4 规范中对应"列表"的编码。邮箱名包围在 " 中，因为名字有可能包括空格或其他导致解析器出错的字符。

代码清单 13-24：imaplib_status.py

```python
import imaplib
import re

from imaplib_connect import open_connection
from imaplib_list_parse import parse_list_response

with open_connection() as c:
    typ, data = c.list()
    for line in data:
        flags, delimiter, mailbox = parse_list_response(line)
        print('Mailbox:', mailbox)
        status = c.status(
            '"{}"'.format(mailbox),
            '(MESSAGES RECENT UIDNEXT UIDVALIDITY UNSEEN)',
        )
        print(status)
```

返回值仍是 `tuple`，其中包含一个响应码和一个来自服务器的信息列表。在这里，列表中包含一个字符串，其格式为首先是邮箱名（用引号包围），然后是状态条件和值（用括号括起）。

```
$ python3 imaplib_status.py

Response code: OK
Server response: b'(\\HasChildren) "." Example'
Parsed response: ('\\HasChildren', '.', 'Example')
Server response: b'(\\HasNoChildren) "." Example.2016'
Parsed response: ('\\HasNoChildren', '.', 'Example.2016')
Server response: b'(\\HasNoChildren) "." Archive'
Parsed response: ('\\HasNoChildren', '.', 'Archive')
Server response: b'(\\HasNoChildren) "." "Deleted Messages"'
Parsed response: ('\\HasNoChildren', '.', 'Deleted Messages')
Server response: b'(\\HasNoChildren) "." INBOX'
Parsed response: ('\\HasNoChildren', '.', 'INBOX')
Mailbox: Example
('OK', [b'Example (MESSAGES 0 RECENT 0 UIDNEXT 2 UIDVALIDITY 1457297771 UNSEEN 0)'])
Mailbox: Example.2016
('OK', [b'Example.2016 (MESSAGES 1 RECENT 0 UIDNEXT 3 UIDVALIDITY 1457297772 UNSEEN 0)'])
Mailbox: Archive
('OK', [b'Archive (MESSAGES 0 RECENT 0 UIDNEXT 1 UIDVALIDITY 1457297770 UNSEEN 0)'])
```

```
Mailbox: Deleted Messages
('OK', [b'"Deleted Messages" (MESSAGES 3 RECENT 0 UIDNEXT 4 UIDV
ALIDITY 1457297773 UNSEEN 0)'])
Mailbox: INBOX
('OK', [b'INBOX (MESSAGES 2 RECENT 0 UIDNEXT 6 UIDVALIDITY 14572
97769 UNSEEN 1)'])
```

13.4.6 选择邮箱

一旦客户得到认证,基本操作模式便为选择一个邮箱,然后向服务器询问邮箱中的消息。这个连接是有状态的,所以选择一个邮箱之后,所有命令都会处理该邮箱中的消息,直至选择一个新邮箱。

代码清单 13-25:**imaplib_select.py**

```
import imaplib
import imaplib_connect

with imaplib_connect.open_connection() as c:
    typ, data = c.select('INBOX')
    print(typ, data)
    num_msgs = int(data[0])
    print('There are {} messages in INBOX'.format(num_msgs))
```

响应数据包含邮箱中的消息总数。

```
$ python3 imaplib_select.py

OK [b'1']
There are 1 messages in INBOX
```

如果指定了一个不合法的邮箱,则响应码为 NO。

代码清单 13-26:**imaplib_select_invalid.py**

```
import imaplib
import imaplib_connect

with imaplib_connect.open_connection() as c:
    typ, data = c.select('Does-Not-Exist')
    print(typ, data)
```

在这个例子中,数据包含一个描述问题的错误消息。

```
$ python3 imaplib_select_invalid.py

NO [b"Mailbox doesn't exist: Does-Not-Exist"]
```

13.4.7 搜索消息

选择邮箱之后,可以使用 search() 来获取邮箱中消息的 ID。

代码清单 13-27:**imaplib_search_all.py**

```
import imaplib
import imaplib_connect
```

```
from imaplib_list_parse import parse_list_response
with imaplib_connect.open_connection() as c:
    typ, mbox_data = c.list()
    for line in mbox_data:
        flags, delimiter, mbox_name = parse_list_response(line)
        c.select('"{}"'.format(mbox_name), readonly=True)
        typ, msg_ids = c.search(None, 'ALL')
        print(mbox_name, typ, msg_ids)
```

消息ID由服务器分配,这要依赖于具体实现。IMAP4协议区分了两种ID,一种是事务期间给定时刻消息的顺序ID,另一种是消息的UID标识符,不过并不是所有服务器都同时实现了这两种ID。

```
$ python3 imaplib_search_all.py

Response code: OK
Server response: b'(\\HasChildren) "." Example'
Parsed response: ('\\HasChildren', '.', 'Example')
Server response: b'(\\HasNoChildren) "." Example.2016'
Parsed response: ('\\HasNoChildren', '.', 'Example.2016')
Server response: b'(\\HasNoChildren) "." Archive'
Parsed response: ('\\HasNoChildren', '.', 'Archive')
Server response: b'(\\HasNoChildren) "." "Deleted Messages"'
Parsed response: ('\\HasNoChildren', '.', 'Deleted Messages')
Server response: b'(\\HasNoChildren) "." INBOX'
Parsed response: ('\\HasNoChildren', '.', 'INBOX')
Example OK [b'']
Example.2016 OK [b'1']
Archive OK [b'']
Deleted Messages OK [b'']
INBOX OK [b'1']
```

在这里,INBOX和Example.2016分别有一个id为1的不同消息。其他邮箱为空。

13.4.8 搜索规则

还可以使用很多其他搜索规则,包括查看消息的日期、标志和其他首部。要全面了解有关详细信息,可以参考**RFC 3501**[⊖]的6.4.4节。

要查找主题中包含`Example message 2`的消息,应当如下构造搜索规则:

```
(SUBJECT "Example message 2")
```

这个例子会查找所有邮箱中标题为"Example message 2"的消息。

代码清单 13-28: `imaplib_search_subject.py`

```
import imaplib
import imaplib_connect
from imaplib_list_parse import parse_list_response

with imaplib_connect.open_connection() as c:
    typ, mbox_data = c.list()
    for line in mbox_data:
        flags, delimiter, mbox_name = parse_list_response(line)
```

⊖ https://tools.ietf.org/html/rfc3501

```
            c.select('"{}"'.format(mbox_name), readonly=True)
            typ, msg_ids = c.search(
                None,
                '(SUBJECT "Example message 2")',
            )
            print(mbox_name, typ, msg_ids)
```

这个账户中只有一个这样的消息,其位于 INBOX 中。

```
$ python3 imaplib_search_subject.py

Response code: OK
Server response: b'(\\HasChildren) "." Example'
Parsed response: ('\\HasChildren', '.', 'Example')
Server response: b'(\\HasNoChildren) "." Example.2016'
Parsed response: ('\\HasNoChildren', '.', 'Example.2016')
Server response: b'(\\HasNoChildren) "." Archive'
Parsed response: ('\\HasNoChildren', '.', 'Archive')
Server response: b'(\\HasNoChildren) "." "Deleted Messages"'
Parsed response: ('\\HasNoChildren', '.', 'Deleted Messages')
Server response: b'(\\HasNoChildren) "." INBOX'
Parsed response: ('\\HasNoChildren', '.', 'INBOX')
Example OK [b'']
Example.2016 OK [b'']
Archive OK [b'']
Deleted Messages OK [b'']
INBOX OK [b'1']
```

还可以组合搜索规则。

代码清单 13-29：imaplib_search_from.py

```
import imaplib
import imaplib_connect
from imaplib_list_parse import parse_list_response

with imaplib_connect.open_connection() as c:
    typ, mbox_data = c.list()
    for line in mbox_data:
        flags, delimiter, mbox_name = parse_list_response(line)
        c.select('"{}"'.format(mbox_name), readonly=True)
        typ, msg_ids = c.search(
            None,
            '(FROM "Doug" SUBJECT "Example message 2")',
        )
        print(mbox_name, typ, msg_ids)
```

这里用一个逻辑与(and)操作来组合搜索规则。

```
$ python3 imaplib_search_from.py

Response code: OK
Server response: b'(\\HasChildren) "." Example'
Parsed response: ('\\HasChildren', '.', 'Example')
Server response: b'(\\HasNoChildren) "." Example.2016'
Parsed response: ('\\HasNoChildren', '.', 'Example.2016')
Server response: b'(\\HasNoChildren) "." Archive'
Parsed response: ('\\HasNoChildren', '.', 'Archive')
Server response: b'(\\HasNoChildren) "." "Deleted Messages"'
Parsed response: ('\\HasNoChildren', '.', 'Deleted Messages')
Server response: b'(\\HasNoChildren) "." INBOX'
```

```
Parsed response: ('\\HasNoChildren', '.', 'INBOX')
Example OK [b'']
Example.2016 OK [b'']
Archive OK [b'']
Deleted Messages OK [b'']
INBOX OK [b'1']
```

13.4.9 获取消息

使用 `fetch()` 方法，可以利用 `search()` 返回的标识符获取消息的内容（或部分内容），以便做进一步处理。这个方法有两个参数：要获取的消息 ID 和所获取消息的（多个）部分。

`message_ids` 参数是一个用逗号分隔的 ID 列表（例如，`"1"`、`"1,2"`）或者是一个 ID 区间（如 `1:2`）。`message_parts` 参数是一个消息段名 IMAP 列表。与 `search()` 的搜索规则类似，IMAP 协议指定了命名消息段，所以客户可以高效地获取他们真正需要的那部分消息。例如，要获取一个邮箱中消息的首部，可以使用 `fetch()` 并指定参数 `BODY.PEEK[HEADER]`。

说明： 还可以使用另一种方法获取首部（`BODY[HEADERS]`），不过这种形式有一个副作用，会隐式地将消息标志为已读，而在很多情况下并不希望如此。

代码清单 13-30：imaplib_fetch_raw.py

```python
import imaplib
import pprint

import imaplib_connect

imaplib.Debug = 4
with imaplib_connect.open_connection() as c:
    c.select('INBOX', readonly=True)
    typ, msg_data = c.fetch('1', '(BODY.PEEK[HEADER] FLAGS)')
    pprint.pprint(msg_data)
```

在这个例子中，`fetch()` 的返回值已经被部分解析，所以与 `list()` 的返回值相比，从某种程度上讲会更难处理。可以打开调试来显示客户与服务器之间完整的交互，以理解为什么会这样。

```
$ python3 imaplib_fetch_raw.py

  19:40.68 imaplib version 2.58
  19:40.68 new IMAP4 connection, tag=b'IIEN'
  19:40.70 < b'* OK [CAPABILITY IMAP4rev1 LITERAL+ SASL-IR LOGIN
-REFERRALS ID ENABLE IDLE AUTH=PLAIN] Dovecot (Ubuntu) ready.'
  19:40.70 > b'IIEN0 CAPABILITY'
  19:40.73 < b'* CAPABILITY IMAP4rev1 LITERAL+ SASL-IR LOGIN-REF
ERRALS ID ENABLE IDLE AUTH=PLAIN'
  19:40.73 < b'IIEN0 OK Pre-login capabilities listed, post-logi
n capabilities have more.'
  19:40.73 CAPABILITIES: ('IMAP4REV1', 'LITERAL+', 'SASL-IR', 'L
OGIN-REFERRALS', 'ID', 'ENABLE', 'IDLE', 'AUTH=PLAIN')
  19:40.73 > b'IIEN1 LOGIN example "TMFw00fpymotw"'
  19:40.79 < b'* CAPABILITY IMAP4rev1 LITERAL+ SASL-IR LOGIN-REF
```

```
ERRALS ID ENABLE IDLE SORT SORT=DISPLAY THREAD=REFERENCES THREAD
=REFS THREAD=ORDEREDSUBJECT MULTIAPPEND URL-PARTIAL CATENATE UNS
ELECT CHILDREN NAMESPACE UIDPLUS LIST-EXTENDED I18NLEVEL=1 CONDS
TORE QRESYNC ESEARCH ESORT SEARCHRES WITHIN CONTEXT=SEARCH LIST-
STATUS SPECIAL-USE BINARY MOVE'
  19:40.79 < b'IIEN1 OK Logged in'
  19:40.79 > b'IIEN2 EXAMINE INBOX'
  19:40.82 < b'* FLAGS (\\Answered \\Flagged \\Deleted \\Seen \\
Draft)'
  19:40.82 < b'* OK [PERMANENTFLAGS ()] Read-only mailbox.'
  19:40.82 < b'* 2 EXISTS'
  19:40.82 < b'* 0 RECENT'
  19:40.82 < b'* OK [UNSEEN 1] First unseen.'
  19:40.82 < b'* OK [UIDVALIDITY 1457297769] UIDs valid'
  19:40.82 < b'* OK [UIDNEXT 6] Predicted next UID'
  19:40.82 < b'* OK [HIGHESTMODSEQ 20] Highest'
  19:40.82 < b'IIEN2 OK [READ-ONLY] Examine completed (0.000 sec
s).'
  19:40.82 > b'IIEN3 FETCH 1 (BODY.PEEK[HEADER] FLAGS)'
  19:40.86 < b'* 1 FETCH (FLAGS () BODY[HEADER] {3108}'
  19:40.86 read literal size 3108
  19:40.86 < b')'
  19:40.89 < b'IIEN3 OK Fetch completed.'
  19:40.89 > b'IIEN4 LOGOUT'
  19:40.93 < b'* BYE Logging out'
  19:40.93 BYE response: b'Logging out'
[(b'1 (FLAGS () BODY[HEADER] {3108}',
  b'Return-Path: <doug@doughellmann.com>\r\nReceived: from compu
te4.internal ('
  b'compute4.nyi.internal [10.202.2.44])\r\n\t by sloti26t01 (Cy
rus 3.0.0-beta1'
  b'-git-fastmail-12410) with LMTPA;\r\n\t Sun, 06 Mar 2016 16:1
6:03 -0500\r'
  b'\nX-Sieve: CMU Sieve 2.4\r\nX-Spam-known-sender: yes, fadd1c
f2-dc3a-4984-a0'
  b'8b-02cef3cf1221="doug",\r\n   ea349ad0-9299-47b5-b632-6ff1e39
4cc7d="both he'
  b'llfly"\r\nX-Spam-score: 0.0\r\nX-Spam-hits: ALL_TRUSTED -1,
BAYES_00 -1.'
  b'9, LANGUAGES unknown, BAYES_USED global,\r\n    SA_VERSION 3.3
.2\r\nX-Spam'
  b'-source: IP=\'127.0.0.1\', Host=\'unk\', Country=\'unk\', FromHead
er=\'com\',\r\n  "
  b'" MailFrom=\'com\'\r\nX-Spam-charsets: plain=\'us-ascii\'\r\nX-Re
solved-to: d"
  b'oughellmann@fastmail.fm\r\nX-Delivered-to: doug@doughellmann
.com\r\nX-Ma'
  b'il-from: doug@doughellmann.com\r\nReceived: from mx5 ([10.20
2.2.204])\r'
  b'\n  by compute4.internal (LMTPProxy); Sun, 06 Mar 2016 16:16
:03 -0500\r\nRe'
  b'ceived: from mx5.nyi.internal (localhost [127.0.0.1])\r\n\tb
y mx5.nyi.inter'
  b'nal (Postfix) with ESMTP id 47CBA280DB3\r\n\tfor <doug@dough
ellmann.com>; S'
  b'un,  6 Mar 2016 16:16:03 -0500 (EST)\r\nReceived: from mx5.n
yi.internal (l'
  b'ocalhost [127.0.0.1])\r\n     by mx5.nyi.internal (Authentica
tion Milter) w'
  b'ith ESMTP\r\n    id A717886846E.30BA4280D81;\r\n    Sun, 6 M
ar 2016 16:1'
  b'6:03 -0500\r\nAuthentication-Results: mx5.nyi.internal;\r\n
```

```
        dkim=pass'
    b' (1024-bit rsa key) header.d=messagingengine.com header.i=@m
essagingengi'
    b'ne.com header.b=Jrsm+pCo;\r\n     x-local-ip=pass\r\nReceived
: from mailo'
    b'ut.nyi.internal (gateway1.nyi.internal [10.202.2.221])\r\n\t
(using TLSv1.2 '
    b'with cipher ECDHE-RSA-AES256-GCM-SHA384 (256/256 bits))\r\n\
t(No client cer'
    b'tificate requested)\r\n\tby mx5.nyi.internal (Postfix) with
ESMTPS id 30BA4'
    b'280D81\r\n\tfor <doug@doughellmann.com>; Sun,  6 Mar 2016 16
:16:03 -0500 (E'
    b'ST)\r\nReceived: from compute2.internal (compute2.nyi.intern
al [10.202.2.4'
    b'2])\r\n\tby mailout.nyi.internal (Postfix) with ESMTP id 174
0420D0A\r\n\tf'
    b'or <doug@doughellmann.com>; Sun,  6 Mar 2016 16:16:03 -0500
(EST)\r\nRecei'
    b'ved: from frontend2 ([10.202.2.161])\r\n     by compute2.intern
al (MEProxy); '
    b'Sun, 06 Mar 2016 16:16:03 -0500\r\nDKIM-Signature: v=1; a=rs
a-sha1; c=rela'
    b'xed/relaxed; d=\r\n\tmessagingengine.com; h=content-transfer
-encoding:conte'
    b'nt-type\r\n\t:date:from:message-id:mime-version:subject:to:x
-sasl-enc\r\n'
    b'\t:x-sasl-enc; s=smtpout; bh=P98NTsEo015suwJ4gk71knAWLa4=; b
=Jrsm+\r\n\t'
    b'pCovRIoQIRyp8Fl0L6JHOI8sbZy2obx7O28JF2iTlTWmX33Rhlq9403XRklw
N3JA\r\n\t7KSPq'
    b'MTp30Qdx6yIUaADwQql0+QMuQq/QxBHdjeebmdhgVfjhqxrzTbSMww/ZNhL\
r\n\tYwv/QM/oDH'
    b'bXiLSUlB3Qrg+9wsE/0jU/E0isiU=\r\nX-Sasl-enc: 8ZJ+4ZRE8AGPzdL
RWQFivGymJb8pa'
    b'4G9JGcb7k4xKn+I 1457298962\r\nReceived: from [192.168.1.14]
(75-137-1-34.d'
    b'hcp.nwnn.ga.charter.com [75.137.1.34])\r\n\tby mail.messagin
gengine.com (Po'
    b'stfix) with ESMTPA id C0B366801CD\r\n\tfor <doug@doughellman
n.com>; Sun,  6'
    b' Mar 2016 16:16:02 -0500 (EST)\r\nFrom: Doug Hellmann <doug@
doughellmann.c'
    b'om>\r\nContent-Type: text/plain; charset=us-ascii\r\nContent
-Transfer-En'
    b'coding: 7bit\r\nSubject: PyMOTW Example message 2\r\nMessage
-Id: <00ABCD'
    b'46-DADA-4912-A451-D27165BC3A2F@doughellmann.com>\r\nDate: Su
n, 6 Mar 2016 '
    b'16:16:02 -0500\r\nTo: Doug Hellmann <doug@doughellmann.com>\
r\nMime-Vers'
    b'ion: 1.0 (Mac OS X Mail 9.2 \\(3112\\))\r\nX-Mailer: Apple M
ail (2.3112)'
    b'\r\n\r\n'),
    b')']
```

FETCH 命令的响应中，首先是标志，然后指示有 595 字节的首部数据。客户用这个消息响应构造一个元组，然后用一个包含右括号（)）的字符串（服务器在获取命令响应的最后会发送这个字符串）结束这个序列。由于采用了这种格式，就能更容易地单独获取信息的不同部分，或者重新组合响应并在客户端解析。

代码清单 13-31：**imaplib_fetch_separately.py**

```python
import imaplib
import pprint
import imaplib_connect

with imaplib_connect.open_connection() as c:
    c.select('INBOX', readonly=True)

    print('HEADER:')
    typ, msg_data = c.fetch('1', '(BODY.PEEK[HEADER])')
    for response_part in msg_data:
        if isinstance(response_part, tuple):
            print(response_part[1])

    print('\nBODY TEXT:')
    typ, msg_data = c.fetch('1', '(BODY.PEEK[TEXT])')
    for response_part in msg_data:
        if isinstance(response_part, tuple):
            print(response_part[1])

    print('\nFLAGS:')
    typ, msg_data = c.fetch('1', '(FLAGS)')
    for response_part in msg_data:
        print(response_part)
        print(imaplib.ParseFlags(response_part))
```

单独地获取值还有一个额外的好处，这样可以很容易地使用 `ParseFlags()` 解析响应中的标志。

```
$ python3 imaplib_fetch_separately.py

HEADER:
b'Return-Path: <doug@doughellmann.com>\r\nReceived: from compute
4.internal (compute4.nyi.internal [10.202.2.44])\r\n\t by sloti2
6t01 (Cyrus 3.0.0-beta1-git-fastmail-12410) with LMTPA;\r\n\t Su
n, 06 Mar 2016 16:16:03 -0500\r\nX-Sieve: CMU Sieve 2.4\r\nX-Spa
m-known-sender: yes, fadd1cf2-dc3a-4984-a08b-02cef3cf1221="doug"
,\r\n  ea349ad0-9299-47b5-b632-6ff1e394cc7d="both hellfly"\r\nX-
Spam-score: 0.0\r\nX-Spam-hits: ALL_TRUSTED -1, BAYES_00 -1.9, L
ANGUAGES unknown, BAYES_USED global,\r\n  SA_VERSION 3.3.2\r\nX-
Spam-source: IP=\'127.0.0.1\', Host=\'unk\', Country=\'unk\', Fr
omHeader=\'com\',\r\n  MailFrom=\'com\'\r\nX-Spam-charsets: plai
n=\'us-ascii\'\r\nX-Resolved-to: doughellmann@fastmail.fm\r\nX-D
elivered-to: doug@doughellmann.com\r\nX-Mail-from: doug@doughell
mann.com\r\nReceived: from mx5 ([10.202.2.204])\r\n  by compute4
.internal (LMTPProxy); Sun, 06 Mar 2016 16:16:03 -0500\r\nReceiv
ed: from mx5.nyi.internal (localhost [127.0.0.1])\r\n\tby mx5.ny
i.internal (Postfix) with ESMTP id 47CBA280DB3\r\n\tfor <doug@do
ughellmann.com>; Sun,  6 Mar 2016 16:16:03 -0500 (EST)\r\nReceiv
ed: from mx5.nyi.internal (localhost [127.0.0.1])\r\n    by mx5.
nyi.internal (Authentication Milter) with ESMTP\r\n    id A71788
6846E.30BA4280D81;\r\n    Sun, 6 Mar 2016 16:16:03 -0500\r\nAuth
entication-Results: mx5.nyi.internal;\r\n    dkim=pass (1024-bit
 rsa key) header.d=messagingengine.com header.i=@messagingengine
.com header.b=Jrsm+pCo;\r\n    x-local-ip=pass\r\nReceived: from
 mailout.nyi.internal (gateway1.nyi.internal [10.202.2.221])\r\n
\t(using TLSv1.2 with cipher ECDHE-RSA-AES256-GCM-SHA384 (256/25
6 bits))\r\n\t(No client certificate requested)\r\n\tby mx5.nyi.
internal (Postfix) with ESMTPS id 30BA4280D81\r\n\tfor <doug@dou
ghellmann.com>; Sun,  6 Mar 2016 16:16:03 -0500 (EST)\r\nReceive
d: from compute2.internal (compute2.nyi.internal [10.202.2.42])\
```

```
r\n\tby mailout.nyi.internal (Postfix) with ESMTP id 1740420D0A\
r\n\tfor <doug@doughellmann.com>; Sun,  6 Mar 2016 16:16:03 -050
0 (EST)\r\nReceived: from frontend2 ([10.202.2.161])\r\n  by com
pute2.internal (MEProxy); Sun, 06 Mar 2016 16:16:03 -0500\r\nDKI
M-Signature: v=1; a=rsa-sha1; c=relaxed/relaxed; d=\r\n\tmessagi
ngengine.com; h=content-transfer-encoding:content-type\r\n\t:dat
e:from:message-id:mime-version:subject:to:x-sasl-enc\r\n\t:x-sas
l-enc; s=smtpout; bh=P98NTsEo015suwJ4gk71knAWLa4=; b=Jrsm+\r\n\t
pCovRIoQIRyp8Fl0L6JHOI8sbZy2obx7O28JF2iTlTWmX33Rhlq9403XRklwN3JA
\r\n\t7KSPqMTp30Qdx6yIUaADwQql0+QMuQq/QxBHdjeebmdhgVfjhqxrzTbSMw
w/ZNhL\r\n\tYwv/QM/oDHbXiLSUlB3Qrg+9wsE/0jU/E0isiU=\r\nX-Sasl-en
c: 8ZJ+4ZRE8AGPzdLRWQFivGymJb8pa4G9JGcb7k4xKn+I 1457298962\r\nRe
ceived: from [192.168.1.14] (75-137-1-34.dhcp.nwnn.ga.charter.co
m [75.137.1.34])\r\n\tby mail.messagingengine.com (Postfix) with
 ESMTPA id C0B366801CD\r\n\tfor <doug@doughellmann.com>; Sun,  6
 Mar 2016 16:16:02 -0500 (EST)\r\nFrom: Doug Hellmann <doug@doug
hellmann.com>\r\nContent-Type: text/plain; charset=us-ascii\r\nC
ontent-Transfer-Encoding: 7bit\r\nSubject: PyMOTW Example messag
e 2\r\nMessage-Id: <00ABCD46-DADA-4912-A451-D27165BC3A2F@doughel
lmann.com>\r\nDate: Sun, 6 Mar 2016 16:16:02 -0500\r\nTo: Doug H
ellmann <doug@doughellmann.com>\r\nMime-Version: 1.0 (Mac OS X M
ail 9.2 \\(3112\\))\r\nX-Mailer: Apple Mail (2.3112)\r\n\r\n'

BODY TEXT:
b'This is the second example message.\r\n'

FLAGS:
b'1 (FLAGS ())'
()
```

13.4.10 完整消息

如前所述，客户可以向服务器单独请求消息中的单个部分。还可以获取整个消息（采用 **RFC 822**[⊖] 规范格式化的邮件消息），并用 email 模块的类进行解析。

代码清单 13-32：**imaplib_fetch_rfc822.py**

```python
import imaplib
import email
import email.parser

import imaplib_connect


with imaplib_connect.open_connection() as c:
    c.select('INBOX', readonly=True)

    typ, msg_data = c.fetch('1', '(RFC822)')
    for response_part in msg_data:
        if isinstance(response_part, tuple):
            email_parser = email.parser.BytesFeedParser()
            email_parser.feed(response_part[1])
            msg = email_parser.close()
            for header in ['subject', 'to', 'from']:
                print('{:^8}: {}'.format(
                    header.upper(), msg[header]))
```

利用 email 模块中的解析器，可以非常容易地访问和处理消息。下面的例子只打印了各个消息的一些首部。

⊖ https://tools.ietf.org/html/rfc822

```
$ python3 imaplib_fetch_rfc822.py

SUBJECT : PyMOTW Example message 2
   TO   : Doug Hellmann <doug@doughellmann.com>
  FROM  : Doug Hellmann <doug@doughellmann.com>
```

13.4.11 上传消息

要向邮箱增加一个新消息，需要构造一个 Message 实例，并把它传递到 append() 方法，同时提供消息的时间戳。

代码清单 13-33：imaplib_append.py

```python
import imaplib
import time
import email.message
import imaplib_connect

new_message = email.message.Message()
new_message.set_unixfrom('pymotw')
new_message['Subject'] = 'subject goes here'
new_message['From'] = 'pymotw@example.com'
new_message['To'] = 'example@example.com'
new_message.set_payload('This is the body of the message.\n')

print(new_message)

with imaplib_connect.open_connection() as c:
    c.append('INBOX', '',
             imaplib.Time2Internaldate(time.time()),
             str(new_message).encode('utf-8'))

    # Show the headers for all messages in the mailbox.
    c.select('INBOX')
    typ, [msg_ids] = c.search(None, 'ALL')
    for num in msg_ids.split():
        typ, msg_data = c.fetch(num, '(BODY.PEEK[HEADER])')
        for response_part in msg_data:
            if isinstance(response_part, tuple):
                print('\n{}:'.format(num))
                print(response_part[1])
```

这个例子中使用的消息内容（payload）是一个简单的纯文本 email 体。Message 还支持 MIME 编码的多部分消息。

```
$ python3 imaplib_append.py

Subject: subject goes here
From: pymotw@example.com
To: example@example.com

This is the body of the message.

b'1':
b'Return-Path: <doug@doughellmann.com>\r\nReceived: from compute
4.internal (compute4.nyi.internal [10.202.2.44])\r\n\t by sloti2
6t01 (Cyrus 3.0.0-beta1-git-fastmail-12410) with LMTPA;\r\n\t Su
n, 06 Mar 2016 16:16:03 -0500\r\nX-Sieve: CMU Sieve 2.4\r\nX-Spa
m-known-sender: yes, fadd1cf2-dc3a-4984-a08b-02cef3cf1221="doug"
```

,\r\n   ea349ad0-9299-47b5-b632-6ff1e394cc7d="both hellfly"\r\nX-Spam-score: 0.0\r\nX-Spam-hits: ALL_TRUSTED -1, BAYES_00 -1.9, LANGUAGES unknown, BAYES_USED global,\r\n   SA_VERSION 3.3.2\r\nX-Spam-source: IP=\'127.0.0.1\', Host=\'unk\', Country=\'unk\', FromHeader=\'com\',\r\n   MailFrom=\'com\'\'\r\nX-Spam-charsets: plain=\'us-ascii\'\'\r\nX-Resolved-to: doughellmann@fastmail.fm\r\nX-Delivered-to: doug@doughellmann.com\r\nX-Mail-from: doug@doughellmann.com\r\nReceived: from mx5 ([10.202.2.204])\r\n by compute4.internal (LMTPProxy); Sun, 06 Mar 2016 16:16:03 -0500\r\nReceived: from mx5.nyi.internal (localhost [127.0.0.1])\r\n\tby mx5.nyi.internal (Postfix) with ESMTP id 47CBA280DB3\r\n\tfor <doug@doughellmann.com>; Sun, 6 Mar 2016 16:16:03 -0500 (EST)\r\nReceived: from mx5.nyi.internal (localhost [127.0.0.1])\r\n   by mx5.nyi.internal (Authentication Milter) with ESMTP\r\n   id A717886846E.30BA4280D81;\r\n   Sun, 6 Mar 2016 16:16:03 -0500\r\nAuthentication-Results: mx5.nyi.internal;\r\n   dkim=pass (1024-bit rsa key) header.d=messagingengine.com header.i=@messagingengine.com header.b=Jrsm+pCo;\r\n   x-local-ip=pass\r\nReceived: from mailout.nyi.internal (gateway1.nyi.internal [10.202.2.221])\r\n\t(using TLSv1.2 with cipher ECDHE-RSA-AES256-GCM-SHA384 (256/256 bits))\r\n\t(No client certificate requested)\r\n\tby mx5.nyi.internal (Postfix) with ESMTPS id 30BA4280D81\r\n\tfor <doug@doughellmann.com>; Sun, 6 Mar 2016 16:16:03 -0500 (EST)\r\nReceived: from compute2.internal (compute2.nyi.internal [10.202.2.42])\r\n\tby mailout.nyi.internal (Postfix) with ESMTP id 1740420D0A\r\n\tfor <doug@doughellmann.com>; Sun, 6 Mar 2016 16:16:03 -0500 (EST)\r\nReceived: from frontend2 ([10.202.2.161])\r\n by compute2.internal (MEProxy); Sun, 06 Mar 2016 16:16:03 -0500\r\nDKIM-Signature: v=1; a=rsa-sha1; c=relaxed/relaxed; d=\r\n\tmessagingengine.com; h=content-transfer-encoding:content-type\r\n\t:date:from:message-id:mime-version:subject:to:x-sasl-enc\r\n\t:x-sasl-enc; s=smtpout; bh=P98NTsEo015suwJ4gk71knAWLa4=; b=Jrsm+\r\n\tpCovRIoQIRyp8Fl0L6JHOI8sbZy2obx7O28JF2iTlTWmX33Rhlq9403XRklwN3JA\r\n\t\r\n\tt7KSPqMTp30Qdx6yIUaADwQqlO+QMuQq/QxBHdjeebmdhgVfjhqxrzTbSMww/ZNhL\r\n\ttYwv/QM/oDHbXiLSUlB3Qrg+9wsE/0jU/EOisiU=\r\nX-Sasl-enc: 8ZJ+4ZRE8AGPzdLRWQFivGymJb8pa4G9JGcb7k4xKn+I 1457298962\r\nReceived: from [192.168.1.14] (75-137-1-34.dhcp.nwnn.ga.charter.com [75.137.1.34])\r\n\tby mail.messagingengine.com (Postfix) with ESMTPA id C0B366801CD\r\n\tfor <doug@doughellmann.com>; Sun, 6 Mar 2016 16:16:02 -0500 (EST)\r\nFrom: Doug Hellmann <doug@doughellmann.com>\r\nContent-Type: text/plain; charset=us-ascii\r\nContent-Transfer-Encoding: 7bit\r\nSubject: PyMOTW Example message 2\r\nMessage-Id: <00ABCD46-DADA-4912-A451-D27165BC3A2F@doughellmann.com>\r\nDate: Sun, 6 Mar 2016 16:16:02 -0500\r\nTo: Doug Hellmann <doug@doughellmann.com>\r\nMime-Version: 1.0 (Mac OS X Mail 9.2 \\(3112\\))\r\nX-Mailer: Apple Mail (2.3112)\r\n\r\n'

b'2':
b'Subject: subject goes here\r\nFrom: pymotw@example.com\r\nTo: example@example.com\r\n\r\n'

13.4.12 移动和复制消息

一旦消息上传到服务器,便可以分别使用 move() 或 copy() 来移动或复制,而无须下载。与 fetch() 一样,这些方法可以处理消息 ID 区间。

代码清单 13-34: **imaplib_archive_read.py**

```
import imaplib
import imaplib_connect
```

```
with imaplib_connect.open_connection() as c:
    # Find the "SEEN" messages in INBOX.
    c.select('INBOX')
    typ, [response] = c.search(None, 'SEEN')
    if typ != 'OK':
        raise RuntimeError(response)
    msg_ids = ','.join(response.decode('utf-8').split(' '))

    # Create a new mailbox, "Example.Today".
    typ, create_response = c.create('Example.Today')
    print('CREATED Example.Today:', create_response)

    # Copy the messages.
    print('COPYING:', msg_ids)
    c.copy(msg_ids, 'Example.Today')

    # Look at the results.
    c.select('Example.Today')
    typ, [response] = c.search(None, 'ALL')
    print('COPIED:', response)
```

这个示例脚本在 `Example` 下创建一个新的邮箱,并把已读消息从 `INBOX` 复制到这个邮箱。

```
$ python3 imaplib_archive_read.py

CREATED Example.Today: [b'Completed']
COPYING: 2
COPIED: b'1'
```

再次运行这个脚本,可以看出检查返回码的重要性。这里调用 `create()` 创建新邮箱时没有产生一个异常,而是会报告这个邮箱已经存在。

```
$ python3 imaplib_archive_read.py

CREATED Example.Today: [b'[ALREADYEXISTS] Mailbox already exists
']
COPYING: 2
COPIED: b'1 2'
```

13.4.13 删除消息

尽管很多现代邮箱客户程序使用一个"垃圾文件夹"模型来处理已删除的消息,但是这些消息往往并没有被移动到一个真正的文件夹中。实际上,删除邮件只会更新它们的标志,以增加一个 `\Deleted` 标志。"清空"垃圾箱的操作是通过 `EXPUNGE` 命令实现的。下面这个示例脚本将查找主题包含" Lorem ipsum "的归档消息,设置已删除标志,然后再次查询服务器,可以看到这些消息仍在文件夹中。

代码清单 13-35: `imaplib_delete_messages.py`

```
import imaplib
import imaplib_connect
from imaplib_list_parse import parse_list_response

with imaplib_connect.open_connection() as c:
    c.select('Example.Today')
```

```python
# Which IDs are in the mailbox?
typ, [msg_ids] = c.search(None, 'ALL')
print('Starting messages:', msg_ids)

# Find the message(s).
typ, [msg_ids] = c.search(
    None,
    '(SUBJECT "subject goes here")',
)
msg_ids = ','.join(msg_ids.decode('utf-8').split(' '))
print('Matching messages:', msg_ids)

# What are the current flags?
typ, response = c.fetch(msg_ids, '(FLAGS)')
print('Flags before:', response)

# Change the Deleted flag.
typ, response = c.store(msg_ids, '+FLAGS', r'(\Deleted)')

# What are the flags now?
typ, response = c.fetch(msg_ids, '(FLAGS)')
print('Flags after:', response)

# Really delete the message.
typ, response = c.expunge()
print('Expunged:', response)

# Which IDs are left in the mailbox?
typ, [msg_ids] = c.search(None, 'ALL')
print('Remaining messages:', msg_ids)
```

显式地调用 `expunge()` 会删除消息，不过调用 `close()` 也能达到同样的效果。区别在于，调用 `close()` 时，客户不会得到删除通知。

```
$ python3 imaplib_delete_messages.py

Response code: OK
Server response: b'(\\HasChildren) "." Example'
Parsed response: ('\\HasChildren', '.', 'Example')
Server response: b'(\\HasNoChildren) "." Example.Today'
Parsed response: ('\\HasNoChildren', '.', 'Example.Today')
Server response: b'(\\HasNoChildren) "." Example.2016'
Parsed response: ('\\HasNoChildren', '.', 'Example.2016')
Server response: b'(\\HasNoChildren) "." Archive'
Parsed response: ('\\HasNoChildren', '.', 'Archive')
Server response: b'(\\HasNoChildren) "." "Deleted Messages"'
Parsed response: ('\\HasNoChildren', '.', 'Deleted Messages')
Server response: b'(\\HasNoChildren) "." INBOX'
Parsed response: ('\\HasNoChildren', '.', 'INBOX')
Starting messages: b'1 2'
Matching messages: 1,2
Flags before: [b'1 (FLAGS (\\Seen))', b'2 (FLAGS (\\Seen))']
Flags after: [b'1 (FLAGS (\\Deleted \\Seen))', b'2 (FLAGS (\\Del
eted \\Seen))']
Expunged: [b'2', b'1']
Remaining messages: b''
```

提示：相关阅读材料

- `imaplib` 的标准库文档[⊖]。

[⊖] https://docs.python.org/3.5/library/imaplib.html

- `rfc822`：rfc822 模块包含一个 RFC 822/RFC 5322 解析器。
- `email`：email 模块用于解析 email 消息。
- `mailbox`：本地邮箱解析器。
- `ConfigParser`：读写配置文件。
- University of Washington IMAP Information Center[①]：提供 IMAP 信息的一个很好的资源，还会提供源代码。
- **RFC 3501**[②]：Internet 消息访问协议。
- **RFC 5322**[③]：Internet 消息格式。
- IMAP Backup Script[④]：这个脚本可以用于备份 IMAP 服务器的邮件。
- IMAPClient[⑤]：这是一个与 IMAP 服务器通信的更高层客户程序，由 Menno Smits 编写。
- `offlineimap`[⑥]：这个 Python 应用用于保证邮箱的本地设置与 IMAP 服务器同步。
- `imaplib` 的 Python 2 到 Python 3 移植说明。

[①] www.washington.edu/imap/
[②] https://tools.ietf.org/html/rfc3501.html
[③] https://tools.ietf.org/html/rfc5322.html
[④] http://snipplr.com/view/7955/imap-backup-script/
[⑤] http://imapclient.freshfoo.com/
[⑥] www.offlineimap.org

第 14 章

应用构建模块

Python 标准库的强大从它的规模就可见一斑。它包括众多程序结构的实现,由于它提供的实现如此丰富,开发人员可以集中精力考虑如何让自己的应用别具一格,而不用反复实现所有这些基本的内容。这一章将介绍一些经常重用的构建模块,它们可以解决大多数应用中常见的问题。

`argparse` 是一个用于解析和验证命令行参数的接口。它支持将参数从字符串转换为整数和其他类型,遇到某个选项时可以运行回调,可以为用户未提供的选项设置默认值,还可以为程序自动生成使用说明。`getopt` 实现了 C 程序和 shell 脚本中可用的底层参数处理模型。与另外几个选项解析库相比,这个库的特性较少,不过由于其简单性和熟悉度,其成了一个流行的选择。

交互式程序应当使用 `readline` 为用户提供一个命令提示窗口。这个模块包括一些工具,可以管理历史、自动完成部分命令,还可以用 `emacs` 和 `vi` 按键绑定交互式地完成编辑。如果要安全地提示用户输入一个密码或其他秘密值,而不要在其键入时将值回显在屏幕上,可以使用 `getpass`。

`cmd` 模块为交互式、命令驱动的 shell 类程序提供了一个框架。它提供了主循环,并处理与用户的交互,所以应用只需要实现各个命令的处理回调。

`shlex` 是一个用于 shell 类语法的解析器,在这种语法中,各行由 token 构成,并用空白符分隔。这个解析器足够 "聪明",可以很好地处理引号和转义序列,所以嵌入空格的文本会被处理为单个 token。`shlex` 很适合作为领域特定语言的词法分析器,如配置文件或编程语言。

用 `configparser` 可以很容易地管理应用配置文件。它能够在程序运行之间保存用户首选项,可以在下一次应用开始时读取用户首选项,或者甚至可以作为一个简单的数据文件格式来提供。

真实世界中部署的应用要为用户提供调试信息。简单的错误消息和 traceback 会有帮助,不过如果很难再生问题,那么完整的活动日志可以直接指向导致失败的事件链。`logging` 模块包含一个功能完备的 API,可以管理日志文件,支持多个线程,甚至可以通过与远程日志守护进程交互来实现集中式日志。

对于 UNIX 环境中的程序,最常用的模式之一是逐行过滤器,即读取数据、修改数据,再将其写回。读取文件非常简单,不过要创建一个过滤器应用,再没有比使用 `fileinput` 模块更简单的方法了。它的 API 是一个行迭代器,会提供各个输入行,所以程序主体是一个简单的 `for` 循环。这个模块会为要处理的文件名解析命令行参数,或者只是直接从标准输入读取，所以基于 `fileinput` 建立的工具可以在文件上直接运行，也可以作为某个管道的一部分。

解释器关闭一个程序时，可以使用 `atexit` 调度要运行的函数。注册退出回调对于通过注销远程服务、关闭文件以及其他方式释放资源很有用。

`sched` 模块实现了一个调度器，用于在将来某些时刻触发事件。这个 API 没有给出 "时间" 的定义，所以可以使用任何时间，从真实时钟时间到解释器步数都是允许的。

14.1 `argparse`：命令行选项和参数解析

`argparse` 模块包含一些工具来构建命令行参数和选项处理器。这是在 Python 2.7 中增加的，以取代 `optparse`。`argparse` 的实现支持一些新特性，其中有些特性可能无法轻松地增加到 `optparse`，有些特性则要求 API 有变化，而这些改变不是向后兼容的。所以，干脆在库中增加了一个新的模块。现在 `optparse` 已经废弃。

14.1.1 建立解析器

使用 `argparse` 的第一步是创建一个解析器对象，并告诉它需要什么参数。程序运行时可以使用这个解析器处理命令行参数。解析器类（`ArgumentParser`）的构造函数可以取多个参数，来建立程序帮助文本中使用的描述以及其他全局行为或设置。

```
import argparse
parser = argparse.ArgumentParser(
    description='This is a PyMOTW sample program',
)
```

14.1.2 定义参数

`argparse` 是一个完整的参数处理库。参数可以触发不同的动作，由 `add_argument()` 的 `action` 参数指定。支持的动作包括存储参数（单独存储，或者作为列表的一部分存储），遇到这个参数时存储一个常量值（包括对 Boolean 分支的 true/false 值的特殊处理），统计遇到一个参数的次数，以及调用一个回调来使用定制处理指令。

默认动作是存储参数值。如果提供了一个类型，那么存储值之前要将值转换为该类型。如果提供了 `dest` 参数，那么解析命令行参数时要用这个名来保存值。

14.1.3 解析命令行

定义了所有参数之后，可以将一个参数串序列传递到 `parse_args()` 以解析命令行。默认地，参数由 `sys.argv[1:]` 得到，不过也可以使用任意的串列表。选项使用 GNU/POSIX 语法处理，所以选项和参数值可以混合出现在序列中。

`parse_args()` 的返回值是一个包含命令参数的 `Namespace`。这个对象会保存参数值（作为属性），所以如果参数的 `dest` 被设置为 `"myoption"`，那么便可以用 `args.myoption` 访问这个值。

14.1.4 简单示例

下面是一个简单的例子,有 3 个不同选项:一个布尔选项(-a),一个简单的串选项(-b),以及一个整数选项(-c)。

代码清单 14-1:**argparse_short.py**

```python
import argparse

parser = argparse.ArgumentParser(description='Short sample app')

parser.add_argument('-a', action="store_true", default=False)
parser.add_argument('-b', action="store", dest="b")
parser.add_argument('-c', action="store", dest="c", type=int)

print(parser.parse_args(['-a', '-bval', '-c', '3']))
```

向单字符选项传值有多种方法。前面的例子使用了两种不同形式,-bval 和 -c val。

$ python3 argparse_short.py

Namespace(a=True, b='val', c=3)

与输出中 'c' 关联的值类型是整数,因为要求在 ArgumentParser 存储参数之前要先完成转换。

"长"选项名(名中包含多个字符)也用同样的方式处理。

代码清单 14-2:**argparse_long.py**

```python
import argparse

parser = argparse.ArgumentParser(
    description='Example with long option names',
)

parser.add_argument('--noarg', action="store_true",
                    default=False)
parser.add_argument('--witharg', action="store",
                    dest="witharg")
parser.add_argument('--witharg2', action="store",
                    dest="witharg2", type=int)

print(
    parser.parse_args(
        ['--noarg', '--witharg', 'val', '--witharg2=3']
    )
)
```

结果也是类似的。

$ python3 argparse_long.py

Namespace(noarg=True, witharg='val', witharg2=3)

argparse 是一个完整的命令行参数解析工具,可选参数和必要参数都可以处理。

代码清单 14-3:**argparse_arguments.py**

```python
import argparse
```

```python
parser = argparse.ArgumentParser(
    description='Example with nonoptional arguments',
)

parser.add_argument('count', action="store", type=int)
parser.add_argument('units', action="store")

print(parser.parse_args())
```

在这个例子中,"count"参数是一个整数,"units"参数被保存为一个字符串。如果命令行中遗漏了其中任何一个参数,或者给定值不能被转换为正确的类型,那么便会报告一个错误。

```
$ python3 argparse_arguments.py 3 inches

Namespace(count=3, units='inches')

$ python3 argparse_arguments.py some inches

usage: argparse_arguments.py [-h] count units
argparse_arguments.py: error: argument count: invalid int value:
'some'

$ python3 argparse_arguments.py

usage: argparse_arguments.py [-h] count units
argparse_arguments.py: error: the following arguments are
required: count, units
```

14.1.4.1 参数动作

遇到一个参数时会触发 6 个内置动作。

`store`:保存值,可能首先要将值转换为一个不同的类型(可选)。如果没有显式指定任何动作,这将是默认动作。

`store_const`:保存参数规范中定义的一个值,而不是来自所解析参数的一个值。这通常用于实现非布尔值的命令行标志。

`store_true`/`store_false`:保存适当的布尔值。这些动作用于实现 Boolean 分支语句。

`append`:将值保存到一个列表。如果参数重复则会保存多个值。

`append_const`:将参数规范中定义的一个值保存到一个列表。

`version`:打印程序的版本详细信息,然后退出。

下面这个示例程序展示了以上每一种动作类型,这里提供了触发各个动作所需的最小配置。

代码清单 14-4:`argparse_action.py`

```python
import argparse

parser = argparse.ArgumentParser()

parser.add_argument('-s', action='store',
                    dest='simple_value',
                    help='Store a simple value')
```

```python
parser.add_argument('-c', action='store_const',
                    dest='constant_value',
                    const='value-to-store',
                    help='Store a constant value')
parser.add_argument('-t', action='store_true',
                    default=False,
                    dest='boolean_t',
                    help='Set a switch to true')
parser.add_argument('-f', action='store_false',
                    default=True,
                    dest='boolean_f',
                    help='Set a switch to false')
parser.add_argument('-a', action='append',
                    dest='collection',
                    default=[],
                    help='Add repeated values to a list')
parser.add_argument('-A', action='append_const',
                    dest='const_collection',
                    const='value-1-to-append',
                    default=[],
                    help='Add different values to list')
parser.add_argument('-B', action='append_const',
                    dest='const_collection',
                    const='value-2-to-append',
                    help='Add different values to list')
parser.add_argument('--version', action='version',
                    version='%(prog)s 1.0')

results = parser.parse_args()
print('simple_value     = {!r}'.format(results.simple_value))
print('constant_value   = {!r}'.format(results.constant_value))
print('boolean_t        = {!r}'.format(results.boolean_t))
print('boolean_f        = {!r}'.format(results.boolean_f))
print('collection       = {!r}'.format(results.collection))
print('const_collection = {!r}'.format(results.const_collection))
```

-t 和 -f 选项被配置为修改不同的选项值，分别存储 True 或 False。-A 和 -B 的 dest 值相同，因此它们的常量值会被追加到同一个列表。

```
$ python3 argparse_action.py -h

usage: argparse_action.py [-h] [-s SIMPLE_VALUE] [-c] [-t] [-f]
                          [-a COLLECTION] [-A] [-B] [--version]

optional arguments:
  -h, --help       show this help message and exit
  -s SIMPLE_VALUE  Store a simple value
  -c               Store a constant value
  -t               Set a switch to true
  -f               Set a switch to false
  -a COLLECTION    Add repeated values to a list
  -A               Add different values to list
  -B               Add different values to list
  --version        show program's version number and exit

$ python3 argparse_action.py -s value

simple_value     = 'value'
```

```
constant_value      = None
boolean_t           = False
boolean_f           = True
collection          = []
const_collection    = []

$ python3 argparse_action.py -c

simple_value        = None
constant_value      = 'value-to-store'
boolean_t           = False
boolean_f           = True
collection          = []
const_collection    = []

$ python3 argparse_action.py -t

simple_value        = None
constant_value      = None
boolean_t           = True
boolean_f           = True
collection          = []
const_collection    = []

$ python3 argparse_action.py -f

simple_value        = None
constant_value      = None
boolean_t           = False
boolean_f           = False
collection          = []
const_collection    = []

$ python3 argparse_action.py -a one -a two -a three

simple_value        = None
constant_value      = None
boolean_t           = False
boolean_f           = True
collection          = ['one', 'two', 'three']
const_collection    = []

$ python3 argparse_action.py -B -A

simple_value        = None
constant_value      = None
boolean_t           = False
boolean_f           = True
collection          = []
const_collection    = ['value-2-to-append', 'value-1-to-append']

$ python3 argparse_action.py --version

argparse_action.py 1.0
```

14.1.4.2 选项前缀

选项的默认语法基于一个 UNIX 约定：使用一个短横线（-）前缀来指示命令行开关。**argparse** 还支持其他前缀，所以程序可以采用本地平台的默认设置（也就是说，在 Windows 上就使用"/"），或者遵循一个不同的约定。

代码清单 14-5：`argparse_prefix_chars.py`

```python
import argparse

parser = argparse.ArgumentParser(
    description='Change the option prefix characters',
    prefix_chars='-+/',
)
parser.add_argument('-a', action="store_false",
                    default=None,
                    help='Turn A off',
                    )
parser.add_argument('+a', action="store_true",
                    default=None,
                    help='Turn A on',
                    )
parser.add_argument('//noarg', '++noarg',
                    action="store_true",
                    default=False)

print(parser.parse_args())
```

将 `ArgumentParser` 的 `prefix_chars` 参数设置为一个字符串，其中包含允许指示选项的所有字符。有一点要了解，尽管 `prefix_chars` 建立了允许的开关字符，但是要由各个参数定义来指定一个给定开关的语法。这种明显的冗余可以显式地控制使用不同前缀的选项究竟是别名（如平台独立的命令行语法就属于这种情况）还是替代选项（例如，使用 + 指示打开一个开关，用 – 关闭开关）。在前面的例子中，`+a` 和 `-a` 是不同的参数，`//noarg` 也可以被指定为 `++noarg`，不过不能指定为 `--noarg`。

```
$ python3 argparse_prefix_chars.py -h

usage: argparse_prefix_chars.py [-h] [-a] [+a] [//noarg]
Change the option prefix characters

optional arguments:
  -h, --help         show this help message and exit
  -a                 Turn A off
  +a                 Turn A on
  //noarg, ++noarg

$ python3 argparse_prefix_chars.py +a

Namespace(a=True, noarg=False)

$ python3 argparse_prefix_chars.py -a

Namespace(a=False, noarg=False)

$ python3 argparse_prefix_chars.py //noarg

Namespace(a=None, noarg=True)

$ python3 argparse_prefix_chars.py ++noarg

Namespace(a=None, noarg=True)

$ python3 argparse_prefix_chars.py --noarg

usage: argparse_prefix_chars.py [-h] [-a] [+a] [//noarg]
argparse_prefix_chars.py: error: unrecognized arguments: --noarg
```

14.1.4.3 参数来源

在目前为止的例子中，提供给解析器的参数列表要么是显式传入的一个列表，要么隐式地从 `sys.argv` 取参数。有一些类命令行指令并非来自命令行（如配置文件中的指令），使用 `argparse` 处理这种指令时，显式地传入列表会很有用。

代码清单 14-6：**argparse_with_shlex.py**

```python
import argparse
from configparser import ConfigParser
import shlex

parser = argparse.ArgumentParser(description='Short sample app')

parser.add_argument('-a', action="store_true", default=False)
parser.add_argument('-b', action="store", dest="b")
parser.add_argument('-c', action="store", dest="c", type=int)

config = ConfigParser()
config.read('argparse_with_shlex.ini')
config_value = config.get('cli', 'options')
print('Config  :', config_value)

argument_list = shlex.split(config_value)
print('Arg List:', argument_list)

print('Results :', parser.parse_args(argument_list))
```

这个例子使用 `configparse` 读取一个配置文件。

```
[cli]
options = -a -b 2
```

利用 `shlex` 可以很容易地分解存储在配置文件中的字符串。

```
$ python3 argparse_with_shlex.py

Config  : -a -b 2
Arg List: ['-a', '-b', '2']
Results : Namespace(a=True, b='2', c=None)
```

在应用代码中处理配置文件还有一种做法，可以使用 `fromfile_prefix_chars` 告诉 `argparse` 如何识别一个指定输入文件的参数（这个文件中包含一组要处理的参数）。

代码清单 14-7：**argparse_fromfile_prefix_chars.py**

```python
import argparse
import shlex

parser = argparse.ArgumentParser(description='Short sample app',
                                 fromfile_prefix_chars='@',
                                 )

parser.add_argument('-a', action="store_true", default=False)
parser.add_argument('-b', action="store", dest="b")
parser.add_argument('-c', action="store", dest="c", type=int)

print(parser.parse_args(['@argparse_fromfile_prefix_chars.txt']))
```

这个例子在发现一个有 @ 前缀的参数时会停止,然后读取指定的文件来查找更多参数。这个文件应当每行包含一个参数,如以下代码清单所示。

代码清单 14-8:**argparse_fromfile_prefix_chars.txt**

```
-a
-b
2
```

处理 `argparse_from_prefix_chars.txt` 文件时生成的输出如下。

```
$ python3 argparse_fromfile_prefix_chars.py

Namespace(a=True, b='2', c=None)
```

14.1.5 帮助输出

14.1.5.1 自动生成的帮助

argparse 会自动增加选项来生成帮助(如果有这个配置)。ArgumentParser 的 add_help 参数会控制与帮助相关的选项。

代码清单 14-9:**argparse_with_help.py**

```
import argparse

parser = argparse.ArgumentParser(add_help=True)

parser.add_argument('-a', action="store_true", default=False)
parser.add_argument('-b', action="store", dest="b")
parser.add_argument('-c', action="store", dest="c", type=int)

print(parser.parse_args())
```

会默认增加帮助选项(`-h` 和 `--help`),不过也可以将 `add_help` 设置为 `false` 以禁用这些帮助选项。

代码清单 14-10:**argparse_without_help.py**

```
import argparse

parser = argparse.ArgumentParser(add_help=False)

parser.add_argument('-a', action="store_true", default=False)
parser.add_argument('-b', action="store", dest="b")
parser.add_argument('-c', action="store", dest="c", type=int)

print(parser.parse_args())
```

尽管 `-h` 和 `--help` 是用于请求帮助的事实上的标准选项名,但是有些应用或 argparse 的某些用法可能不需要提供帮助,或者要用这些选项名来提供其他用途。

```
$ python3 argparse_with_help.py -h

usage: argparse_with_help.py [-h] [-a] [-b B] [-c C]
```

```
optional arguments:
  -h, --help  show this help message and exit
  -a
  -b B
  -c C

$ python3 argparse_without_help.py -h

usage: argparse_without_help.py [-a] [-b B] [-c C]
argparse_without_help.py: error: unrecognized arguments: -h
```

14.1.5.2 定制帮助

对于需要直接处理帮助输出的应用，`ArgumentParser` 提供了一些很有用的工具方法，可以创建定制动作来打印包含额外信息的帮助。

代码清单 14-11: **argparse_custom_help.py**

```python
import argparse

parser = argparse.ArgumentParser(add_help=True)

parser.add_argument('-a', action="store_true", default=False)
parser.add_argument('-b', action="store", dest="b")
parser.add_argument('-c', action="store", dest="c", type=int)

print('print_usage output:')
parser.print_usage()
print()

print('print_help output:')
parser.print_help()
```

`print_usage()` 会为一个参数解析器打印简短的用法消息，`print_help()` 会打印完整的帮助输出。

```
$ python3 argparse_custom_help.py

print_usage output:
usage: argparse_custom_help.py [-h] [-a] [-b B] [-c C]

print_help output:
usage: argparse_custom_help.py [-h] [-a] [-b B] [-c C]

optional arguments:
  -h, --help  show this help message and exit
  -a
  -b B
  -c C
```

`ArgumentParser` 使用一个格式化器类来控制帮助输出的外观。要改变这个类，可以在实例化 `ArgumentParser` 时传入 `formatter_class`。

例如，`RawDescriptionHelpFormatter` 避开了默认格式化器提供的换行。

代码清单 14-12: **argparse_raw_description_help_formatter.py**

```python
import argparse

parser = argparse.ArgumentParser(
```

```
    add_help=True,
    formatter_class=argparse.RawDescriptionHelpFormatter,
    description="""
    description
        not
            wrapped""",
    epilog="""
    epilog
      not
          wrapped""",
)

parser.add_argument(
    '-a', action="store_true",
    help="""argument
    help is
    wrapped
    """,
)

parser.print_help()
```

命令描述和 epilog 中的所有文档都保持不变。

```
$ python3 argparse_raw_description_help_formatter.py

usage: argparse_raw_description_help_formatter.py [-h] [-a]

    description
        not
            wrapped

optional arguments:
  -h, --help  show this help message and exit
  -a          argument help is wrapped

    epilog
      not
          wrapped
```

RawTextHelpFormatter 会把所有帮助文本处理为好像已经预先格式化一样。

代码清单 14-13：**argparse_raw_text_help_formatter.py**

```
import argparse

parser = argparse.ArgumentParser(
    add_help=True,
    formatter_class=argparse.RawTextHelpFormatter,
    description="""
    description
        not
            wrapped""",
    epilog="""
    epilog
      not
          wrapped""",
)

parser.add_argument(
    '-a', action="store_true",
    help="""argument
```

```
        help is not
        wrapped
        """,
)

parser.print_help()
```

对应 -a 参数的帮助文档不会再妥善地换行。

```
$ python3 argparse_raw_text_help_formatter.py

usage: argparse_raw_text_help_formatter.py [-h] [-a]

    description
       not
          wrapped

optional arguments:
  -h, --help  show this help message and exit
  -a          argument
                    help is not
                    wrapped


    epilog
      not
         wrapped
```

有些应用的描述或 epilog 中有一些例子，改变文本的格式可能会使这些例子不再有效，对于这些应用，原始格式化器可能很有用。

MetavarTypeHelpFormatter 会打印每种选项类型的名字，而不是目标变量，这对于有大量不同类型选项的应用可能很有用。

代码清单 14-14：argparse_metavar_type_help_formatter.py

```python
import argparse

parser = argparse.ArgumentParser(
    add_help=True,
    formatter_class=argparse.MetavarTypeHelpFormatter,
)

parser.add_argument('-i', type=int, dest='notshown1')
parser.add_argument('-f', type=float, dest='notshown2')

parser.print_help()
```

并不是显示 dest 值，而是会打印与选项关联的类型名。

```
$ python3 argparse_metavar_type_help_formatter.py

usage: argparse_metavar_type_help_formatter.py [-h] [-i int] [-f
 float]

optional arguments:
  -h, --help  show this help message and exit
  -i int
  -f float
```

14.1.6 解析器组织

argparse 包含很多用于组织参数解析器的特点，以实现或者改善帮助输出的可用性。

14.1.6.1 共享解析器原则

程序员通常需要实现一组命令行工具，它们都取一组参数，然后完成某种特殊化动作。例如，如果程序在采取具体行动之前都需要认证用户，那么它们就都需要支持 --user 和 --password 选项。不必显式地将这些选项增加到每一个 ArgumentParser，完全可以用这些共享选项定义一个父解析器，然后让各个程序的解析器继承这个父解析器的选项。

第一步是用共享参数定义来建立解析器。由于父解析器的各个后续用户会尝试增加相同的帮助选项，而这会导致一个异常，所以要在基解析器中关闭自动帮助生成。

代码清单 14-15：**argparse_parent_base.py**

```python
import argparse

parser = argparse.ArgumentParser(add_help=False)

parser.add_argument('--user', action="store")
parser.add_argument('--password', action="store")
```

接下来，用 **parents** 集合创建另一个解析器。

代码清单 14-16：**argparse_uses_parent.py**

```python
import argparse
import argparse_parent_base

parser = argparse.ArgumentParser(
    parents=[argparse_parent_base.parser],
)

parser.add_argument('--local-arg',
                    action="store_true",
                    default=False)

print(parser.parse_args())
```

得到的程序将有 3 个选项。

```
$ python3 argparse_uses_parent.py -h
usage: argparse_uses_parent.py [-h] [--user USER]
                               [--password PASSWORD]
                               [--local-arg]

optional arguments:
  -h, --help           show this help message and exit
  --user USER
  --password PASSWORD
  --local-arg
```

14.1.6.2 选项冲突

前面的例子指出，在使用相同的参数名向一个解析器增加两个参数处理器时，会导致一个异常。可以传入一个 **conflict_handler** 来改变冲突解决行为。有两个内置的处理器，分别是 error（默认）和 resolve，它们会根据处理器增加的顺序来选择处理器。

代码清单 14-17：argparse_conflict_handler_resolve.py

```
import argparse

parser = argparse.ArgumentParser(conflict_handler='resolve')

parser.add_argument('-a', action="store")
parser.add_argument('-b', action="store", help='Short alone')
parser.add_argument('--long-b', '-b',
                    action="store",
                    help='Long and short together')

print(parser.parse_args(['-h']))
```

在这个例子中，使用了给定参数名的最后一个处理器。因此，独立选项 -b 被 --long-b 的别名屏蔽。

```
$ python3 argparse_conflict_handler_resolve.py

usage: argparse_conflict_handler_resolve.py [-h] [-a A]
[--long-b LONG_B]

optional arguments:
  -h, --help            show this help message and exit
  -a A
  --long-b LONG_B, -b LONG_B
                        Long and short together
```

切换 add_argument() 调用的顺序，将不再屏蔽独立选项。

代码清单 14-18：argparse_conflict_handler_resolve2.py

```
import argparse

parser = argparse.ArgumentParser(conflict_handler='resolve')

parser.add_argument('-a', action="store")
parser.add_argument('--long-b', '-b',
                    action="store",
                    help='Long and short together')
parser.add_argument('-b', action="store", help='Short alone')

print(parser.parse_args(['-h']))
```

现在两个选项可以一起使用。

```
$ python3 argparse_conflict_handler_resolve2.py

usage: argparse_conflict_handler_resolve2.py [-h] [-a A]
                                             [--long-b LONG_B]
                                             [-b B]

optional arguments:
  -h, --help       show this help message and exit
  -a A
  --long-b LONG_B  Long and short together
  -b B             Short alone
```

14.1.6.3 参数组

argparse 将参数定义合并为"组"。默认地,它会使用两个组,一个对应选项,另一个对应必要的基于位置的参数。

代码清单 14-19:`argparse_default_grouping.py`

```python
import argparse

parser = argparse.ArgumentParser(description='Short sample app')

parser.add_argument('--optional', action="store_true",
                    default=False)
parser.add_argument('positional', action="store")

print(parser.parse_args())
```

从帮助输出的"位置参数"(positional argument)和"可选参数"(optional argument)部分可以反映这种分组。

```
$ python3 argparse_default_grouping.py -h

usage: argparse_default_grouping.py [-h] [--optional] positional

Short sample app

positional arguments:
  positional

optional arguments:
  -h, --help  show this help message and exit
  --optional
```

可以调整这种分组,使帮助更有条理,并将相关的选项或值放在一起。例如,可以使用定制组重新编写前面的共享选项示例,使得认证选项在帮助中一起出现。

用 add_argument_group() 创建 "authentication" 组,然后将与认证有关的各个选项增加到这个组,而不是增加到基解析器。

代码清单 14-20:`argparse_parent_with_group.py`

```python
import argparse

parser = argparse.ArgumentParser(add_help=False)

group = parser.add_argument_group('authentication')

group.add_argument('--user', action="store")
group.add_argument('--password', action="store")
```

如果程序使用了基于组的父解析器,那么会像前面一样把它列在 parents 值中。

代码清单 14-21:`argparse_uses_parent_with_group.py`

```python
import argparse
import argparse_parent_with_group

parser = argparse.ArgumentParser(
```

```
                parents=[argparse_parent_with_group.parser],
)
parser.add_argument('--local-arg',
                    action="store_true",
                    default=False)

print(parser.parse_args())
```

现在帮助输出会把认证选项显示在一起。

```
$ python3 argparse_uses_parent_with_group.py -h

usage: argparse_uses_parent_with_group.py [-h] [--user USER]
                                          [--password PASSWORD]
                                          [--local-arg]

optional arguments:
  -h, --help           show this help message and exit
  --local-arg

authentication:
  --user USER
  --password PASSWORD
```

14.1.6.4 互斥选项

定义互斥选项是选项分组特性的一个特殊情况，它依赖于 `add_mutually_exclusive_group()` 而不是 `add_argument_group()`。

代码清单 14-22：`argparse_mutually_exclusive.py`

```
import argparse

parser = argparse.ArgumentParser()

group = parser.add_mutually_exclusive_group()
group.add_argument('-a', action='store_true')
group.add_argument('-b', action='store_true')

print(parser.parse_args())
```

`argparse` 会保证这种互斥性，以便只能指定组中的一个选项。

```
$ python3 argparse_mutually_exclusive.py -h

usage: argparse_mutually_exclusive.py [-h] [-a | -b]

optional arguments:
  -h, --help  show this help message and exit
  -a
  -b

$ python3 argparse_mutually_exclusive.py -a

Namespace(a=True, b=False)

$ python3 argparse_mutually_exclusive.py -b

Namespace(a=False, b=True)
```

```
$ python3 argparse_mutually_exclusive.py -a -b

usage: argparse_mutually_exclusive.py [-h] [-a | -b]
argparse_mutually_exclusive.py: error: argument -b: not allowed
with argument -a
```

14.1.6.5 嵌套解析器

前面介绍的父解析器方法只是在相关命令之间共享选项的一种方法。还可以采用另一种方法，将命令结合到一个程序中，然后使用子解析器处理命令行的各个部分。其做法类似于 svn、hg 和其他有多个命令行动作或子命令的程序。

如果一个程序要处理文件系统上的目录，则可以通过定义命令来创建、删除和列出目录的内容。

代码清单 14-23：`argparse_subparsers.py`

```python
import argparse

parser = argparse.ArgumentParser()

subparsers = parser.add_subparsers(help='commands')

# A list command
list_parser = subparsers.add_parser(
    'list', help='List contents')
list_parser.add_argument(
    'dirname', action='store',
    help='Directory to list')

# A create command
create_parser = subparsers.add_parser(
    'create', help='Create a directory')
create_parser.add_argument(
    'dirname', action='store',
    help='New directory to create')
create_parser.add_argument(
    '--read-only', default=False, action='store_true',
    help='Set permissions to prevent writing to the directory',
)

# A delete command
delete_parser = subparsers.add_parser(
    'delete', help='Remove a directory')
delete_parser.add_argument(
    'dirname', action='store', help='The directory to remove')
delete_parser.add_argument(
    '--recursive', '-r', default=False, action='store_true',
    help='Remove the contents of the directory, too',
)

print(parser.parse_args())
```

帮助输出将指定子解析器显示为"commands"，其可以在命令行上被指定为位置参数。

```
$ python3 argparse_subparsers.py -h

usage: argparse_subparsers.py [-h] {list,create,delete} ...
positional arguments:
  {list,create,delete}  commands
```

```
  list                List contents
  create              Create a directory
  delete              Remove a directory

optional arguments:
  -h, --help          show this help message and exit
```

每个子解析器都有自己的帮助，以描述该命令的参数和选项。

```
$ python3 argparse_subparsers.py create -h

usage: argparse_subparsers.py create [-h] [--read-only] dirname

positional arguments:
  dirname        New directory to create

optional arguments:
  -h, --help     show this help message and exit
  --read-only    Set permissions to prevent writing to the directory
```

解析参数时，`parse_args()` 返回的 `Namespace` 对象只包含与指定命令相关的值。

```
$ python3 argparse_subparsers.py delete -r foo

Namespace(dirname='foo', recursive=True)
```

14.1.7 高级参数处理

到目前为止，前面的例子展示了简单的布尔标志，带字符串或数值参数的选项，以及位置参数。`argparse` 还支持变长参数表、枚举和常量值等复杂的参数规范。

14.1.7.1 可变参数表

参数定义可以被配置为要考虑到命令行上所解析的多个参数。根据必要或期望的参数个数，可以将 `nargs` 设置为表 14-1 所示的某个标志值。

表 14-1　`argparse` 中可变参数定义的标志

值	含义
N	参数的绝对个数（例如 3）
?	0 或 1 个参数
*	0 或所有参数
+	所有（至少 1 个）参数

代码清单 14-24：`argparse_nargs.py`

```python
import argparse

parser = argparse.ArgumentParser()
parser.add_argument('--three', nargs=3)
parser.add_argument('--optional', nargs='?')
parser.add_argument('--all', nargs='*', dest='all')
parser.add_argument('--one-or-more', nargs='+')

print(parser.parse_args())
```

解析器会执行参数统计指令，并生成一个准确的语法图，作为命令帮助文档的一部分。

```
$ python3 argparse_nargs.py -h

usage: argparse_nargs.py [-h] [--three THREE THREE THREE]
                [--optional [OPTIONAL]]
                [--all [ALL [ALL ...]]]
                [--one-or-more ONE_OR_MORE [ONE_OR_MORE ...]]

optional arguments:
  -h, --help            show this help message and exit
  --three THREE THREE THREE
  --optional [OPTIONAL]
  --all [ALL [ALL ...]]
  --one-or-more ONE_OR_MORE [ONE_OR_MORE ...]

$ python3 argparse_nargs.py

Namespace(all=None, one_or_more=None, optional=None, three=None)

$ python3 argparse_nargs.py --three

usage: argparse_nargs.py [-h] [--three THREE THREE THREE]
                [--optional [OPTIONAL]]
                [--all [ALL [ALL ...]]]
                [--one-or-more ONE_OR_MORE [ONE_OR_MORE ...]]
argparse_nargs.py: error: argument --three: expected 3
argument(s)

$ python3 argparse_nargs.py --three a b c

Namespace(all=None, one_or_more=None, optional=None,
three=['a', 'b', 'c'])

$ python3 argparse_nargs.py --optional

Namespace(all=None, one_or_more=None, optional=None, three=None)

$ python3 argparse_nargs.py --optional with_value

Namespace(all=None, one_or_more=None, optional='with_value',
three=None)

$ python3 argparse_nargs.py --all with multiple values

Namespace(all=['with', 'multiple', 'values'], one_or_more=None,
optional=None, three=None)

$ python3 argparse_nargs.py --one-or-more with_value

Namespace(all=None, one_or_more=['with_value'], optional=None,
three=None)

$ python3 argparse_nargs.py --one-or-more with multiple values

Namespace(all=None, one_or_more=['with', 'multiple', 'values'],
optional=None, three=None)

$ python3 argparse_nargs.py --one-or-more

usage: argparse_nargs.py [-h] [--three THREE THREE THREE]
                [--optional [OPTIONAL]]
                [--all [ALL [ALL ...]]]
                [--one-or-more ONE_OR_MORE [ONE_OR_MORE ...]]
```

```
argparse_nargs.py: error: argument --one-or-more: expected
at least one argument
```

14.1.7.2 参数类型

argparse 将所有参数值都处理为字符串，除非明确要求将字符串转换为另一个类型。add_argument() 的 **type** 参数定义了一个转换器函数，ArgumentParser 使用这个函数将参数值从一个字符串转换为另外一种类型。

代码清单 14-25：**argparse_type.py**

```python
import argparse

parser = argparse.ArgumentParser()

parser.add_argument('-i', type=int)
parser.add_argument('-f', type=float)
parser.add_argument('--file', type=open)

try:
    print(parser.parse_args())
except IOError as msg:
    parser.error(str(msg))
```

所有取一个字符串参数的可调用对象（callable）都可以作为 **type** 参数传入，包括内置类型（如 **int** 和 **float**）或者 **open()**。

```
$ python3 argparse_type.py -i 1

Namespace(f=None, file=None, i=1)

$ python3 argparse_type.py -f 3.14

Namespace(f=3.14, file=None, i=None)

$ python3 argparse_type.py --file argparse_type.py

Namespace(f=None, file=<_io.TextIOWrapper
name='argparse_type.py' mode='r' encoding='UTF-8'>, i=None)
```

如果类型转换失败，那么 argparse 会生成一个异常。**TypeError** 和 **ValueError** 异常会被自动截获，并转换为一个简单的错误消息提供给用户。其他异常，如下一个例子中输入文件不存在时产生的 **IOError**，则必须由调用者来处理。

```
$ python3 argparse_type.py -i a

usage: argparse_type.py [-h] [-i I] [-f F] [--file FILE]
argparse_type.py: error: argument -i: invalid int value: 'a'

$ python3 argparse_type.py -f 3.14.15

usage: argparse_type.py [-h] [-i I] [-f F] [--file FILE]
argparse_type.py: error: argument -f: invalid float value:
'3.14.15'

$ python3 argparse_type.py --file does_not_exist.txt

usage: argparse_type.py [-h] [-i I] [-f F] [--file FILE]
argparse_type.py: error: [Errno 2] No such file or directory:
'does_not_exist.txt'
```

如果要将作为输入参数接受的值限制在一个预定义的集合中，则可以使用 choices 参数。

代码清单 14-26：argparse_choices.py

```
import argparse

parser = argparse.ArgumentParser()

parser.add_argument(
    '--mode',
    choices=('read-only', 'read-write'),
)

print(parser.parse_args())
```

如果 --mode 的参数不是所允许的某个值，则会生成一个错误，并停止处理。

```
$ python3 argparse_choices.py -h

usage: argparse_choices.py [-h] [--mode {read-only,read-write}]

optional arguments:
  -h, --help            show this help message and exit
  --mode {read-only,read-write}

$ python3 argparse_choices.py --mode read-only

Namespace(mode='read-only')

$ python3 argparse_choices.py --mode invalid

usage: argparse_choices.py [-h] [--mode {read-only,read-write}]
argparse_choices.py: error: argument --mode: invalid choice:
'invalid' (choose from 'read-only', 'read-write')
```

14.1.7.3 文件参数

尽管可以用一个字符串参数来实例化 file 对象，但是不包括访问模式参数。FileType 提供了一种更为灵活的方式来指定一个参数应当是文件，而且包括模式和缓冲区大小。

代码清单 14-27：argparse_FileType.py

```
import argparse

parser = argparse.ArgumentParser()

parser.add_argument('-i', metavar='in-file',
                    type=argparse.FileType('rt'))
parser.add_argument('-o', metavar='out-file',
                    type=argparse.FileType('wt'))

try:
    results = parser.parse_args()
    print('Input file:', results.i)
    print('Output file:', results.o)
except IOError as msg:
    parser.error(str(msg))
```

与参数名关联的值是打开的文件句柄。不再使用这个文件时，应用要负责关闭文件。

```
$ python3 argparse_FileType.py -h

usage: argparse_FileType.py [-h] [-i in-file] [-o out-file]

optional arguments:
  -h, --help   show this help message and exit
  -i in-file
  -o out-file

$ python3 argparse_FileType.py -i argparse_FileType.py -o tmp_\
file.txt

Input file: <_io.TextIOWrapper name='argparse_FileType.py'
mode='rt' encoding='UTF-8'>
Output file: <_io.TextIOWrapper name='tmp_file.txt' mode='wt'
encoding='UTF-8'>

$ python3 argparse_FileType.py -i no_such_file.txt

usage: argparse_FileType.py [-h] [-i in-file] [-o out-file]
argparse_FileType.py: error: argument -i: can't open
'no_such_file.txt': [Errno 2] No such file or directory:
'no_such_file.txt'
```

14.1.7.4 定制动作

除了前面描述的内置动作之外，还可以提供一个实现了Action API 的对象来定义定制动作。作为动作（action）传入 `add_argument()` 的对象有一些形参描述所定义的参数（为 `add_argument()` 指定的所有参数），并返回一个可调用对象，它的形参包括 `parser`（用来处理参数），`namespace`（包含解析操作），`value`（所处理参数的值），以及触发动作的 `option_string`。

已经提供了类 `Action`，可以把它作为定义新动作的一个很方便的起点。构造函数处理参数定义，所以只需要在子类中覆盖 `__call__()`。

代码清单 14-28：**argparse_custom_action.py**

```python
import argparse


class CustomAction(argparse.Action):
    def __init__(self,
                 option_strings,
                 dest,
                 nargs=None,
                 const=None,
                 default=None,
                 type=None,
                 choices=None,
                 required=False,
                 help=None,
                 metavar=None):
        argparse.Action.__init__(self,
                                 option_strings=option_strings,
                                 dest=dest,
                                 nargs=nargs,
                                 const=const,
                                 default=default,
                                 type=type,
                                 choices=choices,
```

```
                                       required=required,
                                       help=help,
                                       metavar=metavar,
                                       )
        print('Initializing CustomAction')
        for name, value in sorted(locals().items()):
            if name == 'self' or value is None:
                continue
            print('  {} = {!r}'.format(name, value))
        print()
        return

    def __call__(self, parser, namespace, values,
                 option_string=None):
        print('Processing CustomAction for {}'.format(self.dest))
        print('  parser = {}'.format(id(parser)))
        print('  values = {!r}'.format(values))
        print('  option_string = {!r}'.format(option_string))

        # Do some arbitrary processing of the input values.
        if isinstance(values, list):
            values = [v.upper() for v in values]
        else:
            values = values.upper()
        # Save the results in the namespace using the destination
        # variable given to our constructor.
        setattr(namespace, self.dest, values)
        print()

parser = argparse.ArgumentParser()

parser.add_argument('-a', action=CustomAction)
parser.add_argument('-m', nargs='*', action=CustomAction)

results = parser.parse_args(['-a', 'value',
                             '-m', 'multivalue',
                             'second'])
print(results)
```

values 的类型取决于 nargs 的值。如果参数允许多个值，那么 values 就是一个列表（即使其中只包含一项）。

option_string 的值也取决于最初的参数规范。对于必要的位置参数，option_string 总是 None。

```
$ python3 argparse_custom_action.py

Initializing CustomAction
  dest = 'a'
  option_strings = ['-a']
  required = False

Initializing CustomAction
  dest = 'm'
  nargs = '*'
  option_strings = ['-m']
  required = False

Processing CustomAction for a
  parser = 4315836992
  values = 'value'
```

```
    option_string = '-a'
Processing CustomAction for m
  parser = 4315836992
  values = ['multivalue', 'second']
  option_string = '-m'

Namespace(a='VALUE', m=['MULTIVALUE', 'SECOND'])
```

提示：相关阅读材料
- `argparse` 的标准库文档[⊖]。
- `configparser`：读写配置文件。
- `shlex`：解析类 shell 语法。
- `argparse` 的 Python 2 到 Python 3 移植说明。

14.2 getopt：命令行选项解析

getopt 模块是原来的命令行选项解析器，支持 UNIX 函数 getopt 建立的约定。它会解析一个参数序列，如 `sys.argv`，并返回包含（选项，参数）对的一个元组序列和一个非选项参数序列。

目前支持的选项语法包括短格式和长格式选项：

```
-a
-bval
-b val
--noarg
--witharg=val
--witharg val
```

说明：getopt 并没有被废弃，不过 `argparse` 的维护更积极，所以新开发的程序应当使用 `argparse`。

14.2.1 函数参数

`getopt()` 函数有 3 个参数：
- 第一个参数是要解析的参数序列。这个信息通常来自 `sys.argv[1:]`（忽略 `sys.arg[0]` 中的程序名）。
- 第二个参数是对应单字符选项的选项定义串。如果某个选项需要一个参数，那么相应字母后面会有一个冒号。
- 第三个参数（如果使用）应当是一个长格式选项名序列。长格式选项可以包含多个字符，如 `--noarg` 或 `--witharg`。序列中的选项名不包括 `--` 前缀。如果某个长格式选项需要一个参数，那么它的名应当有一个后缀 `=`。

可以在一个调用中结合使用短格式和长格式选项。

⊖ https://docs.python.org/3.5/library/argparse.html

14.2.2 短格式选项

下面的示例程序接收 3 个选项。`-a` 是一个简单标志，`-b` 和 `-c` 需要一个参数。选项定义串为 `"ab:c:"`。

代码清单 14-29：**getopt_short.py**

```python
import getopt
opts, args = getopt.getopt(['-a', '-bval', '-c', 'val'], 'ab:c:')
for opt in opts:
    print(opt)
```

这个程序将一个模拟选项值列表传递到 `getopt()`，以显示如何进行处理。

```
$ python3 getopt_short.py

('-a', '')
('-b', 'val')
('-c', 'val')
```

14.2.3 长格式选项

对于一个有两个选项的程序（`--noarg` 和 `--witharg`），长参数序列应为 `['noarg', 'witharg=']`。

代码清单 14-30：**getopt_long.py**

```python
import getopt
opts, args = getopt.getopt(
    ['--noarg',
     '--witharg', 'val',
     '--witharg2=another'],
    '',
    ['noarg', 'witharg=', 'witharg2='],
)
for opt in opts:
    print(opt)
```

由于这个示例程序没有任何短格式选项，所以 `getopt()` 的第二个参数是一个空串。

```
$ python3 getopt_long.py

('--noarg', '')
('--witharg', 'val')
('--witharg2', 'another')
```

14.2.4 一个完整的例子

以下代码清单中的例子是一个更完整的程序，它有 5 个选项：`-o`、`-v`、`--output`、`--verbose` 和 `--version`。其中 `-o`、`--output` 和 `--version` 选项都需要一个参数。

代码清单 14-31：**getopt_example.py**

```python
import getopt
import sys
```

```
version = '1.0'
verbose = False
output_filename = 'default.out'

print('ARGV     :', sys.argv[1:])

try:
    options, remainder = getopt.getopt(
        sys.argv[1:],
        'o:v',
        ['output=',
         'verbose',
         'version=',
         ])
except getopt.GetoptError as err:
    print('ERROR:', err)
    sys.exit(1)

print('OPTIONS   :', options)

for opt, arg in options:
    if opt in ('-o', '--output'):
        output_filename = arg
    elif opt in ('-v', '--verbose'):
        verbose = True
    elif opt == '--version':
        version = arg

print('VERSION   :', version)
print('VERBOSE   :', verbose)
print('OUTPUT    :', output_filename)
print('REMAINING :', remainder)
```

可以采用多种不同方式来调用这个程序。如果不带任何参数地调用这个程序,那么会使用默认设置。

```
$ python3 getopt_example.py

ARGV      : []
OPTIONS   : []
VERSION   : 1.0
VERBOSE   : False
OUTPUT    : default.out
REMAINING : []
```

单字符选项与其参数可以用空白符分隔。

```
$ python3 getopt_example.py -o foo

ARGV      : ['-o', 'foo']
OPTIONS   : [('-o', 'foo')]
VERSION   : 1.0
VERBOSE   : False
OUTPUT    : foo
REMAINING : []
```

或者,也可以把选项和值结合到一个参数中。

```
$ python3 getopt_example.py -ofoo

ARGV      : ['-ofoo']
OPTIONS   : [('-o', 'foo')]
```

```
VERSION    : 1.0
VERBOSE    : False
OUTPUT     : foo
REMAINING  : []
```

也可以类似地将长格式选项与值分隔。

```
$ python3 getopt_example.py --output foo

ARGV       : ['--output', 'foo']
OPTIONS    : [('--output', 'foo')]
VERSION    : 1.0
VERBOSE    : False
OUTPUT     : foo
REMAINING  : []
```

一个长格式选项与它的值结合时，选项名和值要用一个 = 分隔。

```
$ python3 getopt_example.py --output=foo

ARGV       : ['--output=foo']
OPTIONS    : [('--output', 'foo')]
VERSION    : 1.0
VERBOSE    : False
OUTPUT     : foo
REMAINING  : []
```

14.2.5 缩写长格式选项

只要提供了一个唯一的前缀，就不必在命令行上完整地拼写出长格式选项。

```
$ python3 getopt_example.py --o foo

ARGV       : ['--o', 'foo']
OPTIONS    : [('--output', 'foo')]
VERSION    : 1.0
VERBOSE    : False
OUTPUT     : foo
REMAINING  : []
```

如果没有提供一个唯一的前缀，则会产生一个异常。

```
$ python3 getopt_example.py --ver 2.0

ARGV       : ['--ver', '2.0']
ERROR: option --ver not a unique prefix
```

14.2.6 GNU 式选项解析

正常情况下，一旦遇到第一个非选项参数，选项处理就会停止。

```
$ python3 getopt_example.py -v not_an_option --output foo

ARGV       : ['-v', 'not_an_option', '--output', 'foo']
OPTIONS    : [('-v', '')]
VERSION    : 1.0
VERBOSE    : True
OUTPUT     : default.out
REMAINING  : ['not_an_option', '--output', 'foo']
```

要想以任意顺序混合选项和非选项参数，则要使用 **gnu_getopt()**。

代码清单 14-32：**getopt_gnu.py**

```
import getopt
import sys

version = '1.0'
verbose = False
output_filename = 'default.out'

print('ARGV      :', sys.argv[1:])

try:
    options, remainder = getopt.gnu_getopt(
        sys.argv[1:],
        'o:v',
        ['output=',
         'verbose',
         'version=',
         ])
except getopt.GetoptError as err:
    print('ERROR:', err)
    sys.exit(1)

print('OPTIONS   :', options)

for opt, arg in options:
    if opt in ('-o', '--output'):
        output_filename = arg
    elif opt in ('-v', '--verbose'):
        verbose = True
    elif opt == '--version':
        version = arg

print('VERSION   :', version)
print('VERBOSE   :', verbose)
print('OUTPUT    :', output_filename)
print('REMAINING :', remainder)
```

修改前例中的调用之后，可以清楚地看出两种方法的区别。

```
$ python3 getopt_gnu.py -v not_an_option --output foo

ARGV      : ['-v', 'not_an_option', '--output', 'foo']
OPTIONS   : [('-v', ''), ('--output', 'foo')]
VERSION   : 1.0
VERBOSE   : True
OUTPUT    : foo
REMAINING : ['not_an_option']
```

14.2.7　结束参数处理

如果 getopt() 在输入参数中遇到 --，那么它会停止处理余下的参数（作为选项）。这个特性可以用来传递看上去像选项的参数值，如以一个短横线（-）开头的文件名。

```
$ python3 getopt_example.py -v -- --output foo

ARGV      : ['-v', '--', '--output', 'foo']
OPTIONS   : [('-v', '')]
VERSION   : 1.0
VERBOSE   : True
```

```
OUTPUT    : default.out
REMAINING : ['--output', 'foo']
```

提示：相关阅读材料
- `getopt` 的标准库文档[○]。
- `argparse`：`argparse` 模块取代 `getopt` 在新应用中使用。

14.3 readline：GNU readline 库

`readline` 模块提供了 GNU readline 库的一个接口。它可以用于改进交互式命令行程序，使之更易于使用，例如，增加命令行文本完成，即 "tab 完成" 功能（tab completion）。

说明：由于 `readline` 与控制台内容交互，所以如果打印调试消息，则会很难看出哪些是示例代码完成的工作，而哪些是 `readline` 自动完成的工作。下面的例子使用 `logging` 模块将调试信息写到一个单独的文件。每个示例都会显示日志输出。

说明：默认情况下，并非所有平台都提供 `readline` 所需的 GNU 库。如果你的系统确实不包括这些库，那么在安装依赖库之后可能需要重新编译 Python 解释器以启用这个模块。Python Package Index 在 `gnureadline` 名之下还发布了这个库的一个独立版本[○]。这一节中的例子首先尝试导入 `gnureadline`，然后再使用 `readline`。

14.3.1 配置 readline

有两种方法配置底层 readline 库，可以使用一个配置文件，或者利用 `parse_and_bind()` 函数。配置选项包括调用完成特性的按键绑定，编辑模式（`vi` 或 `emacs`），以及其他一些值。有关的详细信息可以参考 GNU Readline 库的文档。

要启用"tab 完成"功能，最容易的方法就是利用一个 `parse_and_bind()` 调用。其他选项可以同时设置。下面这个例子会改变编辑控制，将使用 `vi` 模式而不是默认的 `emacs`。要编辑当前输入行，可以按下 ESC，然后使用常规的 `vi` 导航键，如 `j`、`k`、`l` 和 `h`。

代码清单 14-33：`readline_parse_and_bind.py`

```
try:
    import gnureadline as readline
except ImportError:
    import readline

readline.parse_and_bind('tab: complete')
readline.parse_and_bind('set editing-mode vi')

while True:
    line = input('Prompt ("stop" to quit): ')
```

○ https://docs.python.org/3.5/library/getopt.html
○ https://pypi.python.org/pypi/gnureadline

```
    if line == 'stop':
        break
    print('ENTERED: {!r}'.format(line))
```

这个配置可以作为指令存储在一个文件中,由库通过一个调用来读取。如果 myreadline.rc 包含以下内容:

代码清单 14-34:myreadline.rc

```
# Turn on tab completion.
tab: complete
# Use vi editing mode instead of emacs.
set editing-mode vi
```

则可以用 read_init_file() 来读取这个文件。

代码清单 14-35:readline_read_init_file.py

```
try:
    import gnureadline as readline
except ImportError:
    import readline

readline.read_init_file('myreadline.rc')

while True:
    line = input('Prompt ("stop" to quit): ')
    if line == 'stop':
        break
    print('ENTERED: {!r}'.format(line))
```

14.3.2 完成文本

下一个程序有一组内置命令,用户输入指令时将使用 tab 完成功能。

代码清单 14-36:readline_completer.py

```
try:
    import gnureadline as readline
except ImportError:
    import readline
import logging

LOG_FILENAME = '/tmp/completer.log'
logging.basicConfig(
    format='%(message)s',
    filename=LOG_FILENAME,
    level=logging.DEBUG,
)


class SimpleCompleter:

    def __init__(self, options):
        self.options = sorted(options)

    def complete(self, text, state):
        response = None
```

```python
            if state == 0:
                # This is the first time for this text,
                # so build a match list.
                if text:
                    self.matches = [
                        s
                        for s in self.options
                        if s and s.startswith(text)
                    ]
                    logging.debug('%s matches: %s',
                                  repr(text), self.matches)
                else:
                    self.matches = self.options[:]
                    logging.debug('(empty input) matches: %s',
                                  self.matches)

            # Return the state'th item from the match list,
            # if that many items are present.
            try:
                response = self.matches[state]
            except IndexError:
                response = None
            logging.debug('complete(%s, %s) => %s',
                          repr(text), state, repr(response))
            return response


def input_loop():
    line = ''
    while line != 'stop':
        line = input('Prompt ("stop" to quit): ')
        print('Dispatch {}'.format(line))

# Register the completer function.
OPTIONS = ['start', 'stop', 'list', 'print']
readline.set_completer(SimpleCompleter(OPTIONS).complete)

# Use the tab key for completion.
readline.parse_and_bind('tab: complete')

# Prompt the user for text.
input_loop()
```

这个程序中的 `input_loop()` 函数进行逐行读取,直至输入值为 `"stop"`。更复杂的程序还可以具体解析输入行,并运行命令。

`SimpleCompleter` 类维护了一个"选项"列表,作为自动完成的候选项。实例的 `complete()` 方法使用 `readline` 注册为完成源。参数是一个要完成的文本串(text)和一个状态值(state),状态值指示对这个文本调用函数的次数。这个函数会反复调用,每次调用将使状态值递增。如果对应这个状态值有一个候选动作,则应当返回一个串,如果没有更多的候选项,则返回 `None`。之前代码清单中的 `complete()` 实现会在 `state` 为 `0` 时查找一组匹配,然后在后续调用时返回所有候选匹配,一次返回一个。

运行前一个代码清单中的代码时,会生成以下初始输出:

```
$ python3 readline_completer.py

Prompt ("stop" to quit):
```

按两次 tab，会显示一个选项列表。

```
$ python3 readline_completer.py

Prompt ("stop" to quit):
list   print  start  stop
Prompt ("stop" to quit):
```

日志文件显示出这里分别利用了两个状态值序列调用 complete()。

```
$ tail -f /tmp/completer.log

(empty input) matches: ['list', 'print', 'start', 'stop']
complete('', 0) => 'list'
complete('', 1) => 'print'
complete('', 2) => 'start'
complete('', 3) => 'stop'
complete('', 4) => None
(empty input) matches: ['list', 'print', 'start', 'stop']
complete('', 0) => 'list'
complete('', 1) => 'print'
complete('', 2) => 'start'
complete('', 3) => 'stop'
complete('', 4) => None
```

第一个序列来自第一次按下 tab 键。完成算法会查询所有候选项，不过并不扩展空的输入行。然后，第二次按下 tab 时，重新计算候选项列表，以便显示给用户。

如果下一个输入为 l，然后是另一个 tab，则会生成以下输出：

```
Prompt ("stop" to quit): list
```

日志反映了 complete() 的不同参数：

```
'l' matches: ['list']
complete('l', 0) => 'list'
complete('l', 1) => None
```

现在按下回车（Enter）会导致 input() 返回这个值，并且 while 循环继续。

```
Dispatch list
Prompt ("stop" to quit):
```

对于以 s 开头的命令，完成这种命令有两种可能的情况。键入 s，然后按下 tab，可以发现 start 和 stop 是候选项，不过自动完成特性只能部分完成屏幕上的文本，即增加一个 t。

日志文件显示了以下信息：

```
's' matches: ['start', 'stop']
complete('s', 0) => 'start'
complete('s', 1) => 'stop'
complete('s', 2) => None
```

屏幕上也会生成输出：

```
Prompt ("stop" to quit): st
```

说明：如果一个完成器函数产生一个异常，那么它会悄无声息地将其忽略，并且 readline 会认为没有匹配的完成选择。

14.3.3 访问完成缓冲区

SimpleCompleter 中的完成算法只查看传入函数的文本参数,而没有使用 readline 内部状态的任何其他信息。还可以使用 readline 函数来管理输入缓冲区的文本。

代码清单 14-37:`readline_buffer.py`

```python
try:
    import gnureadline as readline
except ImportError:
    import readline
import logging

LOG_FILENAME = '/tmp/completer.log'
logging.basicConfig(
    format='%(message)s',
    filename=LOG_FILENAME,
    level=logging.DEBUG,
)


class BufferAwareCompleter:

    def __init__(self, options):
        self.options = options
        self.current_candidates = []

    def complete(self, text, state):
        response = None
        if state == 0:
            # This is the first time for this text,
            # so build a match list.

            origline = readline.get_line_buffer()
            begin = readline.get_begidx()
            end = readline.get_endidx()
            being_completed = origline[begin:end]
            words = origline.split()

            logging.debug('origline=%s', repr(origline))
            logging.debug('begin=%s', begin)
            logging.debug('end=%s', end)
            logging.debug('being_completed=%s', being_completed)
            logging.debug('words=%s', words)

            if not words:
                self.current_candidates = sorted(
                    self.options.keys()
                )
            else:
                try:
                    if begin == 0:
                        # First word
                        candidates = self.options.keys()
                    else:
                        # Later word
                        first = words[0]
                        candidates = self.options[first]

                    if being_completed:
                        # Match options with portion of input
```

```
                            # being completed
                            self.current_candidates = [
                                w for w in candidates
                                if w.startswith(being_completed)
                            ]
                        else:
                            # Matching empty string,
                            # use all candidates
                            self.current_candidates = candidates

                        logging.debug('candidates=%s',
                                      self.current_candidates)

                except (KeyError, IndexError) as err:
                    logging.error('completion error: %s', err)
                    self.current_candidates = []

        try:
            response = self.current_candidates[state]
        except IndexError:
            response = None
        logging.debug('complete(%s, %s) => %s',
                      repr(text), state, response)
        return response


def input_loop():
    line = ''
    while line != 'stop':
        line = input('Prompt ("stop" to quit): ')
        print('Dispatch {}'.format(line))

# Register our completer function.
completer = BufferAwareCompleter({
    'list': ['files', 'directories'],
    'print': ['byname', 'bysize'],
    'stop': [],
})
readline.set_completer(completer.complete)

# Use the tab key for completion.
readline.parse_and_bind('tab: complete')

# Prompt the user for text.
input_loop()
```

这个例子要完成有子选项的命令。complete()方法需要在输入缓冲区中查看完成的位置，来确定它是第一个词的一部分还是后一个词的一部分。如果目标是第一个词，则将使用选项字典的键作为候选项。如果不是第一个词，则使用第一个词在选项字典中查找候选项。

这里有 3 个顶层命令，其中两个有子命令。

- list
  - files
  - directories
- print
  - byname

- bysize
- stop

按照前面同样的动作序列，两次按下 tab 会给出 3 个顶层命令。

```
$ python3 readline_buffer.py

Prompt ("stop" to quit):
list   print  stop
Prompt ("stop" to quit):
```

日志包括以下信息：

```
origline=''
begin=0
end=0
being_completed=
words=[]
complete('', 0) => list
complete('', 1) => print
complete('', 2) => stop
complete('', 3) => None
origline=''
begin=0
end=0
being_completed=
words=[]
complete('', 0) => list
complete('', 1) => print
complete('', 2) => stop
complete('', 3) => None
```

如果第一个词是 `'list'`（单词后面有一个空格），则完成候选项是不同的。

```
Prompt ("stop" to quit): list
directories   files
```

从日志可以看到，所完成的文本不是一整行，则只是 `list` 后的部分。

```
origline='list '
begin=5
end=5
being_completed=
words=['list']
candidates=['files', 'directories']
complete('', 0) => files
complete('', 1) => directories
complete('', 2) => None
origline='list '
begin=5
end=5
being_completed=
words=['list']
candidates=['files', 'directories']
complete('', 0) => files
complete('', 1) => directories
complete('', 2) => None
```

14.3.4 输入历史

readline 会自动跟踪输入历史。有两组不同的函数来处理历史。当前会话的历史

可以用 `get_current_history_length()` 和 `get_history_item()` 访问。这个历史可以保存到一个文件中，以后分别用 `write_history_file()` 和 `read_history_file()` 重新加载。默认地会保存整个历史，不过可以用 `set_history_length()` 设置文件的最大长度。如果值为 –1，则表示长度没有限制。

代码清单 14-38：**readline_history.py**

```python
try:
    import gnureadline as readline
except ImportError:
    import readline
import logging
import os

LOG_FILENAME = '/tmp/completer.log'
HISTORY_FILENAME = '/tmp/completer.hist'

logging.basicConfig(
    format='%(message)s',
    filename=LOG_FILENAME,
    level=logging.DEBUG,
)


def get_history_items():
    num_items = readline.get_current_history_length() + 1
    return [
        readline.get_history_item(i)
        for i in range(1, num_items)
    ]


class HistoryCompleter:

    def __init__(self):
        self.matches = []

    def complete(self, text, state):
        response = None
        if state == 0:
            history_values = get_history_items()
            logging.debug('history: %s', history_values)
            if text:
                self.matches = sorted(
                    h
                    for h in history_values
                    if h and h.startswith(text)
                )
            else:
                self.matches = []
            logging.debug('matches: %s', self.matches)
        try:
            response = self.matches[state]
        except IndexError:
            response = None
        logging.debug('complete(%s, %s) => %s',
                      repr(text), state, repr(response))
        return response


def input_loop():
```

```python
        if os.path.exists(HISTORY_FILENAME):
            readline.read_history_file(HISTORY_FILENAME)
    print('Max history file length:',
          readline.get_history_length())
    print('Startup history:', get_history_items())
    try:
        while True:
            line = input('Prompt ("stop" to quit): ')
            if line == 'stop':
                break
            if line:
                print('Adding {!r} to the history'.format(line))
    finally:
        print('Final history:', get_history_items())
        readline.write_history_file(HISTORY_FILENAME)

# Register our completer function.
readline.set_completer(HistoryCompleter().complete)

# Use the tab key for completion.
readline.parse_and_bind('tab: complete')

# Prompt the user for text.
input_loop()
```

HistoryCompleter 记住键入的所有内容，并在完成后续输入时使用这些值。

```
$ python3 readline_history.py

Max history file length: -1
Startup history: []
Prompt ("stop" to quit): foo
Adding 'foo' to the history
Prompt ("stop" to quit): bar
Adding 'bar' to the history
Prompt ("stop" to quit): blah
Adding 'blah' to the history
Prompt ("stop" to quit): b
bar   blah
Prompt ("stop" to quit): b
Prompt ("stop" to quit): stop
Final history: ['foo', 'bar', 'blah', 'stop']
```

按下 b 后再按下两次 tab，日志会显示以下输出。

```
history: ['foo', 'bar', 'blah']
matches: ['bar', 'blah']
complete('b', 0) => 'bar'
complete('b', 1) => 'blah'
complete('b', 2) => None
history: ['foo', 'bar', 'blah']
matches: ['bar', 'blah']
complete('b', 0) => 'bar'
complete('b', 1) => 'blah'
complete('b', 2) => None
```

第二次运行脚本时，将从文件读取所有历史。

```
$ python3 readline_history.py

Max history file length: -1
Startup history: ['foo', 'bar', 'blah', 'stop']
```

```
Prompt ("stop" to quit):
```

还有一些函数可以删除单个历史项，也可以清除整个历史。

14.3.5 hook

可以使用一些 hook 来触发动作，作为交互序列的一部分。显示提示符之前会调用启动（startup）hook，显示提示符之后但从用户读取文本之前会运行预输入（pre-input）hook。

代码清单 14-39：**readline_hooks.py**

```
try:
    import gnureadline as readline
except ImportError:
    import readline


def startup_hook():
    readline.insert_text('from startup_hook')


def pre_input_hook():
    readline.insert_text(' from pre_input_hook')
    readline.redisplay()

readline.set_startup_hook(startup_hook)
readline.set_pre_input_hook(pre_input_hook)
readline.parse_and_bind('tab: complete')

while True:
    line = input('Prompt ("stop" to quit): ')
    if line == 'stop':
        break
    print('ENTERED: {!r}'.format(line))
```

每个 hook 都可能是一个很好的机会，可以利用这些机会使用 `insert_text()` 修改输入缓冲区。

```
$ python3 readline_hooks.py

Prompt ("stop" to quit): from startup_hook from pre_input_hook
```

如果在预输入 hook 中修改了缓冲区，那么必须调用 `redisplay()` 更新屏幕。

提示：相关阅读材料
- `readline` 的标准库文档[一]。
- GNU readline[二]：GNU readline 库的文档。
- `readline` init file format[三]：初始化和配置文件格式。
- effbot：The `readline` module[四]：effbot 的 `readline` 模块指南。

[一] https://docs.python.org/3.5/library/readline.html
[二] http://tiswww.case.edu/php/chet/readline/readline.html
[三] http://tiswww.case.edu/php/chet/readline/readline.html#SEC10
[四] http://sandbox.effbot.org/librarybook/readline.htm

- gnureadline[①]：readline 的一个静态链接版本，可用于很多平台，而且可以通过 pip 安装。
- pyreadline[②]：作为基于 Python 的 readline 替代库，在 Windows 上使用。
- cmd：cmd 模块在命令接口中大量使用 readline 实现 tab 完成。这一节的一些例子就是根据 cmd 中的代码改写的。
- rlcompleter：使用 readline 为交互式 Python 解释器增加 tab 完成功能。

14.4 getpass：安全密码提示

很多程序通过终端与用户交互，这些程序需要向用户询问密码值，但不在屏幕上显示用户键入的内容。getpass 模块提供了一种可移植的方法，可以安全地处理这种密码提示。

14.4.1 示例

getpass() 函数会显示一个提示符，然后读取用户的输入，直到用户按下回车键。输入会作为一个字符串返回给调用者。

代码清单 14-40：getpass_defaults.py

```
import getpass

try:
    p = getpass.getpass()
except Exception as err:
    print('ERROR:', err)
else:
    print('You entered:', p)
```

如果调用者没有指定其他提示符，则默认的提示符为"Password:"。

```
$ python3 getpass_defaults.py

Password:
You entered: sekret
```

这个提示符可以被改为所需的任何值。

代码清单 14-41：getpass_prompt.py

```
import getpass

p = getpass.getpass(prompt='What is your favorite color? ')
if p.lower() == 'blue':
    print('Right.  Off you go.')
else:
    print('Auuuuugh!')
```

---

① https://pypi.python.org/pypi/gnureadline
② http://ipython.org/pyreadline.html

有些程序要求输入一个"通行短语"（passphrase）来提供更好的安全性，而不是一个简单的密码。

```
$ python3 getpass_prompt.py

What is your favorite color?
Right.  Off you go.

$ python3 getpass_prompt.py

What is your favorite color?
Auuuuugh!
```

默认地，`getpass()`使用`sys.stdout`显示提示符字符串。如果程序会在`sys.stdout`上生成有用的输出，那么通常更好的选择是将提示符发送到另一个流，如`sys.stderr`。

代码清单 14-42：**getpass_stream.py**

```python
import getpass
import sys
p = getpass.getpass(stream=sys.stderr)
print('You entered:', p)
```

对提示符使用`sys.stderr`表示标准输出可以被重定向（到一个管道或一个文件），而不会看到密码提示。用户输入的值仍然不会在屏幕上回显。

```
$ python3 getpass_stream.py >/dev/null

Password:
```

14.4.2 无终端使用 getpass

在 UNIX 下，`getpass()`往往需要一个 tty，它能通过`termios`控制这个 tty，从而禁用输入回显。采用这种方法，不会从一个重定向到标准输入的非终端流读取值。`getpass`会尝试访问 tty 来完成处理，如果函数可以访问 tty，那么便不会产生错误。

```
$ echo "not sekret" | python3 getpass_defaults.py

Password:
You entered: sekret
```

要由调用者负责检测输入流并非 tty 的情况，并在这种情况下使用一个候选方法来读取。

代码清单 14-43：**getpass_noterminal.py**

```python
import getpass
import sys

if sys.stdin.isatty():
    p = getpass.getpass('Using getpass: ')
else:
    print('Using readline')
    p = sys.stdin.readline().rstrip()

print('Read: ', p)
```

如果是 tty，输出为：

```
$ python3 ./getpass_noterminal.py
Using getpass:
Read:  sekret
```

如果不是 tty，输出为：

```
$ echo "sekret" | python3 ./getpass_noterminal.py
Using readline
Read:  sekret
```

提示：相关阅读材料
- getpass 的标准库文档[⊖]。
- readline：交互式提示库。

14.5 cmd：面向行的命令处理器

cmd 模块包含一个公共类 Cmd，这个类被用作交互式 shell 和其他命令解释器的基类。默认地，它使用 readline 完成交互式提示处理、命令行编辑和命令完成。

14.5.1 处理命令

用 cmd 创建的命令解释器使用一个循环从输入读取所有行，进行解析，然后将命令分派到一个适当的命令处理器（command handler）。输入行会被解析为两部分：命令以及该行上的所有其他文本。例如，如果用户输入 foo bar，而且解释器类包含一个名为 do_foo() 的方法，那么便会调用这个方法并以 "bar" 作为它的唯一参数。

文件末尾（end-of-file）标志被分派至 do_EOF()。如果一个命令处理器返回 true 值，那么程序会妥善地退出。所以如果要提供一个简洁的方法退出解释器，就一定要实现 do_EOF()，并让它返回 True。

下面这个简单的示例程序支持"greet"命令。

代码清单 14-44：**cmd_simple.py**

```python
import cmd

class HelloWorld(cmd.Cmd):

    def do_greet(self, line):
        print("hello")
    def do_EOF(self, line):
        return True

if __name__ == '__main__':
    HelloWorld().cmdloop()
```

⊖ https://docs.python.org/3.5/library/getpass.html

交互式地运行这个程序展示了如何分派命令，并显示 Cmd 中包含的一些特性。

```
$ python3 cmd_simple.py

(Cmd)
```

首先要注意的是命令提示符（Cmd）。这个提示符可以通过属性 prompt 来配置。提示符值是动态的；也就是说，如果一个命令处理器改变了提示符属性，就会使用新值来询问下一个命令。

```
Documented commands (type help <topic>):
========================================
help

Undocumented commands:
======================
EOF  greet
```

help 命令被内置在 Cmd 中。如果没有提供参数，则 help 会显示可用命令的列表。如果输入包括一个命令名，则输出会更为详细，并且只显示这个命令的详细信息（如果有）。如果命令是 greet，则会调用 do_greet() 来处理。

```
(Cmd) greet
hello
```

对应一个命令，如果类中没有包含特定的命令处理器，则会调用方法 default()，并以整个输入行作为参数。default() 的内置实现会报告一个错误。

```
(Cmd) foo
*** Unknown syntax: foo
```

由于 do_EOF() 返回 True，键入 Ctrl-D 会使解释器退出。

```
(Cmd) ^D$
```

退出时不会打印换行，所以结果看上去有些乱。

14.5.2 命令参数

下面这个例子做了一些改进来消除存在的一些问题，并为 greet 命令增加帮助。

代码清单 14-45：cmd_arguments.py

```python
import cmd


class HelloWorld(cmd.Cmd):

    def do_greet(self, person):
        """greet [person]
        Greet the named person"""
        if person:
            print("hi,", person)
        else:
            print('hi')

    def do_EOF(self, line):
        return True
```

```
    def postloop(self):
        print()

if __name__ == '__main__':
    HelloWorld().cmdloop()
```

增加到 `do_greet()` 的 docstring 会成为这个命令的帮助文本。

```
$ python3 cmd_arguments.py

(Cmd) help

Documented commands (type help <topic>):
========================================
greet  help

Undocumented commands:
======================
EOF

(Cmd) help greet
greet [person]
        Greet the named person
```

输出显示了 `greet` 的一个可选参数 `person`。尽管这个参数对命令来说是可选的,但是命令和回调方法之间有一个区别。方法总是有参数,不过有时这个值是一个空串。要由命令处理器负责确定空参数是否合法,或者是否要对命令做进一步的解析和处理。在这个例子中,如果提供了一个人名,那么便会提供个性化的欢迎词。

```
(Cmd) greet Alice
hi, Alice
(Cmd) greet
hi
```

不论用户是否指定参数,传递到命令处理器的值都不会包含命令本身。这会简化命令处理器中的解析,特别是如果需要多个参数。

14.5.3 现场帮助

在前面的例子中,帮助文本的格式化还需要有所改进。由于它来自 docstring,所以保留了源文件中的缩进。可以修改源文件,删除多余的空白符,不过这会使应用代码的格式看起来很糟糕。更好的解决方案是为 `greet` 命令实现一个帮助处理器,名为 `help_greet()`。将调用这个帮助处理器为指定的命令生成帮助文本。

代码清单 14-46:**cmd_do_help.py**

```
# Set up gnureadline as readline if installed.
try:
    import gnureadline
    import sys
    sys.modules['readline'] = gnureadline
except ImportError:
    pass

import cmd
```

```python
class HelloWorld(cmd.Cmd):

    def do_greet(self, person):
        if person:
            print("hi,", person)
        else:
            print('hi')

    def help_greet(self):
        print('\n'.join([
            'greet [person]',
            'Greet the named person',
        ]))

    def do_EOF(self, line):
        return True

if __name__ == '__main__':
    HelloWorld().cmdloop()
```

在这个例子中，文本是静态的，不过格式更美观。还可以使用前面的命令状态将帮助文档的内容调整到当前上下文。

```
$ python3 cmd_do_help.py

(Cmd) help greet
greet [person]
Greet the named person
```

要由帮助处理器具体输出帮助消息，而不只是返回帮助文本以便在其他地方处理。

14.5.4 自动完成

Cmd 利用处理器方法支持基于命令名的命令完成。用户在输入提示符处按下 tab 键就会触发这个完成功能。可能有多个完成候选项时，按下两次 tab 会显示一个选项列表。

> 说明：默认地，并不是所有平台上都提供 readline 所需的 GNU 库。在这些情况下，可能不具备 tab 完成功能。如果你的 Python 安装没有这些库，那么可以参见 readline 来了解安装必要的库的有关提示。

```
$ python3 cmd_do_help.py

(Cmd) <tab><tab>
EOF    greet   help
(Cmd) h<tab>
(Cmd) help
```

命令已知时，可以由带 complete_ 前缀的方法来处理参数完成。这就允许新的完成处理器使用任意的规则来组装一个完成候选项列表（也就是说，可以查询数据库，或者查看文件系统上的一个文件或目录）。在这里，程序硬编码编写了一个"朋友"集合，与一些有名字或匿名的陌生人相比，这些朋友会受到更亲热的欢迎。实际的程序可能会在某个地方保存这个列表，读取一次后将内容缓存，以便在需要时查看。

代码清单14-47：cmd_arg_completion.py

```
# Set up gnureadline as readline if installed.
try:
    import gnureadline
    import sys
    sys.modules['readline'] = gnureadline
except ImportError:
    pass

import cmd

class HelloWorld(cmd.Cmd):

    FRIENDS = ['Alice', 'Adam', 'Barbara', 'Bob']

    def do_greet(self, person):
        "Greet the person"
        if person and person in self.FRIENDS:
            greeting = 'hi, {}!'.format(person)
        elif person:
            greeting = 'hello, {}'.format(person)
        else:
            greeting = 'hello'
        print(greeting)

    def complete_greet(self, text, line, begidx, endidx):
        if not text:
            completions = self.FRIENDS[:]
        else:
            completions = [
                f
                for f in self.FRIENDS
                if f.startswith(text)
            ]
        return completions

    def do_EOF(self, line):
        return True

if __name__ == '__main__':
    HelloWorld().cmdloop()
```

如果有输入文本，那么 `complete_greet()` 会返回一个与输入匹配的朋友列表。否则，返回整个朋友列表。

```
$ python3 cmd_arg_completion.py

(Cmd) greet <tab><tab>
Adam     Alice    Barbara  Bob
(Cmd) greet A<tab><tab>
Adam     Alice
(Cmd) greet Ad<tab>
(Cmd) greet Adam
hi, Adam!
```

如果给定的名字没有出现在朋友列表中，那么会给出正式的欢迎词。

```
(Cmd) greet Joe
hello, Joe
```

14.5.5 覆盖基类方法

Cmd 包括的很多方法可以被覆盖为 hook，来采取动作或改变基类行为。下面这个例子并不详尽，不过其中包含很多经常用到的很有用的方法。

代码清单 14-48：**cmd_illustrate_methods.py**

```python
# Set up gnureadline as readline if installed.
try:
    import gnureadline
    import sys
    sys.modules['readline'] = gnureadline
except ImportError:
    pass

import cmd


class Illustrate(cmd.Cmd):
    "Illustrate the base class method use."

    def cmdloop(self, intro=None):
        print('cmdloop({})'.format(intro))
        return cmd.Cmd.cmdloop(self, intro)

    def preloop(self):
        print('preloop()')

    def postloop(self):
        print('postloop()')

    def parseline(self, line):
        print('parseline({!r}) =>'.format(line), end='')
        ret = cmd.Cmd.parseline(self, line)
        print(ret)
        return ret

    def onecmd(self, s):
        print('onecmd({})'.format(s))
        return cmd.Cmd.onecmd(self, s)

    def emptyline(self):
        print('emptyline()')
        return cmd.Cmd.emptyline(self)

    def default(self, line):
        print('default({})'.format(line))
        return cmd.Cmd.default(self, line)

    def precmd(self, line):
        print('precmd({})'.format(line))
        return cmd.Cmd.precmd(self, line)

    def postcmd(self, stop, line):
        print('postcmd({}, {})'.format(stop, line))
        return cmd.Cmd.postcmd(self, stop, line)

    def do_greet(self, line):
        print('hello,', line)

    def do_EOF(self, line):
```

```
            "Exit"
            return True

if __name__ == '__main__':
    Illustrate().cmdloop('Illustrating the methods of cmd.Cmd')
```

`cmdloop()`是解释器的主处理循环。通常没有必要覆盖这个循环,因为可以使用`preloop()`和`postloop()` hook。

每次`cmdloop()`迭代都会调用`onecmd()`,将命令分派到它的处理器。实际输入行用`parseline()`解析来创建一个元组,其中包含命令和该行上的其余部分。

如果这一行为空,则调用`emptyline()`。默认实现会再次运行前面的命令。如果这一行包含一个命令,那么首先调用`precmd()`,然后查看并调用处理器。如果没有找到,则调用`default()`。最后调用`postcmd()`。

以下输出显示增加了`print`语句的一个示例会话。

```
$ python3 cmd_illustrate_methods.py

cmdloop(Illustrating the methods of cmd.Cmd)
preloop()
Illustrating the methods of cmd.Cmd
(Cmd) greet Bob
precmd(greet Bob)
onecmd(greet Bob)
parseline(greet Bob) => ('greet', 'Bob', 'greet Bob')
hello, Bob
postcmd(None, greet Bob)
(Cmd) ^Dprecmd(EOF)
onecmd(EOF)
parseline(EOF) => ('EOF', '', 'EOF')
postcmd(True, EOF)
postloop()
```

14.5.6 通过属性配置 Cmd

除了前面描述的方法,还有很多属性可以用来控制命令解释器。可以把`prompt`设置为一个字符串,每次要求用户输入一个新命令时就会显示这个字符串。`intro`是程序开始时打印的"欢迎"消息。`cmdloop()`取对应这个值的一个参数,或者也可以在类上直接设置。打印帮助时,可以用`doc_header`、`misc_header`、`undoc_header`和`ruler`属性来格式化输出。

代码清单14-49:`cmd_attributes.py`

```
import cmd

class HelloWorld(cmd.Cmd):

    prompt = 'prompt: '
    intro = "Simple command processor example."

    doc_header = 'doc_header'
    misc_header = 'misc_header'
    undoc_header = 'undoc_header'
```

```
        ruler = '-'

    def do_prompt(self, line):
        "Change the interactive prompt"
        self.prompt = line + ': '

    def do_EOF(self, line):
        return True

if __name__ == '__main__':
    HelloWorld().cmdloop()
```

这个示例类显示了一个命令处理器,允许用户控制交互式会话的提示符。

```
$ python3 cmd_attributes.py

Simple command processor example.
prompt: prompt hello
hello: help

doc_header
----------
help   prompt

undoc_header
------------
EOF

hello:
```

14.5.7 运行 shell 命令

作为对标准命令处理的补充,Cmd 包括了两个特殊的命令前缀。问号(?)等价于内置的 help 命令,可以用同样的方式使用。感叹号(!)对应 do_shell(),它要"作为外壳"运行其他命令,就像这个例子中一样。

代码清单 14-50:cmd_do_shell.py

```python
import cmd
import subprocess


class ShellEnabled(cmd.Cmd):

    last_output = ''

    def do_shell(self, line):
        "Run a shell command"
        print("running shell command:", line)
        sub_cmd = subprocess.Popen(line,
                                   shell=True,
                                   stdout=subprocess.PIPE)
        output = sub_cmd.communicate()[0].decode('utf-8')
        print(output)
        self.last_output = output

    def do_echo(self, line):
        """Print the input, replacing '$out' with
        the output of the last shell command
        """
```

```
        # Obviously not robust
        print(line.replace('$out', self.last_output))

    def do_EOF(self, line):
        return True
if __name__ == '__main__':
    ShellEnabled().cmdloop()
```

这个 echo 命令实现将其参数中的串 $out 替换为前面 shell 命令的输出。

```
$ python3 cmd_do_shell.py

(Cmd) ?

Documented commands (type help <topic>):
========================================
echo  help  shell

Undocumented commands:
======================
EOF

(Cmd) ? shell
Run a shell command
(Cmd) ? echo
Print the input, replacing '$out' with
       the output of the last shell command
(Cmd) shell pwd
running shell command: pwd
.../pymotw-3/source/cmd

(Cmd) ! pwd
running shell command: pwd
.../pymotw-3/source/cmd

(Cmd) echo $out
.../pymotw-3/source/cmd
```

14.5.8 候选输入

Cmd() 的默认模式是通过 readline 库与用户交互，不过也可以使用标准 UNIX shell 重定向来为标准输入传递一系列命令。

```
$ echo help | python3 cmd_do_help.py

(Cmd)
Documented commands (type help <topic>):
========================================
greet  help

Undocumented commands:
======================
EOF

(Cmd)
```

要让程序直接读取一个脚本文件，可能还需要另外一些修改。因为 readline 与 terminal/tty 设备交互，而不是与标准输入流交互，从文件读取脚本时应当将其禁用。另外，为了避免打印多余的提示符，可以把提示符设置为一个空串。下面这个例子显示了如何打开一个文件,并将其作为输入传递到 HelloWorld 例子的一个修改版本。

代码清单 14-51:cmd_file.py

```python
import cmd


class HelloWorld(cmd.Cmd):

    # Disable rawinput module use.
    use_rawinput = False

    # Do not show a prompt after each command read.
    prompt = ''

    def do_greet(self, line):
        print("hello,", line)

    def do_EOF(self, line):
        return True

if __name__ == '__main__':
    import sys
    with open(sys.argv[1], 'rt') as input:
        HelloWorld(stdin=input).cmdloop()
```

将 `use_rawinput` 设置为 `False`,`prompt` 设置为一个空串,现在可以对输入文件(每行上有一个命令)调用这个脚本。

代码清单 14-52:cmd_file.txt

```
greet
greet Alice and Bob
```

对示例输入运行这个示例脚本会生成以下输出。

```
$ python3 cmd_file.py cmd_file.txt

hello,
hello, Alice and Bob
```

14.5.9 `sys.argv` 的命令

也可以将程序的命令行参数处理为命令,提供给解释器类,而不是从控制台或文件读取命令。要使用命令行参数,可以直接调用 `onecmd()`,如下例所示。

代码清单 14-53:cmd_argv.py

```python
import cmd


class InteractiveOrCommandLine(cmd.Cmd):
    """Accepts commands via the normal interactive
    prompt or on the command line.
    """

    def do_greet(self, line):
        print('hello,', line)
```

```
        def do_EOF(self, line):
            return True

if __name__ == '__main__':
    import sys
    if len(sys.argv) > 1:
        InteractiveOrCommandLine().onecmd(' '.join(sys.argv[1:]))
    else:
        InteractiveOrCommandLine().cmdloop()
```

由于 onecmd() 取一个字符串作为输入，所以在参数传入之前，需要把程序的参数连接起来。

```
$ python3 cmd_argv.py greet Command-Line User

hello, Command-Line User

$ python3 cmd_argv.py

(Cmd) greet Interactive User
hello, Interactive User
(Cmd)
```

提示：相关阅读材料
- cmd 的标准库文档[一]。
- cmd2[二]：直接替换 cmd，提供了额外的特性。
- GNU readline[三]：GNU readline 库提供了一些函数，允许用户在键入时编辑输入行。
- readline：GNU readline 的 Python 标准库接口。
- subprocess：用于管理其他进程及其输出。

14.6 shlex：解析 shell 类语法

shlex 模块实现了一个类来解析简单的类 shell 语法，其可以用来编写领域特定的语言，或者解析加引号的字符串（这个任务没有表面看起来那么简单）。

14.6.1 解析加引号的字符串

处理输入文本时有一个常见的问题，往往要把一个加引号的单词序列识别为一个实体。根据引号划分文本可能与预想的并不一样，特别是在嵌套有多层引号时。以下面的文本为例。

```
This string has embedded "double quotes" and
'single quotes' in it, and even "a 'nested example'".
```

一种简单的方法是构造一个正则表达式，以查找引号之外的文本部分，并且将它们与引号内的文本分开，或者反之。这个正则表达式可能带来不必要的复杂性，而且很容易

[一] https://docs.python.org/3.5/library/cmd.html
[二] http://pypi.python.org/pypi/cmd2
[三] http://tiswww.case.edu/php/chet/readline/rltop.html

因为边界条件出错，如撇号或者拼写错误。更好的解决方案是使用一个真正的解析器，如 shlex 模块提供的解析器。以下是一个简单的例子，它使用 shlex 类来打印输入文件中找到的 token。

代码清单 14-54：**shlex_example.py**

```
import shlex
import sys

if len(sys.argv) != 2:
    print('Please specify one filename on the command line.')
    sys.exit(1)

filename = sys.argv[1]
with open(filename, 'r') as f:
    body = f.read()
print('ORIGINAL: {!r}'.format(body))
print()

print('TOKENS:')
lexer = shlex.shlex(body)
for token in lexer:
    print('{!r}'.format(token))
```

如果数据中包含嵌入的引号，那么以这样的数据作为输入运行这个解析器时，会得到期望的 token 列表。

```
$ python3 shlex_example.py quotes.txt

ORIGINAL: 'This string has embedded "double quotes" and\n\'singl
e quotes\' in it, and even "a \'nested example\'".\n'

TOKENS:
'This'
'string'
'has'
'embedded'
'"double quotes"'
'and'
"'single quotes'"
'in'
'it'
','
'and'
'even'
'"a \'nested example\'"'
'.'
```

这个解析器还能正确地处理孤立的引号（如撇号）。考虑以下输入文件：

```
This string has an embedded apostrophe, doesn't it?
```

完全可以找出包含嵌入撇号的 token。

```
$ python3 shlex_example.py apostrophe.txt

ORIGINAL: "This string has an embedded apostrophe, doesn't it?"

TOKENS:
'This'
```

```
'string'
'has'
'an'
'embedded'
'apostrophe'
','
"doesn't"
'it'
'?'
```

14.6.2　为 shell 建立安全的字符串

quote() 函数可以完成逆向的操作，对现有的引号转义，并为字符串增加缺少的引号，使它们能够安全地用在 shell 命令中。

代码清单 14-55：**shlex_quote.py**

```
import shlex

examples = [
    "Embedded'SingleQuote",
    'Embedded"DoubleQuote',
    'Embedded Space',
    '~SpecialCharacter',
    r'Back\slash',
]

for s in examples:
    print('ORIGINAL : {}'.format(s))
    print('QUOTED   : {}'.format(shlex.quote(s)))
    print()
```

通常更安全的做法是在用 **subprocess.Popen** 时使用一个参数列表。不过，如果无法做到这一点，那么在这种情况下，**quote()** 可以通过确保对特殊字符和空白符正确地加引号来提供一些保护。

```
$ python3 shlex_quote.py

ORIGINAL : Embedded'SingleQuote
QUOTED   : 'Embedded'"'"'SingleQuote'

ORIGINAL : Embedded"DoubleQuote
QUOTED   : 'Embedded"DoubleQuote'

ORIGINAL : Embedded Space
QUOTED   : 'Embedded Space'

ORIGINAL : ~SpecialCharacter
QUOTED   : '~SpecialCharacter'

ORIGINAL : Back\slash
QUOTED   : 'Back\slash'
```

14.6.3　嵌入注释

由于解析器要用于命令语言，所以还需要处理注释。默认地，# 后面的文本会被认为是注释的一部分，并被忽略。由于解析器的特点，它只支持单字符注释前缀。可以通过

commenters 属性配置所使用的注释字符集。

```
$ python3 shlex_example.py comments.txt

ORIGINAL: 'This line is recognized.\n# But this line is ignored.
\nAnd this line is processed.'

TOKENS:
'This'
'line'
'is'
'recognized'
'.'
'And'
'this'
'line'
'is'
'processed'
'.'
```

14.6.4 将字符串分解为 token

函数 `split()` 是解析器的一个简单包装器，可以用来把一个现有的字符串分解为其组成 token。

代码清单 14-56：`shlex_split.py`

```python
import shlex

text = """This text has "quoted parts" inside it."""
print('ORIGINAL: {!r}'.format(text))
print()

print('TOKENS:')
print(shlex.split(text))
```

结果是一个列表。

```
$ python3 shlex_split.py

ORIGINAL: 'This text has "quoted parts" inside it.'

TOKENS:
['This', 'text', 'has', 'quoted parts', 'inside', 'it.']
```

14.6.5 包含其他 token 源

`shlex` 类包括很多配置属性来控制其行为。`source` 属性支持代码（或配置）重用，允许一个 token 流包含另一个 token 流。这个特性类似于 Bourne shell 的 `source` 操作符——其也因此得名。

代码清单 14-57：`shlex_source.py`

```python
import shlex

text = "This text says to source quotes.txt before continuing."
print('ORIGINAL: {!r}'.format(text))
print()
```

```
lexer = shlex.shlex(text)
lexer.wordchars += '.'
lexer.source = 'source'

print('TOKENS:')
for token in lexer:
    print('{!r}'.format(token))
```

原文本中的字符串 `source quotes.txt` 会得到特殊处理。由于 lexer 的 `source` 属性被设置为 `"source"`，所以遇到这个关键字时，会自动包含下一行上出现的文件名。为了让文件名作为单个 token 出现，需要在单词包含的字符列表中增加 `.` 字符；否则 `quotes.txt` 会变成 3 个 token：`quotes`、`.` 和 `txt`。输出如下。

```
$ python3 shlex_source.py

ORIGINAL: 'This text says to source quotes.txt before
continuing.'

TOKENS:
'This'
'text'
'says'
'to'
'This'
'string'
'has'
'embedded'
'"double quotes"'
'and'
'"\'single quotes\'"'
'in'
'it'
','
'and'
'even'
'"a \\\'nested example\\\'"'
'.'
'before'
'continuing.'
```

`source` 特性使用了一个名为 `sourcehook()` 的方法来加载额外的输入源，所以 shlex 的子类可以提供一个候选实现，从不是文件的其他位置加载数据。

14.6.6 控制解析器

前面的例子展示了可以通过改变 `wordchars` 值来控制单词中包含哪些字符。还可以设置 `quotes` 字符以使用额外或替代引号。每个引号必须是单个字符，所以不可能有不同的开始和结束引号（例如，根据括号解析就是不允许的）。

代码清单 14-58：**shlex_table.py**

```
import shlex

text = """|Col 1||Col 2||Col 3|"""
print('ORIGINAL: {!r}'.format(text))
print()
```

```
lexer = shlex.shlex(text)
lexer.quotes = '|'

print('TOKENS:')
for token in lexer:
    print('{!r}'.format(token))
```

在这个例子中,每个表单元格都用竖线包围。

```
$ python3 shlex_table.py

ORIGINAL: '|Col 1||Col 2||Col 3|'

TOKENS:
'|Col 1|'
'|Col 2|'
'|Col 3|'
```

还可以控制用来分解单词的空白符。

代码清单 14-59:**shlex_whitespace.py**

```
import shlex
import sys

if len(sys.argv) != 2:
    print('Please specify one filename on the command line.')
    sys.exit(1)

filename = sys.argv[1]
with open(filename, 'r') as f:
    body = f.read()
print('ORIGINAL: {!r}'.format(body))
print()

print('TOKENS:')
lexer = shlex.shlex(body)
lexer.whitespace += '.,'
for token in lexer:
    print('{!r}'.format(token))
```

现在,如果 shlex_example.py 中的例子被修改为包含点号和逗号,那么结果也会改变。

```
$ python3 shlex_whitespace.py quotes.txt

ORIGINAL: 'This string has embedded "double quotes" and\n\'singl
e quotes\' in it, and even "a \'nested example\'".\n'

TOKENS:
'This'
'string'
'has'
'embedded'
'"double quotes"'
'and'
"'single quotes'"
'in'
'it'
'and'
```

```
'even'
'"a \'nested example\'"'
```

14.6.7 错误处理

在有引号的串结束之前,如果解析器提前遇到输入末尾,则会产生 `ValueError`。出现这种情况时,可以检查解析器处理输入时维护的一些属性,这很有用。例如,`infile` 指示所处理的文件的名(如果一个文件用 source 包含另一个文件,则可能与原文件不同)。`lineno` 会报告发现错误时正在处理的文本行。`lineno` 通常在文件末尾,这可能与第一个引号相距很远。`token` 属性包含尚未包括在一个合法 token 中的文本缓冲区。`error_leader()` 方法会用类似 UNIX 编译器的方式生成一个消息前缀,这将启用编辑器(如 `emacs`)解析错误,并直接把用户带到有问题的那一行。

代码清单 14-60:**shlex_errors.py**

```
import shlex

text = """This line is OK.
This line has an "unfinished quote.
This line is OK, too.
"""

print('ORIGINAL: {!r}'.format(text))
print()

lexer = shlex.shlex(text)

print('TOKENS:')
try:
    for token in lexer:
        print('{!r}'.format(token))
except ValueError as err:
    first_line_of_error = lexer.token.splitlines()[0]
    print('ERROR: {} {}'.format(lexer.error_leader(), err))
    print('following {!r}'.format(first_line_of_error))
```

这个例子会生成以下输出。

```
$ python3 shlex_errors.py

ORIGINAL: 'This line is OK.\nThis line has an "unfinished quote.
\nThis line is OK, too.\n'

TOKENS:
'This'
'line'
'is'
'OK'
'.'
'This'
'line'
'has'
'an'
ERROR: "None", line 4:  No closing quotation
following '"unfinished quote.'
```

14.6.8 POSIX 与非 POSIX 解析

解析器的默认行为是使用一种向后兼容方式，这不符合 POSIX。对于 POSIX 行为，构造解析器时要设置 `posix` 参数。

代码清单 14-61：`shlex_posix.py`

```
import shlex

examples = [
    'Do"Not"Separate',
    '"Do"Separate',
    'Escaped \e Character not in quotes',
    'Escaped "\e" Character in double quotes',
    "Escaped '\e' Character in single quotes",
    r"Escaped '\'' \"\'\" single quote",
    r'Escaped "\"" \'\"\' double quote',
    "\"'Strip extra layer of quotes'\"",
]

for s in examples:
    print('ORIGINAL : {!r}'.format(s))
    print('non-POSIX: ', end='')

    non_posix_lexer = shlex.shlex(s, posix=False)
    try:
        print('{!r}'.format(list(non_posix_lexer)))
    except ValueError as err:
        print('error({})'.format(err))

    print('POSIX    : ', end='')
    posix_lexer = shlex.shlex(s, posix=True)
    try:
        print('{!r}'.format(list(posix_lexer)))
    except ValueError as err:
        print('error({})'.format(err))

    print()
```

下面几个例子展示了解析行为的差别。

```
$ python3 shlex_posix.py

ORIGINAL : 'Do"Not"Separate'
non-POSIX: ['Do"Not"Separate']
POSIX    : ['DoNotSeparate']

ORIGINAL : '"Do"Separate'
non-POSIX: ['"Do"', 'Separate']
POSIX    : ['DoSeparate']

ORIGINAL : 'Escaped \\e Character not in quotes'
non-POSIX: ['Escaped', '\\', 'e', 'Character', 'not', 'in',
'quotes']
POSIX    : ['Escaped', 'e', 'Character', 'not', 'in', 'quotes']

ORIGINAL : 'Escaped "\\e" Character in double quotes'
non-POSIX: ['Escaped', '"\\e"', 'Character', 'in', 'double',
'quotes']
POSIX    : ['Escaped', '\\e', 'Character', 'in', 'double',
'quotes']
```

```
ORIGINAL : "Escaped '\\e' Character in single quotes"
non-POSIX: ['Escaped', "'\\e'", 'Character', 'in', 'single',
           'quotes']
POSIX    : ['Escaped', '\\e', 'Character', 'in', 'single',
           'quotes']

ORIGINAL : 'Escaped \'\\\'\' \\"\\\'\\" single quote'
non-POSIX: error(No closing quotation)
POSIX    : ['Escaped', '\\ \\"\\"', 'single', 'quote']

ORIGINAL : 'Escaped "\\"" \\\'\\"\\\' double quote'
non-POSIX: error(No closing quotation)
POSIX    : ['Escaped', '"', '\'"\'', 'double', 'quote']

ORIGINAL : '"\'Strip extra layer of quotes\'"'
non-POSIX: ['"\'Strip extra layer of quotes\'"']
POSIX    : ["'Strip extra layer of quotes'"]
```

提示：相关阅读材料
- `shlex` 的标准库文档○。
- `cmd`：构建交互式命令解释器的工具。
- `argparse`：命令行选项解析。
- `subprocess`：解析命令行后运行命令。

14.7 configparser：处理配置文件

使用 `configparser` 模块可以为应用管理用户可编辑的配置文件，这里使用一种与 Windows INI 文件类似的格式。配置文件的内容可以被组织为组，并支持多个选项值类型，包括整数、浮点值和布尔值。可以使用 Python 格式化字符串组合选项值来建立更长的值，如由主机名和端口号等较短的值构造 URL。

14.7.1 配置文件格式

`configparser` 使用的文件格式类似于 Microsoft Windows 较早版本使用的格式。它由一个或多个命名的节（section）组成，每一节包含由名和值构成的选项（option）。

解析器通过查找以"["开头并以"]"结束的行来标识配置文件的节。中括号之间的值是节名，其中可以包含除中括号以外的任何字符。

在一节中，每行列出一个选项。行以选项名开头，选项名与值之间用一个冒号（:）或一个等号（=）分隔。解析文件时，分隔符两边的空白符会被忽略。

以一个分号（;）或井字符（#）开头的行被看作是注释。通过程序访问配置文件的内容时，这些行将被忽略。下面的示例配置文件有一个名为 `bug_tracker` 的节，其中包含 3 个选项：`url`、`username` 和 `password`。

```
# This is a simple example with comments.
[bug_tracker]
```

○ https://docs.python.org/3.5/library/shlex.html

```
url = http://localhost:8080/bugs/
username = dhellmann
; You should not store passwords in plain text
; configuration files.
password = SECRET
```

14.7.2 读取配置文件

用户或系统管理员通常用一个常规的文本编辑器编辑文件，以设置默认的应用行为，然后让应用读取这个文件，进行解析，并根据其内容采取动作。可以使用 ConfigParser 的 read() 方法来读取配置文件。

代码清单 14-62：**configparser_read.py**

```python
from configparser import ConfigParser

parser = ConfigParser()
parser.read('simple.ini')

print(parser.get('bug_tracker', 'url'))
```

这个程序读取上一节的 simple.ini 文件，并打印 bug_tracker 节中 url 选项的值。

```
$ python3 configparser_read.py

http://localhost:8080/bugs/
```

read() 方法还接受一个文件名列表。依次检查这个列表中的各个名，如果文件存在，就打开并读取该文件。

代码清单 14-63：**configparser_read_many.py**

```python
from configparser import ConfigParser
import glob

parser = ConfigParser()

candidates = ['does_not_exist.ini', 'also-does-not-exist.ini',
              'simple.ini', 'multisection.ini']

found = parser.read(candidates)

missing = set(candidates) - set(found)

print('Found config files:', sorted(found))
print('Missing files     :', sorted(missing))
```

read() 返回一个列表，其中包含成功加载的文件的名。通过检查这个列表，程序可以发现缺少哪些配置文件，并确定是将其忽略还是把这个条件当作一个错误。

```
$ python3 configparser_read_many.py

Found config files: ['multisection.ini', 'simple.ini']
Missing files     : ['also-does-not-exist.ini',
'does_not_exist.ini']
```

Unicode 配置数据

包含 Unicode 数据的配置文件应当使用适当的编码值来读取。下面的示例文件将原输入的密码值改为包含 Unicode 字符，并使用 UTF-8 编码。

代码清单 14-64：**unicode.ini**

```
[bug_tracker]
url = http://localhost:8080/bugs/
username = dhellmann
password = †ßéç®é
```

可以用适当的解析器打开这个文件，将 UTF-8 数据转换为原生 Unicode 串。

代码清单 14-65：**configparser_unicode.py**

```python
from configparser import ConfigParser
import codecs

parser = ConfigParser()
# Open the file with the correct encoding.
parser.read('unicode.ini', encoding='utf-8')

password = parser.get('bug_tracker', 'password')

print('Password:', password.encode('utf-8'))
print('Type    :', type(password))
print('repr()  :', repr(password))
```

`get()` 返回的值是一个 Unicode 串。为了安全地打印这个串，必须将它重新编码为 UTF-8。

```
$ python3 configparser_unicode.py

Password: b'\xc3\x9f\xc3\xa9\xc3\xa7\xc2\xae\xc3\xa9\xe2\x80\xa0'
Type    : <class 'str'>
repr()  : '†ßéç®é'
```

14.7.3　访问配置设置

`ConfigParser` 包含一些方法来检查所解析配置的结构，包括列出节和选项，以及得到它们的值。下面这个配置文件包含两个节，分别对应不同的 Web 服务。

```
[bug_tracker]
url = http://localhost:8080/bugs/
username = dhellmann
password = SECRET

[wiki]
url = http://localhost:8080/wiki/
username = dhellmann
password = SECRET
```

下面这个示例程序使用了一些方法来查看配置数据，包括 `sections()`、`options()` 和 `items()`。

代码清单 14-66：configparser_structure.py

```
from configparser import ConfigParser

parser = ConfigParser()
parser.read('multisection.ini')

for section_name in parser.sections():
    print('Section:', section_name)
    print('  Options:', parser.options(section_name))
    for name, value in parser.items(section_name):
        print('  {} = {}'.format(name, value))
    print()
```

sections() 和 options() 会返回字符串列表，而 items() 返回一个元组列表，元组包含名–值对。

```
$ python3 configparser_structure.py

Section: bug_tracker
  Options: ['url', 'username', 'password']
  url = http://localhost:8080/bugs/
  username = dhellmann
  password = SECRET

Section: wiki
  Options: ['url', 'username', 'password']
  url = http://localhost:8080/wiki/
  username = dhellmann
  password = SECRET
```

ConfigParser 还支持与 dict 同样的映射 API，ConfigParser 相当于一个字典，其中包含对应各个节的不同字典。

代码清单 14-67：configparser_structure_dict.py

```
from configparser import ConfigParser

parser = ConfigParser()
parser.read('multisection.ini')

for section_name in parser:
    print('Section:', section_name)
    section = parser[section_name]
    print('  Options:', list(section.keys()))
    for name in section:
        print('  {} = {}'.format(name, section[name]))
    print()
```

使用这个映射 API 访问相同的配置文件会生成相同的输出。

```
$ python3 configparser_structure_dict.py

Section: DEFAULT
  Options: []

Section: bug_tracker
  Options: ['url', 'username', 'password']
  url = http://localhost:8080/bugs/
  username = dhellmann
```

```
    password = SECRET
Section: wiki
  Options: ['url', 'username', 'password']
  url = http://localhost:8080/wiki/
  username = dhellmann
  password = SECRET
```

14.7.3.1 测试值是否存在

要测试一个节是否存在,可以使用 `has_section()`,并传入节名作为方法参数。

代码清单 14-68:**configparser_has_section.py**

```python
from configparser import ConfigParser

parser = ConfigParser()
parser.read('multisection.ini')

for candidate in ['wiki', 'bug_tracker', 'dvcs']:
    print('{:<12}: {}'.format(
        candidate, parser.has_section(candidate)))
```

调用 `get()` 之前先测试一个节是否存在,这样可以避免因为缺少数据而导致产生异常。

```
$ python3 configparser_has_section.py

wiki        : True
bug_tracker : True
dvcs        : False
```

使用 `has_option()` 可以测试一个节中某个选项是否存在。

代码清单 14-69:**configparser_has_option.py**

```python
from configparser import ConfigParser

parser = ConfigParser()
parser.read('multisection.ini')

SECTIONS = ['wiki', 'none']
OPTIONS = ['username', 'password', 'url', 'description']

for section in SECTIONS:
    has_section = parser.has_section(section)
    print('{} section exists: {}'.format(section, has_section))
    for candidate in OPTIONS:
        has_option = parser.has_option(section, candidate)
        print('{}.{:<12}  : {}'.format(
            section, candidate, has_option))
    print()
```

如果节不存在,那么 `has_option()` 会返回 `False`。

```
$ python3 configparser_has_option.py

wiki section exists: True
wiki.username    : True
wiki.password    : True
wiki.url         : True
wiki.description : False
```

```
none section exists: False
none.username       : False
none.password       : False
none.url            : False
none.description    : False
```

14.7.3.2 值类型

所有节和选项名都被处理为字符串，不过选项值可以是字符串、整数、浮点数或者布尔值。可以用多个不同的字符串值表示配置文件中的布尔值；访问时它们会被转换为 True 或 False。下面的文件中包含了一些数值类型的例子，另外还包含所有被解析器识别为布尔值的值。

代码清单 14-70：**types.ini**

```
[ints]
positive = 1
negative = -5

[floats]
positive = 0.2
negative = -3.14

[booleans]
number_true = 1
number_false = 0
yn_true = yes
yn_false = no
tf_true = true
tf_false = false
onoff_true = on
onoff_false = false
```

ConfigParser 不会尝试去了解选项类型，而会希望应用使用正确的方法来获取所需类型的值。get() 总会返回一个字符串。使用 getint() 可以得到整数，getfloat() 得到浮点数，使用 getboolean() 得到布尔值。

代码清单 14-71：**configparser_value_types.py**

```python
from configparser import ConfigParser

parser = ConfigParser()
parser.read('types.ini')

print('Integers:')
for name in parser.options('ints'):
    string_value = parser.get('ints', name)
    value = parser.getint('ints', name)
    print('  {:<12} : {!r:<7} -> {}'.format(
        name, string_value, value))

print('\nFloats:')
for name in parser.options('floats'):
    string_value = parser.get('floats', name)
    value = parser.getfloat('floats', name)
    print('  {:<12} : {!r:<7} -> {:0.2f}'.format(
        name, string_value, value))
```

```
print('\nBooleans:')
for name in parser.options('booleans'):
    string_value = parser.get('booleans', name)
    value = parser.getboolean('booleans', name)
    print('  {:<12} : {!r:<7} -> {}'.format(
        name, string_value, value))
```

使用示例输入运行这个程序,可以生成以下结果。

```
$ python3 configparser_value_types.py

Integers:
  positive     : '1'     -> 1
  negative     : '-5'    -> -5
Floats:
  positive     : '0.2'   -> 0.20
  negative     : '-3.14' -> -3.14
Booleans:
  number_true  : '1'     -> True
  number_false : '0'     -> False
  yn_true      : 'yes'   -> True
  yn_false     : 'no'    -> False
  tf_true      : 'true'  -> True
  tf_false     : 'false' -> False
  onoff_true   : 'on'    -> True
  onoff_false  : 'false' -> False
```

可以在 `ConfigParser` 的 `converters` 参数中传入转换函数来增加定制类型转换器。每个转换器接收一个输入值,然后将它转换为适当的返回类型。

代码清单 14-72: **configparser_custom_types.py**

```
from configparser import ConfigParser
import datetime


def parse_iso_datetime(s):
    print('parse_iso_datetime({!r})'.format(s))
    return datetime.datetime.strptime(s, '%Y-%m-%dT%H:%M:%S.%f')


parser = ConfigParser(
    converters={
        'datetime': parse_iso_datetime,
    }
)
parser.read('custom_types.ini')

string_value = parser['datetimes']['due_date']
value = parser.getdatetime('datetimes', 'due_date')
print('due_date : {!r} -> {!r}'.format(string_value, value))
```

增加转换器会让 `ConfigParser` 自动为这个类型创建一个获取方法,并使用 `converters` 中指定的类型名。在这个例子中,`'datetime'` 转换器会让 `ConfigParser` 增加一个新的 `getdatetime()` 方法。

```
$ python3 configparser_custom_types.py

parse_iso_datetime('2015-11-08T11:30:05.905898')
due_date : '2015-11-08T11:30:05.905898' -> datetime.datetime(201
5, 11, 8, 11, 30, 5, 905898)
```

还可以向 `ConfigParser` 的子类直接增加转换器方法。

14.7.3.3 选项作为标志

通常，解析器要求每个选项都有一个明确的值，不过，如果 `ConfigParser` 参数 `allow_no_value` 被设置为 `True`，那么选项可以在输入文件中单独作为一行，而且可以被用作一个标志。

代码清单 14-73：**configparser_allow_no_value.py**

```python
import configparser

# Require values.
try:
    parser = configparser.ConfigParser()
    parser.read('allow_no_value.ini')
except configparser.ParsingError as err:
    print('Could not parse:', err)

# Allow stand-alone option names.
print('\nTrying again with allow_no_value=True')
parser = configparser.ConfigParser(allow_no_value=True)
parser.read('allow_no_value.ini')
for flag in ['turn_feature_on', 'turn_other_feature_on']:
    print('\n', flag)
    exists = parser.has_option('flags', flag)
    print('  has_option:', exists)
    if exists:
        print('        get:', parser.get('flags', flag))
```

选项没有明确的值时，`has_option()` 会报告这个选项存在，并且 `get()` 返回 `None`。

```
$ python3 configparser_allow_no_value.py

Could not parse: Source contains parsing errors:
'allow_no_value.ini'
        [line  2]: 'turn_feature_on\n'

Trying again with allow_no_value=True

 turn_feature_on
  has_option: True
        get: None

 turn_other_feature_on
  has_option: False
```

14.7.3.4 多行字符串

字符串值可以跨多行，前提是后面的行要缩进。

```
[example]
message = This is a multiline string.
  With two paragraphs.

  They are separated by a completely empty line.
```

在缩进的多行值中，空行会作为值的一部分保留。

```
$ python3 configparser_multiline.py

This is a multiline string.
With two paragraphs.

They are separated by a completely empty line.
```

14.7.4 修改设置

ConfigParser 主要通过从文件读取设置来进行配置，不过也可以填充设置，通过调用 add_section() 来创建一个新的节，另外调用 set() 可以增加或修改一个选项。

代码清单 14-74：**configparser_populate.py**

```
import configparser

parser = configparser.SafeConfigParser()

parser.add_section('bug_tracker')
parser.set('bug_tracker', 'url', 'http://localhost:8080/bugs')
parser.set('bug_tracker', 'username', 'dhellmann')
parser.set('bug_tracker', 'password', 'secret')

for section in parser.sections():
    print(section)
    for name, value in parser.items(section):
        print('  {} = {!r}'.format(name, value))
```

所有选项都必须被设置为字符串，即使它们将被获取为整数、浮点数或布尔值。

```
$ python3 configparser_populate.py

bug_tracker
  url = 'http://localhost:8080/bugs'
  username = 'dhellmann'
  password = 'secret'
```

可以分别用 remove_section() 和 remove_option() 从 ConfigParser 删除节和选项。

代码清单 14-75：**configparser_remove.py**

```
from configparser import ConfigParser

parser = ConfigParser()
parser.read('multisection.ini')

print('Read values:\n')
for section in parser.sections():
    print(section)
    for name, value in parser.items(section):
        print('  {} = {!r}'.format(name, value))

parser.remove_option('bug_tracker', 'password')
parser.remove_section('wiki')

print('\nModified values:\n')
```

```
for section in parser.sections():
    print(section)
    for name, value in parser.items(section):
        print('  {} = {!r}'.format(name, value))
```

删除一节也会删除其中包含的所有选项。

```
$ python3 configparser_remove.py

Read values:

bug_tracker
  url = 'http://localhost:8080/bugs/'
  username = 'dhellmann'
  password = 'SECRET'
wiki
  url = 'http://localhost:8080/wiki/'
  username = 'dhellmann'
  password = 'SECRET'

Modified values:

bug_tracker
  url = 'http://localhost:8080/bugs/'
  username = 'dhellmann'
```

14.7.5 保存配置文件

用所需的数据填充 ConfigParser 后，就可以调用 write() 方法将它保存到一个文件。这种方法可以用来提供一个用于编辑配置设置的用户接口，而不需要编写任何代码来管理文件。

代码清单 14-76：**configparser_write.py**

```
import configparser
import sys

parser = configparser.ConfigParser()

parser.add_section('bug_tracker')
parser.set('bug_tracker', 'url', 'http://localhost:8080/bugs')
parser.set('bug_tracker', 'username', 'dhellmann')
parser.set('bug_tracker', 'password', 'secret')

parser.write(sys.stdout)
```

write() 方法取一个类似文件的对象作为参数。它采用 INI 格式写出数据，以便再由 ConfigParser 解析。

```
$ python3 configparser_write.py

[bug_tracker]
url = http://localhost:8080/bugs
username = dhellmann
password = secret
```

警告：读取、修改和重写配置文件时，原配置文件中的注释不会保留。

14.7.6 选项搜索路径

ConfigParser 查找选项时使用了一个多步搜索过程。开始搜索选项之前,首先会测试节名。如果这个节不存在,而且名不是特殊值 DEFAULT,则产生一个 NoSectionError 异常。

1. 如果选项名出现在传递到 get() 的 vars 字典中,则会返回 vars 的值。
2. 如果选项名出现在指定的节中,则返回该节中的值。
3. 如果选项名出现在 DEFAULT 节中,则会返回相应的值。
4. 如果选项名出现在传递到构造函数的 defaults 字典中,则会返回相应的值。

如果这个名未出现在以上任何位置,则产生 NoOptionError。

可以用以下配置文件来展示这个搜索路径行为。

```
[DEFAULT]
file-only = value from DEFAULT section
init-and-file = value from DEFAULT section
from-section = value from DEFAULT section
from-vars = value from DEFAULT section

[sect]
section-only = value from section in file
from-section = value from section in file
from-vars = value from section in file
```

以下代码清单中的测试程序包括配置文件中未指定的一些默认选项设置,并覆盖了文件中定义的一些值。

代码清单 14-77: **configparser_defaults.py**

```python
import configparser

# Define the names of the options.
option_names = [
    'from-default',
    'from-section', 'section-only',
    'file-only', 'init-only', 'init-and-file',
    'from-vars',
]

# Initialize the parser with some defaults.
DEFAULTS = {
    'from-default': 'value from defaults passed to init',
    'init-only': 'value from defaults passed to init',
    'init-and-file': 'value from defaults passed to init',
    'from-section': 'value from defaults passed to init',
    'from-vars': 'value from defaults passed to init',
}
parser = configparser.ConfigParser(defaults=DEFAULTS)

print('Defaults before loading file:')
defaults = parser.defaults()
for name in option_names:
    if name in defaults:
        print('  {:<15} = {!r}'.format(name, defaults[name]))
# Load the configuration file.
parser.read('with-defaults.ini')
```

```python
    print('\nDefaults after loading file:')
    defaults = parser.defaults()
    for name in option_names:
        if name in defaults:
            print('  {:<15} = {!r}'.format(name, defaults[name]))

    # Define some local overrides.
    vars = {'from-vars': 'value from vars'}

    # Show the values of all the options.
    print('\nOption lookup:')
    for name in option_names:
        value = parser.get('sect', name, vars=vars)
        print('  {:<15} = {!r}'.format(name, value))

    # Show error messages for options that do not exist.
    print('\nError cases:')
    try:
        print('No such option :', parser.get('sect', 'no-option'))
    except configparser.NoOptionError as err:
        print(err)

    try:
        print('No such section:', parser.get('no-sect', 'no-option'))
    except configparser.NoSectionError as err:
        print(err)
```

输出显示了各个选项值的来源，并展示了不同来源的默认值是如何覆盖现有值的。

```
$ python3 configparser_defaults.py

Defaults before loading file:
  from-default    = 'value from defaults passed to init'
  from-section    = 'value from defaults passed to init'
  init-only       = 'value from defaults passed to init'
  init-and-file   = 'value from defaults passed to init'
  from-vars       = 'value from defaults passed to init'

Defaults after loading file:
  from-default    = 'value from defaults passed to init'
  from-section    = 'value from DEFAULT section'
  file-only       = 'value from DEFAULT section'
  init-only       = 'value from defaults passed to init'
  init-and-file   = 'value from DEFAULT section'
  from-vars       = 'value from DEFAULT section'

Option lookup:
  from-default    = 'value from defaults passed to init'
  from-section    = 'value from section in file'
  section-only    = 'value from section in file'
  file-only       = 'value from DEFAULT section'
  init-only       = 'value from defaults passed to init'
  init-and-file   = 'value from DEFAULT section'
  from-vars       = 'value from vars'

Error cases:
No option 'no-option' in section: 'sect'
No section: 'no-sect'
```

14.7.7 用拼接合并值

ConfigParser 提供了一个特性，名为拼接（interpolation），可以将值结合在一起。

如果值包含标准 Python 格式串，那么获取这个值时就会触发拼接特性。获取的值中指定的各个选项会按顺序依次被替换为相应的值，直到再不需要更多替换。

可以重写本节前面的 URL 例子，改为使用拼接，从而可以更容易地只改变部分值。例如，下面的配置文件将 URL 的协议、主机名和端口分开，作为单独的选项。

```
[bug_tracker]
protocol = http
server = localhost
port = 8080
url = %(protocol)s://%(server)s:%(port)s/bugs/
username = dhellmann
password = SECRET
```

每次调用 `get()` 时会默认地完成拼接。通过在 `raw` 参数中传入一个 `true` 值，可以获取未拼接的原值。

代码清单 14-78：**configparser_interpolation.py**

```python
from configparser import ConfigParser

parser = ConfigParser()
parser.read('interpolation.ini')

print('Original value       :', parser.get('bug_tracker', 'url'))

parser.set('bug_tracker', 'port', '9090')
print('Altered port value   :', parser.get('bug_tracker', 'url'))

print('Without interpolation:', parser.get('bug_tracker', 'url',
                                           raw=True))
```

由于值由 `get()` 计算，所以改变 `url` 值所用的某个设置也会改变返回值。

```
$ python3 configparser_interpolation.py

Original value       : http://localhost:8080/bugs/
Altered port value   : http://localhost:9090/bugs/
Without interpolation: %(protocol)s://%(server)s:%(port)s/bugs/
```

14.7.7.1 使用默认值

并不要求拼接的值出现在原选项所在的同一节中。默认值可以与覆盖值混合使用。

```
[DEFAULT]
url = %(protocol)s://%(server)s:%(port)s/bugs/
protocol = http
server = bugs.example.com
port = 80

[bug_tracker]
server = localhost
port = 8080
username = dhellmann
password = SECRET
```

对于这个配置，`url` 的值来自 `DEFAULT` 节，替换从查看 `bug_tracker` 开始，一开始没有找到的值再在 `DEFAULT` 中查找。

代码清单 14-79：**configparser_interpolation_defaults.py**

```
from configparser import ConfigParser

parser = ConfigParser()
parser.read('interpolation_defaults.ini')

print('URL:', parser.get('bug_tracker', 'url'))
```

`hostname` 和 `port` 值来自 `bug_tracker` 节，但是 `protocol` 来自 `DEFAULT`。

```
$ python3 configparser_interpolation_defaults.py

URL: http://localhost:8080/bugs/
```

14.7.7.2 替换错误

`MAX_INTERPOLATION_DEPTH` 步骤之后替换停止，以避免递归引用导致的问题。

代码清单 14-80：**configparser_interpolation_recursion.py**

```
import configparser

parser = configparser.ConfigParser()

parser.add_section('sect')
parser.set('sect', 'opt', '%(opt)s')

try:
    print(parser.get('sect', 'opt'))
except configparser.InterpolationDepthError as err:
    print('ERROR:', err)
```

如果有过多替换步骤，则会产生一个 `InterpolationDepthError` 异常。

```
$ python3 configparser_interpolation_recursion.py

ERROR: Recursion limit exceeded in value substitution: option 'o
pt' in section 'sect' contains an interpolation key which cannot
 be substituted in 10 steps. Raw value: '%(opt)s'
```

缺少值会导致一个 `InterpolationMissingOptionError` 异常。

代码清单 14-81：**configparser_interpolation_error.py**

```
import configparser

parser = configparser.ConfigParser()

parser.add_section('bug_tracker')
parser.set('bug_tracker', 'url',
           'http://%(server)s:%(port)s/bugs')

try:
    print(parser.get('bug_tracker', 'url'))
except configparser.InterpolationMissingOptionError as err:
    print('ERROR:', err)
```

由于没有定义 `server` 值，所以无法构造 `url`。

```
$ python3 configparser_interpolation_error.py

ERROR: Bad value substitution: option 'url' in section
'bug_tracker' contains an interpolation key 'server' which is
not a valid option name. Raw value:
'http://%(server)s:%(port)s/bugs'
```

14.7.7.3 转义特殊字符

由于拼接指令以 `%` 开始，值中的字面量 `%` 必须转义为 `%%`。

```
[escape]
value = a literal %% must be escaped
```

读取这个值并不需要任何特殊的考虑。

代码清单 14-82：configparser_escape.py

```python
from configparser import ConfigParser
import os

filename = 'escape.ini'
config = ConfigParser()
config.read([filename])

value = config.get('escape', 'value')

print(value)
```

读取这个值时，`%%` 会自动转换为 `%`。

```
$ python3 configparser_escape.py

a literal % must be escaped
```

14.7.7.4 扩展拼接

`ConfigParser` 通过 `interpolation` 参数来支持候选的拼接实现。`interpolation` 参数给定的对象要实现 `Interpolation` 类定义的 API。例如，使用 `ExtendedInterpolation` 而不是默认的 `BasicInterpolation` 会支持一种不同的语法：使用 `${}` 指示变量。

代码清单 14-83：configparser_extendedinterpolation.py

```python
from configparser import ConfigParser, ExtendedInterpolation

parser = ConfigParser(interpolation=ExtendedInterpolation())
parser.read('extended_interpolation.ini')

print('Original value       :', parser.get('bug_tracker', 'url'))
parser.set('intranet', 'port', '9090')
print('Altered port value   :', parser.get('bug_tracker', 'url'))
print('Without interpolation:', parser.get('bug_tracker', 'url',
                                           raw=True))
```

利用扩展拼接访问配置文件中其他节的值时，可以在变量名前加节名和一个冒号（`:`）作为前缀。

```
[intranet]
server = localhost
port = 8080

[bug_tracker]
url = http://${intranet:server}:${intranet:port}/bugs/
username = dhellmann
password = SECRET
```

通过引用文件中其他节的值,可以共享一个值层次结构,而不用把所有默认值都放在 DEFAULTS 节中。

```
$ python3 configparser_extendedinterpolation.py

Original value       : http://localhost:8080/bugs/
Altered port value   : http://localhost:9090/bugs/
Without interpolation: http://${intranet:server}:${intranet:port
}/bugs/
```

14.7.7.5 禁用拼接

如果要禁用拼接,则应传入 None 而不是一个 Interpolation 对象。

代码清单 14-84:configparser_nointerpolation.py

```python
from configparser import ConfigParser

parser = ConfigParser(interpolation=None)
parser.read('interpolation.ini')

print('Without interpolation:', parser.get('bug_tracker', 'url'))
```

通过禁用拼接,原先拼接对象能处理的所有语法都会被安全地忽略。

```
$ python3 configparser_nointerpolation.py

Without interpolation: %(protocol)s://%(server)s:%(port)s/bugs/
```

提示:相关阅读材料

- configparser 的标准库文档[⊖]。
- ConfigObj[⊖]:一个高级配置文件解析器,支持内容验证等特性。
- configparser 的 Python 2 到 Python 3 移植说明。

14.8 logging:报告状态、错误和信息消息

logging 模块定义了一个标准 API,用来报告应用和库的错误及状态信息。由一个标准库模块提供日志 API 的主要好处在于:所有 Python 模块都可以参与日志记录,所以应用的日志还可以包含来自第三方模块的消息。

⊖ https://docs.python.org/3.5/library/configparser.html

⊖ http://configobj.readthedocs.org/en/latest/configobj.html

14.8.1 日志系统的组成

日志系统包括4类相互交互的对象。任何一个模块或应用如果希望把某个活动记入日志，都可以使用 Logger 实例来向日志增加消息。调用日志记录器会创建一个 LogRecord，它把信息保存在内存中，直到得到处理。一个 Logger 可以有多个 Handler 对象，其被配置为要接收和处理日志记录。Handler 使用一个 Formatter 将日志记录转换为输出消息。

14.8.2 应用与库中的日志记录

应用开发人员和库作者都可以使用 logging，不过分别有不同的考虑。

应用开发人员要配置 logging 模块，将消息定向到适当的输出通道。例如，可以采用不同的详细程度记录消息，或者记录到不同的目标。可以把日志消息写入文件，写入 HTTP GET/POST 位置，通过 SMTP 写入 email 邮件，写入通用套接字或采用操作系统特定的日志机制，logging 模块中提供了所有这些处理器，但是对于内置类未处理的特殊需求，还可以创建定制的日志目标类。

库开发人员也可以使用 logging 来完成自己的工作，而且要使用这个模块，需要的工作甚至更少。只需要为各个上下文分别创建一个日志记录器实例，使用一个适当的名字，然后使用标准级别记录日志。只要库使用 logging API，并提供一致命名和级别选择，便可以根据需要配置应用来显示或隐藏库的消息。

14.8.3 记入文件

大多数应用都被配置为将日志记入文件。使用 basicConfig() 函数建立默认处理器，将调试消息写至一个文件。

代码清单 14-85：logging_file_example.py

```
import logging

LOG_FILENAME = 'logging_example.out'
logging.basicConfig(
    filename=LOG_FILENAME,
    level=logging.DEBUG,
)

logging.debug('This message should go to the log file')

with open(LOG_FILENAME, 'rt') as f:
    body = f.read()

print('FILE:')
print(body)
```

运行这个脚本时，日志消息被写至 logging_example.out。

```
$ python3 logging_file_example.py

FILE:
DEBUG:root:This message should go to the log file
```

14.8.4 旋转日志文件

反复运行以上代码清单中的脚本会向这个文件追加更多消息。要想在每次程序运行时创建一个新文件，可以向 `basicConfig()` 的参数 `filemode` 传入值 `'w'`。不过，最好不要采用这种方式管理文件的创建，更好的做法是使用一个 RotatingFileHandler，它会自动创建新文件，同时保留原来的日志文件。

代码清单 14-86：`logging_rotatingfile_example.py`

```python
import glob
import logging
import logging.handlers

LOG_FILENAME = 'logging_rotatingfile_example.out'

# Set up a specific logger with the desired output level.
my_logger = logging.getLogger('MyLogger')
my_logger.setLevel(logging.DEBUG)
# Add the log message handler to the logger.
handler = logging.handlers.RotatingFileHandler(
    LOG_FILENAME,
    maxBytes=20,
    backupCount=5,
)
my_logger.addHandler(handler)

# Log some messages.
for i in range(20):
    my_logger.debug('i = %d' % i)

# See which files are created.
logfiles = glob.glob('%s*' % LOG_FILENAME)
for filename in logfiles:
    print(filename)
```

最后会得到 6 个单独的文件，分别包含应用的部分日志历史。

```
$ python3 logging_rotatingfile_example.py

logging_rotatingfile_example.out
logging_rotatingfile_example.out.1
logging_rotatingfile_example.out.2
logging_rotatingfile_example.out.3
logging_rotatingfile_example.out.4
logging_rotatingfile_example.out.5
```

在这个例子中，最新的文件总是 `logging_rotatingfile_example.out`，每次达到大小限制时，就会加后缀 `.1` 进行重命名。现有的各个备份文件也会被重命名，使后缀递增（`.1` 变成 `.2`，等等），并且 `.5` 文件会被删除。

说明：显然，这个例子将日志长度设置得太小，只可作为一个极端的例子。在实际程序中要把 `maxBytes` 设置为一个更合适的值。

14.8.5 详细级别

`logging` API 还有一个有用的特性，即能够采用不同的日志级别（log level）生成不同的消息。例如，这说明代码可以附带调试消息，并且可以适当地设置日志级别，使其不会在生产系统中写出这些调试消息。表 14-2 列出了 `logging` 定义的日志级别。

表 14-2　日志级别

级别	值
CRITICAL	50
ERROR	40
WARNING	30
INFO	20
DEBUG	10
UNSET	0

对于某个级别的日志消息，只有当处理器和日志记录器被配置为可以发布该级别（或更高级别）的消息时，才会显示这个日志消息。例如，如果一个消息的级别是 CRITICAL，而日志记录器被设置为 ERROR，那么就会生成这个消息（50 > 40）。如果消息是 WARNING，而日志记录器被设置为只生成已设置为 ERROR 的消息，那么就不会生成这个消息（30 < 40）。

代码清单 14-87：**logging_level_example.py**

```python
import logging
import sys

LEVELS = {
    'debug': logging.DEBUG,
    'info': logging.INFO,
    'warning': logging.WARNING,
    'error': logging.ERROR,
    'critical': logging.CRITICAL,
}

if len(sys.argv) > 1:
    level_name = sys.argv[1]
    level = LEVELS.get(level_name, logging.NOTSET)
    logging.basicConfig(level=level)

logging.debug('This is a debug message')
logging.info('This is an info message')
logging.warning('This is a warning message')
logging.error('This is an error message')
logging.critical('This is a critical error message')
```

运行这个脚本并提供参数（如 debug 或 warning），以查看在不同级别会显示哪些消息。

```
$ python3 logging_level_example.py debug

DEBUG:root:This is a debug message
INFO:root:This is an info message
WARNING:root:This is a warning message
ERROR:root:This is an error message
CRITICAL:root:This is a critical error message

$ python3 logging_level_example.py info

INFO:root:This is an info message
WARNING:root:This is a warning message
```

```
ERROR:root:This is an error message
CRITICAL:root:This is a critical error message
```

14.8.6 命名日志记录器实例

前面的所有日志消息都内嵌有一个单词"root",因为这个代码使用了根日志记录器。如果要指出一个特定的日志消息来自哪里,那么一种容易的方法是对各个模块使用一个单独的日志记录器对象;发送到一个日志记录器的日志消息会包含这个日志记录器的名。下面的例子展示了如何记录来自不同模块的日志,以跟踪消息的来源。

代码清单 14-88:`logging_modules_example.py`

```python
import logging

logging.basicConfig(level=logging.WARNING)

logger1 = logging.getLogger('package1.module1')
logger2 = logging.getLogger('package2.module2')

logger1.warning('This message comes from one module')
logger2.warning('This comes from another module')
```

输出显示了各个输出行中不同的模块名。

```
$ python3 logging_modules_example.py

WARNING:package1.module1:This message comes from one module
WARNING:package2.module2:This comes from another module
```

14.8.7 日志树

`Logger` 实例采用一种树结构配置(根据它们的名字),如图 14-1 所示。通常每个应用或库都会定义一个基名,各个模块的日志记录器被设置为子节点。根日志记录器没有名字。

树结构对于配置日志记录很有用,因为这样各个日志记录器就不需要有自己的一组处理器。如果一个日志记录器没有任何处理器,那么消息会传给它的父日志记录器来处理。因此,对于大多数应用,只需要在根日志记录器上配置处理器,所有日志信息都将被收集和发送到相同的地方,如图 14-2 所示。

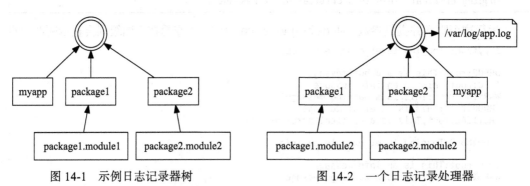

图 14-1 示例日志记录器树   图 14-2 一个日志记录处理器

树结构还允许为应用或库的不同部分设置不同的详细级别、处理器和格式化器。这种灵活性使得程序员能控制要记录哪些消息,以及要把它们记录到哪里,如图 14-3 所示。

14.8.8 与 warnings 模块集成

logging 模块通过 captureWarnings() 函数与 warnings 集成,这会配置 warnings 以便通过日志系统发送消息而不是直接输出。

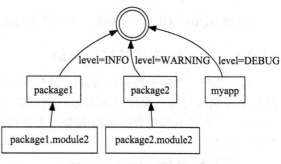

图 14-3 不同级别和处理器

代码清单 14-89:logging_capture_warnings.py

```
import logging
import warnings
logging.basicConfig(
    level=logging.INFO,
)

warnings.warn('This warning is not sent to the logs')

logging.captureWarnings(True)

warnings.warn('This warning is sent to the logs')
```

这个警告消息使用 WARNING 级别被发送到一个名为 py.warnings 的日志记录器。

```
$ python3 logging_capture_warnings.py

logging_capture_warnings.py:13: UserWarning: This warning is not
 sent to the logs
  warnings.warn('This warning is not sent to the logs')
WARNING:py.warnings:logging_capture_warnings.py:17: UserWarning:
 This warning is sent to the logs
  warnings.warn('This warning is sent to the logs')
```

提示:相关阅读材料

- logging 的标准库文档[一]:logging 的文档内容相当多,还包括很多教程和参考材料,而不只是这里给出的示例。
- logging 的 Python 2 到 Python 3 移植说明。
- warnings:非致命警告。
- logging_tree[二]:Brandon Rhodes 提供的一个第三方包,用于显示一个应用的日志记录器树。
- Logging Cookbook[三]:标准库文档的一部分,提供了使用 logging 完成不同任务的例子。

[一] https://docs.python.org/3.5/library/logging.html
[二] https://pypi.python.org/pypi/logging_tree
[三] https://docs.python.org/3.5/howto/logging-cookbook.html

14.9 `fileinput`：命令行过滤器框架

`fileinput`模块是一个框架，可以作为过滤器用来创建用于处理文本文件的命令行程序。

14.9.1 将 m3u 文件转换为 RSS

过滤器的一个例子是 m3utorss[⊖]，这个程序可以将一组 MP3 文件转换为一个可以作为播客共享的 RSS 提要。程序的输入是一个或多个 m3u 文件，其中列出要发布的 MP3 文件。输出是一个打印到控制台的 RSS 提要。要处理输入，程序需要迭代处理文件名列表，并完成以下步骤：

1. 打开各个文件。
2. 读取文件的各行。
3. 明确这一行是否指示一个 MP3 文件。
4. 如果是，则向 RSS 提要增加一个新元素。
5. 打印输出。

所有这些文件处理都可以通过手工编写代码完成。这并不是太复杂，只是利用了一些测试，甚至也可以自行编写错误处理。不过`fileinput`可以处理所有这些细节，能使程序大为简化。

```
for line in fileinput.input(sys.argv[1:]):
    mp3filename = line.strip()
    if not mp3filename or mp3filename.startswith('#'):
        continue
    item = SubElement(rss, 'item')
    title = SubElement(item, 'title')
    title.text = mp3filename
    encl = SubElement(item, 'enclosure',
                     {'type': 'audio/mpeg',
                      'url': mp3filename})
```

`input()`函数取要检查的文件名列表作为参数。如果这个列表为空，则模块会从标准输入读取数据。这个函数会返回一个迭代器，从被处理的文本文件生成各个文本行。调用者只需循环处理各行，跳过空格和注释，查找指向 MP3 文件的引用。

下面的代码清单提供了完整的程序。

代码清单 14-90：**`fileinput_example.py`**

```
import fileinput
import sys
import time
from xml.etree.ElementTree import Element, SubElement, tostring
from xml.dom import minidom
# Establish the RSS and channel nodes.
rss = Element('rss',
              {'xmlns:dc': "http://purl.org/dc/elements/1.1/",
               'version': '2.0'})
channel = SubElement(rss, 'channel')
title = SubElement(channel, 'title')
```

⊖ https://pypi.python.org/pypi/m3utorss

```
title.text = 'Sample podcast feed'
desc = SubElement(channel, 'description')
desc.text = 'Generated for PyMOTW'
pubdate = SubElement(channel, 'pubDate')
pubdate.text = time.asctime()
gen = SubElement(channel, 'generator')
gen.text = 'https://pymotw.com/'

for line in fileinput.input(sys.argv[1:]):
    mp3filename = line.strip()
    if not mp3filename or mp3filename.startswith('#'):
        continue
    item = SubElement(rss, 'item')
    title = SubElement(item, 'title')
    title.text = mp3filename
    encl = SubElement(item, 'enclosure',
                     {'type': 'audio/mpeg',
                      'url': mp3filename})

rough_string = tostring(rss)
reparsed = minidom.parseString(rough_string)
print(reparsed.toprettyxml(indent="  "))
```

以下代码清单中的示例输入文件包含多个 MP3 文件的文件名。

代码清单 14-91：sample_data.m3u

```
# This is a sample m3u file.
episode-one.mp3
episode-two.mp3
```

利用以上示例输入，运行 `fileinput_example.py` 可以生成 RSS 格式的 XML 数据。

```
$ python3 fileinput_example.py sample_data.m3u

<?xml version="1.0" ?>
<rss version="2.0" xmlns:dc="http://purl.org/dc/elements/1.1/">
  <channel>
    <title>Sample podcast feed</title>
    <description>Generated for PyMOTW</description>
    <pubDate>Sun Jul 10 10:45:01 2016</pubDate>
    <generator>https://pymotw.com/</generator>
  </channel>
  <item>
    <title>episode-one.mp3</title>
    <enclosure type="audio/mpeg" url="episode-one.mp3"/>
  </item>
  <item>
    <title>episode-two.mp3</title>
    <enclosure type="audio/mpeg" url="episode-two.mp3"/>
  </item>
</rss>
```

14.9.2 进度元数据

在前面的例子中，文件名和正在处理的行号并不重要。但有时其他工具（如类 grep 的搜索工具）可能需要这个信息。`fileinput` 包含一些函数来访问有关当前行的所有元数据（`filename()`、`filelineno()` 和 `lineno()`）。

代码清单 14-92：`fileinput_grep.py`

```python
import fileinput
import re
import sys

pattern = re.compile(sys.argv[1])

for line in fileinput.input(sys.argv[2:]):
    if pattern.search(line):
        if fileinput.isstdin():
            fmt = '{lineno}:{line}'
        else:
            fmt = '{filename}:{lineno}:{line}'
        print(fmt.format(filename=fileinput.filename(),
                         lineno=fileinput.filelineno(),
                         line=line.rstrip()))
```

可以用一个基本的模式匹配循环来查找串 "`fileinput`" 在这些示例源文件中的出现。

```
$ python3 fileinput_grep.py fileinput *.py

fileinput_change_subnet.py:10:import fileinput
fileinput_change_subnet.py:17:for line in fileinput.input(files,
 inplace=True):
fileinput_change_subnet_noisy.py:10:import fileinput
fileinput_change_subnet_noisy.py:18:for line in fileinput.input(
files, inplace=True):
fileinput_change_subnet_noisy.py:19:        if fileinput.isfirstline
():
fileinput_change_subnet_noisy.py:21:            fileinput.filena
me()))
fileinput_example.py:6:"""Example for fileinput module.
fileinput_example.py:10:import fileinput
fileinput_example.py:30:for line in fileinput.input(sys.argv[1:]
):
fileinput_grep.py:10:import fileinput
fileinput_grep.py:16:for line in fileinput.input(sys.argv[2:]):
fileinput_grep.py:18:    if fileinput.isstdin():
fileinput_grep.py:22:        print(fmt.format(filename=fileinput
.filename(),
fileinput_grep.py:23:                         lineno=fileinput.f
ilelineno(),
```

还可以从标准输入读取文本。

```
$ cat *.py | python fileinput_grep.py fileinput

10:import fileinput
17:for line in fileinput.input(files, inplace=True):
29:import fileinput
37:for line in fileinput.input(files, inplace=True):
38:        if fileinput.isfirstline():
40:            fileinput.filename()))
54:"""Example for fileinput module.
58:import fileinput
78:for line in fileinput.input(sys.argv[1:]):
101:import fileinput
107:for line in fileinput.input(sys.argv[2:]):
109:    if fileinput.isstdin():
113:        print(fmt.format(filename=fileinput.filename(),
114:                         lineno=fileinput.filelineno(),
```

14.9.3 原地过滤

另一种常见的文件处理操作是原地（in-place）修改一个文件的内容，而不是创建一个新文件（其中包含修改后的内容）。例如，如果一个子网范围改变，那么 UNIX 主机文件就可能需要更新。

代码清单 14-93：`etc_hosts.txt before modifications`

```
##
# Host Database
#
# localhost is used to configure the loopback interface
# when the system is booting.  Do not change this entry.
##
127.0.0.1       localhost
255.255.255.255 broadcasthost
::1             localhost
fe80::1%lo0     localhost
10.16.177.128   hubert hubert.hellfly.net
10.16.177.132   cubert cubert.hellfly.net
10.16.177.136   zoidberg zoidberg.hellfly.net
```

要自动完成这个修改，安全的做法是根据输入创建一个新文件，然后用编辑后的副本替换原来的文件。`fileinput` 使用 `inplace` 选项支持这种方法。

代码清单 14-94：`fileinput_change_subnet.py`

```python
import fileinput
import sys

from_base = sys.argv[1]
to_base = sys.argv[2]
files = sys.argv[3:]

for line in fileinput.input(files, inplace=True):
    line = line.rstrip().replace(from_base, to_base)
    print(line)
```

尽管前面的脚本使用了 `print()`，但是由于 `fileinput` 将标准输出重定向到所覆盖的文件，所以不生成任何输出。

```
$ python3 fileinput_change_subnet.py 10.16 10.17 etc_hosts.txt
```

更新后的文件包含了 `10.16.0.0/16` 网络上所有服务器更改后的 IP 地址。

代码清单 14-95：`etc_hosts.txt after modifications`

```
##
# Host Database
#
# localhost is used to configure the loopback interface
# when the system is booting.  Do not change this entry.
##
127.0.0.1       localhost
255.255.255.255 broadcasthost
::1             localhost
fe80::1%lo0     localhost
```

```
10.17.177.128   hubert hubert.hellfly.net
10.17.177.132   cubert cubert.hellfly.net
10.17.177.136   zoidberg zoidberg.hellfly.net
```

处理开始之前，会使用原来的文件名加上 .bak 扩展名来创建一个备份文件。

代码清单 14-96：fileinput_change_subnet_noisy.py

```python
import fileinput
import glob
import sys

from_base = sys.argv[1]
to_base = sys.argv[2]
files = sys.argv[3:]

for line in fileinput.input(files, inplace=True):
    if fileinput.isfirstline():
        sys.stderr.write('Started processing {}\n'.format(
            fileinput.filename()))
        sys.stderr.write('Directory contains: {}\n'.format(
            glob.glob('etc_hosts.txt*')))
    line = line.rstrip().replace(from_base, to_base)
    print(line)

sys.stderr.write('Finished processing\n')
sys.stderr.write('Directory contains: {}\n'.format(
    glob.glob('etc_hosts.txt*')))
```

输入结束时会删除这个备份文件。

```
$ python3 fileinput_change_subnet_noisy.py 10.16. 10.17. etc_hosts.txt

Started processing etc_hosts.txt
Directory contains: ['etc_hosts.txt', 'etc_hosts.txt.bak']
Finished processing
Directory contains: ['etc_hosts.txt']
```

提示：相关阅读材料

- `fileinput` 的标准库文档[⊖]。
- m3utorss[⊖]：这个脚本可以把包含 MP3 列表的 m3u 文件转换为一个适合用作播客提要的 RSS 文件。
- `xml.etree`：提供了使用 ElementTree 生成 XML 的更多详细内容。

14.10 atexit：程序关闭回调

atexit 模块提供了一个接口，可以注册程序正常关闭时调用的函数。

14.10.1 注册退出回调

下面的例子通过调用 register() 注册了一个函数。

⊖ https://docs.python.org/3.5/library/fileinput.html
⊖ https://pypi.python.org/pypi/m3utorss

第 14 章 应用构建模块

代码清单 14-97：**atexit_simple.py**

```
import atexit

def all_done():
    print('all_done()')

print('Registering')
atexit.register(all_done)
print('Registered')
```

由于程序不做其他事情，所以会立即调用 all_done()。

```
$ python3 atexit_simple.py

Registering
Registered
all_done()
```

还可以注册多个函数，并向注册的函数传递参数。这种方法对于妥善地断开数据库连接、删除临时文件等可能很有用。不用为需要释放的资源维护一个特殊的列表，完全可以对每个资源注册一个单独的清理函数。

代码清单 14-98：**atexit_multiple.py**

```
import atexit

def my_cleanup(name):
    print('my_cleanup({})'.format(name))

atexit.register(my_cleanup, 'first')
atexit.register(my_cleanup, 'second')
atexit.register(my_cleanup, 'third')
```

退出函数会按注册的逆序来调用。这个方法以模块导入顺序（相应地，也就是注册 atexit 函数的顺序）的逆序来完成模块清理，这样能减少依赖冲突。

```
$ python3 atexit_multiple.py

my_cleanup(third)
my_cleanup(second)
my_cleanup(first)
```

14.10.2 修饰符语法

不需要任何参数的函数可以使用 register() 被注册为一个修饰符。这种候选语法对于处理模块级全局数据的清理函数很方便。

代码清单 14-99：**atexit_decorator.py**

```
import atexit

@atexit.register
```

```
def all_done():
    print('all_done()')

print('starting main program')
```

因为会在定义函数的同时注册这个函数，所以还要确保即使模块不做任何其他工作这个函数也要能正常工作，这很重要。如果它要清理的资源从未被初始化过，那么调用退出回调不应产生错误。

```
$ python3 atexit_decorator.py

starting main program
all_done()
```

14.10.3 撤销回调

要撤销一个退出回调，可以使用 unregister() 把它从注册表删除。

代码清单 14-100：atexit_unregister.py

```python
import atexit

def my_cleanup(name):
    print('my_cleanup({})'.format(name))

atexit.register(my_cleanup, 'first')
atexit.register(my_cleanup, 'second')
atexit.register(my_cleanup, 'third')

atexit.unregister(my_cleanup)
```

对同一个回调的所有调用都会被撤销，而不论它注册了多少次。

```
$ python3 atexit_unregister.py
```

删除原来未注册的回调不会被视为一个错误。

代码清单 14-101：atexit_unregister_not_registered.py

```python
import atexit

def my_cleanup(name):
    print('my_cleanup({})'.format(name))

if False:
    atexit.register(my_cleanup, 'never registered')

atexit.unregister(my_cleanup)
```

因为其会悄悄地忽略未知的回调，所以即使在注册序列未知的情况下也可以使用 unregister()。

```
$ python3 atexit_unregister_not_registered.py
```

14.10.4 什么情况下不调用 atexit 函数

如果满足以下任意一个条件,就不会调用为 atexit 注册的回调。
- 程序由于一个信号而中止。
- 直接调用了 os._exit()。
- 检测到解释器中的一个致命错误。

可以更新 subprocess 一节中的例子,显示程序因为一个信号而中止时会发生什么。这里涉及两个文件,父程序和子程序。父程序启动子程序,暂停,然后中止子程序。

代码清单 14-102: **atexit_signal_parent.py**

```
import os
import signal
import subprocess
import time

proc = subprocess.Popen('./atexit_signal_child.py')
print('PARENT: Pausing before sending signal...')
time.sleep(1)
print('PARENT: Signaling child')
os.kill(proc.pid, signal.SIGTERM)
```

子程序建立一个 atexit 回调,然后休眠,直至信号到来。

代码清单 14-103: **atexit_signal_child.py**

```
import atexit
import time
import sys


def not_called():
    print('CHILD: atexit handler should not have been called')

print('CHILD: Registering atexit handler')
sys.stdout.flush()
atexit.register(not_called)

print('CHILD: Pausing to wait for signal')
sys.stdout.flush()
time.sleep(5)
```

运行这个脚本时,会生成以下输出。

```
$ python3 atexit_signal_parent.py

CHILD: Registering atexit handler
CHILD: Pausing to wait for signal
PARENT: Pausing before sending signal...
PARENT: Signaling child
```

子程序不会打印嵌在 not_called() 中的消息。

通过使用 os._exit(),程序员就可以避免调用 atexit 回调。

代码清单 14-104：**atexit_os_exit.py**

```python
import atexit
import os


def not_called():
    print('This should not be called')

print('Registering')
atexit.register(not_called)
print('Registered')

print('Exiting...')
os._exit(0)
```

由于这个例子绕过了正常的退出路径，所以没有运行回调。另外，打印输出不会刷新，所以要提供 -u 选项运行这个示例来启用无缓冲的 I/O。

```
$ python3 -u atexit_os_exit.py

Registering
Registered
Exiting...
```

为了确保运行回调，可以在要执行的语句外运行或者通过调用 **sys.exit()** 使程序中止。

代码清单 14-105：**atexit_sys_exit.py**

```python
import atexit
import sys


def all_done():
    print('all_done()')

print('Registering')
atexit.register(all_done)
print('Registered')

print('Exiting...')
sys.exit()
```

这个例子调用了 **sys.exit()**，所以会调用注册的回调。

```
$ python3 atexit_sys_exit.py

Registering
Registered
Exiting...
all_done()
```

14.10.5 处理异常

对于 **atexit** 回调中产生的异常，会在控制台上打印这些异常的 Traceback，最后产生的异常会被重新抛出，并作为程序的最后一个错误消息。

代码清单 14-106：**atexit_exception.py**

```
import atexit

def exit_with_exception(message):
    raise RuntimeError(message)

atexit.register(exit_with_exception, 'Registered first')
atexit.register(exit_with_exception, 'Registered second')
```

注册顺序会控制执行顺序。如果一个回调中的某个错误引入了另一个回调中的一个错误（较早注册，但较后调用），那么向用户显示时，最后的错误消息可能并不是最有用的错误消息。

```
$ python3 atexit_exception.py

Error in atexit._run_exitfuncs:
Traceback (most recent call last):
  File "atexit_exception.py", line 11, in exit_with_exception
    raise RuntimeError(message)
RuntimeError: Registered second
Error in atexit._run_exitfuncs:
Traceback (most recent call last):
  File "atexit_exception.py", line 11, in exit_with_exception
    raise RuntimeError(message)
RuntimeError: Registered first
```

通常最好的办法是在清理函数中处理异常并悄悄地记入日志，因为程序退出时显示一大堆错误会很乱。

提示：相关阅读材料
- `atexit` 的标准库文档[⊖]。
- 17.2.4 节：未捕获异常的全局处理。
- `atexit` 的 Python 2 到 Python 3 移植说明。

14.11 `sched`：定时事件调度器

`sched` 模块实现了一个通用事件调度器，可以在指定时刻运行任务。调度器类使用一个 `time` 函数来掌握当前时间，另外利用一个 `delay` 函数来等待一个指定时间段。具体的时间单位并不重要，所以接口足够灵活，可以用于很多用途。

调用 `time` 函数时不带任何参数，并且它会返回一个表示当前时间的数。调用 `delay` 函数要提供一个整数参数，使用的单位与 `time` 函数相同，返回之前会等待指定数目的时间单位。例如，可以使用 `time` 模块的 `monotonic()` 和 `sleep()`，不过这一节中的例子使用了 `time.time()`（也满足这些需求），因为这样输出更容易理解。

要支持多线程应用，生成各事件之后可以调用 `delay` 函数并提供参数 0，以确保其他线程也有机会运行。

⊖ https://docs.python.org/3.5/library/atexit.html

14.11.1 有延迟地运行事件

可以调度事件在一个延迟之后运行或者在一个指定时间运行。要有延迟地调度事件，可以使用 enter() 方法，它有 4 个参数：
- 表示延迟的一个数
- 一个优先级值
- 一个要调用的函数
- 函数参数的一个元组

这个例子调度两个不同的事件分别在 2 秒和 3 秒后运行。一旦达到事件的时间，便会调用 print_event()，打印当前时间和传至事件的 name 参数。

代码清单 14-107：**sched_basic.py**

```
import sched
import time

scheduler = sched.scheduler(time.time, time.sleep)


def print_event(name, start):
    now = time.time()
    elapsed = int(now - start)
    print('EVENT: {} elapsed={} name={}'.format(
        time.ctime(now), elapsed, name))


start = time.time()
print('START:', time.ctime(start))
scheduler.enter(2, 1, print_event, ('first', start))
scheduler.enter(3, 1, print_event, ('second', start))

scheduler.run()
```

运行这个程序会生成以下输出。

```
$ python3 sched_basic.py

START: Sun Sep  4 16:21:01 2016
EVENT: Sun Sep  4 16:21:03 2016 elapsed=2 name=first
EVENT: Sun Sep  4 16:21:04 2016 elapsed=3 name=second
```

为第一个事件打印的时间是开始时间（start）后的 2 秒，第二个事件的时间则为开始时间（start）后的 3 秒。

14.11.2 重叠事件

run() 调用会阻塞，直至所有事件都已经被处理。每个事件都在相同的线程中运行，所以如果一个事件需要很长时间运行，并且超出了事件之间的延迟，那么就会出现重叠。重叠可以通过推迟后面的事件来解决。这样不会丢失事件，但有些事件可能比其调度时间更晚调用。在下面的例子中，long_event() 调用 sleep 休眠，不过也可以完成一个长时间的计算或者在 I/O 上阻塞，这样也能很容易地延迟处理。

代码清单 14-108：**sched_overlap.py**

```python
import sched
import time

scheduler = sched.scheduler(time.time, time.sleep)

def long_event(name):
    print('BEGIN EVENT :', time.ctime(time.time()), name)
    time.sleep(2)
    print('FINISH EVENT:', time.ctime(time.time()), name)

print('START:', time.ctime(time.time()))
scheduler.enter(2, 1, long_event, ('first',))
scheduler.enter(3, 1, long_event, ('second',))

scheduler.run()
```

其结果是第一个事件一旦完成就会立即运行第二个事件，因为第一个事件花费的时间足够长，使时钟超过了第二个事件期望的开始时间。

```
$ python3 sched_overlap.py

START: Sun Sep  4 16:21:04 2016
BEGIN EVENT : Sun Sep  4 16:21:06 2016 first
FINISH EVENT: Sun Sep  4 16:21:08 2016 first
BEGIN EVENT : Sun Sep  4 16:21:08 2016 second
FINISH EVENT: Sun Sep  4 16:21:10 2016 second
```

14.11.3 事件优先级

如果调度多个事件在同一时间运行，那么就要使用事件的优先级来确定它们以什么顺序运行。

代码清单 14-109：**sched_priority.py**

```python
import sched
import time

scheduler = sched.scheduler(time.time, time.sleep)

def print_event(name):
    print('EVENT:', time.ctime(time.time()), name)

now = time.time()
print('START:', time.ctime(now))
scheduler.enterabs(now + 2, 2, print_event, ('first',))
scheduler.enterabs(now + 2, 1, print_event, ('second',))

scheduler.run()
```

这个例子要确保将事件调度为在完全相同的时间运行，所以要使用 enterabs() 方法而不是 enter()。enterabs() 的第一个参数是运行事件的时间，而不是要延迟的时间。

```
$ python3 sched_priority.py

START: Sun Sep  4 16:21:10 2016
```

```
EVENT: Sun Sep  4 16:21:12 2016 second
EVENT: Sun Sep  4 16:21:12 2016 first
```

14.11.4 取消事件

enter()和enterabs()都会返回事件的一个引用,以后可以用这个引用来取消事件。由于run()会阻塞,所以必须在一个不同的线程中取消这个事件。在这个例子中,线程开始运行调度器,并用主处理线程取消事件。

代码清单 14-110:**sched_cancel.py**

```python
import sched
import threading
import time

scheduler = sched.scheduler(time.time, time.sleep)

# Set up a global to be modified by the threads.
counter = 0


def increment_counter(name):
    global counter
    print('EVENT:', time.ctime(time.time()), name)
    counter += 1
    print('NOW:', counter)

print('START:', time.ctime(time.time()))
e1 = scheduler.enter(2, 1, increment_counter, ('E1',))
e2 = scheduler.enter(3, 1, increment_counter, ('E2',))

# Start a thread to run the events.
t = threading.Thread(target=scheduler.run)
t.start()

# Back in the main thread, cancel the first scheduled event.
scheduler.cancel(e1)

# Wait for the scheduler to finish running in the thread.
t.join()

print('FINAL:', counter)
```

这里调度了两个事件,不过第一个事件随后被取消了。只运行了第二个事件,所以counter变量只递增一次。

```
$ python3 sched_cancel.py

START: Sun Sep  4 16:21:13 2016
EVENT: Sun Sep  4 16:21:16 2016 E2
NOW: 1
FINAL: 1
```

提示:相关阅读材料
- sched 的标准库文档⊖。
- time:time 模块。

⊖ https://docs.python.org/3.5/library/sched.html

第 15 章 国际化和本地化

Python 提供了两个模块来支持应用处理多种自然语言和文化设置。gettext 采用不同语言创建消息编目，从而能以用户能够理解的语言显示提示符和错误消息。locale 会考虑到文化差异（例如如何指示负数，本地货币符号是什么，等等），改变数字、货币、日期和时间的格式化方式。这两个模块都会与其他工具和操作环境交互，使 Python 应用能够与系统上的所有其他程序很好地配合。

15.1 gettext：消息编目

gettext 模块提供了一个纯 Python 实现，与 GNU gettext 库兼容，用于完成消息转换和编目管理。利用 Python 源代码发布版提供的工具，可以从一组源文件中抽取消息，构建一个包含转换的消息编目，并使用这个消息编目在运行时为用户显示一个适当的消息。

消息编目可以用来为程序提供国际化接口，使用适合用户的语言来显示消息。还可以用于其他消息定制，包括为不同包装器或合作伙伴的界面"换肤"。

说明：尽管标准库文档声称 Python 已经包含所有必要的工具，但是即使提供了适当的命令行选项，pygettext.py 也无法抽取包装在 ngettext 调用中的消息。这些例子使用了 GNU gettext 工具集的 xgettext，而不是 pygettext.py。

15.1.1 转换工作流概述

建立和使用转换的过程包括 5 个步骤。

1. 标识并标记源代码中包含待转换消息的字面量串。

首先在程序源代码中标识需要转换的消息，并标记字面量串，以便抽取程序发现这些字面量串。

2. 抽取消息。

标识源代码中可转换的串之后，使用 xgettext 抽取出这些串，并创建一个 .pot 文件，或转换模板（translation template）。这个模板是一个文本文件，包含所有已标识串的副本及对应其转换的占位符。

3. 转换消息。

将 .pot 文件的一个副本提供给转换器，将扩展名改为 .po。这个 .po 文件是一个可编辑的源文件，被用作下一步编译的输入。转换器要更新这个文件中的首部文本，提供所有串的转换。

4. 由转换"编译"消息编目。

转换器发回完整的 .po 文件时，使用 msgfmt 将这个文本文件编译为二进制编目格式。运行时编目的查找代码将使用这个二进制格式。

5. 运行时加载并启动适当的消息编目。

最后一步是向应用增加几行代码，配置和加载消息编目，并安装转换函数。对此有几种方法，各有优缺点。

这一节余下的内容将更详细地介绍这些步骤，首先从需要完成的代码修改开始。

15.1.2 由源代码创建消息编目

gettext 首先在一个转换数据库中查找字面量串，并取出适当的转换串。常用模式是将适当的查找函数与名"_"（单个下划线字符）绑定，使得代码中不会堆积大量长名函数调用。

消息抽取程序 xgettext 会查找嵌入在编目查找函数（catalog lookup function）调用中的消息。它知道不同的源语言，并分别使用适当的解析器。如果查找函数有别名，或者增加了额外的函数，那么便要为 xgettext 提供这些额外符号的名，从而能够在抽取消息时考虑到。

以下脚本提供了一个消息，可以完成转换。

代码清单 15-1：**gettext_example.py**

```
import gettext

# Set up message catalog access.
t = gettext.translation(
    'example_domain', 'locale',
    fallback=True,
)
_ = t.gettext

print(_('This message is in the script.'))
```

文本 "This message is in the script." 是将要由编目替换的消息。这里启用了 Fallback 模式，所以如果运行脚本时没有一个消息编目，则会打印内联的消息。

```
$ python3 gettext_example.py

This message is in the script.
```

下一步是抽取消息，并创建 .pot 文件，这里可以使用 pygettext.py 或 xgettext。

```
$ xgettext -o example.pot gettext_example.py
```

生成的输出文件包含以下内容。

代码清单 15-2：**example.pot**

```
# SOME DESCRIPTIVE TITLE.
# Copyright (C) YEAR THE PACKAGE'S COPYRIGHT HOLDER
# This file is distributed under the same license as the PACKAGE package.
```

```
# FIRST AUTHOR <EMAIL@ADDRESS>, YEAR.
#
#, fuzzy
msgid ""
msgstr ""
"Project-Id-Version: PACKAGE VERSION\n"
"Report-Msgid-Bugs-To: \n"
"POT-Creation-Date: 2016-07-10 10:45-0400\n"
"PO-Revision-Date: YEAR-MO-DA HO:MI+ZONE\n"
"Last-Translator: FULL NAME <EMAIL@ADDRESS>\n"
"Language-Team: LANGUAGE <LL@li.org>\n"
"Language: \n"
"MIME-Version: 1.0\n"
"Content-Type: text/plain; charset=CHARSET\n"
"Content-Transfer-Encoding: 8bit\n"

#: gettext_example.py:19
msgid "This message is in the script."
msgstr ""
```

消息编目被安装到按域（domain）和语言（language）组织的目录中。域由应用或库提供，通常是一个唯一值，如应用名。在这里，`gettext_example.py` 中的域是 `example_domain`。语言值则由用户环境在运行时通过某个环境变量（LANGUAGE、LC_ALL、LC_MESSAGES 或 LANG）提供，这取决于其配置和平台。这一章中的例子在运行时都将语言设置为 `en_US`。

既然模板已经准备好，那么下一步便是创建必要的目录结构，并把模板复制到适当的位置。PyMOTW 源码树中的 `locale` 目录可以作为这些示例的消息编目目录的根，不过通常最好使用全系统都可以访问的一个目录，使所有用户都能访问消息编目。这个编目输入源文件的完整路径为 `$localedir/$language/LC_MESSAGES/$domain.po`，并且实际编目的文件扩展名为 `.mo`。

要创建编目，应将 `example.pot` 复制到 `locale/en_US/LC_MESSAGES/example.po`，编辑这个文件，改变首部中的值，并设置替换消息。结果如以下代码清单所示。

代码清单 15-3：`locale/en_US/LC_MESSAGES/example.po`

```
# Messages from gettext_example.py.
# Copyright (C) 2009 Doug Hellmann
# Doug Hellmann <doug@doughellmann.com>, 2016.
#
msgid ""
msgstr ""
"Project-Id-Version: PyMOTW-3\n"
"Report-Msgid-Bugs-To: Doug Hellmann <doug@doughellmann.com>\n"
"POT-Creation-Date: 2016-01-24 13:04-0500\n"
"PO-Revision-Date: 2016-01-24 13:04-0500\n"
"Last-Translator: Doug Hellmann <doug@doughellmann.com>\n"
"Language-Team: US English <doug@doughellmann.com>\n"
"MIME-Version: 1.0\n"
"Content-Type: text/plain; charset=UTF-8\n"
"Content-Transfer-Encoding: 8bit\n"


#: gettext_example.py:16
msgid "This message is in the script."
msgstr "This message is in the en_US catalog."
```

使用 msgformat 从 .po 文件构建这个编目。

```
$ cd locale/en_US/LC_MESSAGES; msgfmt -o example.mo example.po
```

gettext_example.py 中的域是 example_domain，但是文件名为 example.pot。要让 gettext 找到正确的转换文件，名字必须匹配。

代码清单 15-4：**gettext_example_corrected.py**

```
t = gettext.translation(
    'example', 'locale',
    fallback=True,
)
```

现在运行脚本时，会打印编目中的消息而不是内联字符串。

```
$ python3 gettext_example_corrected.py

This message is in the en_US catalog.
```

15.1.3 运行时查找消息编目

如前所述，包含消息编目的 locale 目录根据语言来组织，编目按程序的域命名。不同的操作系统分别定义了自己的默认值，不过 gettext 并不知道所有这些默认值。它使用一个默认的 locale 目录 sys.prefix + '/share/locale'，但是大多数情况下，更安全的做法是显式地提供一个 localedir 值而不是指望这个默认值总是合法的。find() 函数负责在运行时找到一个合适的消息编目。

代码清单 15-5：**gettext_find.py**

```
import gettext

catalogs = gettext.find('example', 'locale', all=True)
print('Catalogs:', catalogs)
```

路径的语言部分由某个环境变量（LANGUAGE、LC_ALL、LC_MESSAGES 和 LANG）得到，这些环境变量可以用于配置本地化特性。总是使用找到的第一个变量。要选择多种语言，可以将值用冒号（:）分隔。要了解这是如何做到的，来看下面的例子，这里使用 gettext_find.py 来运行一些试验。

```
$ cd locale/en_CA/LC_MESSAGES; msgfmt -o example.mo example.po
$ cd ../../..
$ python3 gettext_find.py

Catalogs: ['locale/en_US/LC_MESSAGES/example.mo']

$ LANGUAGE=en_CA python3 gettext_find.py

Catalogs: ['locale/en_CA/LC_MESSAGES/example.mo']

$ LANGUAGE=en_CA:en_US python3 gettext_find.py

Catalogs: ['locale/en_CA/LC_MESSAGES/example.mo',
'locale/en_US/LC_MESSAGES/example.mo']
```

```
$ LANGUAGE=en_US:en_CA python3 gettext_find.py

Catalogs: ['locale/en_US/LC_MESSAGES/example.mo',
'locale/en_CA/LC_MESSAGES/example.mo']
```

尽管 `find()` 显示了完整的编目列表，但实际上却只为消息查找加载了这个序列中的第一个编目。

```
$ python3 gettext_example_corrected.py

This message is in the en_US catalog.

$ LANGUAGE=en_CA python3 gettext_example_corrected.py

This message is in the en_CA catalog.

$ LANGUAGE=en_CA:en_US python3 gettext_example_corrected.py

This message is in the en_CA catalog.

$ LANGUAGE=en_US:en_CA python3 gettext_example_corrected.py

This message is in the en_US catalog.
```

15.1.4 复数值

简单的消息替换可以处理大多数转换需求，不过 `gettext` 将复数处理为一种特殊情况。一个消息的单数和复数形式之间会有差别，而且取决于具体语言，这种差别可能也有所不同，有些只是某个单词的末尾不同，有些则是整个句子结构都不同。根据复数的层次，可能还会有不同的形式。为了更容易地管理复数（在某些情况下，甚至是为了能管理复数），模块提供了一组单独的函数来询问一个消息的复数形式。

代码清单 15-6：**gettext_plural.py**

```python
from gettext import translation
import sys

t = translation('plural', 'locale', fallback=False)
num = int(sys.argv[1])
msg = t.ngettext('{num} means singular.',
                 '{num} means plural.',
                 num)

# Still need to add the values to the message ourselves
print(msg.format(num=num))
```

使用 `ungettext()` 来访问一个消息的复数版本。参数是要转换的消息和项数。

```
$ xgettext -L Python -o plural.pot gettext_plural.py
```

由于有一些候选的转换形式，所以这些替换形式会被列在一个数组中。通过使用数组，就可以对有多种复数形式的语言完成转换（例如，波兰语就用不同的形式来表示相对数量）。

代码清单 15-7：**plural.pot**

```
# SOME DESCRIPTIVE TITLE.
# Copyright (C) YEAR THE PACKAGE'S COPYRIGHT HOLDER
```

```
# This file is distributed under the same license as the PACKAGE package.
# FIRST AUTHOR <EMAIL@ADDRESS>, YEAR.
#
#, fuzzy
msgid ""
msgstr ""
"Project-Id-Version: PACKAGE VERSION\n"
"Report-Msgid-Bugs-To: \n"
"POT-Creation-Date: 2016-07-10 10:45-0400\n"
"PO-Revision-Date: YEAR-MO-DA HO:MI+ZONE\n"
"Last-Translator: FULL NAME <EMAIL@ADDRESS>\n"
"Language-Team: LANGUAGE <LL@li.org>\n"
"Language: \n"
"MIME-Version: 1.0\n"
"Content-Type: text/plain; charset=CHARSET\n"
"Content-Transfer-Encoding: 8bit\n"
"Plural-Forms: nplurals=INTEGER; plural=EXPRESSION;\n"

#: gettext_plural.py:15
#, python-brace-format
msgid "{num} means singular."
msgid_plural "{num} means plural."
msgstr[0] ""
msgstr[1] ""
```

除了填入转换串之外，还需要告诉库采用哪种复数形式，让它知道对应给定的数量值如何在数组中索引。行 "Plural-Forms: nplurals=INTEGER; plural=EXPRESSION;\n" 包含两个值，需要手工替换：nplurals 是一个整数，指示数组的大小（使用的转换数）；plural 是一个 C 语言表达式，用于将得到的数量转换为查找转换时需要的数组索引。字面量串 n 会被替换为传递给 ungettext() 的数量。

例如，英语包括两种复数形式。数量 0 会被处理为复数（"0 bananas"）。Plural-Forms 项如下。

```
Plural-Forms: nplurals=2; plural=n != 1;
```

单数转换位于位置 0，复数转换在位置 1。

代码清单 15-8：locale/en_US/LC_MESSAGES/plural.po

```
# Messages from gettext_plural.py
# Copyright (C) 2009 Doug Hellmann
# This file is distributed under the same license
# as the PyMOTW package.
# Doug Hellmann <doug@doughellmann.com>, 2016.
#
#, fuzzy
msgid ""
msgstr ""
"Project-Id-Version: PyMOTW-3\n"
"Report-Msgid-Bugs-To: Doug Hellmann <doug@doughellmann.com>\n"
"POT-Creation-Date: 2016-01-24 13:04-0500\n"
"PO-Revision-Date: 2016-01-24 13:04-0500\n"
"Last-Translator: Doug Hellmann <doug@doughellmann.com>\n"
"Language-Team: en_US <doug@doughellmann.com>\n"
"MIME-Version: 1.0\n"
"Content-Type: text/plain; charset=UTF-8\n"
"Content-Transfer-Encoding: 8bit\n"
```

```
"Plural-Forms: nplurals=2; plural=n != 1;"

#: gettext_plural.py:15
#, python-format
msgid "{num} means singular."
msgid_plural "{num} means plural."
msgstr[0] "In en_US, {num} is singular."
msgstr[1] "In en_US, {num} is plural."
```

编译编目之后,运行几次测试脚本,展示不同的 N 值如何被转换为对应转换字符串的索引。

```
$ cd locale/en_US/LC_MESSAGES/; msgfmt -o plural.mo plural.po
$ cd ../../..
$ python3 gettext_plural.py 0

In en_US, 0 is plural.

$ python3 gettext_plural.py 1

In en_US, 1 is singular.

$ python3 gettext_plural.py 2

In en_US, 2 is plural.
```

15.1.5 应用与模块本地化

转换的作用域定义了如何安装 `gettext`,以及如何将 `gettext` 用于一个代码体。

15.1.5.1 应用本地化

对于全应用范围的转换,可以让作者使用 `__builtins__` 命名空间全局地安装类似 `ungettext()` 的函数,这是可以接受的,因为作者可以控制应用的顶层代码,并理解完整的需求。

代码清单 15-9:**gettext_app_builtin.py**

```python
import gettext

gettext.install(
    'example',
    'locale',
    names=['ngettext'],
)

print(_('This message is in the script.'))
```

`install()` 函数将 `gettext()` 绑定到 `__builtins__` 命名空间中的名 `_()`。它还增加了 `ngettext()` 和 `names` 中列出的其他函数。

15.1.5.2 模块本地化

对于一个库或单个模块,修改 `__builtins__` 并不是一个好主意,因为这可能会与某个应用全局值冲突。实际上,应当在模块前面手工地导入或重新绑定转换函数名。

代码清单 15-10:**gettext_module_global.py**

```python
import gettext
```

```
t = gettext.translation(
    'example',
    'locale',
    fallback=False,
)
_ = t.gettext
ngettext = t.ngettext

print(_('This message is in the script.'))
```

15.1.6 切换转换

前面的例子在整个程序期间都使用了一个转换。有些情况下，特别是对于 web 应用，还需要在不同时间使用不同的消息编目，而不要退出和重新设置环境。对于这些情况，`gettext` 中提供的基于类的 API 会更为方便。这些 API 调用实际上与这一节中介绍的全局调用相同，不过会发布消息编目对象，而且可以直接管理，因此可以使用多个编目。

提示：相关阅读材料
- `gettext` 的标准库文档[一]。
- `locale`：其他本地化工具。
- GNU gettext[二]：这个模块的消息编目格式、API 等都是基于 GNU 原来的 gettext 包。编目文件格式是兼容的，命令行脚本有类似的选项（甚至完全相同）。GNU gettext 手册[三]中对文件格式做了详细的描述，并且还介绍了处理这些文件格式的工具的 GNU 版本。
- Plural forms[四]：处理不同语言中单词和句子的复数形式。
- Internationalizing Python[五]：Martin von Löwis 撰写的关于 Python 应用国际化技术的一篇文章。
- Django Internationalization[六]：另一个关于使用 `gettext` 的很好的信息来源，还包括一些真实的示例。

15.2 `locale`：文化本地化 API

`locale` 模块是 Python 国际化和本地化支持库的一部分。它提供了一种标准方法，可以处理可能依赖于语言或用户位置的操作。例如，它会将数字格式化为货币形式，比较字符串以完成排序，并且还会处理日期。它不涉及转换（见 `gettext` 模块）或 Unicode 编码（见 `codecs` 模块）。

[一] https://docs.python.org/3.5/library/gettext.html
[二] www.gnu.org/software/gettext/
[三] www.gnu.org/software/gettext/manual/gettext.html
[四] www.gnu.org/software/gettext/manual/gettext.html#Plural-forms
[五] http://legacy.python.org/workshops/1997-10/proceedings/loewis.html
[六] https://docs.djangoproject.com/en/dev/topics/i18n/

说明：改变本地化环境可能会对整个应用带来影响，所以推荐的实践做法是避免改变库中的值，而是让应用一次性设置。在这一节的例子中，会在一个小程序中多次改变本地化环境，以强调不同本地化环境设置的差别。通常更常见的情况是由应用在启动时设置一次本地化环境，或者在接收到一个 Web 请求时设置一次本地化环境，而不会反复改变。

这一节将介绍 locale 模块中的一些高级函数。其他函数则为低级函数（format_string()）或者与管理应用的本地化环境有关（resetlocale()）。

15.2.1 探查当前本地化环境

要让用户改变应用的本地化环境设置，最常见的方式是通过一个环境变量（LC_ALL、LC_CTYPE、LANG 或 LANGUAGE，这取决于使用哪个平台）。然后应用调用 setlocale() 而不是指定硬编码的值，并且还会使用环境值。

代码清单 15-11：**locale_env.py**

```python
import locale
import os
import pprint

# Default settings based on the user's environment
locale.setlocale(locale.LC_ALL, '')

print('Environment settings:')
for env_name in ['LC_ALL', 'LC_CTYPE', 'LANG', 'LANGUAGE']:
    print('  {} = {}'.format(
        env_name, os.environ.get(env_name, ''))
    )

# What is the locale?
print('\nLocale from environment:', locale.getlocale())

template = """
Numeric formatting:

  Decimal point      : "{decimal_point}"
  Grouping positions : {grouping}
  Thousands separator: "{thousands_sep}"

Monetary formatting:

  International currency symbol   : "{int_curr_symbol!r}"
  Local currency symbol           : {currency_symbol!r}
  Symbol precedes positive value  : {p_cs_precedes}
  Symbol precedes negative value  : {n_cs_precedes}
  Decimal point                   : "{mon_decimal_point}"
  Digits in fractional values     : {frac_digits}
  Digits in fractional values,
                    international : {int_frac_digits}
  Grouping positions              : {mon_grouping}
  Thousands separator             : "{mon_thousands_sep}"
  Positive sign                   : "{positive_sign}"
  Positive sign position          : {p_sign_posn}
  Negative sign                   : "{negative_sign}"
  Negative sign position          : {n_sign_posn}

"""
```

```
    sign_positions = {
        0: 'Surrounded by parentheses',
        1: 'Before value and symbol',
        2: 'After value and symbol',
        3: 'Before value',
        4: 'After value',
        locale.CHAR_MAX: 'Unspecified',
    }

    info = {}
    info.update(locale.localeconv())
    info['p_sign_posn'] = sign_positions[info['p_sign_posn']]
    info['n_sign_posn'] = sign_positions[info['n_sign_posn']]

    print(template.format(**info))
```

localeconv()方法返回一个字典,其中包含本地化环境的约定。标准库文档中给出了完整的值名和定义。

在运行OS X 10.11.6的Mac上(所有变量都未设置),会生成以下输出。

```
$ export LANG=; export LC_CTYPE=; python3 locale_env.py

Environment settings:
  LC_ALL =
  LC_CTYPE =
  LANG =
  LANGUAGE =

Locale from environment: (None, None)

Numeric formatting:

  Decimal point      : "."
  Grouping positions : []
  Thousands separator: ""

Monetary formatting:

  International currency symbol   : "'"
  Local currency symbol           : ''
  Symbol precedes positive value  : 127
  Symbol precedes negative value  : 127
  Decimal point                   : ""
  Digits in fractional values     : 127
  Digits in fractional values,
                    international : 127
  Grouping positions              : []
  Thousands separator             : ""
  Positive sign                   : ""
  Positive sign position          : Unspecified
  Negative sign                   : ""
  Negative sign position          : Unspecified
```

运行同样的脚本,但设置了LANG变量,会显示本地化环境和默认编码如何改变。

美式英语(en_US):

```
$ LANG=en_US LC_CTYPE=en_US LC_ALL=en_US python3 locale_env.py

Environment settings:
  LC_ALL = en_US
```

```
    LC_CTYPE = en_US
    LANG = en_US
    LANGUAGE =

Locale from environment: ('en_US', 'ISO8859-1')

Numeric formatting:

    Decimal point      : "."
    Grouping positions : [3, 3, 0]
    Thousands separator: ","

Monetary formatting:

    International currency symbol   : "'USD '"
    Local currency symbol           : '$'
    Symbol precedes positive value  : 1
    Symbol precedes negative value  : 1
    Decimal point                   : "."
    Digits in fractional values     : 2
    Digits in fractional values,
                      international : 2
    Grouping positions              : [3, 3, 0]
    Thousands separator             : ","
    Positive sign                   : ""
    Positive sign position          : Before value and symbol
    Negative sign                   : "-"
    Negative sign position          : Before value and symbol
```

法语（fr_FR）:

```
$ LANG=fr_FR LC_CTYPE=fr_FR LC_ALL=fr_FR python3 locale_env.py

Environment settings:
    LC_ALL = fr_FR
    LC_CTYPE = fr_FR
    LANG = fr_FR
    LANGUAGE =

Locale from environment: ('fr_FR', 'ISO8859-1')

Numeric formatting:

    Decimal point      : ","
    Grouping positions : [127]
    Thousands separator: ""

Monetary formatting:

    International currency symbol   : "'EUR '"
    Local currency symbol           : 'Eu'
    Symbol precedes positive value  : 0
    Symbol precedes negative value  : 0
    Decimal point                   : ","
    Digits in fractional values     : 2
    Digits in fractional values,
                      international : 2
    Grouping positions              : [3, 3, 0]
    Thousands separator             : " "
    Positive sign                   : ""
    Positive sign position          : Before value and symbol
    Negative sign                   : "-"
    Negative sign position          : After value and symbol
```

西班牙语(es_ES):

```
$ LANG=es_ES LC_CTYPE=es_ES LC_ALL=es_ES python3 locale_env.py

Environment settings:
  LC_ALL = es_ES
  LC_CTYPE = es_ES
  LANG = es_ES
  LANGUAGE =

Locale from environment: ('es_ES', 'ISO8859-1')

Numeric formatting:
  Decimal point        : ","
  Grouping positions : [127]
  Thousands separator: ""

Monetary formatting:

  International currency symbol   : "'EUR '"
  Local currency symbol           : 'Eu'
  Symbol precedes positive value  : 0
  Symbol precedes negative value  : 0
  Decimal point                   : ","
  Digits in fractional values     : 2
  Digits in fractional values,
              international       : 2
  Grouping positions              : [3, 3, 0]
  Thousands separator             : "."
  Positive sign                   : ""
  Positive sign position          : Before value and symbol
  Negative sign                   : "-"
  Negative sign position          : Before value and symbol
```

葡萄牙语(pt_PT):

```
$ LANG=pt_PT LC_CTYPE=pt_PT LC_ALL=pt_PT python3 locale_env.py

Environment settings:
  LC_ALL = pt_PT
  LC_CTYPE = pt_PT
  LANG = pt_PT
  LANGUAGE =

Locale from environment: ('pt_PT', 'ISO8859-1')

Numeric formatting:

  Decimal point        : ","
  Grouping positions : []
  Thousands separator: " "

Monetary formatting:

  International currency symbol   : "'EUR '"
  Local currency symbol           : 'Eu'
  Symbol precedes positive value  : 0
  Symbol precedes negative value  : 0
  Decimal point                   : "."
  Digits in fractional values     : 2
  Digits in fractional values,
              international       : 2
  Grouping positions              : [3, 3, 0]
```

```
  Thousands separator          : "."
  Positive sign                : ""
  Positive sign position       : Before value and symbol
  Negative sign                : "-"
  Negative sign position       : Before value and symbol
```

波兰语(pl_PL):

```
$ LANG=pl_PL LC_CTYPE=pl_PL LC_ALL=pl_PL python3 locale_env.py
Environment settings:
  LC_ALL = pl_PL
  LC_CTYPE = pl_PL
  LANG = pl_PL
  LANGUAGE =

Locale from environment: ('pl_PL', 'ISO8859-2')

Numeric formatting:

  Decimal point     : ","
  Grouping positions : [3, 3, 0]
  Thousands separator: " "

Monetary formatting:

  International currency symbol    : "'PLN '"
  Local currency symbol            : 'z'
  Symbol precedes positive value   : 1
  Symbol precedes negative value   : 1
  Decimal point                    : ","
  Digits in fractional values      : 2
  Digits in fractional values,
                     international : 2
  Grouping positions               : [3, 3, 0]
  Thousands separator              : " "
  Positive sign                    : ""
  Positive sign position           : After value
  Negative sign                    : "-"
  Negative sign position           : After value
```

15.2.2 货币

从前面的示例输出可以看到，改变本地化环境会更新货币符号设置，还会改变分隔整数和小数部分的字符。这个例子循环处理多个不同的本地化环境，针对各个本地化环境，分别打印一个格式化的正货币值和负货币值。

代码清单 15-12：locale_currency.py

```python
import locale

sample_locales = [
    ('USA', 'en_US'),
    ('France', 'fr_FR'),
    ('Spain', 'es_ES'),
    ('Portugal', 'pt_PT'),
    ('Poland', 'pl_PL'),
]

for name, loc in sample_locales:
```

```
    locale.setlocale(locale.LC_ALL, loc)
    print('{:>10}: {:>10}   {:>10}'.format(
        name,
        locale.currency(1234.56),
        locale.currency(-1234.56),
    ))
```

输出为下面这个小表。

```
$ python3 locale_currency.py

      USA:    $1234.56     -$1234.56
   France: 1234,56 Eu    1234,56 Eu-
    Spain: 1234,56 Eu    -1234,56 Eu
 Portugal: 1234.56 Eu    -1234.56 Eu
   Poland: ł z 1234,56   ł z 1234,56-
```

15.2.3 格式化数字

与货币无关的数字也会根据本地化环境以不同方式格式化。具体来讲，数字中会有一些分组（grouping）字符，用于将大数字分隔为可读的小块。

代码清单 15-13：**locale_grouping.py**

```
import locale

sample_locales = [
    ('USA', 'en_US'),
    ('France', 'fr_FR'),
    ('Spain', 'es_ES'),
    ('Portugal', 'pt_PT'),
    ('Poland', 'pl_PL'),
]

print('{:>10} {:>10} {:>15}'.format(
    'Locale', 'Integer', 'Float')
)
for name, loc in sample_locales:
    locale.setlocale(locale.LC_ALL, loc)

    print('{:>10}'.format(name), end=' ')
    print(locale.format('%10d', 123456, grouping=True), end=' ')
    print(locale.format('%15.2f', 123456.78, grouping=True))
```

要格式化不带货币符号的数字，可以使用 `format()` 而不是 `currency()`。

```
$ python3 locale_grouping.py

    Locale    Integer           Float
       USA    123,456       123,456.78
    France     123456       123456,78
     Spain     123456       123456,78
  Portugal     123456       123456,78
    Poland    123 456       123 456,78
```

要把根据本地化环境格式化的数字转换为规范化的数字（格式与本地化环境无关），可以使用 `delocalize()`。

代码清单 15-14：**locale_delocalize.py**

```
import locale

sample_locales = [
    ('USA', 'en_US'),
    ('France', 'fr_FR'),
    ('Spain', 'es_ES'),
    ('Portugal', 'pt_PT'),
    ('Poland', 'pl_PL'),
]

for name, loc in sample_locales:
    locale.setlocale(locale.LC_ALL, loc)
    localized = locale.format('%0.2f', 123456.78, grouping=True)
    delocalized = locale.delocalize(localized)
    print('{:>10}: {:>10}   {:>10}'.format(
        name,
        localized,
        delocalized,
    ))
```

这会删除分组符号，另外小数分隔符总是被转换为一个点号（.）。

```
$ python3 locale_delocalize.py

       USA: 123,456.78   123456.78
    France: 123456,78    123456.78
     Spain: 123456,78    123456.78
  Portugal: 123456,78    123456.78
    Poland: 123 456,78   123456.78
```

15.2.4 解析数字

除了以不同格式生成输出，locale 模块还可以帮助解析输入。它包含 atoi() 和 atof() 函数，可以根据本地化环境的数值格式约定将字符串转换为整数和浮点值。

代码清单 15-15：**locale_atof.py**

```
import locale

sample_data = [
    ('USA', 'en_US', '1,234.56'),
    ('France', 'fr_FR', '1234,56'),
    ('Spain', 'es_ES', '1234,56'),
    ('Portugal', 'pt_PT', '1234.56'),
    ('Poland', 'pl_PL', '1 234,56'),
]

for name, loc, a in sample_data:
    locale.setlocale(locale.LC_ALL, loc)
    print('{:>10}: {:>9} => {:f}'.format(
        name,
        a,
        locale.atof(a),
    ))
```

解析器会识别本地化环境的分组和小数分隔符值。

```
$ python3 locale_atof.py

    USA:  1,234.56 => 1234.560000
 France:  1234,56  => 1234.560000
  Spain:  1234,56  => 1234.560000
Portugal: 1234.56  => 1234.560000
 Poland:  1 234,56 => 1234.560000
```

15.2.5 日期和时间

本地化的另一个重要方面是日期和时间格式化。

代码清单 15-16：**locale_date.py**

```python
import locale
import time

sample_locales = [
    ('USA', 'en_US'),
    ('France', 'fr_FR'),
    ('Spain', 'es_ES'),
    ('Portugal', 'pt_PT'),
    ('Poland', 'pl_PL'),
]

for name, loc in sample_locales:
    locale.setlocale(locale.LC_ALL, loc)
    format = locale.nl_langinfo(locale.D_T_FMT)
    print('{:>10}: {}'.format(name, time.strftime(format)))
```

这个例子使用本地化环境的日期格式化串来打印当前日期和时间。

```
$ python3 locale_date.py

     USA: Fri Aug  5 17:33:31 2016
  France: Ven  5 aoû 17:33:31 2016
   Spain: vie  5 ago 17:33:31 2016
Portugal: Sex  5 Ago 17:33:31 2016
  Poland: ptk  5 sie 17:33:31 2016
```

提示：相关阅读材料
- `locale` 的标准库文档[○]。
- `locale` 的 Python 2 到 Python 3 移植说明。
- `gettext`：完成转换的消息编目。

[○] https://docs.python.org/3.5/library/locale.html

第16章
开发工具

在Python的发展历程中，它已经演化为一个庞大的模块生态系统，这些模块的作用就是让Python开发人员的日子更好过，有了它们，便不再需要开发人员对一切都从头开始构建。开发人员完成工作所依赖的工具也是如此（即使有些工具并不真正用于程序的最后版本）。本章将涵盖Python包含的常用模块，这些模块可以为常见的开发任务（如测试、调试和性能分析）提供便利。

对开发人员来说，最基本的帮助形式就是所使用的代码的文档。`pydoc`模块可以从任何可移植模块源代码中包含的docstring生成格式化的参考文档。

Python包含两个测试框架，可以自动地执行代码，并验证代码是否正常工作。`doctest`可以从文档中包含的示例抽取测试场景，这些示例可能在源代码中，也可能作为独立的文件。`unittest`是一个功能完备的自动化测试框架，支持固件、预定义测试套件和测试发现。

`trace`模块会监视Python如何执行一个程序，生成一个报告并显示每行运行多少次。这个信息可以用于查找自动化测试套件未测试的代码路径，并研究函数调用图来查找模块之间的依赖性。

编写和运行测试可以发现大多数程序中的问题。Python可以更容易地完成调试，因为通常未处理的错误会作为traceback打印到控制台上。如果不是在文本控制台环境中运行程序，则可以用`traceback`为日志文件或消息对话框准备类似的输出。有些情况下，标准traceback没有提供足够的信息，对于这些情况，可以使用`cgitb`查看详细信息，如栈中每一级的局部变量设置和源代码上下文。`cgitb`还可以将traceback格式化为HTML形式，用于在Web应用中报告错误。

一旦找出问题的位置，便可以使用`pdb`模块中的交互式调试工具单步调试代码，通过显示导致出现错误情况的代码路径，可以更迅速地得到解决方案。这个模块还有利于使用现场（live）对象和代码来尝试修改，这会减少迭代次数，可以更快地找出去除一个错误所需的正确修改。

测试并调试程序使它能正常工作之后，下一步就是研究性能。通过使用`profile`和`timeit`，开发人员可以测量一个程序的运行速度，找出速度慢的部分，将它们隔离出来并加以改进。

在类似Python的语言中，要以一致的方式缩进源代码，这很重要，在这里空白符也是语法的一部分。`tabnanny`模块提供了一个扫描器，会报告有歧义的缩进使用；这个模块还可以在测试中用于确保代码满足最低标准，然后再签入源代码存储库。

运行 Python 程序时，会向解释器提供原程序源代码的一个字节编译版本。字节编译版本可以动态创建，也可以在程序打包时创建。`compileall` 模块提供了安装程序使用的界面，另外还提供了打包工具，可以创建包含一个模块字节代码的文件。它可以在开发环境中用来确保一个文件没有语法错误，并在发布程序时构建可用于打包的字节编译文件。

在源代码级，`pyclbr` 模块提供了一个类浏览器，文本编辑器或其他程序可以使用这个类浏览器扫描 Python 源文件，查找感兴趣的符号，如函数和类；这一步不必导入代码，也不会引发副作用。

Python 虚拟环境由 `venv` 管理，会通过定义隔离的环境来安装包和运行程序。这样可以很容易地用不同版本的依赖库测试相同的程序，还可以在同一个计算机上安装依赖库有冲突的不同程序。

要充分利用扩展模块、框架和通过 Python Package Index 提供的庞大的工具生态系统，这需要一个包安装工具。Python 的包安装工具 pip 不随解释器发布，因为与工具所需的更新相比，Python 语言的发布周期太长。`ensurepip` 模块可以用来安装最新版本的 pip。

16.1 `pydoc`：模块的联机帮助

`pydoc` 模块导入了一个 Python 模块，并使用它的内容在运行时生成帮助文本。只要对象中包含 docstring，输出便会包括所有这些对象的 docstring。这里会描述模块的所有类、方法和函数。

16.1.1 纯文本帮助

作为一个命令行程序运行 `pydoc` 并传入一个模块的名字，会在控制台上生成这个模块的帮助文本和它的内容，这里可能会使用一个分页程序（如果已配置）。例如，要查看 `atexit` 模块的帮助文本，可以运行 `pydoc atexit`。

```
$ pydoc atexit

Help on built-in module atexit:

NAME
    atexit - allow programmer to define multiple exit functions
to be executed upon normal program termination.
DESCRIPTION
    Two public functions, register and unregister, are defined.

FUNCTIONS
    register(...)
        register(func, *args, **kwargs) -> func

        Register a function to be executed upon normal program
termination.

            func - function to be called at exit
            args - optional arguments to pass to func
            kwargs - optional keyword arguments to pass to func

            func is returned to facilitate usage as a decorator.
```

```
        unregister(...)
            unregister(func) -> None

            Unregister an exit function which was previously
registered using
            atexit.register

                func - function to be unregistered
FILE
    (built-in)
```

16.1.2　HTML 帮助

`pydoc` 还可以生成 HTML 输出，可能将一个静态文件写至一个本地目录，或者启动一个 Web 服务器在线浏览文档。

```
$ pydoc -w atexit
```

前面的代码将在当前目录创建 `atexit.html`。

```
$ pydoc -p 5000
Server ready at http://localhost:5000/
Server commands: [b]rowser, [q]uit
server> q
Server stopped
```

以上代码将启动一个 Web 服务器监听 `http://localhost:5000/`。这个服务器会在你浏览时动态生成文档。使用 b 命令可以自动打开一个浏览器窗口，使用 q 会停止服务器。

16.1.3　交互式帮助

`pydoc` 还为 `__builtins__` 增加了一个函数 `help()`，可以从 Python 解释器提示窗口访问同样的信息。

```
$ python

Python 3.5.2 (v3.5.2:4def2a2901a5, Jun 26 2016, 10:47:25)
[GCC 4.2.1 (Apple Inc. build 5666) (dot 3)] on darwin
Type "help", "copyright", "credits" or "license" for more
information.
>>> help('atexit')
Help on module atexit:

NAME
    atexit - allow programmer to define multiple exit functions
to be executed upon normal program termination.

...
```

提示：相关阅读材料

- `pydoc` 的标准库文档[○]。
- `inspect`：可以通过编程用 `inspect` 模块获取一个对象的 docstring。

[○] https://docs.python.org/3.5/library/pydoc.html

16.2 doctest：通过文档完成测试

doctest 会运行文档中嵌入的例子，并验证它们是否能生成所期望的结果，来对源代码进行测试。它的做法是解析帮助文档，找到例子，运行这些例子，然后将输出文本与所期望的值进行比较。很多开发人员发现 doctest 比 unittest 更易于使用，因为如果采用最简单的形式，那么使用 doctest 之前无须学习新的 API。不过，随着例子变得越来越复杂，由于缺乏固件管理，编写 doctest 测试可能会比使用 unittest 更麻烦。

16.2.1 起步

建立 doctest 的第一步是使用交互式解释器创建例子，然后把这些例子复制粘贴到模块的 docstring 中。在这里，my_function() 给出了两个例子。

代码清单 16-1：doctest_simple.py

```
def my_function(a, b):
    """
    >>> my_function(2, 3)
    6
    >>> my_function('a', 3)
    'aaa'
    """
    return a * b
```

运行这些测试时，通过指定 -m 选项将 doctest 作为主程序。运行测试时通常不会生成输出，所以下面的例子包含了 -v 选项来得到更详细的输出。

```
$ python3 -m doctest -v doctest_simple.py

Trying:
    my_function(2, 3)
Expecting:
    6
ok
Trying:
    my_function('a', 3)
Expecting:
    'aaa'
ok
1 items had no tests:
    doctest_simple
1 items passed all tests:
   2 tests in doctest_simple.my_function
2 tests in 2 items.
2 passed and 0 failed.
Test passed.
```

例子并不总能独立地作为一个函数的解释，所以 doctest 还允许有包围文本。它会查找以解释器提示符（>>>）开头的行，找出测试用例的开始位置；用例以一个空行结束，或者以下一个解释器提示符结束。介于中间的文本会被忽略，并且它们可以采用任何格式（只要看上去不像是一个测试用例即可）。

代码清单 16-2: **doctest_simple_with_docs.py**

```
def my_function(a, b):
    """Returns a * b.

    Works with numbers:

    >>> my_function(2, 3)
    6

    and strings:

    >>> my_function('a', 3)
    'aaa'
    """
    return a * b
```

更新的 docstring 中如果有包围文本，那么这些包围文本对人类读者更有用。由于它会被 doctest 忽略，所以结果是一样的。

```
$ python3 -m doctest -v doctest_simple_with_docs.py

Trying:
    my_function(2, 3)
Expecting:
    6
ok
Trying:
    my_function('a', 3)
Expecting:
    'aaa'
ok
1 items had no tests:
    doctest_simple_with_docs
1 items passed all tests:
   2 tests in doctest_simple_with_docs.my_function
2 tests in 2 items.
2 passed and 0 failed.
Test passed.
```

16.2.2 处理不可预测的输出

还有一些情况可能无法预测准确的输出，不过仍可以测试。例如，每次运行测试时本地日期和时间值以及对象 ID 会改变，浮点值表示中使用的默认精度取决于编译器选项，另外容器对象（如字典）的串表示可能是不确定的。尽管这些条件可能无法控制，但是确实有一些技术可以处理。

例如，在 CPython 中，对象标识符是基于保存这个对象的数据结构的内存地址。

代码清单 16-3: **doctest_unpredictable.py**

```
class MyClass:
    pass
def unpredictable(obj):
    """Returns a new list containing obj.

    >>> unpredictable(MyClass())
    [<doctest_unpredictable.MyClass object at 0x10055a2d0>]
    """
    return [obj]
```

每次程序运行时，这些 ID 值都会改变，因为这些值会被加载到内存的不同部分。

```
$ python3 -m doctest -v doctest_unpredictable.py

Trying:
    unpredictable(MyClass())
Expecting:
    [<doctest_unpredictable.MyClass object at 0x10055a2d0>]
**********************************************************************
File ".../doctest_unpredictable.py", line 17, in doctest_unpredi
ctable.unpredictable
Failed example:
    unpredictable(MyClass())
Expected:
    [<doctest_unpredictable.MyClass object at 0x10055a2d0>]
Got:
    [<doctest_unpredictable.MyClass object at 0x1016a4160>]
2 items had no tests:
    doctest_unpredictable
    doctest_unpredictable.MyClass
**********************************************************************
1 items had failures:
   1 of   1 in doctest_unpredictable.unpredictable
1 tests in 3 items.
0 passed and 1 failed.
***Test Failed*** 1 failures.
```

测试的值可能会以不可预测的方式改变时，如果具体值对于测试结果并不重要，则可以使用 ELLIPSIS 选项告诉 doctest 忽略验证值的某些部分。

代码清单 16-4：doctest_ellipsis.py

```python
class MyClass:
    pass


def unpredictable(obj):
    """Returns a new list containing obj.
    >>> unpredictable(MyClass()) #doctest: +ELLIPSIS
    [<doctest_ellipsis.MyClass object at 0x...>]
    """
    return [obj]
```

unpredictable() 调用后面的 #doctest: +ELLIPSIS 注释告诉 doctest 打开这个测试的 ELLIPSIS 选项。... 将替换对象 ID 中的内存地址，这样就会忽略期望值中的这一部分。实际输出将成功匹配，并且通过测试。

```
$ python3 -m doctest -v doctest_ellipsis.py

Trying:
    unpredictable(MyClass()) #doctest: +ELLIPSIS
Expecting:
    [<doctest_ellipsis.MyClass object at 0x...>]
ok
2 items had no tests:
    doctest_ellipsis
    doctest_ellipsis.MyClass
1 items passed all tests:
   1 tests in doctest_ellipsis.unpredictable
1 tests in 3 items.
```

```
1 passed and 0 failed.
Test passed.
```

有些情况下,不能忽略不可预测的值,因为这会让测试不完备或不准确。例如,处理一些数据类型时,如果其数据的串表示不一致,那么简单的测试很快会变得越来越复杂。举例来说,字典的串形式可能会根据增加键的顺序不同而改变。

代码清单 16-5:doctest_hashed_values.py

```python
keys = ['a', 'aa', 'aaa']

print('dict:', {k: len(k) for k in keys})
print('set :', set(keys))
```

由于散列随机化和键冲突,每次运行这个脚本时,字典的内部键列表的顺序都会有所不同。集合(set)会使用相同的散列算法,并提供相同的行为。

```
$ python3 doctest_hashed_values.py

dict: {'aa': 2, 'a': 1, 'aaa': 3}
set : {'aa', 'a', 'aaa'}

$ python3 doctest_hashed_values.py

dict: {'a': 1, 'aa': 2, 'aaa': 3}
set : {'a', 'aa', 'aaa'}
```

要处理这些潜在的差异,最好的方法是创建测试来生成不太可能改变的值。对于字典和集合,这可能意味着需要分别查找特定的键,并生成数据结构内容的一个有序列表,或者与一个字面值比较相等性而不依赖于串表示。

代码清单 16-6:doctest_hashed_values_tests.py

```python
import collections


def group_by_length(words):
    """Returns a dictionary grouping words into sets by length.

    >>> grouped = group_by_length([ 'python', 'module', 'of',
    ... 'the', 'week' ])
    >>> grouped == { 2:set(['of']),
    ...              3:set(['the']),
    ...              4:set(['week']),
    ...              6:set(['python', 'module']),
    ...              }
    True

    """
    d = collections.defaultdict(set)
    for word in words:
        d[len(word)].add(word)
    return d
```

以上代码中的这个例子实际上会被解释为两个单独的测试,第一个没有任何控制台输出,第二个则会得到比较操作的布尔结果。

```
$ python3 -m doctest -v doctest_hashed_values_tests.py

Trying:
    grouped = group_by_length([ 'python', 'module', 'of',
    'the', 'week' ])
Expecting nothing
ok
Trying:
    grouped == { 2:set(['of']),
                 3:set(['the']),
                 4:set(['week']),
                 6:set(['python', 'module']),
                 }
Expecting:
    True
ok
1 items had no tests:
    doctest_hashed_values_tests
1 items passed all tests:
   2 tests in doctest_hashed_values_tests.group_by_length
2 tests in 2 items.
2 passed and 0 failed.
Test passed.
```

16.2.3 traceback

traceback 是不断变化的数据的一个特殊情况。由于 traceback 中的路径取决于模块安装在文件系统上的具体位置，如果像其他输出一样处理，则可能无法编写可移植的测试。

代码清单 16-7：**doctest_tracebacks.py**

```
def this_raises():
    """This function always raises an exception.

    >>> this_raises()
    Traceback (most recent call last):
      File "<stdin>", line 1, in <module>
      File "/no/such/path/doctest_tracebacks.py", line 14, in
      this_raises
        raise RuntimeError('here is the error')
    RuntimeError: here is the error
    """
    raise RuntimeError('here is the error')
```

doctest 做了一些特殊工作来识别 traceback，并忽略可能因系统不同而改变的部分。

```
$ python3 -m doctest -v doctest_tracebacks.py

Trying:
    this_raises()
Expecting:
    Traceback (most recent call last):
      File "<stdin>", line 1, in <module>
      File "/no/such/path/doctest_tracebacks.py", line 14, in
      this_raises
        raise RuntimeError('here is the error')
    RuntimeError: here is the error
ok
1 items had no tests:
    doctest_tracebacks
1 items passed all tests:
```

```
    1 tests in doctest_tracebacks.this_raises
1 tests in 2 items.
1 passed and 0 failed.
Test passed.
```

实际上，整个 traceback 体都将被忽略，可以略去。

代码清单 16-8：doctest_tracebacks_no_body.py

```
def this_raises():
    """This function always raises an exception.

    >>> this_raises()
    Traceback (most recent call last):
    RuntimeError: here is the error

    >>> this_raises()
    Traceback (innermost last):
    RuntimeError: here is the error
    """
    raise RuntimeError('here is the error')
```

doctest 看到一个 traceback 首部行时（可能是 "Traceback (most recent call last):" 或 "Traceback (innermost last):"，这取决于所用的 Python 版本），它会跳过去继续查找异常类型和消息，完全忽略中间的各行。

```
$ python3 -m doctest -v doctest_tracebacks_no_body.py

Trying:
    this_raises()
Expecting:
    Traceback (most recent call last):
    RuntimeError: here is the error
ok
Trying:
    this_raises()
Expecting:
    Traceback (innermost last):
    RuntimeError: here is the error
ok
1 items had no tests:
    doctest_tracebacks_no_body
1 items passed all tests:
    2 tests in doctest_tracebacks_no_body.this_raises
2 tests in 2 items.
2 passed and 0 failed.
Test passed.
```

16.2.4 避开空白符

在实际应用中，输出通常包括空白符，如空行、tab 和多余的间隔，目的是使输出更可读。尤其是空行会导致 doctest 出现问题，因为一般会用空行作为测试的分界线。

代码清单 16-9：doctest_blankline_fail.py

```
def double_space(lines):
    """Prints a list of double-spaced lines.

    >>> double_space(['Line one.', 'Line two.'])
```

```
    Line one.

    Line two.
    """
    for l in lines:
        print(l)
        print()
```

double_space()取一个输入行列表,在输入行之间加入空行,打印时行之间会有两个空行间隔。

```
$ python3 -m doctest -v doctest_blankline_fail.py

Trying:
    double_space(['Line one.', 'Line two.'])
Expecting:
    Line one.
**********************************************************************
File ".../doctest_blankline_fail.py", line 12, in doctest_blankl
ine_fail.double_space
Failed example:
    double_space(['Line one.', 'Line two.'])
Expected:
    Line one.
Got:
    Line one.
    <BLANKLINE>
    Line two.
    <BLANKLINE>
1 items had no tests:
    doctest_blankline_fail
**********************************************************************
1 items had failures:
   1 of   1 in doctest_blankline_fail.double_space
1 tests in 2 items.
0 passed and 1 failed.
***Test Failed*** 1 failures.
```

在前面的例子中,测试会失败,因为它把docstring中包含Line one.的那一行后面的空行解释为示例输出的末尾。为了与空行匹配,要把示例输入中的空行替换为<BLANKLINE>。

代码清单16-10:**doctest_blankline.py**

```
def double_space(lines):
    """Prints a list of double-spaced lines.

    >>> double_space(['Line one.', 'Line two.'])
    Line one.
    <BLANKLINE>
    Line two.
    <BLANKLINE>
    """
    for l in lines:
        print(l)
        print()
```

在这个例子中,完成比较之前,doctest将具体的空行替换为相同的字面量,所以现在具体值和期望值匹配，并且测试通过。

```
$ python3 -m doctest -v doctest_blankline.py

Trying:
    double_space(['Line one.', 'Line two.'])
Expecting:
    Line one.
    <BLANKLINE>
    Line two.
    <BLANKLINE>
ok
1 items had no tests:
    doctest_blankline
1 items passed all tests:
   1 tests in doctest_blankline.double_space
1 tests in 2 items.
1 passed and 0 failed.
Test passed.
```

行中包含的空白符也可能导致测试出问题。下面这个例子在 6 后面有一个额外的空格。

代码清单 16-11：doctest_extra_space.py

```
def my_function(a, b):
    """
    >>> my_function(2, 3)
    6
    >>> my_function('a', 3)
    'aaa'
    """
    return a * b
```

这个额外的空格可能是因为复制粘贴错误而引入代码，不过由于它们位于行末尾，所以其可能在源文件中不被注意，在测试失败报告中也看不到。

```
$ python3 -m doctest -v doctest_extra_space.py

Trying:
    my_function(2, 3)
Expecting:
    6
**********************************************************
File ".../doctest_extra_space.py", line 15, in doctest_extra_spa
ce.my_function
Failed example:
    my_function(2, 3)
Expected:
    6
Got:
    6
Trying:
    my_function('a', 3)
Expecting:
    'aaa'
ok
1 items had no tests:
    doctest_extra_space
**********************************************************
1 items had failures:
   1 of   2 in doctest_extra_space.my_function
```

```
2 tests in 2 items.
1 passed and 1 failed.
***Test Failed*** 1 failures.
```

使用某个基于差异的报告选项，如 REPORT_NDIFF，可以更详细地显示实际值和期望值之间的差异，这样就会看到这个额外的空格。

代码清单 16-12：doctest_ndiff.py

```python
def my_function(a, b):
    """
    >>> my_function(2, 3) #doctest: +REPORT_NDIFF
    6
    >>> my_function('a', 3)
    'aaa'
    """
    return a * b
```

也可以使用统一 diff (REPORT_UDIFF) 和上下文 diff (REPORT_CDIFF)。

```
$ python3 -m doctest -v doctest_ndiff.py

Trying:
    my_function(2, 3) #doctest: +REPORT_NDIFF
Expecting:
    6
**********************************************************************
File ".../doctest_ndiff.py", line 16, in doctest_ndiff.my_functi
on
Failed example:
    my_function(2, 3) #doctest: +REPORT_NDIFF
Differences (ndiff with -expected +actual):
    - 6
    ?  -
    + 6
Trying:
    my_function('a', 3)
Expecting:
    'aaa'
ok
1 items had no tests:
    doctest_ndiff
**********************************************************************
1 items had failures:
   1 of   2 in doctest_ndiff.my_function
2 tests in 2 items.
1 passed and 1 failed.
***Test Failed*** 1 failures.
```

有些情况下，可能要在测试的示例输出中增加额外的空白符，而让 doctest 忽略这些空白符。例如，尽管有些数据结构的表示可以显示在一行上，不过多行显示可能更易读。

```python
def my_function(a, b):
    """Returns a * b.

    >>> my_function(['A', 'B'], 3) #doctest: +NORMALIZE_WHITESPACE
    ['A', 'B',
     'A', 'B',
     'A', 'B']
    This does not match because of the extra space after the [ in
    the list.
```

```
    >>> my_function(['A', 'B'], 2) #doctest: +NORMALIZE_WHITESPACE
    [ 'A', 'B',
      'A', 'B', ]
    """
    return a * b
```

NORMALIZE_WHITESPACE 被打开时，实际值和期望值中的空白符会被认为是匹配的。如果输出中不存在空白符，那么期望值中就不能增加空白符，不过空白符序列的长度和实际的空白字符不需要一致。第一个测试示例满足这个原则，并且测试通过（尽管输入包含额外的空格和换行）。第二个测试示例在 [与] 之间有额外的空白符，所以测试失败。

```
$ python3 -m doctest -v doctest_normalize_whitespace.py

Trying:
    my_function(['A', 'B'], 3) #doctest: +NORMALIZE_WHITESPACE
Expecting:
    ['A', 'B',
     'A', 'B',
     'A', 'B',]
**********************************************************************
File "doctest_normalize_whitespace.py", line 13, in doctest_nor
malize_whitespace.my_function
Failed example:
    my_function(['A', 'B'], 3) #doctest: +NORMALIZE_WHITESPACE
Expected:
    ['A', 'B',
     'A', 'B',
     'A', 'B',]
Got:
    ['A', 'B', 'A', 'B', 'A', 'B']
Trying:
    my_function(['A', 'B'], 2) #doctest: +NORMALIZE_WHITESPACE
Expecting:
    [ 'A', 'B',
      'A', 'B', ]
**********************************************************************
File "doctest_normalize_whitespace.py", line 21, in doctest_nor
malize_whitespace.my_function
Failed example:
    my_function(['A', 'B'], 2) #doctest: +NORMALIZE_WHITESPACE
Expected:
    [ 'A', 'B',
      'A', 'B', ]
Got:
    ['A', 'B', 'A', 'B']
1 items had no tests:
    doctest_normalize_whitespace
**********************************************************************
1 items had failures:
   2 of   2 in doctest_normalize_whitespace.my_function
2 tests in 2 items.
0 passed and 2 failed.
***Test Failed*** 2 failures.
```

16.2.5 测试位置

目前为止的例子中，所有测试都写在所测试的函数的 docstring 中。对于查看 docstring 的用户来说这很方便，可以帮助他们使用这个函数（特别是对于 **pydoc**），不过 **doctest** 还会在其他地方查找测试。最可能完成测试的地方就是在模块中其他位置找到的 docstring。

代码清单 16-13：`doctest_docstrings.py`

```
"""Tests can appear in any docstring within the module.

Module-level tests cross class and function boundaries.

>>> A('a') == B('b')
False
"""


class A:
    """Simple class.

    >>> A('instance_name').name
    'instance_name'
    """

    def __init__(self, name):
        self.name = name

    def method(self):
        """Returns an unusual value.

        >>> A('name').method()
        'eman'
        """
        return ''.join(reversed(self.name))


class B(A):
    """Another simple class.

    >>> B('different_name').name
    'different_name'
    """
```

模块级、类级和函数级的 docstring 都可以包含测试。

```
$ python3 -m doctest -v doctest_docstrings.py

Trying:
    A('a') == B('b')
Expecting:
    False
ok
Trying:
    A('instance_name').name
Expecting:
    'instance_name'
ok
Trying:
    A('name').method()
Expecting:
    'eman'
ok
Trying:
    B('different_name').name
Expecting:
    'different_name'
ok
1 items had no tests:
    doctest_docstrings.A.__init__
4 items passed all tests:
   1 tests in doctest_docstrings
```

```
        1 tests in doctest_docstrings.A
        1 tests in doctest_docstrings.A.method
        1 tests in doctest_docstrings.B
    4 tests in 5 items.
    4 passed and 0 failed.
    Test passed.
```

有些情况下，模块的测试应当包含在源代码中，而不是放在模块的帮助文本中。在这种情况下，需要把测试放在 docstring 以外的位置。doctest 还会查找一个模块级变量，名为 __test__，用它来找到其他测试。__test__ 的值应当是一个字典，将测试集合名（字符串）映射为字符串、模块、类或函数。

代码清单 16-14：doctest_private_tests.py

```
import doctest_private_tests_external

__test__ = {
    'numbers': """
>>> my_function(2, 3)
6

>>> my_function(2.0, 3)
6.0
""",

    'strings': """
>>> my_function('a', 3)
'aaa'

>>> my_function(3, 'a')
'aaa'
""",

    'external': doctest_private_tests_external,
}

def my_function(a, b):
    """Returns a * b
    """
    return a * b
```

如果与一个键关联的值是字符串，那么其会被处理为一个 docstring，并在其中扫描测试。如果值是一个类或函数，则 doctest 会递归地搜索 docstring，然后在其中扫描测试。在下面这个例子中，模块 doctest_private_tests_external 的 docstring 内有一个测试。

代码清单 16-15：doctest_private_tests_external.py

```
"""External tests associated with doctest_private_tests.py.

>>> my_function(['A', 'B', 'C'], 2)
['A', 'B', 'C', 'A', 'B', 'C']
"""
```

扫描示例文件之后，doctest 总共找到了 5 个要运行的测试。

```
$ python3 -m doctest -v doctest_private_tests.py
Trying:
    my_function(['A', 'B', 'C'], 2)
Expecting:
    ['A', 'B', 'C', 'A', 'B', 'C']
ok
Trying:
    my_function(2, 3)
Expecting:
    6
ok
Trying:
    my_function(2.0, 3)
Expecting:
    6.0
ok
Trying:
    my_function('a', 3)
Expecting:
    'aaa'
ok
Trying:
    my_function(3, 'a')
Expecting:
    'aaa'
ok
2 items had no tests:
    doctest_private_tests
    doctest_private_tests.my_function
3 items passed all tests:
   1 tests in doctest_private_tests.__test__.external
   2 tests in doctest_private_tests.__test__.numbers
   2 tests in doctest_private_tests.__test__.strings
5 tests in 5 items.
5 passed and 0 failed.
Test passed.
```

16.2.6 外部文档

将测试混合在常规代码中并不是使用 **doctest** 的唯一办法。也可以使用外部工程文档文件（如 reStructuredText 文件）中嵌入的示例。

代码清单 16-16：doctest_in_help.py

```
def my_function(a, b):
    """Returns a*b
    """
    return a * b
```

这个示例模块的帮助被保存在一个单独的文件 **doctest_in_help.txt** 中。展示如何使用模块的例子包含在帮助文本中，可以使用 **doctest** 查找并运行。

代码清单 16-17：doctest_in_help.txt

```
==============================
How to Use doctest_in_help.py
==============================

This library is very simple, since it only has one function called
```

```
''my_function()''.

Numbers
=======

''my_function()'' returns the product of its arguments.  For numbers,
that value is equivalent to using the ''*'' operator.

::

    >>> from doctest_in_help import my_function
    >>> my_function(2, 3)
    6

It also works with floating-point values.

::

    >>> my_function(2.0, 3)
    6.0

Non-Numbers
===========

Because ''*'' is also defined on data types other than numbers,
''my_function()'' works just as well if one of the arguments is a
string, a list, or a tuple.

::

    >>> my_function('a', 3)
    'aaa'

    >>> my_function(['A', 'B', 'C'], 2)
    ['A', 'B', 'C', 'A', 'B', 'C']
```

文本文件中的测试可以从命令行运行，类似于 Python 源模块。

```
$ python3 -m doctest -v doctest_in_help.txt

Trying:
    from doctest_in_help import my_function
Expecting nothing
ok
Trying:
    my_function(2, 3)
Expecting:
    6
ok
Trying:
    my_function(2.0, 3)
Expecting:
    6.0
ok
Trying:
    my_function('a', 3)
Expecting:
    'aaa'
ok
Trying:
    my_function(['A', 'B', 'C'], 2)
Expecting:
```

```
            ['A', 'B', 'C', 'A', 'B', 'C']
ok
1 items passed all tests:
   5 tests in doctest_in_help.txt
5 tests in 1 items.
5 passed and 0 failed.
Test passed.
```

正常情况下，doctest 会建立测试执行环境来包含所测试模块的成员，这样测试就不需要再显式地导入这个模块。不过，在这里，测试不在 Python 模块中定义，doctest 不知道如何建立全局命名空间，所以这些例子需要自行完成导入工作。一个给定文件中的所有测试都共享相同的执行上下文，所以在文件最前面导入一次模块就足够了。

16.2.7 运行测试

前面的例子都使用 doctest 内置的命令行测试运行工具。测试单个模块时，这很容易也很方便，不过随着包划分到多个文件，这很快会变得很麻烦。对于这些情况，很多其他方法会更高效。

16.2.7.1 由模块运行

可以在模块最下面包含相应指令来对源代码运行 doctest。

代码清单 16-18：doctest_testmod.py

```python
def my_function(a, b):
    """
    >>> my_function(2, 3)
    6
    >>> my_function('a', 3)
    'aaa'
    """
    return a * b

if __name__ == '__main__':
    import doctest
    doctest.testmod()
```

只有当前模块名是 __main__ 时才会调用 testmod()，这可以确保仅当模块作为主程序调用时才运行测试。

```
$ python3 doctest_testmod.py -v

Trying:
    my_function(2, 3)
Expecting:
    6
ok
Trying:
    my_function('a', 3)
Expecting:
    'aaa'
ok
1 items had no tests:
    __main__
1 items passed all tests:
   2 tests in __main__.my_function
```

```
2 tests in 2 items.
2 passed and 0 failed.
Test passed.
```

`testmod()`的第一个参数是一个模块，包含需要扫描的代码（检查其中是否有测试）。其他测试脚本可以使用这个特性导入实际代码，并依次运行各个模块中的测试。

代码清单 16-19：**doctest_testmod_other_module.py**

```python
import doctest_simple

if __name__ == '__main__':
    import doctest
    doctest.testmod(doctest_simple)
```

通过导入各个模块并运行它们的测试，可以为工程构造一个测试套件。

```
$ python3 doctest_testmod_other_module.py -v

Trying:
    my_function(2, 3)
Expecting:
    6
ok
Trying:
    my_function('a', 3)
Expecting:
    'aaa'
ok
1 items had no tests:
    doctest_simple
1 items passed all tests:
   2 tests in doctest_simple.my_function
2 tests in 2 items.
2 passed and 0 failed.
Test passed.
```

16.2.7.2　由文件运行

`testfile()`的做法类似于`testmod()`，允许在测试程序中从一个外部文件显式调用测试。

代码清单 16-20：**doctest_testfile.py**

```python
import doctest

if __name__ == '__main__':
    doctest.testfile('doctest_in_help.txt')
```

`testmod()`和`testfile()`包括一些可选的参数，能通过`doctest`选项控制测试的行为。关于这些特性的更多详细信息可以参考标准库文档——不过要注意，大多数情况下并不需要这些特性。

```
$ python3 doctest_testfile.py -v

Trying:
    from doctest_in_help import my_function
Expecting nothing
```

```
ok
Trying:
    my_function(2, 3)
Expecting:
    6
ok
Trying:
    my_function(2.0, 3)
Expecting:
    6.0
ok
Trying:
    my_function('a', 3)
Expecting:
    'aaa'
ok
Trying:
    my_function(['A', 'B', 'C'], 2)
Expecting:
    ['A', 'B', 'C', 'A', 'B', 'C']
ok
1 items passed all tests:
   5 tests in doctest_in_help.txt
5 tests in 1 items.
5 passed and 0 failed.
Test passed.
```

16.2.7.3 unittest 套件

unittest 和 doctest 用于在不同情况测试相同的代码，此时可以使用 doctest 中的 unittest 集成一起运行测试。可以应用两个类（DocTestSuite 和 DocFileSuite）来创建与 unittest 的测试运行工具 API 兼容的测试套件。

代码清单 16-21：doctest_unittest.py

```python
import doctest
import unittest

import doctest_simple

suite = unittest.TestSuite()
suite.addTest(doctest.DocTestSuite(doctest_simple))
suite.addTest(doctest.DocFileSuite('doctest_in_help.txt'))

runner = unittest.TextTestRunner(verbosity=2)
runner.run(suite)
```

各个源文件的测试会合并到一个结果，而不是单独报告。

```
$ python3 doctest_unittest.py

my_function (doctest_simple)
Doctest: doctest_simple.my_function ... ok
doctest_in_help.txt
Doctest: doctest_in_help.txt ... ok

----------------------------------------------------------------------
Ran 2 tests in 0.002s

OK
```

16.2.8 测试上下文

doctest 运行测试时创建的执行上下文包含测试模块中模块级全局变量的一个副本。每个测试源（如函数、类、模块）都有自己的一组全局值，让测试在某种程度上相互隔离，使它们不太可能相互干扰。

代码清单 16-22：doctest_test_globals.py

```
class TestGlobals:

    def one(self):
        """
        >>> var = 'value'
        >>> 'var' in globals()
        True
        """

    def two(self):
        """
        >>> 'var' in globals()
        False
        """
```

TestGlobals 有两个方法：`one()` 和 `two()`。`one()` 的 docstring 中的测试设置了一个全局变量，而 `two()` 的测试则要查找这个变量（但是应该找不到）。

```
$ python3 -m doctest -v doctest_test_globals.py

Trying:
    var = 'value'
Expecting nothing
ok
Trying:
    'var' in globals()
Expecting:
    True
ok
Trying:
    'var' in globals()
Expecting:
    False
ok
2 items had no tests:
    doctest_test_globals
    doctest_test_globals.TestGlobals
2 items passed all tests:
   2 tests in doctest_test_globals.TestGlobals.one
   1 tests in doctest_test_globals.TestGlobals.two
3 tests in 4 items.
3 passed and 0 failed.
Test passed.
```

不过，这并不表示测试不能相互干扰，如果测试要改变模块中定义的可变变量的内容，那么反而要希望它们能交互。

代码清单 16-23：doctest_mutable_globals.py

```
_module_data = {}


class TestGlobals:
```

```
    def one(self):
        """
        >>> TestGlobals().one()
        >>> 'var' in _module_data
        True
        """
        _module_data['var'] = 'value'

    def two(self):
        """
        >>> 'var' in _module_data
        False
        """
```

one() 的测试改变了模块变量 _module_data，导致 two() 的测试失败。

```
$ python3 -m doctest -v doctest_mutable_globals.py

Trying:
    TestGlobals().one()
Expecting nothing
ok
Trying:
    'var' in _module_data
Expecting:
    True
ok
Trying:
    'var' in _module_data
Expecting:
    False
**********************************************************************
File ".../doctest_mutable_globals.py", line 25, in doctest_mutab
le_globals.TestGlobals.two
Failed example:
    'var' in _module_data
Expected:
    False
Got:
    True
2 items had no tests:
    doctest_mutable_globals
    doctest_mutable_globals.TestGlobals
1 items passed all tests:
   2 tests in doctest_mutable_globals.TestGlobals.one
**********************************************************************
1 items had failures:
   1 of   1 in doctest_mutable_globals.TestGlobals.two
3 tests in 4 items.
2 passed and 1 failed.
***Test Failed*** 1 failures.
```

如果测试需要全局值，例如对应一个环境进行参数化，则可以将值传递到 `testmod()` 和 `testfile()`，以使用调用者控制的数据建立上下文。

提示：相关阅读材料

- `doctest` 的标准库文档[一]。

[一] https://docs.python.org/3.5/library/doctest.html

- The Mighty Dictionary[1]：Brandon Rhodes 在 PyCon 2010 中关于 dict 内部操作的演示文稿。
- difflib：Python 的顺序差异计算库，用于生成 ndiff 输出。
- Sphinx[2]：除了作为 Python 标准库的文档处理工具，Sphinx 已经被很多第三方项目所采用，因为它不仅易于使用，还可以采用多种数字和打印格式生成简洁的输出。Sphinx 包括一个扩展包，可以运行 doctest（作为其进程的文档源文件），所以例子总是正确的。
- py.test[3]：提供 doctest 支持的第三方测试运行工具。
- nose2[4]：提供 doctest 支持的第三方测试运行工具。
- Manuel[5]：基于文档的第三方测试运行工具，提供更高级的测试用例抽取以及与 Sphinx 的集成。

16.3 unittest：自动测试框架

unittest 中的自动测试框架基于 Kent Beck 和 Erich Gamma 提出的 XUnit 框架设计。同样的模式在很多其他语言中都有出现，包括 C、Perl、Java 和 Smalltalk。unittest 实现的框架支持固件和测试套件，还提供了一个测试运行工具来完成自动测试。

16.3.1 基本测试结构

按照 unittest 的定义，测试有两个部分：管理测试依赖库的代码（名为"固件"）和测试本身。各个测试通过派生 TestCase 并覆盖或增加适当的方法来创建。在下面的例子中，SimplisticTest 有一个 test() 方法，如果 a 不同于 b，那么这个测试便会失败。

代码清单 16-24：unittest_simple.py

```
import unittest


class SimplisticTest(unittest.TestCase):

    def test(self):
        a = 'a'
        b = 'a'
        self.assertEqual(a, b)
```

16.3.2 运行测试

运行 unittest 测试时，最容易的方法是通过命令行接口使用自动发现。

---

[1] www.youtube.com/watch?v=C4Kc8xzcA68
[2] www.sphinx-doc.org
[3] http://doc.pytest.org/en/latest/
[4] https://nose2.readthedocs.io/en/latest/
[5] https://pythonhosted.org/manuel/

```
$ python3 -m unittest unittest_simple.py
.
----------------------------------------------------------------------
Ran 1 test in 0.000s

OK
```

这里的简略输出包括测试花费的时间，并为每个测试提供了一个状态指示符（输出第一行上的"."表示测试通过）。要得到更详细的测试结果，可以包括 -v 选项。

```
$ python3 -m unittest -v unittest_simple.py

test (unittest_simple.SimplisticTest) ... ok

----------------------------------------------------------------------
Ran 1 test in 0.000s

OK
```

16.3.3 测试结果

测试有 3 种可能的结果，如表 16-1 所述。没有明确的方法让一个测试"通过"，所以一个测试的状态取决于是否出现异常。

表 16-1 测试用例结果

结果	描述
ok	测试通过
FAIL	测试没有通过，产生一个 AssertionError 异常
ERROR	测试产生 AssertionError 以外的某个异常

代码清单 16-25：**unittest_outcomes.py**

```
import unittest


class OutcomesTest(unittest.TestCase):

    def testPass(self):
        return

    def testFail(self):
        self.assertFalse(True)

    def testError(self):
        raise RuntimeError('Test error!')
```

当一个测试失败或生成一个错误时，输出中会包含 traceback。

```
$ python3 -m unittest unittest_outcomes.py

EF.
======================================================================
ERROR: testError (unittest_outcomes.OutcomesTest)
----------------------------------------------------------------------
Traceback (most recent call last):
  File ".../unittest_outcomes.py", line 18, in testError
```

```
        raise RuntimeError('Test error!')
RuntimeError: Test error!

======================================================================
FAIL: testFail (unittest_outcomes.OutcomesTest)
----------------------------------------------------------------------
Traceback (most recent call last):
  File ".../unittest_outcomes.py", line 15, in testFail
    self.assertFalse(True)
AssertionError: True is not false

----------------------------------------------------------------------
Ran 3 tests in 0.001s

FAILED (failures=1, errors=1)
```

在前面的例子中，testFail() 失败，traceback 显示了失败代码所在的那一行。不过，读测试输出的人要查看代码，并明确失败测试的含义。

代码清单 16-26：unittest_failwithmessage.py

```
import unittest

class FailureMessageTest(unittest.TestCase):

    def testFail(self):
        self.assertFalse(True, 'failure message goes here')
```

为了更容易地理解一个测试失败的实质，fail*() 和 assert*() 方法都接受一个参数 msg，可以用来生成一个更详细的错误消息。

```
$ python3 -m unittest -v unittest_failwithmessage.py

testFail (unittest_failwithmessage.FailureMessageTest) ... FAIL

======================================================================
FAIL: testFail (unittest_failwithmessage.FailureMessageTest)
----------------------------------------------------------------------
Traceback (most recent call last):
  File ".../unittest_failwithmessage.py", line 12, in testFail
    self.assertFalse(True, 'failure message goes here')
AssertionError: True is not false : failure message goes here
----------------------------------------------------------------------
Ran 1 test in 0.000s

FAILED (failures=1)
```

16.3.4 断言真值

大多数测试会断言某个条件的真值。编写真值检查测试有两种不同的方法，这取决于测试作者的偏好以及所测试代码的预期结果。

代码清单 16-27：unittest_truth.py

```
import unittest

class TruthTest(unittest.TestCase):
```

```python
    def testAssertTrue(self):
        self.assertTrue(True)

    def testAssertFalse(self):
        self.assertFalse(False)
```

如果代码生成一个可能为 true 的值，则应当使用方法 assertTrue()。如果代码生成一个 false 值，则方法 assertFalse() 更有意义。

```
$ python3 -m unittest -v unittest_truth.py

testAssertFalse (unittest_truth.TruthTest) ... ok
testAssertTrue (unittest_truth.TruthTest) ... ok

----------------------------------------------------------------------
Ran 2 tests in 0.000s

OK
```

16.3.5 测试相等性

作为一种特殊情况，unittest 还包括测试两个值相等性的方法。

代码清单 16-28：**unittest_equality.py**

```python
import unittest


class EqualityTest(unittest.TestCase):

    def testExpectEqual(self):
        self.assertEqual(1, 3 - 2)

    def testExpectEqualFails(self):
        self.assertEqual(2, 3 - 2)

    def testExpectNotEqual(self):
        self.assertNotEqual(2, 3 - 2)

    def testExpectNotEqualFails(self):
        self.assertNotEqual(1, 3 - 2)
```

如果失败，则这些特殊的测试方法会生成错误消息，其中包括所比较的值。

```
$ python3 -m unittest -v unittest_equality.py

testExpectEqual (unittest_equality.EqualityTest) ... ok
testExpectEqualFails (unittest_equality.EqualityTest) ... FAIL
testExpectNotEqual (unittest_equality.EqualityTest) ... ok
testExpectNotEqualFails (unittest_equality.EqualityTest) ...
FAIL

======================================================================
FAIL: testExpectEqualFails (unittest_equality.EqualityTest)
----------------------------------------------------------------------
Traceback (most recent call last):
  File ".../unittest_equality.py", line 15, in
testExpectEqualFails
    self.assertEqual(2, 3 - 2)
AssertionError: 2 != 1
```

```
==============================================================
FAIL: testExpectNotEqualFails (unittest_equality.EqualityTest)
--------------------------------------------------------------
Traceback (most recent call last):
  File ".../unittest_equality.py", line 21, in
testExpectNotEqualFails
    self.assertNotEqual(1, 3 - 2)
AssertionError: 1 == 1

--------------------------------------------------------------
Ran 4 tests in 0.001s

FAILED (failures=2)
```

16.3.6 几乎相等?

除了严格相等性,还可以使用 assertAlmostEqual() 和 assertNotAlmostEqual() 测试浮点数的近似相等性。

代码清单 16-29: **unittest_almostequal.py**

```
import unittest

class AlmostEqualTest(unittest.TestCase):
    def testEqual(self):
        self.assertEqual(1.1, 3.3 - 2.2)

    def testAlmostEqual(self):
        self.assertAlmostEqual(1.1, 3.3 - 2.2, places=1)

    def testNotAlmostEqual(self):
        self.assertNotAlmostEqual(1.1, 3.3 - 2.0, places=1)
```

参数是要比较的值以及测试所用的小数位数。

```
$ python3 -m unittest unittest_almostequal.py

.F.
==============================================================
FAIL: testEqual (unittest_almostequal.AlmostEqualTest)
--------------------------------------------------------------
Traceback (most recent call last):
  File ".../unittest_almostequal.py", line 12, in testEqual
    self.assertEqual(1.1, 3.3 - 2.2)
AssertionError: 1.1 != 1.0999999999999996

--------------------------------------------------------------
Ran 3 tests in 0.001s

FAILED (failures=1)
```

16.3.7 容器

除了通用的 assertEqual() 和 assertNotEqual() 方法,还有一些特殊方法可以用来比较如 list、dict 和 set 对象等的容器。

代码清单 16-30: **unittest_equality_container.py**

```python
import textwrap
import unittest


class ContainerEqualityTest(unittest.TestCase):

    def testCount(self):
        self.assertCountEqual(
            [1, 2, 3, 2],
            [1, 3, 2, 3],
        )

    def testDict(self):
        self.assertDictEqual(
            {'a': 1, 'b': 2},
            {'a': 1, 'b': 3},
        )

    def testList(self):
        self.assertListEqual(
            [1, 2, 3],
            [1, 3, 2],
        )

    def testMultiLineString(self):
        self.assertMultiLineEqual(
            textwrap.dedent("""
            This string
            has more than one
            line.
            """),
            textwrap.dedent("""
            This string has
            more than two
            lines.
            """),
        )

    def testSequence(self):
        self.assertSequenceEqual(
            [1, 2, 3],
            [1, 3, 2],
        )

    def testSet(self):
        self.assertSetEqual(
            set([1, 2, 3]),
            set([1, 3, 2, 4]),
        )

    def testTuple(self):
        self.assertTupleEqual(
            (1, 'a'),
            (1, 'b'),
        )
```

这些方法分别使用对应输入类型的格式来报告是否不相等,以便更容易地理解和修正测试失败。

```
$ python3 -m unittest unittest_equality_container.py

FFFFFFF
======================================================================
FAIL: testCount
(unittest_equality_container.ContainerEqualityTest)
----------------------------------------------------------------------
Traceback (most recent call last):
  File ".../unittest_equality_container.py", line 15, in testCount
    [1, 3, 2, 3],
AssertionError: Element counts were not equal:
First has 2, Second has 1:  2
First has 1, Second has 2:  3

======================================================================
FAIL: testDict
(unittest_equality_container.ContainerEqualityTest)
----------------------------------------------------------------------
Traceback (most recent call last):
  File ".../unittest_equality_container.py", line 21, in testDict
    {'a': 1, 'b': 3},
AssertionError: {'b': 2, 'a': 1} != {'b': 3, 'a': 1}
- {'a': 1, 'b': 2}
?               ^

+ {'a': 1, 'b': 3}
?               ^


======================================================================
FAIL: testList
(unittest_equality_container.ContainerEqualityTest)
----------------------------------------------------------------------
Traceback (most recent call last):
  File ".../unittest_equality_container.py", line 27, in testList
    [1, 3, 2],
AssertionError: Lists differ: [1, 2, 3] != [1, 3, 2]

First differing element 1:
2
3

- [1, 2, 3]
+ [1, 3, 2]

======================================================================
FAIL: testMultiLineString
(unittest_equality_container.ContainerEqualityTest)
----------------------------------------------------------------------
Traceback (most recent call last):
  File ".../unittest_equality_container.py", line 41, in testMultiLineString
    """),
AssertionError: '\nThis string\nhas more than one\nline.\n' != '\nThis string has\nmore than two\nlines.\n'

- This string
+ This string has
?            ++++
- has more than one
```

```
 ? ----              --
 + more than two
 ?               ++
 - line.
 + lines.
 ?      +


 ======================================================================
 FAIL: testSequence
 (unittest_equality_container.ContainerEqualityTest)
 ----------------------------------------------------------------------
 Traceback (most recent call last):
   File ".../unittest_equality_container.py", line 47, in
 testSequence
     [1, 3, 2]),
 AssertionError: Sequences differ: [1, 2, 3] != [1, 3, 2]

 First differing element 1:
 2
 3

 - [1, 2, 3]
 + [1, 3, 2]


 ======================================================================
 FAIL: testSet
 (unittest_equality_container.ContainerEqualityTest)
 ----------------------------------------------------------------------
 Traceback (most recent call last):
   File ".../unittest_equality_container.py", line 53, in testSet
     set([1, 3, 2, 4]),
 AssertionError: Items in the second set but not the first:
 4


 ======================================================================
 FAIL: testTuple
 (unittest_equality_container.ContainerEqualityTest)
 ----------------------------------------------------------------------
 Traceback (most recent call last):
   File ".../unittest_equality_container.py", line 59, in
 testTuple
     (1, 'b'),
 AssertionError: Tuples differ: (1, 'a') != (1, 'b')

 First differing element 1:
 'a'
 'b'

 - (1, 'a')
 ?     ^
 + (1, 'b')
 ?     ^


 ----------------------------------------------------------------------
 Ran 7 tests in 0.004s

 FAILED (failures=7)
```

可以使用assertIn()来测试容器的成员关系。

代码清单 16-31：**unittest_in.py**

```python
import unittest

class ContainerMembershipTest(unittest.TestCase):

    def testDict(self):
        self.assertIn(4, {1: 'a', 2: 'b', 3: 'c'})

    def testList(self):
        self.assertIn(4, [1, 2, 3])

    def testSet(self):
        self.assertIn(4, set([1, 2, 3]))
```

支持 in 操作符或容器 API 的所有对象都可以使用 assertIn() 来测试。

```
$ python3 -m unittest unittest_in.py

FFF
======================================================================
FAIL: testDict (unittest_in.ContainerMembershipTest)
----------------------------------------------------------------------
Traceback (most recent call last):
  File ".../unittest_in.py", line 12, in testDict
    self.assertIn(4, {1: 'a', 2: 'b', 3: 'c'})
AssertionError: 4 not found in {1: 'a', 2: 'b', 3: 'c'}

======================================================================
FAIL: testList (unittest_in.ContainerMembershipTest)
----------------------------------------------------------------------
Traceback (most recent call last):
  File ".../unittest_in.py", line 15, in testList
    self.assertIn(4, [1, 2, 3])
AssertionError: 4 not found in [1, 2, 3]

======================================================================
FAIL: testSet (unittest_in.ContainerMembershipTest)
----------------------------------------------------------------------
Traceback (most recent call last):
  File ".../unittest_in.py", line 18, in testSet
    self.assertIn(4, set([1, 2, 3]))
AssertionError: 4 not found in {1, 2, 3}

----------------------------------------------------------------------
Ran 3 tests in 0.001s

FAILED (failures=3)
```

16.3.8 测试异常

正如前面提到的，如果测试产生 AssertionError 以外的一个异常，则会将其处理为一个错误。这种行为对于发现错误很有用，还有助于修改现有测试覆盖的代码。不过，有些情况下，测试要验证代码确实会产生一个异常。例如，如果为对象的属性提供一个非法值，那么与在测试中捕获异常相比，assertRaises() 可以得到更简洁的代码。下面的例子包含两个测试，可以在这个基础上进行比较。

代码清单 16-32: **unittest_exception.py**

```python
import unittest


def raises_error(*args, **kwds):
    raise ValueError('Invalid value: ' + str(args) + str(kwds))


class ExceptionTest(unittest.TestCase):

    def testTrapLocally(self):
        try:
            raises_error('a', b='c')
        except ValueError:
            pass
        else:
            self.fail('Did not see ValueError')

    def testAssertRaises(self):
        self.assertRaises(
            ValueError,
            raises_error,
            'a',
            b='c',
        )
```

这两个测试的结果是相同的，不过第二个使用 `assertRaises()` 的测试更为简洁。

```
$ python3 -m unittest -v unittest_exception.py

testAssertRaises (unittest_exception.ExceptionTest) ... ok
testTrapLocally (unittest_exception.ExceptionTest) ... ok

----------------------------------------------------------------
Ran 2 tests in 0.000s

OK
```

16.3.9 测试固件

固件是测试所需的外部资源。例如，一个类的测试可能都需要另一个类的实例（用来提供配置设置）或者另一个共享资源。其他测试固件包括数据库连接和临时文件。(很多人可能对此有争议，认为使用外部资源会使这些测试不再是"单元"测试，不过它们仍是测试，而且仍然很有用）。

unittest 包括一个特殊的 hook，用来配置和清理测试所需的所有固件。要为各个单独的测试用例建立固件，需要覆盖 TestCase 上的 `setUp()`。要完成清理，则要覆盖 `tearDown()`。要为一个测试类的所有实例管理一组固件，则需要覆盖 TestCase 的类方法 `setUpClass()` 和 `tearDownClass()`。最后如果要为一个模块中的所有测试处理代价特别昂贵的建立操作，则可以使用模块级函数 `setUpModule()` 和 `tearDownModule()`。

代码清单 16-33: **unittest_fixtures.py**

```python
import random
import unittest
```

```python
def setUpModule():
    print('In setUpModule()')


def tearDownModule():
    print('In tearDownModule()')


class FixturesTest(unittest.TestCase):

    @classmethod
    def setUpClass(cls):
        print('In setUpClass()')
        cls.good_range = range(1, 10)

    @classmethod
    def tearDownClass(cls):
        print('In tearDownClass()')
        del cls.good_range

    def setUp(self):
        super().setUp()
        print('\nIn setUp()')
        # Pick a number sure to be in the range. The range is
        # defined as not including the "stop" value, so this
        # value should not be included in the set of allowed
        # values for our choice.
        self.value = random.randint(
            self.good_range.start,
            self.good_range.stop - 1,
        )

    def tearDown(self):
        print('In tearDown()')
        del self.value
        super().tearDown()

    def test1(self):
        print('In test1()')
        self.assertIn(self.value, self.good_range)

    def test2(self):
        print('In test2()')
        self.assertIn(self.value, self.good_range)
```

运行这个示例测试时，固件和测试方法执行的顺序很明显。

```
$ python3 -u -m unittest -v unittest_fixtures.py

In setUpModule()
In setUpClass()
test1 (unittest_fixtures.FixturesTest) ...
In setUp()
In test1()
In tearDown()
ok
test2 (unittest_fixtures.FixturesTest) ...
In setUp()
In test2()
In tearDown()
ok
In tearDownClass()
```

```
In tearDownModule()

----------------------------------------------------------------
Ran 2 tests in 0.001s

OK
```

如果清理固件的过程中出现错误,那么 tearDown 方法可能不会都被调用。为了确保总能正确地释放固件,要使用 addCleanup()。

代码清单 16-34: unittest_addcleanup.py

```python
import random
import shutil
import tempfile
import unittest


def remove_tmpdir(dirname):
    print('In remove_tmpdir()')
    shutil.rmtree(dirname)


class FixturesTest(unittest.TestCase):

    def setUp(self):
        super().setUp()
        self.tmpdir = tempfile.mkdtemp()
        self.addCleanup(remove_tmpdir, self.tmpdir)

    def test1(self):
        print('\nIn test1()')

    def test2(self):
        print('\nIn test2()')
```

这个示例测试会创建一个临时目录,然后在测试完成时使用 shutil 清理这个目录。

```
$ python3 -u -m unittest -v unittest_addcleanup.py

test1 (unittest_addcleanup.FixturesTest) ...
In test1()
In remove_tmpdir()
ok
test2 (unittest_addcleanup.FixturesTest) ...
In test2()
In remove_tmpdir()
ok

----------------------------------------------------------------
Ran 2 tests in 0.003s

OK
```

16.3.10 用不同输入重复测试

用不同的输入运行相同的测试逻辑通常很有用。不是为每个小用例定义一个单独的测试方法,常用的一种技术是创建一个测试方法,其中包含多个相关的断言调用。这种方法的问题在于，一旦一个断言失败，就会跳过其他断言。一种更好的方法是使用 subTest() 在一个测试方法中为一个测试创建一个上下文。如果这个测试失败，则报告失败，其余的测试仍继续。

代码清单 16-35：unittest_subtest.py

```python
import unittest

class SubTest(unittest.TestCase):

    def test_combined(self):
        self.assertRegex('abc', 'a')
        self.assertRegex('abc', 'B')
        # The next assertions are not verified!
        self.assertRegex('abc', 'c')
        self.assertRegex('abc', 'd')

    def test_with_subtest(self):
        for pat in ['a', 'B', 'c', 'd']:
            with self.subTest(pattern=pat):
                self.assertRegex('abc', pat)
```

在这个例子中，test_combined() 方法永远不会运行对应模式 'c' 和 'd' 的断言。但 test_with_subtest() 方法会运行这两个断言，它能正确地报告各个失败。需要说明，测试运行工具仍认为只存在两个测试用例，尽管报告了 3 个失败。

```
$ python3 -m unittest -v unittest_subtest.py

test_combined (unittest_subtest.SubTest) ... FAIL
test_with_subtest (unittest_subtest.SubTest) ...
======================================================================
FAIL: test_combined (unittest_subtest.SubTest)
----------------------------------------------------------------------
Traceback (most recent call last):
  File ".../unittest_subtest.py", line 13, in test_combined
    self.assertRegex('abc', 'B')
AssertionError: Regex didn't match: 'B' not found in 'abc'

======================================================================
FAIL: test_with_subtest (unittest_subtest.SubTest) (pattern='B')
----------------------------------------------------------------------
Traceback (most recent call last):
  File ".../unittest_subtest.py", line 21, in test_with_subtest
    self.assertRegex('abc', pat)
AssertionError: Regex didn't match: 'B' not found in 'abc'

======================================================================
FAIL: test_with_subtest (unittest_subtest.SubTest) (pattern='d')
----------------------------------------------------------------------
Traceback (most recent call last):
  File ".../unittest_subtest.py", line 21, in test_with_subtest
    self.assertRegex('abc', pat)
AssertionError: Regex didn't match: 'd' not found in 'abc'

----------------------------------------------------------------------
Ran 2 tests in 0.001s

FAILED (failures=3)
```

16.3.11 跳过测试

倘若一些外部条件不满足，那么跳过测试通常会很有用。例如，如果编写测试来检查一个特定 Python 版本中某个库的行为，那么就没有必要在 Python 的其他版本下运行这些测试。测试类和方法可以用 `skip()` 修饰，这样就会跳过测试。修饰符 `skipIf()` 和 `skipUnless()` 可以在跳过测试之前检查一个条件。

代码清单 16-36：**unittest_skip.py**

```python
import sys
import unittest

class SkippingTest(unittest.TestCase):

    @unittest.skip('always skipped')
    def test(self):
        self.assertTrue(False)

    @unittest.skipIf(sys.version_info[0] > 2,
                    'only runs on python 2')
    def test_python2_only(self):
        self.assertTrue(False)

    @unittest.skipUnless(sys.platform == 'Darwin',
                        'only runs on macOS')
    def test_macos_only(self):
        self.assertTrue(True)

    def test_raise_skiptest(self):
        raise unittest.SkipTest('skipping via exception')
```

如果条件很复杂，很难用一个表达式来表述并传递到 `skipIf()` 或 `skipUnless()`，那么在这种情况下，测试用例可以直接产生 `SkipTest` 来跳过测试。

```
$ python3 -m unittest -v unittest_skip.py

test (unittest_skip.SkippingTest) ... skipped 'always skipped'
test_macos_only (unittest_skip.SkippingTest) ... skipped 'only
runs on macOS'
test_python2_only (unittest_skip.SkippingTest) ... skipped 'only
runs on python 2'
test_raise_skiptest (unittest_skip.SkippingTest) ... skipped
'skipping via exception'

----------------------------------------------------------------
Ran 4 tests in 0.000s

OK (skipped=4)
```

16.3.12 忽略失败测试

不用删除永久失败的测试，可以用 `expectedFailure()` 修饰符标志这些测试，从而忽略它们的失败。

代码清单 16-37：**unittest_expectedfailure.py**

```
import unittest

class Test(unittest.TestCase):

    @unittest.expectedFailure
    def test_never_passes(self):
        self.assertTrue(False)

    @unittest.expectedFailure
    def test_always_passes(self):
        self.assertTrue(True)
```

如果一个预计失败的测试确实通过了，那么这个条件会作为一种特殊类型的失败，并报告为"意外成功"（unexpected success）。

```
$ python3 -m unittest -v unittest_expectedfailure.py

test_always_passes (unittest_expectedfailure.Test) ...
unexpected success
test_never_passes (unittest_expectedfailure.Test) ... expected
failure

----------------------------------------------------------------------
Ran 2 tests in 0.001s

FAILED (expected failures=1, unexpected successes=1)
```

提示：相关阅读材料
- **unittest** 的标准库文档[一]。
- **doctest**：提供了另一种候选方式来运行嵌入到 docstring 或外部文档文件中的测试。
- **nose**[二]：这个第三方测试运行工具提供了复杂的发现特性。
- **pytest**[三]：一个流行的第三方测试运行工具，支持分布式执行和一个候选的固件管理系统。
- **testrepository**[四]：OpenStack 项目使用的第三方测试运行工具，支持并行执行和跟踪失败。

16.4 trace：执行程序流

trace 模块对于了解程序以何种方式运行很有用。它会监视所执行的语句，生成覆盖报告，并有助于研究相互调用的函数之间的关系。

16.4.1 示例程序

这一节余下的例子中都会使用这个程序。它导入另一个名为 recurse 的模块，然后运

[一] https://docs.python.org/3/5/library/unittest.html
[二] https://nose.readthedocs.io/en/latest/
[三] http://doc.pytest.org/en/latest/
[四] http://testrepository.readthedocs.io/en/latest/

行其中的一个函数。

代码清单 16-38：`trace_example/main.py`

```python
from recurse import recurse


def main():
    print('This is the main program.')
    recurse(2)

if __name__ == '__main__':
    main()
```

`recurse()` 函数会调用其自身，直至 level 参数达到 0。

代码清单 16-39：`trace_example/recurse.py`

```python
def recurse(level):
    print('recurse({})'.format(level))
    if level:
        recurse(level - 1)


def not_called():
    print('This function is never called.')
```

16.4.2 跟踪执行

可以从命令行直接使用 `trace`，这很容易。给定 `--trace` 选项时，会打印程序运行时执行的语句。这个例子还会忽略 Python 标准库的位置，避免跟踪进入 `importlib` 和其他模块（在另一个例子中这可能比较有意思，不过对于这个简单的例子，可能会让输出很混乱）。

```
$ python3 -m trace --ignore-dir=.../lib/python3.5 \
--trace trace_example/main.py

 --- modulename: main, funcname: <module>
main.py(7): """
main.py(10): from recurse import recurse
 --- modulename: recurse, funcname: <module>
recurse.py(7): """
recurse.py(11): def recurse(level):
recurse.py(17): def not_called():
main.py(13): def main():
main.py(17): if __name__ == '__main__':
main.py(18):     main()
 --- modulename: main, funcname: main
main.py(14):     print('This is the main program.')
This is the main program.
main.py(15):     recurse(2)
 --- modulename: recurse, funcname: recurse
recurse.py(12):     print('recurse({})'.format(level))
recurse(2)
recurse.py(13):     if level:
recurse.py(14):         recurse(level - 1)
 --- modulename: recurse, funcname: recurse
recurse.py(12):     print('recurse({})'.format(level))
recurse(1)
```

```
recurse.py(13):         if level:
recurse.py(14):             recurse(level - 1)
 --- modulename: recurse, funcname: recurse
recurse.py(12):     print('recurse({})'.format(level))
recurse(0)
recurse.py(13):         if level:
 --- modulename: trace, funcname: _unsettrace
trace.py(77):           sys.settrace(None)
```

输出的第一部分显示了 `trace` 完成的建立操作。输出的余下部分显示了每个函数的入口，包括函数所在的模块，以及执行时源文件的代码行。不出所料，根据 `main()` 中调用的方式，会进入 3 次 `recurse()` 函数。

16.4.3 代码覆盖

给定 `--count` 选项时，从命令行运行 `trace` 会生成代码覆盖报告信息，详细列出运行了哪些代码行，而哪些行被跳过。由于复杂的程序通常由多个文件组成，所以会为各个文件分别生成一个单独的覆盖报告。默认地，覆盖报告文件会写至模块所在的目录，根据模块来命名，不过扩展名是 `.cover` 而不是 `.py`。

```
$ python3 -m trace --count trace_example/main.py

This is the main program.
recurse(2)
recurse(1)
recurse(0)
```

这里会生成两个输出文件。以下是 `trace_example/main.cover`。

代码清单 16-40: **trace_example/main.cover**

```
1: from recurse import recurse

1: def main():
1:     print 'This is the main program.'
1:     recurse(2)
1:     return

1: if __name__ == '__main__':
1:     main()
```

以下是 `trace_example/recurse.cover`。

代码清单 16-41: **trace_example/recurse.cover**

```
1: def recurse(level):
3:     print 'recurse(%s)' % level
3:     if level:
2:         recurse(level-1)
3:     return

1: def not_called():
       print 'This function is never called.'
```

说明：尽管代码行 `def recurse(level)`: 计数为 1，但这并不表示这个函数只运行了一次。实际上，它表示只执行了一次函数定义。这同样适用于 `def not_called():`，因为即使这个函数本身从未调用，也会评估函数定义。

还可以提供不同的选项让这个程序运行多次，来保存覆盖数据并生成一个合并的报告。第一次运行 `trace` 并提供一个输出文件时，如果创建文件之前试图加载现有数据与新结果合并，那么便会报告一个错误。

```
$ python3 -m trace --coverdir coverdir1 --count \
--file coverdir1/coverage_report.dat trace_example/main.py

This is the main program.
recurse(2)
recurse(1)
recurse(0)
Skipping counts file 'coverdir1/coverage_report.dat': [Errno 2]
No such file or directory: 'coverdir1/coverage_report.dat'

$ python3 -m trace --coverdir coverdir1 --count \
--file coverdir1/coverage_report.dat trace_example/main.py

This is the main program.
recurse(2)
recurse(1)
recurse(0)

$ python3 -m trace --coverdir coverdir1 --count \
--file coverdir1/coverage_report.dat trace_example/main.py

This is the main program.
recurse(2)
recurse(1)
recurse(0)

$ ls coverdir1

coverage_report.dat
```

一旦覆盖信息被记录到 `.cover` 文件，那么要生成报告，可以使用 `--report` 选项。

```
$ python3 -m trace --coverdir coverdir1 --report --summary \
--missing --file coverdir1/coverage_report.dat \
trace_example/main.py

lines   cov%   module              (path)
  537     0%   trace               (.../lib/python3.5/trace.py)
    7   100%   trace_example.main  (trace_example/main.py)
    7    85%   trace_example.recurse
(trace_example/recurse.py)
```

由于这个程序运行了 3 次，所以覆盖报告显示出，值比第一个报告中大 3 倍。`--summary` 选项会在输出中增加百分比覆盖信息。`recurse` 模块只覆盖了 87%。查看 `recurse` 的覆盖文件可以看到，如 `>>>>>>` 前缀所指示的，`not_called()` 的体实际上从未运行。

代码清单 16-42：**coverdir1/trace_example.recurse.cover**

```
    3: def recurse(level):
    9:     print('recurse({})'.format(level))
    9:     if level:
```

```
  6:            recurse(level - 1)

  3: def not_called():
>>>>>>         print('This function is never called.')
```

16.4.4 调用关系

除了覆盖信息，trace 还会收集和报告相互调用的函数之间的关系。要得到所调用函数的一个简单列表，可以使用 --listfuncs。

```
$ python3 -m trace --listfuncs trace_example/main.py | \
grep -v importlib

This is the main program.
recurse(2)
recurse(1)
recurse(0)

functions called:
filename: .../lib/python3.5/trace.py, modulename: trace,
funcname: _unsettrace
filename: trace_example/main.py, modulename: main, funcname:
<module>
filename: trace_example/main.py, modulename: main, funcname:
main
filename: trace_example/recurse.py, modulename: recurse,
funcname: <module>
filename: trace_example/recurse.py, modulename: recurse,
funcname: recurse
```

要得到调用者的更多详细信息，可以使用 --trackcalls。

```
$ python3 -m trace --listfuncs --trackcalls \
trace_example/main.py | grep -v importlib

This is the main program.
recurse(2)
recurse(1)
recurse(0)

calling relationships:

*** .../lib/python3.5/trace.py ***
   trace.Trace.runctx -> trace._unsettrace
 --> trace_example/main.py
   trace.Trace.runctx -> main.<module>

 --> trace_example/recurse.py

*** trace_example/main.py ***
   main.<module> -> main.main
 --> trace_example/recurse.py
   main.main -> recurse.recurse

*** trace_example/recurse.py ***
   recurse.recurse -> recurse.recurse
```

说明： --listfuncs 和 --trackcalls 都不考虑 --ignore-dirs 或 --ignore-mods 参数，所以使用 grep 去除了这个例子的部分输出。

16.4.5　编程接口

要想更多地控制 trace 接口，可以在程序中使用一个 Trace 对象调用这个接口。Trace 支持在运行一个函数或执行一个要跟踪的 Python 命令之前先建立固件和其他依赖对象。

代码清单 16-43：**trace_run.py**

```
import trace
from trace_example.recurse import recurse

tracer = trace.Trace(count=False, trace=True)
tracer.run('recurse(2)')
```

由于这个例子只跟踪到 recurse() 函数，所以输出中不包含 main.py 的任何信息。

```
$ python3 trace_run.py

 --- modulename: trace_run, funcname: <module>
<string>(1):  --- modulename: recurse, funcname: recurse
recurse.py(12):     print('recurse({})'.format(level))
recurse(2)
recurse.py(13):     if level:
recurse.py(14):         recurse(level - 1)
 --- modulename: recurse, funcname: recurse
recurse.py(12):     print('recurse({})'.format(level))
recurse(1)
recurse.py(13):     if level:
recurse.py(14):         recurse(level - 1)
 --- modulename: recurse, funcname: recurse
recurse.py(12):     print('recurse({})'.format(level))
recurse(0)
recurse.py(13):     if level:
 --- modulename: trace, funcname: _unsettrace
trace.py(77):           sys.settrace(None)
```

用 runfunc() 方法也可以生成同样的输出。

代码清单 16-44：**trace_runfunc.py**

```
import trace
from trace_example.recurse import recurse

tracer = trace.Trace(count=False, trace=True)
tracer.runfunc(recurse, 2)
```

runfunc() 接收任意的位置和关键字参数，由 tracer 调用时这些参数会被传递到函数。

```
$ python3 trace_runfunc.py

 --- modulename: recurse, funcname: recurse
recurse.py(12):     print('recurse({})'.format(level))
recurse(2)
recurse.py(13):     if level:
recurse.py(14):         recurse(level - 1)
 --- modulename: recurse, funcname: recurse
recurse.py(12):     print('recurse({})'.format(level))
recurse(1)
recurse.py(13):     if level:
recurse.py(14):         recurse(level - 1)
```

```
--- modulename: recurse, funcname: recurse
recurse.py(12):     print('recurse({})'.format(level))
recurse(0)
recurse.py(13):     if level:
```

16.4.6 保存结果数据

类似于命令行接口，还可以记录统计和覆盖信息。这些数据必须使用 Trace 对象的 CoverageResults 实例显式保存。

代码清单 16-45：**trace_CoverageResults.py**

```
import trace
from trace_example.recurse import recurse

tracer = trace.Trace(count=True, trace=False)
tracer.runfunc(recurse, 2)

results = tracer.results()
results.write_results(coverdir='coverdir2')
```

这个例子将覆盖结果保存到目录 coverdir2。

```
$ python3 trace_CoverageResults.py

recurse(2)
recurse(1)
recurse(0)

$ find coverdir2

coverdir2
coverdir2/trace_example.recurse.cover
```

输出文件包含以下内容。

```
       #!/usr/bin/env python
       # encoding: utf-8
       #
       # Copyright (c) 2008 Doug Hellmann. All rights reserved.
       #
       """
>>>>>> """

       #end_pymotw_header

>>>>>> def recurse(level):
    3:     print('recurse({})'.format(level))
    3:     if level:
    2:         recurse(level - 1)
>>>>>> def not_called():
>>>>>>     print('This function is never called.')
```

要保存统计数据来生成报告，可以对 Trace 使用 infile 和 outfile 参数。

代码清单 16-46：**trace_report.py**

```
import trace
from trace_example.recurse import recurse
```

```
tracer = trace.Trace(count=True,
                     trace=False,
                     outfile='trace_report.dat')
tracer.runfunc(recurse, 2)

report_tracer = trace.Trace(count=False,
                            trace=False,
                            infile='trace_report.dat')
results = tracer.results()
results.write_results(summary=True, coverdir='/tmp')
```

将一个文件名传至 `infile` 来读取先前存储的数据，将一个文件名传至 `outfile` 可以在跟踪之后将新结果写入该文件。如果 `infile` 和 `outfile` 相同，则以上代码的效果就是用累积的数据更新文件。

```
$ python3 trace_report.py

recurse(2)
recurse(1)
recurse(0)
lines   cov%   module        (path)
    7    42%   trace_example.recurse
(.../trace_example/recurse.py)
```

16.4.7 选项

`Trace` 的构造函数有多个可选的参数，用来控制运行时行为。

`count` Boolean。打开行号统计；默认为 `True`。
`countfuncs` Boolean。打开运行期间调用的函数列表；默认为 `False`。
`countcallers` Boolean。打开跟踪调用者和被调用者；默认为 `False`。
`ignoremods` Sequence。跟踪覆盖情况时所要忽略的模块或包列表；默认为一个空元组。
`ignoredirs` Sequence。这个目录列表包含要忽略的模块或包；默认为一个空元组。
`infile`：包含缓存计数值的文件的文件名；默认为 `None`。
`outfile`：这个文件用来存储缓存计数文件；默认为 `None`，数据未存储。

提示：相关阅读材料
- `trace` 的标准库文档⊖。
- 17.2.7 节：`sys` 模块包含一些工具，可以在运行时为解释器增加一个定制跟踪函数。
- coverage.py⊖：Ned Batchelder 编写的覆盖模块。
- figleaf⊜：Titus Brown 编写的覆盖应用。

16.5 traceback：异常和栈轨迹

`traceback` 模块处理调用栈来生成错误消息。traceback 是指从异常处理器沿调用链

⊖ https://docs.python.org/3.5/library/trace.html
⊖ http://nedbatchelder.com/code/modules/coverage.html
⊜ http://darcs.idyll.org/~t/projects/figleaf/doc/

向下直到产生异常的那一点的栈轨迹。也可以在当前调用栈从调用位置（没有错误上下文）向上访问 traceback，这对于确定进入函数的路径很有用。

traceback 中的高层 API 使用 StackSummary 和 FrameSummary 实例保存栈的表示。可以从一个 traceback 或当前执行栈构造这些类，然后用同样的方式处理。

traceback 中的底层函数通常可以分为几类。有些函数用于从当前运行时环境（可能是一个 traceback 的异常处理器或者是常规的栈）抽取原始 traceback。所抽取的栈轨迹是一个元组序列，元组中包含文件名、行号、函数名和源代码行文本。

一旦抽取了栈轨迹，便可以使用类似 format_exception() 和 format_stack() 等函数格式化。格式化函数会返回一个字符串列表，其中对消息进行格式化以便打印。还有一些用于打印格式化值的简写函数。

traceback 中的函数默认地会模拟交互式解释器的行为，不仅如此，它们对另外一些情况下异常的处理也很有用，这些情况下并不需要将完整的栈轨迹转储到控制台。例如，一个 Web 应用可能需要格式化 traceback，以便用 HTML 格式很好地显示，或者一个 IDE 可以将栈轨迹的元素转换为一个可点击的列表，以便用户浏览源文本。

16.5.1 支持函数

这一节中的例子使用了模块 **traceback_example.py**。

代码清单 16-47：**traceback_example.py**

```python
import traceback
import sys


def produce_exception(recursion_level=2):
    sys.stdout.flush()
    if recursion_level:
        produce_exception(recursion_level - 1)
    else:
        raise RuntimeError()


def call_function(f, recursion_level=2):
    if recursion_level:
        return call_function(f, recursion_level - 1)
    else:
        return f()
```

16.5.2 检查栈

要检查当前栈，可从 **walk_stack()** 构造一个 StackSummary。

代码清单 16-48：**traceback_stacksummary.py**

```python
import traceback
import sys

from traceback_example import call_function


def f():
```

```python
    summary = traceback.StackSummary.extract(
        traceback.walk_stack(None)
    )
    print(''.join(summary.format()))


print('Calling f() directly:')
f()

print()
print('Calling f() from 3 levels deep:')
call_function(f)
```

format() 方法生成一个可以打印的格式化字符串序列。

```
$ python3 traceback_stacksummary.py

Calling f() directly:
  File "traceback_stacksummary.py", line 18, in f
    traceback.walk_stack(None)
  File "traceback_stacksummary.py", line 24, in <module>
    f()


Calling f() from 3 levels deep:
  File "traceback_stacksummary.py", line 18, in f
    traceback.walk_stack(None)
  File ".../traceback_example.py", line 26, in call_function
    return f()
  File ".../traceback_example.py", line 24, in call_function
    return call_function(f, recursion_level - 1)
  File ".../traceback_example.py", line 24, in call_function
    return call_function(f, recursion_level - 1)
  File "traceback_stacksummary.py", line 28, in <module>
    call_function(f)
```

StackSummary 是一个包含 **FrameSummary** 实例的可迭代的容器。

代码清单 16-49：**traceback_framesummary.py**

```python
import traceback
import sys

from traceback_example import call_function

template = (
    '{fs.filename:<26}:{fs.lineno}:{fs.name}:\n'
    '    {fs.line}'
)


def f():
    summary = traceback.StackSummary.extract(
        traceback.walk_stack(None)
    )
    for fs in summary:
        print(template.format(fs=fs))


print('Calling f() directly:')
f()

print()
```

```
print('Calling f() from 3 levels deep:')
call_function(f)
```

每个 FrameSummary 都描述了一个栈帧，包括程序源文件中执行上下文的位置。

```
$ python3 traceback_framesummary.py

Calling f() directly:
traceback_framesummary.py :23:f:
    traceback.walk_stack(None)
traceback_framesummary.py :30:<module>:
    f()

Calling f() from 3 levels deep:
traceback_framesummary.py :23:f:
    traceback.walk_stack(None)
.../traceback_example.py:26:call_function:
    return f()
.../traceback_example.py:24:call_function:
    return call_function(f, recursion_level - 1)
.../traceback_example.py:24:call_function:
    return call_function(f, recursion_level - 1)
traceback_framesummary.py :34:<module>:
    call_function(f)
```

16.5.3 traceback 异常

TracebackException 类是高层接口，可以在处理 traceback 时构建一个 StackSummary。

代码清单 16-50：`traceback_tracebackexception.py`

```python
import traceback
import sys

from traceback_example import produce_exception

print('with no exception:')
exc_type, exc_value, exc_tb = sys.exc_info()
tbe = traceback.TracebackException(exc_type, exc_value, exc_tb)
print(''.join(tbe.format()))

print('\nwith exception:')
try:
    produce_exception()
except Exception as err:
    exc_type, exc_value, exc_tb = sys.exc_info()
    tbe = traceback.TracebackException(
        exc_type, exc_value, exc_tb,
    )
    print(''.join(tbe.format()))

    print('\nexception only:')
    print(''.join(tbe.format_exception_only()))
```

format() 方法生成完整 traceback 的格式化版本，而 format_exception_only() 只显示异常消息。

```
$ python3 traceback_tracebackexception.py

with no exception:
```

```
None

with exception:
Traceback (most recent call last):
  File "traceback_tracebackexception.py", line 22, in <module>
    produce_exception()
  File ".../traceback_example.py", line 17, in produce_exception
    produce_exception(recursion_level - 1)
  File ".../traceback_example.py", line 17, in produce_exception
    produce_exception(recursion_level - 1)
  File ".../traceback_example.py", line 19, in produce_exception
    raise RuntimeError()
RuntimeError


exception only:
RuntimeError
```

16.5.4 底层异常 API

处理异常报告的另一种方法是利用 `print_exc()`。这种方法使用 `sys.exc_info()` 得到当前线程的异常信息，格式化这个结构，并把文本输出到一个文件句柄（默认为 `sys.stder`）。

代码清单 16-51：**traceback_print_exc.py**

```
import traceback
import sys

from traceback_example import produce_exception

print('print_exc() with no exception:')
traceback.print_exc(file=sys.stdout)
print()

try:
    produce_exception()
except Exception as err:
    print('print_exc():')
    traceback.print_exc(file=sys.stdout)
    print()
    print('print_exc(1):')
    traceback.print_exc(limit=1, file=sys.stdout)
```

在这个例子中替换了 `sys.stdout` 文件句柄，使信息消息和 traceback 消息能正确混合。

```
$ python3 traceback_print_exc.py

print_exc() with no exception:
NoneType

print_exc():
Traceback (most recent call last):
  File "traceback_print_exc.py", line 20, in <module>
    produce_exception()
  File ".../traceback_example.py", line 17, in produce_exception
    produce_exception(recursion_level - 1)
  File ".../traceback_example.py", line 17, in produce_exception
    produce_exception(recursion_level - 1)
```

```
    File ".../traceback_example.py", line 19, in produce_exception
        raise RuntimeError()
RuntimeError

print_exc(1):
Traceback (most recent call last):
  File "traceback_print_exc.py", line 20, in <module>
    produce_exception()
RuntimeError
```

print_exc() 就是 print_exception() 的快捷方式，print_exception() 需要显式参数。

代码清单 16-52：traceback_print_exception.py

```
import traceback
import sys

from traceback_example import produce_exception

try:
    produce_exception()
except Exception as err:
    print('print_exception():')
    exc_type, exc_value, exc_tb = sys.exc_info()
    traceback.print_exception(exc_type, exc_value, exc_tb)
```

print_exception() 的参数由 sys.exc_info() 生成。

```
$ python3 traceback_print_exception.py

Traceback (most recent call last):
  File "traceback_print_exception.py", line 16, in <module>
    produce_exception()
  File ".../traceback_example.py", line 17, in produce_exception
    produce_exception(recursion_level - 1)
  File ".../traceback_example.py", line 17, in produce_exception
    produce_exception(recursion_level - 1)
  File ".../traceback_example.py", line 19, in produce_exception
    raise RuntimeError()
RuntimeError
print_exception():
```

print_exception() 使用 format_exception() 准备文本。

代码清单 16-53：traceback_format_exception.py

```
import traceback
import sys
from pprint import pprint

from traceback_example import produce_exception

try:
    produce_exception()
except Exception as err:
    print('format_exception():')
    exc_type, exc_value, exc_tb = sys.exc_info()
    pprint(
        traceback.format_exception(exc_type, exc_value, exc_tb),
        width=65,
    )
```

`format_exception()` 中使用同样的 3 个参数——异常类型、异常值和 traceback。

```
$ python3 traceback_format_exception.py

format_exception():
['Traceback (most recent call last):\n',
 '  File "traceback_format_exception.py", line 17, in '
'<module>\n'
 '    produce_exception()\n',
 '  File '
'".../traceback_example.py", '
'line 17, in produce_exception\n'
 '    produce_exception(recursion_level - 1)\n',
 '  File '
'".../traceback_example.py", '
'line 17, in produce_exception\n'
 '    produce_exception(recursion_level - 1)\n',
 '  File '
'".../traceback_example.py", '
'line 19, in produce_exception\n'
 '    raise RuntimeError()\n',
 'RuntimeError\n']
```

要以另外某种方式处理 traceback，如以不同的方式格式化，可以使用 `extract_tb()` 以一种可用的形式获取数据。

代码清单 16-54：traceback_extract_tb.py

```python
import traceback
import sys
import os
from traceback_example import produce_exception

template = '{filename:<23}:{linenum}:{funcname}:\n    {source}'

try:
    produce_exception()
except Exception as err:
    print('format_exception():')
    exc_type, exc_value, exc_tb = sys.exc_info()
    for tb_info in traceback.extract_tb(exc_tb):
        filename, linenum, funcname, source = tb_info
        if funcname != '<module>':
            funcname = funcname + '()'
        print(template.format(
            filename=os.path.basename(filename),
            linenum=linenum,
            source=source,
            funcname=funcname)
        )
```

返回值是由 traceback 表示的每一层栈元素的一个列表。每个元素是一个元组，包括 4 个部分：源文件的名，文件中的行号，函数名，以及该行去除了所有空白符的源文本（如果可以得到源文本）。

```
$ python3 traceback_extract_tb.py

format_exception():
traceback_extract_tb.py:18:<module>:
```

```
    produce_exception()
traceback_example.py     :17:produce_exception():
    produce_exception(recursion_level - 1)
traceback_example.py     :17:produce_exception():
    produce_exception(recursion_level - 1)
traceback_example.py     :19:produce_exception():
    raise RuntimeError()
```

16.5.5 底层栈 API

还有一组类似的函数可以用来对当前调用栈而不是 traceback 完成同样的操作。`print_stack()` 会打印当前栈,而不生成异常。

代码清单 16-55:`traceback_print_stack.py`

```
import traceback
import sys

from traceback_example import call_function


def f():
    traceback.print_stack(file=sys.stdout)


print('Calling f() directly:')
f()

print()
print('Calling f() from 3 levels deep:')
call_function(f)
```

输出看起来就像一个 traceback,只是没有错误消息。

```
$ python3 traceback_print_stack.py

Calling f() directly:
  File "traceback_print_stack.py", line 21, in <module>
    f()
  File "traceback_print_stack.py", line 17, in f
    traceback.print_stack(file=sys.stdout)

Calling f() from 3 levels deep:
  File "traceback_print_stack.py", line 25, in <module>
    call_function(f)
  File ".../traceback_example.py", line 24, in call_function
    return call_function(f, recursion_level - 1)
  File ".../traceback_example.py", line 24, in call_function
    return call_function(f, recursion_level - 1)
  File ".../traceback_example.py", line 26, in call_function
    return f()
  File "traceback_print_stack.py", line 17, in f
    traceback.print_stack(file=sys.stdout)
```

就像 `format_exception()` 准备 traceback 一样,`format_stack()` 以同样的方式准备栈轨迹。

代码清单 16-56:`traceback_format_stack.py`

```
import traceback
import sys
```

```python
from pprint import pprint

from traceback_example import call_function


def f():
    return traceback.format_stack()


formatted_stack = call_function(f)
pprint(formatted_stack)
```

它会返回一个字符串列表,每个字符串构成输出的一行。

```
$ python3 traceback_format_stack.py

['  File "traceback_format_stack.py", line 21, in <module>\n'
 '    formatted_stack = call_function(f)\n',
 '  File '
 '".../traceback_example.py", '
 'line 24, in call_function\n'
 '    return call_function(f, recursion_level - 1)\n',
 '  File '
 '".../traceback_example.py", '
 'line 24, in call_function\n'
 '    return call_function(f, recursion_level - 1)\n',
 '  File '
 '".../traceback_example.py", '
 'line 26, in call_function\n'
 '    return f()\n',
 '  File "traceback_format_stack.py", line 18, in f\n'
 '    return traceback.format_stack()\n']
```

extract_stack() 函数的做法与 extract_tb() 类似。

代码清单 16-57:**traceback_extract_stack.py**

```python
import traceback
import sys
import os

from traceback_example import call_function

template = '{filename:<26}:{linenum}:{funcname}:\n    {source}'


def f():
    return traceback.extract_stack()


stack = call_function(f)
for filename, linenum, funcname, source in stack:
    if funcname != '<module>':
        funcname = funcname + '()'
    print(template.format(
        filename=os.path.basename(filename),
        linenum=linenum,
        source=source,
        funcname=funcname)
    )
```

它也接受参数（不过这里没有显示），从而可以从栈帧中一个候选位置开始，或者用来限制遍历的深度。

```
$ python3 traceback_extract_stack.py

traceback_extract_stack.py:23:<module>:
    stack = call_function(f)
traceback_example.py        :24:call_function():
    return call_function(f, recursion_level - 1)
traceback_example.py        :24:call_function():
    return call_function(f, recursion_level - 1)
traceback_example.py        :26:call_function():
    return f()
traceback_extract_stack.py:20:f():
    return traceback.extract_stack()
```

提示：相关阅读材料
- `traceback` 的标准库文档⊖。
- `sys`：`sys` 模块包括一些保存当前异常的单例对象。
- `inspect`：`inspect` 模块包括另外一些函数来探查栈中的帧。
- `cgitb`：这是另一个用于美观地格式化 traceback 的模块。

16.6 `cgitb`：详细的 traceback 报告

`cgitb` 是标准库中一个很有价值的调试工具。它原来是被设计来显示 Web 应用中的错误和调试信息的。尽管后来得到更新，包含了纯文本输出，不过遗憾的是，更新后并没有相应地改名。这带来了模糊性，以至于这个模块没有得到应有的关注，本来它应该更为常用（它比 `traceback` 包含更详细的 traceback 信息）。

16.6.1 标准 traceback 转储

Python 的默认异常处理行为是向标准错误输出流打印一个 traceback，并提供直至错误位置的调用栈。这个基本输出包含的信息通常足以了解异常的原因并做出修正。

代码清单 16-58：`cgitb_basic_traceback.py`

```
def func2(a, divisor):
    return a / divisor


def func1(a, b):
    c = b - 5
    return func2(a, c)

func1(1, 5)
```

这个示例程序在 `func2()` 中有一个小错误。

⊖ https://docs.python.org/3.5/library/traceback.html

```
$ python3 cgitb_basic_traceback.py

Traceback (most recent call last):
  File "cgitb_basic_traceback.py", line 18, in <module>
    func1(1, 5)
  File "cgitb_basic_traceback.py", line 16, in func1
    return func2(a, c)
  File "cgitb_basic_traceback.py", line 11, in func2
    return a / divisor
ZeroDivisionError: division by zero
```

16.6.2 启用详细的 traceback

尽管基本 traceback 包括了足够的信息来发现错误，不过启用 `cgitb` 将给出更多详细信息。`cgitb` 将 `sys.excepthook` 替换为另一个函数，它能提供更丰富的 traceback。

代码清单 16-59：`cgitb_local_vars.py`

```
import cgitb
cgitb.enable(format='text')
```

这个例子生成的错误报告比原先要丰富得多。会列出栈的每一帧，并提供以下信息：
- 源文件的完整路径，而不只是基名。
- 栈中各个函数的参数值。
- 错误路径中当前行周围的几行源代码上下文。
- 导致错误的表达式中的变量值。

由于能够访问错误栈中涉及的变量，这有助于程序员找到出现在栈中更高位置的逻辑错误，而不只是生成具体异常的那一行代码。

```
$ python3 cgitb_local_vars.py

ZeroDivisionError
Python 3.5.2: .../bin/python3
Thu Dec 29 09:30:37 2016

A problem occurred in a Python script.  Here is the sequence of
function calls leading up to the error, in the order they
occurred.

 .../cgitb_local_vars.py in <module>()
   18 def func1(a, b):
   19     c = b - 5
   20     return func2(a, c)
   21
   22 func1(1, 5)
func1 = <function func1>

 .../cgitb_local_vars.py in func1(a=1, b=5)
   18 def func1(a, b):
   19     c = b - 5
   20     return func2(a, c)
   21
   22 func1(1, 5)
global func2 = <function func2>
a = 1
c = 0
```

```
.../cgitb_local_vars.py in func2(a=1, divisor=0)
   13
   14 def func2(a, divisor):
   15     return a / divisor
   16
   17
a = 1
divisor = 0
ZeroDivisionError: division by zero
    __cause__ = None
    __class__ = <class 'ZeroDivisionError'>
    __context__ = None
    __delattr__ = <method-wrapper '__delattr__' of
ZeroDivisionError object>
    __dict__ = {}
    __dir__ = <built-in method __dir__ of ZeroDivisionError
object>
    __doc__ = 'Second argument to a division or modulo operation
was zero.'
    __eq__ = <method-wrapper '__eq__' of ZeroDivisionError
object>
    __format__ = <built-in method __format__ of
ZeroDivisionError object>
    __ge__ = <method-wrapper '__ge__' of ZeroDivisionError
object>
    __getattribute__ = <method-wrapper '__getattribute__' of
ZeroDivisionError object>
    __gt__ = <method-wrapper '__gt__' of ZeroDivisionError
object>
    __hash__ = <method-wrapper '__hash__' of ZeroDivisionError
object>
    __init__ = <method-wrapper '__init__' of ZeroDivisionError
object>
    __le__ = <method-wrapper '__le__' of ZeroDivisionError
object>
    __lt__ = <method-wrapper '__lt__' of ZeroDivisionError
object>
    __ne__ = <method-wrapper '__ne__' of ZeroDivisionError
object>
    __new__ = <built-in method __new__ of type object>
    __reduce__ = <built-in method __reduce__ of
ZeroDivisionError object>
    __reduce_ex__ = <built-in method __reduce_ex__ of
ZeroDivisionError object>
    __repr__ = <method-wrapper '__repr__' of ZeroDivisionError
object>
    __setattr__ = <method-wrapper '__setattr__' of
ZeroDivisionError object>
    __setstate__ = <built-in method __setstate__ of
ZeroDivisionError object>
    __sizeof__ = <built-in method __sizeof__ of
ZeroDivisionError object>
    __str__ = <method-wrapper '__str__' of ZeroDivisionError
object>
    __subclasshook__ = <built-in method __subclasshook__ of type
object>
    __suppress_context__ = False
    __traceback__ = <traceback object>
    args = ('division by zero',)
    with_traceback = <built-in method with_traceback of
ZeroDivisionError object>
```

The above is a description of an error in a Python program.

```
Here is
the original traceback:

Traceback (most recent call last):
  File "cgitb_local_vars.py", line 22, in <module>
    func1(1, 5)
  File "cgitb_local_vars.py", line 20, in func1
    return func2(a, c)
  File "cgitb_local_vars.py", line 15, in func2
    return a / divisor
ZeroDivisionError: division by zero
```

对于这个存在 `ZeroDivisionError` 异常的代码,显然是因为 `func1()` 中 c 值的计算带来的问题,而不是因为 `func2()` 中在哪里使用这个值。

输出的最后还包括了异常对象的完整细节(除了 `message` 以外可能还有其他属性,对调试会很有用),以及 traceback 转储的原始形式。

16.6.3 traceback 中的局部变量

`cgitb` 中的代码会检查栈帧中使用的导致错误的变量,这些代码足够聪明,还可以计算对象属性并显示。

代码清单 16-60:`cgitb_with_classes.py`

```
import cgitb
cgitb.enable(format='text', context=12)


class BrokenClass:
    """This class has an error.
    """

    def __init__(self, a, b):
        """Be careful passing arguments in here.
        """
        self.a = a
        self.b = b
        self.c = self.a * self.b
        # Really
        # long
        # comment
        # goes
        # here.
        self.d = self.a / self.b
        return

o = BrokenClass(1, 0)
```

如果一个函数或方法包括大量内联注释、空白符或其他代码,使它篇幅很长,那么倘若只有默认的 5 行上下文,则可能无法提供足够的指示。如果将函数体从代码窗口取出,在屏幕上不再可见,那么就没有足够的上下文来了解出现错误的位置。可以对 `cgitb` 使用一个更大的上下文值来解决这个问题。向 `enable()` 传入一个整数作为上下文(`context`)参数,控制其为 traceback 的各行显示多少代码。

下面的输出显示,容易出错的代码与 `self.a` 和 `self.b` 有关。

```
$ python3 cgitb_with_classes.py

ZeroDivisionError
Python 3.5.2: .../bin/python3
Thu Dec 29 09:30:37 2016

A problem occurred in a Python script.  Here is the sequence of
function calls leading up to the error, in the order they
occurred.

 .../cgitb_with_classes.py in <module>()
    21          self.a = a
    22          self.b = b
    23          self.c = self.a * self.b
    24          # Really
    25          # long
    26          # comment
    27          # goes
    28          # here.
    29          self.d = self.a / self.b
    30          return
    31
    32 o = BrokenClass(1, 0)
o undefined
BrokenClass = <class '__main__.BrokenClass'>

 .../cgitb_with_classes.py in
 __init__(self=<__main__.BrokenClass object>, a=1, b=0)
    21          self.a = a
    22          self.b = b
    23          self.c = self.a * self.b
    24          # Really
    25          # long
    26          # comment
    27          # goes
    28          # here.
    29          self.d = self.a / self.b
    30          return
    31
    32 o = BrokenClass(1, 0)
self = <__main__.BrokenClass object>
self.d undefined
self.a = 1
self.b = 0
ZeroDivisionError: division by zero
    __cause__ = None
    __class__ = <class 'ZeroDivisionError'>
    __context__ = None
    __delattr__ = <method-wrapper '__delattr__' of
ZeroDivisionError object>
    __dict__ = {}
    __dir__ = <built-in method __dir__ of ZeroDivisionError
object>
    __doc__ = 'Second argument to a division or modulo operation
was zero.'
    __eq__ = <method-wrapper '__eq__' of ZeroDivisionError
object>
    __format__ = <built-in method __format__ of
ZeroDivisionError object>
    __ge__ = <method-wrapper '__ge__' of ZeroDivisionError
object>
    __getattribute__ = <method-wrapper '__getattribute__' of
```

```
           ZeroDivisionError object>
       __gt__ = <method-wrapper '__gt__' of ZeroDivisionError
       object>
       __hash__ = <method-wrapper '__hash__' of ZeroDivisionError
       object>
       __init__ = <method-wrapper '__init__' of ZeroDivisionError
       object>
       __le__ = <method-wrapper '__le__' of ZeroDivisionError
       object>
       __lt__ = <method-wrapper '__lt__' of ZeroDivisionError
       object>
       __ne__ = <method-wrapper '__ne__' of ZeroDivisionError
       object>
       __new__ = <built-in method __new__ of type object>
       __reduce__ = <built-in method __reduce__ of
       ZeroDivisionError object>
       __reduce_ex__ = <built-in method __reduce_ex__ of
       ZeroDivisionError object>
       __repr__ = <method-wrapper '__repr__' of ZeroDivisionError
       object>
       __setattr__ = <method-wrapper '__setattr__' of
       ZeroDivisionError object>
       __setstate__ = <built-in method __setstate__ of
       ZeroDivisionError object>
       __sizeof__ = <built-in method __sizeof__ of
       ZeroDivisionError object>
       __str__ = <method-wrapper '__str__' of ZeroDivisionError
       object>
       __subclasshook__ = <built-in method __subclasshook__ of type
       object>
       __suppress_context__ = False
       __traceback__ = <traceback object>
       args = ('division by zero',)
       with_traceback = <built-in method with_traceback of
       ZeroDivisionError object>

The above is a description of an error in a Python program.
Here is
the original traceback:

Traceback (most recent call last):
  File "cgitb_with_classes.py", line 32, in <module>
    o = BrokenClass(1, 0)
  File "cgitb_with_classes.py", line 29, in __init__
    self.d = self.a / self.b
ZeroDivisionError: division by zero
```

16.6.4 异常属性

除了显示每个栈帧局部变量的属性，cgitb 还会显示异常对象的所有属性。定制异常类型的额外属性会作为错误报告的一部分打印。

代码清单 16-61：`cgitb_exception_properties.py`

```
import cgitb
cgitb.enable(format='text')


class MyException(Exception):
    """Add extra properties to a special exception
```

```
    """

    def __init__(self, message, bad_value):
        self.bad_value = bad_value
        Exception.__init__(self, message)
        return

raise MyException('Normal message', bad_value=99)
```

在这个例子中，除了标准的 `message` 和 `args` 值，还包含 `bad_value` 属性。

```
$ python3 cgitb_exception_properties.py

MyException
Python 3.5.2: .../bin/python3
Thu Dec 29 09:30:37 2016

A problem occurred in a Python script.  Here is the sequence of
function calls leading up to the error, in the order they
occurred.

 .../cgitb_exception_properties.py in <module>()
   19         self.bad_value = bad_value
   20         Exception.__init__(self, message)
   21         return
   22
   23 raise MyException('Normal message', bad_value=99)
MyException = <class '__main__.MyException'>
bad_value undefined
MyException: Normal message
    __cause__ = None
    __class__ = <class '__main__.MyException'>
    __context__ = None
    __delattr__ = <method-wrapper '__delattr__' of MyException
object>
    __dict__ = {'bad_value': 99}
    __dir__ = <built-in method __dir__ of MyException object>
    __doc__ = 'Add extra properties to a special exception\n
'
    __eq__ = <method-wrapper '__eq__' of MyException object>
    __format__ = <built-in method __format__ of MyException
object>
    __ge__ = <method-wrapper '__ge__' of MyException object>
    __getattribute__ = <method-wrapper '__getattribute__' of
MyException object>
    __gt__ = <method-wrapper '__gt__' of MyException object>
    __hash__ = <method-wrapper '__hash__' of MyException object>
    __init__ = <bound method MyException.__init__ of
MyException('Normal message',)>
    __le__ = <method-wrapper '__le__' of MyException object>
    __lt__ = <method-wrapper '__lt__' of MyException object>
    __module__ = '__main__'
    __ne__ = <method-wrapper '__ne__' of MyException object>
    __new__ = <built-in method __new__ of type object>
    __reduce__ = <built-in method __reduce__ of MyException
object>
    __reduce_ex__ = <built-in method __reduce_ex__ of
MyException object>
    __repr__ = <method-wrapper '__repr__' of MyException object>
    __setattr__ = <method-wrapper '__setattr__' of MyException
object>
    __setstate__ = <built-in method __setstate__ of MyException
```

```
      object>
    __sizeof__ = <built-in method __sizeof__ of MyException
      object>
    __str__ = <method-wrapper '__str__' of MyException object>
    __subclasshook__ = <built-in method __subclasshook__ of type
      object>
    __suppress_context__ = False
    __traceback__ = <traceback object>
    __weakref__ = None
    args = ('Normal message',)
    bad_value = 99
    with_traceback = <built-in method with_traceback of
      MyException object>

The above is a description of an error in a Python program.
Here is
the original traceback:

Traceback (most recent call last):
  File "cgitb_exception_properties.py", line 23, in <module>
    raise MyException('Normal message', bad_value=99)
MyException: Normal message
```

16.6.5 HTML 输出

由于 cgitb 原来是为了处理 Web 应用中的异常而开发的,所以如果不介绍原来的 HTML 输出格式,就不能算是完整的讨论。前面的例子都显示了纯文本输出。为了生成 HTML 输出,要省略格式(format)参数(或者指定 "html")。大多数现代 Web 应用都使用一个包含错误报告功能的框架来构造,所以某种程度上讲,HTML 格式已经过时了。

16.6.6 记录 traceback

很多情况下,将 traceback 细节打印到标准错误输出是最好的解决方案。不过,在生产系统中,更好的做法是记录错误日志。enable() 函数包括一个可选参数 logdir,用来启用错误日志。为这个方法提供一个目录名时,每个异常都会被记入指定目录内该异常自己的文件中。

代码清单 16-62:cgitb_log_exception.py

```python
import cgitb
import os

LOGDIR = os.path.join(os.path.dirname(__file__), 'LOGS')

if not os.path.exists(LOGDIR):
    os.makedirs(LOGDIR)

cgitb.enable(
    logdir=LOGDIR,
    display=False,
    format='text',
)


def func(a, divisor):
    return a / divisor

func(1, 0)
```

尽管"抑制"了错误显示，但会打印一个消息来指出错误日志的位置。

```
$ python3 cgitb_log_exception.py

<p>A problem occurred in a Python script.
.../LOGS/tmptxqq_6yx.txt contains the description of this error.

$ ls LOGS

tmptxqq_6yx.txt

$ cat LOGS/*.txt

ZeroDivisionError
Python 3.5.2: .../bin/python3
Thu Dec 29 09:30:38 2016

A problem occurred in a Python script.  Here is the sequence of
function calls leading up to the error, in the order they
occurred.

 .../cgitb_log_exception.py in <module>()
   24
   25 def func(a, divisor):
   26     return a / divisor
   27
   28 func(1, 0)
func = <function func>

 .../cgitb_log_exception.py in func(a=1, divisor=0)
   24
   25 def func(a, divisor):
   26     return a / divisor
   27
   28 func(1, 0)
a = 1
divisor = 0
ZeroDivisionError: division by zero
    __cause__ = None
    __class__ = <class 'ZeroDivisionError'>
    __context__ = None
    __delattr__ = <method-wrapper '__delattr__' of
ZeroDivisionError object>
    __dict__ = {}
    __dir__ = <built-in method __dir__ of ZeroDivisionError
object>
    __doc__ = 'Second argument to a division or modulo operation
was zero.'
    __eq__ = <method-wrapper '__eq__' of ZeroDivisionError
object>
    __format__ = <built-in method __format__ of
ZeroDivisionError object>
    __ge__ = <method-wrapper '__ge__' of ZeroDivisionError
object>
    __getattribute__ = <method-wrapper '__getattribute__' of
ZeroDivisionError object>
    __gt__ = <method-wrapper '__gt__' of ZeroDivisionError
object>
    __hash__ = <method-wrapper '__hash__' of ZeroDivisionError
object>
    __init__ = <method-wrapper '__init__' of ZeroDivisionError
object>
    __le__ = <method-wrapper '__le__' of ZeroDivisionError
```

```
         object>
         __lt__ = <method-wrapper '__lt__' of ZeroDivisionError
         object>
         __ne__ = <method-wrapper '__ne__' of ZeroDivisionError
         object>
         __new__ = <built-in method __new__ of type object>
         __reduce__ = <built-in method __reduce__ of
         ZeroDivisionError object>
         __reduce_ex__ = <built-in method __reduce_ex__ of
         ZeroDivisionError object>
         __repr__ = <method-wrapper '__repr__' of ZeroDivisionError
         object>
         __setattr__ = <method-wrapper '__setattr__' of
         ZeroDivisionError object>
         __setstate__ = <built-in method __setstate__ of
         ZeroDivisionError object>
         __sizeof__ = <built-in method __sizeof__ of
         ZeroDivisionError object>
         __str__ = <method-wrapper '__str__' of ZeroDivisionError
         object>
         __subclasshook__ = <built-in method __subclasshook__ of type
         object>
         __suppress_context__ = False
         __traceback__ = <traceback object>
         args = ('division by zero',)
         with_traceback = <built-in method with_traceback of
         ZeroDivisionError object>

The above is a description of an error in a Python program.
Here is
the original traceback:

Traceback (most recent call last):
  File "cgitb_log_exception.py", line 28, in <module>
    func(1, 0)
  File "cgitb_log_exception.py", line 26, in func
    return a / divisor
ZeroDivisionError: division by zero
```

提示：相关阅读材料

- `cgitb` 的标准库文档[⊖]。
- `traceback`：处理 traceback 的标准库模块。
- `inspect`：`inspect` 模块包含更多用于检查栈的函数。
- `sys`：`sys` 模块允许访问当前异常值和出现异常时所调用的 `excepthook` 处理器。
- 改进的 `traceback` 模块[⊖]：Python 开发邮件列表中关于 `traceback` 模块的改进以及其他开发人员本地使用的相关改进的讨论。

16.7 pdb：交互式调试工具

pdb 为 Python 程序实现了一个交互式调试环境。它包括一些特性，可以暂停程序，查

⊖ https://docs.python.org/3.5/library/cgitb.html
⊖ https://lists.gt.net/python/dev/802870

看变量值，以及逐步监视程序执行，从而能了解程序具体做了什么，并查找逻辑中存在的 bug。

16.7.1 启动调试工具

使用 pdb 的第一步是让解释器在适当的时候进入调试工具。可以采用很多不同的方法达到这个目的，具体取决于起始条件和所要调试的内容。

16.7.1.1 从命令行运行

使用调试工具最直接的方法是从命令行运行调试工具，提供程序作为输入，使它知道要运行什么。

代码清单 16-63：**pdb_script.py**

```
 1  #!/usr/bin/env python3
 2  # encoding: utf-8
 3  #
 4  # Copyright (c) 2010 Doug Hellmann.  All rights reserved.
 5  #
 6
 7
 8  class MyObj:
 9
10      def __init__(self, num_loops):
11          self.count = num_loops
12
13      def go(self):
14          for i in range(self.count):
15              print(i)
16          return
17
18  if __name__ == '__main__':
19      MyObj(5).go()
```

从命令行运行调试工具时，它会加载源文件，并在找到的第一条语句处停止执行。在这里，它会在第 8 行评估类 **MyObj** 的定义之前停止。

```
$ python3 -m pdb pdb_script.py

> .../pdb_script.py(8)<module>()
-> class MyObj(object):
(Pdb)
```

说明：通常，pdb 打印一个文件名时会在输出中包含各模块的完整路径。为了简化这一节中的例子，示例输出中的路径被替换为一个省略号（...）。

16.7.1.2 在解释器中运行

很多 Python 开发人员在开发模块的较早版本时会使用交互式解释器，因为这样他们能反复试验，而不用像创建独立脚本时那样，需要完整地保存/运行/重复周期。要在一个交互式解释器中运行调试工具，可以使用 run() 或 runeval()。

```
$ python3
Python 3.5.1 (v3.5.1:37a07cee5969, Dec  5 2015, 21:12:44)
```

```
[GCC 4.2.1 (Apple Inc. build 5666) (dot 3)] on darwin
Type "help", "copyright", "credits" or "license" for more information.
>>> import pdb_script
>>> import pdb
>>> pdb.run('pdb_script.MyObj(5).go()')
> <string>(1)<module>()
(Pdb)
```

`run()` 的参数是一个串表达式，可以由 Python 解释器估算。调试工具会进行解析，然后在估算第一个表达式之前暂停执行。这里介绍的调试工具命令可以用来导航和控制执行。

16.7.1.3 在程序中运行

前面的两个例子都是从程序一开始就启动调试工具。对于一个长时间运行的进程，问题可能出现在程序执行比较靠后的时刻，更方便的做法是在程序中使用 `set_trace()` 启用调试工具。

代码清单 16-64：**pdb_set_trace.py**

```
 1  #!/usr/bin/env python3
 2  # encoding: utf-8
 3  #
 4  # Copyright (c) 2010 Doug Hellmann.  All rights reserved.
 5  #
 6
 7  import pdb
 8
 9
10  class MyObj:
11
12      def __init__(self, num_loops):
13          self.count = num_loops
14
15      def go(self):
16          for i in range(self.count):
17              pdb.set_trace()
18              print(i)
19          return
20
21  if __name__ == '__main__':
22      MyObj(5).go()
```

示例脚本的第 17 行在执行到该点时触发调试工具，使它在第 18 行暂停。

```
$ python3 ./pdb_set_trace.py

> .../pdb_set_trace.py(18)go()
-> print(i)
(Pdb)
```

`set_trace()` 只是一个 Python 函数，所以可以在程序中任意位置调用。这样就可以根据程序中的条件进入调试工具，包括从一个异常处理器进入，或者通过一个控制语句的特定分支进入。

16.7.1.4 失败后运行

在程序终止后调试失败被称为事后剖析调试（post-mortem debugging）。`pdb` 通过 `pm()` 和 `post_mortem()` 函数支持事后剖析调试。

代码清单 16-65：`pdb_post_mortem.py`

```python
#!/usr/bin/env python3
# encoding: utf-8
#
# Copyright (c) 2010 Doug Hellmann.  All rights reserved.
#


class MyObj:

    def __init__(self, num_loops):
        self.count = num_loops

    def go(self):
        for i in range(self.num_loops):
            print(i)
        return
```

在这个例子中，第 14 行上不正确的属性名触发了一个 `AttributeError` 异常，导致执行停止，`pm()` 查找活动 traceback，并在调用栈中出现异常的位置启动调试工具。

```
$ python3
Python 3.5.1 (v3.5.1:37a07cee5969, Dec  5 2015, 21:12:44)
[GCC 4.2.1 (Apple Inc. build 5666) (dot 3)] on darwin
Type "help", "copyright", "credits" or "license" for more information.
>>> from pdb_post_mortem import MyObj
>>> MyObj(5).go()
Traceback (most recent call last):
  File "<stdin>", line 1, in <module>
  File ".../pdb_post_mortem.py", line 14, in go
    for i in range(self.num_loops):
AttributeError: 'MyObj' object has no attribute 'num_loops'
>>> import pdb
>>> pdb.pm()
> .../pdb/pdb_post_mortem.py(14)go()
-> for i in range(self.num_loops):
(Pdb)
```

16.7.2 控制调试工具

调试工具的接口是一个很小的命令语言，允许你在调用栈中移动，检查和修改变量的值，以及控制调试工具如何执行程序。这个交互式调试工具使用 `readline` 接受命令，而且支持对命令、文件名和函数名的 tab 完成特性。输入一个空行会使其重新运行之前的命令，除非是一个 `list` 操作。

16.7.2.1 导航执行栈

调试工具运行的任何时刻，使用 `where`（缩写为 `w`）可以得出正在执行哪一行，以及程序在调用栈的哪个位置。在下述情况下，模块 `pdb_set_trace.py` 的 `go()` 方法中的第 18 行将停止执行。

```
$ python3 pdb_set_trace.py
> .../pdb_set_trace.py(18)go()
-> print(i)
(Pdb) where
  .../pdb_set_trace.py(22)<module>()
-> MyObj(5).go()
```

```
> .../pdb_set_trace.py(18)go()
-> print(i)
(Pdb)
```

要增加当前位置的更多上下文，可以使用 list (l)。

```
(Pdb) l
 13            self.count = num_loops
 14
 15        def go(self):
 16            for i in range(self.count):
 17                pdb.set_trace()
 18 ->             print(i)
 19            return
 20
 21    if __name__ == '__main__':
 22        MyObj(5).go()
[EOF]
(Pdb)
```

默认地会列出包括当前行在内的周围 11 行（前面 5 行，后面 5 行）。如果使用 list 并提供一个数值参数，则会列出指定行（而不是当前行）周围的 11 行。

```
(Pdb) list 14
  9
 10    class MyObj(object):
 11
 12        def __init__(self, num_loops):
 13            self.count = num_loops
 14
 15        def go(self):
 16            for i in range(self.count):
 17                pdb.set_trace()
 18 ->             print(i)
 19            return
```

如果 list 接收两个参数，则会把它们分别解释为第一行和最后一行，包含在其输出中。

```
(Pdb) list 7, 19
  7    import pdb
  8
  9
 10    class MyObj(object):
 11
 12        def __init__(self, num_loops):
 13            self.count = num_loops
 14
 15        def go(self):
 16            for i in range(self.count):
 17                pdb.set_trace()
 18 ->             print(i)
 19            return
```

longlist (ll) 命令打印当前函数或帧的源码，而不必提前确定行号。这个命令名为 "longlist"，这是因为对于长函数，与 list 的默认输出相比，它会生成更多的输出。

```
(Pdb) longlist
 15        def go(self):
 16            for i in range(self.count):
 17                pdb.set_trace()
 18 ->             print(i)
 19            return
```

source 命令加载和打印一个任意的类、函数或模块的完整源代码。

```
(Pdb) source MyObj
 10  class MyObj:
 11
 12      def __init__(self, num_loops):
 13          self.count = num_loops
 14
 15      def go(self):
 16          for i in range(self.count):
 17              pdb.set_trace()
 18              print(i)
 19          return
```

可以使用 up 和 down 在当前调用栈的帧之间移动。up（缩写为 u）向栈中较旧的帧移动。down（缩写为 d）则移向较新的帧。每次在栈中上移或下移时，调试工具都会打印当前位置，格式与 where 生成的输出格式相同。

```
(Pdb) up
> .../pdb_set_trace.py(22)<module>()
-> MyObj(5).go()

(Pdb) down
> .../pdb_set_trace.py(18)go()
-> print(i)
```

向 up 或 down 传入一个数值参数以便一次在栈中上移或下移指定的步数。

16.7.2.2　检查栈中的变量

栈中的各个帧会维护一组变量，包括所执行函数的局部值和全局状态信息。pdb 提供了多种方法来检查这些变量的内容。

代码清单 16-66：pdb_function_arguments.py

```
 1  #!/usr/bin/env python3
 2  # encoding: utf-8
 3  #
 4  # Copyright (c) 2010 Doug Hellmann.  All rights reserved.
 5  #
 6
 7  import pdb
 8
 9
10  def recursive_function(n=5, output='to be printed'):
11      if n > 0:
12          recursive_function(n - 1)
13      else:
14          pdb.set_trace()
15          print(output)
16      return
17
18  if __name__ == '__main__':
19      recursive_function()
```

args 命令（缩写为 a）会打印当前帧中活动函数的所有参数。这个例子还使用了一个递归函数，可以显示 where 打印内容时的一个更深的栈。

```
$ python3 pdb_function_arguments.py
> .../pdb_function_arguments.py(15)recursive_function()
```

```
    -> print(output)
(Pdb) where
  .../pdb_function_arguments.py(19)<module>()
-> recursive_function()
  .../pdb_function_arguments.py(12)recursive_function()
-> recursive_function(n - 1)
  .../pdb_function_arguments.py(12)recursive_function()
-> recursive_function(n - 1)
  .../pdb_function_arguments.py(12)recursive_function()
-> recursive_function(n - 1)
  .../pdb_function_arguments.py(12)recursive_function()
-> recursive_function(n - 1)
> .../pdb_function_arguments.py(15)recursive_function()
-> print(output)

(Pdb) args
n = 0
output = to be printed

(Pdb) up
> .../pdb_function_arguments.py(12)recursive_function()
-> recursive_function(n - 1)

(Pdb) args
n = 1
output = to be printed
```

p命令会评估作为参数给定的一个表达式，并打印结果。也可以使用Python的 `print()` 函数，不过要把它传递到解释器执行，而不是在调试工具中作为一个命令运行。

```
(Pdb) p n
1

(Pdb) print(n)
1
```

类似地，在一个表达式前面加上前缀！就会把它传递到Python解释器进行评估。这个特性可以用来执行任意的Python语句，包括修改变量。下面这个例子在允许调试工具继续运行程序之前会修改 output 的值。set_trace() 调用后的下一条语句在打印 output 的值时会显示修改后的值。

```
$ python3 pdb_function_arguments.py

> .../pdb_function_arguments.py(14)recursive_function()
-> print(output)

(Pdb) !output
'to be printed'

(Pdb) !output='changed value'

(Pdb) continue
changed value
```

对于更复杂的值，如嵌套数据结构或大型数据结构，要使用 pp 以"完美打印"的格式来打印。下面这个程序从一个文件读取多个文本行。

代码清单 16-67：**pdb_pp.py**

```
1  #!/usr/bin/env python3
2  # encoding: utf-8
3  #
4  # Copyright (c) 2010 Doug Hellmann.  All rights reserved.
5  #
6
7  import pdb
8
9  with open('lorem.txt', 'rt') as f:
10     lines = f.readlines()
11
12 pdb.set_trace()
```

用 `p` 打印变量 `lines` 时，得到的输出很难读，因为它的换行很糟糕。`pp` 使用 `pprint` 格式化值，从而能更美观地打印。

```
$ python3 pdb_pp.py

> .../pdb_pp.py(12)<module>()->None
-> pdb.set_trace()
(Pdb) p lines
['Lorem ipsum dolor sit amet, consectetuer adipiscing elit.
\n', 'Donec egestas, enim et consecte tuer ullamcorper, lect
us \n', 'ligula rutrum leo, a elementum el it tortor eu quam
.\n']

(Pdb) pp lines
['Lorem ipsum dolor sit amet, consectetuer adipiscing elit. \n',
 'Donec egestas, enim et consectetuer ullamcorper, lectus \n',
 'ligula rutrum leo, a elementum elit tortor eu quam.\n']

(Pdb)
```

对于交互式探索和试验，可以从调试工具回到标准的 Python 交互式提示窗口（已经填充当前帧的全局和局部值）。

```
$ python3 -m pdb pdb_interact.py
> .../pdb_interact.py(7)<module>()
-> import pdb
(Pdb) break 14
Breakpoint 1 at .../pdb_interact.py:14

(Pdb) continue
> .../pdb_interact.py(14)f()
-> print(l, m, n)

(Pdb) p l
['a', 'b']

(Pdb) p m
9

(Pdb) p n
5

(Pdb) interact
*interactive*

>>> l
```

```
['a', 'b']
>>> m
9
>>> n
5
```

交互式解释器可以改变类似列表等可变对象。与之相反，不可变的对象是不能改变的，它们的名字不能被重新绑定到新值。

```
>>> l.append('c')
>>> m += 7
>>> n = 3
>>> l
['a', 'b', 'c']
>>> m
16
>>> n
3
```

使用文件结束序列 Ctrl-D 退出交互式提示窗口，并返回到调试工具。在这个例子中，列表 l 已经改变，但是 m 和 n 的值没有改变。

```
>>> ^D
(Pdb) p l
['a', 'b', 'c']
(Pdb) p m
9
(Pdb) p n
5
(Pdb)
```

16.7.2.3 单步执行程序

除了程序暂停时在调用栈中上下导航外，还可以在进入调试工具那一点之后单步执行程序。

代码清单 16-68：**pdb_step.py**

```
1  #!/usr/bin/env python3
2  # encoding: utf-8
3  #
4  # Copyright (c) 2010 Doug Hellmann.  All rights reserved.
5  #
6
7  import pdb
8
9
10 def f(n):
11     for i in range(n):
12         j = i * n
13         print(i, j)
14     return
```

```
15
16  if __name__ == '__main__':
17      pdb.set_trace()
18      f(5)
```

使用 step（缩写为 s）执行当前行，然后在下一个执行点停止，这可能是所调用函数中的第一条语句，也可能是当前函数的下一行语句。

```
$ python3 pdb_step.py

> .../pdb_step.py(18)<module>()
-> f(5)
```

解释器会在 set_trace() 调用处暂停，并将控制交给调试工具。第一个 step 会执行进入 f()。

```
(Pdb) step
--Call--
> .../pdb_step.py(10)f()
-> def f(n):
```

下一个 step 会执行到 f() 的第一行，并开始循环。

```
(Pdb) step
> .../pdb_step.py(11)f()
-> for i in range(n):
```

执行下一步会移动到循环中的第一行，这里定义了 j。

```
(Pdb) step
> .../pdb_step.py(12)f()
-> j = i * n

(Pdb) p i
0
```

i 的值为 0，所以再执行一步后，j 的值应当也是 0。

```
(Pdb) step
> .../pdb_step.py(13)f()
-> print(i, j)

(Pdb) p j
0

(Pdb)
```

像这样一次执行一步，如果在出现错误那一点之前需要执行大量代码，或者如果需要反复调用相同的函数，那么这会变得很麻烦。

代码清单 16-69：pdb_next.py

```
1  #!/usr/bin/env python3
2  # encoding: utf-8
3  #
4  # Copyright (c) 2010 Doug Hellmann.  All rights reserved.
5  #
6
7  import pdb
8
9
```

```
10  def calc(i, n):
11      j = i * n
12      return j
13
14
15  def f(n):
16      for i in range(n):
17          j = calc(i, n)
18          print(i, j)
19      return
20
21  if __name__ == '__main__':
22      pdb.set_trace()
23      f(5)
```

在这个例子中，`calc()`没有错误，所以如果每次在`f()`的循环中调用这个函数时都进行单步跟踪，那么执行时会显示`calc()`的所有代码行，这就对真正有用的输出造成了干扰。

```
$ python3 pdb_next.py

> .../pdb_next.py(23)<module>()
-> f(5)
(Pdb) step
--Call--
> .../pdb_next.py(15)f()
-> def f(n):

(Pdb) step
> .../pdb_next.py(16)f()
-> for i in range(n):

(Pdb) step
> .../pdb_next.py(17)f()
-> j = calc(i, n)

(Pdb) step
--Call--
> .../pdb_next.py(10)calc()
-> def calc(i, n):

(Pdb) step
> .../pdb_next.py(11)calc()
-> j = i * n

(Pdb) step
> .../pdb_next.py(12)calc()
-> return j

(Pdb) step
--Return--
> .../pdb_next.py(12)calc()->0
-> return j

(Pdb) step
> .../pdb_next.py(18)f()
-> print(i, j)

(Pdb) step
0 0
```

```
> .../pdb_next.py(16)f()
-> for i in range(n):
(Pdb)
```

next命令（缩写为n）有些类似step，不过不会从正在执行的语句进入所调用的函数。实际上，它会用一个操作完成整个函数调用，直接进入当前函数的下一条语句。

```
> .../pdb_next.py(16)f()
-> for i in range(n):
(Pdb) step
> .../pdb_next.py(17)f()
-> j = calc(i, n)

(Pdb) next
> .../pdb_next.py(18)f()
-> print(i, j)

(Pdb)
```

until命令类似于next，只不过它会继续执行，直至执行到同一个函数中行号大于当前值的一行。这说明，可以用until跳过循环末尾。

```
$ python3 pdb_next.py

> .../pdb_next.py(23)<module>()
-> f(5)
(Pdb) step
--Call--
> .../pdb_next.py(15)f()
-> def f(n):

(Pdb) step
> .../pdb_next.py(16)f()
-> for i in range(n):

(Pdb) step
> .../pdb_next.py(17)f()
-> j = calc(i, n)

(Pdb) next
> .../pdb_next.py(18)f()
-> print(i, j)

(Pdb) until
0 0
1 5
2 10
3 15
4 20
> .../pdb_next.py(19)f()
-> return

(Pdb)
```

运行until命令之前，当前行为18，即循环的最后一行。运行until之后，位于第19行，并且循环已经退出。

要想一直执行到一个特定的行，可以把这个行号传递给until命令。与设置断点时不同，传入until的行号必须大于当前行号，所以这个命令对于通过跳过很长的代码块来在函数中导航而言最为有用。

```
$ python3 pdb_next.py
> .../pdb_next.py(23)<module>()
-> f(5)
(Pdb) list
 18         print(i, j)
 19     return
 20
 21 if __name__ == '__main__':
 22     pdb.set_trace()
 23 ->     f(5)
[EOF]

(Pdb) until 18
*** "until" line number is smaller than current line number

(Pdb) step
--Call--
> .../pdb_next.py(15)f()
-> def f(n):

(Pdb) step
> .../pdb_next.py(16)f()
-> for i in range(n):

(Pdb) list
 11         j = i * n
 12         return j
 13
 14
 15 def f(n):
 16 ->     for i in range(n):
 17         j = calc(i, n)
 18         print(i, j)
 19     return
 20
 21 if __name__ == '__main__':

(Pdb) until 19
0 0
1 5
2 10
3 15
4 20
> .../pdb_next.py(19)f()
-> return

(Pdb)
```

return命令也是绕开函数部分的一个捷径。它会继续执行,直至函数准备执行一个 return 语句,然后会暂停,使得在函数返回之前有时间查看返回值。

```
$ python3 pdb_next.py

> .../pdb_next.py(23)<module>()
-> f(5)
(Pdb) step
--Call--
> .../pdb_next.py(15)f()
-> def f(n):

(Pdb) step
> .../pdb_next.py(16)f()
```

```
-> for i in range(n):
(Pdb) return
0 0
1 5
2 10
3 15
4 20
--Return--
> .../pdb_next.py(19)f()->None
-> return

(Pdb)
```

16.7.3 断点

随着程序越来越长，即使使用 `next` 和 `until` 也会变得很慢，很烦琐。不用手动地单步跟踪程序，一种更好的解决方案是让它正常运行，直至达到某一点，并且调试工具要在这一点中断执行。`set_trace()` 可以启动调试工具，不过只有当程序中有一个要暂停的点时这才适用。更方便的做法是通过调试工具运行程序，使用断点（breakpoint）提前告诉调试工具在哪里停止。调试工具会监视程序，并且在其到达断点描述的位置时，程序会在执行那一行之前暂停。

代码清单 16-70：**pdb_break.py**

```
 1  #!/usr/bin/env python3
 2  # encoding: utf-8
 3  #
 4  # Copyright (c) 2010 Doug Hellmann.  All rights reserved.
 5  #
 6
 7
 8  def calc(i, n):
 9      j = i * n
10      print('j =', j)
11      if j > 0:
12          print('Positive!')
13      return j
14
15
16  def f(n):
17      for i in range(n):
18          print('i =', i)
19          j = calc(i, n)  # noqa
20      return
21
22  if __name__ == '__main__':
23      f(5)
```

`break` 命令（缩写为 `b`）在设置断点时可以使用很多选项，包括要暂停处理的行号、文件和函数。要在当前文件的一个特定行设置断点，可以使用 `break lineno`。

```
$ python3 -m pdb pdb_break.py

> .../pdb_break.py(8)<module>()
-> def calc(i, n):
(Pdb) break 12
```

```
Breakpoint 1 at .../pdb_break.py:12

(Pdb) continue
i = 0
j = 0
i = 1
j = 5
> .../pdb_break.py(12)calc()
-> print('Positive!')

(Pdb)
```

命令 continue（缩写为 c）告诉调试工具继续运行程序，直到到达下一个断点。在这个例子中，它会运行完 f() 中 for 循环的第一次迭代，并在第二次迭代期间于 calc() 中停止。

还可以指定函数名而不是一个行号，把断点设置到一个函数的第一行。下面这个例子显示了为 calc() 函数增加一个断点会发生什么情况。

```
$ python3 -m pdb pdb_break.py

> .../pdb_break.py(8)<module>()
-> def calc(i, n):
(Pdb) break calc
Breakpoint 1 at .../pdb_break.py:8

(Pdb) continue
i = 0
> .../pdb_break.py(9)calc()
-> j = i * n

(Pdb) where
  .../pdb_break.py(23)<module>()
-> f(5)
  .../pdb_break.py(19)f()
-> j = calc(i, n)
> .../pdb_break.py(9)calc()
-> j = i * n

(Pdb)
```

要在另一个文件中指定一个断点，可以在行或函数参数前加一个文件名前缀。

代码清单 16-71：pdb_break_remote.py

```
1  #!/usr/bin/env python3
2  # encoding: utf-8
3
4  from pdb_break import f
5
6  f(5)
```

在这里，启动主程序 pdb_break_remote.py 之后，为 pdb_break.py 的第 12 行设置了一个断点。

```
$ python3 -m pdb pdb_break_remote.py

> .../pdb_break_remote.py(4)<module>()
-> from pdb_break import f
```

```
(Pdb) break pdb_break.py:12
Breakpoint 1 at .../pdb_break.py:12

(Pdb) continue
i = 0
j = 0
i = 1
j = 5
> .../pdb_break.py(12)calc()
-> print('Positive!')

(Pdb)
```

文件名可以是源文件的完整路径，也可以是相对于 sys.path 上某个文件的相对路径。

要列出当前设置的断点，可以使用 break 而不带任何参数。输出的信息包括文件和各个断点的行号，以及这个断点被遇到了多少次。

```
$ python3 -m pdb pdb_break.py

> .../pdb_break.py(8)<module>()
-> def calc(i, n):

(Pdb) break 12
Breakpoint 1 at .../pdb_break.py:12

(Pdb) break
Num Type         Disp Enb   Where
1   breakpoint   keep yes   at .../pdb_break.py:12

(Pdb) continue
i = 0
j = 0
i = 1
j = 5
> .../pdb/pdb_break.py(12)calc()
-> print('Positive!')

(Pdb) continue
Positive!
i = 2
j = 10
> .../pdb_break.py(12)calc()
-> print('Positive!')

(Pdb) break
Num Type         Disp Enb   Where
1   breakpoint   keep yes   at .../pdb_break.py:12
        breakpoint already hit 2 times

(Pdb)
```

16.7.3.1 管理断点

在增加各个新断点时，会为它指定一个数值标识符。这些 ID 号用于交互式地启用、禁用和删除断点。用 disable 关闭一个断点时，会告诉调试工具到达这一行时不要停止。在这种情况下，会记住这个断点，但是会将其忽略。

```
$ python3 -m pdb pdb_break.py

> .../pdb_break.py(8)<module>()
-> def calc(i, n):
```

```
(Pdb) break calc
Breakpoint 1 at .../pdb_break.py:8

(Pdb) break 12
Breakpoint 2 at .../pdb_break.py:12

(Pdb) break
Num Type         Disp Enb   Where
1   breakpoint   keep yes   at .../pdb_break.py:8
2   breakpoint   keep yes   at .../pdb_break.py:12

(Pdb) disable 1

(Pdb) break
Num Type         Disp Enb   Where
1   breakpoint   keep no    at .../pdb_break.py:8
2   breakpoint   keep yes   at .../pdb_break.py:12

(Pdb) continue
i = 0
j = 0
i = 1
j = 5
> .../pdb_break.py(12)calc()
-> print('Positive!')

(Pdb)
```

下一个调试会话在程序中设置了两个断点，然后禁用了其中一个。程序会一直运行，直到遇到剩下的那个断点，然后在继续执行之前用 **enable** 把另一个断点打开。

```
$ python3 -m pdb pdb_break.py

> .../pdb_break.py(8)<module>()
-> def calc(i, n):
(Pdb) break calc
Breakpoint 1 at .../pdb_break.py:8

(Pdb) break 18
Breakpoint 2 at .../pdb_break.py:18

(Pdb) disable 1

(Pdb) continue
> .../pdb_break.py(18)f()
-> print('i =', i)

(Pdb) list
 13             return j
 14
 15
 16     def f(n):
 17         for i in range(n):
 18 B->         print('i =', i)
 19             j = calc(i, n)
 20         return
 21
 22     if __name__ == '__main__':
 23         f(5)

(Pdb) continue
i = 0
```

```
    j = 0
> .../pdb_break.py(18)f()
-> print('i =', i)

(Pdb) list
 13         return j
 14
 15
 16   def f(n):
 17       for i in range(n):
 18 B->        print('i =', i)
 19            j = calc(i, n)
 20       return
 21
 22   if __name__ == '__main__':
 23       f(5)

(Pdb) p i
 1

(Pdb) enable 1
Enabled breakpoint 1 at .../pdb_break.py:8

(Pdb) continue
i = 1
> .../pdb_break.py(9)calc()
-> j = i * n

(Pdb) list
  4   # Copyright (c) 2010 Doug Hellmann.  All rights reserved.
  5   #
  6
  7
  8 B   def calc(i, n):
  9 ->      j = i * n
 10         print('j =', j)
 11         if j > 0:
 12             print('Positive!')
 13         return j
 14

(Pdb)
```

list 的输出中带 B 前缀的行显示，程序中哪里设置了断点（第 8 行和第 18 行）。可以使用 clear 完全删除一个断点。

```
$ python3 -m pdb pdb_break.py

> .../pdb_break.py(8)<module>()
-> def calc(i, n):
(Pdb) break calc
Breakpoint 1 at .../pdb_break.py:8

(Pdb) break 12
Breakpoint 2 at .../pdb_break.py:12

(Pdb) break 18
Breakpoint 3 at .../pdb_break.py:18

(Pdb) break
Num Type         Disp Enb   Where
1   breakpoint   keep yes   at .../pdb_break.py:8
```

```
2   breakpoint    keep yes   at .../pdb_break.py:12
3   breakpoint    keep yes   at .../pdb_break.py:18

(Pdb) clear 2
Deleted breakpoint 2

(Pdb) break
Num Type         Disp Enb   Where
1   breakpoint   keep yes   at .../pdb_break.py:8
3   breakpoint   keep yes   at .../pdb_break.py:18

(Pdb)
```

其他断点仍保留原来的标识符，不会重新编号。

16.7.3.2 临时断点

程序第一次执行到临时断点时会将它自动清除。通过使用临时断点，可以很快到达程序流中的特定位置，这与常规断点一样。不过，因为它会立即清除，所以如果这部分程序反复运行，则临时断点不会干扰后续执行。

```
$ python3 -m pdb pdb_break.py

> .../pdb_break.py(8)<module>()
-> def calc(i, n):
(Pdb) tbreak 12
Breakpoint 1 at .../pdb_break.py:12

(Pdb) continue
i = 0
j = 0
i = 1
j = 5
Deleted breakpoint 1 at .../pdb_break.py:12
> .../pdb_break.py(12)calc()
-> print('Positive!')

(Pdb) break

(Pdb) continue
Positive!
i = 2
j = 10
Positive!
i = 3
j = 15
Positive!
i = 4
j = 20
Positive!
The program finished and will be restarted
> .../pdb_break.py(8)<module>()
-> def calc(i, n):

(Pdb)
```

程序第一次到达第 12 行时，会把这个断点删除，在程序完成之前不会再停止执行。

16.7.3.3 条件断点

可以对断点应用一些规则，以便其仅当条件满足时才停止执行。相对于手动地启用和禁用断点，使用条件断点可以对调试工具如何暂停程序提供更精细的控制。可以用两种方式设置条件断点。第一种是使用 break 在设置断点时指定条件。

```
$ python3 -m pdb pdb_break.py

> .../pdb_break.py(8)<module>()
-> def calc(i, n):
(Pdb) break 10, j>0
Breakpoint 1 at .../pdb_break.py:10

(Pdb) break
Num Type         Disp Enb   Where
1   breakpoint   keep yes   at .../pdb_break.py:10
        stop only if j>0

(Pdb) continue
i = 0
j = 0
i = 1
> .../pdb_break.py(10)calc()
-> print('j =', j)

(Pdb)
```

条件参数必须是一个表达式,要使用定义断点的栈帧中可见的值。如果表达式估算为 true,则在断点处停止执行。

还可以使用 condition 命令对一个现有的断点应用条件。这个命令的参数包括断点 ID 和表达式。

```
$ python3 -m pdb pdb_break.py

> .../pdb_break.py(8)<module>()
-> def calc(i, n):
(Pdb) break 10
Breakpoint 1 at .../pdb_break.py:10

(Pdb) break
Num Type         Disp Enb   Where
1   breakpoint   keep yes   at .../pdb_break.py:10

(Pdb) condition 1 j>0

(Pdb) break
Num Type         Disp Enb   Where
1   breakpoint   keep yes   at .../pdb_break.py:10
        stop only if j>0

(Pdb)
```

16.7.3.4 忽略断点

有些程序包含有循环,或者使用了大量相同函数的递归调用,通过在执行中"前跳"(skipping ahead)可以更容易地调试这些程序,而不是监视每一个调用或断点。ignore 命令告诉调试工具跳过一个断点而不停止。每次处理遇到这个断点时,它都会递减"忽略计数器"。当这个计数器为 0 时,会再次启用这个断点。

```
$ python3 -m pdb pdb_break.py

> .../pdb_break.py(8)<module>()
-> def calc(i, n):
```

```
(Pdb) break 19
Breakpoint 1 at .../pdb_break.py:19

(Pdb) continue
i = 0
> .../pdb_break.py(19)f()
-> j = calc(i, n)

(Pdb) next
j = 0
> .../pdb_break.py(17)f()
-> for i in range(n):

(Pdb) ignore 1 2
Will ignore next 2 crossings of breakpoint 1.

(Pdb) break
Num Type         Disp Enb   Where
1   breakpoint   keep yes   at .../pdb_break.py:19
        ignore next 2 hits
        breakpoint already hit 1 time

(Pdb) continue
i = 1
j = 5
Positive!
i = 2
j = 10
Positive!
i = 3
> .../pdb_break.py(19)f()
-> j = calc(i, n)

(Pdb) break
Num Type         Disp Enb   Where
1   breakpoint   keep yes   at .../pdb_break.py:19
        breakpoint already hit 4 times
```

显式地重新设置忽略计数器为 0，可以立即再次启用这个断点。

```
$ python3 -m pdb pdb_break.py

> .../pdb_break.py(8)<module>()
-> def calc(i, n):
(Pdb) break 19
Breakpoint 1 at .../pdb_break.py:19

(Pdb) ignore 1 2
Will ignore next 2 crossings of breakpoint 1.

(Pdb) break
Num Type         Disp Enb   Where
1   breakpoint   keep yes   at .../pdb_break.py:19
        ignore next 2 hits

(Pdb) ignore 1 0
Will stop next time breakpoint 1 is reached.

(Pdb) break
Num Type         Disp Enb   Where
1   breakpoint   keep yes   at .../pdb_break.py:19
```

16.7.3.5　在断点触发动作

除了纯交互式模式，pdb 还支持基本脚本模式。通过使用 commands，可以在遇到一个特定断点时执行一系列解释器命令，也包括 Python 语句。当运行 commands 并提供断点号作为参数时，调试工具提示符会变为（com）。一次输入一个指定的命令，用 end 结束命令列表以保存脚本，然后返回到主调试工具提示符。

```
$ python3 -m pdb pdb_break.py

> .../pdb_break.py(8)<module>()
-> def calc(i, n):
(Pdb) break 10
Breakpoint 1 at .../pdb_break.py:10

(Pdb) commands 1
(com) print('debug i =', i)
(com) print('debug j =', j)
(com) print('debug n =', n)
(com) end

(Pdb) continue
i = 0
debug i = 0
debug j = 0
debug n = 5
> .../pdb_break.py(10)calc()
-> print('j =', j)

(Pdb) continue
j = 0
i = 1
debug i = 1
debug j = 5
debug n = 5
> .../pdb_break.py(10)calc()
-> print 'j =', j

(Pdb)
```

如果代码使用了大量数据结构或变量，那么这个特性对调试这些代码尤其有用。可以让调试工具自动地输出所有值，而不是每次遇到断点时手动地输出值。

16.7.3.6 监视数据改变

还可以在程序执行期间监视值的改变，而不需要写显式的 print 命令。为此，要使用 display 命令。

```
$ python3 -m pdb pdb_break.py
> .../pdb_break.py(8)<module>()
-> def calc(i, n):
(Pdb) break 18
Breakpoint 1 at .../pdb_break.py:18

(Pdb) continue
> .../pdb_break.py(18)f()
-> print('i =', i)

(Pdb) display j
display j: ** raised NameError: name 'j' is not defined **

(Pdb) next
i = 0
> .../pdb_break.py(19)f()
-> j = calc(i, n)   # noqa
```

```
(Pdb) next
j = 0
> .../pdb_break.py(17)f()
-> for i in range(n):
display j: 0  [old: ** raised NameError: name 'j' is not defined **]

(Pdb)
```

每次在帧中停止执行时,都会计算这个表达式。如果有变化,会输出这个结果以及原来的值。如果没有参数,`display`命令会输出当前帧中活动的显示值列表。

```
(Pdb) display
Currently displaying:
j: 0

(Pdb) up
> .../pdb_break.py(23)<module>()
-> f(5)

(Pdb) display
Currently displaying:

(Pdb)
```

要删除显示表达式,可以使用

```
(Pdb) display
Currently displaying:
j: 0

(Pdb) undisplay j

(Pdb) display
Currently displaying:

(Pdb)
```

16.7.4 改变执行流

`jump`命令可以在运行时改变程序流,而不修改代码。它可以向前跳以避免运行某些代码,或者向后跳以再次运行某些代码。下面的示例程序会生成一个数字列表。

代码清单 16-72:**pdb_jump.py**

```
 1  #!/usr/bin/env python3
 2  # encoding: utf-8
 3  #
 4  # Copyright (c) 2010 Doug Hellmann.  All rights reserved.
 5  #
 6
 7
 8  def f(n):
 9      result = []
10      j = 0
11      for i in range(n):
12          j = i * n + j
13          j += n
14          result.append(j)
15      return result
16
```

```
17  if __name__ == '__main__':
18      print(f(5))
```

不加干扰地运行时,这个例子的输出是整除 5 的递增数字组成的一个序列。

```
$ python3 pdb_jump.py

[5, 15, 30, 50, 75]
```

16.7.4.1 前跳

前跳(jump ahead)会把执行点移至当前位置之后,而不再执行老位置和新位置之间的任何语句。这个例子中,由于跳过了第 13 行,j 的值并不递增,所以后面所有依赖于它的值都会稍小一点。

```
$ python3 -m pdb pdb_jump.py
> .../pdb_jump.py(8)<module>()
-> def f(n):
(Pdb) break 13
Breakpoint 1 at .../pdb_jump.py:13

(Pdb) continue
> .../pdb_jump.py(13)f()
-> j += n

(Pdb) p j
0

(Pdb) step
> .../pdb_jump.py(14)f()
-> result.append(j)

(Pdb) p j
5

(Pdb) continue
> .../pdb_jump.py(13)f()
-> j += n

(Pdb) jump 14
> .../pdb_jump.py(14)f()
-> result.append(j)

(Pdb) p j
10

(Pdb) disable 1

(Pdb) continue
[5, 10, 25, 45, 70]
The program finished and will be restarted
> .../pdb_jump.py(8)<module>()
-> def f(n):
(Pdb)
```

16.7.4.2 后跳

也可以后跳,让程序执行之前已经执行过的语句,使代码可以再次运行。在这里,j 值多递增一次,所以结果序列中的数都比原本的值大。

```
$ python3 -m pdb pdb_jump.py

> .../pdb_jump.py(8)<module>()
-> def f(n):
(Pdb) break 14
Breakpoint 1 at .../pdb_jump.py:14

(Pdb) continue
> .../pdb_jump.py(14)f()
-> result.append(j)

(Pdb) p j
5

(Pdb) jump 13
> .../pdb_jump.py(13)f()
-> j += n

(Pdb) continue
> .../pdb_jump.py(14)f()
-> result.append(j)

(Pdb) p j
10

(Pdb) disable 1

(Pdb) continue
[10, 20, 35, 55, 80]

The program finished and will be restarted
> .../pdb_jump.py(8)<module>()
-> def f(n):
(Pdb)
```

16.7.4.3 非法跳转

跳入和跳出某些流控制语句会很危险，而且不确定，因此调试工具不允许这些行为。

代码清单 16-73：pdb_no_jump.py

```
 1  #!/usr/bin/env python3
 2  # encoding: utf-8
 3  #
 4  # Copyright (c) 2010 Doug Hellmann.  All rights reserved.
 5  #
 6
 7
 8  def f(n):
 9      if n < 0:
10          raise ValueError('Invalid n: {}'.format(n))
11      result = []
12      j = 0
13      for i in range(n):
14          j = i * n + j
15          j += n
16          result.append(j)
17      return result
18
19
20  if __name__ == '__main__':
21      try:
22          print(f(5))
```

```
23      finally:
24          print('Always printed')
25
26      try:
27          print(f(-5))
28      except:
29          print('There was an error')
30      else:
31          print('There was no error')
32
33      print('Last statement')
```

尽管可以使用 jump 进入一个函数，但是参数未定义，代码也不能正常工作。

```
$ python3 -m pdb pdb_no_jump.py

> .../pdb_no_jump.py(8)<module>()
-> def f(n):
(Pdb) break 22
Breakpoint 1 at .../pdb_no_jump.py:22

(Pdb) jump 9
> .../pdb_no_jump.py(9)<module>()
-> if n < 0:

(Pdb) p n
*** NameError: NameError("name 'n' is not defined",)

(Pdb) args

(Pdb)
```

jump 不会进入类似 for 循环或 try:except 语句等代码块的中间。

```
$ python3 -m pdb pdb_no_jump.py

> .../pdb_no_jump.py(8)<module>()
-> def f(n):
(Pdb) break 22
Breakpoint 1 at .../pdb_no_jump.py:22

(Pdb) continue
> .../pdb_no_jump.py(22)<module>()
-> print(f(5))

(Pdb) jump 27
*** Jump failed: can't jump into the middle of a block

(Pdb)
```

finally 块中的代码必须全部执行，所以 jump 不会离开 finally 块。

```
$ python3 -m pdb pdb_no_jump.py

> .../pdb_no_jump.py(8)<module>()
-> def f(n):
(Pdb) break 24
Breakpoint 1 at .../pdb_no_jump.py:24

(Pdb) continue
[5, 15, 30, 50, 75]
> .../pdb_no_jump.py(24)<module>()
```

```
-> print 'Always printed'

(Pdb) jump 26
*** Jump failed: can't jump into or out of a 'finally' block

(Pdb)
```

最基本的限制是，跳转要受调用栈底帧的约束。如果已经使用 **up** 命令改变了调试上下文，那么执行流就不能改变。

```
$ python3 -m pdb pdb_no_jump.py

> .../pdb_no_jump.py(8)<module>()
-> def f(n):
(Pdb) break 12
Breakpoint 1 at .../pdb_no_jump.py:12

(Pdb) continue
> .../pdb_no_jump.py(12)f()
-> j = 0

(Pdb) where
  .../lib/python3.5/bdb.py(
431)run()
-> exec cmd in globals, locals
  <string>(1)<module>()
  .../pdb_no_jump.py(22)<module>()
-> print(f(5))
> .../pdb_no_jump.py(12)f()
-> j = 0

(Pdb) up
> .../pdb_no_jump.py(22)<module>()
-> print(f(5))

(Pdb) jump 25
*** You can only jump within the bottom frame

(Pdb)
```

16.7.4.4 重启程序

当调试工具到达程序末尾时，程序会自行重新启动，不过也可以显式地重启而不用退出调试工具，这样就不会丢失当前的断点或其他设置。

代码清单 16-74：**pdb_run.py**

```python
1  #!/usr/bin/env python3
2  # encoding: utf-8
3  #
4  # Copyright (c) 2010 Doug Hellmann.  All rights reserved.
5  #
6
7  import sys
8
9
10 def f():
11     print('Command-line args:', sys.argv)
12     return
13
14 if __name__ == '__main__':
15     f()
```

在调试工具中运行前面的程序直至结束，其会输出脚本文件的名，因为命令行上没有提供任何其他参数。

```
$ python3 -m pdb pdb_run.py

> .../pdb_run.py(7)<module>()
-> import sys
(Pdb) continue

Command line args: ['pdb_run.py']
The program finished and will be restarted
> .../pdb_run.py(7)<module>()
-> import sys

(Pdb)
```

还可以使用 `run` 重启程序。传入 `run` 的参数将由 `shlex` 解析并传递到程序，就好像它们是命令行参数一样，所以可以用不同的设置重启程序。

```
(Pdb) run a b c "this is a long value"
Restarting pdb_run.py with arguments:
        a b c this is a long value
> .../pdb_run.py(7)<module>()
-> import sys

(Pdb) continue
Command line args: ['pdb_run.py', 'a', 'b', 'c',
'this is a long value']
The program finished and will be restarted
> .../pdb_run.py(7)<module>()
-> import sys

(Pdb)
```

还可以在处理中的任何其他位置使用 `run` 重启程序。

```
$ python3 -m pdb pdb_run.py

> .../pdb_run.py(7)<module>()
-> import sys
(Pdb) break 11
Breakpoint 1 at .../pdb_run.py:11

(Pdb) continue
> .../pdb_run.py(11)f()
-> print('Command line args:', sys.argv)

(Pdb) run one two three
Restarting pdb_run.py with arguments:
        one two three
> .../pdb_run.py(7)<module>()
-> import sys

(Pdb)
```

16.7.5 用别名定制调试工具

可以使用 `alias` 定义一个快捷方式，从而无须反复键入复杂的命令。别名扩展被应用于每个命令的第一个词。别名的体由可在调试工具提示窗口合法键入的任何命令组成，包括其他调试工具命令和纯 Python 表达式。别名定义中允许递归，所以一个别名甚至可以调用另一个别名。

```
$ python3 -m pdb pdb_function_arguments.py

> .../pdb_function_arguments.py(7)<module>()
-> import pdb
(Pdb) break 11
Breakpoint 1 at .../pdb_function_arguments.py:11

(Pdb) continue
> .../pdb_function_arguments.py(11)recursive_function()
-> if n > 0:

(Pdb) pp locals().keys()
dict_keys(['output', 'n'])

(Pdb) alias pl pp locals().keys()

(Pdb) pl
dict_keys(['output', 'n'])
```

运行 `alias` 而不带任何参数时，会显示已定义的别名的一个列表。如果给出一个参数，则认为这是一个别名的名，并且将打印这个别名的定义。

```
(Pdb) alias
pl = pp locals().keys()

(Pdb) alias pl
pl = pp locals().keys()

(Pdb)
```

`alias` 的参数使用 `%n` 来引用，其中 n 会被替换为指示参数位置的一个数，从 1 开始。要消费所有参数，可以使用 `%*`。

```
$ python3 -m pdb pdb_function_arguments.py

> .../pdb_function_arguments.py(7)<module>()
-> import pdb
(Pdb) alias ph !help(%1)

(Pdb) ph locals
Help on built-in function locals in module builtins:

locals()
    Return a dictionary containing the current scope's local
    variables.

    NOTE: Whether or not updates to this dictionary will affect
    name lookups in the local scope and vice-versa is
    *implementation dependent* and not covered by any backwards
    compatibility guarantees.
```

可以用 `unalias` 删除一个别名的定义。

```
(Pdb) unalias ph

(Pdb) ph locals
*** SyntaxError: invalid syntax (<stdin>, line 1)

(Pdb)
```

16.7.6 保存配置设置

调试一个程序需要做大量重复：运行代码，观察输出，调整代码或输入，以及再次运行代码。pdb 努力减少所需的重复以控制调试体验，使你能集中精力考虑代码而不是调试工具。为了减少让调试工具执行相同命令的次数，pdb 可以从启动时解释的文本文件读取已保存的配置。

首先读取文件 ~/.pdbrc，它会建立所有调试会话的全局个人首选项。然后从当前工作目录读取 ./.pdbrc，以设置特定项目的局部首选项。

```
$ cat ~/.pdbrc

# Show python help
alias ph !help(%1)
# Overridden alias
alias redefined p 'home definition'

$ cat .pdbrc

# Breakpoints
break 11
# Overridden alias
alias redefined p 'local definition'

$ python3 -m pdb pdb_function_arguments.py

Breakpoint 1 at .../pdb_function_arguments.py:11
> .../pdb_function_arguments.py(7)<module>()
-> import pdb
(Pdb) alias
ph = !help(%1)
redefined = p 'local definition'

(Pdb) break
Num Type         Disp Enb   Where
1   breakpoint   keep yes   at .../pdb_function_arguments.py:11

(Pdb)
```

能够在调试工具提示窗口键入的所有配置命令都可以被保存到某个启动文件中。有些控制执行的命令（例如 continue、next）也可以用相同的方式保存。

```
$ cat .pdbrc
break 11
continue
list

$ python3 -m pdb pdb_function_arguments.py
Breakpoint 1 at .../pdb_function_arguments.py:11
  6
  7     import pdb
  8
  9
 10     def recursive_function(n=5, output='to be printed'):
 11 B->     if n > 0:
 12             recursive_function(n - 1)
 13         else:
 14             pdb.set_trace()
 15             print(output)
```

```
    16      return
> .../pdb_function_arguments.py(11)recursive_function()
-> if n > 0:
(Pdb)
```

保存 `run` 命令尤其有用。这么做意味着可以在 `./.pdbrc` 中设置一个调试会话的命令行参数，从而在多次运行时保持一致。

```
$ cat .pdbrc
run a b c "long argument"

$ python3 -m pdb pdb_run.py
Restarting pdb_run.py with arguments:
    a b c "long argument"
> .../pdb_run.py(7)<module>()
-> import sys

(Pdb) continue
Command-line args: ['pdb_run.py', 'a', 'b', 'c',
'long argument']
The program finished and will be restarted
> .../pdb_run.py(7)<module>()
-> import sys

(Pdb)
```

提示：相关阅读材料
- `pdb` 的标准库文档[一]。
- `readline`：交互式提示符编辑库。
- `cmd`：构建交互式程序。
- `shlex`：shell 命令行解析。
- Python issue 26053[二]：如果 `run` 的输出与这里提供的值不匹配，那么可以参考这个 bug 来了解 2.7 和 3.5 版本之间 pdb 输出中一个回归的详细内容。

16.8　`profile` 和 `pstats`：性能分析

`profile` 模块提供了一些 API，可以用来收集和分析有关 Python 源代码消耗处理器资源的统计信息。

说明：这一节的输出报告已经重新调整了格式，以便在这里显示。以反斜线（\）结尾的行表示未完，下一行继续。

16.8.1　运行性能分析工具

`profile` 模块中最基本的起点是 `run()`。它以一个字符串语句为参数，并且会创建一个报告，指出在运行这个语句时执行不同代码行所花费的时间。

[一] https://docs.python.org/3.5/library/pdb.html
[二] http://bugs.python.org/issue26053

代码清单 16-75： `profile_fibonacci_raw.py`

```python
import profile


def fib(n):
    # from literateprograms.org
    # http://bit.ly/hlOQ5m
    if n == 0:
        return 0
    elif n == 1:
        return 1
    else:
        return fib(n - 1) + fib(n - 2)


def fib_seq(n):
    seq = []
    if n > 0:
        seq.extend(fib_seq(n - 1))
    seq.append(fib(n))
    return seq


profile.run('print(fib_seq(20)); print()')
```

这是一个递归版本的 Fibonacci 序列计算器，它对于展示 `profile` 模块尤其有用，因为这个程序的性能可以得到显著改善。标准报告格式会显示一个总结，然后给出所执行的各个函数的详细信息。

```
$ python3 profile_fibonacci_raw.py

[0, 1, 1, 2, 3, 5, 8, 13, 21, 34, 55, 89, 144, 233, 377, 610, 98\
7, 1597, 2584, 4181, 6765]

         57359 function calls (69 primitive calls) in 0.127 seco\
nds

   Ordered by: standard name

   ncalls  tottime  percall  cumtime  percall filename:lineno(fu\
nction)
       21    0.000    0.000    0.000    0.000 :0(append)
        1    0.000    0.000    0.127    0.127 :0(exec)
       20    0.000    0.000    0.000    0.000 :0(extend)
        2    0.000    0.000    0.000    0.000 :0(print)
        1    0.001    0.001    0.001    0.001 :0(setprofile)
        1    0.000    0.000    0.127    0.127 <string>:1(<module\
>)
        1    0.000    0.000    0.127    0.127 profile:0(print(fi\
b_seq(20)); print())
        0    0.000             0.000          profile:0(profiler\
)
 57291/21    0.126    0.000    0.126    0.006 profile_fibonacci_\
raw.py:11(fib)
     21/1    0.000    0.000    0.127    0.127 profile_fibonacci_\
raw.py:22(fib_seq)
```

原始版本有 57 359 个不同的函数调用，并且运行时间为 0.127 秒。实际上这里只有 69 个原始调用，这个事实说明这 57 359 个调用中大部分都是递归调用。所用时间的详细信息按代码清单中的函数分解，显示了调用数，函数花费的总时间，每个调用花费的时间（`tottime/ncalls`），一个函数花费的累积时间，以及累积时间与基本调用之比。

并不奇怪，这里大部分时间都花费在反复调用 `fib()` 上。增加一个缓存修饰符可以减少递归调用数，这会对这个函数的性能产生显著影响。

代码清单 16-76：`profile_fibonacci_memoized.py`

```python
import functools
import profile


@functools.lru_cache(maxsize=None)
def fib(n):
    # from literateprograms.org
    # http://bit.ly/hlOQ5m
    if n == 0:
        return 0
    elif n == 1:
        return 1
    else:
        return fib(n - 1) + fib(n - 2)


def fib_seq(n):
    seq = []
    if n > 0:
        seq.extend(fib_seq(n - 1))
    seq.append(fib(n))
    return seq


if __name__ == '__main__':
    profile.run('print(fib_seq(20)); print()')
```

通过记住各层的 Fibonacci 值，大多数调用都可以避免，并且将运行的调用减至 89 个，这只需要 0.001 秒。`fib()` 的 `ncalls` 数显示出它完全没有递归。

```
$ python3 profile_fibonacci_memoized.py

[0, 1, 1, 2, 3, 5, 8, 13, 21, 34, 55, 89, 144, 233, 377, 610, 98\
7, 1597, 2584, 4181, 6765]

         89 function calls (69 primitive calls) in 0.001 seconds

   Ordered by: standard name

   ncalls  tottime  percall  cumtime  percall filename:lineno(fu\
nction)
       21    0.000    0.000    0.000    0.000 :0(append)
        1    0.000    0.000    0.000    0.000 :0(exec)
       20    0.000    0.000    0.000    0.000 :0(extend)
        2    0.000    0.000    0.000    0.000 :0(print)
        1    0.001    0.001    0.001    0.001 :0(setprofile)
        1    0.000    0.000    0.000    0.000 <string>:1(<module\
>)
        1    0.000    0.000    0.001    0.001 profile:0(print(fi\
b_seq(20)); print())
        0    0.000             0.000          profile:0(profiler\
)
```

```
          21    0.000    0.000    0.000    0.000 profile_fibonacci_\
memoized.py:12(fib)
        21/1    0.000    0.000    0.000    0.000 profile_fibonacci_\
memoized.py:24(fib_seq)
```

16.8.2 在上下文中运行

有时，不用为 run() 构造一个复杂的表达式，更容易的做法是构建一个简单的表达式，并使用 runctx() 通过一个上下文为它传递参数。

代码清单 16-77：profile_runctx.py

```
import profile
from profile_fibonacci_memoized import fib, fib_seq

if __name__ == '__main__':
    profile.runctx(
        'print(fib_seq(n)); print()',
        globals(),
        {'n': 20},
    )
```

在这个例子中，n 的值通过局部变量上下文传递，而不是直接嵌入到传至 runctx() 的语句中。

```
$ python3 profile_runctx.py

[0, 1, 1, 2, 3, 5, 8, 13, 21, 34, 55, 89, 144, 233, 377, 610,
987, 1597, 2584, 4181, 6765]

         148 function calls (90 primitive calls) in 0.002 seconds

   Ordered by: standard name

   ncalls  tottime  percall  cumtime  percall filename:lineno(\
function)
       21    0.000    0.000    0.000    0.000 :0(append)
        1    0.000    0.000    0.001    0.001 :0(exec)
       20    0.000    0.000    0.000    0.000 :0(extend)
        2    0.000    0.000    0.000    0.000 :0(print)
        1    0.001    0.001    0.001    0.001 :0(setprofile)
        1    0.000    0.000    0.001    0.001 <string>:1(<module\
>)
        1    0.000    0.000    0.002    0.002 profile:0(print(fi\
b_seq(n)); print())
        0    0.000             0.000          profile:0(profiler\
)
    59/21    0.000    0.000    0.000    0.000 profile_fibonacci_\
memoized.py:19(__call__)
       21    0.000    0.000    0.000    0.000 profile_fibonacci_\
memoized.py:27(fib)
     21/1    0.000    0.000    0.001    0.001 profile_fibonacci_\
memoized.py:39(fib_seq)
```

16.8.3 pstats：保存和处理统计信息

profile 函数创建的标准报告不太灵活。不过，可以保存 run() 和 runctx() 的原始性能数据并用 pstats.Stats 类单独处理，以生成定制报告。

下面的例子会多次迭代运行同一个测试，并且合并该结果。

代码清单 16-78：**profile_stats.py**

```
import cProfile as profile
import pstats
from profile_fibonacci_memoized import fib, fib_seq

# Create 5 sets of stats.
for i in range(5):
    filename = 'profile_stats_{}.stats'.format(i)
    profile.run('print({}, fib_seq(20))'.format(i), filename)

# Read all 5 stats files into a single object.
stats = pstats.Stats('profile_stats_0.stats')
for i in range(1, 5):
    stats.add('profile_stats_{}.stats'.format(i))

# Clean up filenames for the report.
stats.strip_dirs()

# Sort the statistics by the cumulative time spent
# in the function.
stats.sort_stats('cumulative')

stats.print_stats()
```

输出报告按其在函数中所花费的累积时间的降序排序，另外打印的文件名中去掉了目录名，以节省页面上的水平空间。

```
$ python3 profile_stats.py

0 [0, 1, 1, 2, 3, 5, 8, 13, 21, 34, 55, 89, 144, 233, 377, 610, \
987, 1597, 2584, 4181, 6765]
1 [0, 1, 1, 2, 3, 5, 8, 13, 21, 34, 55, 89, 144, 233, 377, 610, \
987, 1597, 2584, 4181, 6765]
2 [0, 1, 1, 2, 3, 5, 8, 13, 21, 34, 55, 89, 144, 233, 377, 610, \
987, 1597, 2584, 4181, 6765]
3 [0, 1, 1, 2, 3, 5, 8, 13, 21, 34, 55, 89, 144, 233, 377, 610, \
987, 1597, 2584, 4181, 6765]
4 [0, 1, 1, 2, 3, 5, 8, 13, 21, 34, 55, 89, 144, 233, 377, 610, \
987, 1597, 2584, 4181, 6765]
Sat Dec 31 07:46:22 2016    profile_stats_0.stats
Sat Dec 31 07:46:22 2016    profile_stats_1.stats
Sat Dec 31 07:46:22 2016    profile_stats_2.stats
Sat Dec 31 07:46:22 2016    profile_stats_3.stats
Sat Dec 31 07:46:22 2016    profile_stats_4.stats

         351 function calls (251 primitive calls) in 0.000 secon\
ds

   Ordered by: cumulative time

   ncalls  tottime  percall  cumtime  percall filename:lineno(fu\
nction)
        5    0.000    0.000    0.000    0.000 {built-in method b\
uiltins.exec}
        5    0.000    0.000    0.000    0.000 <string>:1(<module\
>)
    105/5    0.000    0.000    0.000    0.000 profile_fibonacci_\
memoized.py:24(fib_seq)
```

```
       5    0.000    0.000    0.000    0.000 {built-in method b\
uiltins.print}
     100    0.000    0.000    0.000    0.000 {method 'extend' o\
f 'list' objects}
      21    0.000    0.000    0.000    0.000 profile_fibonacci_\
memoized.py:12(fib)
     105    0.000    0.000    0.000    0.000 {method 'append' o\
f 'list' objects}
       5    0.000    0.000    0.000    0.000 {method 'disable' \
of '_lsprof.Profiler' objects}
```

16.8.4 限制报告内容

可以按函数限制输出。下面这个版本只显示 `fib()` 和 `fib_seq()` 性能的有关信息，这里使用一个正则表达式来匹配所需的 `filename:lineno(function)` 值。

代码清单 16-79：`profile_stats_restricted.py`

```python
import profile
import pstats
from profile_fibonacci_memoized import fib, fib_seq

# Read all 5 stats files into a single object.
stats = pstats.Stats('profile_stats_0.stats')
for i in range(1, 5):
    stats.add('profile_stats_{}.stats'.format(i))
stats.strip_dirs()
stats.sort_stats('cumulative')

# Limit output to lines with "(fib" in them.
stats.print_stats('\(fib')
```

这个正则表达式包含一个字面量左括号（`(`），以匹配位置值的函数名部分。

```
$ python3 profile_stats_restricted.py

Sat Dec 31 07:46:22 2016    profile_stats_0.stats
Sat Dec 31 07:46:22 2016    profile_stats_1.stats
Sat Dec 31 07:46:22 2016    profile_stats_2.stats
Sat Dec 31 07:46:22 2016    profile_stats_3.stats
Sat Dec 31 07:46:22 2016    profile_stats_4.stats

         351 function calls (251 primitive calls) in 0.000 secon\
ds

   Ordered by: cumulative time
   List reduced from 8 to 2 due to restriction <'\\(fib'>

   ncalls  tottime  percall  cumtime  percall filename:lineno(fu\
nction)
    105/5    0.000    0.000    0.000    0.000 profile_fibonacci_\
memoized.py:24(fib_seq)
       21    0.000    0.000    0.000    0.000 profile_fibonacci_\
memoized.py:12(fib)
```

16.8.5 调用者 / 被调用者图

`stats` 还包括一些方法可以打印函数的调用者和被调用者。

代码清单 16-80：`profile_stats_callers.py`

```python
import cProfile as profile
import pstats
from profile_fibonacci_memoized import fib, fib_seq

# Read all 5 stats files into a single object.
stats = pstats.Stats('profile_stats_0.stats')
for i in range(1, 5):
    stats.add('profile_stats_{}.stats'.format(i))
stats.strip_dirs()
stats.sort_stats('cumulative')

print('INCOMING CALLERS:')
stats.print_callers('\(fib')

print('OUTGOING CALLEES:')
stats.print_callees('\(fib')
```

`print_callers()` 和 `print_callees()` 的参数与 `print_stats()` 的限制参数类似。输出会显示调用者、被调用者、调用数以及累积时间。

```
$ python3 profile_stats_callers.py

INCOMING CALLERS:
   Ordered by: cumulative time
   List reduced from 8 to 2 due to restriction <'\\(fib'>

Function                                    was called by...
                                                ncalls  tottime  \
cumtime
profile_fibonacci_memoized.py:24(fib_seq)   <-       5    0.000  \
  0.000   <string>:1(<module>)
                                                   100/5    0.000  \
  0.000   profile_fibonacci_memoized.py:24(fib_seq)
profile_fibonacci_memoized.py:12(fib)       <-      21    0.000  \
  0.000   profile_fibonacci_memoized.py:24(fib_seq)


OUTGOING CALLEES:
   Ordered by: cumulative time
   List reduced from 8 to 2 due to restriction <'\\(fib'>

Function                                    called...
                                                ncalls  tottime  \
cumtime
profile_fibonacci_memoized.py:24(fib_seq)   ->      21    0.000  \
  0.000   profile_fibonacci_memoized.py:12(fib)
                                                   100/5    0.000  \
  0.000   profile_fibonacci_memoized.py:24(fib_seq)
                                                     105    0.000  \
  0.000   {method 'append' of 'list' objects}
                                                     100    0.000  \
  0.000   {method 'extend' of 'list' objects}
profile_fibonacci_memoized.py:12(fib)       ->
```

提示：相关阅读材料

- `profile` 的标准库文档[⊖]。

⊖ https://docs.python.org/3.5/library/profile.html

- functools.lru_cache()：缓存修饰符，用来改进这个例子的性能。
- The Stats Class[⊖]：pstats.Stats 的标准库文档。
- Gprof2Dot[⊖]：性能输出数据的可视化工具。
- Smiley[⊖]：Python 应用跟踪器。

16.9 timeit：测量小段 Python 代码执行的时间

timeit 模块提供了一个简单的接口来确定小段 Python 代码的执行时间。它使用一个平台特定的时间函数，尽可能提供最准确的时间计算，并减少反复执行代码时启动或关闭开销对时间计算的影响。

16.9.1 模块内容

timeit 定义了一个公共类 Timer。Timer 的构造函数有两个参数，一个是要测量时间的语句，另一个是"建立"语句（例如，用来初始化变量）。Python 语句应当是字符串，可以包含嵌入的换行符。

timeit() 方法会运行一次建立语句，然后反复执行主语句。它会返回过去了多少时间。timeit() 的参数控制要运行多少次语句；默认为 1 000 000。

16.9.2 基本示例

为了展示如何使用 Timer 的各个参数，下面给出一个简单的例子，执行各个语句时会打印一个标识值。

代码清单 16-81：timeit_example.py

```
import timeit

# Using setitem
t = timeit.Timer("print('main statement')", "print('setup')")

print('TIMEIT:')
print(t.timeit(2))

print('REPEAT:')
print(t.repeat(3, 2))
```

输出显示了反复调用 print() 的结果。

```
$ python3 timeit_example.py

TIMEIT:
setup
main statement
```

⊖ https://docs.python.org/3/5/library/profile.html#the-stats-class
⊖ http://code.google.com/p/jrfonseca/wiki/Gprof2Dot
⊖ https://github.com/dhellmann/smiley

```
main statement
3.7070130929350853e-06
REPEAT:
setup
main statement
main statement
setup
main statement
main statement
setup
main statement
main statement
[1.4499528333544731e-06, 1.1939555406570435e-06,
1.1870870366692543e-06]
```

`timeit()` 运行一次建立语句，然后调用 `count` 次主语句。它返回一个浮点值表示运行主语句花费的累积时间。

使用 `repeat()` 时，它会多次调用 `timeit()`（在这里是 3 次），所有响应都返回到一个列表中。

16.9.3 将值存储在字典中

下一个例子更为复杂，它比较了使用不同方法以大量值填充一个字典所需的时间。首先，需要一些常量来配置 `Timer`。`setup_statement` 变量初始化一个元组列表，这些元组中包含主语句用来构建字典的字符串和整数，字符串被用作键，整数被存储为关联的值。

```python
# A few constants
range_size = 1000
count = 1000
setup_statement = ';'.join([
    "l = [(str(x), x) for x in range(1000)]",
    "d = {}",
])
```

这里定义了一个工具函数 `show_results()`，它采用一种有用的格式来打印结果。`timeit()` 方法返回反复执行这个语句所花费的时间。`show_results()` 的输出将这个值转换为每次迭代花费的时间，然后进一步将这个值缩减为在字典中存储一项所花费的平均时间。

```python
def show_results(result):
    "Print microseconds per pass and per item."
    global count, range_size
    per_pass = 1000000 * (result / count)
    print('{:6.2f} usec/pass'.format(per_pass), end=' ')
    per_item = per_pass / range_size
    print('{:6.2f} usec/item'.format(per_item))


print("{} items".format(range_size))
print("{} iterations".format(count))
print()
```

为了建立一个基准，测试的第一个配置使用了 `__setitem__()`。所有其他版本都不会覆盖字典中已经有的值，所以这个简单版本应该是最快的。

`Timer` 的第一个参数是一个多行的字符串，这里保留了空白符，以确保运行时函数能正确地解析这个字符串。第二个参数是一个常量，用来初始化值列表和字典。

```
# Using __setitem__ without checking for existing values first
print('__setitem__:', end=' ')
t = timeit.Timer(
    textwrap.dedent(
        """
        for s, i in l:
            d[s] = i
        """),
    setup_statement,
)
show_results(t.timeit(number=count))
```

下一个版本使用 `setdefault()` 确保字典中已有的值不会被覆盖。

```
# Using setdefault
print('setdefault :', end=' ')
t = timeit.Timer(
    textwrap.dedent(
        """
        for s, i in l:
            d.setdefault(s, i)
        """),
    setup_statement,
)
show_results(t.timeit(number=count))
```

如果查找现有值时产生了一个 `KeyError` 异常，则这个方法会增加值。

```
# Using exceptions
print('KeyError   :', end=' ')
t = timeit.Timer(
    textwrap.dedent(
        """
        for s, i in l:
            try:
                existing = d[s]
            except KeyError:
                d[s] = i
        """),
    setup_statement,
)
show_results(t.timeit(number=count))
```

最后一个方法使用"in"来确定字典是否有某个特定的键。

```
# Using "in"
print('"not in"   :', end=' ')
t = timeit.Timer(
    textwrap.dedent(
        """
        for s, i in l:
            if s not in d:
                d[s] = i
        """),
    setup_statement,
)
show_results(t.timeit(number=count))
```

运行时，脚本会生成以下输出。

```
$ python3 timeit_dictionary.py

1000 items
```

```
1000 iterations

__setitem__ :   91.79 usec/pass    0.09 usec/item
setdefault   :  182.85 usec/pass    0.18 usec/item
KeyError     :   80.87 usec/pass    0.08 usec/item
"not in"     :   66.77 usec/pass    0.07 usec/item
```

这个输出中的时间是在 MacMini 上得到的。当然，取决于使用什么硬件以及系统上正在运行着哪些其他程序，结果可能会有所不同。可以尝试不同的 range_size 和 count 变量来试验，因为不同的组合可能会生成不同的结果。

16.9.4 从命令行执行

除了编程接口，timeit 还提供了一个命令行接口来测试模块，而不需要自动化测试 (instrumentation)。

要运行模块，可以对 Python 解释器使用 -m 选项以查找模块，并把它作为主程序。

```
$ python3 -m timeit
```

例如，使用以下命令来获得帮助。

```
$ python3 -m timeit -h

Tool for measuring execution time of small code snippets.

This module avoids a number of common traps for measuring execution
times.  See also Tim Peters' introduction to the Algorithms chapter in
the Python Cookbook, published by O'Reilly.

...
```

命令行上的 statement 参数与 Timer 的参数稍有不同。并不是传入一个长字符串，而是要将每行指令作为一个单独的命令行参数传递给这个方法。如果需要缩进行（如在一个循环中），则可以用引号包围代码行从而在字符串中嵌入空格。

```
$ python3 -m timeit -s \
"d={}" \
"for i in range(1000):" \
"    d[str(i)] = i"

1000 loops, best of 3: 306 usec per loop
```

还可以用更复杂的代码来定义一个函数，然后从命令行调用这个函数。

代码清单 16-82：**timeit_setitem.py**

```
def test_setitem(range_size=1000):
    l = [(str(x), x) for x in range(range_size)]
    d = {}
    for s, i in l:
        d[s] = i
```

要运行测试，可以传入代码导入模块并运行测试函数。

```
$ python3 -m timeit \
"import timeit_setitem; timeit_setitem.test_setitem()"

1000 loops, best of 3: 401 usec per loop
```

提示：相关阅读材料
- `timeit` 的标准库文档[⊖]。
- `profile`：`profile` 模块对于性能分析也很有用。
- 单调时钟：对 `time` 模块中单调时钟的讨论。

16.10 tabnanny：缩进验证工具

对于像 Python 这样的语言，一致地使用缩进非常重要，并且在这里空白符也是很重要的。`tabnanny` 模块提供了一个扫描器，会报告存在二义性的缩进使用。

从命令行运行

使用 `tabnanny` 最简单的方法是从命令行运行，传入要检查的文件的名。如果传入目录名，则会递归地扫描这些目录以查找要检查的 `.py` 文件。

通过对 PyMOTW 源代码运行 `tabnanny`，可以发现一个老模块使用了 tab 缩进而没有使用空格。

```
$ python3 -m tabnanny .
./source/queue/fetch_podcasts.py 65 "    \t\tparsed_url = \
urlparse(enclosure['url'])\n"
```

`fetch_podcasts.py` 的第 65 行包含两个 tab 而不是 8 个空格。在文本编辑器中，并不能清楚地看出 tab，如果其被配置为对应 4 个空格的制表符，那么看起来两个 tab 和 8 个空格并没有区别。

```
        for enclosure in entry.get('enclosures', []):
            parsed_url = urlparse(enclosure['url'])
            message('queuing {}'.format(
                parsed_url.path.rpartition('/')[-1]))
            enclosure_queue.put(enclosure['url'])
```

修正第 65 行，再运行 tabnanny，显示出第 66 行存在另一个错误。最后一个问题出现在第 67 行上。

如果想扫描文件，但是不想看错误的有关详细信息，则可以使用 `-q` 选项来抑制信息，除了文件名之外不显示其他信息。

```
$ python3 -m tabnanny -q .
./source/queue/fetch_podcasts.py
```

要查看所扫描文件的更多信息，可以使用 `-v` 选项。

```
'source/queue/': listing directory
'source/queue/fetch_podcasts.py': *** Line 65: trouble in tab
city! ***
offending line: "    \t\tparsed_url = urlparse(enclosure['url'])
\n"
indent not greater e.g. at tab sizes 1, 2
```

⊖ https://docs.python.org/3.5/library/timeit.html

```
'source/queue/queue_fifo.py': Clean bill of health.
'source/queue/queue_lifo.py': Clean bill of health.
'source/queue/queue_priority.py': Clean bill of health.
```

说明：对 PyMOTW 源文件运行这些示例不会报告相同错误，因为这些问题已经得到了修正。

提示：相关阅读材料
- `tabnanny` 的标准库文档[一]。
- `tokenize`：Python 源代码的词法扫描工具。
- `flake8`[二]：模块化源代码检查工具。
- `pycodestyle`[三]：Python 样式检查工具。
- `pylint`[四]：Python 代码静态检查工具。

16.11 compileall：字节编译源文件

`compileall` 模块查找 Python 源文件，并把它们编译为字节码表示，将结果保存在 `.pyc` 文件中。

16.11.1 编译一个目录

`compile_dir()` 用于递归地扫描一个目录，并对其中的文件完成字节编译。

代码清单 16-83：**compileall_compile_dir.py**

```python
import compileall
import glob


def show(title):
    print(title)
    for filename in glob.glob('examples/**',
                              recursive=True):
        print('  {}'.format(filename))
    print()

show('Before')

compileall.compile_dir('examples')

show('\nAfter')
```

默认地，会扫描所有子目录，直至深度达到 10。输出文件被写至一个 `__pycache__` 目录，并基于 Python 解释器版本命名。

[一] https://docs.python.org/3/5/library/tabnanny.html
[二] https://pypi.python.org/pypi/flake8
[三] https://pycodestyle.readthedocs.io/en/latest/
[四] https://pypi.python.org/pypi/pylint

```
$ python3 compileall_compile_dir.py

Before
  examples/
  examples/README
  examples/a.py
  examples/subdir
  examples/subdir/b.py

Listing 'examples'...
Compiling 'examples/a.py'...
Listing 'examples/subdir'...
Compiling 'examples/subdir/b.py'...

After
  examples/
  examples/README
  examples/__pycache__
  examples/__pycache__/a.cpython-35.pyc
  examples/a.py
  examples/subdir
  examples/subdir/__pycache__
  examples/subdir/__pycache__/b.cpython-35.pyc
  examples/subdir/b.py
```

16.11.2　忽略文件

要筛除目录，可以使用 rx 参数提供一个正则表达式来匹配要排除的目录名。

代码清单 16-84：**compileall_exclude_dirs.py**

```python
import compileall
import re

compileall.compile_dir(
    'examples',
    rx=re.compile(r'/subdir'),
)
```

这个版本会排除 subdir 子目录中的文件。

```
$ python3 compileall_exclude_dirs.py

Listing 'examples'...
Compiling 'examples/a.py'...
Listing 'examples/subdir'...
```

maxlevels 参数控制递归深度。例如，要完全避免递归，可以传入 0 作为这个参数的值。

代码清单 16-85：**compileall_recursion_depth.py**

```python
import compileall
import re

compileall.compile_dir(
    'examples',
    maxlevels=0,
)
```

在这里，只会编译传递到 `compile_dir()` 的目录中的文件。

```
$ python3 compileall_recursion_depth.py

Listing 'examples'...
Compiling 'examples/a.py'...
```

16.11.3 编译 `sys.path`

只需一个 `compile_path()` 调用，就可以编译 `sys.path` 中找到的所有 Python 源文件。

代码清单 16-86：**compileall_path.py**

```python
import compileall
import sys

sys.path[:] = ['examples', 'notthere']
print('sys.path =', sys.path)
compileall.compile_path()
```

这个例子替换了 `sys.path` 的默认内容，以避免运行脚本时的权限错误，不过仍然能很好地展示默认行为。注意 `maxlevels` 值默认为 `0`。

```
$ python3 compileall_path.py

sys.path = ['examples', 'notthere']
Listing 'examples'...
Compiling 'examples/a.py'...
Listing 'notthere'...
Can't list 'notthere'
```

16.11.4 编译单个文件

要编译一个文件而不是一个完整的文件目录，可以使用 `compile_file()`。

代码清单 16-87：**compileall_compile_file.py**

```python
import compileall
import glob


def show(title):
    print(title)
    for filename in glob.glob('examples/**',
                              recursive=True):
        print('  {}'.format(filename))
    print()
show('Before')

compileall.compile_file('examples/a.py')

show('\nAfter')
```

第一个参数应当是这个文件的名，可以采用完整路径形式，也可以使用相对路径。

```
$ python3 compileall_compile_file.py

Before
  examples/
```

```
examples/README
examples/a.py
examples/subdir
examples/subdir/b.py

Compiling 'examples/a.py'...

After
examples/
examples/README
examples/__pycache__
examples/__pycache__/a.cpython-35.pyc
examples/a.py
examples/subdir
examples/subdir/b.py
```

16.11.5 从命令行运行

也可以从命令行调用 `compileall`，从而能通过一个 Makefile 与构建系统集成。下面给出一个例子。

```
$ python3 -m compileall -h

usage: compileall.py [-h] [-l] [-r RECURSION] [-f] [-q] [-b] [-d DESTDIR]
                    [-x REGEXP] [-i FILE] [-j WORKERS]
                    [FILE|DIR [FILE|DIR ...]]

Utilities to support installing Python libraries.

positional arguments:
  FILE|DIR              zero or more file and directory names to compile; if
                        no arguments given, defaults to the
                        equivalent of -l
                        sys.path

optional arguments:
  -h, --help            show this help message and exit
  -l                    don't recurse into subdirectories
  -r RECURSION          control the maximum recursion level. if
'-l' and '-r'
                        options are specified, then '-r' takes
precedence.
  -f                    force rebuild even if timestamps are up
to date
  -q                    output only error messages; -qq will
suppress the
                        error messages as well.
  -b                    use legacy (pre-PEP3147) compiled file
locations
  -d DESTDIR            directory to prepend to file paths for
use in compile-
                        time tracebacks and in runtime
tracebacks in cases
                        where the source file is unavailable
  -x REGEXP             skip files matching the regular
expression; the regexp
                        is searched for in the full path of each
file
                        considered for compilation
  -i FILE               add all the files and directories listed
```

```
  in FILE to
                          the list considered for compilation; if
"-", names are
                          read from stdin
  -j WORKERS, --workers WORKERS
                          Run compileall concurrently
```

要重新创建之前的例子，跳过 subdir 目录，可以运行以下命令。

```
$ python3 -m compileall -x '/subdir' examples

Listing 'examples'...
Compiling 'examples/a.py'...
Listing 'examples/subdir'...
```

提示：相关阅读材料

- compileall 的标准库文档[⊖]。

16.12 pyclbr：类浏览器

pyclbr 可以扫描 Python 源代码以查找类和独立的函数。可以使用 tokenize 收集类、方法和函数名及行号的有关信息，而无须导入代码。

这一节中的例子都使用以下源文件作为输入。

代码清单 16-88：pyclbr_example.py

```
"""Example source for pyclbr.
"""


class Base:
    """This is the base class.
    """

    def method1(self):
        return


class Sub1(Base):
    """This is the first subclass.
    """


class Sub2(Base):
    """This is the second subclass.
    """


class Mixin:
    """A mixin class.
    """

    def method2(self):
        return
```

⊖ https://docs.python.org/3.5/library/compileall.html

```python
class MixinUser(Sub2, Mixin):
    """Overrides method1 and method2
    """

    def method1(self):
        return

    def method2(self):
        return
    def method3(self):
        return


def my_function():
    """Stand-alone function.
    """
    return
```

16.12.1 扫描类

`pyclbr` 公布了两个公共函数。第一个是 `readmodule()`，它以模块名作为参数，并返回一个映射，将类名映射到 `Class` 对象（其中包含有关类源代码的元数据）。

代码清单 16-89：**pyclbr_readmodule.py**

```python
import pyclbr
import os
from operator import itemgetter


def show_class(name, class_data):
    print('Class:', name)
    filename = os.path.basename(class_data.file)
    print('  File: {0} [{1}]'.format(
        filename, class_data.lineno))
    show_super_classes(name, class_data)
    show_methods(name, class_data)
    print()


def show_methods(class_name, class_data):
    for name, lineno in sorted(class_data.methods.items(),
                               key=itemgetter(1)):
        print('  Method: {0} [{1}]'.format(name, lineno))


def show_super_classes(name, class_data):
    super_class_names = []
    for super_class in class_data.super:
        if super_class == 'object':
            continue
        if isinstance(super_class, str):
            super_class_names.append(super_class)
        else:
            super_class_names.append(super_class.name)
    if super_class_names:
        print('  Super classes:', super_class_names)


example_data = pyclbr.readmodule('pyclbr_example')
```

```
for name, class_data in sorted(example_data.items(),
                               key=lambda x: x[1].lineno):
    show_class(name, class_data)
```

类的源数据包括定义这个类的文件及所在的行号, 还包括超类的类名。类的方法被保存为方法名与行号之间的一个映射。输出显示了这些类和方法 (按其在源文件中行号的顺序列出)。

```
$ python3 pyclbr_readmodule.py

Class: Base
  File: pyclbr_example.py [11]
  Method: method1 [15]

Class: Sub1
  File: pyclbr_example.py [19]
  Super classes: ['Base']

Class: Sub2
  File: pyclbr_example.py [24]
  Super classes: ['Base']

Class: Mixin
  File: pyclbr_example.py [29]
  Method: method2 [33]

Class: MixinUser
  File: pyclbr_example.py [37]
  Super classes: ['Sub2', 'Mixin']
  Method: method1 [41]
  Method: method2 [44]
  Method: method3 [47]
```

16.12.2 扫描函数

`pyclbr` 中的另一个公共函数是 `readmodule_ex()`。它能完成 `readmodule()` 所做的全部工作, 还为结果集增加了一些函数。

代码清单 16-90: **pyclbr_readmodule_ex.py**

```
import pyclbr
import os
from operator import itemgetter

example_data = pyclbr.readmodule_ex('pyclbr_example')

for name, data in sorted(example_data.items(),
                         key=lambda x: x[1].lineno):
    if isinstance(data, pyclbr.Function):
        print('Function: {0} [{1}]'.format(name, data.lineno))
```

`Function` 对象的属性与 `Class` 对象很类似。

```
$ python3 pyclbr_readmodule_ex.py

Function: my_function [51]
```

提示: 相关阅读材料

- `pyclbr` 的标准库文档[⊖]。

⊖ https://docs.python.org/3.5/library/pyclbr.html

- inspect：inspect 模块可以发现有关类和函数的更多元数据，不过需要导入代码。
- tokenize：tokenize 模块将 Python 源代码解析为 token。

16.13 venv：创建虚拟环境

Python 虚拟环境由 venv 管理，建立虚拟环境用于以一种隔离的方式安装包和运行程序，使它们与系统上安装的其他包隔离。由于每个环境都有自己的解释器和安装包的目录，所以可以很容易地在同一个计算机上创建多个环境，分别配置有不同的 Python 和包版本。

16.13.1 创建环境

venv 的主命令行接口依赖于 Python 能够使用 -m 选项运行一个模块中的"主"函数。

```
$ python3 -m venv /tmp/demoenv
```

取决于如何构建和打包 Python 解释器，可以安装一个单独的 pyvenv 命令行应用。以下命令与前例中的命令有相同的效果。

```
$ pyvenv /tmp/demoenv
```

我们更倾向于使用 -m venv，因为这需要显式地选择一个 Python 解释器。这种方法可以确保版本号或与虚拟环境关联的导入路径不会有混淆。

16.13.2 虚拟环境的内容

每个虚拟环境都包含一个 bin 目录，本地解释器和所有可执行的脚本都安装在这里；另外包含一个 include 目录，其中包括构建 C 扩展包的相关文件；还有一个 lib 目录，有一个单独的 site-packages 位置来安装包。

```
$ ls -F /tmp/demoenv

bin/
include/
lib/
pyvenv.cfg
```

默认的 bin 目录中对应多种 UNIX shell 版本分别包含一个"激活"脚本。可以用这些脚本在 shell 的搜索路径上安装虚拟环境，从而确保 shell 选择这个环境中安装的程序。尽管并不是必须激活一个环境才能使用其中安装的程序，但这种技术可能更方便。

```
$ ls -F /tmp/demoenv/bin

activate
activate.csh
activate.fish
easy_install*
easy_install-3.5*
pip*
pip3*
pip3.5*
python@
python3@
```

在支持虚拟环境的平台上，会使用符号链接而不是复制类似 Python 解释器的可执行文件。在这种环境中，`pip` 被安装为一个本地副本，不过解释器是一个符号链接。

最后，环境还包含一个 `pyvenv.cfg` 文件，其中的设置描述了如何配置这个环境，以及会有怎样的行为。`home` 变量指示 Python 解释器的位置，并且将在这里运行 `venv` 以创建环境。`include-system-site-packages` 是一个布尔变量，指示包是在虚拟环境之外安装、在系统级安装还是在虚拟环境内部可见。`version` 是创建这个环境所用的 Python 版本。

代码清单 16-91：`pyvenv.cfg`

```
home = /Library/Frameworks/Python.framework/Versions/3.5/bin
include-system-site-packages = false
version = 3.5.2
```

结合 `pip` 和 `setuptools` 等工具时，虚拟环境会更有用，可以用这些工具来安装其他包，所以 `pyvenv` 默认地安装了这些工具。要创建一个不带这些工具的基本环境，可以在命令行传入 `--without-pip`。

16.13.3 使用虚拟环境

虚拟环境通常用来运行程序的不同版本，或者用程序依赖库的不同版本测试某个给定版本的程序。例如，从 Sphinx 的一个版本升级到另一个版本之前，可能有必要使用老版本和新版本测试输入文档文件。为此，首先创建两个虚拟环境。

```
$ python3 -m venv /tmp/sphinx1
$ python3 -m venv /tmp/sphinx2
```

然后安装要测试的工具的版本。

```
$ /tmp/sphinx1/bin/pip install Sphinx==1.3.6

Collecting Sphinx==1.3.6
  Using cached Sphinx-1.3.6-py2.py3-none-any.whl
Collecting Jinja2>=2.3 (from Sphinx==1.3.6)
  Using cached Jinja2-2.8-py2.py3-none-any.whl
Collecting Pygments>=2.0 (from Sphinx==1.3.6)
  Using cached Pygments-2.1.3-py2.py3-none-any.whl
Collecting babel!=2.0,>=1.3 (from Sphinx==1.3.6)
  Using cached Babel-2.3.4-py2.py3-none-any.whl
Collecting snowballstemmer>=1.1 (from Sphinx==1.3.6)
  Using cached snowballstemmer-1.2.1-py2.py3-none-any.whl
Collecting alabaster<0.8,>=0.7 (from Sphinx==1.3.6)
  Using cached alabaster-0.7.9-py2.py3-none-any.whl
Collecting six>=1.4 (from Sphinx==1.3.6)
  Using cached six-1.10.0-py2.py3-none-any.whl
Collecting sphinx-rtd-theme<2.0,>=0.1 (from Sphinx==1.3.6)
  Using cached sphinx_rtd_theme-0.1.9-py2.py3-none-any.whl
Collecting docutils>=0.11 (from Sphinx==1.3.6)
  Using cached docutils-0.13.1-py3-none-any.whl
Collecting MarkupSafe (from Jinja2>=2.3->Sphinx==1.3.6)
Collecting pytz>=0a (from babel!=2.0,>=1.3->Sphinx==1.3.6)
  Using cached pytz-2016.10-py2.py3-none-any.whl
Installing collected packages: MarkupSafe, Jinja2, Pygments,
pytz, babel, snowballstemmer, alabaster, six, sphinx-rtd-theme,
```

```
docutils, Sphinx
Successfully installed Jinja2-2.8 MarkupSafe-0.23 Pygments-2.1.3
Sphinx-1.3.6 alabaster-0.7.9 babel-2.3.4 docutils-0.13.1
pytz-2016.10 six-1.10.0 snowballstemmer-1.2.1 sphinx-rtd-
theme-0.1.9

$ /tmp/sphinx2/bin/pip install Sphinx==1.4.4

Collecting Sphinx==1.4.4
  Using cached Sphinx-1.4.4-py2.py3-none-any.whl
Collecting Jinja2>=2.3 (from Sphinx==1.4.4)
  Using cached Jinja2-2.8-py2.py3-none-any.whl
Collecting imagesize (from Sphinx==1.4.4)
  Using cached imagesize-0.7.1-py2.py3-none-any.whl
Collecting Pygments>=2.0 (from Sphinx==1.4.4)
  Using cached Pygments-2.1.3-py2.py3-none-any.whl
Collecting babel!=2.0,>=1.3 (from Sphinx==1.4.4)
  Using cached Babel-2.3.4-py2.py3-none-any.whl
Collecting snowballstemmer>=1.1 (from Sphinx==1.4.4)
  Using cached snowballstemmer-1.2.1-py2.py3-none-any.whl
Collecting alabaster<0.8,>=0.7 (from Sphinx==1.4.4)
  Using cached alabaster-0.7.9-py2.py3-none-any.whl
Collecting six>=1.4 (from Sphinx==1.4.4)
  Using cached six-1.10.0-py2.py3-none-any.whl
Collecting docutils>=0.11 (from Sphinx==1.4.4)
  Using cached docutils-0.13.1-py3-none-any.whl
Collecting MarkupSafe (from Jinja2>=2.3->Sphinx==1.4.4)
Collecting pytz>=0a (from babel!=2.0,>=1.3->Sphinx==1.4.4)
  Using cached pytz-2016.10-py2.py3-none-any.whl
Installing collected packages: MarkupSafe, Jinja2, imagesize,
Pygments, pytz, babel, snowballstemmer, alabaster, six,
docutils, Sphinx
Successfully installed Jinja2-2.8 MarkupSafe-0.23 Pygments-2.1.3
Sphinx-1.4.4 alabaster-0.7.9 babel-2.3.4 docutils-0.13.1
imagesize-0.7.1 pytz-2016.10 six-1.10.0 snowballstemmer-1.2.1
```

在这里，两个虚拟环境中不同的 Sphinx 版本可以单独运行，这样就可以利用相同的输入文件完成测试。

```
$ /tmp/sphinx1/bin/sphinx-build --version

Sphinx (sphinx-build) 1.3.6

$ /tmp/sphinx2/bin/sphinx-build --version

Sphinx (sphinx-build) 1.4.4
```

提示：相关阅读材料

- `venv` 的标准库文档⊖。
- **PEP 405**⊜：Python 虚拟环境。
- virtualenv⊛：Python 虚拟环境的一个版本，适用于 Python 2 和 Python 3。
- virtualenvwrapper⊕：virtualenv 的一组 shell 包装器，以便更容易地管理大量环境。

⊖ https://docs.python.org/3.5/library/venv.html
⊜ www.python.org/dev/peps/pep-0405
⊛ https://pypi.python.org/pypi/virtualenv
⊕ https://pypi.python.org/pypi/virtualenvwrapper

- Sphinx[⊖]：将 reStructuredText 输入文件转换为 HTML、LaTeX 和其他格式来进行处理的工具。

16.14 `ensurepip`：安装 Python 包安装工具

Python 是一种"内含动力"的编程语言，它的标准库中提供了丰富的模块，甚至还可以从 Python Package Index[⊖]安装更多的库、框架和工具。要安装这些包，开发人员需要安装工具 `pip`。要安装一个工具，而且这个工具要用来安装其他工具，这就带来一个有意思的自引导问题，`ensurepip` 可以解决这个问题。

安装 `pip`

这个例子使用了一个虚拟环境，这里其被配置为未安装 `pip`。

```
$ python3 -m venv --without-pip /tmp/demoenv
$ ls -F /tmp/demoenv/bin

activate
activate.csh
activate.fish
python@
python3@
```

从命令行对 Python 解释器使用 `-m` 选项运行 `ensurepip`。默认地，会安装标准库提供的一个 `pip` 副本。然后，可以用这个版本安装 `pip` 的一个更新版本。要想直接安装 `pip` 的一个最新版本，可以对 `ensurepip` 使用 `--upgrade` 选项。

```
$ /tmp/demoenv/bin/python3 -m ensurepip --upgrade

Ignoring indexes: https://pypi.python.org/simple
Collecting setuptools
Collecting pip
Installing collected packages: setuptools, pip
Successfully installed pip-8.1.1 setuptools-20.10.1
```

这个命令会在这个虚拟环境中分别安装 `pip3` 和 `pip3.5`（作为单独的程序），还包括所需的 `setuptools` 依赖库。

```
$ ls -F /tmp/demoenv/bin

activate
activate.csh
activate.fish
easy_install-3.5*
pip3*
pip3.5*
python@
python3@
```

⊖ www.sphinx-doc.org/en/stable/

⊖ https://pypi.python.org/pypi

提示：相关阅读材料
- `ensurepip` 的标准库文档[一]。
- `venv`：虚拟环境。
- **PEP 453**[二]：Python 安装中 `pip` 的自引导问题。
- Installing Python modules[三]：安装用于 Python 的其他包的有关说明。
- **Python Package Index**[四]：为 Python 程序员提供的扩展模块托管网站。
- `pip`[五]：安装 Python 包的工具。

[一] https://docs.python.org/3.5/library/ensurepip.html
[二] www.python.org/dev/peps/pep-0453
[三] https://docs.python.org/3.5/installing/index.html#installing-index
[四] https://pypi.python.org/pypi
[五] https://pypi.python.org/pypi/pip

第 17 章
运行时特性

这一章将介绍 Python 标准库中的一些运行时特性,利用这些特性,程序可以与解释器或它所在的运行环境交互。

启动时,解释器会加载 `site` 模块来配置特定于当前安装的设置。通过结合环境设置、解释器构建参数以及配置文件来构造导入路径。

`sys` 模块是标准库中最大的模块之一。它包含大量用于访问各种解释器和系统设置的函数,包括解释器构建设置和限制,命令行参数和程序退出码,异常处理,线程调试和控制,导入机制和导入模块,运行时控制流跟踪,以及进程的标准输入和输出流。

`sys` 的重点是解释器设置,`os` 模块则允许访问操作系统信息。`os` 模块可以作为系统调用的可移植接口,返回正在运行的进程的有关详细信息,如进程所有者和环境变量。`os` 模块还包括一些用于处理系统和进程管理的函数。

Python 通常作为一个跨平台语言用来创建可移植程序。即使一个程序可能需要在任意环境中都能运行,有时也有必要知道当前系统的操作系统或硬件体系结构。`platform` 模块提供了一些函数来获取这些设置。

可以通过 `resource` 模块探查和修改系统资源的限制,如最大进程栈大小或打开的文件数目。它还会报告当前消耗率,从而能监视进程的资源泄漏情况。

`gc` 模块允许访问 Python 垃圾回收系统的内部状态。它包括一些有用的信息来检测和打破对象引用环,打开和关闭垃圾收集器,以及调整自动触发垃圾回收清扫的阈值。

`sysconfig` 模块包含构建脚本的编译时变量。构建和打包工具可以用这个模块动态生成路径和其他设置。

17.1 `site`:全站点配置

`site` 模块处理站点特定的配置,特别是导入路径。

17.1.1 导入路径

每次解释器启动时会自动导入 `site`。导入时,会用站点特定的名扩展 `sys.path`,可以通过组合前缀值(`sys.prefix` 和 `sys.exec_prefix`)和一些后缀来构造这个名。使用的前缀值将被保存在模块级变量 `PREFIXES` 中,以便以后引用。在 Windows 下,后缀是一个空串和 `lib/site-packages`。对于类 UNIX 平台,后缀值是 `lib/python$version/site-packages`(其中 `$version` 被替换为解释器的主版本号和次版本号，如 3.5）和 `lib/site-python`。

代码清单 17-1：`site_import_path.py`

```python
import sys
import os
import site

if 'Windows' in sys.platform:
    SUFFIXES = [
        '',
        'lib/site-packages',
    ]
else:
    SUFFIXES = [
        'lib/python{}/site-packages'.format(sys.version[:3]),
        'lib/site-python',
    ]

print('Path prefixes:')
for p in site.PREFIXES:
    print('  ', p)

for prefix in sorted(set(site.PREFIXES)):
    print()
    print(prefix)
    for suffix in SUFFIXES:
        print()
        print(' ', suffix)
        path = os.path.join(prefix, suffix).rstrip(os.sep)
        print('    exists :', os.path.exists(path))
        print('    in path:', path in sys.path)
```

对通过组合得到的各个路径进行测试，将确实存在的路径增加到 `sys.path`。下面这个输出显示了安装在一个 Mac OS X 系统上的 Python 框架版本。

```
$ python3 site_import_path.py

Path prefixes:
   /Library/Frameworks/Python.framework/Versions/3.5
   /Library/Frameworks/Python.framework/Versions/3.5

/Library/Frameworks/Python.framework/Versions/3.5

  lib/python3.5/site-packages
   exists : True
   in path: True

  lib/site-python
   exists : False
   in path: False
```

17.1.2 用户目录

除了全局的 `site-packages` 路径，`site` 还负责向导入路径增加用户特定的位置。用户特定的路径都基于 `USER_BASE` 目录，通常位于文件系统中当前用户拥有（而且可写）的一部分。`USER_BASE` 目录中有一个 `site-packages` 目录，其中包含路径可以用 `USER_SITE` 访问。

代码清单 17-2：**site_user_base.py**

```
import site

print('Base:', site.USER_BASE)
print('Site:', site.USER_SITE)
```

USER_SITE 路径名是使用同样的平台特定后缀（如前所述）来创建的。

```
$ python3 site_user_base.py

Base: /Users/dhellmann/.local
Site: /Users/dhellmann/.local/lib/python3.5/site-packages
```

用户基目录可以通过 PYTHONUSERBASE 环境变量设置，而且有平台特定的默认值（对于 Windows 是 ~/Python$version/site-packages，对于非 Windows 平台为 ~/.local）。

```
$ PYTHONUSERBASE=/tmp/$USER python3 site_user_base.py

Base: /tmp/dhellmann
Site: /tmp/dhellmann/lib/python3.5/site-packages
```

在一些可能出现安全问题的情况下会禁用用户目录（例如，如果运行进程的用户或组 ID 与原先启动这个进程的实际用户不同）。应用可以通过查看 ENABLE_USER_SITE 来检查这个设置。

代码清单 17-3：**site_enable_user_site.py**

```
import site

status = {
    None: 'Disabled for security',
    True: 'Enabled',
    False: 'Disabled by command-line option',
}

print('Flag   :', site.ENABLE_USER_SITE)
print('Meaning:', status[site.ENABLE_USER_SITE])
```

还可以在命令行上用 -S 显式禁用用户目录。

```
$ python3 site_enable_user_site.py

Flag   : True
Meaning: Enabled

$ python3 -s site_enable_user_site.py

Flag   : False
Meaning: Disabled by command-line option
```

17.1.3 路径配置文件

路径被增加到导入路径时，还会在这些路径中扫描路径配置文件（path configuration file）。路径配置文件是一个纯文本文件，扩展名为 .pth。文件中的每一行可以有以下 4 种形式：

- 要增加到导入路径的另一个位置的完全或相对路径。

- 一个要执行的 Python 语句。所有这样的行都必须以一个 `import` 语句开头。
- 空行，这些行会被忽略。
- 以 `#` 开头的行，被处理为注释并被忽略。

可以用路径配置文件扩展导入路径，以查看不能自动增加的路径。例如，`setuptools` 包使用 `python setup.py develop` 以开发模式安装一个包时会向 `easy-install.pth` 增加一个路径。

扩展 `sys.path` 的函数是公共的，可以在示例程序中用来显示路径配置文件如何工作。给定一个名为 `with_modules` 的目录，其中包含文件 `mymodule.py`，以下 `print` 语句显示了这个模块如何导入。

代码清单 17-4：**with_modules/mymodule.py**

```python
import os
print('Loaded {} from {}'.format(
    __name__, __file__[len(os.getcwd()) + 1:])
)
```

下面这个脚本显示了 `addsitedir()` 如何扩展导入路径，使得解释器能够找到所需的模块。

代码清单 17-5：**site_addsitedir.py**

```python
import site
import os
import sys

script_directory = os.path.dirname(__file__)
module_directory = os.path.join(script_directory, sys.argv[1])

try:
    import mymodule
except ImportError as err:
    print('Could not import mymodule:', err)

print()
before_len = len(sys.path)
site.addsitedir(module_directory)
print('New paths:')
for p in sys.path[before_len:]:
    print(p.replace(os.getcwd(), '.'))  # Shorten dirname

print()
import mymodule
```

将包含模块的目录增加到 `sys.path` 之后，这个脚本可以顺利地导入 `mymodule`。

```
$ python3 site_addsitedir.py with_modules

Could not import mymodule: No module named 'mymodule'

New paths:
./with_modules

Loaded mymodule from with_modules/mymodule.py
```

addsitedir() 完成的路径调整不只是向 sys.path 追加参数。如果为 addsitedir() 提供的目录包括与模式 *.pth 匹配的文件，那么它们也会作为路径配置文件被加载。例如，对于以下目录结构：

```
with_pth
  pymotw.pth
  subdir
      mymodule.py
```

如果 with_pth/pymotw.pth 包含以下内容：

```
# Add a single subdirectory to the path.
./subdir
```

通过增加 with_pth（作为一个站点目录），可以导入 with_pth/subdir/mymodule.py，尽管模块不在那个目录中，这是因为 with_pth 和 with_pth/subdir 都已经增加到导入路径。

```
$ python3 site_addsitedir.py with_pth

Could not import mymodule: No module named 'mymodule'

New paths:
./with_pth
./with_pth/subdir

Loaded mymodule from with_pth/subdir/mymodule.py
```

如果一个站点目录包含多个 .pth 文件，会按字母顺序进行处理。

```
$ ls -F multiple_pth

a.pth
b.pth
from_a/
from_b/

$ cat multiple_pth/a.pth

./from_a

$ cat multiple_pth/b.pth

./from_b
```

在这里，会在 multiple_pth/from_a 中找到模块，因为 a.pth 比 b.pth 先读取。

```
$ python3 site_addsitedir.py multiple_pth

Could not import mymodule: No module named 'mymodule'

New paths:
./multiple_pth
./multiple_pth/from_a
./multiple_pth/from_b

Loaded mymodule from multiple_pth/from_a/mymodule.py
```

17.1.4 定制站点配置

site 模块还负责加载整个站点的定制设置，这个设置在 sitecustomize 模块中由本地站点所有者定义。sitecustomize 可以用来扩展导入路径，以及启用覆盖、性能分析或其他开发工具。

例如，以下代码清单中的 sitecustomize.py 脚本用基于当前平台的一个目录扩展了导入路径。/opt/python 中的平台特定路径会增加到导入路径，从而可以导入其中安装的所有包。如果网络中不同主机之间要通过一个共享文件系统来共享包含编译扩展模块的包，那么这就很有用。只需要在各个主机上安装 sitecustomize.py 脚本，其他包都可以从文件服务器访问。

代码清单 17-6：**with_sitecustomize/sitecustomize.py**

```python
print('Loading sitecustomize.py')

import site
import platform
import os
import sys

path = os.path.join('/opt',
                    'python',
                    sys.version[:3],
                    platform.platform(),
                    )
print('Adding new path', path)

site.addsitedir(path)
```

可以用一个简单的脚本来展示在 Python 开始运行你自己的代码之前会先导入 sitecustomize.py。

代码清单 17-7：**with_sitecustomize/site_sitecustomize.py**

```python
import sys

print('Running main program from\n{}'.format(sys.argv[0]))

print('End of path:', sys.path[-1])
```

由于 sitecustomize 被用来建立整个系统的配置，所以应当将其安装在默认路径上的某个位置（通常在 site-packages 目录中）。下面这个例子显式地设置了 PYTHONPATH，以确保可以导入这个模块。

```
$ PYTHONPATH=with_sitecustomize python3 with_sitecustomize/sit\
e_sitecustomize.py

Loading sitecustomize.py
Adding new path /opt/python/3.5/Darwin-15.6.0-x86_64-i386-64bit
Running main program from
with_sitecustomize/site_sitecustomize.py
End of path: /opt/python/3.5/Darwin-15.6.0-x86_64-i386-64bit
```

17.1.5 定制用户配置

类似于 sitecustomize，每次解释器启动时都可以用 usercustomize 模块建立用户特定设置。`usercustomize`在`sitecustomize`之后加载,所以可以覆盖整个站点的定制设置。

有些环境下,运行不同操作系统或不同版本的多个服务器会共享用户的主目录,标准用户目录机制可能不适用于安装用户特定的包。在这些情况下,可以使用一个平台特定的目录树。

代码清单 17-8: with_usercustomize/usercustomize.py

```
print('Loading usercustomize.py')

import site
import platform
import os
import sys

path = os.path.expanduser(os.path.join('~',
                                       'python',
                                       sys.version[:3],
                                       platform.platform(),
                                       ))
print('Adding new path', path)

site.addsitedir(path)
```

类似于介绍`sitecustomize`时所用的脚本,可以用另一个简单的脚本来展示在Python开始运行其他代码之前会先导入`usercustomize.py`。

代码清单 17-9: with_usercustomize/site_usercustomize.py

```
import sys

print('Running main program from\n{}'.format(sys.argv[0]))

print('End of path:', sys.path[-1])
```

由于`usercustomize`被用来建立一个用户的用户特定配置,所以应当将其安装在用户默认路径上的某个位置,而不是放在整个站点路径上。默认的`USER_BASE`目录就是一个很合适的位置。这个例子显式地设置了`PYTHONPATH`,以确保可以导入这个模块。

```
$ PYTHONPATH=with_usercustomize python3 with_usercustomize/site\
_usercustomize.py

Loading usercustomize.py
Adding new path /Users/dhellmann/python/3.5/Darwin-15.5.0-x86_64\
-i386-64bit
Running main program from
with_usercustomize/site_usercustomize.py
End of path: /Users/dhellmann/python/3.5/Darwin-15.5.0-x86_64\
-i386-64bit
```

当用户站点目录特性被禁用时,不会导入`usercustomize`,不论它位于用户站点目录还是在其他位置。

```
$ PYTHONPATH=with_usercustomize python3 -s with_usercustomize/s\
ite_usercustomize.py
```

```
Running main program from
with_usercustomize/site_usercustomize.py
End of path: /Users/dhellmann/Envs/pymotw35/lib/python3.5/site-
packages
```

17.1.6 禁用 `site` 模块

之前的 Python 版本没有增加自动导入特性，为了维护与这些版本的向后兼容性，解释器还接受一个 `-S` 选项。

```
$ python3 -S site_import_path.py

Path prefixes:
   /Users/dhellmann/Envs/pymotw35/bin/..
   /Users/dhellmann/Envs/pymotw35/bin/..

/Users/dhellmann/Envs/pymotw35/bin/..

   lib/python3.5/site-packages
     exists : True
     in path: False

   lib/site-python
     exists : False
     in path: False
```

提示：相关阅读材料
- `site` 的标准库文档[⊖]。
- 17.2.6 节：描述了在 `sys` 中定义的导入路径如何工作。
- `setuptools`[⊖]：打包库和安装工具 `easy_install`。
- Running code at Python startup[⊜]：Ned Batchelder 的一个帖子，讨论了多种方法可以让 Python 解释器在启动主程序执行之前先运行定制代码。

17.2 `sys`：系统特定配置

`sys` 模块包括一组服务，可以探查或修改解释器的运行时配置以及资源，与当前程序之外的操作环境交互。

17.2.1 解释器设置

`sys` 包含有一些属性和函数，可以访问解释器的编译时或运行时配置设置。

17.2.1.1 构建时版本信息

构建 C 解释器所用的版本可以有多种形式。`sys.version` 是人类可读的一个串，通常包含完整的版本号，以及有关构建日期、编译器和平台的信息。`sys.hexversion` 可以

⊖ https://docs.python.org/3.5/library/site.html
⊖ https://setuptools.readthedocs.io/en/latest/index.html
⊜ http://nedbatchelder.com/blog/201001/running_code_at_python_startup.html

更容易地用来检查解释器版本，因为它是一个简单的整数。使用 hex() 格式化时，可以清楚地看出，sys.hexversion 的某些部分来自更可读的 sys.version_info 中的版本信息（这是一个包含 5 部分的命名元组，只表示版本号）。当前解释器使用的 C API 版本保存在 sys.api_version 中。

代码清单 17-10：**sys_version_values.py**

```
import sys

print('Version info:')
print()
print('sys.version      =', repr(sys.version))
print('sys.version_info =', sys.version_info)
print('sys.hexversion   =', hex(sys.hexversion))
print('sys.api_version  =', sys.api_version)
```

所有这些值都依赖于运行示例程序的具体解释器。

```
$ python3 sys_version_values.py

Version info:

sys.version      = '3.5.2 (v3.5.2:4def2a2901a5, Jun 26 2016,
10:47:25) \n[GCC 4.2.1 (Apple Inc. build 5666) (dot 3)]'
sys.version_info = sys.version_info(major=3, minor=5, micro=2,
releaselevel='final', serial=0)
sys.hexversion   = 0x30502f0
sys.api_version  = 1013
```

用来构建解释器的操作系统平台被保存为 sys.platform。

代码清单 17-11：**sys_platform.py**

```
import sys

print('This interpreter was built for:', sys.platform)
```

对于大多数 UNIX 系统，这个值由命令 uname -s 的输出与 uname -r 中版本的第一部分组合而成。对于其他操作系统，会使用一个硬编码的值表。

```
$ python3 sys_platform.py

This interpreter was built for: darwin
```

提示：相关阅读材料
- platform values[○]：sys.platform 为系统提供的硬编码值，但不包括 uname。

17.2.1.2 解释器实现

CPython 解释器是 Python 语言的实现之一。sys.implementation 可以检测处理解释器差异所需的库的当前实现。

[○] https://docs.python.org/3/library/sys.html#sys.platform

代码清单17-12：**sys_implementation.py**

```
import sys

print('Name:', sys.implementation.name)
print('Version:', sys.implementation.version)
print('Cache tag:', sys.implementation.cache_tag)
```

对于 CPython，`sys.implementation.version` 与 `sys.version_info` 相同，不过对于其他解释器而言，它们却是不同的。

```
$ python3 sys_implementation.py

Name: cpython
Version: sys.version_info(major=3, minor=5, micro=2, releaseleve
l='final', serial=0)
Cache tag: cpython-35
```

提示：相关阅读材料

- PEP 421[⊖]：增加 `sys.implementation`。

17.2.1.3 命令行选项

CPython 解释器接受一些命令行选项以控制解释器的行为，这些选项如表 17-1 所列。其中一些选项可以在程序中通过 `sys.flags` 检查。

表 17-1　CPython 命令行选项标志

选项	含义
-B	导入时不写 .py[co] 文件
-b	发出警告，指出在将字节转换为字符串时没有正确地解码和比较字节与字符串
-bb	将处理字节对象的警告转换为错误
-d	调试解析器的输出
-E	忽略 PYTHON* 环境变量（如 PYTHONPATH）
-i	在运行脚本后进行交互式检查
-O	对生成的字节码稍做优化
-OO	除了完成 -O 优化外，还会删除 docstring
-s	不向 `sys.path` 增加用户站点目录
-S	不在初始化时运行"import site"
-t	发出警告，指出 tab 使用不一致
-tt	发出错误，指出 tab 使用不一致
-v	详细显示

代码清单17-13：**sys_flags.py**

```
import sys

if sys.flags.bytes_warning:
```

[⊖] www.python.org/dev/peps/pep-0421

```
        print('Warning on bytes/str errors')
if sys.flags.debug:
    print('Debuging')
if sys.flags.inspect:
    print('Will enter interactive mode after running')
if sys.flags.optimize:
    print('Optimizing byte-code')
if sys.flags.dont_write_bytecode:
    print('Not writing byte-code files')
if sys.flags.no_site:
    print('Not importing "site"')
if sys.flags.ignore_environment:
    print('Ignoring environment')
if sys.flags.verbose:
    print('Verbose mode')
```

可以尝试使用 `sys_flags.py` 来了解命令行选项如何映射到标志设置。

```
$ python3 -S -E -b sys_flags.py

Warning on bytes/str errors
Not importing "site"
Ignoring environment
```

17.2.1.4 Unicode 默认编码

要得到解释器使用的默认 Unicode 编码名,可以调用 `getdefaultencoding()`。这个值在启动时设置,在会话期间不能改变。

对于某些操作系统,内部编码默认设置与文件系统编码可能不同,所以要有另外一种方法来获取文件系统设置。`getfilesystemencoding()` 会返回操作系统特定的一个值(而不是文件系统特定的值)。

代码清单 17-14:`sys_unicode.py`

```
import sys

print('Default encoding     :', sys.getdefaultencoding())
print('File system encoding :', sys.getfilesystemencoding())
```

大多数 Unicode 专家并不建议依赖全局默认编码,而是推荐让应用显式地设置 Unicode。这种方法提供了两个好处:不同的数据源有不同的 Unicode 编码,这样可以更简洁地处理;而且其可以减少对应用代码中编码的假设。

```
$ python3 sys_unicode.py

Default encoding     : utf-8
File system encoding : utf-8
```

17.2.1.5 交互式提示符

交互式解释器使用两种不同的提示符来指示默认输入层次(`ps1`)和多行语句的"继续"(`ps2`)。这些值只用于交互式解释器。

```
>>> import sys
>>> sys.ps1
'>>> '
>>> sys.ps2
```

```
'... '
>>>
```

这两个提示符中任意一个（或者二者）都可以被改为不同的串。

```
>>> sys.ps1 = '::: '
::: sys.ps2 = '~~~ '
::: for i in range(3):
~~~     print i
~~~
0
1
2
:::
```

或者，只要一个对象可以转换为字符串（通过 `__str__`），就可以作为提示符。

代码清单 17-15：sys_ps1.py

```python
import sys

class LineCounter:

    def __init__(self):
        self.count = 0

    def __str__(self):
        self.count += 1
        return '({:3d})> '.format(self.count)
```

LineCounter 会记录它使用了多少次，所以提示符中的数字每次都会增加。

```
$ python
Python 3.4.2 (v3.4.2:ab2c023a9432, Oct  5 2014, 20:42:22)
[GCC 4.2.1 (Apple Inc. build 5666) (dot 3)] on darwin
Type "help", "copyright", "credits" or "license" for more information.
>>> from sys_ps1 import LineCounter
>>> import sys
>>> sys.ps1 = LineCounter()
(  1)>
(  2)>
(  3)>
```

17.2.1.6 显示 hook

每次用户进入一个表达式时，交互式解释器都会调用 `sys.displayhook`。这个表达式的计算结果将作为唯一的参数传至函数。

代码清单 17-16：sys_displayhook.py

```python
import sys

class ExpressionCounter:

    def __init__(self):
        self.count = 0
        self.previous_value = self

    def __call__(self, value):
```

```
            print()
            print('  Previous:', self.previous_value)
            print('  New     :', value)
            print()
            if value != self.previous_value:
                self.count += 1
                sys.ps1 = '({:3d})> '.format(self.count)
            self.previous_value = value
            sys.__displayhook__(value)


print('installing')
sys.displayhook = ExpressionCounter()
```

默认值（保存在 sys.__displayhook__ 中）将结果打印到标准输出（stdout），并把它保存到 _ 以便以后引用。

```
$ python3
Python 3.4.2 (v3.4.2:ab2c023a9432, Oct  5 2014, 20:42:22)
[GCC 4.2.1 (Apple Inc. build 5666) (dot 3)] on darwin
Type "help", "copyright", "credits" or "license" for more
information.
>>> import sys_displayhook
installing
>>> 1 + 2

  Previous: <sys_displayhook.ExpressionCounter
object at 0x1021035f8>
  New     : 3

3
(  1)> 'abc'

  Previous: 3
  New     : abc

'abc'
(  2)> 'abc'

  Previous: abc
  New     : abc

'abc'
(  2)> 'abc' * 3

  Previous: abc
  New     : abcabcabc

'abcabcabc'
(  3)>
```

17.2.1.7 安装位置

只要有合适的解释器路径，所有系统便都可以由 sys.executable 得到具体解释器程序的路径。可以检查这个信息来确保使用了正确的解释器，而且还能对基于解释器位置设置的路径给出线索。

sys.prefix 指示解释器安装的父目录。它通常包括 bin 和 lib 目录，分别存放可执行文件和已安装模块。

代码清单 17-17：`sys_locations.py`

```
import sys

print('Interpreter executable:')
print(sys.executable)
print('\nInstallation prefix:')
print(sys.prefix)
```

下面的示例输出是在 Mac 上生成的，它运行了一个从 python.org 安装的框架。

```
$ python3 sys_locations.py

Interpreter executable:
/Library/Frameworks/Python.framework/Versions/3.5/bin/python3

Installation prefix:
/Library/Frameworks/Python.framework/Versions/3.5
```

17.2.2 运行时环境

`sys` 提供了一些底层 API 与应用的外部系统交互，可以接受命令行参数，访问用户输入，以及向用户传递消息和状态值。

17.2.2.1 命令行参数

解释器捕获的参数将由解释器处理，不会传递到它运行的程序。其余的所有选项和参数（包括脚本名本身）都保存到 `sys.argv`，因为程序有可能确实需要用到。

代码清单 17-18：`sys_argv.py`

```
import sys

print('Arguments:', sys.argv)
```

这里显示的第 3 个例子中，`-u` 选项由解释器处理，不会传递到它运行的程序。

```
$ python3 sys_argv.py

Arguments: ['sys_argv.py']

$ python3 sys_argv.py -v foo blah

Arguments: ['sys_argv.py', '-v', 'foo', 'blah']

$ python3 -u sys_argv.py

Arguments: ['sys_argv.py']
```

提示：相关阅读材料
- `argparse`：解析命令行参数的模块。

17.2.2.2 输入和输出流

按照 UNIX 编程范式，默认地，Python 程序可以访问 3 个文件描述符。

代码清单 17-19：**sys_stdio.py**

```
import sys

print('STATUS: Reading from stdin', file=sys.stderr)

data = sys.stdin.read()

print('STATUS: Writing data to stdout', file=sys.stderr)

sys.stdout.write(data)
sys.stdout.flush()

print('STATUS: Done', file=sys.stderr)
```

stdin 是读取输入的标准方法，通常从控制台读取，不过也可以通过管道从其他程序读取。stdout 是为用户写输出的标准方法（写至控制台），或者发送到管线中的下一个程序。stderr 用于写警告或错误消息。

```
$ cat sys_stdio.py | python3 -u sys_stdio.py

STATUS: Reading from stdin
STATUS: Writing data to stdout
#!/usr/bin/env python3

#end_pymotw_header
import sys

print('STATUS: Reading from stdin', file=sys.stderr)

data = sys.stdin.read()

print('STATUS: Writing data to stdout', file=sys.stderr)

sys.stdout.write(data)
sys.stdout.flush()

print('STATUS: Done', file=sys.stderr)
STATUS: Done
```

提示：相关阅读材料

- **subprocess** 和 **pipes**：这两个模块都提供了将程序通过管线相结合的特性。

17.2.2.3 返回状态

要从一个程序返回一个退出码，需要向 `sys.exit()` 传递一个整数值。

代码清单 17-20：**sys_exit.py**

```
import sys

exit_code = int(sys.argv[1])
sys.exit(exit_code)
```

非 0 值表示程序退出时有一个错误。

```
$ python3 sys_exit.py 0 ; echo "Exited $?"

Exited 0
```

```
$ python3 sys_exit.py 1 ; echo "Exited $?"

Exited 1
```

17.2.3 内存管理和限制

sys 包含一些用于了解和控制内存使用的函数。

17.2.3.1 引用数

Python 的主实现（CPython）使用引用计数（reference counting）和垃圾回收（garbage collection）来完成自动内存管理。当一个对象的引用数降至 0 时，它会自动标记为回收。要查看一个现有对象的引用数，可以使用 **getrefcount()**。

代码清单 17-21：sys_getrefcount.py

```
import sys

one = []
print('At start          :', sys.getrefcount(one))

two = one

print('Second reference :', sys.getrefcount(one))

del two

print('After del         :', sys.getrefcount(one))
```

报告的这个数实际上比期望的计数多 1，因为 **getrefcount()** 本身会维护对象的一个临时引用。

```
$ python3 sys_getrefcount.py

At start          : 2
Second reference : 3
After del         : 2
```

提示：相关阅读材料

- gc：通过 gc 中提供的函数来控制垃圾回收器。

17.2.3.2 对象大小

了解一个对象有多少引用可以帮助开发人员发现环或内存泄漏，不过还不足以确定哪些对象消耗的内存最多。这需要知道对象有多大。

代码清单 17-22：sys_getsizeof.py

```
import sys

class MyClass:
    pass
objects = [
    [], (), {}, 'c', 'string', b'bytes', 1, 2.3,
    MyClass, MyClass(),
]
```

```
    for obj in objects:
        print('{:>10} : {}'.format(type(obj).__name__,
                                   sys.getsizeof(obj)))
```

getsizeof() 会报告一个对象的大小（单位为字节）。

```
$ python3 sys_getsizeof.py

      list : 64
     tuple : 48
      dict : 288
       str : 50
       str : 55
     bytes : 38
       int : 28
     float : 24
      type : 1016
   MyClass : 56
```

报告的定制类大小不包括属性值的大小。

代码清单 17-23：`sys_getsizeof_object.py`

```
import sys


class WithoutAttributes:
    pass


class WithAttributes:
    def __init__(self):
        self.a = 'a'
        self.b = 'b'
        return


without_attrs = WithoutAttributes()
print('WithoutAttributes:', sys.getsizeof(without_attrs))

with_attrs = WithAttributes()
print('WithAttributes:', sys.getsizeof(with_attrs))
```

这可能会让人对实际消耗的内存量有一个错误的印象。

```
$ python3 sys_getsizeof_object.py

WithoutAttributes: 56
WithAttributes: 56
```

对于一个类所用的空间，为了得到更全面的估计，这个模块提供了一个 `__sizeof__()` 方法来计算这个值，它会累计一个对象各个属性的大小。

代码清单 17-24：`sys_getsizeof_custom.py`

```
import sys


class WithAttributes:
    def __init__(self):
```

```python
        self.a = 'a'
        self.b = 'b'
        return

    def __sizeof__(self):
        return object.__sizeof__(self) + \
            sum(sys.getsizeof(v) for v in self.__dict__.values())


my_inst = WithAttributes()
print(sys.getsizeof(my_inst))
```

这个版本在计算对象大小时,会在对象基本大小的基础上加上存储在内部 __dict__ 中的所有属性的大小。

```
$ python3 sys_getsizeof_custom.py

156
```

17.2.3.3 递归

Python 应用中允许无限递归,这可能会引入解释器本身的栈溢出,导致崩溃。为了消除这种情况,解释器提供了一种方法,可以使用 setrecursionlimit() 和 getrecursionlimit() 来控制最大递归深度。

代码清单 17-25: **sys_recursionlimit.py**

```python
import sys

print('Initial limit:', sys.getrecursionlimit())
sys.setrecursionlimit(10)

print('Modified limit:', sys.getrecursionlimit())


def generate_recursion_error(i):
    print('generate_recursion_error({})'.format(i))
    generate_recursion_error(i + 1)


try:
    generate_recursion_error(1)
except RuntimeError as err:
    print('Caught exception:', err)
```

一旦栈大小达到递归限制,解释器便会产生一个 RuntimeError 异常,使程序有机会处理这种情况。

```
$ python3 sys_recursionlimit.py

Initial limit: 1000
Modified limit: 10
generate_recursion_error(1)
generate_recursion_error(2)
generate_recursion_error(3)
generate_recursion_error(4)
generate_recursion_error(5)
generate_recursion_error(6)
generate_recursion_error(7)
```

```
generate_recursion_error(8)
Caught exception: maximum recursion depth exceeded while calling
a Python object
```

17.2.3.4 最大值

除了运行时可配置的值，`sys` 还包括一些变量，可以用来定义随系统不同而变化的一些类型的最大值。

代码清单 17-26：sys_maximums.py

```python
import sys

print('maxsize   :', sys.maxsize)
print('maxunicode:', sys.maxunicode)
```

`maxsize` 是 C 解释器中 `size` 类型指示的列表、字典、串或其他数据结构的最大大小。`maxunicode` 是当前配置的解释器支持的最大整数 Unicode 值。

```
$ python3 sys_maximums.py

maxsize   : 9223372036854775807
maxunicode: 1114111
```

17.2.3.5 浮点值

结构 `float_info` 包含解释器所用的浮点类型表示的有关信息（基于底层系统的 `float` 实现）。

代码清单 17-27：sys_float_info.py

```python
import sys

print('Smallest difference (epsilon):', sys.float_info.epsilon)
print()
print('Digits (dig)                 :', sys.float_info.dig)
print('Mantissa digits (mant_dig):', sys.float_info.mant_dig)
print()
print('Maximum (max):', sys.float_info.max)
print('Minimum (min):', sys.float_info.min)
print()
print('Radix of exponents (radix):', sys.float_info.radix)
print()
print('Maximum exponent for radix (max_exp):',
      sys.float_info.max_exp)
print('Minimum exponent for radix (min_exp):',
      sys.float_info.min_exp)
print()
print('Max. exponent power of 10 (max_10_exp):',
      sys.float_info.max_10_exp)
print('Min. exponent power of 10 (min_10_exp):',
      sys.float_info.min_10_exp)
print()
print('Rounding for addition (rounds):', sys.float_info.rounds)
```

这些值依赖于编译器和底层系统。下面这些例子是在 Intel Core i7 上的 OS X 10.9.5 上生成的。

```
$ python3 sys_float_info.py

Smallest difference (epsilon): 2.220446049250313e-16

Digits (dig)              : 15
Mantissa digits (mant_dig): 53

Maximum (max): 1.7976931348623157e+308
Minimum (min): 2.2250738585072014e-308

Radix of exponents (radix): 2

Maximum exponent for radix (max_exp): 1024
Minimum exponent for radix (min_exp): -1021

Max. exponent power of 10 (max_10_exp): 308
Min. exponent power of 10 (min_10_exp): -307

Rounding for addition (rounds): 1
```

提示：相关阅读材料
- 本地编译器的 float.h C 头文件包含这些设置的更多详细信息。

17.2.3.6 整数值
结构 int_info 包含解释器中所用整数内部表示的有关信息。

代码清单 17-28：sys_int_info.py

```python
import sys

print('Number of bits used to hold each digit:',
      sys.int_info.bits_per_digit)
print('Size in bytes of C type used to hold each digit:',
      sys.int_info.sizeof_digit)
```

下面的输出是在 Intel Core i7 上的 OS X 10.9.5 上生成的。

```
$ python3 sys_int_info.py

Number of bits used to hold each digit: 30
Size in bytes of C type used to hold each digit: 4
```

会在构建解释器时确定在内部用来存储整数的 C 类型。默认地，64 位体系结构会自动使用 30 位整数，不过通过设置配置标志 --enable-big-digits，32 位体系结构也可以启用 30 位整数。

提示：相关阅读材料
- "What's New in Python 3.1" 中构建和 C API 的改变[⊖]。

17.2.3.7 字节序
byteorder 被设置为内置字节序。

⊖ https://docs.python.org/3.1/whatsnew/3.1.html#build-and-c-api-changes

代码清单 17-29：`sys_byteorder.py`

```
import sys

print(sys.byteorder)
```

这个值可以是 `big` 表示大端（big endian），或者是 `little` 表示小端（little endian）。

```
$ python3 sys_byteorder.py

little
```

提示：相关阅读材料
- Wikipedia：Endianness[○]：大端和小端内存系统的描述。
- `array` 和 `struct`：依赖于数据字节序的其他模块。
- `float.h`：本地编译器的 C 头文件包含有关这些设置的更多详细信息。

17.2.4 异常处理

`sys` 包含一些用于捕获和处理异常的特性。

17.2.4.1 未处理异常

很多应用的结构都包括一个主循环，将执行包围在一个全局异常处理器中，以捕获较低层次未处理的错误。要达到同样的目的，还有一种方法是将 `sys.excepthook` 设置为一个函数，它有 3 个参数（错误类型、错误值和 traceback），由这个函数处理未处理的错误。

代码清单 17-30：`sys_excepthook.py`

```
import sys

def my_excepthook(type, value, traceback):
    print('Unhandled error:', type, value)

sys.excepthook = my_excepthook
print('Before exception')

raise RuntimeError('This is the error message')

print('After exception')
```

由于产生异常的代码行未包围在 `try:except` 块中，所以不会运行后面的 `print()` 调用（尽管设置了 `excepthook`）。

```
$ python3 sys_excepthook.py

Before exception
Unhandled error: <class 'RuntimeError'> This is the error
message
```

17.2.4.2 当前异常

有些情况下，不论是出于代码简洁性考虑，还是为了避免与试图安装 `excepthook` 的

[○] https://en.wikipedia.org/wiki/Byte_order

库发生冲突，使用一个显式的异常处理器是更合适的。在这些情况下，程序员可以创建一个处理函数，通过调用 `exc_info()` 来获取线程的当前异常，因此不需要显式地为它传递异常对象。

`exc_info()` 的返回值是一个包含 3 个成员的元组，其中包含异常类、异常实例和 traceback。使用 `exc_info()` 要优于原来的形式（使用 `exc_type`、`exc_value` 和 `exc_traceback`），因为它是线程安全的。

代码清单 17-31：**sys_exc_info.py**

```python
import sys
import threading
import time


def do_something_with_exception():
    exc_type, exc_value = sys.exc_info()[:2]
    print('Handling {} exception with message "{}" in {}'.format(
        exc_type.__name__, exc_value,
        threading.current_thread().name))


def cause_exception(delay):
    time.sleep(delay)
    raise RuntimeError('This is the error message')


def thread_target(delay):
    try:
        cause_exception(delay)
    except:
        do_something_with_exception()


threads = [
    threading.Thread(target=thread_target, args=(0.3,)),
    threading.Thread(target=thread_target, args=(0.1,)),
]

for t in threads:
    t.start()
for t in threads:
    t.join()
```

这个例子忽略了 `exc_info()` 的部分返回值，从而避免在 traceback 对象和当前帧中的一个局部变量之间引入循环引用。如果需要 traceback（例如，以便记入日志），则可以显式地删除局部变量（使用 `del`）以避免环。

```
$ python3 sys_exc_info.py

Handling RuntimeError exception with message "This is the error
message" in Thread-2
Handling RuntimeError exception with message "This is the error
message" in Thread-1
```

17.2.4.3 以前的交互式异常

交互式解释器中只有一个交互线程。该线程中的未处理异常会保存到 `sys` 的 3 个变量

（`last_type`、`last_value`和`last_traceback`）中，从而能轻松地获取以完成调试。通过使用`pdb`中的事后剖析调试工具，不再需要直接使用这些值。

```
$ python3
Python 3.4.2 (v3.4.2:ab2c023a9432, Oct  5 2014, 20:42:22)
[GCC 4.2.1 (Apple Inc. build 5666) (dot 3)] on darwin
Type "help", "copyright", "credits" or "license" for more information.
>>> def cause_exception():
...     raise RuntimeError('This is the error message')
...
>>> cause_exception()
Traceback (most recent call last):
  File "<stdin>", line 1, in <module>
  File "<stdin>", line 2, in cause_exception
RuntimeError: This is the error message
>>> import pdb
>>> pdb.pm()
> <stdin>(2)cause_exception()
(Pdb) where
  <stdin>(1)<module>()
> <stdin>(2)cause_exception()
(Pdb)
```

提示：相关阅读材料

- `exceptions`：内置错误。
- `pdb`：Python调试工具。
- `traceback`：处理traceback的模块。

17.2.5 底层线程支持

`sys`包括一些用于控制和调试线程行为的底层函数。

17.2.5.1 切换间隔

Python 3使用一个全局锁来防止单个线程破坏解释器状态。一个时间间隔之后（这个时间间隔是可配置的），字节码执行会暂停，解释器检查是否需要执行某个信号处理器。在这个间隔检查期间，当前线程还会释放全局解释器锁（Global Interpreter Lock，GIL），然后重新请求，使其他线程比释放这个锁的线程更优先。

默认的切换间隔是5毫秒，可以用`sys.getswitchinterval()`获取当前值。用`sys.getswitchinterval()`改变这个间隔可能会对应用的性能产生影响，这取决于所完成的操作。

代码清单17-32：`sys_switchinterval.py`

```
import sys
import threading
from queue import Queue


def show_thread(q):
    for i in range(5):
        for j in range(1000000):
            pass
        q.put(threading.current_thread().name)
```

```
        return

def run_threads():
    interval = sys.getswitchinterval()
    print('interval = {:0.3f}'.format(interval))
    q = Queue()
    threads = [
        threading.Thread(target=show_thread,
                         name='T{}'.format(i),
                         args=(q,))
        for i in range(3)
    ]
    for t in threads:
        t.setDaemon(True)
        t.start()
    for t in threads:
        t.join()
    while not q.empty():
        print(q.get(), end=' ')
    print()
    return


for interval in [0.001, 0.1]:
    sys.setswitchinterval(interval)
    run_threads()
    print()
```

当切换间隔小于线程完成运行所要花费的时间时,解释器会把控制权交给另一个线程,让它运行一段时间。这种行为从下面的第一组输出情况可以看到,其中间隔被设置为 1 毫秒。

如果有更长的间隔,那么在强制交出控制之前,活动线程能完成更多工作。这种情况可以用第二个例子中队列内的 name 值顺序来说明,这里使用的间隔为 10 毫秒。

```
$ python3 sys_switchinterval.py

interval = 0.001
T0 T1 T2 T1 T0 T2 T0 T1 T2 T1 T0 T2 T1 T0 T2

interval = 0.100
T0 T0 T0 T0 T0 T1 T1 T1 T1 T1 T2 T2 T2 T2 T2
```

除了切换间隔,还有很多其他因素也会控制 Python 线程的上下文切换。例如,如果一个线程完成 I/O,则它会释放 GIL,并且可能因此允许另一个线程接管执行。

17.2.5.2 调试

找出死锁可能是处理线程最困难的方面之一。`sys._current_frames()` 会有所帮助,它能准确地显示出线程在哪里停止。

代码清单 17-33: `sys_current_frames.py`

```
1  import sys
2  import threading
3  import time
4
5  io_lock = threading.Lock()
```

```
 6  blocker = threading.Lock()
 7
 8
 9  def block(i):
10      t = threading.current_thread()
11      with io_lock:
12          print('{} with ident {} going to sleep'.format(
13              t.name, t.ident))
14      if i:
15          blocker.acquire()  # Acquired but never released
16          time.sleep(0.2)
17      with io_lock:
18          print(t.name, 'finishing')
19      return
20
21  # Create and start several threads that "block."
22  threads = [
23      threading.Thread(target=block, args=(i,))
24      for i in range(3)
25  ]
26  for t in threads:
27      t.setDaemon(True)
28      t.start()
29
30  # Map the threads from their identifier to the thread object.
31  threads_by_ident = dict((t.ident, t) for t in threads)
32
33  # Show where each thread is "blocked."
34  time.sleep(0.01)
35  with io_lock:
36      for ident, frame in sys._current_frames().items():
37          t = threads_by_ident.get(ident)
38          if not t:
39              # Main thread
40              continue
41          print('{} stopped in {} at line {} of {}'.format(
42              t.name, frame.f_code.co_name,
43              frame.f_lineno, frame.f_code.co_filename))
```

sys._current_frames()返回的字典以线程标识符为键，而不是线程名。需要稍做一点工作将这些标识符映射为线程对象。

由于线程-1没有休眠，在检查它的状态之前它就已经完成了。由于这个线程不再是活动的，所以不会出现在输出中。线程-2请求锁阻塞器（blocker），然后休眠很短一段时间。与此同时，线程-3尝试请求阻塞器，不过无法得到，因为已经被线程-2占用。

```
$ python3 sys_current_frames.py

Thread-1 with ident 123145307557888 going to sleep
Thread-1 finishing
Thread-2 with ident 123145307557888 going to sleep
Thread-3 with ident 123145312813056 going to sleep
Thread-3 stopped in block at line 18 of sys_current_frames.py
Thread-2 stopped in block at line 19 of sys_current_frames.py
```

提示：相关阅读材料

- threading：threading模块包括一些用于创建Python线程的类。
- Queue：Queue模块提供了一个FIFO数据结构的线程安全实现。

- Reworking the GIL[①]：Antoine Pitrou 向 python-dev 邮件列表发出的一封邮件，描述了引入交换间隔的 GIL 实现变更。

17.2.6 模块和导入

大多数 Python 程序最后都会是一个组合，其中包括多个模块以及导入这些模块的一个主应用。无论是使用标准库的特性，还是将定制代码组织到单独的文件中以便于维护，理解和管理程序的依赖关系都是开发的一个重要方面。sys 包含了应用可用模块的有关信息，这些模块可能是内置模块，也可能是导入的模块。sys 还定义了一些 hook，为特殊情况覆盖标准导入行为。

17.2.6.1 导入的模块

sys.modules 是一个字典，将所导入模块的名映射到包含具体代码的模块对象。

代码清单 17-34：sys_modules.py

```
import sys
import textwrap
names = sorted(sys.modules.keys())
name_text = ', '.join(names)

print(textwrap.fill(name_text, width=64))
```

随着新模块的导入，sys.modules 的内容会改变。

```
$ python3 sys_modules.py

__main__, _bootlocale, _codecs, _collections_abc,
_frozen_importlib, _frozen_importlib_external, _imp, _io,
_locale, _signal, _sre, _stat, _thread, _warnings, _weakref,
_weakrefset, abc, builtins, codecs, copyreg, encodings,
encodings.aliases, encodings.latin_1, encodings.utf_8, errno,
genericpath, io, marshal, os, os.path, posix, posixpath, re,
site, sre_compile, sre_constants, sre_parse, stat, sys,
textwrap, zipimport
```

17.2.6.2 内置模块

Python 解释器在编译时可以内置一些 C 模块，因此不需要作为单独的共享库发布这些 C 模块。这些模块不会出现在 sys.modules 管理的导入模块列表中，因为从理论上讲，它们并不是导入的模块。要查找这些可用的内置模块，唯一的方法就是通过 sys.builtin_module_names。

代码清单 17-35：sys_builtins.py

```
import sys
import textwrap

name_text = ', '.join(sorted(sys.builtin_module_names))

print(textwrap.fill(name_text, width=64))
```

[①] https://mail.python.org/pipermail/python-dev/2009-October/093321.html

这个脚本的输出可能会有变化，特别是用解释器的一个定制构建版本来运行时。以下输出是通过从 OS X 的标准 python.org 安装程序安装的一个解释器副本来创建的。

```
$ python3 sys_builtins.py

_ast, _codecs, _collections, _functools, _imp, _io, _locale,
_operator, _signal, _sre, _stat, _string, _symtable, _thread,
_tracemalloc, _warnings, _weakref, atexit, builtins, errno,
faulthandler, gc, itertools, marshal, posix, pwd, sys, time,
xxsubtype, zipimport
```

提示：相关阅读材料
- Build Instructions[⊖]：构建 Python 的说明，摘自随源代码发布的 README 文件。

17.2.6.3 导入路径

模块的搜索路径作为一个 Python 列表保存在 `sys.path` 中。这个路径的默认内容包括启动应用所使用脚本的目录和当前工作目录。

代码清单 17-36：**sys_path_show.py**

```python
import sys

for d in sys.path:
    print(d)
```

搜索路径中的第一个目录是示例脚本本身的主目录。后面是一系列平台特定的路径，指示已编译的扩展模块（用 C 编写）可能安装在哪些位置。最后列出全局 `site-packages` 目录。

```
$ python3 sys_path_show.py

/Users/dhellmann/Documents/PyMOTW/pymotw-3/source/sys
.../python35.zip
.../lib/python3.5
.../lib/python3.5/plat-darwin
.../python3.5/lib-dynload
.../lib/python3.5/site-packages
```

通过将 shell 变量 **PYTHONPATH** 设置为一个用冒号分隔的目录列表，可以在启动解释器之前修改导入搜索路径列表。

```
$ PYTHONPATH=/my/private/site-packages:/my/shared/site-packages \
> python3 sys_path_show.py

/Users/dhellmann/Documents/PyMOTW/pymotw-3/source/sys
/my/private/site-packages
/my/shared/site-packages
.../python35.zip
.../lib/python3.5
.../lib/python3.5/plat-darwin
.../python3.5/lib-dynload
.../lib/python3.5/site-packages
```

[⊖] https://hg.python.org/cpython/file/tip/README

程序还可以直接向 `sys.path` 增加元素以修改路径。

代码清单 17-37：sys_path_modify.py

```python
import imp
import os
import sys

base_dir = os.path.dirname(__file__) or '.'
print('Base directory:', base_dir)

# Insert the package_dir_a directory at the front of the path.
package_dir_a = os.path.join(base_dir, 'package_dir_a')
sys.path.insert(0, package_dir_a)

# Import the example module.
import example
print('Imported example from:', example.__file__)
print('   ', example.DATA)

# Make package_dir_b the first directory in the search path.
package_dir_b = os.path.join(base_dir, 'package_dir_b')
sys.path.insert(0, package_dir_b)

# Reload the module to get the other version.
imp.reload(example)
print('Reloaded example from:', example.__file__)
print('   ', example.DATA)
```

重新加载一个已经导入的模块时，会重新导入这个文件，并使用相同的 `module` 对象来保存结果。如果在第一次导入和 `reload()` 调用之间改变了路径，这意味着第二次可能加载一个不同的模块。

```
$ python3 sys_path_modify.py

Base directory: .
Imported example from: ./package_dir_a/example.py
    This is example A
Reloaded example from: ./package_dir_b/example.py
    This is example B
```

17.2.6.4 定制导入工具

通过修改搜索路径，程序员可以控制如何找到标准 Python 模块。不过，如果一个程序需要导入其他地方的代码，而不是从文件系统上常规的 `.py` 或 `.pyc` 文件导入，那么又该怎么做呢？PEP 302[⊖] 解决了这个问题，通过引入导入 hook 的思想，它会捕获到在搜索路径上查找一个模块的意图，并采用候选策略从其他位置加载代码或者对它做预处理。

可以通过两个不同阶段来实现定制导入工具。查找工具（finder）负责找到一个模块，并提供一个加载工具（loader）来管理具体的导入。可以向 `sys.path_hooks` 列表追加一个工厂来增加定制模块查找工具。导入时，会把路径的各个部分提供给一个查找工具，直到一个部分声称提供支持（不产生 `ImportError`）。然后，查找工具负责搜索数据存储（由对应命名模块的路径入口表示）。

⊖ www.python.org/dev/peps/pep-0302

代码清单 17-38: **sys_path_hooks_noisy.py**

```python
import sys


class NoisyImportFinder:

    PATH_TRIGGER = 'NoisyImportFinder_PATH_TRIGGER'

    def __init__(self, path_entry):
        print('Checking {}:'.format(path_entry), end=' ')
        if path_entry != self.PATH_TRIGGER:
            print('wrong finder')
            raise ImportError()
        else:
            print('works')
            return

    def find_module(self, fullname, path=None):
        print('Looking for {!r}'.format(fullname))
        return None


sys.path_hooks.append(NoisyImportFinder)

for hook in sys.path_hooks:
    print('Path hook: {}'.format(hook))

sys.path.insert(0, NoisyImportFinder.PATH_TRIGGER)

try:
    print('importing target_module')
    import target_module
except Exception as e:
    print('Import failed:', e)
```

这个例子展示了如何实例化和查询查找工具。如果提供的路径入口与查找工具的特殊触发值不匹配，这显然不是文件系统上的一个真正的路径，那么实例化时 **NoisyImportFinder** 会产生一个 **ImportError**。这个测试可以避免 NoisyImportFinder 破坏真正模块的导入。

```
$ python3 sys_path_hooks_noisy.py

Path hook: <class 'zipimport.zipimporter'>
Path hook: <function
FileFinder.path_hook.<locals>.path_hook_for_FileFinder at
0x100734950>
Path hook: <class '__main__.NoisyImportFinder'>
importing target_module
Checking NoisyImportFinder_PATH_TRIGGER: works
Looking for 'target_module'
Import failed: No module named 'target_module'
```

17.2.6.5 从 shelf 导入

查找工具找到一个模块时，它要负责返回一个能够导入该模块的加载工具（loader）。这一节的例子展示了一个定制导入工具，它会将它的模块内容保存到由 **shelve** 创建的一个数据库中。

首先，使用一个脚本以一个包（其中包含一个子模块和子包）来填充这个 shelf。

代码清单 17-39：sys_shelve_importer_create.py

```
import shelve
import os

filename = '/tmp/pymotw_import_example.shelve'
if os.path.exists(filename + '.db'):
    os.unlink(filename + '.db')
with shelve.open(filename) as db:
    db['data:README'] = b"""
==============
package README
==============

This is the README for "package".
"""
    db['package.__init__'] = b"""
print('package imported')
message = 'This message is in package.__init__'
"""
    db['package.module1'] = b"""
print('package.module1 imported')
message = 'This message is in package.module1'
"""
    db['package.subpackage.__init__'] = b"""
print('package.subpackage imported')
message = 'This message is in package.subpackage.__init__'
"""
    db['package.subpackage.module2'] = b"""
print('package.subpackage.module2 imported')
message = 'This message is in package.subpackage.module2'
"""
    db['package.with_error'] = b"""
print('package.with_error being imported')
raise ValueError('raising exception to break import')
"""
    print('Created {} with:'.format(filename))
    for key in sorted(db.keys()):
        print(' ', key)
```

真正的打包脚本会从文件系统读取内容，不过对于像这样一个简单的例子来说，使用硬编码的值就足够了。

```
$ python3 sys_shelve_importer_create.py

Created /tmp/pymotw_import_example.shelve with:
    data:README
    package.__init__
    package.module1
    package.subpackage.__init__
    package.subpackage.module2
    package.with_error
```

这个定制导入工具需要提供查找工具和加载工具，它们要知道如何在 shelf 中查找模块或包的源代码。

代码清单 17-40：sys_shelve_importer.py

```
import imp
import os
```

```python
import shelve
import sys


def _mk_init_name(fullname):
    """Return the name of the __init__ module
    for a given package name.
    """
    if fullname.endswith('.__init__'):
        return fullname
    return fullname + '.__init__'


def _get_key_name(fullname, db):
    """Look in an open shelf for fullname or
    fullname.__init__, and return the name found.
    """
    if fullname in db:
        return fullname
    init_name = _mk_init_name(fullname)
    if init_name in db:
        return init_name
    return None


class ShelveFinder:
    """Find modules collected in a shelve archive."""

    _maybe_recursing = False

    def __init__(self, path_entry):
        # Loading shelve creates an import recursive loop when it
        # imports dbm, and we know we will not load the
        # module being imported. Thus, when we seem to be
        # recursing, just ignore the request so another finder
        # will be used.
        if ShelveFinder._maybe_recursing:
            raise ImportError
        try:
            # Test the path_entry to see if it is a valid shelf.
            try:
                ShelveFinder._maybe_recursing = True
                with shelve.open(path_entry, 'r'):
                    pass
            finally:
                ShelveFinder._maybe_recursing = False
        except Exception as e:
            print('shelf could not import from {}: {}'.format(
                path_entry, e))
            raise
        else:
            print('shelf added to import path:', path_entry)
            self.path_entry = path_entry
        return

    def __str__(self):
        return '<{} for {!r}>'.format(self.__class__.__name__,
                                      self.path_entry)

    def find_module(self, fullname, path=None):
        path = path or self.path_entry
        print('\nlooking for {!r}\n  in {}'.format(
            fullname, path))
```

```python
            with shelve.open(self.path_entry, 'r') as db:
                key_name = _get_key_name(fullname, db)
                if key_name:
                    print('  found it as {}'.format(key_name))
                    return ShelveLoader(path)
            print('  not found')
            return None

class ShelveLoader:
    """Load source for modules from shelve databases."""

    def __init__(self, path_entry):
        self.path_entry = path_entry
        return

    def _get_filename(self, fullname):
        # Make up a fake filename that starts with the path entry
        # so pkgutil.get_data() works correctly.
        return os.path.join(self.path_entry, fullname)

    def get_source(self, fullname):
        print('loading source for {!r} from shelf'.format(
            fullname))
        try:
            with shelve.open(self.path_entry, 'r') as db:
                key_name = _get_key_name(fullname, db)
                if key_name:
                    return db[key_name]
                raise ImportError(
                    'could not find source for {}'.format(
                        fullname)
                )
        except Exception as e:
            print('could not load source:', e)
            raise ImportError(str(e))

    def get_code(self, fullname):
        source = self.get_source(fullname)
        print('compiling code for {!r}'.format(fullname))
        return compile(source, self._get_filename(fullname),
                       'exec', dont_inherit=True)

    def get_data(self, path):
        print('looking for data\n  in {}\n  for {!r}'.format(
            self.path_entry, path))
        if not path.startswith(self.path_entry):
            raise IOError
        path = path[len(self.path_entry) + 1:]
        key_name = 'data:' + path
        try:
            with shelve.open(self.path_entry, 'r') as db:
                return db[key_name]
        except Exception:
            # Convert all errors to IOError.
            raise IOError()

    def is_package(self, fullname):
        init_name = _mk_init_name(fullname)
        with shelve.open(self.path_entry, 'r') as db:
            return init_name in db

    def load_module(self, fullname):
```

```python
        source = self.get_source(fullname)

        if fullname in sys.modules:
            print('reusing module from import of {!r}'.format(
                fullname))
            mod = sys.modules[fullname]
        else:
            print('creating a new module object for {!r}'.format(
                fullname))
            mod = sys.modules.setdefault(
                fullname,
                imp.new_module(fullname)
            )

        # Set a few properties required by PEP 302.
        mod.__file__ = self._get_filename(fullname)
        mod.__name__ = fullname
        mod.__path__ = self.path_entry
        mod.__loader__ = self
        # PEP-366 specifies that packages set __package__ to
        # their name, and modules have it set to their parent
        # package (if any).
        if self.is_package(fullname):
            mod.__package__ = fullname
        else:
            mod.__package__ = '.'.join(fullname.split('.')[:-1])

        if self.is_package(fullname):
            print('adding path for package')
            # Set __path__ for packages
            # so we can find the submodules.
            mod.__path__ = [self.path_entry]
        else:
            print('imported as regular module')
        print('execing source...')
        exec(source, mod.__dict__)
        print('done')
        return mod
```

现在可以用 ShelveFinder 和 ShelveLoader 从一个 shelf 导入代码。例如，可以用以下代码导入刚创建的 package。

代码清单 17-41: **sys_shelve_importer_package.py**

```python
import sys
import sys_shelve_importer


def show_module_details(module):
    print('  message    :', module.message)
    print('  __name__   :', module.__name__)
    print('  __package__:', module.__package__)
    print('  __file__   :', module.__file__)
    print('  __path__   :', module.__path__)
    print('  __loader__ :', module.__loader__)


filename = '/tmp/pymotw_import_example.shelve'
sys.path_hooks.append(sys_shelve_importer.ShelveFinder)
sys.path.insert(0, filename)

print('Import of "package":')
```

```python
import package

print()
print('Examine package details:')
show_module_details(package)

print()
print('Global settings:')
print('sys.modules entry:')
print(sys.modules['package'])
```

修改路径之后，第一次出现导入时会把这个 shelf 增加到导入路径。查找工具找到这个 shelf，并返回一个加载工具，这个加载工具将用于完成来自该 shelf 的所有导入。初始的包级别导入会创建一个新的模块对象，然后使用 **exec** 来运行从 shelf 加载的源代码。它将这个新模块作为命名空间，所以源代码中定义的名字可以作为模块级属性保留。

```
$ python3 sys_shelve_importer_package.py

Import of "package":
shelf added to import path: /tmp/pymotw_import_example.shelve

looking for 'package'
  in /tmp/pymotw_import_example.shelve
  found it as package.__init__
loading source for 'package' from shelf
creating a new module object for 'package'
adding path for package
execing source...
package imported
done

Examine package details:
  message    : This message is in package.__init__
  __name__   : package
  __package__: package
  __file__   : /tmp/pymotw_import_example.shelve/package
  __path__   : ['/tmp/pymotw_import_example.shelve']
  __loader__ : <sys_shelve_importer.ShelveLoader object at
0x101467860>

Global settings:
sys.modules entry:
<module 'package' (<sys_shelve_importer.ShelveLoader object at
0x101467860>)>
```

17.2.6.6　定制包导入

可以采用同样的方式加载其他模块和子包。

代码清单 17-42：`sys_shelve_importer_module.py`

```python
import sys
import sys_shelve_importer

def show_module_details(module):
    print('  message    :', module.message)
    print('  __name__   :', module.__name__)
    print('  __package__:', module.__package__)
    print('  __file__   :', module.__file__)
```

```
    print('  __path__   :', module.__path__)
    print('  __loader__ :', module.__loader__)

filename = '/tmp/pymotw_import_example.shelve'
sys.path_hooks.append(sys_shelve_importer.ShelveFinder)
sys.path.insert(0, filename)
print('Import of "package.module1":')
import package.module1

print()
print('Examine package.module1 details:')
show_module_details(package.module1)

print()
print('Import of "package.subpackage.module2":')
import package.subpackage.module2

print()
print('Examine package.subpackage.module2 details:')
show_module_details(package.subpackage.module2)
```

查找工具接收要加载的模块的完整名（有点号），并返回一个 **ShelveLoader**，其被配置为从指向这个 shelf 文件的路径入口加载模块。模块的完全限定名被传递到加载工具的 **load_module()** 方法，它会构造并返回一个 **module** 实例。

```
$ python3 sys_shelve_importer_module.py

Import of "package.module1":
shelf added to import path: /tmp/pymotw_import_example.shelve

looking for 'package'
  in /tmp/pymotw_import_example.shelve
  found it as package.__init__
loading source for 'package' from shelf
creating a new module object for 'package'
adding path for package
execing source...
package imported
done

looking for 'package.module1'
  in /tmp/pymotw_import_example.shelve
  found it as package.module1
loading source for 'package.module1' from shelf
creating a new module object for 'package.module1'
imported as regular module
execing source...
package.module1 imported
done

Examine package.module1 details:
  message    : This message is in package.module1
  __name__   : package.module1
  __package__: package
  __file__   : /tmp/pymotw_import_example.shelve/package.module1
  __path__   : /tmp/pymotw_import_example.shelve
  __loader__ : <sys_shelve_importer.ShelveLoader object at
0x101376e10>

Import of "package.subpackage.module2":
```

```
looking for 'package.subpackage'
  in /tmp/pymotw_import_example.shelve
  found it as package.subpackage.__init__
loading source for 'package.subpackage' from shelf
creating a new module object for 'package.subpackage'
adding path for package
execing source...
package.subpackage imported
done

looking for 'package.subpackage.module2'
  in /tmp/pymotw_import_example.shelve
  found it as package.subpackage.module2
loading source for 'package.subpackage.module2' from shelf
creating a new module object for 'package.subpackage.module2'
imported as regular module
execing source...
package.subpackage.module2 imported
done

Examine package.subpackage.module2 details:
  message    : This message is in package.subpackage.module2
  __name__   : package.subpackage.module2
  __package__: package.subpackage
  __file__   :
/tmp/pymotw_import_example.shelve/package.subpackage.module2
  __path__   : /tmp/pymotw_import_example.shelve
  __loader__ : <sys_shelve_importer.ShelveLoader object at
0x1013a6c88>
```

17.2.6.7 在定制导入工具中重新加载模块

关于重新加载一个模块的处理稍有不同。不是创建一个新的模块对象，而是重用现有的模块。

代码清单 17-43：sys_shelve_importer_reload.py

```python
import importlib
import sys
import sys_shelve_importer

filename = '/tmp/pymotw_import_example.shelve'
sys.path_hooks.append(sys_shelve_importer.ShelveFinder)
sys.path.insert(0, filename)

print('First import of "package":')
import package

print()
print('Reloading "package":')
importlib.reload(package)
```

通过重用相同的对象，会保留这个模块现有的引用，即使重新加载步骤修改了类或函数定义。

```
$ python3 sys_shelve_importer_reload.py

First import of "package":
shelf added to import path: /tmp/pymotw_import_example.shelve

looking for 'package'
```

```
    in /tmp/pymotw_import_example.shelve
    found it as package.__init__
loading source for 'package' from shelf
creating a new module object for 'package'
adding path for package
execing source...
package imported
done

Reloading "package":

looking for 'package'
    in /tmp/pymotw_import_example.shelve
    found it as package.__init__
loading source for 'package' from shelf
reusing module from import of 'package'
adding path for package
execing source...
package imported
done
```

17.2.6.8 处理导入错误

当所有查找工具都无法找到一个模块时,主导入代码会产生一个 `ImportError`。

代码清单 17-44:`sys_shelve_importer_missing.py`

```
import sys
import sys_shelve_importer
filename = '/tmp/pymotw_import_example.shelve'
sys.path_hooks.append(sys_shelve_importer.ShelveFinder)
sys.path.insert(0, filename)

try:
    import package.module3
except ImportError as e:
    print('Failed to import:', e)
```

会传播导入期间的其他错误。

```
$ python3 sys_shelve_importer_missing.py

shelf added to import path: /tmp/pymotw_import_example.shelve

looking for 'package'
    in /tmp/pymotw_import_example.shelve
    found it as package.__init__
loading source for 'package' from shelf
creating a new module object for 'package'
adding path for package
execing source...
package imported
done

looking for 'package.module3'
    in /tmp/pymotw_import_example.shelve
    not found
Failed to import: No module named 'package.module3'
```

17.2.6.9 包数据

除了定义 API 来加载可执行的 Python 代码,PEP 302 还定义了一个可选的 API 以获取包数据，用于发布数据文件、文档和包所需的其他非代码资源。通过实现 `get_data()`，加载工具允许调用应用以获取与包关联的数据，而不用考虑这个包具体如何安装（特别是不用假设这个包作为文件存储在一个文件系统上）。

代码清单 17-45：`sys_shelve_importer_get_data.py`

```
import sys
import sys_shelve_importer
import os
import pkgutil

filename = '/tmp/pymotw_import_example.shelve'
sys.path_hooks.append(sys_shelve_importer.ShelveFinder)
sys.path.insert(0, filename)

import package

readme_path = os.path.join(package.__path__[0], 'README')

readme = pkgutil.get_data('package', 'README')
# Equivalent to:
#   readme = package.__loader__.get_data(readme_path)
print(readme.decode('utf-8'))

foo_path = os.path.join(package.__path__[0], 'foo')
try:
    foo = pkgutil.get_data('package', 'foo')
    # Equivalent to:
    #   foo = package.__loader__.get_data(foo_path)
except IOError as err:
    print('ERROR: Could not load "foo"', err)
else:
    print(foo)
```

`get_data()` 根据拥有数据的模块或包来取一个路径。它将资源"文件"的内容作为一个字节串返回，或者如果这个资源不存在，则产生一个 `IOError`。

```
$ python3 sys_shelve_importer_get_data.py

shelf added to import path: /tmp/pymotw_import_example.shelve

looking for 'package'
  in /tmp/pymotw_import_example.shelve
  found it as package.__init__
loading source for 'package' from shelf
creating a new module object for 'package'
adding path for package
execing source...
package imported
done
looking for data
  in /tmp/pymotw_import_example.shelve
  for '/tmp/pymotw_import_example.shelve/README'

==============
package README
==============

This is the README for "package".
```

```
looking for data
  in /tmp/pymotw_import_example.shelve
  for '/tmp/pymotw_import_example.shelve/foo'
ERROR: Could not load "foo"
```

提示：相关阅读材料

- pkgutil：包含 get_data()，用于从一个包获取数据。

17.2.6.10　导入工具缓存

每次导入一个模块时都要搜索所有 hook，这样可能很耗费时间。为了节省时间，会维护一个 sys.path_importer_cache，作为路径入口与加载工具（可以使用值查找模块）之间的一个映射。

代码清单 17-46：sys_path_importer_cache.py

```
import os
import sys

prefix = os.path.abspath(sys.prefix)

print('PATH:')
for name in sys.path:
    name = name.replace(prefix, '...')
    print(' ', name)

print()
print('IMPORTERS:')
for name, cache_value in sys.path_importer_cache.items():
    if '..' in name:
        name = os.path.abspath(name)
    name = name.replace(prefix, '...')
    print('  {}: {!r}'.format(name, cache_value))
```

使用一个 FileFinder 来标识文件系统上找到的路径位置。对于所有查找工具都不支持的路径，其中的位置与一个 None 值关联，因为它们不能用来导入模块。以下输出显示了缓存导入，不过由于版面限制，这里做了删减。

```
$ python3 sys_path_importer_cache.py

PATH:
  /Users/dhellmann/Documents/PyMOTW/Python3/pymotw-3/source/sys
  .../lib/python35.zip
  .../lib/python3.5
  .../lib/python3.5/plat-darwin
  .../lib/python3.5/lib-dynload
  .../lib/python3.5/site-packages

IMPORTERS:
  sys_path_importer_cache.py: None
  .../lib/python3.5/encodings: FileFinder(
'.../lib/python3.5/encodings')
  .../lib/python3.5/lib-dynload: FileFinder(
'.../lib/python3.5/lib-dynload')
  .../lib/python3.5/lib-dynload: FileFinder(
'.../lib/python3.5/lib-dynload')
  .../lib/python3.5/site-packages: FileFinder(
```

```
'.../lib/python3.5/site-packages')
.../lib/python3.5: FileFinder(
'.../lib/python3.5/')
.../lib/python3.5/plat-darwin: FileFinder(
'.../lib/python3.5/plat-darwin')
.../lib/python3.5: FileFinder(
'.../lib/python3.5')
.../lib/python35.zip: None
.../lib/python3.5/plat-darwin: FileFinder(
'.../lib/python3.5/plat-darwin')
```

17.2.6.11 元路径

`sys.meta_path` 进一步扩展了可能的导入来源，允许在扫描常规的 `sys.path` 之前先搜索查找工具。元路径上查找工具的 API 与常规路径上查找工具的 API 是一样的，只不过元查找工具不限于 `sys.path` 中的一项，它可以搜索任何地方。

代码清单 17-47：sys_meta_path.py

```python
import sys
import imp


class NoisyMetaImportFinder:

    def __init__(self, prefix):
        print('Creating NoisyMetaImportFinder for {}'.format(
            prefix))
        self.prefix = prefix
        return

    def find_module(self, fullname, path=None):
        print('looking for {!r} with path {!r}'.format(
            fullname, path))
        name_parts = fullname.split('.')
        if name_parts and name_parts[0] == self.prefix:
            print(' ... found prefix, returning loader')
            return NoisyMetaImportLoader(path)
        else:
            print(' ... not the right prefix, cannot load')
        return None


class NoisyMetaImportLoader:

    def __init__(self, path_entry):
        self.path_entry = path_entry
        return

    def load_module(self, fullname):
        print('loading {}'.format(fullname))
        if fullname in sys.modules:
            mod = sys.modules[fullname]
        else:
            mod = sys.modules.setdefault(
                fullname,
                imp.new_module(fullname))

        # Set a few properties required by PEP 302.
        mod.__file__ = fullname
        mod.__name__ = fullname
```

```python
        # Always looks like a package
        mod.__path__ = ['path-entry-goes-here']
        mod.__loader__ = self
        mod.__package__ = '.'.join(fullname.split('.')[:-1])

        return mod


# Install the meta-path finder.
sys.meta_path.append(NoisyMetaImportFinder('foo'))

# Import some modules that are "found" by the meta-path finder.
print()
import foo

print()
import foo.bar

# Import a module that is not found.
print()
try:
    import bar
except ImportError as e:
    pass
```

搜索 sys.path 之前，会询问元路径上的各个查找工具，所以总是可以让一个中心导入工具加载模块，而不必显式地修改 sys.path。一旦"找到"模块，加载工具 API 就会像常规加载工具一样正常工作（不过为简单起见，这个例子有所删减）。

```
$ python3 sys_meta_path.py

Creating NoisyMetaImportFinder for foo

looking for 'foo' with path None
 ... found prefix, returning loader
loading foo

looking for 'foo.bar' with path ['path-entry-goes-here']
 ... found prefix, returning loader
loading foo.bar

looking for 'bar' with path None
 ... not the right prefix, cannot load
```

提示：相关阅读材料

- **importlib**：创建定制导入工具的基类和其他工具。
- **zipimport**：从 ZIP 归档实现导入 Python 模块。
- The Internal Structure of Python Eggs[1]：egg 格式的 setuptools 文档。
- Wheel[2]：可安装 Python 代码的 wheel 归档格式的文档。
- **PEP 302**[3]：导入 hook。

---

[1] http://setuptools.readthedocs.io/en/latest/formats.html?highlight=egg

[2] http://wheel.readthedocs.org/en/latest/

[3] www.python.org/dev/peps/pep-0302

- **PEP 366**⊖：主模块显式相对导入。
- **PEP 427**⊖：Wheel 二进制包格式 1.0。
- Import this, that, and the other thing：custom importers⊜：Brett Cannon 的 PyCon 2010 演示文稿。

17.2.7 跟踪程序运行情况

有两种方法注入代码以监视一个程序的运行，包括跟踪（tracing）和性能分析（profiling）。它们很相似，不过分别有不同的用途，所以也有不同的约束。监视一个程序最容易也最低效的方法是通过一个跟踪 hook，可以用它来编写一个调试工具，监视代码覆盖，或者达到其他目的。

可以向 sys.settrace() 传递一个回调函数以修改跟踪 hook。这个回调接收 3 个参数：所运行代码的栈帧，命名通知类型的串，以及一个事件特定的参数值。表 17-2 列出了程序执行时出现的不同层次信息所对应的 7 个不同事件类型。

表 17-2　settrace() 的事件 hook

事件	何时出现	参数值
call	调用函数之前	None
line	执行一行代码之前	None
return	函数返回之前	返回的值
exception	出现一个异常之后	（异常，值，traceback）元组
c_call	调用一个 C 函数之前	C 函数对象
c_return	C 函数返回之后	None
c_exception	C 函数抛出一个错误之后	None

17.2.7.1　跟踪函数调用

在每个函数调用之前都会生成一个 call 事件。传入回调的帧可以用来查找其在调用哪个函数，以及从哪里调用。

代码清单 17-48：**sys_settrace_call.py**

```
1  #!/usr/bin/env python3
2  # encoding: utf-8
3
4  import sys
5
6
7  def trace_calls(frame, event, arg):
8      if event != 'call':
9          return
```

⊖ www.python.org/dev/peps/pep-0366
⊖ www.python.org/dev/peps/pep-0427
⊜ http://pyvideo.org/pycon-us-2010/pycon-2010--import-this--that--and-the-other-thin.html

```
10      co = frame.f_code
11      func_name = co.co_name
12      if func_name == 'write':
13          # Ignore write() calls from printing.
14          return
15      func_line_no = frame.f_lineno
16      func_filename = co.co_filename
17      caller = frame.f_back
18      caller_line_no = caller.f_lineno
19      caller_filename = caller.f_code.co_filename
20      print('* Call to', func_name)
21      print('*  on line {} of {}'.format(
22          func_line_no, func_filename))
23      print('*  from line {} of {}'.format(
24          caller_line_no, caller_filename))
25      return
26
27
28  def b():
29      print('inside b()\n')
30
31
32  def a():
33      print('inside a()\n')
34      b()
35
36  sys.settrace(trace_calls)
37  a()
```

这个例子忽略了 write() 调用,因为 print 会使用 write() 写至 sys.stdout。

```
$ python3 sys_settrace_call.py

* Call to a
*  on line 32 of sys_settrace_call.py
*  from line 37 of sys_settrace_call.py
inside a()

* Call to b
*  on line 28 of sys_settrace_call.py
*  from line 34 of sys_settrace_call.py
inside b()
```

17.2.7.2 跟踪内部函数

跟踪 hook 可以返回一个新 hook,并在新作用域中使用(局部跟踪函数)。例如,可以控制跟踪,使其只在某些模块或函数中逐行运行。

代码清单 17-49:`sys_settrace_line.py`

```
1  #!/usr/bin/env python3
2  # encoding: utf-8
3
4  import functools
5  import sys
6
7
8  def trace_lines(frame, event, arg):
9      if event != 'line':
10         return
11     co = frame.f_code
```

```
12      func_name = co.co_name
13      line_no = frame.f_lineno
14      print('*  {} line {}'.format(func_name, line_no))
15
16
17  def trace_calls(frame, event, arg, to_be_traced):
18      if event != 'call':
19          return
20      co = frame.f_code
21      func_name = co.co_name
22      if func_name == 'write':
23          # Ignore write() calls from printing.
24          return
25      line_no = frame.f_lineno
26      filename = co.co_filename
27      print('* Call to {} on line {} of {}'.format(
28          func_name, line_no, filename))
29      if func_name in to_be_traced:
30          # Trace into this function.
31          return trace_lines
32      return
33
34
35  def c(input):
36      print('input =', input)
37      print('Leaving c()')
38
39
40  def b(arg):
41      val = arg * 5
42      c(val)
43      print('Leaving b()')
44
45
46  def a():
47      b(2)
48      print('Leaving a()')
49
50
51  tracer = functools.partial(trace_calls, to_be_traced=['b'])
52  sys.settrace(tracer)
53  a()
```

在这个例子中，函数列表被保存在变量 `to_be_traced` 内，所以在 `trace_calls()` 运行时，它能返回 `trace_lines()` 以启用 `b()` 内部的跟踪。

```
$ python3 sys_settrace_line.py

* Call to a on line 46 of sys_settrace_line.py
* Call to b on line 40 of sys_settrace_line.py
*  b line 41
*  b line 42
* Call to c on line 35 of sys_settrace_line.py
input = 10
Leaving c()
*  b line 43
Leaving b()
Leaving a()
```

17.2.7.3 监视栈

使用 hook 的另一种方法是跟踪正在调用哪些函数，以及它们的返回值是什么。为了监视返回值，可以监视 return 事件。

代码清单 17-50：sys_settrace_return.py

```
1  #!/usr/bin/env python3
2  # encoding: utf-8
3
4  import sys
5
6
7  def trace_calls_and_returns(frame, event, arg):
8      co = frame.f_code
9      func_name = co.co_name
10     if func_name == 'write':
11         # Ignore write() calls from printing.
12         return
13     line_no = frame.f_lineno
14     filename = co.co_filename
15     if event == 'call':
16         print('* Call to {} on line {} of {}'.format(
17             func_name, line_no, filename))
18         return trace_calls_and_returns
19     elif event == 'return':
20         print('* {} => {}'.format(func_name, arg))
21     return
22
23
24 def b():
25     print('inside b()')
26     return 'response_from_b '
27
28
29 def a():
30     print('inside a()')
31     val = b()
32     return val * 2
33
34
35 sys.settrace(trace_calls_and_returns)
36 a()
```

局部跟踪函数用于监视返回事件，这说明调用一个函数时 trace_calls_and_returns() 需要返回自身的一个引用，从而能监视返回值。

```
$ python3 sys_settrace_return.py

* Call to a on line 29 of sys_settrace_return.py
inside a()
* Call to b on line 24 of sys_settrace_return.py
inside b()
* b => response_from_b
* a => response_from_b response_from_b
```

17.2.7.4 异常传播

可以在一个局部跟踪函数中查找 exception 事件以监视异常。出现异常时，会调用跟踪 hook 并提供一个元组，其中包含异常类型、异常对象和一个 traceback 对象。

代码清单 17-51：sys_settrace_exception.py

```
1  #!/usr/bin/env python3
```

```python
 2  # encoding: utf-8
 3
 4  import sys
 5
 6
 7  def trace_exceptions(frame, event, arg):
 8      if event != 'exception':
 9          return
10      co = frame.f_code
11      func_name = co.co_name
12      line_no = frame.f_lineno
13      exc_type, exc_value, exc_traceback = arg
14      print(('* Tracing exception:\n'
15             '* {} "{}"\n'
16             '* on line {} of {}\n').format(
17                 exc_type.__name__, exc_value, line_no,
18                 func_name))
19
20
21  def trace_calls(frame, event, arg):
22      if event != 'call':
23          return
24      co = frame.f_code
25      func_name = co.co_name
26      if func_name in TRACE_INTO:
27          return trace_exceptions
28
29
30  def c():
31      raise RuntimeError('generating exception in c()')
32
33
34  def b():
35      c()
36      print('Leaving b()')
37
38
39  def a():
40      b()
41      print('Leaving a()')
42
43
44  TRACE_INTO = ['a', 'b', 'c']
45
46  sys.settrace(trace_calls)
47  try:
48      a()
49  except Exception as e:
50      print('Exception handler:', e)
```

要注意应用局部函数的限制,因为格式化错误消息的一些内部函数会生成自己的异常并将其忽略。不论调用者是否捕获和忽略异常,每个异常都会被跟踪 hook 看到。

```
$ python3 sys_settrace_exception.py

* Tracing exception:
* RuntimeError "generating exception in c()"
* on line 31 of c

* Tracing exception:
* RuntimeError "generating exception in c()"
```

```
	* on line 35 of b
	* Tracing exception:
	* RuntimeError "generating exception in c()"
	* on line 40 of a
Exception handler: generating exception in c()
```

> **提示：相关阅读材料**
> - `profile`：`profile` 模块文档介绍了如何使用一个现成的性能分析工具。
> - `trace`：`trace` 模块实现了多个代码分析特性。
> - Types and members[一]：帧和代码对象及其属性的描述。
> - Tracing Python code[二]：另一个 `settrace()` 教程。
> - Wicked hack：Python bytecode tracing[三]：Ned Batchelder 完成的试验，使用比源代码行级更细的粒度进行跟踪。
> - smiley[四]：Python 应用跟踪工具。

> **提示：`sys` 模块的相关阅读材料**
> - `sys` 的标准库文档[五]。
> - `sys` 的 Python 2 到 Python 3 移植说明。

17.3 os：可移植访问操作系统特定特性

os 模块为平台特定的模块（如 posix、nt 和 mac）提供了一个包装器。所有平台上函数的 API 都应当是相同的，所以使用 os 模块可以提供一定的可移植性。不过，并不是所有函数在每一个平台上都可用。具体来说，这个总结中介绍的许多进程管理函数就对 Windows 不适用。

os 模块的 Python 文档的子标题是"杂类操作系统接口"。这个模块主要包括用于创建和管理运行进程或文件系统内容（文件和目录）的函数，只有很少一部分涉及其他功能。

17.3.1 检查文件系统内容

要准备文件系统上一个目录的内容列表，可以使用 `listdir()`。

代码清单 17-52：**os_listdir.py**

```
import os
import sys

print(os.listdir(sys.argv[1]))
```

[一] https://docs.python.org/3/library/inspect.html#types-and-members
[二] www.dalkescientific.com/writings/diary/archive/2005/04/20/tracing_python_code.html
[三] http://nedbatchelder.com/blog/200804/wicked_hack_python_bytecode_tracing.html
[四] https://pypi.python.org/pypi/smiley
[五] https://docs.python.org/3.5/library/sys.html

返回值是一个列表,其中包含给定目录中的所有命名成员,这里不会区分文件、子目录和符号链接。

```
$ python3 os_listdir.py .

['index.rst', 'os_access.py', 'os_cwd_example.py',
'os_directories.py', 'os_environ_example.py',
'os_exec_example.py', 'os_fork_example.py',
'os_kill_example.py', 'os_listdir.py', 'os_listdir.py~',
'os_process_id_example.py', 'os_process_user_example.py',
'os_rename_replace.py', 'os_rename_replace.py~',
'os_scandir.py', 'os_scandir.py~', 'os_spawn_example.py',
'os_stat.py', 'os_stat_chmod.py', 'os_stat_chmod_example.txt',
'os_strerror.py', 'os_strerror.py~', 'os_symlinks.py',
'os_system_background.py', 'os_system_example.py',
'os_system_shell.py', 'os_wait_example.py',
'os_waitpid_example.py', 'os_walk.py']
```

函数 walk() 会递归地遍历一个目录。对于每个子目录,它会生成一个 tuple,其中包含目录路径、该路径的直接子目录以及一个列表,这个列表包含该目录中所有文件的文件名。

代码清单 17-53:os_walk.py

```python
import os
import sys

# If we are not given a path to list, use /tmp.
if len(sys.argv) == 1:
    root = '/tmp'
else:
    root = sys.argv[1]

for dir_name, sub_dirs, files in os.walk(root):
    print(dir_name)
    # Make the subdirectory names stand out with /.
    sub_dirs = [n + '/' for n in sub_dirs]
    # Mix the directory contents together.
    contents = sub_dirs + files
    contents.sort()
    # Show the contents.
    for c in contents:
        print('  {}'.format(c))
    print()
```

该示例展示了一个递归目录列表。

```
$ python3 os_walk.py ../zipimport

../zipimport
  __init__.py
  example_package/
  index.rst
  zipimport_example.zip
  zipimport_find_module.py
  zipimport_get_code.py
  zipimport_get_data.py
  zipimport_get_data_nozip.py
  zipimport_get_data_zip.py
```

```
zipimport_get_source.py
zipimport_is_package.py
zipimport_load_module.py
zipimport_make_example.py

../zipimport/example_package
    README.txt
    __init__.py
```

如果除了文件名外还需要更多信息，那么 scandir() 可能比 listdir() 更高效：扫描目录时一个系统调用可以收集更多信息。

代码清单 17-54：**os_scandir.py**

```
import os
import sys

for entry in os.scandir(sys.argv[1]):
    if entry.is_dir():
        typ = 'dir'
    elif entry.is_file():
        typ = 'file'
    elif entry.is_symlink():
        typ = 'link'
    else:
        typ = 'unknown'
    print('{name} {typ}'.format(
        name=entry.name,
        typ=typ,
    ))
```

scandir() 会为目录中的元素返回一个 DirEntry 实例序列。这个对象提供了很多属性和方法来访问文件的元数据。

```
$ python3 os_scandir.py .

index.rst file
os_access.py file
os_cwd_example.py file
os_directories.py file
os_environ_example.py file
os_exec_example.py file
os_fork_example.py file
os_kill_example.py file
os_listdir.py file
os_listdir.py~ file
os_process_id_example.py file
os_process_user_example.py file
os_rename_replace.py file
os_rename_replace.py~ file
os_scandir.py file
os_scandir.py~ file
os_spawn_example.py file
os_stat.py file
os_stat_chmod.py file
os_stat_chmod_example.txt file
os_strerror.py file
os_strerror.py~ file
os_symlinks.py file
os_system_background.py file
```

```
os_system_example.py file
os_system_shell.py file
os_wait_example.py file
os_waitpid_example.py file
os_walk.py file
```

17.3.2 管理文件系统权限

可以使用 stat() 或 lstat() 访问关于文件的详细信息（lstat() 方法用于检查一个可能是符号链接的对象的状态）。

代码清单 17-55：os_stat.py

```python
import os
import sys
import time

if len(sys.argv) == 1:
    filename = __file__
else:
    filename = sys.argv[1]

stat_info = os.stat(filename)

print('os.stat({}):'.format(filename))
print('  Size:', stat_info.st_size)
print('  Permissions:', oct(stat_info.st_mode))
print('  Owner:', stat_info.st_uid)
print('  Device:', stat_info.st_dev)
print('  Created      :', time.ctime(stat_info.st_ctime))
print('  Last modified:', time.ctime(stat_info.st_mtime))
print('  Last accessed:', time.ctime(stat_info.st_atime))
```

取决于示例代码如何安装，输出可能有变化。可以试用这个函数，尝试在命令行上向 os_stat.py 传递不同的文件名。

```
$ python3 os_stat.py

os.stat(os_stat.py):
  Size: 593
  Permissions: 0o100644
  Owner: 527
  Device: 16777218
  Created      : Sat Dec 17 12:09:51 2016
  Last modified: Sat Dec 17 12:09:51 2016
  Last accessed: Sat Dec 31 12:33:19 2016

$ python3 os_stat.py index.rst

os.stat(index.rst):
  Size: 26878
  Permissions: 0o100644
  Owner: 527
  Device: 16777218
  Created      : Sat Dec 31 12:33:10 2016
  Last modified: Sat Dec 31 12:33:10 2016
  Last accessed: Sat Dec 31 12:33:19 2016
```

在类 UNIX 的系统上，可以使用 chmod() 改变文件权限，只需传入模式（一个整数）。

模式值可以使用 stat 模块中定义的常量来构造。下面这个例子会来回切换用户的执行权限位。

代码清单 17-56：**os_stat_chmod.py**

```
import os
import stat

filename = 'os_stat_chmod_example.txt'
if os.path.exists(filename):
    os.unlink(filename)
with open(filename, 'wt') as f:
    f.write('contents')

# Determine which permissions are already set using stat.
existing_permissions = stat.S_IMODE(os.stat(filename).st_mode)

if not os.access(filename, os.X_OK):
    print('Adding execute permission')
    new_permissions = existing_permissions | stat.S_IXUSR
else:
    print('Removing execute permission')
    # Use xor to remove the user execute permission.
    new_permissions = existing_permissions ^ stat.S_IXUSR

os.chmod(filename, new_permissions)
```

这个脚本假设其有所需的权限，可以在运行时修改文件的模式。

```
$ python3 os_stat_chmod.py

Adding execute permission
```

可以使用函数 `access()` 来测试一个进程对文件的访问权限。

代码清单 17-57：**os_access.py**

```
import os

print('Testing:', __file__)
print('Exists:', os.access(__file__, os.F_OK))
print('Readable:', os.access(__file__, os.R_OK))
print('Writable:', os.access(__file__, os.W_OK))
print('Executable:', os.access(__file__, os.X_OK))
```

取决于示例代码如何安装，结果会有所变化，不过输出可能与下述类似。

```
$ python3 os_access.py

Testing: os_access.py
Exists: True
Readable: True
Writable: True
Executable: False
```

`access()` 的库文档给出了两个特殊的警告。首先，对一个文件具体调用 `open()` 之前，调用 `access()` 检查这个文件是否能够打开并没有太大意义。这两个调用之间有一个很小（但确实存在）的时间窗，在此期间文件权限可能改变。另一个警告主要适用于扩展了

POSIX 权限语义的网络文件系统。有些类型的文件系统响应 POSIX 调用时，可能会指出一个进程有权限访问一个文件，但是当试图使用 `open()` 打开文件时，却由于未通过 POSIX 调用测试的原因报告一个失败。更好的策略是基于所需的模式调用 `open()`，如果出现问题就捕获产生的 `IOError`。

17.3.3 创建和删除目录

有很多函数可以用来处理文件系统上的目录，包括创建内容、列出内容和删除目录。

代码清单 17-58: **os_directories.py**

```
import os

dir_name = 'os_directories_example'

print('Creating', dir_name)
os.makedirs(dir_name)

file_name = os.path.join(dir_name, 'example.txt')
print('Creating', file_name)
with open(file_name, 'wt') as f:
    f.write('example file')

print('Cleaning up')
os.unlink(file_name)
os.rmdir(dir_name)
```

有两组用于创建和删除目录的函数。当使用 `mkdir()` 创建一个新目录时，所有父目录都必须存在。用 `rmdir()` 删除一个目录时，实际上只会删除叶子目录（路径的最后一部分）。与此相反，`makedirs()` 和 `removedirs()` 会处理路径中的所有节点。`makedirs()` 会创建路径上所有不存在的部分，`removedirs()` 将删除所有父目录（只要它们为空）。

```
$ python3 os_directories.py

Creating os_directories_example
Creating os_directories_example/example.txt
Cleaning up
```

17.3.4 处理符号链接

对于支持符号链接（symlink）的平台和文件系统，还有一些相应的处理函数。

代码清单 17-59: **os_symlinks.py**

```
import os

link_name = '/tmp/' + os.path.basename(__file__)

print('Creating link {} -> {}'.format(link_name, __file__))
os.symlink(__file__, link_name)

stat_info = os.lstat(link_name)
print('Permissions:', oct(stat_info.st_mode))

print('Points to:', os.readlink(link_name))
```

```
# Clean up.
os.unlink(link_name)
```

使用 symlink() 可以创建一个符号链接，使用 readlink() 可以读取符号链接以确定该链接指示的原始文件。lstat() 函数类似于 stat()，不过它处理的是符号链接。

```
$ python3 os_symlinks.py

Creating link /tmp/os_symlinks.py -> os_symlinks.py
Permissions: 0o120755
Points to: os_symlinks.py
```

17.3.5 安全地替换现有文件

替换或重命名一个现有文件不是幂等操作（安全操作），可能会导致应用出现竞态条件。rename() 和 replace() 函数为这些动作实现了安全算法，会尽可能使用 POSIX 兼容系统上的原子操作。

代码清单 17-60：**os_rename_replace.py**

```python
import glob
import os

with open('rename_start.txt', 'w') as f:
    f.write('starting as rename_start.txt')

print('Starting:', glob.glob('rename*.txt'))
os.rename('rename_start.txt', 'rename_finish.txt')

print('After rename:', glob.glob('rename*.txt'))

with open('rename_finish.txt', 'r') as f:
    print('Contents:', repr(f.read()))

with open('rename_new_contents.txt', 'w') as f:
    f.write('ending with contents of rename_new_contents.txt')

os.replace('rename_new_contents.txt', 'rename_finish.txt')

with open('rename_finish.txt', 'r') as f:
    print('After replace:', repr(f.read()))

for name in glob.glob('rename*.txt'):
    os.unlink(name)
```

大多数情况下，rename() 和 replace() 函数都可以跨文件系统使用。如果文件移动到一个新的文件系统，或者目标已经存在，那么重命名文件可能失败。

```
$ python3 os_rename_replace.py

Starting: ['rename_start.txt']
After rename: ['rename_finish.txt']
Contents: 'starting as rename_start.txt'
After replace: 'ending with contents of rename_new_contents.txt'
```

17.3.6 检测和改变进程所有者

os 提供的下一组函数用于确定和改变进程所有者 ID。守护进程或一些特殊系统程序（需要改变权限级别而不是作为 root 运行）的作者最常使用这些函数。这一节不打算解释 UNIX 安全、进程所有者和其他进程有关问题的所有复杂细节。有关的更多详细内容请参考本节最后的参考文献列表。

下面的例子显示了一个进程的有效用户和组信息，然后改变这些有效值。这类似于系统自引导期间一个守护进程作为 root 启动时可能要做的工作，比如降低权限等级，以及作为一个不同的用户运行。

说明：运行这个例子之前，要改变 TEST_GID 和 TEST_UID 值，从而与系统上定义的一个真实用户一致。

代码清单 17-61：os_process_user_example.py

```python
import os

TEST_GID = 502
TEST_UID = 502


def show_user_info():
    print('User  (actual/effective)  : {} / {}'.format(
        os.getuid(), os.geteuid()))
    print('Group (actual/effective) : {} / {}'.format(
        os.getgid(), os.getegid()))
    print('Actual Groups   :', os.getgroups())


print('BEFORE CHANGE:')
show_user_info()
print()

try:
    os.setegid(TEST_GID)
except OSError:
    print('ERROR: Could not change effective group. '
          'Rerun as root.')
else:
    print('CHANGE GROUP:')
    show_user_info()
    print()

try:
    os.seteuid(TEST_UID)
except OSError:
    print('ERROR: Could not change effective user. '
          'Rerun as root.')
else:
    print('CHANGE USER:')
    show_user_info()
    print()
```

在 OS X 上，作为 ID 为 502、组为 502 的用户运行时，这个代码会生成以下输出。

```
$ python3 os_process_user_example.py

BEFORE CHANGE:
User (actual/effective)  : 527 / 527
Group (actual/effective) : 501 / 501
Actual Groups   : [501, 701, 402, 702, 500, 12, 61, 80, 98, 398,
399, 33, 100, 204, 395]

ERROR: Could not change effective group. Rerun as root.
ERROR: Could not change effective user. Rerun as root.
```

这些值不会改变,因为只要不作为 root 运行,进程就不能改变其有效的所有者值。试图将有效用户 ID 或组 ID 设置为非当前用户的其他值时,会导致一个 OSError。使用 sudo 运行同样的脚本,使它启动时有 root 权限,结果则会完全不同。

```
$ sudo python3 os_process_user_example.py

BEFORE CHANGE:

User (actual/effective)  : 0 / 0
Group (actual/effective) : 0 / 0
Actual Groups : [0, 1, 2, 3, 4, 5, 8, 9, 12, 20, 29, 61, 80,
702, 33, 98, 100, 204, 395, 398, 399, 701]

CHANGE GROUP:
User (actual/effective)  : 0 / 0
Group (actual/effective) : 0 / 502
Actual Groups   : [0, 1, 2, 3, 4, 5, 8, 9, 12, 20, 29, 61, 80,
702, 33, 98, 100, 204, 395, 398, 399, 701]

CHANGE USER:
User (actual/effective)  : 0 / 502
Group (actual/effective) : 0 / 502
Actual Groups   : [0, 1, 2, 3, 4, 5, 8, 9, 12, 20, 29, 61, 80,
702, 33, 98, 100, 204, 395, 398, 399, 701]
```

在这种情况下,由于它作为 root 启动,所以脚本可以改变进程的有效用户和组。一旦改变了有效 UID,进程则被限于该用户的权限。由于非根用户不能改变其有效组,所以程序在改变用户之前需要先改变组。

17.3.7 管理进程环境

操作系统通过 os 模块为程序提供的另一个特性是环境。环境中设置的变量作为字符串可见,可以通过 os.environ 或 getenv() 来读取这些字符串。环境变量通常用于配置值,如搜索路径、文件位置和调试标志。下面这个例子显示了如何获取一个环境变量,并将一个值传递到一个子进程。

代码清单 17-62:os_environ_example.py

```python
import os

print('Initial value:', os.environ.get('TESTVAR', None))
print('Child process:')
os.system('echo $TESTVAR')

os.environ['TESTVAR'] = 'THIS VALUE WAS CHANGED'
```

```
print()
print('Changed value:', os.environ['TESTVAR'])
print('Child process:')
os.system('echo $TESTVAR')

del os.environ['TESTVAR']

print()
print('Removed value:', os.environ.get('TESTVAR', None))
print('Child process:')
os.system('echo $TESTVAR')
```

`os.environ` 对象使用标准 Python 映射 API 来获取和设置值。对 `os.environ` 的改变会导出到子进程。

```
$ python3 -u os_environ_example.py

Initial value: None
Child process:

Changed value: THIS VALUE WAS CHANGED
Child process:
THIS VALUE WAS CHANGED

Removed value: None
Child process:
```

17.3.8 管理进程工作目录

如果操作系统拥有具有层次结构的文件系统，那么其会有一个"当前工作目录"（current working directory）的概念，也就是说，在使用相对路径访问文件时，进程将使用文件系统上的这个目录作为起始位置。当前工作目录可以用 `getcwd()` 获取，用 `chdir()` 改变。

代码清单 17-63：os_cwd_example.py

```
import os

print('Starting:', os.getcwd())

print('Moving up one:', os.pardir)
os.chdir(os.pardir)

print('After move:', os.getcwd())
```

利用 `os.curdir` 和 `os.pardir` 可以分别采用一种可移植的方式来指示当前目录和父目录。

```
$ python3 os_cwd_example.py

Starting: .../pymotw-3/source/os
Moving up one: ..
After move: .../pymotw-3/source
```

17.3.9 运行外部命令

> **警告**：对于处理进程的很多函数，其可移植性都很有限。要采用一种更一致的方法以一种平台独立的方式处理进程，请参考 `subprocess` 模块。

要运行一个单独的命令而不与之交互,最基本的方法就是使用 `system()` 函数。它有一个字符串参数,也就是命令行,其将由运行 shell 的子进程执行。

代码清单 17-64:**os_system_example.py**

```
import os

# Simple command
os.system('pwd')
```

`system()` 的返回值是运行这个程序的 shell 的退出值,被包装为一个 16 位数字,这个数的高字节为退出状态,低字节是导致进程结束的信号数或 0。

```
$ python3 -u os_system_example.py

.../pymotw-3/source/os
```

由于命令被直接传递到 shell 来处理,所以它可以包含 shell 语法,如文件名模式匹配或环境变量。

代码清单 17-65:**os_system_shell.py**

```
import os

# Command with shell expansion
os.system('echo $TMPDIR')
```

shell 运行命令行时,这个串中的环境变量 $TMPDIR 会展开。

```
$ python3 -u os_system_shell.py

/var/folders/5q/8gk0wq888xlggz008k8dr7180000hg/T/
```

除非显式地在后台运行命令,否则 `system()` 调用会阻塞,直至完成。子进程的标准输入、输出和错误通道默认被绑定到调用者拥有的适当的流中,不过可以使用 shell 语法重定向。

代码清单 17-66:**os_system_background.py**

```
import os
import time

print('Calling...')
os.system('date; (sleep 3; date) &')

print('Sleeping...')
time.sleep(5)
```

但是这会让 shell 很麻烦,还有更好的方法来达到同样的目的。

```
$ python3 -u os_system_background.py

Calling...
Sat Dec 31 12:33:20 EST 2016
Sleeping...
Sat Dec 31 12:33:23 EST 2016
```

17.3.10 用 `os.fork()` 创建进程

通过 os 模块提供了 POSIX 函数 `fork()` 和 `exec()`（在 Mac OS X、Linux 和其他 UNIX 系统上都可用）。有一些书用所有内容来介绍如何可靠地使用这些函数，所以不要仅限于这里给出的介绍，请参考库或者去书店了解更多细节。

要创建一个新进程作为当前进程的一个克隆，可以使用 `fork()`。

代码清单 17-67：`os_fork_example.py`

```python
import os

pid = os.fork()

if pid:
    print('Child process id:', pid)
else:
    print('I am the child')
```

输出会根据每次运行例子时的系统状态而改变，不过基本上类似下面的输出。

```
$ python3 -u os_fork_example.py

Child process id: 29190
I am the child
```

创建子进程之后，这两个进程运行同样的代码。程序要想分辨是在父进程还是在子进程中运行，需要检查 `fork()` 的返回值。如果值是 0，则当前进程是子进程，如果不是 0，则说明程序在父进程中运行，返回值就是子进程的进程 ID。

代码清单 17-68：`os_kill_example.py`

```python
import os
import signal
import time

def signal_usr1(signum, frame):
    "Callback invoked when a signal is received"
    pid = os.getpid()
    print('Received USR1 in process {}'.format(pid))

print('Forking...')
child_pid = os.fork()
if child_pid:
    print('PARENT: Pausing before sending signal...')
    time.sleep(1)
    print('PARENT: Signaling {}'.format(child_pid))
    os.kill(child_pid, signal.SIGUSR1)
else:
    print('CHILD: Setting up signal handler')
    signal.signal(signal.SIGUSR1, signal_usr1)
    print('CHILD: Pausing to wait for signal')
    time.sleep(5)
```

父进程可以使用 `kill()` 和 `signal` 模块向子进程发送信号。首先，定义一个信号处理器，接收到这个信号时会调用这个处理器。然后调用 `fork()`，在父进程中，使用

kill()发送一个USR1信号之前会暂停很短一段时间。这个例子使用了这个短暂的暂停，使子进程有时间建立信号处理器。实际应用不需要（或不希望）调用sleep()。在子进程中，要建立信号处理器，并休眠一段时间，使父进程有足够的时间发送信号。

```
$ python3 -u os_kill_example.py

Forking...
PARENT: Pausing before sending signal...
CHILD: Setting up signal handler
CHILD: Pausing to wait for signal
PARENT: Signaling 29193
Received USR1 in process 29193
```

要在子进程中处理不同的行为，一种简单的方法是检查fork()的返回值并建立分支。要实现更复杂的行为，可能需要完成更多代码分离而不只是一个简单的分支。另外一些情况下，还可能需要包装现有的程序。对于这两类情况，可以用exec*()系列的函数运行另一个程序。

代码清单17-69：os_exec_example.py

```
import os

child_pid = os.fork()
if child_pid:
    os.waitpid(child_pid, 0)
else:
    os.execlp('pwd', 'pwd', '-P')
```

由exec()运行程序时，这个程序的代码会替换现有进程的代码。

```
$ python3 os_exec_example.py

.../pymotw-3/source/os
```

exec()有很多变种，使用哪个变种取决于参数的形式、父进程的路径和环境是否要复制到子进程以及其他因素。对于所有这些变种，第一个参数都是路径或文件名，其余的参数将控制程序如何运行。参数可能作为命令行参数传递，或者会覆盖进程"环境"（参见os.environ和os.getenv）。请参考库文档来全面了解完整的详细信息。

17.3.11 等待子进程

很多包含大量计算的程序会使用多个进程，以绕开Python的线程限制和全局解释器锁。当启动多个进程来运行不同任务时，主进程开始新的子进程之前需要等待一个或多个子进程完成，以避免服务器负载过大。为此，取决于具体情况，可以使用wait()和相关函数来实现。

如果哪个子进程最先退出并不重要，那么可以使用wait()。一旦有子进程退出它就会返回。

代码清单17-70：os_wait_example.py

```
import os
import sys
import time
```

```
for i in range(2):
    print('PARENT {}: Forking {}'.format(os.getpid(), i))
    worker_pid = os.fork()
    if not worker_pid:
        print('WORKER {}: Starting'.format(i))
        time.sleep(2 + i)
        print('WORKER {}: Finishing'.format(i))
        sys.exit(i)

for i in range(2):
    print('PARENT: Waiting for {}'.format(i))
    done = os.wait()
    print('PARENT: Child done:', done)
```

`wait()` 的返回值是一个元组，其中包含进程 ID 和退出状态（被组合为一个 16 位值）。低字节是结束进程的信号编号，高字节是退出时进程返回的状态码。

```
$ python3 -u os_wait_example.py

PARENT 29202: Forking 0
PARENT 29202: Forking 1
PARENT: Waiting for 0
WORKER 0: Starting
WORKER 1: Starting
WORKER 0: Finishing
PARENT: Child done: (29203, 0)
PARENT: Waiting for 1
WORKER 1: Finishing
PARENT: Child done: (29204, 256)
```

要等待一个特定的进程，可以使用 `waitpid()`。

代码清单 17-71: **os_waitpid_example.py**

```
import os
import sys
import time

workers = []
for i in range(2):
    print('PARENT {}: Forking {}'.format(os.getpid(), i))
    worker_pid = os.fork()
    if not worker_pid:
        print('WORKER {}: Starting'.format(i))
        time.sleep(2 + i)
        print('WORKER {}: Finishing'.format(i))
        sys.exit(i)
    workers.append(worker_pid)

for pid in workers:
    print('PARENT: Waiting for {}'.format(pid))
    done = os.waitpid(pid, 0)
    print('PARENT: Child done:', done)
```

传入目标进程的进程 ID。`waitpid()` 会阻塞，直至该进程退出。

```
$ python3 -u os_waitpid_example.py

PARENT 29211: Forking 0
PARENT 29211: Forking 1
```

```
PARENT: Waiting for 29212
WORKER 0: Starting
WORKER 1: Starting
WORKER 0: Finishing
PARENT: Child done: (29212, 0)
PARENT: Waiting for 29213
WORKER 1: Finishing
PARENT: Child done: (29213, 256)
```

`wait3()` 和 `wait4()` 采用类似的方式，不过会返回关于子进程的更详细的信息，包括进程 ID、退出状态和资源使用情况。

17.3.12 Spawn 创建新进程

为提供便利，`spawn()` 系列函数可以在一个语句中一次完成 `fork()` 和 `exec()` 处理。

代码清单 17-72：`os_spawn_example.py`

```python
import os

os.spawnlp(os.P_WAIT, 'pwd', 'pwd', '-P')
```

第一个参数是一个模式，指示返回之前是否等待进程完成。这个例子会等待。使用 `P_NOWAIT` 允许另一个进程开始，不过之后会恢复执行当前进程。

```
$ python3 os_spawn_example.py

.../pymotw-3/source/os
```

17.3.13 操作系统错误码

操作系统定义的错误码（在 `errno` 模块中管理）可以使用 `strerror()` 转换为消息字符串。

代码清单 17-73：`os_strerror.py`

```python
import errno
import os

for num in [errno.ENOENT, errno.EINTR, errno.EBUSY]:
    name = errno.errorcode[num]
    print('[{num:>2}] {name:<6}: {msg}'.format(
        name=name, num=num, msg=os.strerror(num)))
```

下面的输出显示了常见错误码的相关消息。

```
$ python3 os_strerror.py

[ 2] ENOENT: No such file or directory
[ 4] EINTR : Interrupted system call
[16] EBUSY : Resource busy
```

提示：相关阅读材料

- `os` 的标准库文档[⊖]。

⊖ https://docs.python.org/3.5/library/os.html

- os 的 Python 2 到 Python 3 移植说明。
- signal：关于 signal 模块的部分更详细地介绍信号处理技术。
- subprocess：subprocess 模块取代了 os.popen()。
- multiprocessing：multiprocessing 模块使额外进程的处理更为容易。
- tempfile：tempfile 模块用于处理临时文件。
- 6.7.3 节：shutil 模块还包括一些处理目录树的函数。
- Speaking UNIX, Part 8[①]：了解 UNIX 如何实现多任务。
- Wikipedia：Standard streams[②]：对 stdin、stdout 和 stderr 的更多讨论。
- Delve into Unix Process Creation[③]：解释 UNIX 进程的生命周期。
- 《Advanced Programming in the UNIX Environment》是由 W. Richard Stevens 和 Stephen A. Rago 所著，由 Addison-Wesley Professional 于 2005 出版（ISBN-10:0201433079）。这本书介绍了如何使用多进程，如处理信号、关闭重复的文件描述符等。

17.4 platform：系统版本信息

尽管 Python 通常被用作一个跨平台的语言，但有时还是有必要知道程序在哪种系统上运行。构建工具需要这个信息，另外应用可能也需要知道它使用的一些库或外部命令在不同操作系统上有不同的接口。例如，一个管理操作系统网络配置的工具可能对网络接口、别名、IP 地址和其他 OS 特定的信息定义了可移植的表示。不过，在编辑配置文件时，它必须对主机有更多了解，从而能使用正确的操作系统配置命令和文件。platform 模块提供了一些工具来了解运行程序的解释器、操作系统和硬件平台。

说明：这一节的示例输出是在 3 个系统上生成的：一个是运行 OS X 10.11.6 的 Mac mini；一个是运行 Ubuntu Linux 14.04 的 Dell PC；还有一个是运行 Windows 10 的 VirtualBox VM。在 OS X 和 Windows 系统上使用 python.org 的预编译安装工具安装 Python。Linux 系统运行系统包中的一个版本。

17.4.1 解释器

有 4 个函数可以得到当前 Python 解释器的有关信息。python_version() 和 python_version_tuple() 可以返回不同形式的解释器版本，包括主版本、次版本和补丁级组件。python_compiler() 会报告构建解释器所用的编译器。python_build() 将给出解释器构建的版本串。

代码清单 17-74：platform_python.py

```
import platform
```

[①] www.ibm.com/developerworks/aix/library/au-speakingunix8/index.html
[②] https://en.wikipedia.org/wiki/Standard_streams
[③] www.ibm.com/developerworks/aix/library/au-unixprocess.html

```
print('Version        :', platform.python_version())
print('Version tuple:', platform.python_version_tuple())
print('Compiler       :', platform.python_compiler())
print('Build          :', platform.python_build())
```

OS X：

```
$ python3 platform_python.py

Version       : 3.5.2
Version tuple: ('3', '5', '2')
Compiler      : GCC 4.2.1 (Apple Inc. build 5666) (dot 3)
Build         : ('v3.5.2:4def2a2901a5', 'Jun 26 2016 10:47:25')
```

Linux：

```
$ python3 platform_python.py

Version       : 3.5.2
Version tuple: ('3', '5', '2')
Compiler      : GCC 4.8.4
Build         : ('default', 'Jul 17 2016 00:00:00')
```

Windows：

```
C:\>Desktop\platform_python.py

Version       : 3.5.1
Version tuple: ('3', '5', '1')
Compiler      : MSC v.1900 64 bit (AMD64)
Build         : ('v3.5.1:37a07cee5969', 'Dec  6 2015 01:54:25')
```

17.4.2 平台

platform() 函数返回一个字符串，其中包含一个通用的平台标识符。这个函数接受两个可选的布尔参数。如果 aliased 为 True，则返回值中的名会从一个正式名转换为更常用的格式。如果 terse 为 true，则会返回一个最小值，即去除某些部分，而不是返回完整的串。

代码清单 17-75：platform_platform.py

```
import platform

print('Normal :', platform.platform())
print('Aliased:', platform.platform(aliased=True))
print('Terse  :', platform.platform(terse=True))
```

OS X：

```
$ python3 platform_platform.py

Normal : Darwin-15.6.0-x86_64-i386-64bit
Aliased: Darwin-15.6.0-x86_64-i386-64bit
Terse  : Darwin-15.6.0
```

Linux：

```
$ python3 platform_platform.py

Normal : Linux-3.13.0-55-generic-x86_64-with-Ubuntu-14.04-trusty
Aliased: Linux-3.13.0-55-generic-x86_64-with-Ubuntu-14.04-trusty
Terse  : Linux-3.13.0-55-generic-x86_64-with-glibc2.9
```

Windows：

```
C:\>platform_platform.py

Normal  : Windows-10-10.0.10240-SP0
Aliased : Windows-10-10.0.10240-SP0
Terse   : Windows-10
```

17.4.3 操作系统和硬件信息

还可以得到运行解释器的操作系统和硬件的更多详细信息。uname()返回一个元组，其中包含系统、节点、发行号、版本、机器和处理器值。可以通过同名的函数访问各个值，如表17-3所列。

表 17-3 平台信息函数

函　　数	返回值
system()	操作系统名
node()	服务器主机名，不是完全限定名
release()	操作系统发行号
version()	更详细的系统版本信息
machine()	硬件类型标识符，如 'i386'
processor()	处理器实际标识符（有些情况下与 machine() 值相同）

代码清单 17-76：platform_os_info.py

```python
import platform

print('uname:', platform.uname())

print()
print('system   :', platform.system())
print('node     :', platform.node())
print('release  :', platform.release())
print('version  :', platform.version())
print('machine  :', platform.machine())
print('processor:', platform.processor())
```

OS X：

```
$ python3 platform_os_info.py

uname: uname_result(system='Darwin', node='hubert.local',
release='15.6.0', version='Darwin Kernel Version 15.6.0: Thu Jun
23 18:25:34 PDT 2016; root:xnu-3248.60.10~1/RELEASE_X86_64',
machine='x86_64', processor='i386')

system    : Darwin
node      : hubert.local
release   : 15.6.0
version   : Darwin Kernel Version 15.6.0: Thu Jun 23 18:25:34 PDT
2016; root:xnu-3248.60.10~1/RELEASE_X86_64
machine   : x86_64
processor : i386
```

Linux：

```
$ python3 platform_os_info.py

uname: uname_result(system='Linux', node='apu',
release='3.13.0-55-generic', version='#94-Ubuntu SMP Thu Jun 18
00:27:10 UTC 2015', machine='x86_64', processor='x86_64')

system   : Linux
node     : apu
release  : 3.13.0-55-generic
version  : #94-Ubuntu SMP Thu Jun 18 00:27:10 UTC 2015
machine  : x86_64
processor: x86_64
```

Windows：

```
C:\>Desktop\platform_os_info.py

uname: uname_result(system='Windows', node='IE11WIN10',
release='10', version='10.0.10240', machine='AMD64',
processor='Intel64 Family 6 Model 70 Stepping 1, GenuineIntel')

system   : Windows
node     : IE11WIN10
release  : 10
version  : 10.0.10240
machine  : AMD64
processor: Intel64 Family 6 Model 70 Stepping 1, GenuineIntel
```

17.4.4 可执行程序体系结构

可以使用 `architecture()` 函数查看程序的体系结构信息。第一个参数是可执行程序的路径（默认为 `sys.executable`，即 Python 解释器）。返回值是一个元组，包含位体系结构和使用的链接格式。

代码清单 17-77：**platform_architecture.py**

```
import platform

print('interpreter:', platform.architecture())
print('/bin/ls    :', platform.architecture('/bin/ls'))
```

OS X：

```
$ python3 platform_architecture.py

interpreter: ('64bit', '')
/bin/ls    : ('64bit', '')
```

Linux：

```
$ python3 platform_architecture.py

interpreter: ('64bit', 'ELF')
/bin/ls    : ('64bit', 'ELF')
```

Windows：

```
C:\>Desktop\platform_architecture.py
```

```
interpreter: ('64bit', 'WindowsPE')
/bin/ls    : ('64bit', '')
```

提示：相关阅读材料
- platform 的标准库文档[⊖]。
- platform 的 Python 2 到 Python 3 移植说明。

17.5 resource：系统资源管理

resource 中的函数可以检查一个进程消耗的当前系统资源，并做出限制，以控制一个程序对系统可能增加的负载。

17.5.1 当前使用情况

使用 getrusage() 来查看当前进程和其子进程使用的资源。返回值是一个数据结构，其中包含当前系统状态下的一些资源度量。

说明：这里并没有显示所收集的全部资源值。更完整的列表可以参考 resource 的标准库文档。

代码清单 17-78：resource_getrusage.py

```python
import resource
import time

RESOURCES = [
    ('ru_utime', 'User time'),
    ('ru_stime', 'System time'),
    ('ru_maxrss', 'Max. Resident Set Size'),
    ('ru_ixrss', 'Shared Memory Size'),
    ('ru_idrss', 'Unshared Memory Size'),
    ('ru_isrss', 'Stack Size'),
    ('ru_inblock', 'Block inputs'),
    ('ru_oublock', 'Block outputs'),
]

usage = resource.getrusage(resource.RUSAGE_SELF)

for name, desc in RESOURCES:
    print('{:<25} ({:<10}) = {}'.format(
        desc, name, getattr(usage, name)))
```

由于这个测试程序极其简单，所以它没有使用太多资源。

```
$ python3 resource_getrusage.py

User time                 (ru_utime  ) = 0.021876
System time               (ru_stime  ) = 0.006726999999999995
Max. Resident Set Size    (ru_maxrss ) = 6479872
Shared Memory Size        (ru_ixrss  ) = 0
Unshared Memory Size      (ru_idrss  ) = 0
```

[⊖] https://docs.python.org/3.5/library/platform.html

```
Stack Size              (ru_isrss  ) = 0
Block inputs            (ru_inblock) = 0
Block outputs           (ru_oublock) = 0
```

17.5.2 资源限制

除了当前实际使用的资源外,还可以检查对应用的限制,然后做出修改。

代码清单 17-79: `resource_getrlimit.py`

```python
import resource

LIMITS = [
    ('RLIMIT_CORE', 'core file size'),
    ('RLIMIT_CPU', 'CPU time'),
    ('RLIMIT_FSIZE', 'file size'),
    ('RLIMIT_DATA', 'heap size'),
    ('RLIMIT_STACK', 'stack size'),
    ('RLIMIT_RSS', 'resident set size'),
    ('RLIMIT_NPROC', 'number of processes'),
    ('RLIMIT_NOFILE', 'number of open files'),
    ('RLIMIT_MEMLOCK', 'lockable memory address'),
]

print('Resource limits (soft/hard):')
for name, desc in LIMITS:
    limit_num = getattr(resource, name)
    soft, hard = resource.getrlimit(limit_num)
    print('{:<23} {}/{}'.format(desc, soft, hard))
```

对应各个限制的返回值分别是一个元组,包含当前配置的软(soft)限制,以及操作系统的硬(hard)限制。

```
$ python3 resource_getrlimit.py

Resource limits (soft/hard):
core file size          0/9223372036854775807
CPU time                9223372036854775807/9223372036854775807
file size               9223372036854775807/9223372036854775807
heap size               9223372036854775807/9223372036854775807
stack size              8388608/67104768
resident set size       9223372036854775807/9223372036854775807
number of processes     709/1064
number of open files    7168/9223372036854775807
lockable memory address 9223372036854775807/9223372036854775807
```

可以用 `setrlimit()` 改变限制。

代码清单 17-80: `resource_setrlimit_nofile.py`

```python
import resource
import os

soft, hard = resource.getrlimit(resource.RLIMIT_NOFILE)
print('Soft limit starts as  :', soft)

resource.setrlimit(resource.RLIMIT_NOFILE, (4, hard))

soft, hard = resource.getrlimit(resource.RLIMIT_NOFILE)
print('Soft limit changed to :', soft)
```

```python
random = open('/dev/random', 'r')
print('random has fd =', random.fileno())
try:
    null = open('/dev/null', 'w')
except IOError as err:
    print(err)
else:
    print('null has fd =', null.fileno())
```

这个例子使用 **RLIMIT_NOFILE** 来控制允许打开的文件数，将它改为比默认值小的一个软限制。

```
$ python3 resource_setrlimit_nofile.py

Soft limit starts as   : 7168
Soft limit changed to  : 4
random has fd = 3
[Errno 24] Too many open files: '/dev/null'
```

限制一个进程消耗的 CPU 时间可能也很有用，可以避免一个进程耗费太多 CPU 时间。当进程运行的时间超过所分配的时间时，会向它发送一个 **SIGXCPU** 信号。

代码清单 17-81：**resource_setrlimit_cpu.py**

```python
import resource
import sys
import signal
import time


# Set up a signal handler to notify us
# when we run out of time.
def time_expired(n, stack):
    print('EXPIRED :', time.ctime())
    raise SystemExit('(time ran out)')
signal.signal(signal.SIGXCPU, time_expired)

# Adjust the CPU time limit.
soft, hard = resource.getrlimit(resource.RLIMIT_CPU)
print('Soft limit starts as  :', soft)

resource.setrlimit(resource.RLIMIT_CPU, (1, hard))

soft, hard = resource.getrlimit(resource.RLIMIT_CPU)
print('Soft limit changed to :', soft)
print()

# Consume some CPU time in a pointless exercise.
print('Starting:', time.ctime())
for i in range(200000):
    for i in range(200000):
        v = i * i

# We should never make it this far.
print('Exiting :', time.ctime())
```

正常情况下，信号处理器应当刷新输出所有打开的文件，然后将其关闭，不过在这里，它只是打印一个消息然后退出。

```
$ python3 resource_setrlimit_cpu.py

Soft limit starts as   : 9223372036854775807
Soft limit changed to : 1

Starting: Sun Aug 21 19:18:51 2016
EXPIRED : Sun Aug 21 19:18:52 2016
(time ran out)
```

提示：相关阅读材料

- resource 的标准库文档[⊖]。
- signal：提供了注册信号处理器的详细内容。

17.6 gc：垃圾回收器

gc 提供了 Python 的底层内存管理机制，即自动垃圾回收器。这个模块包括一些函数，可以控制回收器如何操作，以及如何检查系统已知的对象，这些对象可能在等待收集，也可能已陷入引用环而无法释放。

17.6.1 跟踪引用

利用 gc，可以使用对象之间来回的引用来查找复杂数据结构中的环。如果已知一个数据结构中存在环，那么可以使用定制代码来检查它的属性。如果环在未知代码中，则可以使用 get_referents() 和 get_referrers() 函数建立通用的调试工具。

例如，get_referents() 显示了输入参数引用的对象。

代码清单 17-82：gc_get_referents.py

```
import gc
import pprint


class Graph:

    def __init__(self, name):
        self.name = name
        self.next = None

    def set_next(self, next):
        print('Linking nodes {}.next = {}'.format(self, next))
        self.next = next

    def __repr__(self):
        return '{}({})'.format(
            self.__class__.__name__, self.name)


# Construct a graph cycle.
one = Graph('one')
two = Graph('two')
three = Graph('three')
```

⊖ https://docs.python.org/3.5/library/resource.html

```
one.set_next(two)
two.set_next(three)
three.set_next(one)

print()
print('three refers to:')
for r in gc.get_referents(three):
    pprint.pprint(r)
```

在这里，Graph 实例 three 包含其实例字典（在 __dict__ 属性中）以及其类的引用。

```
$ python3 gc_get_referents.py

Linking nodes Graph(one).next = Graph(two)
Linking nodes Graph(two).next = Graph(three)
Linking nodes Graph(three).next = Graph(one)

three refers to:
{'name': 'three', 'next': Graph(one)}
<class '__main__.Graph'>
```

下一个例子使用一个 Queue 来对所有对象引用完成一个广度优先遍历，以查找环。插入到队列中的元素为元组，其中包含目前为止的引用链和下一个要检查的对象。检查从 three 开始，要查看它引用的所有对象。跳过类意味着不检查它的方法、模块和其他组件。

代码清单 17-83：gc_get_referents_cycles.py

```
import gc
import pprint
import queue

class Graph:

    def __init__(self, name):
        self.name = name
        self.next = None

    def set_next(self, next):
        print('Linking nodes {}.next = {}'.format(self, next))
        self.next = next

    def __repr__(self):
        return '{}({})'.format(
            self.__class__.__name__, self.name)

# Construct a graph cycle.
one = Graph('one')
two = Graph('two')
three = Graph('three')
one.set_next(two)
two.set_next(three)
three.set_next(one)

print()

seen = set()
to_process = queue.Queue()

# Start with an empty object chain and Graph three.
```

```
        to_process.put(([], three))

    # Look for cycles, building the object chain for each object
    # found in the queue so the full cycle can be printed at the
    # end.
    while not to_process.empty():
        chain, next = to_process.get()
        chain = chain[:]
        chain.append(next)
        print('Examining:', repr(next))
        seen.add(id(next))
        for r in gc.get_referents(next):
            if isinstance(r, str) or isinstance(r, type):
                # Ignore strings and classes.
                pass
            elif id(r) in seen:
                print()
                print('Found a cycle to {}:'.format(r))
                for i, link in enumerate(chain):
                    print('  {}: '.format(i), end=' ')
                    pprint.pprint(link)
            else:
                to_process.put((chain, r))
```

通过监视已处理的对象，可以很容易地发现节点中的环。所以这些对象的引用不会被收集，它们的 id() 值会缓存在一个集合中。环中找到的字典对象是 Graph 实例的 __dict__ 值，其包含它们的实例属性。

```
$ python3 gc_get_referents_cycles.py

Linking nodes Graph(one).next = Graph(two)
Linking nodes Graph(two).next = Graph(three)
Linking nodes Graph(three).next = Graph(one)

Examining: Graph(three)
Examining: {'next': Graph(one), 'name': 'three'}
Examining: Graph(one)
Examining: {'next': Graph(two), 'name': 'one'}
Examining: Graph(two)
Examining: {'next': Graph(three), 'name': 'two'}

Found a cycle to Graph(three):
  0:  Graph(three)
  1:  {'name': 'three', 'next': Graph(one)}
  2:  Graph(one)
  3:  {'name': 'one', 'next': Graph(two)}
  4:  Graph(two)
  5:  {'name': 'two', 'next': Graph(three)}
```

17.6.2 强制垃圾回收

尽管在解释器执行一个程序时会自动运行垃圾回收器，但是如果需要释放大量对象，或者如果当时没有太多工作，那么相应地回收器在不影响应用性能的情况下也可以触发垃圾回收器，让它在一个特定的时间运行。可以使用 collect() 触发垃圾回收。

代码清单 17-84： gc_collect.py

```
import gc
import pprint
```

```python
class Graph:

    def __init__(self, name):
        self.name = name
        self.next = None

    def set_next(self, next):
        print('Linking nodes {}.next = {}'.format(self, next))
        self.next = next

    def __repr__(self):
        return '{}({})'.format(
            self.__class__.__name__, self.name)


# Construct a graph cycle.
one = Graph('one')
two = Graph('two')
three = Graph('three')
one.set_next(two)
two.set_next(three)
three.set_next(one)

# Remove references to the graph nodes in this module's namespace.
one = two = three = None

# Show the effect of garbage collection.
for i in range(2):
    print('\nCollecting {} ...'.format(i))
    n = gc.collect()
    print('Unreachable objects:', n)
    print('Remaining Garbage:', end=' ')
    pprint.pprint(gc.garbage)
```

在这个例子中，回收第一次运行时就会清除环，因为除了自身以外，Graph 节点不再有其他引用。collect() 会返回它找到的"不可达的"对象数。在这里，这个值是 6，表示有 3 个对象，它们分别有自己的实例属性字典。

```
$ python3 gc_collect.py

Linking nodes Graph(one).next = Graph(two)
Linking nodes Graph(two).next = Graph(three)
Linking nodes Graph(three).next = Graph(one)

Collecting 0 ...
Unreachable objects: 34
Remaining Garbage: []

Collecting 1 ...
Unreachable objects: 0
Remaining Garbage: []
```

17.6.3 查找无法回收的对象引用

要在垃圾列表中查找哪个对象包含另一个对象的引用，这比查看一个对象引用了什么要麻烦一些。因为查找引用的代码本身也需要包含引用，所以需要忽略一些包含引用的对象（引用者）。下面这个例子创建了一个图环，然后处理 Graph 实例，并删除"父"节点中的引用。

代码清单 17-85：gc_get_referrers.py

```python
import gc
import pprint


class Graph:

    def __init__(self, name):
        self.name = name
        self.next = None

    def set_next(self, next):
        print('Linking nodes {}.next = {}'.format(self, next))
        self.next = next

    def __repr__(self):
        return '{}({})'.format(
            self.__class__.__name__, self.name)

    def __del__(self):
        print('{}.__del__()'.format(self))


# Construct a graph cycle.
one = Graph('one')
two = Graph('two')
three = Graph('three')
one.set_next(two)
two.set_next(three)
three.set_next(one)

# Collecting now keeps the objects as uncollectable,
# but not garbage.
print()
print('Collecting...')
n = gc.collect()
print('Unreachable objects:', n)
print('Remaining Garbage:', end=' ')
pprint.pprint(gc.garbage)

# Ignore references from local variables in this module, global
# variables, and from the garbage collector's bookkeeping.
REFERRERS_TO_IGNORE = [locals(), globals(), gc.garbage]


def find_referring_graphs(obj):
    print('Looking for references to {!r}'.format(obj))
    referrers = (r for r in gc.get_referrers(obj)
                 if r not in REFERRERS_TO_IGNORE)
    for ref in referrers:
        if isinstance(ref, Graph):
            # A graph node
            yield ref
        elif isinstance(ref, dict):
            # An instance or other namespace dictionary
            for parent in find_referring_graphs(ref):
                yield parent


# Look for objects that refer to the objects in the graph.
print()
print('Clearing referrers:')
```

```
    for obj in [one, two, three]:
        for ref in find_referring_graphs(obj):
            print('Found referrer:', ref)
            ref.set_next(None)
            del ref  # Remove reference so the node can be deleted.
        del obj  # Remove reference so the node can be deleted.

# Clear references held by gc.garbage.
print()
print('Clearing gc.garbage:')
del gc.garbage[:]
# Everything should have been freed this time.
print()
print('Collecting...')
n = gc.collect()
print('Unreachable objects:', n)
print('Remaining Garbage:', end=' ')
pprint.pprint(gc.garbage)
```

如果环是已知的,那么这种逻辑有些"大材小用",不过对于数据中不可解释的环,使用 `get_referrers()` 可以发现未曾预料到的关系。

```
$ python3 gc_get_referrers.py

Linking nodes Graph(one).next = Graph(two)
Linking nodes Graph(two).next = Graph(three)
Linking nodes Graph(three).next = Graph(one)

Collecting...
Unreachable objects: 28
Remaining Garbage: []

Clearing referrers:
Looking for references to Graph(one)
Looking for references to {'next': Graph(one), 'name': 'three'}
Found referrer: Graph(three)
Linking nodes Graph(three).next = None
Looking for references to Graph(two)
Looking for references to {'next': Graph(two), 'name': 'one'}
Found referrer: Graph(one)
Linking nodes Graph(one).next = None
Looking for references to Graph(three)
Looking for references to {'next': Graph(three), 'name': 'two'}
Found referrer: Graph(two)
Linking nodes Graph(two).next = None

Clearing gc.garbage:

Collecting...
Unreachable objects: 0
Remaining Garbage: []
Graph(one).__del__()
Graph(two).__del__()
Graph(three).__del__()
```

17.6.4 回收阈值和代

垃圾回收器会维护它运行时看到的 3 个对象列表,分别对应回收器跟踪的各"代"(generation)。在各代中检查对象时,这些对象要么被回收,要么变老进入下一代,直至最终达到永久保存的阶段。

根据回收器运行之间对象分配数和撤销数之差，可以调整回收器例程，按不同的频率运行。如果分配数减去撤销数大于当前这一代的阈值，则运行垃圾回收器。当前阈值可以用 `get_threshold()` 检查。

代码清单 17-86：**gc_get_threshold.py**

```python
import gc

print(gc.get_threshold())
```

返回值是一个元组，包含各代的阈值。

```
$ python3 gc_get_threshold.py

(700, 10, 10)
```

可以用 `set_threshold()` 改变阈值。下面这个示例程序使用一个命令行参数来设置第 0 代的阈值，然后分配一系列对象。

代码清单 17-87：**gc_threshold.py**

```python
import gc
import pprint
import sys

try:
    threshold = int(sys.argv[1])
except (IndexError, ValueError, TypeError):
    print('Missing or invalid threshold, using default')
    threshold = 5


class MyObj:

    def __init__(self, name):
        self.name = name
        print('Created', self.name)


gc.set_debug(gc.DEBUG_STATS)

gc.set_threshold(threshold, 1, 1)
print('Thresholds:', gc.get_threshold())
print('Clear the collector by forcing a run')
gc.collect()
print()

print('Creating objects')
objs = []
for i in range(10):
    objs.append(MyObj(i))
print('Exiting')

# Turn off debugging.
gc.set_debug(0)
```

不同阈值会导致在不同的时间完成垃圾回收清扫。下面显示的值是因为启用了调试。

```
$ python3 -u gc_threshold.py 5

gc: collecting generation 1...
gc: objects in each generation: 240 1439 4709
gc: done, 0.0013s elapsed
Thresholds: (5, 1, 1)
Clear the collector by forcing a run
gc: collecting generation 2...
gc: objects in each generation: 1 0 6282
gc: done, 0.0025s elapsed

gc: collecting generation 0...
gc: objects in each generation: 5 0 6275
gc: done, 0.0000s elapsed
Creating objects
gc: collecting generation 0...
gc: objects in each generation: 8 0 6275
gc: done, 0.0000s elapsed
Created 0
Created 1
Created 2
gc: collecting generation 1...
gc: objects in each generation: 9 2 6275
gc: done, 0.0000s elapsed
Created 3
Created 4
Created 5
gc: collecting generation 0...
gc: objects in each generation: 9 0 6280
gc: done, 0.0000s elapsed
Created 6
Created 7
Created 8
gc: collecting generation 0...
gc: objects in each generation: 9 3 6280
gc: done, 0.0000s elapsed
Created 9
Exiting
```

较小的阈值会导致更频繁地运行清扫。

```
$ python3 -u gc_threshold.py 2

gc: collecting generation 1...
gc: objects in each generation: 240 1439 4709
gc: done, 0.0003s elapsed
Thresholds: (2, 1, 1)
Clear the collector by forcing a run
gc: collecting generation 2...
gc: objects in each generation: 1 0 6282
gc: done, 0.0010s elapsed
gc: collecting generation 0...
gc: objects in each generation: 3 0 6275
gc: done, 0.0000s elapsed

Creating objects
gc: collecting generation 0...
gc: objects in each generation: 6 0 6275
gc: done, 0.0000s elapsed
gc: collecting generation 1...
gc: objects in each generation: 3 4 6275
```

```
gc: done, 0.0000s elapsed
Created 0
Created 1
gc: collecting generation 0...
gc: objects in each generation: 4 0 6277
gc: done, 0.0000s elapsed
Created 2
gc: collecting generation 0...
gc: objects in each generation: 8 1 6277
gc: done, 0.0000s elapsed
Created 3
Created 4
gc: collecting generation 1...
gc: objects in each generation: 4 3 6277
gc: done, 0.0000s elapsed
Created 5
gc: collecting generation 0...
gc: objects in each generation: 8 0 6281
gc: done, 0.0000s elapsed
Created 6
Created 7
gc: collecting generation 0...
gc: objects in each generation: 4 2 6281
gc: done, 0.0000s elapsed
Created 8
gc: collecting generation 1...
gc: objects in each generation: 8 3 6281
gc: done, 0.0000s elapsed
Created 9
Exiting
```

17.6.5 调试

调试内存泄漏可能很有难度。gc 包括有一些选项，可以提供代码的内部工作情况，使这个任务更为简单。这些选项都是位标志，可以组合传递到 set_debug()，在程序运行时配置垃圾回收器。调试信息被打印到 sys.stderr。

DEBUG_STATS 标志打开统计报告。这会使垃圾回收器报告它何时运行、每一代跟踪的对象数以及完成清扫花费的时间。

代码清单 17-88：gc_debug_stats.py

```python
import gc

gc.set_debug(gc.DEBUG_STATS)

gc.collect()
print('Exiting')
```

这个示例输出显示运行了两次回收器。第一次是在显式调用时运行，第二次是在解释器退出时运行。

```
$ python3 gc_debug_stats.py

gc: collecting generation 2...
gc: objects in each generation: 123 1063 4711
gc: done, 0.0008s elapsed
Exiting
```

```
gc: collecting generation 2...
gc: objects in each generation: 1 0 5880
gc: done, 0.0007s elapsed
gc: collecting generation 2...
gc: objects in each generation: 99 0 5688
gc: done, 2114 unreachable, 0 uncollectable, 0.0011s elapsed
gc: collecting generation 2...
gc: objects in each generation: 0 0 3118
gc: done, 292 unreachable, 0 uncollectable, 0.0003s elapsed
```

启用 DEBUG_COLLECTABLE 和 DEBUG_UNCOLLECTABLE 会让回收器报告它检查的各个对象能不能回收。如果看到不能回收的对象，但这还不能提供足够的信息来了解数据保留在哪里，则可以启用 DEBUG_SAVEALL，这会让 gc 保留它找到的所有对象，但在 garbage 列表中没有任何引用。

代码清单 17-89：**gc_debug_saveall.py**

```python
import gc

flags = (gc.DEBUG_COLLECTABLE |
         gc.DEBUG_UNCOLLECTABLE |
         gc.DEBUG_SAVEALL
         )

gc.set_debug(flags)

class Graph:

    def __init__(self, name):
        self.name = name
        self.next = None

    def set_next(self, next):
        self.next = next

    def __repr__(self):
        return '{}({})'.format(
            self.__class__.__name__, self.name)


class CleanupGraph(Graph):

    def __del__(self):
        print('{}.__del__()'.format(self))


# Construct a graph cycle.
one = Graph('one')
two = Graph('two')
one.set_next(two)
two.set_next(one)

# Construct another node that stands on its own.
three = CleanupGraph('three')
# Construct a graph cycle with a finalizer.
four = CleanupGraph('four')
five = CleanupGraph('five')
four.set_next(five)
five.set_next(four)
```

```python
# Remove references to the graph nodes in this module's namespace.
one = two = three = four = five = None

# Force a sweep.
print('Collecting')
gc.collect()
print('Done')

# Report on what was left.
for o in gc.garbage:
    if isinstance(o, Graph):
        print('Retained: {} 0x{:x}'.format(o, id(o)))

# Reset the debug flags before exiting to avoid dumping a lot
# of extra information and making the example output more
# confusing.
gc.set_debug(0)
```

这个代码允许在垃圾回收后检查对象,有些情况下这会很有用,例如,不能把构造函数改为在创建各个对象时打印对象 ID。

```
$ python3 -u gc_debug_saveall.py

CleanupGraph(three).__del__()
Collecting
gc: collectable <Graph 0x101be7240>
gc: collectable <Graph 0x101be72e8>
gc: collectable <dict 0x101994108>
gc: collectable <dict 0x101994148>
gc: collectable <CleanupGraph 0x101be73c8>
gc: collectable <CleanupGraph 0x101be7400>
gc: collectable <dict 0x101bee548>
gc: collectable <dict 0x101bee488>
CleanupGraph(four).__del__()
CleanupGraph(five).__del__()
Done
Retained: Graph(one) 0x101be7240
Retained: Graph(two) 0x101be72e8
Retained: CleanupGraph(four) 0x101be73c8
Retained: CleanupGraph(five) 0x101be7400
```

为简单起见,DEBUG_LEAK 被定义为所有其他选项的一个组合。

代码清单 17-90:gc_debug_leak.py

```python
import gc

flags = gc.DEBUG_LEAK

gc.set_debug(flags)


class Graph:

    def __init__(self, name):
        self.name = name
        self.next = None

    def set_next(self, next):
        self.next = next
```

```python
    def __repr__(self):
        return '{}({})'.format(
            self.__class__.__name__, self.name)


class CleanupGraph(Graph):

    def __del__(self):
        print('{}.__del__()'.format(self))


# Construct a graph cycle.
one = Graph('one')
two = Graph('two')
one.set_next(two)
two.set_next(one)

# Construct another node that stands on its own.
three = CleanupGraph('three')

# Construct a graph cycle with a finalizer.
four = CleanupGraph('four')
five = CleanupGraph('five')
four.set_next(five)
five.set_next(four)

# Remove references to the graph nodes in this module's namespace.
one = two = three = four = five = None

# Force a sweep.
print('Collecting')
gc.collect()
print('Done')

# Report on what was left.
for o in gc.garbage:
    if isinstance(o, Graph):
        print('Retained: {} 0x{:x}'.format(o, id(o)))

# Reset the debug flags before exiting to avoid dumping a lot
# of extra information and making the example output more
# confusing.
gc.set_debug(0)
```

要记住，由于DEBUG_SAVEALL由DEBUG_LEAK启用，甚至正常情况下已经被回收和删除的无引用的对象也会保留。

```
$ python3 -u gc_debug_leak.py

CleanupGraph(three).__del__()
Collecting
gc: collectable <Graph 0x1013e7240>
gc: collectable <Graph 0x1013e72e8>
gc: collectable <dict 0x101194108>
gc: collectable <dict 0x101194148>
gc: collectable <CleanupGraph 0x1013e73c8>
gc: collectable <CleanupGraph 0x1013e7400>
gc: collectable <dict 0x1013ee548>
gc: collectable <dict 0x1013ee488>
CleanupGraph(four).__del__()
CleanupGraph(five).__del__()
```

```
Done
Retained: Graph(one) 0x1013e7240
Retained: Graph(two) 0x1013e72e8
Retained: CleanupGraph(four) 0x1013e73c8
Retained: CleanupGraph(five) 0x1013e7400
```

提示：相关阅读材料

- gc 的标准库文档[⊖]。
- gc 的 Python 2 到 Python 3 移植说明。
- weakref：weakref 模块提供了一种创建对象引用的方法，而不会增加其引用数，使对象仍能被垃圾回收。
- Supporting Cyclic Garbage Collection[⊖]：Python C API 文档的背景资料。
- How does Python manage memory?[⊖]：Fredrik Lundh 写的关于 Python 内存管理的一篇文章。

17.7 sysconfig：解释器编译时配置

sysconfig 的特性已经从 distutils 中抽取出来，以创建一个独立的模块。这个模块包括一些函数来确定编译和安装当前解释器所用的设置。

17.7.1 配置变量

可以通过两个函数来访问构建时配置设置：get_config_vars() 和 get_config_var()。其中 get_config_vars() 返回一个字典，将配置变量名映射到值。

代码清单 17-91：sysconfig_get_config_vars.py

```
import sysconfig

config_values = sysconfig.get_config_vars()
print('Found {} configuration settings'.format(
    len(config_values.keys())))

print('\nSome highlights:\n')

print(' Installation prefixes:')
print('  prefix={prefix}'.format(**config_values))
print('  exec_prefix={exec_prefix}'.format(**config_values))

print('\n Version info:')
print('  py_version={py_version}'.format(**config_values))
print('  py_version_short={py_version_short}'.format(
    **config_values))
print('  py_version_nodot={py_version_nodot}'.format(
    **config_values))

print('\n Base directories:')
print('  base={base}'.format(**config_values))
```

⊖ https://docs.python.org/3.5/library/gc.html
⊖ https://docs.python.org/3/c-api/gcsupport.html
⊖ http://effbot.org/pyfaq/how-does-python-manage-memory.htm

```
print('  platbase={platbase}'.format(**config_values))
print('  userbase={userbase}'.format(**config_values))
print('  srcdir={srcdir}'.format(**config_values))
print('\n Compiler and linker flags:')
print('  LDFLAGS={LDFLAGS}'.format(**config_values))
print('  BASECFLAGS={BASECFLAGS}'.format(**config_values))
print('  Py_ENABLE_SHARED={Py_ENABLE_SHARED}'.format(
    **config_values))
```

sysconfig API 提供的详细级别取决于运行程序所在的平台。在 POSIX 系统（如 Linux 和 OS X）上，用来构建解释器的 Makefile 以及为构建生成的 config.h 头文件都会被解析，其中找到的所有变量都可用。在非 POSIX 系统（如 Windows）上，设置则仅限于一些路径、文件名扩展和版本详细信息。

```
$ python3 sysconfig_get_config_vars.py

Found 665 configuration settings

Some highlights:

  Installation prefixes:
   prefix=/Library/Frameworks/Python.framework/Versions/3.5
   exec_prefix=/Library/Frameworks/Python.framework/Versions/3.5

  Version info:
   py_version=3.5.2
   py_version_short=3.5
   py_version_nodot=35

  Base directories:
   base=/Users/dhellmann/Envs/pymotw35
   platbase=/Users/dhellmann/Envs/pymotw35
   userbase=/Users/dhellmann/Library/Python/3.5
   srcdir=/Library/Frameworks/Python.framework/Versions/3.5/lib/p
ython3.5/config-3.5m

  Compiler and linker flags:
   LDFLAGS=-arch i386 -arch x86_64  -g
   BASECFLAGS=-fno-strict-aliasing -Wsign-compare -fno-common
-dynamic
   Py_ENABLE_SHARED=0
```

向 get_config_vars() 传递变量名会把返回值改为一个 list，这是将所有这些变量的值追加在一起创建的。

代码清单 17-92：sysconfig_get_config_vars_by_name.py

```
import sysconfig

bases = sysconfig.get_config_vars('base', 'platbase', 'userbase')
print('Base directories:')
for b in bases:
    print('  ', b)
```

这个例子建立了所有安装基目录的一个列表（可以在这些目录上找到当前系统的模块）。

```
$ python3 sysconfig_get_config_vars_by_name.py

Base directories:
```

```
/Users/dhellmann/Envs/pymotw35
/Users/dhellmann/Envs/pymotw35
/Users/dhellmann/Library/Python/3.5
```

只需要一个配置值时，可以使用 `get_config_var()` 来获取。

代码清单 17-93：sysconfig_get_config_var.py

```python
import sysconfig

print('User base directory:',
      sysconfig.get_config_var('userbase'))
print('Unknown variable   :',
      sysconfig.get_config_var('NoSuchVariable'))
```

如果没有找到变量，则 `get_config_var()` 会返回 `None` 而不是产生一个异常。

```
$ python3 sysconfig_get_config_var.py

User base directory: /Users/dhellmann/Library/Python/3.5
Unknown variable   : None
```

17.7.2 安装路径

`sysconfig` 主要由安装和打包工具使用。因此，尽管可以用来访问通用的配置设置，如解释器版本，但它主要用来访问另外一些信息，即查找一个系统上当前安装的 Python 发布中各个部分所在位置所需的信息。安装一个包所用的位置依赖于使用的方案（scheme）。

方案是平台特定的一组默认目录，根据平台的打包标准和原则来组织。安装到一个全站点位置或是用户所有的一个私有目录时，分别有不同的方案。可以用 `get_scheme_names()` 访问完整的方案集。

代码清单 17-94：sysconfig_get_scheme_names.py

```python
import sysconfig

for name in sysconfig.get_scheme_names():
    print(name)
```

这里没有"当前方案"的概念。默认方案取决于平台，使用的具体方案取决于为安装程序提供的选项。如果当前系统运行一个符合 POSIX 的操作系统，则默认方案为 `posix_prefix`。否则，按照 `os.name` 的定义，默认为操作系统名。

```
$ python3 sysconfig_get_scheme_names.py

nt
nt_user
osx_framework_user
posix_home
posix_prefix
posix_user
```

每个方案都定义了一组用于安装包的路径。要得到路径名列表，可以使用 `get_path_names()`。

代码清单 17-95：sysconfig_get_path_names.py

```python
import sysconfig

for name in sysconfig.get_path_names():
    print(name)
```

对于一个给定方案，有些路径可能是相同的，不过安装工具对于哪些是真正的路径不能做任何假设。每个名都有一个特定的语义含义，所以应当在安装期间用正确的名来查找给定文件的路径。路径名及其含义的完整列表见表 17-4。

表 17-4　sysconfig 中使用的路径名

路径名	描述
stdlib	标准 Python 库文件，非平台特定
platstdlib	标准 Python 库文件，平台特定
platlib	站点特定、平台特定文件
purelib	站点特定、非平台特定文件
include	头文件，非平台特定
platinclude	头文件，平台特定
scripts	可执行脚本文件
data	数据文件

```
$ python3 sysconfig_get_path_names.py

stdlib
platstdlib
purelib
platlib
include
scripts
data
```

可以使用 get_paths() 来获取与一个方案关联的具体目录。

代码清单 17-96：sysconfig_get_paths.py

```python
import sysconfig
import pprint
import os

for scheme in ['posix_prefix', 'posix_user']:
    print(scheme)
    print('=' * len(scheme))
    paths = sysconfig.get_paths(scheme=scheme)
    prefix = os.path.commonprefix(paths.values())
    print('prefix = {}\n'.format(prefix))
    for name, path in sorted(paths.items()):
        print('{}\n  .{}'.format(name, path[len(prefix):]))
    print()
```

这个例子显示了对应 posix_prefix 的全系统路径（在 Mac OS X 上构建的一个框架下）和对应 posix_user 的用户特定值之间的差别。

```
$ python3 sysconfig_get_paths.py

posix_prefix
============
prefix = /Users/dhellmann/Envs/pymotw35

data
   .
include
   ./include/python3.5m
platinclude
   ./include/python3.5m
platlib
   ./lib/python3.5/site-packages
platstdlib
   ./lib/python3.5
purelib
   ./lib/python3.5/site-packages
scripts
   ./bin
stdlib
   ./lib/python3.5

posix_user
==========
prefix = /Users/dhellmann/Library/Python/3.5

data
   .
include
   ./include/python3.5
platlib
   ./lib/python3.5/site-packages
platstdlib
   ./lib/python3.5
purelib
   ./lib/python3.5/site-packages
scripts
   ./bin
stdlib
   ./lib/python3.5
```

要得到单个路径，可以调用 get_path()。

代码清单 17-97：sysconfig_get_path.py

```
import sysconfig
import pprint

for scheme in ['posix_prefix', 'posix_user']:
    print(scheme)
    print('=' * len(scheme))
    print('purelib =', sysconfig.get_path(name='purelib',
                                          scheme=scheme))
    print()
```

使用 get_path() 等价于保存 get_paths() 的值并在字典中查找单个键。如果需要多个路径，则 get_paths() 更为高效，因为它不会每次都重新计算所有路径。

```
$ python3 sysconfig_get_path.py

posix_prefix
============
purelib = /Users/dhellmann/Envs/pymotw35/lib/python3.5/site-pack
ages
posix_user
==========
purelib = /Users/dhellmann/Library/Python/3.5/lib/python3.5/site
-packages
```

17.7.3 Python 版本和平台

尽管 sys 包含一些基本平台标识（见 17.2.1.1 节），但还不够特定，不足以用来安装二进制包，因为 sys.platform 并不总包括有关硬件体系结构、指令大小或影响二进制库兼容性的其他值的信息。要想更准确地指示平台信息，可以使用 get_platform()。

代码清单 17-98: **sysconfig_get_platform.py**

```python
import sysconfig

print(sysconfig.get_platform())
```

用来提供这个示例输出的解释器编译时考虑了 OS X 10.6 兼容性，所以平台串中包含了版本号。

```
$ python3 sysconfig_get_platform.py

macosx-10.6-intel
```

作为一种便利方法，还可以通过 sysconfig 中的 get_python_version() 从 sys.version_info 得到解释器版本。

代码清单 17-99: **sysconfig_get_python_version.py**

```python
import sysconfig
import sys

print('sysconfig.get_python_version():',
      sysconfig.get_python_version())
print('\nsys.version_info:')
print('  major        :', sys.version_info.major)
print('  minor        :', sys.version_info.minor)
print('  micro        :', sys.version_info.micro)
print('  releaselevel:', sys.version_info.releaselevel)
print('  serial       :', sys.version_info.serial)
```

get_python_version() 会返回一个串，在构建版本特定的路径时很适用。

```
$ python3 sysconfig_get_python_version.py

sysconfig.get_python_version(): 3.5

sys.version_info:
  major        : 3
  minor        : 5
```

```
micro       : 2
releaselevel: final
serial      : 0
```

提示：相关阅读材料
- `sysconfig` 的标准库文档[⊖]。
- `distutils`：`sysconfig` 作为 `distutils` 包的一部分。
- `site`：`site` 模块更详细地描述了导入时搜索的路径。
- `os`：包括 `os.name`，当前操作系统的名。
- `sys`：包括其他构建时信息，如平台。

⊖ https://docs.python.org/3.5/library/sysconfig.html

第 18 章 语言工具

除了上一章介绍的开发工具，Python 还包括一些模块，可以用来访问其内部特性。这一章将介绍 Python 中使用的一些工具，这里不区分应用领域。

warnings 模块用于报告非致命条件或可恢复的错误。如果标准库的一个特性被一个新类、接口或模块取代，那么便会生成 DeprecationWarning，这就是警告的一个常见例子。使用 warnings 会报告可能需要用户注意的条件，但不是致命的。

定义符合一个公共 API 的一组类时，如果这个 API 由另外某个人定义或者使用了大量方法，那么这可能很有难度。要解决这个问题，一种常用方法是从一个公共基类派生所有新类。不过，哪些方法应当被覆盖而哪些可以采用默认行为，这一点并不总是很明显。abc 模块的抽象基类提供了 API 的形式化定义，显式地标记某些方法必须由类以某种方式提供，从而避免在类未完全实现时实例化。例如，很多 Python 的容器类型的抽象基类都在 abc 或 collections 中定义。

dis 模块可以用于反汇编程序的字节码版本，以了解解释器运行这个程序时的步骤。调试性能问题或并发问题时，查看反汇编的代码可能很有用，因为它能给出解释器为程序中各个语句执行的原子操作。

inspect 模块为当前进程中的所有对象提供了自省支持。这包括导入的模块、类和函数定义，以及由这些定义所实例化的对象。可以利用自省为源代码生成文档，动态地调整运行时行为，或者检查一个程序的执行环境。

18.1 warnings：非致命警告

warnings 模块由 **PEP 230**[一]引入，由于预见到 Python 3.0 中会出现不能向后兼容的改变，所以以这种方式来警告程序员，指出语言或库特性已出现改变。这个模块还可以用来报告可恢复的配置错误或者因为缺少库而出现的特性降级。不过，最好通过 logging 模块提供用户可见的消息，因为发送到控制台的警告可能丢失。

由于警告不是致命的，所以程序在运行过程中可能会多次遇到相同的警告情况。warnings 模块会抑制相同来源的重复消息，避免因为反复看到同样的警告而厌烦。可以逐情况地控制输出，为此可以使用解释器的命令行选项，也可以调用 warnings 中的函数。

[一] www.python.org/dev/peps/pep-0230

18.1.1 分类和过滤

警告使用内置异常类 `Warning` 的子类进行分类。`exceptions` 模块的联机文档中描述了很多标准值,还可以通过派生 `Warning` 增加定制警告。

警告要根据过滤器(filter)设置来处理。过滤器包括5个部分:动作(`action`)、消息(`message`)、类别(`category`)、模块(`module`)和行号(`line number`)。过滤器的消息部分是一个正则表达式,用来匹配警告文本。类别是一个异常类的类名。模块包含一个正则表达式,要与生成警告的模块名匹配。行号可以用来改变警告出现时的处理。

在生成一个警告时,要与所有注册的过滤器比较。第一个匹配的过滤器将控制对这个警告采取的动作。如果没有匹配的过滤器,则会采取默认动作。过滤机制支持的动作如表18-1所示。

表 18-1 警告过滤器动作

动 作	含 义
error	将警告变成一个异常
ignore	删除警告
always	总是发出一个警告
default	从各个位置第一次生成警告时打印警告
module	从各个模块第一次生成警告时打印警告
once	第一次生成警告时打印警告

18.1.2 生成警告

要发出一个警告,最简单的方法是调用 `warn()`,并提供消息作为参数。

代码清单 18-1:`warnings_warn.py`

```
import warnings

print('Before the warning')
warnings.warn('This is a warning message')
print('After the warning')
```

程序运行时,会打印这个消息。

```
$ python3 -u warnings_warn.py

Before the warning
warnings_warn.py:13: UserWarning: This is a warning message
  warnings.warn('This is a warning message')
After the warning
```

尽管打印了警告,但默认行为是经过这一点继续运行余下的程序。可以用过滤器改变这个行为。

代码清单 18-2:`warnings_warn_raise.py`

```
import warnings

warnings.simplefilter('error', UserWarning)
```

```
print('Before the warning')
warnings.warn('This is a warning message')
print('After the warning')
```

在这个例子中，simplefilter() 函数为内部过滤器列表增加了一项，以告诉 warnings 模块当发出一个 UserWarning 警告时要产生一个异常。

```
$ python3 -u warnings_warn_raise.py

Before the warning
Traceback (most recent call last):
  File "warnings_warn_raise.py", line 15, in <module>
    warnings.warn('This is a warning message')
UserWarning: This is a warning message
```

也可以使用解释器的 -W 选项从命令行控制过滤器行为。只需指定过滤器属性，这是一个包含 5 部分的串（动作、消息、类别、模块和行号），各部分之间用冒号（:）分隔。例如，如果运行 warnings_warn.py，并且过滤器被设置为在出现 UserWarning 时产生一个错误，那么便会生成一个异常。

```
$ python3 -u -W "error::UserWarning::0" warnings_warn.py

Before the warning
Traceback (most recent call last):
  File "warnings_warn.py", line 13, in <module>
    warnings.warn('This is a warning message')
UserWarning: This is a warning message
```

当消息（message）和模块（module）字段为空时，这被解释为与所有内容匹配。

18.1.3　用模式过滤

要通过编程按照更复杂的规则进行过滤，可以使用 filterwarnings()。例如，要根据消息文本的内容过滤，可以提供一个正则表达式作为 message 参数。

代码清单 18-3：**warnings_filterwarnings_message.py**

```
import warnings

warnings.filterwarnings('ignore', '.*do not.*',)

warnings.warn('Show this message')
warnings.warn('Do not show this message')
```

模式中包含 do not，具体的消息使用了 Do not。这个模式会匹配，因为正则表达式被编译为查找不区分大小写的匹配。

```
$ python3 warnings_filterwarnings_message.py

warnings_filterwarnings_message.py:14: UserWarning: Show this message
  warnings.warn('Show this message')
```

下一个示例程序会生成两个警告。

代码清单 18-4：`warnings_filter.py`

```
import warnings

warnings.warn('Show this message')
warnings.warn('Do not show this message')
```

可以在命令行上使用过滤器参数忽略其中一个警告。

```
$ python3 -W "ignore:do not:UserWarning::0" warnings_filter.py

warnings_filter.py:12: UserWarning: Show this message
  warnings.warn('Show this message')
```

同样的模式匹配规则也应用于源模块名（该模块中包含生成警告的调用）。可以将模块名作为模式传至 `module` 参数，抑制来自 `warnings_filtering` 模块的所有消息。

代码清单 18-5：`warnings_filterwarnings_module.py`

```
import warnings

warnings.filterwarnings(
    'ignore',
    '.*',
    UserWarning,
    'warnings_filter',
)

import warnings_filter
```

由于有这个过滤器，所以导入 `warnings_filtering` 时不会发出任何警告。

```
$ python3 warnings_filterwarnings_module.py
```

如果只抑制 `warnings_filtering` 第 13 行上的消息，则可以将这个行号作为 `filterwarnings()` 的最后一个参数。可以使用源文件中的实际行号来限制过滤器，也可以使用 0，对消息的所有出现都应用过滤器。

代码清单 18-6：`warnings_filterwarnings_lineno.py`

```
import warnings

warnings.filterwarnings(
    'ignore',
    '.*',
    UserWarning,
    'warnings_filter',
    13,
)

import warnings_filter
```

这个模式与所有消息都匹配，所以重要的参数是模块名和行号。

```
$ python3 warnings_filterwarnings_lineno.py

.../warnings_filter.py:12: UserWarning: Show this message
  warnings.warn('Show this message')
```

18.1.4 重复的警告

默认地，大多数警告只是在一个给定位置第一次出现时才会打印，这里所说的"位置"是由模块和生成警告的相应行号的组合来定义的。

代码清单 18-7：**warnings_repeated.py**

```
import warnings

def function_with_warning():
    warnings.warn('This is a warning!')

function_with_warning()
function_with_warning()
function_with_warning()
```

这个例子多次调用同一个函数，不过只生成一个警告。

```
$ python3 warnings_repeated.py

warnings_repeated.py:14: UserWarning: This is a warning!
  warnings.warn('This is a warning!')
```

`"once"`动作可以用来抑制相同消息在不同位置多次出现。

代码清单 18-8：**warnings_once.py**

```
import warnings

warnings.simplefilter('once', UserWarning)

warnings.warn('This is a warning!')
warnings.warn('This is a warning!')
warnings.warn('This is a warning!')
```

所有警告的消息文本会被保存，并且只打印一个消息。

```
$ python3 warnings_once.py

warnings_once.py:14: UserWarning: This is a warning!
  warnings.warn('This is a warning!')
```

类似地，`"module"`可以抑制来自相同模块的重复消息，而不论具体行号是什么。

18.1.5 候选消息传送函数

正常情况下，警告会打印到`sys.stderr`。可以通过替换`warnings`模块中的`showwarning()`函数来改变这个行为。例如，要把警告发送到一个日志文件而不是标准错误输出，可以把`showwarning()`替换为将警告记入日志的一个函数。

代码清单 18-9：**warnings_showwarning.py**

```
import warnings
import logging
```

```
def send_warnings_to_log(message, category, filename, lineno,
                         file=None):
    logging.warning(
        '%s:%s: %s:%s',
        filename, lineno,
        category.__name__, message,
    )
logging.basicConfig(level=logging.INFO)
old_showwarning = warnings.showwarning
warnings.showwarning = send_warnings_to_log

warnings.warn('message')
```

调用 `warn()` 时,警告会随其余日志消息发出。

```
$ python3 warnings_showwarning.py

WARNING:root:warnings_showwarning.py:28: UserWarning:message
```

18.1.6 格式化

如果警告要发送到标准错误输出,但是需要重新格式化,那么可以替换 `formatwarning()`。

代码清单 18-10:**warnings_formatwarning.py**

```
import warnings


def warning_on_one_line(message, category, filename, lineno,
                        file=None, line=None):
    return '-> {}:{}: {}:{}'.format(
        filename, lineno, category.__name__, message)

warnings.warn('Warning message, before')
warnings.formatwarning = warning_on_one_line
warnings.warn('Warning message, after')
```

格式化函数必须返回一个串,其中包含要显示给用户的警告的适当表示。

```
$ python3 -u warnings_formatwarning.py

warnings_formatwarning.py:18: UserWarning: Warning message,
before
  warnings.warn('Warning message, before')
-> warnings_formatwarning.py:20: UserWarning:Warning message,
after
```

18.1.7 警告中的栈层次

默认地,警告消息包括生成该消息的源代码行(如果有)。不过,只是看到具体警告消息和代码行并不太有用。实际上,可以告诉 `warn()` 要在栈中上行多远才能找到调用相应函数(即包含这个警告的函数)的代码行。这样一来,使用了过期函数的用户就可以看到函数在哪里调用,而不是看到函数的实现。

代码清单 18-11:**warnings_warn_stacklevel.py**

```
1  #!/usr/bin/env python3
```

```
 2  # encoding: utf-8
 3
 4  import warnings
 5
 6
 7  def old_function():
 8      warnings.warn(
 9          'old_function() is deprecated, use new_function()',
10          stacklevel=2)
11
12
13  def caller_of_old_function():
14      old_function()
15
16
17  caller_of_old_function()
```

在这个例子中,warn()需要在栈中上行两层,一层对应它自身,另一层对应old_function()。

```
$ python3 warnings_warn_stacklevel.py

warnings_warn_stacklevel.py:14: UserWarning: old_function() is deprecated,
 use new_function()
   old_function()
```

提示:相关阅读材料

- warnings 的标准库文档[○]。
- PEP 230[○]:警告框架。
- exceptions:异常和警告的基类。
- logging:传送警告的一种候选机制是写入日志。

18.2 abc:抽象基类

抽象基类是一种接口,与单个 hasattr() 检查特定方法相比,抽象基类的检查更为严格。通过定义一个抽象基类,可以为一组子类建立一个公共 API。有些情况下,可能需要一个对应用源代码不太熟悉的人提供插件扩展,这种情况下这个功能就特别有用,另外对于大型团队合作或者处理一个很大的代码基(同时跟踪所有类很困难,甚至不可能)也很有帮助。

18.2.1 ABC 如何工作

abc 的做法是,将基类的方法标记为抽象,然后注册具体类作为这个抽象基类的实现。如果应用或库需要一个特定的 API,则可以用 issubclass() 或 isinstance() 根据抽象类检查对象。

要使用 abc 模块,首先,定义一个抽象基类来表示一组插件的 API,用于保存和加载数据。设置新基类的 metaclass 为 ABCMeta,并使用修饰符为这个类建立公共 API。下面的例子使用了 abc_base.py。

○ https://docs.python.org/3.5/library/warnings.html
○ www.python.org/dev/peps/pep-0230

代码清单 18-12：**abc_base.py**

```python
import abc


class PluginBase(metaclass=abc.ABCMeta):

    @abc.abstractmethod
    def load(self, input):
        """Retrieve data from the input source
        and return an object.
        """

    @abc.abstractmethod
    def save(self, output, data):
        """Save the data object to the output."""
```

18.2.2 注册一个具体类

有两种方法指示一个具体类实现了一个抽象 API：可以显式地注册这个类，或者直接从抽象基类创建新的子类。当类提供了所需的 API 时，可以使用 `register()` 类方法作为一个具体类的修饰符显式地增加这个具体类，不过它不属于抽象基类继承树。

代码清单 18-13：**abc_register.py**

```python
import abc
from abc_base import PluginBase


class LocalBaseClass:
    pass


@PluginBase.register
class RegisteredImplementation(LocalBaseClass):

    def load(self, input):
        return input.read()

    def save(self, output, data):
        return output.write(data)


if __name__ == '__main__':
    print('Subclass:', issubclass(RegisteredImplementation,
                                  PluginBase))
    print('Instance:', isinstance(RegisteredImplementation(),
                                  PluginBase))
```

在这个例子中，`RegisteredImplementation` 派生自 `LocalBaseClass`，但是它被注册为实现 `PluginBase` API。这说明 `issubclass()` 和 `isinstance()` 会把它看作是从 `PluginBase` 派生的。

```
$ python3 abc_register.py

Subclass: True
Instance: True
```

18.2.3 通过派生实现

直接从基类派生子类可以避免显式地注册类。

代码清单 18-14：**abc_subclass.py**

```python
import abc
from abc_base import PluginBase


class SubclassImplementation(PluginBase):
    def load(self, input):
        return input.read()

    def save(self, output, data):
        return output.write(data)


if __name__ == '__main__':
    print('Subclass:', issubclass(SubclassImplementation,
                                  PluginBase))
    print('Instance:', isinstance(SubclassImplementation(),
                                  PluginBase))
```

在这里，使用了常规的 Python 类管理特性来识别 **SubclassImplementation** 实现了抽象基类 **PluginBase**。

```
$ python3 abc_subclass.py

Subclass: True
Instance: True
```

使用直接派生有一个副作用：向基类询问由其派生的类的列表时，可以找到一个插件的所有实现（这不是 abc 特有的一个特性，所有类都有这个特性）。

代码清单 18-15：**abc_find_subclasses.py**

```python
import abc
from abc_base import PluginBase
import abc_subclass
import abc_register

for sc in PluginBase.__subclasses__():
    print(sc.__name__)
```

尽管导入了 **abc_register()**，但 **RegisteredImplementation** 并不在子类列表中，因为它并非真正派生自这个基类。

```
$ python3 abc_find_subclasses.py

SubclassImplementation
```

18.2.4 辅助基类

如果没有适当地设置元类（metaclass），那么便不会强制要求具体实现 API。为了更容易地正确建立抽象类，可以提供一个基类，由它自动地指定元类。

代码清单 18-16：**abc_abc_base.py**

```python
import abc

class PluginBase(abc.ABC):

    @abc.abstractmethod
    def load(self, input):
        """Retrieve data from the input source
        and return an object.
        """

    @abc.abstractmethod
    def save(self, output, data):
        """Save the data object to the output."""

class SubclassImplementation(PluginBase):

    def load(self, input):
        return input.read()

    def save(self, output, data):
        return output.write(data)

if __name__ == '__main__':
    print('Subclass:', issubclass(SubclassImplementation,
                                  PluginBase))
    print('Instance:', isinstance(SubclassImplementation(),
                                  PluginBase))
```

To create a new abstract class, simply inherit from ABC.

```
$ python3 abc_abc_base.py

Subclass: True
Instance: True
```

18.2.5 不完整的实现

直接从抽象基类派生子类还有一个好处：除非子类完全实现了 API 的抽象部分，否则子类不能被实例化。

代码清单 18-17：**abc_incomplete.py**

```python
import abc
from abc_base import PluginBase
@PluginBase.register
class IncompleteImplementation(PluginBase):

    def save(self, output, data):
        return output.write(data)


if __name__ == '__main__':
    print('Subclass:', issubclass(IncompleteImplementation,
                                  PluginBase))
    print('Instance:', isinstance(IncompleteImplementation(),
                                  PluginBase))
```

这会避免不完整的实现在运行时触发预料之外的错误。

```
$ python3 abc_incomplete.py

Subclass: True
Traceback (most recent call last):
  File "abc_incomplete.py", line 24, in <module>
    print('Instance:', isinstance(IncompleteImplementation(),
TypeError: Can't instantiate abstract class
IncompleteImplementation with abstract methods load
```

18.2.6 ABC 中的具体方法

具体类必须提供所有抽象方法的实现，不仅如此，抽象基类也可以提供实现，可以通过 `super()` 来调用。这就允许将公共逻辑放在基类中来实现重用，而要求子类用（可能）定制的逻辑提供一个覆盖方法。

代码清单 18-18：**abc_concrete_method.py**

```python
import abc
import io


class ABCWithConcreteImplementation(abc.ABC):

    @abc.abstractmethod
    def retrieve_values(self, input):
        print('base class reading data')
        return input.read()


class ConcreteOverride(ABCWithConcreteImplementation):

    def retrieve_values(self, input):
        base_data = super(ConcreteOverride,
                          self).retrieve_values(input)
        print('subclass sorting data')
        response = sorted(base_data.splitlines())
        return response


input = io.StringIO("""line one
line two
line three
""")

reader = ConcreteOverride()
print(reader.retrieve_values(input))
print()
```

由于 `ABCWithConcreteImplementation()` 是一个抽象基类，所以不能实例化这个类直接使用。子类必须为 `retrieve_values()` 提供一个覆盖方法，在这里，具体类要在返回数据之前先整理数据。

```
$ python3 abc_concrete_method.py

base class reading data
subclass sorting data
['line one', 'line three', 'line two']
```

18.2.7 抽象属性

如果一个 API 规范除了方法外还包括属性,那么便可以结合 abstractmethod() 和 property() 来要求具体类中要包括这些属性。

代码清单 18-19:abc_abstractproperty.py

```
import abc

class Base(abc.ABC):

    @property
    @abc.abstractmethod
    def value(self):
        return 'Should never reach here'

    @property
    @abc.abstractmethod
    def constant(self):
        return 'Should never reach here'
class Implementation(Base):

    @property
    def value(self):
        return 'concrete property'

    constant = 'set by a class attribute'


try:
    b = Base()
    print('Base.value:', b.value)
except Exception as err:
    print('ERROR:', str(err))

i = Implementation()
print('Implementation.value   :', i.value)
print('Implementation.constant:', i.constant)
```

这个例子中的 Base 类不能被实例化,因为它对于 value 和 constant 只有一个抽象的属性获取方法。Implementation 中为 value 属性提供了一个具体的获取方法,另外还使用了一个类属性来定义 constant。

```
$ python3 abc_abstractproperty.py

ERROR: Can't instantiate abstract class Base with abstract
methods constant, value
Implementation.value   : concrete property
Implementation.constant: set by a class attribute
```

还可以定义抽象的读写属性。

代码清单 18-20:abc_abstractproperty_rw.py

```
import abc

class Base(abc.ABC):
```

```python
        @property
        @abc.abstractmethod
        def value(self):
            return 'Should never reach here'

        @value.setter
        @abc.abstractmethod
        def value(self, new_value):
            return
class PartialImplementation(Base):

        @property
        def value(self):
            return 'Read-only'


class Implementation(Base):

        _value = 'Default value'

        @property
        def value(self):
            return self._value

        @value.setter
        def value(self, new_value):
            self._value = new_value


try:
    b = Base()
    print('Base.value:', b.value)
except Exception as err:
    print('ERROR:', str(err))

p = PartialImplementation()
print('PartialImplementation.value:', p.value)

try:
    p.value = 'Alteration'
    print('PartialImplementation.value:', p.value)
except Exception as err:
    print('ERROR:', str(err))

i = Implementation()
print('Implementation.value:', i.value)

i.value = 'New value'
print('Changed value:', i.value)
```

与抽象属性一致，具体属性必须用同样的方式定义，可以是读写属性或只读属性。如果试图用一个只读属性覆盖 `PartialImplementation` 中的一个读写属性，则会使属性变成只读的；也就是说，不能重用基类中这个属性的设置方法。

```
$ python3 abc_abstractproperty_rw.py

ERROR: Can't instantiate abstract class Base with abstract
methods value
PartialImplementation.value: Read-only
ERROR: can't set attribute
Implementation.value: Default value
Changed value: New value
```

要对读写抽象属性使用修饰符语法，获取和设置值的方法必须同名。

18.2.8 抽象类和静态方法

类和静态方法还可以被标记为抽象。

代码清单 18-21：abc_class_static.py

```
import abc


class Base(abc.ABC):

    @classmethod
    @abc.abstractmethod
    def factory(cls, *args):
        return cls()

    @staticmethod
    @abc.abstractmethod
    def const_behavior():
        return 'Should never reach here'


class Implementation(Base):

    def do_something(self):
        pass

    @classmethod
    def factory(cls, *args):
        obj = cls(*args)
        obj.do_something()
        return obj

    @staticmethod
    def const_behavior():
        return 'Static behavior differs'


try:
    o = Base.factory()
    print('Base.value:', o.const_behavior())
except Exception as err:
    print('ERROR:', str(err))

i = Implementation.factory()
print('Implementation.const_behavior :', i.const_behavior())
```

尽管类方法是在类上调用而不是在实例上调用，但如果未定义类方法，则还是会导致类无法实例化。

```
$ python3 abc_class_static.py

ERROR: Can't instantiate abstract class Base with abstract
methods const_behavior, factory
Implementation.const_behavior : Static behavior differs
```

提示：相关阅读材料

- abc 的标准库文档[⊖]。

⊖ https://docs.python.org/3.5/library/abc.html

- **PEP 3119**[一]：介绍抽象基类。
- **collections**：**collections** 模块包含很多用于集合类型的抽象基类。
- **PEP 3141**[二]：数字的一个类型层次结构。
- Wikipedia：Strategy pattern[三]：策略模式的描述和示例，这是一个常用的插件实现模式。
- Dynamic Code Patterns：Extending Your Applications with Plugins[四]：Doug Hellmann 提供的 PyCon 2013 演示文稿。
- abc 的 Python 2 到 Python 3 移植说明。

18.3 `dis`：Python 字节码反汇编工具

`dis` 模块包括一些用于处理 Python 字节码的函数，可以将字节码"反汇编"为更便于人阅读的形式。查看解释器执行的字节码是一种手控调整紧密循环（tight loop）的好办法，还有助于完成其他优化。这个模块对于查找多线程应用中的竞态条件也很有用，因为可以用它来估计代码中哪一点线程控制可能切换。

警告：字节码的使用是 CPython 解释器的一个实现细节，特定于具体版本。针对你使用的解释器版本，参考源代码中的 `Include/opcode.h`，可以查找字节码的正式名列表（canonical list）。

18.3.1 基本反汇编

函数 `dis()` 会打印一个 Python 代码源（模块、类、方法、函数或代码对象）的反汇编表示。可以通过从命令行运行 `dis` 来对类似 `dis_simple.py` 的模块进行反汇编。

代码清单 18-22：**dis_simple.py**

```
1  #!/usr/bin/env python3
2  # encoding: utf-8
3
4  my_dict = {'a': 1}
```

输出按列组织，包括源代码行号、代码对象中的指令地址、操作码名，以及传至操作码的所有参数。

```
$ python3 -m dis dis_simple.py

  4           0 LOAD_CONST               0 ('a')
              3 LOAD_CONST               1 (1)
              6 BUILD_MAP                1
              9 STORE_NAME               0 (my_dict)
             12 LOAD_CONST               2 (None)
             15 RETURN_VALUE
```

[一] www.python.org/dev/peps/pep-3119
[二] www.python.org/dev/peps/pep-3141
[三] https://en.wikipedia.org/wiki/Strategy_pattern
[四] http://pyvideo.org/pycon-us-2013/dynamic-code-patterns-extending-your-application.html

在这里，源代码转换为 4 个不同的操作来创建和填充字典，然后将结果保存到一个局部变量。由于 Python 解释器是基于栈的，所以前几步是用 **LOAD_CONST** 将常量按正确的顺序放入栈中，然后使用 **BUILD_MAP** 弹出要增加到字典的新键和值。用 **STORE_NAME** 将所得到的 `dict` 对象绑定到名 `my_dict`。

18.3.2 反汇编函数

遗憾的是，反汇编整个模块时不会自动地递归反汇编函数。

代码清单 18-23：**dis_function.py**

```
1  #!/usr/bin/env python3
2  # encoding: utf-8
3
4
5  def f(*args):
6      nargs = len(args)
7      print(nargs, args)
8
9
10 if __name__ == '__main__':
11     import dis
12     dis.dis(f)
```

反汇编 `dis_function.py` 的结果显示了将函数代码对象加载到栈然后转换为一个函数的操作（**LOAD_CONST**，**MAKE_FUNCTION**），不过没有函数体。

```
$ python3 -m dis dis_function.py

  5           0 LOAD_CONST               0 (<code object f at
0x10141ba50, file "dis_function.py", line 5>)
              3 LOAD_CONST               1 ('f')
              6 MAKE_FUNCTION            0
              9 STORE_NAME               0 (f)

 10          12 LOAD_NAME                1 (__name__)
             15 LOAD_CONST               2 ('__main__')
             18 COMPARE_OP               2 (==)
             21 POP_JUMP_IF_FALSE       49

 11          24 LOAD_CONST               3 (0)
             27 LOAD_CONST               4 (None)
             30 IMPORT_NAME              2 (dis)
             33 STORE_NAME               2 (dis)

 12          36 LOAD_NAME                2 (dis)
             39 LOAD_ATTR                2 (dis)
             42 LOAD_NAME                0 (f)
             45 CALL_FUNCTION            1 (1 positional, 0 keyword pair)
             48 POP_TOP
        >>   49 LOAD_CONST               4 (None)
             52 RETURN_VALUE
```

要查看函数内部，必须把函数传递到 `dis()`。

```
$ python3 dis_function.py

  6           0 LOAD_GLOBAL              0 (len)
```

```
                 3 LOAD_FAST              0 (args)
                 6 CALL_FUNCTION          1 (1 positional, 0
keyword pair)
                 9 STORE_FAST             1 (nargs)
  7             12 LOAD_GLOBAL            1 (print)
                15 LOAD_FAST              1 (nargs)
                18 LOAD_FAST              0 (args)
                21 CALL_FUNCTION          2 (2 positional, 0
keyword pair)
                24 POP_TOP
                25 LOAD_CONST             0 (None)
                28 RETURN_VALUE
```

要打印函数的一个总结，包括它使用的参数和名的有关信息，可以调用 show_code()，并传入这个函数作为第一个参数。

```
#!/usr/bin/env python3
# encoding: utf-8


def f(*args):
    nargs = len(args)
    print(nargs, args)


if __name__ == '__main__':
    import dis
    dis.show_code(f)
```

show_code() 的参数会传递到 code_info()，它会返回函数、方法、代码串或其他代码对象的一个格式美观的总结，可供打印输出。

```
$ python3 dis_show_code.py

Name:              f
Filename:          dis_show_code.py
Argument count:    0
Kw-only arguments: 0
Number of locals:  2
Stack size:        3
Flags:             OPTIMIZED, NEWLOCALS, VARARGS, NOFREE
Constants:
   0: None
Names:
   0: len
   1: print
Variable names:
   0: args
   1: nargs
```

18.3.3 类

可以把类传递到 dis()，在这种情况下，会依序反汇编类的所有方法。

代码清单 18-24：dis_class.py

```
1  #!/usr/bin/env python3
2  # encoding: utf-8
3
```

```
 4  import dis
 5
 6
 7  class MyObject:
 8      """Example for dis."""
 9
10      CLASS_ATTRIBUTE = 'some value'
11
12      def __str__(self):
13          return 'MyObject({})'.format(self.name)
14
15      def __init__(self, name):
16          self.name = name
17
18
19  dis.dis(MyObject)
```

方法以字母顺序列出,而不是按它们在文件中出现的顺序。

```
$ python3 dis_class.py

Disassembly of __init__:
 16           0 LOAD_FAST                1 (name)
              3 LOAD_FAST                0 (self)
              6 STORE_ATTR               0 (name)
              9 LOAD_CONST               0 (None)
             12 RETURN_VALUE

Disassembly of __str__:
 13           0 LOAD_CONST               1 ('MyObject({})')
              3 LOAD_ATTR                0 (format)
              6 LOAD_FAST                0 (self)
              9 LOAD_ATTR                1 (name)
             12 CALL_FUNCTION            1 (1 positional, 0 keyword pair)
             15 RETURN_VALUE
```

18.3.4 源代码

处理一个程序的源代码通常比处理代码对象本身更便利。dis 中的函数接受包含源代码的字符串参数,并且在生成反汇编或其他输出之前先将它们转换为代码对象。

代码清单 18-25:**dis_string.py**

```
import dis

code = """
my_dict = {'a': 1}
"""

print('Disassembly:\n')
dis.dis(code)

print('\nCode details:\n')
dis.show_code(code)
```

传入一个字符串意味着跳过了编译代码和保存结果引用的相关步骤。在检查函数之外的语句时,这种方法更方便。

```
$ python3 dis_string.py
Disassembly:

  2           0 LOAD_CONST               0 ('a')
              3 LOAD_CONST               1 (1)
              6 BUILD_MAP                1
              9 STORE_NAME               0 (my_dict)
             12 LOAD_CONST               2 (None)
             15 RETURN_VALUE

Code details:

Name:              <module>
Filename:          <disassembly>
Argument count:    0
Kw-only arguments: 0
Number of locals:  0
Stack size:        2
Flags:             NOFREE
Constants:
   0: 'a'
   1: 1
   2: None
Names:
   0: my_dict
```

18.3.5 使用反汇编调试

调试一个异常时，有时要查看是哪个字节码带来了问题，这可能很有用。要对一个错误周围的代码反汇编，有多种方法。第一种策略是在交互式解释器中使用 dis() 报告最后一个异常。如果没有向 dis() 传入任何参数，那么它会查找一个异常，并显示导致这个异常的栈顶元素的反汇编结果。

```
$ python3
Python 3.5.1 (v3.5.1:37a07cee5969, Dec  5 2015, 21:12:44)
[GCC 4.2.1 (Apple Inc. build 5666) (dot 3)] on darwin
Type "help", "copyright", "credits" or "license" for more information.
>>> import dis
>>> j = 4
>>> i = i + 4
Traceback (most recent call last):
  File "<stdin>", line 1, in <module>
NameError: name 'i' is not defined
>>> dis.dis()
  1 -->       0 LOAD_NAME                0 (i)
              3 LOAD_CONST               0 (4)
              6 BINARY_ADD
              7 STORE_NAME               0 (i)
             10 LOAD_CONST               1 (None)
             13 RETURN_VALUE
>>>
```

行号后面的 --> 指示了导致错误的操作码（opcode）。由于没有定义 i 变量，所以无法将与这个名关联的值加载到栈中。

程序还可以打印一个活动 traceback 的有关信息，将它直接传递到 distb()。在下面的例子中，有一个 DivideByZero 异常；不过由于这个公式有两个除法，所以可能不清楚哪一部分为 0。

代码清单 18-26：**dis_traceback.py**

```
 1  #!/usr/bin/env python3
 2  # encoding: utf-8
 3
 4  i = 1
 5  j = 0
 6  k = 3
 7
 8  try:
 9      result = k * (i / j) + (i / k)
10  except:
11      import dis
12      import sys
13      exc_type, exc_value, exc_tb = sys.exc_info()
14      dis.distb(exc_tb)
```

在反汇编版本中，当值加载到栈时便会很容易发现错误。会用 **-->** 突出显示出错的操作，前一行将 j 的值压入栈。

```
$ python3 dis_traceback.py

  4           0 LOAD_CONST               0 (1)
              3 STORE_NAME               0 (i)

  5           6 LOAD_CONST               1 (0)
              9 STORE_NAME               1 (j)

  6          12 LOAD_CONST               2 (3)
             15 STORE_NAME               2 (k)

  8          18 SETUP_EXCEPT            26 (to 47)

  9          21 LOAD_NAME                2 (k)
             24 LOAD_NAME                0 (i)
             27 LOAD_NAME                1 (j)
    -->      30 BINARY_TRUE_DIVIDE
             31 BINARY_MULTIPLY
             32 LOAD_NAME                0 (i)
             35 LOAD_NAME                2 (k)
             38 BINARY_TRUE_DIVIDE
             39 BINARY_ADD
             40 STORE_NAME               3 (result)
...trimmed...
```

18.3.6 循环的性能分析

除了调试错误，`dis` 还有助于发现性能问题。检查反汇编的代码对于紧密循环尤其有用，在这些循环中，Python 指令很少，但是这些指令会转换为一组效率很低的字节码。可以通过查看一个类 Dictionary 的不同实现来了解反汇编提供的帮助，这个类会读取一个单词列表，然后按其首字母分组。

代码清单 18-27：**dis_test_loop.py**

```
import dis
import sys
import textwrap
import timeit
```

```
module_name = sys.argv[1]
module = __import__(module_name)
Dictionary = module.Dictionary

dis.dis(Dictionary.load_data)
print()
t = timeit.Timer(
    'd = Dictionary(words)',
    textwrap.dedent("""
    from {module_name} import Dictionary
    words = [
        l.strip()
        for l in open('/usr/share/dict/words', 'rt')
    ]
    """).format(module_name=module_name)
)
iterations = 10
print('TIME: {:0.4f}'.format(t.timeit(iterations) / iterations))
```

可以用测试驱动应用 dis_test_loop.py 来运行 Dictionary 类的各个实现，首先是一个简单但很慢的实现。

代码清单 18-28：dis_slow_loop.py

```
 1  #!/usr/bin/env python3
 2  # encoding: utf-8
 3
 4
 5  class Dictionary:
 6
 7      def __init__(self, words):
 8          self.by_letter = {}
 9          self.load_data(words)
10
11      def load_data(self, words):
12          for word in words:
13              try:
14                  self.by_letter[word[0]].append(word)
15              except KeyError:
16                  self.by_letter[word[0]] = [word]
```

用这个版本运行测试程序时，会显示反汇编的程序，以及运行所花费的时间。

```
$ python3 dis_test_loop.py dis_slow_loop

 12           0 SETUP_LOOP              83 (to 86)
              3 LOAD_FAST                1 (words)
              6 GET_ITER
        >>    7 FOR_ITER                75 (to 85)
             10 STORE_FAST               2 (word)

 13          13 SETUP_EXCEPT            28 (to 44)

 14          16 LOAD_FAST                0 (self)
             19 LOAD_ATTR                0 (by_letter)
             22 LOAD_FAST                2 (word)
             25 LOAD_CONST               1 (0)
             28 BINARY_SUBSCR
             29 BINARY_SUBSCR
             30 LOAD_ATTR                1 (append)
```

```
                    33 LOAD_FAST                2 (word)
                    36 CALL_FUNCTION            1 (1 positional, 0
keyword pair)
                    39 POP_TOP
                    40 POP_BLOCK
                    41 JUMP_ABSOLUTE            7

    15      >>      44 DUP_TOP
                    45 LOAD_GLOBAL              2 (KeyError)
                    48 COMPARE_OP              10 (exception match)
                    51 POP_JUMP_IF_FALSE       81
                    54 POP_TOP
                    55 POP_TOP
                    56 POP_TOP

    16              57 LOAD_FAST                2 (word)
                    60 BUILD_LIST               1
                    63 LOAD_FAST                0 (self)
                    66 LOAD_ATTR                0 (by_letter)
                    69 LOAD_FAST                2 (word)
                    72 LOAD_CONST               1 (0)
                    75 BINARY_SUBSCR
                    76 STORE_SUBSCR
                    77 POP_EXCEPT
                    78 JUMP_ABSOLUTE            7
            >>      81 END_FINALLY
                    82 JUMP_ABSOLUTE            7
            >>      85 POP_BLOCK
            >>      86 LOAD_CONST               0 (None)
                    89 RETURN_VALUE

TIME: 0.0568
```

前面的输出显示, `dis_slow_loop.py` 花费了 0.0568 秒来加载 OS X 上 /usr/share/dict/words 副本中的 235 886 个单词。这个性能不算太坏, 不过相应的反汇编结果显示出循环做了很多不必要的工作。它在操作码 13 处进入循环时, 程序建立了一个异常上下文 (SETUP_EXCEPT)。然后在将 word 追加到列表之前, 使用了 6 个操作码来查找 self.by_letter[word[0]]。如果由于 word[0] 还不在字典中而生成一个异常, 那么异常处理器会做完全相同的工作来确定 word[0] (3 个操作码), 并把 self.by_letter[word[0]] 设置为包含这个单词的一个新列表。

要避免建立这个异常, 一种技术是对应字母表中的各个字母分别用一个列表来预填充字典 self.by_letter。这意味着总会找到新单词相应的列表, 可以在查找之后保存值。

代码清单 18-29: **dis_faster_loop.py**

```
 1  #!/usr/bin/env python3
 2  # encoding: utf-8
 3
 4  import string
 5
 6
 7  class Dictionary:
 8
 9      def __init__(self, words):
10          self.by_letter = {
11              letter: []
12              for letter in string.ascii_letters
13          }
```

```
14          self.load_data(words)
15
16      def load_data(self, words):
17          for word in words:
18              self.by_letter[word[0]].append(word)
```

这个修改将操作码数减至一半，不过时间只减少到 0.0567 秒。显然，异常处理有一些开销，但并不多。

```
$ python3 dis_test_loop.py dis_faster_loop

 17           0 SETUP_LOOP              38 (to 41)
              3 LOAD_FAST                1 (words)
              6 GET_ITER
        >>    7 FOR_ITER                30 (to 40)
             10 STORE_FAST               2 (word)

 18          13 LOAD_FAST                0 (self)
             16 LOAD_ATTR                0 (by_letter)
             19 LOAD_FAST                2 (word)
             22 LOAD_CONST               1 (0)
             25 BINARY_SUBSCR
             26 BINARY_SUBSCR
             27 LOAD_ATTR                1 (append)
             30 LOAD_FAST                2 (word)
             33 CALL_FUNCTION            1 (1 positional, 0 keyword pair)
             36 POP_TOP
             37 JUMP_ABSOLUTE            7
        >>   40 POP_BLOCK
        >>   41 LOAD_CONST               0 (None)
             44 RETURN_VALUE

TIME: 0.0567
```

将 `self.by_letter` 的查找移到循环之外（毕竟值没有改变），可以进一步提高性能。

代码清单 18-30：`dis_fastest_loop.py`

```
1  #!/usr/bin/env python3
2  # encoding: utf-8
3
4  import collections
5
6
7  class Dictionary:
8
9      def __init__(self, words):
10         self.by_letter = collections.defaultdict(list)
11         self.load_data(words)
12
13     def load_data(self, words):
14         by_letter = self.by_letter
15         for word in words:
16             by_letter[word[0]].append(word)
```

现在操作码 0～6 会查找 `self.by_letter` 的值，并把它保存为一个局部变量 `by_letter`。使用局部变量只需要一个操作码，而不是两个（语句 22 使用 LOAD_FAST 将字典放在栈中）。做了这个修改之后，运行时间降至 0.0473 秒。

```
$ python3 dis_test_loop.py dis_fastest_loop

 14           0 LOAD_FAST                0 (self)
              3 LOAD_ATTR                0 (by_letter)
              6 STORE_FAST               2 (by_letter)

 15           9 SETUP_LOOP              35 (to 47)
             12 LOAD_FAST                1 (words)
             15 GET_ITER
        >>   16 FOR_ITER                27 (to 46)
             19 STORE_FAST               3 (word)

 16          22 LOAD_FAST                2 (by_letter)
             25 LOAD_FAST                3 (word)
             28 LOAD_CONST               1 (0)
             31 BINARY_SUBSCR
             32 BINARY_SUBSCR
             33 LOAD_ATTR                1 (append)
             36 LOAD_FAST                3 (word)
             39 CALL_FUNCTION            1 (1 positional, 0
keyword pair)
             42 POP_TOP
             43 JUMP_ABSOLUTE           16
        >>   46 POP_BLOCK
        >>   47 LOAD_CONST               0 (None)
             50 RETURN_VALUE

TIME: 0.0473
```

Brandon Rhodes 还建议了进一步的优化，可以完全消除 Python 版本的 `for` 循环。如果使用 `itertools.groupby()` 来整理输入，那么将把迭代处理移至 C。这个转移很安全，因为输入已经是有序的。如果并非如此，则程序需要先进行排序。

代码清单 18-31：**dis_eliminate_loop.py**

```
 1  #!/usr/bin/env python3
 2  # encoding: utf-8
 3
 4  import operator
 5  import itertools
 6
 7
 8  class Dictionary:
 9
10      def __init__(self, words):
11          self.by_letter = {}
12          self.load_data(words)
13
14      def load_data(self, words):
15          # Arrange by letter.
16          grouped = itertools.groupby(
17              words,
18              key=operator.itemgetter(0),
19          )
20          # Save arranged sets of words.
21          self.by_letter = {
22              group[0][0]: group
23              for group in grouped
24          }
```

这个 `itertools` 版本运行只需要 0.0332 秒，仅为原程序运行时间的约 60%。

```
$ python3 dis_test_loop.py dis_eliminate_loop

 16           0 LOAD_GLOBAL              0 (itertools)
              3 LOAD_ATTR                1 (groupby)
 17           6 LOAD_FAST                1 (words)
              9 LOAD_CONST               1 ('key')
 18          12 LOAD_GLOBAL              2 (operator)
             15 LOAD_ATTR                3 (itemgetter)
             18 LOAD_CONST               2 (0)
             21 CALL_FUNCTION            1 (1 positional, 0 keyword pair)
             24 CALL_FUNCTION            257 (1 positional, 1 keyword pair)
             27 STORE_FAST               2 (grouped)
 21          30 LOAD_CONST               3 (<code object <dictcomp> at 0x101517930, file ".../dis_eliminate_loop.py", line 21>)
             33 LOAD_CONST               4 ('Dictionary.load_data.<locals>.<dictcomp>')
             36 MAKE_FUNCTION            0
 23          39 LOAD_FAST                2 (grouped)
             42 GET_ITER
             43 CALL_FUNCTION            1 (1 positional, 0 keyword pair)
             46 LOAD_FAST                0 (self)
             49 STORE_ATTR               4 (by_letter)
             52 LOAD_CONST               0 (None)
             55 RETURN_VALUE

TIME: 0.0332
```

18.3.7 编译器优化

对编译的源代码进行反汇编还能发现一些编译器优化。例如，如果可能，字面量表达式会在编译时尽可能折叠。

代码清单 18-32：`dis_constant_folding.py`

```
 1  #!/usr/bin/env python3
 2  # encoding: utf-8
 3
 4  # Folded
 5  i = 1 + 2
 6  f = 3.4 * 5.6
 7  s = 'Hello,' + ' World!'
 8
 9  # Not folded
10  I = i * 3 * 4
11  F = f / 2 / 3
12  S = s + '\n' + 'Fantastic!'
```

5～7 行上表达式中的值不会改变完成操作的方式，所以可以在编译时计算表达式的结果，并将其压缩为单个 `LOAD_CONST` 指令。10～12 行则不同。因为这些行上的表达式涉及一个变量，而且这个变量指示的对象可能会重载涉及的操作符，所以计算必须延迟到运行时。

```
$ python3 -m dis dis_constant_folding.py

  5           0 LOAD_CONST              11 (3)
              3 STORE_NAME               0 (i)

  6           6 LOAD_CONST              12 (19.04)
              9 STORE_NAME               1 (f)

  7          12 LOAD_CONST              13 ('Hello, World!')
             15 STORE_NAME               2 (s)

 10          18 LOAD_NAME                0 (i)
             21 LOAD_CONST               6 (3)
             24 BINARY_MULTIPLY
             25 LOAD_CONST               7 (4)
             28 BINARY_MULTIPLY
             29 STORE_NAME               3 (I)

 11          32 LOAD_NAME                1 (f)
             35 LOAD_CONST               1 (2)
             38 BINARY_TRUE_DIVIDE
             39 LOAD_CONST               6 (3)
             42 BINARY_TRUE_DIVIDE
             43 STORE_NAME               4 (F)

 12          46 LOAD_NAME                2 (s)
             49 LOAD_CONST               8 ('\n')
             52 BINARY_ADD
             53 LOAD_CONST               9 ('Fantastic!')
             56 BINARY_ADD
             57 STORE_NAME               5 (S)
             60 LOAD_CONST              10 (None)
             63 RETURN_VALUE
```

提示：相关阅读材料

- `dis` 的标准库文档[一]：包括字节码指令列表[二]。
- `Include/opcode.h`：CPython 解释器的源代码在 `opcode.h` 中定义了字节码。
- 《Python Essential Reference》，第 4 版，作者 David M. Beazley。
- thomas.apestaart.org：Python Disassembly[三]：简短讨论了 Python 2.5 和 2.6 中在字典内存储值的差别。
- Why is looping over `range()` in Python faster than using a `while` loop?[四]：Stack Overflow 上对于通过反汇编字节码比较两个循环示例的讨论。
- Decorator for binding constants at compile time[五]：Raymond Hettinger 和 Skip Montanaro 提供的 Python 实用技巧，包含一个函数修饰符，它会重写一个函数的字节码，插入全局常量以避免在运行时查找名。

[一] https://docs.python.org/3.5/library/dis.html
[二] https://docs.python.org/3.5/library/dis.html#python-bytecode-instructions
[三] http://thomas.apestaart.org/log/?p=927
[四] http://stackoverflow.com/questions/869229/why-is-looping-over-range-in-python-faster-than-using-a-while-loop
[五] http://code.activestate.com/recipes/277940/

18.4 inspect：检查现场对象

inspect 模块提供了一些函数来了解现场对象，包括模块、类、实例、函数和方法。这个模块中的函数可以用来获取一个函数的原始源代码，查看栈中一个方法的参数，以及抽取对生成源代码库文档有用的信息。

18.4.1 示例模块

这一节余下的例子都会使用这个示例文件 example.py。

代码清单 18-33：**example.py**

```python
# This comment appears first
# and spans 2 lines.

# This comment does not show up in the output of getcomments().

"""Sample file to serve as the basis for inspect examples.
"""


def module_level_function(arg1, arg2='default', *args, **kwargs):
    """This function is declared in the module."""
    local_variable = arg1 * 2
    return local_variable


class A(object):
    """The A class."""

    def __init__(self, name):
        self.name = name

    def get_name(self):
        "Returns the name of the instance."
        return self.name

instance_of_a = A('sample_instance')


class B(A):
    """This is the B class.
    It is derived from A.
    """

    # This method is not part of A.
    def do_something(self):
        """Does some work"""

    def get_name(self):
        "Overrides version from A"
        return 'B(' + self.name + ')'
```

18.4.2 检查模块

第一种对现场对象的检查是为了了解这些对象。可以使用 getmembers() 发现对象的成员属性。返回的成员类型取决于所扫描对象的类型。模块可能包含类和函数；类可能包含方法和属性等。

getmembers()函数的参数是一个待扫描的对象（模块、类或实例）和一个可选的谓词函数，这个谓词函数被用来过滤返回的对象。返回值是一个元组列表，元组中包含两个值：成员名和成员的类型。inspect模块包括很多名为 ismodule()、isclass()等的谓词函数。

代码清单 18-34：**inspect_getmembers_module.py**

```
import inspect

import example

for name, data in inspect.getmembers(example):
    if name.startswith('__'):
        continue
    print('{} : {!r}'.format(name, data))
```

这个例子会打印 example 模块的成员。这个模块有一些私有属性作为导入实现的一部分，另外还包含一组 __builtins__。对于这个例子，其输出中的所有这些成员都将被忽略，因为它们不能真正算是模块的一部分，并且这个列表很长。

```
$ python3 inspect_getmembers_module.py

A : <class 'example.A'>
B : <class 'example.B'>
instance_of_a : <example.A object at 0x1014814a8>
module_level_function : <function module_level_function at
0x10148bc80>
```

可以用谓词（predicate）参数过滤返回对象的类型。

代码清单 18-35：**inspect_getmembers_module_class.py**

```
import inspect

import example

for name, data in inspect.getmembers(example, inspect.isclass):
    print('{} : {!r}'.format(name, data))
```

现在输出中只包括类。

```
$ python3 inspect_getmembers_module_class.py

A : <class 'example.A'>
B : <class 'example.B'>
```

18.4.3 检查类

类似于检查模块，可以采用同样的方式使用 getmembers() 扫描类，不过成员的类型不同。

代码清单 18-36：**inspect_getmembers_class.py**

```
import inspect
from pprint import pprint
```

```
import example

pprint(inspect.getmembers(example.A), width=65)
```

由于没有应用过滤，所以输出显示了属性、方法、槽，以及类的其他成员。

```
$ python3 inspect_getmembers_class.py

[('__class__', <class 'type'>),
 ('__delattr__',
  <slot wrapper '__delattr__' of 'object' objects>),
 ('__dict__',
  mappingproxy({'__dict__': <attribute '__dict__' of 'A'
objects>,
                '__doc__': 'The A class.',
                '__init__': <function A.__init__ at
0x101c99510>,
                '__module__': 'example',
                '__weakref__': <attribute '__weakref__' of 'A'
objects>,
                'get_name': <function A.get_name at
0x101c99598>})),
 ('__dir__', <method '__dir__' of 'object' objects>),
 ('__doc__', 'The A class.'),
 ('__eq__', <slot wrapper '__eq__' of 'object' objects>),
 ('__format__', <method '__format__' of 'object' objects>),
 ('__ge__', <slot wrapper '__ge__' of 'object' objects>),
 ('__getattribute__',
  <slot wrapper '__getattribute__' of 'object' objects>),
 ('__gt__', <slot wrapper '__gt__' of 'object' objects>),
 ('__hash__', <slot wrapper '__hash__' of 'object' objects>),
 ('__init__', <function A.__init__ at 0x101c99510>),
 ('__le__', <slot wrapper '__le__' of 'object' objects>),
 ('__lt__', <slot wrapper '__lt__' of 'object' objects>),
 ('__module__', 'example'),
 ('__ne__', <slot wrapper '__ne__' of 'object' objects>),
 ('__new__',
  <built-in method __new__ of type object at 0x10022bb20>),
 ('__reduce__', <method '__reduce__' of 'object' objects>),
 ('__reduce_ex__', <method '__reduce_ex__' of 'object'
objects>),
 ('__repr__', <slot wrapper '__repr__' of 'object' objects>),
 ('__setattr__',
  <slot wrapper '__setattr__' of 'object' objects>),
 ('__sizeof__', <method '__sizeof__' of 'object' objects>),
 ('__str__', <slot wrapper '__str__' of 'object' objects>),
 ('__subclasshook__',
  <built-in method __subclasshook__ of type object at
0x10061fba8>),
 ('__weakref__', <attribute '__weakref__' of 'A' objects>),
 ('get_name', <function A.get_name at 0x101c99598>)]
```

要查找一个类的成员，可以使用谓词 isfunction()。ismethod() 谓词只识别实例的绑定方法。

代码清单 18-37: **inspect_getmembers_class_methods.py**

```
import inspect
from pprint import pprint

import example

pprint(inspect.getmembers(example.A, inspect.isfunction))
```

Now only unbound methods are returned.

```
$ python3 inspect_getmembers_class_methods.py

[('__init__', <function A.__init__ at 0x10139d510>),
 ('get_name', <function A.get_name at 0x10139d598>)]
```

B的输出包括覆盖的get_name()方法、新方法，以及从A继承的__init__()方法。

代码清单18-38：**inspect_getmembers_class_methods_b.py**

```
import inspect
from pprint import pprint

import example

pprint(inspect.getmembers(example.B, inspect.isfunction))
```

从A继承的方法（如__init__()）会被标识为B的方法。

```
$ python3 inspect_getmembers_class_methods_b.py

[('__init__', <function A.__init__ at 0x10129d510>),
 ('do_something', <function B.do_something at 0x10129d620>),
 ('get_name', <function B.get_name at 0x10129d6a8>)]
```

18.4.4 检查实例

检查实例的方法与检查其他对象相同。

代码清单18-39：**inspect_getmembers_instance.py**

```
import inspect
from pprint import pprint

import example

a = example.A(name='inspect_getmembers')
pprint(inspect.getmembers(a, inspect.ismethod))
```

谓词ismethod()找出示例实例中A的两个绑定方法。

```
$ python3 inspect_getmembers_instance.py

[('__init__', <bound method A.__init__ of <example.A object at 0x101ab1ba8>>),
 ('get_name', <bound method A.get_name of <example.A object at 0x101ab1ba8>>)]
```

18.4.5 文档串

可以用getdoc()获取一个对象的docstring。返回值是__doc__属性，其中的tab被扩展为空格，并且其缩进保持一致。

代码清单18-40：**inspect_getdoc.py**

```
import inspect
import example
```

```
print('B.__doc__:')
print(example.B.__doc__)
print()
print('getdoc(B):')
print(inspect.getdoc(example.B))
```

通过这个属性直接获取 docstring 的第二行时，它会缩进，不过 getdoc() 会把它移至左边界。

```
$ python3 inspect_getdoc.py

B.__doc__:
This is the B class.
    It is derived from A.


getdoc(B):
This is the B class.
It is derived from A.
```

除了具体的 docstring，还可以从实现对象的源文件获取注释（如果可以得到源文件）。getcomments() 函数会查看对象的源文件，并查找实现代码前面的注释。

代码清单 18-41：**inspect_getcomments_method.py**

```
import inspect
import example

print(inspect.getcomments(example.B.do_something))
```

返回的行中包括注释前缀，这里会去除空白符前缀。

```
$ python3 inspect_getcomments_method.py

# This method is not part of A.
```

将一个模块传入 getcomments() 时，返回值总是模块中的第一个注释。

代码清单 18-42：**inspect_getcomments_module.py**

```
import inspect
import example

print(inspect.getcomments(example))
```

示例文件中邻接的行会作为一个注释，不过一旦出现一个空行，注释就会停止。

```
$ python3 inspect_getcomments_module.py

# This comment appears first
# and spans 2 lines.
```

18.4.6 获取源代码

如果可以得到一个模块的 .py 文件，则可以使用 getsource() 和 getsourcelines() 获取类或方法的原始源代码。

代码清单 18-43：**inspect_getsource_class.py**

```
import inspect
import example

print(inspect.getsource(example.A))
```

传入一个类时，输出中会包含这个类的所有方法。

```
$ python3 inspect_getsource_class.py

class A(object):
    """The A class."""

    def __init__(self, name):
        self.name = name

    def get_name(self):
        "Returns the name of the instance."
        return self.name
```

要获取一个方法的源代码，可以将这个方法引用传入 getsource()。

代码清单 18-44：**inspect_getsource_method.py**

```
import inspect
import example

print(inspect.getsource(example.A.get_name))
```

在这里，会保留原来的缩进层次。

```
$ python3 inspect_getsource_method.py

    def get_name(self):
        "Returns the name of the instance."
        return self.name
```

可以使用 getsourcelines() 而不是 getsource() 来获取源文件中的代码行，并将其分解为单独的字符串。

代码清单 18-45：**inspect_getsourcelines_method.py**

```
import inspect
import pprint
import example

pprint.pprint(inspect.getsourcelines(example.A.get_name))
```

getsourcelines() 的返回值是一个 tuple，其中包含一个字符串列表（源文件中的代码行）和文件中源代码出现的起始行号。

```
$ python3 inspect_getsourcelines_method.py

(['    def get_name(self):\n',
  '        "Returns the name of the instance."\n',
  '        return self.name\n'],
 23)
```

如果得不到源文件，则 `getsource()` 和 `getsourcelines()` 会产生一个 `IOError`。

18.4.7 方法和函数签名

除了函数或方法的文档，还可以得到可调用方法参数的完整规范，包括默认值。`signature()` 函数会返回一个 `Signature` 实例，其中包含函数参数的有关信息。

代码清单18-46：**inspect_signature_function.py**

```
import inspect
import example

sig = inspect.signature(example.module_level_function)
print('module_level_function{}'.format(sig))

print('\nParameter details:')
for name, param in sig.parameters.items():
    if param.kind == inspect.Parameter.POSITIONAL_ONLY:
        print('  {} (positional-only)'.format(name))
    elif param.kind == inspect.Parameter.POSITIONAL_OR_KEYWORD:
        if param.default != inspect.Parameter.empty:
            print('  {}={!r}'.format(name, param.default))
        else:
            print('  {}'.format(name))
    elif param.kind == inspect.Parameter.VAR_POSITIONAL:
        print('  *{}'.format(name))
    elif param.kind == inspect.Parameter.KEYWORD_ONLY:
        if param.default != inspect.Parameter.empty:
            print('  {}={!r} (keyword-only)'.format(
                name, param.default))
        else:
            print('  {} (keyword-only)'.format(name))
    elif param.kind == inspect.Parameter.VAR_KEYWORD:
        print('  **{}'.format(name))
```

可以通过 `Signature` 的 `parameters` 属性来得到函数参数。`parameters` 是一个有序字典，将参数名映射到描述这个参数的 `Parameter` 实例。在这个例子中，函数的第一个参数 `arg1` 没有默认值，而 `arg2` 有默认值。

```
$ python3 inspect_signature_function.py

module_level_function(arg1, arg2='default', *args, **kwargs)

Parameter details:
  arg1
  arg2='default'
  *args
  **kwargs
```

修饰符或其他函数可以用一个函数的 `Signature` 来验证输入、提供不同的默认值以及完成其他任务。不过，要编写一个通用而且可重用的验证修饰符，会带来一个特殊的挑战，因为对于同时接收命名参数和位置参数的函数来说，要把收到的参数与相应的参数名对应，这可能会很复杂。`bind()` 和 `bind_partial()` 方法提供了处理这种映射所需的逻辑。它们会返回一个 `BoundArguments` 实例，其中填充与指定函数参数名关联的实参。

代码清单 18-47：**inspect_signature_bind.py**

```
import inspect
import example

sig = inspect.signature(example.module_level_function)

bound = sig.bind(
    'this is arg1',
    'this is arg2',
    'this is an extra positional argument',
    extra_named_arg='value',
)

print('Arguments:')
for name, value in bound.arguments.items():
    print('{} = {!r}'.format(name, value))

print('\nCalling:')
print(example.module_level_function(*bound.args, **bound.kwargs))
```

BoundArguments 实例包含属性 `args` 和 `kwargs`，可以使用这种语法调用函数，以便将元组和字典扩展到参数所在的栈。

```
$ python3 inspect_signature_bind.py

Arguments:
arg1 = 'this is arg1'
arg2 = 'this is arg2'
args = ('this is an extra positional argument',)
kwargs = {'extra_named_arg': 'value'}

Calling:
this is arg1this is arg1
```

如果只能得到一些参数，则 `bind_partial()` 将仍能创建一个 BoundArguments 实例。在增加其余参数之前，这个实例可能还无法正常使用。

代码清单 18-48：**inspect_signature_bind_partial.py**

```
import inspect
import example

sig = inspect.signature(example.module_level_function)

partial = sig.bind_partial(
    'this is arg1',
)

print('Without defaults:')
for name, value in partial.arguments.items():
    print('{} = {!r}'.format(name, value))

print('\nWith defaults:')
partial.apply_defaults()
for name, value in partial.arguments.items():
    print('{} = {!r}'.format(name, value))
```

`apply_defaults()` 会增加参数默认值中的值。

```
$ python3 inspect_signature_bind_partial.py

Without defaults:
arg1 = 'this is arg1'

With defaults:
arg1 = 'this is arg1'
arg2 = 'default'
args = ()
kwargs = {}
```

18.4.8 类层次体系

inspect 提供了两个方法来直接处理类层次体系。第一个方法是 getclasstree()，它会基于给定的类及其基类创建一个类似树的数据结构。返回的列表中，各个元素可能是一个包含类及其基类的元组，也可能是另一个包含子类元组的列表。

代码清单 18-49：**inspect_getclasstree.py**

```python
import inspect
import example


class C(example.B):
    pass


class D(C, example.A):
    pass


def print_class_tree(tree, indent=-1):
    if isinstance(tree, list):
        for node in tree:
            print_class_tree(node, indent + 1)
    else:
        print('  ' * indent, tree[0].__name__)
    return


if __name__ == '__main__':
    print('A, B, C, D:')
    print_class_tree(inspect.getclasstree(
        [example.A, example.B, C, D])
    )
```

这个例子的输出是对应 A、B、C 和 D 类的继承树。D 出现了两次，因为它同时继承了 C 和 A。

```
$ python3 inspect_getclasstree.py

A, B, C, D:
 object
   A
     D
   B
     C
       D
```

如果调用 getclasstree() 时将 unique 设置为值 true，则输出会有所不同。

代码清单 18-50：`inspect_getclasstree_unique.py`

```python
import inspect
import example
from inspect_getclasstree import *

print_class_tree(inspect.getclasstree(
    [example.A, example.B, C, D],
    unique=True,
))
```

这一次，D 在输出中只出现一次。

```
$ python3 inspect_getclasstree_unique.py

object
  A
    B
      C
        D
```

18.4.9 方法解析顺序

处理类层次体系的另一个函数是 `getmro()`，它会返回类的一个 `tuple`，其中类的顺序就是扫描这些类的顺序，即使用方法解析顺序（Method Resolution Order，MRO）解析从基类继承的一个属性时要按这个顺序扫描各个类。每个类在这个序列中只出现一次。

代码清单 18-51：`inspect_getmro.py`

```python
import inspect
import example

class C(object):
    pass

class C_First(C, example.B):
    pass

class B_First(example.B, C):
    pass

print('B_First:')
for c in inspect.getmro(B_First):
    print('   {}'.format(c.__name__))
print()
print('C_First:')
for c in inspect.getmro(C_First):
    print('   {}'.format(c.__name__))
```

这个例子的输出展示了 MRO 搜索的"深度优先"特性。对于 `B_First`，搜索顺序中 A 也在 C 之前，因为 B 派生自 A。

```
$ python3 inspect_getmro.py

B_First:
   B_First
   B
   A
   C
   object
```

```
C_First:
  C_First
  C
  B
  A
  object
```

18.4.10 栈与帧

除了用于检查代码对象的函数，`inspect` 还包括一些函数可以检查程序执行时的运行时环境。其中大多数函数都处理调用栈，并且操作对象为"调用帧"。帧对象包含当前执行上下文，包括正在运行的代码的引用，正在执行的操作，以及局部和全局变量的值。一般地，这些信息会在产生异常时用来建立 traceback；当然，其对于记录日志或调试程序也很有用，因为可以通过查看栈帧来发现传入函数的参数值。

`currentframe()` 会返回位于栈顶的帧（对应当前函数）。

代码清单 18-52：**inspect_currentframe.py**

```python
import inspect
import pprint
def recurse(limit, keyword='default', *, kwonly='must be named'):
    local_variable = '.' * limit
    keyword = 'changed value of argument'
    frame = inspect.currentframe()
    print('line {} of {}'.format(frame.f_lineno,
                                 frame.f_code.co_filename))
    print('locals:')
    pprint.pprint(frame.f_locals)
    print()
    if limit <= 0:
        return
    recurse(limit - 1)
    return local_variable

if __name__ == '__main__':
    recurse(2)
```

`recurse()` 的参数值被包含在帧的局部变量字典中。

```
$ python3 inspect_currentframe.py

line 14 of inspect_currentframe.py
locals:
{'frame': <frame object at 0x1022a7b88>,
 'keyword': 'changed value of argument',
 'kwonly': 'must be named',
 'limit': 2,
 'local_variable': '..'}

line 14 of inspect_currentframe.py
locals:
{'frame': <frame object at 0x102016b28>,
 'keyword': 'changed value of argument',
 'kwonly': 'must be named',
 'limit': 1,
 'local_variable': '.'}

line 14 of inspect_currentframe.py
locals:
{'frame': <frame object at 0x1020176b8>,
 'keyword': 'changed value of argument',
 'kwonly': 'must be named',
```

```
    'limit': 0,
    'local_variable': ''}
```

使用 stack()，还可以访问当前帧到第一个调用者的所有栈帧。这个例子与前面的例子类似，只不过它会一直等待，直到递归结束，再打印栈信息。

代码清单 18-53：inspect_stack.py

```
import inspect
import pprint


def show_stack():
    for level in inspect.stack():
        print('{}[{}]\n  -> {}'.format(
            level.frame.f_code.co_filename,
            level.lineno,
            level.code_context[level.index].strip(),
        ))
        pprint.pprint(level.frame.f_locals)
        print()


def recurse(limit):
    local_variable = '.' * limit
    if limit <= 0:
        show_stack()
        return
    recurse(limit - 1)
    return local_variable


if __name__ == '__main__':
    recurse(2)
```

输出的最后一部分表示主程序，这在 recurse() 函数之外。

```
$ python3 inspect_stack.py

inspect_stack.py[11]
  -> for level in inspect.stack():
{'level': FrameInfo(frame=<frame object at 0x10127e5d0>,
filename='inspect_stack.py', lineno=11, function='show_stack',
code_context=['    for level in inspect.stack():\n'], index=0)}

inspect_stack.py[24]
  -> show_stack()
{'limit': 0, 'local_variable': ''}

inspect_stack.py[26]
  -> recurse(limit - 1)
{'limit': 1, 'local_variable': '.'}

inspect_stack.py[26]
  -> recurse(limit - 1)
{'limit': 2, 'local_variable': '..'}

inspect_stack.py[30]
  -> recurse(2)
{'__builtins__': <module 'builtins' (built-in)>,
 '__cached__': None,
 '__doc__': 'Inspecting the call stack.\n',
 '__file__': 'inspect_stack.py',
 '__loader__': <_frozen_importlib_external.SourceFileLoader
object at 0x1007a97f0>,
```

```
'__name__': '__main__',
'__package__': None,
'__spec__': None,
'inspect': <module 'inspect' from
'.../lib/python3.5/inspect.py'>,
'pprint': <module 'pprint' from '.../lib/python3.5/pprint.py'>,
'recurse': <function recurse at 0x1012aa400>,
'show_stack': <function show_stack at 0x1007a6a60>}
```

还有一些函数可以构建不同上下文中的帧列表，如在处理一个异常时。相关的更多细节可参见 `trace()`、`getouterframes()` 和 `getinnerframes()` 的文档。

18.4.11 命令行接口

`inspect` 模块还包括一个命令行接口，可以利用这个接口得到对象的相关细节，而不必在一个单独的 Python 程序中编写调用。输入是一个模块名，以及输入模块中的对象（可选）。默认输出是指定对象的源代码。如果使用 `--details` 参数，则会打印元数据而不是源代码。

```
$ python3 -m inspect -d example

Target: example
Origin: .../example.py
Cached: .../__pycache__/example.cpython-35.pyc
Loader: <_frozen_importlib_external.SourceFileLoader object at 0
x101527860>


$ python3 -m inspect -d example:A

Target: example:A
Origin: .../example.py
Cached: .../__pycache__/example.cpython-35.pyc
Line: 16


$ python3 -m inspect example:A.get_name

    def get_name(self):
        "Returns the name of the instance."
        return self.name
```

提示：相关阅读材料

- `inspect` 的标准库文档[⊖]。
- `inspect` 的 Python 2 到 Python 3 移植说明。
- Python 2.3 Method Resolution Order[⊜]：Python 2.3 及以后版本使用的 C3 方法解析顺序的文档。
- `pyclbr`：通过解析模块而不是导入，`pyclbr` 模块也可以像 `inspect` 一样访问同样的一些信息。
- **PEP 362**[⊝]：函数签名对象。

⊖ https://docs.python.org/3.5/library/inspect.html
⊜ www.python.org/download/releases/2.3/mro/
⊝ www.python.org/dev/peps/pep-0362

第 19 章
模块和包

Python 的主要扩展机制使用了保存到模块的源代码，并通过 `import` 语句将其结合到程序中。大多数开发人员所认为的"Python"特性实际上是作为模块集合实现的，这个模块集合被称为标准库，这也是本书讨论的主题。尽管导入特性是解释器本身内置的，但库中还有很多与导入过程有关的模块。

`importlib` 模块提供了解释器使用的导入机制的底层实现。它可以用来在运行时动态地导入模块，而不是使用 `import` 语句在启动时加载。如果不能提前知道需要导入的模块名，那么动态地加载模块就很有用，比如一个程序的插件或扩展包。

Python 包除了包括源代码，还可以包括支持资源文件，如模板、默认配置文件、图像和其他数据。`pkgutil` 模块中实现了一个以可移植方式访问资源文件的接口。它还支持修改一个包的导入路径，从而可以将内容作为同一个包的不同部分安装到多个目录上。

`zipimport` 为保存到 ZIP 归档中的模块和包提供了一个定制导入工具。例如，它可以用来加载 Python EGG 文件，还可以作为一种便利方法用来打包和发布应用。

19.1 `importlib`：Python 的导入机制

`importlib` 模块包括一些函数，这些函数实现了 Python 的导入机制，可以加载包和模块中的代码。可以利用它动态导入模块，另外一些情况下，如果需要导入的模块名在编写代码时是未知的（例如，应用的插件或扩展包），那么此时这个包也很有用。

19.1.1 示例包

本节中的例子使用了一个名为 example 的包，其中包含 `__init__.py`。

代码清单 19-1：example/__init__.py

```
print('Importing example package')
```

这个包还包含 `submodule.py`。

代码清单 19-2：example/submodule.py

```
print('Importing submodule')
```

导入这个包或模块时，注意示例输出中 `print()` 调用的文本。

19.1.2 模块类型

Python 支持多种类型的模块。打开模块并将其增加到命名空间时，每种类型的模块都需要它自己的处理，另外不同平台对格式的支持也有所不同。例如，在 Microsoft Windows 下，共享库从扩展名为 `.dll` 或 `.pyd` 的文件加载，而不是 `.so`。使用解释器的调试（debug）构建版本而不是普通的发行（release）构建版本时，C 模块的扩展包也可能改变，因为这些扩展包在编译时可能还包含调试信息。如果一个 C 扩展库或其他模块不能按预想的那样正常加载，则可以使用 `importlib.machinery` 中定义的常量来查找当前平台支持的类型，以及相应的加载参数。

代码清单 19-3：**importlib_suffixes.py**

```python
import importlib.machinery

SUFFIXES = [
    ('Source:', importlib.machinery.SOURCE_SUFFIXES),
    ('Debug:',
     importlib.machinery.DEBUG_BYTECODE_SUFFIXES),
    ('Optimized:',
     importlib.machinery.OPTIMIZED_BYTECODE_SUFFIXES),
    ('Bytecode:', importlib.machinery.BYTECODE_SUFFIXES),
    ('Extension:', importlib.machinery.EXTENSION_SUFFIXES),
]


def main():
    tmpl = '{:<10}  {}'
    for name, value in SUFFIXES:
        print(tmpl.format(name, value))


if __name__ == '__main__':
    main()
```

返回值是一个元组序列，其中包含文件扩展名，打开文件（包含有模块）所用的模式，以及一个类型码（来自模块中定义的一个常量）。下面的表并不完备，因为有些可导入的模块或包类型并不对应单个文件。

```
$ python3 importlib_suffixes.py

Source:     ['.py']
Debug:      ['.pyc']
Optimized:  ['.pyc']
Bytecode:   ['.pyc']
Extension:  ['.cpython-35m-darwin.so', '.abi3.so', '.so']
```

19.1.3 导入模块

给定一个绝对名或相对名，`importlib` 中的高层 API 便可以简化导入一个模块的过程。使用一个相对模块名时，可以指定包含这个模块的包作为一个单独的参数。

代码清单 19-4：**importlib_import_module.py**

```python
import importlib
```

```
m1 = importlib.import_module('example.submodule')
print(m1)

m2 = importlib.import_module('.submodule', package='example')
print(m2)

print(m1 is m2)
```

import_module()的返回值是这个导入创建的模块对象。

```
$ python3 importlib_import_module.py

Importing example package
Importing submodule
<module 'example.submodule' from '.../example/submodule.py'>
<module 'example.submodule' from '.../example/submodule.py'>
True
```

如果无法导入这个模块，则import_module()会产生一个ImportError。

代码清单19-5：**importlib_import_module_error.py**

```
import importlib


try:
    importlib.import_module('example.nosuchmodule')
except ImportError as err:
    print('Error:', err)
```

错误消息包含缺少的模块的名。

```
$ python3 importlib_import_module_error.py

Importing example package
Error: No module named 'example.nosuchmodule'
```

要重新加载一个现有的模块，可以使用reload()。

代码清单19-6：**importlib_reload.py**

```
import importlib


m1 = importlib.import_module('example.submodule')
print(m1)

m2 = importlib.reload(m1)
print(m1 is m2)
```

reload()的返回值是新模块。取决于所用加载工具的类型，这可能是相同的模块实例。

```
$ python3 importlib_reload.py

Importing example package
Importing submodule
<module 'example.submodule' from '.../example/submodule.py'>
Importing submodule
True
```

19.1.4 加载工具

`importlib` 中的底层 API 允许访问加载工具对象,参见 17.2.6 节中(在关于 `sys` 模块的一节中)的介绍。要为一个模块获得一个加载工具,可以使用 `find_loader()`。然后,要获取这个模块,可以使用加载工具的 `load_module()` 方法。

代码清单 19-7:**importlib_find_loader.py**

```
import importlib

loader = importlib.find_loader('example')
print('Loader:', loader)

m = loader.load_module()
print('Module:', m)
```

这个例子加载 `example` 包的顶层。

```
$ python3 importlib_find_loader.py

Loader: <_frozen_importlib_external.SourceFileLoader object at
0x101be0da0>
Importing example package
Module: <module 'example' from '.../example/__init__.py'>
```

包中的子模块需要使用包中的路径单独加载。在下面的例子中,首先加载这个包,然后把它的路径传递到 `find_loader()` 以创建一个能加载子模块的加载工具。

代码清单 19-8:**importlib_submodule.py**

```
import importlib

pkg_loader = importlib.find_loader('example')
pkg = pkg_loader.load_module()

loader = importlib.find_loader('submodule', pkg.__path__)
print('Loader:', loader)

m = loader.load_module()
print('Module:', m)
```

与 `import_module()` 不同,指定子模块的名时没有相对路径前缀,因为加载工具已经被这个包的路径限度。

```
$ python3 importlib_submodule.py

Importing example package
Loader: <_frozen_importlib_external.SourceFileLoader object at
0x1012e5390>
Importing submodule
Module: <module 'submodule' from '.../example/submodule.py'>
```

提示:相关阅读材料

- `importlib` 的标准库文档[⊖]。
- 17.2.6 节:`sys` 模块中的导入 hook、模块搜索路径以及其他相关的机制。

⊖ https://docs.python.org/3.5/library/importlib.html

- **inspect**：通过编程从一个模块加载信息。
- **PEP 302**[⊖]：New-import hook。
- **PEP 369**[⊜]：Post-import hook。
- **PEP 488**[⊟]：PYO 文件消除。

19.2 pkgutil：包工具

pkgutil 模块包含一些函数，可以改变 Python 包的导入规则，以及从包中发布的文件加载非代码资源。

19.2.1 包导入路径

extend_path() 函数可以用来修改搜索路径，并改变从包导入子模块的方式，这样能结合多个不同的目录，使它们就好像是一个目录一样。借助这个函数，可以用包的开发版本覆盖已安装的版本，或者将平台特定的模块与共享模块结合到一个包命名空间。

调用 extend_path() 最常用的方式是在包的 __init__.py 中增加下面两行代码。

```
import pkgutil
__path__ = pkgutil.extend_path(__path__, __name__)
```

extend_path() 会扫描 sys.path 来查找目录，其中包括一个子目录，它的名字基于第二个参数指定的包。这个目录列表与作为第一个参数传入并作为一个列表返回的路径值结合，很适合作为包导入路径。

名为 demopkg 的示例包中包括两个文件，__init__.py 和 shared.py。demopkg1 中的 __init__.py 文件包含 print 语句，以显示修改前和修改后的搜索路径，强调这两个路径之间的区别。

代码清单 19-9：demopkg1/__init__.py

```
import pkgutil
import pprint

print('demopkg1.__path__ before:')
pprint.pprint(__path__)
print()

__path__ = pkgutil.extend_path(__path__, __name__)

print('demopkg1.__path__ after:')
pprint.pprint(__path__)
print()
```

extension 目录（包含对应 demopkg 的特性）还包含另外 3 个源文件。每个目录层次上都有 __init__.py，另外这里还有一个 not_shared.py。

⊖ www.python.org/dev/peps/pep-0302
⊜ www.python.org/dev/peps/pep-0369
⊟ www.python.org/dev/peps/pep-0488

```
$ find extension -name '*.py'

extension/__init__.py
extension/demopkg1/__init__.py
extension/demopkg1/not_shared.py
```

下面这个简单的测试程序会导入 demopkg1 包。

代码清单 19-10：**pkgutil_extend_path.py**

```python
import demopkg1
print('demopkg1                :', demopkg1.__file__)

try:
    import demopkg1.shared
except Exception as err:
    print('demopkg1.shared         : Not found ({})'.format(err))
else:
    print('demopkg1.shared         :', demopkg1.shared.__file__)

try:
    import demopkg1.not_shared
except Exception as err:
    print('demopkg1.not_shared: Not found ({})'.format(err))
else:
    print('demopkg1.not_shared:', demopkg1.not_shared.__file__)
```

直接从命令行运行这个测试程序时，找不到 not_shared 模块。

说明：这些例子中没有给出完整的文件系统路径，其有所简化，以重点强调有变化的部分。

```
$ python3 pkgutil_extend_path.py

demopkg1.__path__ before:
['.../demopkg1']

demopkg1.__path__ after:
['.../demopkg1']

demopkg1            : .../demopkg1/__init__.py
demopkg1.shared     : .../demopkg1/shared.py
demopkg1.not_shared: Not found (No module named 'demopkg1.not_sh
ared')
```

不过，如果将 extension 目录增加到 PYTHONPATH，并且再次运行这个程序，那么便会生成不同的结果。

```
$ PYTHONPATH=extension python3 pkgutil_extend_path.py

demopkg1.__path__ before:
['.../demopkg1']

demopkg1.__path__ after:
['.../demopkg1',
 '.../extension/demopkg1']

demopkg1            : .../demopkg1/__init__.py
demopkg1.shared     : .../demopkg1/shared.py
demopkg1.not_shared: .../extension/demopkg1/not_shared.py
```

extension 目录中的 demopkg1 已经被增加到搜索路径，所以可以在那里找到 not_

shared 模块。

以这种方式扩展路径对于结合平台特定的包和公共包会很有用，特别是在平台特定的包中包含 C 扩展模块时。

19.2.2　包的开发版本

改进一个项目时，开发人员通常需要测试对已安装包的修改。将已安装的版本替换为开发版本可能是个糟糕的想法，因为开发版本不一定正确，而且系统上的其他工具可能会依赖于已安装的包。

可以使用 virtualenv 或 venv 在开发环境中配置包的一个完全独立的副本，不过对于小的修改，建立这样一个包含所有依赖包的虚拟环境开销可能太大。

还有另一种选择，对于正在开发的包，可以使用 pkgutil 修改其中模块的搜索路径。不过，在这种情况下，路径必须逆向设置，以使开发版本会覆盖已安装的版本。

给定一个包 demopkg2，其中包含一个 __init__.py 和 overloaded.py，正在开发的函数位于 demopkg2/overloaded.py，已安装的版本包含：

代码清单 19-11：**demopkg2/overloaded.py**

```
def func():
    print('This is the installed version of func().')
```

另外，demopkg2/__init__.py 包含：

代码清单 19-12：**demopkg2/__init__.py**

```
import pkgutil

__path__ = pkgutil.extend_path(__path__, __name__)
__path__.reverse()
```

reverse() 用来确保扫描默认位置之前先扫描由 pkgutil 增加到搜索路径的目录，以完成导入。

下一个程序会导入 demopkg2.overloaded 并调用 func()。

代码清单 19-13：**pkgutil_devel.py**

```
import demopkg2
print('demopkg2                :', demopkg2.__file__)

import demopkg2.overloaded
print('demopkg2.overloaded:', demopkg2.overloaded.__file__)

print()
demopkg2.overloaded.func()
```

如果运行时没有做任何特殊的路径处理，则会从 func() 的已安装版本生成输出。

```
$ python3 pkgutil_devel.py

demopkg2                : .../demopkg2/__init__.py
```

```
demopkg2.overloaded: .../demopkg2/overloaded.py

This is the installed version of func().
```

开发目录包含以下内容：

```
$ find develop/demopkg2 -name '*.py'

develop/demopkg2/__init__.py
develop/demopkg2/overloaded.py
```

overloaded 的修改版本如下：

代码清单 19-14：develop/demopkg2/overloaded.py

```python
def func():
    print('This is the development version of func().')
```

如果 develop 目录在搜索路径中，则运行测试程序时就会加载这个版本。

```
$ PYTHONPATH=develop python3 pkgutil_devel.py

demopkg2            : .../demopkg2/__init__.py
demopkg2.overloaded: .../develop/demopkg2/overloaded.py

This is the development version of func().
```

19.2.3 用 PKG 文件管理路径

第一个例子展示了如何使用 PYTHONPATH 中包含的额外目录来扩展搜索路径。此外，还可以使用包含目录名的 *.pkg 文件来扩展搜索路径。PKG 文件类似于 site 模块使用的 PTH 文件，其中可以包含要增加到包搜索路径的目录名，每行一个目录名。

对于第一个例子中的应用，要为特定于平台的部分建立结构，还有一种方法：对于各个操作系统分别使用一个单独的目录，并包含一个 .pkg 文件来扩展搜索路径。

下面的例子使用了同样的 demopkg1 文件，还包括以下文件。

```
$ find os_* -type f

os_one/demopkg1/__init__.py
os_one/demopkg1/not_shared.py
os_one/demopkg1.pkg
os_two/demopkg1/__init__.py
os_two/demopkg1/not_shared.py
os_two/demopkg1.pkg
```

PKG 文件被命名为 demopkg1.pkg，与所扩展的包匹配。包名和 PKG 文件名中都包含一行。

```
demopkg
```

这个演示程序显示了导入的模块。

代码清单 19-15：pkgutil_os_specific.py

```python
import demopkg1
print('demopkg1:', demopkg1.__file__)
```

```python
import demopkg1.shared
print('demopkg1.shared:', demopkg1.shared.__file__)

import demopkg1.not_shared
print('demopkg1.not_shared:', demopkg1.not_shared.__file__)
```

可以用一个简单的包装器脚本在这两个包之间切换。

代码清单 19-16：**with_os.sh**

```sh
#!/bin/sh

export PYTHONPATH=os_${1}
echo "PYTHONPATH=$PYTHONPATH"
echo

python3 pkgutil_os_specific.py
```

在指定 "one" 或 "two" 作为参数运行这个脚本时，路径会被调整。

```
$ ./with_os.sh one

PYTHONPATH=os_one

demopkg1.__path__ before:
['.../demopkg1']

demopkg1.__path__ after:
['.../demopkg1',
 '.../os_one/demopkg1',
 'demopkg']

demopkg1: .../demopkg1/__init__.py
demopkg1.shared: .../demopkg1/shared.py
demopkg1.not_shared: .../os_one/demopkg1/not_shared.py

$ ./with_os.sh two

PYTHONPATH=os_two

demopkg1.__path__ before:
['.../demopkg1']

demopkg1.__path__ after:
['.../demopkg1',
 '.../os_two/demopkg1',
 'demopkg']

demopkg1: .../demopkg1/__init__.py
demopkg1.shared: .../demopkg1/shared.py
demopkg1.not_shared: .../os_two/demopkg1/not_shared.py
```

PKG 文件可以出现在正常搜索路径的任何位置，所以还可以用当前工作目录中的一个 PKG 文件来包含一个开发树。

19.2.4 嵌套包

对于嵌套包，只需要修改顶级包的路径。例如，考虑以下目录结构：

```
$ find nested -name '*.py'
```

```
nested/__init__.py
nested/second/__init__.py
nested/second/deep.py
nested/shallow.py
```

其中 `nested/__init__.py` 包含：

代码清单 19-17：nested/__init__.py

```python
import pkgutil

__path__ = pkgutil.extend_path(__path__, __name__)
__path__.reverse()
```

开发树如下：

```
$ find develop/nested -name '*.py'
```

```
develop/nested/__init__.py
develop/nested/second/__init__.py
develop/nested/second/deep.py
develop/nested/shallow.py
```

shallow 和 deep 模块都包含一个函数来打印一条消息，指示消息来自已安装版本还是来自开发版本。使用以下测试程序测试这些新的包。

代码清单 19-18：pkgutil_nested.py

```python
import nested

import nested.shallow
print('nested.shallow:', nested.shallow.__file__)
nested.shallow.func()

print()
import nested.second.deep
print('nested.second.deep:', nested.second.deep.__file__)
nested.second.deep.func()
```

运行 `pkgutil_nested.py` 时，如果未对路径做任何处理，则会使用这两个函数的已安装版本。

```
$ python3 pkgutil_nested.py

nested.shallow: .../nested/shallow.py
This func() comes from the installed version of nested.shallow

nested.second.deep: .../nested/second/deep.py
This func() comes from the installed version of nested.second.de
ep
```

将 develop 目录增加到路径时，这两个函数的开发版本会覆盖已安装版本。

```
$ PYTHONPATH=develop python3 pkgutil_nested.py

nested.shallow: .../develop/nested/shallow.py
This func() comes from the development version of nested.shallow
```

```
nested.second.deep: .../develop/nested/second/deep.py
This func() comes from the development version of nested.second.
deep
```

19.2.5 包数据

除了代码之外,Python 包还可以包含数据文件,如模板、默认配置文件、图像以及包中代码使用的其他支持文件。利用 get_data() 函数,可以采用一种无关格式的方式来访问文件中的数据,所以不论包是作为一个 EGG 发布,还是作为一个冰冻二进制包的一部分,或者是作为文件系统上的常规文件,都没有任何影响。

假设有一个包 pkgwithdata,其中包含一个 templates 目录。

```
$ find pkgwithdata -type f

pkgwithdata/__init__.py
pkgwithdata/templates/base.html
```

其中文件 pkgwithdata/templates/base.html 包含一个简单的 HTML 模板。

代码清单 19-19:**pkgwithdata/templates/base.html**

```
<!DOCTYPE HTML PUBLIC "-//IETF//DTD HTML//EN">
<html> <head>
<title>PyMOTW Template</title>
</head>

<body>
<h1>Example Template</h1>
<p>This is a sample data file.</p>

</body>
</html>
```

下面的程序使用 get_data() 获取模板内容,并打印出来。

代码清单 19-20:**pkgutil_get_data.py**

```
import pkgutil

template = pkgutil.get_data('pkgwithdata', 'templates/base.html')
print(template.decode('utf-8'))
```

get_data() 有两个参数,一个是包的加点名,一个是相对于包顶级目录的文件名。返回值是一个字节序列,所以在打印之前会编码为 UTF-8。

```
$ python3 pkgutil_get_data.py

<!DOCTYPE HTML PUBLIC "-//IETF//DTD HTML//EN">
<html> <head>
<title>PyMOTW Template</title>
</head>

<body>
<h1>Example Template</h1>

<p>This is a sample data file.</p>

</body>
</html>
```

`get_data()` 与发布格式无关,因为它使用 **PEP 302** 中定义的导入 hook 来访问包内容。可以使用任何提供了 hook 的加载工具,包括 `zipfile` 中的 ZIP 归档导入工具。

代码清单 19-21:**pkgutil_get_data_zip.py**

```python
import pkgutil
import zipfile
import sys

# Create a ZIP file with code from the current directory
# and the template using a name that does not appear on the
# local file system.
with zipfile.PyZipFile('pkgwithdatainzip.zip', mode='w') as zf:
    zf.writepy('.')
    zf.write('pkgwithdata/templates/base.html',
             'pkgwithdata/templates/fromzip.html',
             )

# Add the ZIP file to the import path.
sys.path.insert(0, 'pkgwithdatainzip.zip')

# Import pkgwithdata to show that it comes from the ZIP archive.
import pkgwithdata
print('Loading pkgwithdata from', pkgwithdata.__file__)

# Print the template body.
print('\nTemplate:')
data = pkgutil.get_data('pkgwithdata', 'templates/fromzip.html')
print(data.decode('utf-8'))
```

这个例子使用 `PyZipFile.writepy()` 创建一个 ZIP 归档,其中包含 `pkgwithdata` 包的一个副本,它包括模板文件的一个重命名的版本。在使用 `pkgutil` 加载模板并打印之前,将这个 ZIP 归档增加到导入路径。关于如何使用 `writepy()` 的更多细节,可以参考有关 `zipfile` 的讨论。

```
$ python3 pkgutil_get_data_zip.py

Loading pkgwithdata from
pkgwithdatainzip.zip/pkgwithdata/__init__.pyc

Template:
<!DOCTYPE HTML PUBLIC "-//IETF//DTD HTML//EN">
<html> <head>
<title>PyMOTW Template</title>
</head>

<body>
<h1>Example Template</h1>

<p>This is a sample data file.</p>

</body>
</html>
```

提示:相关阅读材料

- `pkgutil` 的标准库文档[1]。
- virtualenv[2]:Ian Bicking 提供的虚拟环境脚本。

---

[1] https://docs.python.org/3.5/library/pkgutil.html

[2] http://pypi.python.org/pypi/virtualenv

- **distutils**：Python 标准库的打包工具。
- **setuptools**[1]：下一代打包工具。
- **PEP 302**[2]：导入 hook。
- **zipfile**：创建可导入的 ZIP 归档。
- **zipimport**：这个导入工具可以导入 ZIP 归档中的包。

19.3 `zipimport`：从 ZIP 归档加载 Python 代码

`zipimport` 模块实现了 `zipimporter` 类，这个类可以用来查找和加载 ZIP 归档中的 Python 模块。`zipimporter` 支持 **PEP 302** 中指定的导入 hook API；Python Eggs 就采用这种方式。

通常没有必要直接使用 `zipimport` 模块，因为只要归档出现在 `sys.path` 中，就可以直接从 ZIP 归档导入。不过，研究如何使用导入工具 API 对于程序员学习可用的特性以及了解如何完成模块导入很有意义。另外如果用 `zipfile.PyZipFile` 创建作为 ZIP 归档打包的应用，那么了解 ZIP 导入工具如何工作便也有助于调试发布这些应用时可能出现的问题。

19.3.1 示例

这些例子重用了讨论 `zipfile` 时的一些代码，以创建一个示例 ZIP 归档，其中包含一些 Python 模块。

代码清单 19-22：`zipimport_make_example.py`

```
import sys
import zipfile

if __name__ == '__main__':
    zf = zipfile.PyZipFile('zipimport_example.zip', mode='w')
    try:
        zf.writepy('.')
        zf.write('zipimport_get_source.py')
        zf.write('example_package/README.txt')
    finally:
        zf.close()
    for name in zf.namelist():
        print(name)
```

运行其余例子之前先运行 `zipimport_make_example.py`，创建一个 ZIP 归档，其内包含示例目录中的所有模块，以及这一节中例子所需的一些测试数据。

```
$ python3 zipimport_make_example.py

__init__.pyc
example_package/__init__.pyc
zipimport_find_module.pyc
zipimport_get_code.pyc
```

---

[1] https://setuptools.readthedocs.io/en/latest/
[2] www.python.org/dev/peps/pep-0302

```
zipimport_get_data.pyc
zipimport_get_data_nozip.pyc
zipimport_get_data_zip.pyc
zipimport_get_source.pyc
zipimport_is_package.pyc
zipimport_load_module.pyc
zipimport_make_example.pyc
zipimport_get_source.py
example_package/README.txt
```

19.3.2 查找模块

给定模块的全名，`find_module()`会尝试在 ZIP 归档中查找这个模块。

代码清单 19-23：**zipimport_find_module.py**

```
import zipimport

importer = zipimport.zipimporter('zipimport_example.zip')

for module_name in ['zipimport_find_module', 'not_there']:
    print(module_name, ':', importer.find_module(module_name))
```

如果找到了这个模块，则会返回`zipimporter`实例。否则，返回`None`。

```
$ python3 zipimport_find_module.py

zipimport_find_module : <zipimporter object
"zipimport_example.zip">
not_there : None
```

19.3.3 访问代码

`get_code()`方法从归档中加载一个模块的代码对象。

代码清单 19-24：**zipimport_get_code.py**

```
import zipimport

importer = zipimport.zipimporter('zipimport_example.zip')
code = importer.get_code('zipimport_get_code')
print(code)
```

代码对象与模块对象不同，不过可以用代码对象创建一个模块对象。

```
$ python3 zipimport_get_code.py

<code object <module> at 0x1012b4ae0, file
"./zipimport_get_code.py", line 6>
```

要加载这个代码作为一个可用的模块，可以使用`load_module()`。

代码清单 19-25：**zipimport_load_module.py**

```
import zipimport

importer = zipimport.zipimporter('zipimport_example.zip')
module = importer.load_module('zipimport_get_code')
print('Name    :', module.__name__)
```

```
print('Loader :', module.__loader__)
print('Code   :', module.code)
```

结果会得到一个模块对象,就好像是从常规的 import 语句中加载代码一样。

```
$ python3 zipimport_load_module.py

<code object <module> at 0x1007b4c00, file
"./zipimport_get_code.py", line 6>
Name   : zipimport_get_code
Loader : <zipimporter object "zipimport_example.zip">
Code   : <code object <module> at 0x1007b4c00, file
"./zipimport_get_code.py", line 6>
```

19.3.4 源代码

类似于 inspect 模块,可以用 zipimport 模块从 ZIP 归档获取一个模块的源代码(如果归档中包含这个源代码)。在下面的例子中,zipimport_example.zip 内只增加了 zipimport_get_source.py;其余模块都是作为 .pyc 文件被增加的。

代码清单 19-26:zipimport_get_source.py

```
import zipimport

modules = [
    'zipimport_get_code',
    'zipimport_get_source',
]

importer = zipimport.zipimporter('zipimport_example.zip')
for module_name in modules:
    source = importer.get_source(module_name)
    print('=' * 80)
    print(module_name)
    print('=' * 80)
    print(source)
    print()
```

如果不能得到一个模块的源代码,则 get_source() 会返回 None。

```
$ python3 zipimport_get_source.py

============================================================
zipimport_get_code
============================================================
None

============================================================
zipimport_get_source
============================================================
#!/usr/bin/env python3
#
# Copyright 2007 Doug Hellmann.
#
"""Retrieving the source code for a module within a zip archive.
"""

#end_pymotw_header
import zipimport
```

```
modules = [
    'zipimport_get_code',
    'zipimport_get_source',
]

importer = zipimport.zipimporter('zipimport_example.zip')
for module_name in modules:
    source = importer.get_source(module_name)
    print('=' * 80)
    print(module_name)
    print('=' * 80)
    print(source)
    print()
```

19.3.5 包

要确定一个名指示的是一个包还是一个常规的模块，可以使用 is_package()。

代码清单 19-27：zipimport_is_package.py

```
import zipimport

importer = zipimport.zipimporter('zipimport_example.zip')
for name in ['zipimport_is_package', 'example_package']:
    print(name, importer.is_package(name))
```

在这里，zipimport_is_package 来自一个模块，example_package 是一个包。

```
$ python3 zipimport_is_package.py

zipimport_is_package False
example_package True
```

19.3.6 数据

有些情况下，源模块或包发布时还需要提供非代码的数据。图像、配置文件、默认数据和测试固件就是数据的例子。通常，可以用模块的 __path__ 或 __file__ 属性来查找这些数据文件（相对于代码安装目录）。

例如，对于一个"正常的"模块，可以由导入的包的 __file__ 属性来构造文件系统路径，如以下代码所示。

代码清单 19-28：zipimport_get_data_nozip.py

```
import os
import example_package

# Find the directory containing the imported
# package and build the data filename from it.
pkg_dir = os.path.dirname(example_package.__file__)
data_filename = os.path.join(pkg_dir, 'README.txt')

# Read the file and show its contents.
print(data_filename, ':')
print(open(data_filename, 'r').read())
```

输出取决于示例代码位于文件系统的哪个位置。

```
$ python3 zipimport_get_data_nozip.py

.../example_package/README.txt :
This file represents sample data which could be embedded in the
ZIP archive.  You could include a configuration file, images, or
any other sort of noncode data.
```

如果 example_package 是从 ZIP 归档导入而不是从文件系统加载,则无法使用 __file__。

代码清单 19-29:**zipimport_get_data_zip.py**

```python
import sys
sys.path.insert(0, 'zipimport_example.zip')

import os
import example_package
print(example_package.__file__)
data_filename = os.path.join(
    os.path.dirname(example_package.__file__),
    'README.txt',
)
print(data_filename, ':')
print(open(data_filename, 'rt').read())
```

包的 __file__ 指向 ZIP 归档,而不是一个目录,所以在构建指向 README.txt 文件的路径时,会给出错误的值。

```
$ python3 zipimport_get_data_zip.py

zipimport_example.zip/example_package/__init__.pyc
zipimport_example.zip/example_package/README.txt :
Traceback (most recent call last):
  File "zipimport_get_data_zip.py", line 20, in <module>
    print(open(data_filename, 'rt').read())
NotADirectoryError: [Errno 20] Not a directory:
'zipimport_example.zip/example_package/README.txt'
```

要获取文件,一种更可靠的方式是使用 get_data() 方法。可以通过已导入模块的 __loader__ 属性来访问加载模块的 zipimporter 实例。

代码清单 19-30:**zipimport_get_data.py**

```python
import sys
sys.path.insert(0, 'zipimport_example.zip')

import os
import example_package
print(example_package.__file__)
data = example_package.__loader__.get_data(
    'example_package/README.txt')
print(data.decode('utf-8'))
```

pkgutil.get_data() 使用这个接口来访问包中的数据。返回的值是一个字节串,打印之前需要将其解码为一个 Unicode 串。

```
$ python3 zipimport_get_data.py

zipimport_example.zip/example_package/__init__.pyc
This file represents sample data which could be embedded in the
ZIP archive.  You could include a configuration file, images, or
any other sort of noncode data.
```

对于不是通过 `zipimport` 导入的模块，不会设置 `__loader__`。

提示：相关阅读材料
- `zipimport` 的标准库文档[一]。
- `zipimport` 的 Python 2 到 Python 3 移植说明。
- `imp`：其他与导入相关的函数。
- `pkgutil`：提供到 `get_data()` 的一个更通用的接口。
- `zipfile`：读写 ZIP 归档文件。
- **PEP 302**[二]：新的导入 hook。

[一] https://docs.python.org/3.5/library/zipimport.html
[二] www.python.org/dev/peps/pep-0302

附录 A
移植说明

本附录介绍从 Python 2 更新到 Python 3 的说明和提示，包括各个模块中变更的相关总结和参考资料。

A.1 参考资料

本节中的说明主要基于 Python 开发团队和发行经理为每个版本准备的系列"新增内容"文档。

- Python 3.0 新增内容[一]
- Python 3.1 新增内容[二]
- Python 3.2 新增内容[三]
- Python 3.3 新增内容[四]
- Python 3.4 新增内容[五]
- Python 3.5 新增内容[六]

关于移植到 Python 3 的更多信息，参见以下文档：

- Porting Python 2 Code to Python 3[七]
- Porting to Python 3[八]，作者 Lennart Regebro
- python-porting[九]邮件列表

A.2 新增模块

Python 3 包含很多新模块，提供了 Python 2 中没有的一些特性：
`asyncio`：异步 I/O、事件循环和其他并发工具。

[一] https://docs.python.org/3.0/whatsnew/3.0.html
[二] https://docs.python.org/3.1/whatsnew/3.1.html
[三] https://docs.python.org/3.2/whatsnew/3.2.html
[四] https://docs.python.org/3.3/whatsnew/3.3.html
[五] https://docs.python.org/3.4/whatsnew/3.4.html
[六] https://docs.python.org/3.5/whatsnew/3.5.html
[七] https://docs.python.org/3/howto/pyporting.html
[八] http://python3porting.com/
[九] http://mail.python.org/mailman/listinfo/python-porting

concurrent.futures：并发任务的管理池。
ensurepip：安装 Python 包安装工具（Python Package Installer，pip）。
enum：定义枚举类型。
ipaddress：处理 Internet Protocol (IP) 地址的类。
pathlib：这是一个处理文件系统路径的面向对象 API。
selectors：I/O 多路复用抽象。
statistics：统计计算。
venv：创建隔离抽象和执行上下文。

A.3 重命名的模块

很多标准库模块已经在 Python 2 和 Python 3 之间重命名，见 **PEP 3108** 说明。所有新模块名都一致地使用小写，有一些模块移到了包中，以便更好地组织相关的模块。使用这些模块的代码通常只需要调整 import 语句就可以被更新为使用 Python 3。重命名模块的完整列表见字典 lib2to3.fixes.fix_imports.MAPPING（键是 Python 2 名，值是 Python 3 名），另外表 A-1 也列出了这些重命名的模块。

提示：相关阅读材料

- Six[1] 包可以用于编写可同时运行于 Python 2 和 Python 3 的代码。具体地，six.moves 模块允许代码使用一个 import 语句来导入重命名的模块，它会根据 Python 的版本自动将 import 重定向到名字的正确版本。
- **PEP 3108**[2]：标准库重组。

表 A-1 重命名的模块

Python 2 名	Python 3 名
__builtin__	builtins
_winreg	winreg
BaseHTTPServer	http.server
CGIHTTPServer	http.server
commands	subprocess
ConfigParser	configparser
Cookie	http.cookies
cookielib	http.cookiejar
copy_reg	copyreg
cPickle	pickle
cStringIO	io
dbhash	dbm.bsd

---

[1] http://pythonhosted.org/six/

[2] www.python.org/dev/peps/pep-3108

（续）

Python 2 名	Python 3 名
dbm	dbm.ndbm
Dialog	tkinter.dialog
DocXMLRPCServer	xmlrpc.server
dumbdbm	dbm.dumb
FileDialog	tkinter.filedialog
gdbm	dbm.gnu
htmlentitydefs	html.entities
HTMLParser	html.parser
httplib	http.client
Queue	queue
repr	reprlib
robotparser	urllib.robotparser
ScrolledText	tkinter.scrolledtext
SimpleDialog	tkinter.simpledialog
SimpleHTTPServer	http.server
SimpleXMLRPCServer	xmlrpc.server
SocketServer	socketserver
StringIO	io
Tix	tkinter.tix
tkColorChooser	tkinter.colorchooser
tkCommonDialog	tkinter.commondialog
Tkconstants	tkinter.constants
Tkdnd	tkinter.dnd
tkFileDialog	tkinter.filedialog
tkFont	tkinter.font
Tkinter	tkinter
tkMessageBox	tkinter.messagebox
tkSimpleDialog	tkinter.simpledialog
ttk	tkinter.ttk
urlparse	urllib.parse
UserList	collections
UserString	collections
xmlrpclib	xmlrpc.client

A.4 删除的模块

以下这些模块可能不再提供，或者它们的特性已经被合并到其他现有模块中。

A.4.1 bsddb

`bsddb` 和 `dbm.bsd` 模块已经被删除。与 Berkeley DB 的绑定现在会在标准库之外作为 bsddb3[①] 维护。

A.4.2 commands

`commands` 模块在 Python 2.6 中废弃,并在 Python 3.0 中被完全删除。参见替代的 `subprocess`。

A.4.3 compiler

`compiler` 模块已经被删除。参见替代的 `ast`。

A.4.4 dircache

`dircache` 模块已经被删除,没有替代模块。

A.4.5 EasyDialogs

`EasyDialogs` 模块已经被删除。参见替代的 `tkinter`。

A.4.6 exceptions

`exceptions` 模块已经被删除,因为这里定义的所有异常都已经作为内置类提供。

A.4.7 htmllib

`htmllib` 模块已经被删除。参见替代的 `html.parser`。

A.4.8 md5

MD5 消息摘要算法的实现已经移到 `hashlib`。

A.4.9 mimetools、MimeWriter、mimify、multifile 和 rfc822

`mimetools`、`MimeWriter`、`mimify`、`multifile` 和 `rfc822` 模块已经被删除。参见替代的 `email`。

A.4.10 popen2

`popen2` 模块已经被删除。参见替代的 `subprocess`。

A.4.11 posixfile

`posixfile` 模块已经被删除。参见替代的 `io`。

[①] https://pypi.python.org/pypi/bsddb3

A.4.12 sets

`sets` 模块在 Python 2.6 中已经废弃，并在 Python 3.0 中被完全删除。应用内置类型 `set` 和 `orderedset`。

A.4.13 sha

SHA-1 消息摘要算法的实现已经移至 `hashlib`。

A.4.14 sre

`sre` 模块在 Python 2.5 中已经废弃，并在 Python 3.0 中被完全删除。使用替代的 `re`。

A.4.15 statvfs

`statvfs` 模块在 Python 2.6 中已经废弃，并在 Python 3.0 中被完全删除。参见 `os` 模块中的 `os.statvfs()`。

A.4.16 thread

`thread` 模块已经被删除。应使用 `threading` 中的更高层 API。

A.4.17 user

`user` 模块在 Python 2.6 中已经废弃，并在 Python 3.0 中被完全删除。参见 `site` 模块中提供的用户定制特性。

A.5 废弃的模块

下面这些模块仍被包含在标准库中，不过已经废弃，在新的 Python 3 程序中不应使用。

A.5.1 asyncore 和 asynchat

异步 I/O 和协议处理器。参见替代的 `asyncio`。

A.5.2 formatter

通用输出格式化器和设备接口。相关详细信息参见 Python issue 18716[⊖]。

A.5.3 imp

访问 import 语句的实现。参见替代的 `importlib`。

A.5.4 optparse

命令行选项解析库。`argparse` 的 API 类似于 `optparse` 提供的 API，在很多情况下，

[⊖] http://bugs.python.org/issue18716

通过更新使用的类和方法名,可以用 `argparse` 作为简单替代。

A.6 模块变更小结

A.6.1 abc

`abstractproperty()`、`abstractclassmethod()` 和 `abstractstaticmethod()` 修饰符已经废弃。将 `abstractmethod()` 与 `property()`、`classmethod()` 以及 `staticmethod()` 修饰符结合使用可以完成预想的工作(Python issue 11610[①])。

A.6.2 anydbm

`anydbm` 模块在 Python 3 中被重命名为 `dbm`。

A.6.3 argparse

`ArgumentParser` 的 `version` 参数已经去除,而代之以一个特殊的 `action` 类型(Python issue 13248[②])。

原来的形式会传入 `version` 作为参数。

```
parser = argparse.ArgumentParser(version='1.0')
```

新形式要求增加一个显式的参数定义。

```
parser = argparse.ArgumentParser()
parser.add_argument('--version', action='version',
                    version='%(prog)s 1.0')
```

可以修改选项名和版本格式串以满足应用的需要。

在 Python 3.4 中,`version` 动作被改为向 stdout 而不是 stderr 打印版本串(Python issue 18920[③])。

A.6.4 array

Python 2 较早版本中用于字符字节的 `'c'` 类型已经被删除。对于字节要使用 `'b'` 或 `'B'`。

对应 Unicode 字符串中字符的 `'u'` 类型已经废弃,并将在 Python 4.0 中被完全删除。

方法 `tostring()` 和 `fromstring()` 已经分别改名为 `tobytes()` 和 `frombytes()`,以消除二义性(Python issue 8990[④])。

A.6.5 atexit

更新 atexit 来包含一个 C 实现时(Python issue 1680961[⑤]),在错误处理逻辑中引

[①] http://bugs.python.org/issue11610
[②] http://bugs.python.org/issue13248
[③] http://bugs.python.org/issue18920
[④] http://bugs.python.org/issue8990
[⑤] http://bugs.python.org/issue1680961

入了一个回归错误,这个错误导致只显示异常的小结,而没有 traceback。这个回归错误在 Python 3.3 中得到了修正(Python issue 18776[一])。

A.6.6 base64

encodestring() 和 decodestring() 函数已经分别改名为 encodebytes() 和 decodebytes()。原来的名字仍可以作为别名,不过已经废弃(Python issue 3613[二])。

增加了两个使用 85 字符字母表的新编码。b85encode() 实现了 Mercurial 和 git 中使用的编码,而 a85encode() 实现了 PDF 文件使用的 Ascii85 格式(Python issue 17618[三])。

A.6.7 bz2

BZ2File 实例现在支持上下文管理器协议,不需要用 contextlib.closing() 包装。

A.6.8 collections

原先在 collections 中定义的抽象基类现在移至 collections.abc,collections 中目前还提供向后兼容的导入(Python issue 11085[四])。

A.6.9 commands

函数 getoutput() 和 getstatusoutput() 已经移至 subprocess,另外 commands 已经被删除。

A.6.10 configparser

老的 ConfigParser 模块已经改名为 configparser。

老的 ConfigParser 类已经被删除,而代之以 SafeConfigParser,而它已经改名为 ConfigParser。可以通过 LegacyInterpolation 得到已废弃的拼接行为。

read() 方法现在支持一个 encoding 参数,所以不再需要使用 codecs 读取包含 Unicode 值的配置文件。

不鼓励使用老的 RawConfigParser。新项目应当使用 ConfigParser(interpolation=None) 来得到同样的行为。

A.6.11 contextlib

contextlib.nested() 已经被删除。应当向同样的 with 语句传入多个上下文管理器。

A.6.12 csv

不要直接使用一个阅读器的 next() 方法,应当使用内置的 next() 函数正确地调用

[一] http://bugs.python.org/issue18776
[二] http://bugs.python.org/issue3613
[三] http://bugs.python.org/issue17618
[四] http://bugs.python.org/issue11085

迭代器。

A.6.13 datetime

从 Python 3.3 开始，本地和考虑时区的 `datetime` 实例之间的相等性比较会返回 `False`，而不是产生 `TypeError`（Python issue 15006①）。

在 Python 3.5 之前，表示午夜的 `datetime.time` 对象在转换为一个布尔值时会被估算为 `False`。这种行为在 Python 3.5 中已经被删除（Python issue 13936②）。

A.6.14 decimal

Python 3.3 加入了基于 `libmpdec` 的 `decimal` 的一个 C 表示。这个变更可以提高性能，不过也包含不同于纯 Python 实现的一些 API 变更和行为差异。相关的详细内容参见 Python 3.3 发布说明③。

A.6.15 fractions

不再需要 `from_float()` 和 `from_decimal()` 类方法。浮点数和 `Decimal` 值可以直接传入 `Fraction` 构造函数。

A.6.16 gc

标志 `DEBUG_OBJECT` 和 `DEBUG_INSTANCE` 已经被删除。不再需要它们来区分新式和老式的类。

A.6.17 gettext

`gettext` 中的所有转换函数都假设有 Unicode 输入和输出，类似 `ugettext()` 的 Unicode 版本已经被删除。

A.6.18 glob

新函数 `escape()` 实现了一种方法来搜索名字中包含元字符的文件（Python issue 8402④）。

A.6.19 http.cookies

除了转义引号，SimpleCookie 还会对值中的逗号和分号编码，从而更好地反映真实浏览器的行为（Python issue 9824⑤）。

---

① http://bugs.python.org/issue15006
② http://bugs.python.org/issue13936
③ https://docs.python.org/3.3/whatsnew/3.3.html#decimal
④ bugs.python.org/issue8402
⑤ http://bugs.python.org/issue9824

A.6.20 imaplib

在 Python 3 下，`imaplib` 返回编码为 UTF-8 的字节串。它接受 Unicode 串并在发出命令时自动编码，或者作为登录服务器的用户名和密码。

A.6.21 inspect

函数 `getargspec()`、`getfullargspec()`、`getargvalues()`、`getcallargs()`、`getargvalues()`、`formatargspec()` 和 `formatargvalues()` 已经废弃，代之以 `signature()`（Python issue 20438[一]）。

A.6.22 itertools

函数 `imap()`、`izip()` 和 `ifilter()` 已经替换为内置函数版本（分别是 `map()`、`zip()` 和 `filter:()`），它们会返回可迭代对象而不是 `list` 对象。

函数 `ifilterfalse()` 已经改名为 `filterfalse()`。

A.6.23 json

`json` API 已经更新为只支持 `str` 而不支持 `bytes`，因为 JSON 规范是使用 Unicode 定义的。

A.6.24 locale

UTF-8 编码的标准名字从"UTF8"改为"UTF-8"，因为 Mac OS X 和 OpenBSD 不支持使用"UTF8"（Python issue 10154[二]和 Python issue 10090[三]）。

A.6.25 logging

`logging` 模块现在包含一个 `lastResort` 日志记录器，如果一个应用没有完成其他日志配置，就使用这个日志记录器。这样如果只是为了避免用户看到错误消息，则将不再需要应用配置 `logging`（有可能应用导入的一个库使用了 `logging`，但是应用本身并没有使用）。

A.6.26 mailbox

`mailbox` 以二进制模式读写邮箱文件，依赖于 email 包来解析消息。`StringIO` 和文本文件输入已经废弃（Python issue 9124[四]）。

A.6.27 mmap

读取 API 返回的值是字节串，在将其处理为文本之前需要先解码。

[一] bugs.python.org/issue20438
[二] http://bugs.python.org/issue10154
[三] http://bugs.python.org/issue10090
[四] http://bugs.python.org/issue9124

A.6.28 operator

`div()` 函数已经被删除。取决于所需的语义，应当使用 `floordiv()` 或 `truediv()`。
`repeat()` 函数已经被删除。应当使用 `mul()`。

函数 `getslice()`、`setslice()` 和 `delslice()` 已经被删除，应当分别使用 `getitem()`、`setitem()` 和 `delitem()`，并使用分片索引。

`isCallable()` 已经被删除。应当使用抽象基类 `collections.Callable`。

```
isinstance(obj, collections.Callable)
```

类型检查函数 `isMappingType()`、`isSequenceType()` 和 `isNumberType()` 已经被删除。应当使用 `collections` 或 `numbers` 中的相应抽象基类。

```
isinstance(obj, collections.Mapping)
isinstance(obj, collections.Sequence)
isinstance(obj, numbers.Number)
```

`sequenceIncludes()` 函数已经被删除。应当使用替代的 `contains()`。

A.6.29 os

函数 `popen2()`、`popen3()` 和 `popen4()` 已经被删除。`popen()` 仍可用，不过已经废弃，如果使用它则会得到一个警告。使用这些函数的代码应当被重写为使用 `subprocess`，从而在不同操作系统之间更可移植。

函数 `os.tmpnam()`、`os.tempnam()` 和 `os.tmpfile()` 已经删除。应当使用 `tempfile` 模块。

函数 `os.stat_float_times()` 已经废弃（Python issue 14711[一]）。`os.unsetenv()` 不再忽略错误（Python issue 13415[二]）。

A.6.30 os.path

`os.path.walk()` 已经被删除。应当使用替代的 `os.walk()`。

A.6.31 pdb

`print` 命令别名已经被删除，不再遮蔽 `print()` 函数（Python issue 18764[三]）。p 快捷方式仍保留。

A.6.32 pickle

Python 2 中 `pickle` 模块的 C 实现已经移至一个新模块，会尽可能自动使用这个实现来替代 Python 实现。原来的导入方式（即在 `pickle` 之前先寻找 `cPickle` 的做法）已经

[一] http://bugs.python.org/issue14711
[二] http://bugs.python.org/issue13415
[三] http://bugs.python.org/issue18764

不再需要。

```
try:
    import cPickle as pickle
except:
    import pickle
```

基于 C 实现的自动导入，只需要直接导入 `pickle` 模块。

```
import pickle
```

对于腌制数据，Python 2.x 和 3.x 之间的互操作性有所改进，这里使用 level 2 或更低级协议来解决由于移植到 Python 3 时大量标准库模块重命名所带来的问题。由于腌制数据包含类和类型名的引用，而那些名字已经改变，所以很难在 Python 2 和 Python 3 程序之间交换腌制数据。现在，对于使用 level 2 或更老协议腌制的数据，读写 `pickle` 流时会自动使用这些类原来的名字。

这种行为是默认的，不过可以使用 `fix_imports` 选项关闭这种行为。这个变化会改善情况，不过不会完全消除不兼容性。具体地，在 Python 3.1 下腌制的数据可能在 Python 3.0 下就不可读。为了确保 Python 3 应用之间最大的可移植性，要使用 level 3 协议（其中不包含这种兼容性特性）。

默认的协议版本已经从 0（人类可读的版本）变为 3，即在 Python 3 应用之间共享时可以提供最佳互操作性的二进制格式。

读回由一个 Python 2.x 应用写至一个 pickle 的字节串数据来创建一个 Unicode 字符串对象时，该数据会被解码。转换的编码默认为 ASCII，可以向 `Unpickler` 传入值来改变编码。

A.6.33　pipes

`pipes.quote()` 已经移至 `shlex`（Python issue 9723[1]）。

A.6.34　platform

`platform.popen()` 已经废弃。应当使用替代的 `subprocess.popen()`（Python issue 11377[2]）。

`platform.uname()` 现在返回一个 `namedtupl`。

由于 Linux 发布版本没有提供一种一致的描述方法，所得到描述的函数（`platform.dist()` 和 `platform.linux_distribution()`）已经废弃，并计划在 Python3.7 中被完全删除（Python issue 1322[3]）。

A.6.35　random

函数 `jumpahead()` 在 Python 3.0 已经被删除。

---

[1] http://bugs.python.org/issue9723
[2] http://bugs.python.org/issue11377
[3] http://bugs.python.org/issue1322

A.6.36 re

`UNICODE` 标志表示默认行为。要恢复 Python 2 中特定于 ASCII 的行为，应当使用 `ASCII` 标志。

A.6.37 shelve

`shelve` 的默认输出格式会创建一个文件，在提供给 `shelve.open()` 的名字后面增加 `.db` 扩展名。

A.6.38 signal

PEP 475[1]强制要求系统调用中断，并重试 EINTR 返回。这会改变信号处理器和其他系统调用的行为。现在，信号处理返回后，会重试中断的调用，除非信号处理产生一个异常。参考 PEP 文档来了解完整的详细内容。

A.6.39 socket

在 Python 2 中，通常可以在 socket 上直接发送 `str` 对象。因为 `str` 取代了 `unicode`，在 Python 3 中，值在发送前必须先编码。socket 一节中的例子使用了已经编码的字节串。

A.6.40 socketserver

`socketserver` 模块在 Python 2 中名为 `SocketServer`。

A.6.41 string

`string` 模块中同时也是 `str` 对象方法的所有函数已经被删除。

常量 `letters`、`lowercase` 和 `uppercase` 已经被删除。有类似名字的新常量仅限于 ASCII 字符集。

`maketrans()` 函数已经被 `str`、`bytes` 和 `bytearra` 上的方法取代，以明确各个转换表支持哪些输入类型。

A.6.42 struct

`struct.pack()` 使用 s 串包装码时只支持字节串，不再隐式地将字符串对象编码为 UTF-8（Python issue 10783[2]）。

A.6.43 subprocess

`subprocess.Popen` 的 `close_fds` 参数的默认值不再总是 `False`。在 UNIX 下它总是默认为 `True`。在 Windows 下，如果标准 I/O 流参数被设置为 `None`，那么默认为 `True`；

---

[1] www.python.org/dev/peps/pep-0475
[2] http://bugs.python.org/issue10783

否则，默认为 False。

A.6.44 sys

变量 `sys.exitfunc` 不再在程序退出时检查是否运行一个清理动作。应当使用 `atexit`。

不再定义变量 `sys.subversion`。

不再定义标志 `sys.flags.py3k_warning`、`sys.flags.division_warning`、`sys.flags.division_new`、`sys.flags.tabcheck` 和 `sys.flags.unicode`。

不再定义 `sys.maxint`；应当使用 `sys.maxsize`。参见 **PEP 237**[1]（统一长整型和整型）。

全局异步跟踪变量 `sys.exc_type`、`sys.exc_value` 和 `sys.exc_traceback` 已经被删除。函数 `sys.exc_clear()` 也已经被删除。

变量 `sys.version_info` 现在是一个 `namedtuple` 实例，包含属性 `major`、`minor`、`micro`、`releaselevel` 和 `serial`（Python issue 4285[2]）。

检查间隔特性（在允许一个线程上下文切换之前控制可执行的操作码数）被替换为一个绝对时间值，这个值由 `sys.setswitchinterval()` 管理。管理检查间隔的老函数 `sys.getcheckinterval()` 和 `sys.setcheckinterval()` 已经废弃。

`sys.meta_path` 和 `sys.path_hooks` 变量现在会提供导入模块的所有路径查找工具和入口 hook。在较早的版本中，只会提供显式增加到路径的查找工具和 hook，C 导入在实现中使用的值无法从外部修改。

对于 Linux 系统，`sys.platform` 不再包含版本号。现在值只是 `linux`，而不是 `linux2` 或 `linux3`。

A.6.45 threading

`thread` 模块已经废弃，代之以 `threading` 中的 API。

`threading` 的调试特性（包括 "verbose" 参数）已经从 API 中删除（Python issue 13550[3]）。

`threading` 较早的实现为一些类使用了工厂函数，因为它们在 C 中被实现为扩展类型，而不能派生子类。这个限制已经去除，所以很多老的工厂函数已经被转换为标准类，从而允许派生子类（Python issue 10968[4]）。

`threading` 导出的公共符号已经被重命名为符合 **PEP 8**[5]。原来的名字仍保留以支持向后兼容，不过在将来的版本中会删除。

A.6.46 time

`time.asctime()` 和 `time.ctime()` 已经改为不使用相同时间的系统函数，从而

---

[1] www.python.org/dev/peps/pep-0237
[2] http://bugs.python.org/issue4285
[3] http://bugs.python.org/issue13550
[4] http://bugs.python.org/issue10968
[5] www.python.org/dev/peps/pep-0008

允许使用更大的年份。`time.ctime()` 现在支持从 1900 到 `maxint` 的年份，不过对于大于 9999 的值，输出串不会超过标准的 24 个字符，使得年份可以不只是 4 位（Python issue 8013[①]）。

A.6.47 unittest

以 "fail" 开头的 `TestCase` 方法（例如 `failIf()`、`failUnless()`）已经废弃。要使用断言方法的替代形式。

很多较老的方法别名已经废弃，代之以现用名。如果使用废弃的名字，则会产生一个警告（Python issue 9424[②]）。参见表 A-2，这里给出了老名字与新名字的一个映射。

表 A-2 废弃的 `unittest.TestCase` 方法

废弃名字	现用名
`assert_()`	`assertTrue()`
`assertEquals()`	`assertEqual()`
`assertNotEquals()`	`assertNotEqual()`
`assertAlmostEquals()`	`assertAlmostEqual()`
`assertNotAlmostEquals()`	`assertNotAlmostEqual()`

A.6.48 UserDict、UserList 和 UserString

`UserDict`、`UserList` 和 `UserString` 类都已经从各自的模块中移出，移到了 `collections` 模块中。`dict`、`list` 和 `str` 可以直接派生，不过 `collections` 中的类会让实现子类更简单，因为可以直接通过实例属性得到容器的内容。`collections.abc` 中的抽象类还可以用来创建满足内置类型 API 的定制容器。

A.6.49 uuid

`uuid.getnode()` 现在使用 PATH 环境变量来查找可以报告 UNIX 主机 MAC 地址的程序（Python issue 19855[③]）。如果在搜索路径上没有找到这样的程序，那么它会在 /sbin 和 /usr/sbin 中查找。如果有 `netstat`、`ifconfig`、`ip` 和 `arp` 之类程序的其他版本，那么这种搜索行为得到的结果可能与 Python 的较早版本不同。

A.6.50 whichdb

`whichdb` 的功能已经移至 `dbm` 模块。

A.6.51 xml.etree.ElementTree

`XMLTreeBuilder` 已经改名为 `TreeBuilder`，而且 API 也经过了多次修改。

[①] http://bugs.python.org/issue8013
[②] http://bugs.python.org/issue9424
[③] http://bugs.python.org/issue19855

ElementTree.getchildren() 已经废弃。要使用 list(elem) 来构建子元素的列表。

ElementTree.getiterator() 已经废弃。要使用 iter() 来创建一个使用普通迭代器协议的迭代器。

解析失败时，不是产生 xml.parsers.expat.ExpatError，现在 XMLParser 会产生 xml.etree.ElementTree.ParseError。

A.6.52 zipimport

get_data() 返回的数据是一个字节串，在被用作 Unicode 串之前需要先解码。

附录 B
标准库之外

Python 标准库相当丰富，除此以外，作为补充还有一个由第三方开发人员提供的健壮的模块生态系统，可以从 Python Package Index①获得。本附录描述了这样一些模块，另外还将介绍会在什么情况下使用这些模块来补充甚至替代标准库。

B.1 文本

string 模块提供了一个非常基本的模板工具。很多 Web 框架都提供了更强大的模板工具，不过 Jinja②和 Mako③是很流行的独立工具。这两个工具都支持循环和条件控制结构，另外还支持其他一些特性，可以结合数据和模板生成文本输出。

re 模块包含一些函数可以使用形式化描述的模式（称为正则表达式）来搜索和解析文本。不过这不是解析文本的唯一途径。

PLY④包支持以 GNU 工具 lexx 和 yacc 的方式构建解析器，这通常用于构建语言编译器。通过提供描述合法 token 的输入、语法工具和遇到各个 token 时要采取的动作，可以构建功能完备的编译器和解释器以及更直接的数据解析器。

PyParsing⑤是构建解析器的另一个工具。其输入是类实例，这些类实例可以使用操作符和方法调用串链在一起来构建一个语法工具。

最后，NLTK⑥是一个处理自然语言文本的包——也就是说，处理人类语言而不是计算机语言。它支持解析句子（解析为语音部分）、找出单词词根和基本的语义处理。

B.2 算法

functools 模块包含一些创建修饰符的工具，修饰符也是函数，它们可以包装其他函数来改变这些函数的行为。wrapt⑦包比 functools.wrap() 更进一步，可以确保适当地构造一个修饰符并适用于所有边界情况。

---

① https://pypi.python.org/pypi
② http://jinja.pocoo.org
③ http://docs.makotemplates.org/en/latest/
④ www.dabeaz.com/ply/
⑤ http://pyparsing.wikispaces.com
⑥ www.nltk.org
⑦ http://wrapt.readthedocs.org/

B.3 日期和时间

`time` 和 `datetime` 模块提供了处理时间和日期值的函数和类。这两个模块都包含一些函数来解析字符串，将它们转换为内部表示。dateutil[①]包包含一个更灵活的解析器，可以更容易地构建健壮的应用，能更好地支持不同的输入格式。

`datetime` 模块包含一个考虑时区的类来表示某一天的某个特定时间。不过，它不包含完备的时区数据库。pytz[②]包则提供了这样一个数据库。这个包未包含在标准库中而是单独发布，因为它由其他作者维护，另外这个数据库需要在控制时区和夏令时的行政机构改变这些值时频繁地更新。

B.4 数学

`math` 模块包含一些高级数学函数的快速实现。NumPy[③]扩展了支持的函数集，包含了线性几何和傅立叶变换函数。它还包含一个快速多维数组实现，改进了 `array` 中的数组版本。

B.5 数据持久存储和交换

`sqlite3` 一节中的例子直接运行 SQL 语句并处理底层数据结构。对于大型应用，通常需要使用一个对象关系映射工具（Object Relational Mapper，ORM）来将类映射到数据库中的表。SQLAlchemy[④] ORM 库提供了相应的 API，可以将类与数据库表关联，建立查询以及连接到不同类型的生产级关系数据库。

lxml[⑤]包包装了 libxml2 和 libxslt 库，以创建替代 `xml.etree.ElementTree` 中 XML 解析器的候选工具。如果开发人员已经熟悉如何使用其他语言中的这些库，那么会发现在 Python 中使用 lxml 更为容易。

defusedxml[⑥]包提供了对"billion laughs"[⑦]以及 Python XML 库中其他实体扩展拒绝服务攻击漏洞的修正，与单独使用标准库相比，可以更安全地处理不可信的 XML。

B.6 密码

构建 cryptography[⑧]包的团队指出，"我们的目标是让它成为你的'密码标准库'"。

---
① https://dateutil.readthedocs.io/
② http://pythonhosted.org/pytz/
③ www.numpy.org
④ www.sqlalchemy.org
⑤ http://lxml.de
⑥ https://pypi.python.org/pypi/defusedxml
⑦ http://en.wikipedia.org/wiki/Billion_laughs
⑧ https://cryptography.io/en/latest/

cryptography 包提供了一些高层 API，可以很容易地向应用增加加密特性。这个包得到了积极的维护，频繁发布的新版本可用来解决底层库（如 OpenSSL）的脆弱性。

B.7 进程、线程和协程的并发

asyncio 中内建的事件循环是基于这个模块定义的抽象 API 的一个参考实现。可以用一个类似 uvloop[一]的库来替代这个事件循环，它能提供更好的性能，但代价是会增加额外的应用依赖库。

curio[二]包是另一个并发包，它类似于 asyncio，不过有一个更小的 API，把所有一切都当作协程。它不像 asyncio 那样支持回调。

Twisted[三]库提供了一个可扩展的框架来完成 Python 编程，特别强调基于事件的网络编程和多协议集合。这个库很成熟、很健壮，而且提供了丰富的文档。

B.8 互联网

requests[四]包是替代 urllib.request 的一个非常流行的工具。它提供了一致的 API 来处理可以通过 HTTP 访问的远程资源（包括健壮的 SSL 支持），而且在多线程应用中可以使用连接池来得到更好的性能。这个包提供的另外一些特性使它很适合访问 REST API，如内置的 JSON 解析。

Python 的 html 模块包括一个基本解析工具来解析良构的 HTML 数据。不过，真实世界的数据很少是良构的，所以解析这些数据很成问题。BeautifulSoup[五]和 PyQuery[六]库是 html 的替代工具，对于杂乱的数据，它们更为健壮。这两个库都定义了解析、修改和构造 HTML 的 API。

内置 http.server 包提供了一些类可以从头创建简单的 HTTP 服务器。不过，除了构建基于 Web 的应用，它并没有提供太多其他的支持。Django[七]和 Pyramid[八]包是两个流行的Web应用框架，对请求解析、URL路由和 cookie 处理等高级特性提供了更多支持。

很多现有的库不能使用 asyncio，因为这些库没有被更新为处理事件循环。作为 aio-libs[九]

[一] http://uvloop.readthedocs.io
[二] https://github.com/dabeaz/curio
[三] https://twistedmatrix.com/
[四] http://docs.python-requests.org/
[五] www.crummy.com/software/BeautifulSoup/
[六] http://pyquery.rtfd.org/
[七] www.djangoproject.com/
[八] https://trypyramid.com/
[九] https://github.com/aio-libs

项目的一部分，已经创建了一组新的库（包括 aiohttp[1]等库）来填补这个空白。

B.9 Email

`imaplib` 的 API 是相当底层的，要求调用者理解 IMAP 协议才能构建查询和解析结果。imapclient[2]包提供了一个更高层的 API，如果构建需要处理 IMAP 邮箱的应用，则这个 API 更易于使用。

B.10 应用构建模块

构建命令行接口的两个标准库模块 `argparse` 和 `getopt` 都把命令行参数的定义与其解析和值处理相分离。作为替代，click[3]（"Command-Line Interface Construction Kit"）定义了命令处理函数，然后使用修饰符将选项和提示定义与这些命令关联。

cliff[4]（"Command-Line Interface Formulation Framework"）提供了一组基类来定义命令和一个插件系统，这个插件系统用于扩展有多个子命令的应用，这些子命令可能分布在单独的包中。它使用 `argparse` 构建帮助文本和参数解析器，所以我们很熟悉它的命令行处理。

docopt[5]包与正常流程相反，开发人员要先为程序写帮助文本，然后再解析这个文本来了解选项和子命令的合法组合。

对于基于终端的交互式程序，prompt_toolkit[6]提供了一些高级特性，如颜色支持、语法突出显示、输入编辑、鼠标支持和可搜索的历史。可以像 `cmd` 模块一样用来构建面向命令的程序，或者可以构建类似文本编辑器的全屏应用。

`configparser` 使用的那种 INI 文件在应用配置中仍然很流行，不过，在配置方面，YAML[7]文件格式也得到了广泛使用。YAML 采用一种更易于人阅读的形式提供了 JSON 的很多数据结构特性。可以利用 PyYAML[8]库访问一个 YAML 解析器和串行化器。

B.11 开发工具

标准库模块 `venv` 是 Python 3 中新增的模块，如果要在 Python 2 和 Python 3 下实现类似的应用隔离，则应当使用 virtualenv[9]。

---

[1] http://aiohttp.readthedocs.io/
[2] http://imapclient.freshfoo.com/
[3] http://click.pocoo.org
[4] http://docs.openstack.org/developer/cliff/
[5] http://docopt.org
[6] http://python-prompt-toolkit.readthedocs.io/en/stable/
[7] http://yaml.org
[8] http://pyyaml.org
[9] https://virtualenv.pypa.io/

fixtures[一]包提供了一些测试资源管理类，专门设计为使用 `unittest` 模块中测试用例的 `addCleanup()` 方法。fixture 类采用一种一致而安全的方式来管理日志记录器、环境变量、临时文件等，确保每个测试用例与套件中的其他测试用例完全隔离。

标准库中原先用 `distutils` 模块打包 Python 模块来进行发布和实现重用，现在 `distutils` 模块已经废弃。它的替代工具 setuptools[二]单独打包，并不包含在标准库中，这样就能更频繁地发布新版本。setuptools 的 API 提供了一些工具来构建包含在包中的文件列表。有一些扩展包可以由一个版本控制系统管理的文件集自动构建这个列表。例如，使用 setuptools-git[三]并结合一个 git[四]存储库中的源代码，就会默认地在包中包含跟踪到的所有文件。构建一个包之后，twine[五]应用将把它上传到 package index，从而可以与其他开发人员共享。

`tabnanny` 之类的工具很适合查找 Python 代码中的常见格式化错误。Python Code Quality Authority[六]维护了一组丰富的更高级的静态分析工具，这些工具可以强制风格要求、查找常见编程错误甚至帮助避免过度的复杂性。

提示：相关阅读材料

- Python Package Index[七] (PyPI)：这个网站用于查找和下载 Python 运行时库以外单独发布的扩展模块。

[一] https://pypi.python.org/pypi/fixtures
[二] https://setuptools.readthedocs.io/en/latest/
[三] https://pypi.python.org/pypi/setuptools-git
[四] https://git-scm.com
[五] https://pypi.python.org/pypi/twine
[六] http://meta.pycqa.org/en/latest/
[七] https://pypi.python.org/pypi

推荐阅读

Effective系列